CAMBRIDGE LIBRARY COLLECTION

Books of enduring scholarly value

Mathematical Sciences

From its pre-historic roots in simple counting to the algorithms powering modern desktop computers, from the genius of Archimedes to the genius of Einstein, advances in mathematical understanding and numerical techniques have been directly responsible for creating the modern world as we know it. This series will provide a library of the most influential publications and writers on mathematics in its broadest sense. As such, it will show not only the deep roots from which modern science and technology have grown, but also the astonishing breadth of application of mathematical techniques in the humanities and social sciences, and in everyday life.

The Scientific Papers of Sir George Darwin

Sir George Darwin (1845-1912) was the second son and fifth child of Charles Darwin. After studying mathematics at Cambridge he read for the Bar, but soon returned to science and to Cambridge, where in 1883 he was appointed Plumian Professor of Astronomy and Experimental Philosophy. His family home is now the location of Darwin College. His work was concerned primarily with the effect of the sun and moon on tidal forces on Earth, and with the theoretical cosmogony which evolved from practical observation: he formulated the fission theory of the formation of the moon (that the moon was formed from still molten matter pulled away from the Earth by solar tides). He also developed a theory of evolution for the Sun–Earth–Moon system based on mathematical analysis in geophysical theory. This volume of his collected papers covers tidal friction and cosmogony.

Cambridge University Press has long been a pioneer in the reissuing of out-of-print titles from its own backlist, producing digital reprints of books that are still sought after by scholars and students but could not be reprinted economically using traditional technology. The Cambridge Library Collection extends this activity to a wider range of books which are still of importance to researchers and professionals, either for the source material they contain, or as landmarks in the history of their academic discipline.

Drawing from the world-renowned collections in the Cambridge University Library, and guided by the advice of experts in each subject area, Cambridge University Press is using state-of-the-art scanning machines in its own Printing House to capture the content of each book selected for inclusion. The files are processed to give a consistently clear, crisp image, and the books finished to the high quality standard for which the Press is recognised around the world. The latest print-on-demand technology ensures that the books will remain available indefinitely, and that orders for single or multiple copies can quickly be supplied.

The Cambridge Library Collection will bring back to life books of enduring scholarly value (including out-of-copyright works originally issued by other publishers) across a wide range of disciplines in the humanities and social sciences and in science and technology.

The Scientific Papers of Sir George Darwin

Volume 2: Tidal Friction and Cosmogony

George Howard Darwin

Cambridge University Press

CAMBRIDGE UNIVERSITY PRESS

Cambridge New York Melbourne Madrid Cape Town Singapore São Paolo Delhi

Published in the United States of America by Cambridge University Press, New York

www.cambridge.org
Information on this title: www.cambridge.org/9781108004435

This edition first published 1908
This digitally printed version 2009

ISBN 978-1-108-00443-5

SCIENTIFIC PAPERS

CAMBRIDGE UNIVERSITY PRESS WAREHOUSE,
C. F. CLAY, MANAGER.
London: FETTER LANE, E.C.
Edinburgh: 100, PRINCES STREET.

Leipzig: F. A. BROCKHAUS.
Berlin: A. ASHER AND CO.
New York: G. P. PUTNAM'S SONS.
Bombay and Calcutta: MACMILLAN AND CO., LTD.

SCIENTIFIC PAPERS

BY

SIR GEORGE HOWARD DARWIN,

K.C.B., F.R.S.

FELLOW OF TRINITY COLLEGE

PLUMIAN PROFESSOR IN THE UNIVERSITY OF CAMBRIDGE

VOLUME II

TIDAL FRICTION

AND

COSMOGONY

CAMBRIDGE:

AT THE UNIVERSITY PRESS

1908

Cambridge:
PRINTED BY JOHN CLAY, M.A.
AT THE UNIVERSITY PRESS.

PREFACE

THE papers contained in the present volume form in effect a single investigation in speculative astronomy. The tidal oscillations of the mobile parts of a planet must be subject to frictional resistance, and this simple cause gives rise to a diversity of astronomical effects worthy of examination.

The earlier portion of the investigation was undertaken with the object of explaining, if possible, the obliquity of the earth's equator to the ecliptic, and the results attained were so fruitful and promising that it seemed well to examine the whole subject with the closest attention, and to discuss the various collateral points which arose in the course of the work.

It is the experience of every investigator that he reaches his goal by a devious path, and this, at least, has been the case in the present group of papers. If then the whole field were now to be retraversed, it is almost certain that the results might be obtained more shortly. Then, again, there is another cause which precludes brevity, for when an entirely new subject is under consideration every branch road must be examined with care. By far the greater number of the forks in the road lead only to blind alleys; and it is often impossible to foresee, at the cross roads, which will be the main highway, and which a blind alley. Clearness of view is only reached by the pioneer after much labour, and as he first passes along his path he has to grope his way in the dark without the help of any sign-post.

This may be illustrated by what actually occurred to me, for when I first found the quartic equation (p. 102) which expresses the identity between the lengths of the day and of the month, I only regarded it as giving the configuration towards which the retrospective integration was leading back. I well remember thinking that it was just as well to find the other roots of the equation, although I had no suspicion that anything of interest would be discovered thereby. As of course I ought to have foreseen, the result threw a flood of light on the whole subject, for it showed how the system must have degraded, through loss of energy, from a configuration represented by the first real root to another represented by the second. Moreover the motion in the first configuration was found to be unstable whilst that in the second was stable. Thus this quartic equation led to the remarkably simple and illuminating view of the theory of tidal friction contained in the fifth paper (p. 195); and yet all this arose from a point which appeared at first sight barely worth examining.

I wish now, after the lapse of more than twenty years, to avail myself of this opportunity of commenting on some portions of the work and of reviewing the theory as a whole.

The observations of Dr Hecker* and of others do not afford evidence of any considerable amount of retardation in the tidal oscillations of the solid earth, for, within the limits of error of observation, the phase of the oscillation appears to be the same as if the earth were purely elastic. Then again modern researches in the lunar theory show that the secular acceleration of the moon's mean motion is so nearly explained by means of pure gravitation as to leave but a small residue to be referred to the effects of tidal friction. We are thus driven to believe that at present tidal friction is producing its inevitable effects with extreme slowness. But we need not therefore hold that the march of events was always so leisurely, and if the earth was ever wholly or in large part molten, it cannot have been the case.

In any case frictional resistance, whether it be much or little and whether applicable to the solid planet or to the superincumbent ocean, is a true cause of change, and it remains desirable that its effects should be investigated. Now for this end it was necessary to adopt some consistent theory of frictionally resisted tides, and the hypothesis of the earth's viscosity afforded the only available theory of the kind. Thus the first paper in the present volume is devoted to the theory of the tides of a viscous spheroid. It may be that nothing material is added by solving the problem also for the case of elastico-viscosity, but it was well that that hypothesis should also be examined.

I had at a previous date endeavoured to determine the amount of modification to which Lord Kelvin's theory of the tides of an elastic globe must be subject in consequence of the heterogeneity of the earth's density, and this investigation is reproduced in the second paper. Dr Herglotz has also treated the problem by means of some laborious analysis, and finds the change due to heterogeneity somewhat greater than I had done. But we both base our conclusions on assumptions which seem to be beyond the reach of verification, and the probability of correctness in the results can only be estimated by means of the plausibility of the assumptions.

The differential equations which specify the rates of change in the various elements of the motions of the moon and the earth were found to be too

* *Veröffentl. d. K. Preuss. Geodät. Inst.*, Neue Folge, No. 32, Potsdam, 1907.

complex to admit of analytical integration, and it therefore became necessary to solve the problem numerically. It was intended to draw conclusions as to the history of the earth and moon, and accordingly the true values of the mass, size and speed of rotation of the earth were taken as the basis of computation. But the earth was necessarily treated as being homogeneous, and thus erroneous values were involved for the ellipticity, for the precessional constant and for the inequalities in the moon's motion due to the oblateness of the earth. It was not until the whole of the laborious integrations had been completed that it occurred to me that an appropriate change in the linear dimensions of the homogeneous earth might afford approximately correct values for every other element. Such a mechanically equivalent substitute for the earth is determined on p. 439, and if my integrations should ever be repeated I suggest that it would be advantageous to adopt the numerical values there specified as the foundation for the computations.

The third paper contains the investigation of the secular changes in the motions of the earth and moon, due to tidal friction, when the lunar orbit is treated as circular and coincident with the ecliptic. The differential equations are obtained by means of the disturbing forces, but the method of the disturbing function is much more elegant. The latter method is used in the sixth paper (p. 208), which is devoted especially to finding the changes in the eccentricity and the inclination of the orbit. However the analysis is so complicated that I do not regret having obtained the equations in two independent ways. As the sixth paper was intended to be supplementary to the third, the disturbing function is developed with the special object of finding the equations for the eccentricity and the inclination, but an artifice is devised whereby it may also be made to furnish the equations for the other elements. It would only need a slight amount of modification to obtain the equations for all the elements simultaneously by straightforward analysis.

This paper also contains an investigation of the motion of a satellite moving about an oblate planet by means of equations, which give simultaneously the nutations of the planet and the corresponding inequalities in the motion of the satellite. The equations are afterwards extended so as to include the effects of tidal friction. I found this portion of the work far more arduous than anything else in the whole series of researches.

The developments and integrations in all these papers are carried out with what may perhaps be regarded as an unnecessary degree of elaboration, but it was impossible to foresee what terms might become important. It does not, however, seem worth while to comment further on minor points such as these.

For the astronomer who is interested in cosmogony the important point is the degree of applicability of the theory as a whole to celestial evolution. To me it seems that the theory has rather gained than lost in the esteem of men of science during the last 25 years, and I observe that several writers are disposed to accept it as an established acquisition to our knowledge of cosmogony.

Undue weight has sometimes been laid on the exact numerical values assigned for defining the primitive configuration of the earth and moon. In so speculative a matter close accuracy is unattainable, for a different theory of frictionally retarded tides would inevitably lead to a slight difference in the conclusion; moreover such a real cause as the secular increase in the masses of the earth and moon through the accumulation of meteoric dust, and possibly other causes are left out of consideration.

The exact nature of the process by which the moon was detached from the earth must remain even more speculative. I suggested that the fission of the primitive planet may have been brought about by the synchronism of the solar tide with the period of the fundamental free oscillation of the planet, and the suggestion has received a degree of attention which I never anticipated. It may be that we shall never attain to a higher degree of certainty in these obscure questions than we now possess, but I would maintain that we may now hold with confidence that the moon originated by a process of fission from the primitive planet, that at first she revolved in an orbit close to the present surface of the earth, and that tidal friction has been the principal agent which transformed the system to its present configuration.

The theory for a long time seemed to lie open to attack on the ground that it made too great demands on time, and this has always appeared to me the greatest difficulty in the way of its acceptance. If we were still compelled to assent to the justice of Lord Kelvin's views as to the period of time which has elapsed since the earth solidified, and as to the age of the solar system, we should also have to admit that the theory of evolution under tidal influence is inapplicable to its full extent. Lord Kelvin's contributions to cosmogony have been of the first order of importance, but his arguments on these points no longer carry conviction with them. Lord Kelvin contended that the actual distribution of land and sea proves that the planet solidified at a time when the day had nearly its present length. If this were true the effects of tidal friction relate to a period antecedent to the solidification. But I have always felt convinced that the earth would adjust its ellipticity to its existing speed of rotation with close approximation. The calculations contained in Paper 9, the plasticity of even the most refractory forms of matter under great stresses, and the contortions of geological strata appear to me, at least, conclusive against Lord Kelvin's view.

The researches of Mr Strutt on the radio-activity of rocks prove that we cannot regard the earth simply as a cooling globe, and therefore Lord Kelvin's argument as to the age of the earth as derived from the observed gradient of temperature must be illusory. Indeed even without regard to the initial temperature of the earth acquired by means of secular contraction, it is hard to understand why the earth is not hotter inside than it is.

It seems probable that Mr Strutt may be able to obtain a rough numerical scale of geological time by means of his measurements of the radio-activity of rocks, and although he has not yet been able to formulate such a scale with any degree of accuracy, he is already confident that the periods involved must be measured in hundreds or perhaps even thousands of millions of years*. The evidence, taken at its lowest, points to a period many times as great as was admitted by Lord Kelvin for the whole history of the solar system.

Lastly the recent discovery of the colossal internal energy resident in the atom shows that it is unsafe to calculate the age of the sun merely from mechanical energy, as did Helmholtz and Kelvin. It is true that the time has not yet arrived at which we can explain exactly the manner in which the atomic energy may be available for maintaining the sun's heat, but when the great age of the earth is firmly established the insufficiency of the supply of heat to the sun by means of purely mechanical energy will prove that atomic energy does become available in some way. On the whole then it may be maintained that deficiency of time does not, according to our present state of knowledge, form a bar to the full acceptability of the theory of terrestrial evolution under the influence of tidal friction.

It is very improbable that tidal friction has been the dominant cause of change in any of the other planetary sub-systems or in the solar system itself, yet it seems to throw light on the distribution of the satellites amongst the several planets. It explains the identity of the rotation of the moon with her orbital motion, as was long ago pointed out by Kant and Laplace, and it tends to confirm the correctness of the observations according to which Venus always presents the same face to the sun. Finally it has been held by Dr See and by others to explain some of the peculiarities of the orbits of double stars.

Lord Kelvin's determination of the strain of an elastic sphere and the solution of the corresponding problem of the tides of a viscous spheroid suggested another interesting question with respect to the earth. This problem is to find the strength of the materials of which the earth must be

* Some of Mr Strutt's preliminary computations are given in *Proc. Roy. Soc.* A, Vol. 81, p. 272 (1908).

built so as to prevent the continents from sinking and the sea bed from rising; this question is treated in Paper 9 (p. 459). The existence of an isostatic layer, at which the hydrostatic pressure is uniform, at no great depth below the earth's surface, is now well established. This proves that I have underestimated in my paper the strength of the superficial layers necessary to prevent subsidence and elevation. The strength of granite and of other rocks is certainly barely adequate to sustain the continents in position, and Mr Hayford* seeks to avoid the difficulty by arguing that the earth is actually 'a failing structure,' and that the subsidence of the continents is only prevented by the countervailing effects of the gradually increasing weight of sedimentation on the adjoining sea-beds.

In his address to the Geological Section of the British Association at Dublin (1908) Professor Joly makes an interesting suggestion which bears on this subject. He supposes that the heat generated by the radio-active materials in sediment has exercised an important influence in bringing about the elevation of mountain ranges and of the adjoining continents.

A subsidiary outcome of this same investigation was given in Vol. I. of these papers, when I attempted to determine the elastic oscillations of the superficial layers of the earth under the varying pressures of the tides and of the atmosphere. Dr Hecker may perhaps be able to verify or disprove these theoretical calculations when he makes the final reduction of his valuable observations with horizontal pendulums at Potsdam.

When the first volume of these papers was published Lord Kelvin was still alive, and I had the pleasure of receiving from him a cordial letter of thanks for my acknowledgement of the deep debt I owe him. His name also occurs frequently in the present volume, and if I dissent from some of his views, I none the less regard him as amongst the greatest of those who have tried to guess the riddle of the history of the universe.

The chronological list of my papers is repeated in this second volume, together with a column showing in which volume they are or will be reproduced.

In conclusion I wish to thank the printers and readers of the Cambridge University Press for their marvellous accuracy and care in setting up the type and in detecting some mistakes in the complicated analysis contained in these papers.

G. H. DARWIN.

October, 1908.

* *Phil. Soc. Washington*, Vol. 15 (1907), p. 57.

CONTENTS

PAGE

Chronological List of Papers with References to the Volumes in which they are or probably will be contained xii

Erratum in Vol. I xvi

TIDAL FRICTION AND COSMOGONY.

1. On the Bodily Tides of Viscous and Semi-elastic Spheroids, and on the Ocean Tides upon a Yielding Nucleus 1
[*Philosophical Transactions of the Royal Society*, Part I. Vol. 170 (1879), pp. 1—35.]

2. Note on Thomson's Theory of the Tides of an Elastic Sphere . 33
[*Messenger of Mathematics*, VIII. (1879), pp. 23—26.]

3. On the Precession of a Viscous Spheroid, and on the Remote History of the Earth 36
[*Philosophical Transactions of the Royal Society*, Part II. Vol. 170 (1879), pp. 447—530.]

4. Problems connected with the Tides of a Viscous Spheroid . 140
[*Philosophical Transactions of the Royal Society*, Part II. Vol. 170 (1879), pp. 539—593.]

5. The Determination of the Secular Effects of Tidal Friction by a Graphical Method 195
[*Proceedings of the Royal Society of London*, XXIX. (1879), pp. 168—181.]

6. On the Secular Changes in the Elements of the Orbit of a Satellite revolving about a Tidally Distorted Planet . 208
[*Philosophical Transactions of the Royal Society*, Vol. 171 (1880), pp. 713—891.]

7. On the Analytical Expressions which give the History of a Fluid Planet of Small Viscosity, attended by a Single Satellite . 383
[*Proceedings of the Royal Society*, Vol. XXX. (1880), pp. 255—278.]

8. On the Tidal Friction of a Planet attended by Several Satellites, and on the Evolution of the Solar System 406
[*Philosophical Transactions of the Royal Society*, Vol. 172 (1881), pp. 491—535.]

9. On the Stresses caused in the Interior of the Earth by the Weight of Continents and Mountains 459
[*Philosophical Transactions of the Royal Society*, Vol. 173 (1882), pp. 187—230, with which is incorporated "Note on a previous paper," *Proc. Roy. Soc.* Vol. 38 (1885), pp. 322—328.]

INDEX 515

CHRONOLOGICAL LIST OF PAPERS WITH REFERENCES TO THE VOLUMES IN WHICH THEY ARE OR PROBABLY WILL BE CONTAINED.

YEAR	TITLE AND REFERENCE	Probable volume in collected papers
1875	On two applications of Peaucellier's cells. London Math. Soc. Proc., 6, 1875, pp. 113, 114.	IV
1875	On some proposed forms of slide-rule. London Math. Soc. Proc., 6, 1875, p. 113.	IV
1875	The mechanical description of equipotential lines. London Math. Soc. Proc., 6, 1875, pp. 115—117.	IV
1875	On a mechanical representation of the second elliptic integral. Messenger of Math., 4, 1875, pp. 113—115.	IV
1875	On maps of the World. Phil. Mag., 50, 1875, pp. 431—444.	IV
1876	On graphical interpolation and integration. Brit. Assoc. Rep., 1876, p. 13.	IV
1876	On the influence of geological changes on the Earth's axis of rotation. Roy. Soc. Proc., 25, 1877, pp. 328—332; Phil. Trans., 167, 1877, pp. 271—312.	III
1876	On an oversight in the *Mécanique Céleste*, and on the internal densities of the planets. Astron. Soc. Month. Not., 37, 1877, pp. 77—89.	III
1877	A geometrical puzzle. Messenger of Math., 6, 1877, p. 87.	IV
1877	A geometrical illustration of the potential of a distant centre of force. Messenger of Math., 6, 1877, pp. 97, 98.	IV
1877	Note on the ellipticity of the Earth's strata. Messenger of Math., 6, 1877, pp. 109, 110.	III
1877	On graphical interpolation and integration. Messenger of Math., 6, 1877, pp. 134—136.	IV
1877	On a theorem in spherical harmonic analysis. Messenger of Math., 6, 1877, pp. 165—168.	IV
1877	On a suggested explanation of the obliquity of planets to their orbits. Phil. Mag., 3, 1877, pp. 188—192.	III
1877	On fallible measures of variable quantities, and on the treatment of meteorological observations. Phil. Mag., 4, 1877, pp. 1—14.	IV

YEAR	TITLE AND REFERENCE	Probable volume in collected papers
1878	On Professor Haughton's estimate of geological time. Roy. Soc. Proc., 27, 1878, pp. 179—183.	III
1878	On the bodily tides of viscous and semi-elastic spheroids, and on the Ocean tides on a yielding nucleus. Roy. Soc. Proc., 27, 1878, pp. 419—424; Phil. Trans., 170, 1879, pp. 1—35.	II
1878	On the precession of a viscous spheroid. Brit. Assoc. Rep., 1878, pp. 482—485.	omitted
1879	On the precession of a viscous spheroid, and on the remote history of the Earth. Roy. Soc. Proc., 28, 1879, pp. 184—194; Phil. Trans., 170, 1879, pp. 447—538.	II
1879	Problems connected with the tides of a viscous spheroid. Roy. Soc. Proc., 28, 1879, pp. 194—199; Phil. Trans., 170, 1879, pp. 539—593.	II
1879	Note on Thomson's theory of the tides of an elastic sphere. Messenger of Math., 8, 1879, pp. 23—26.	II
1879	The determination of the secular effects of tidal friction by a graphical method. Roy. Soc. Proc., 29, 1879, pp. 168—181.	II
1880	On the secular changes in the elements of the orbit of a satellite revolving about a tidally distorted planet. Roy. Soc. Proc., 30, 1880, pp. 1—10; Phil. Trans., 171, 1880, pp. 713—891.	II
1880	On the analytical expressions which give the history of a fluid planet of small viscosity, attended by a single satellite. Roy. Soc. Proc., 30, 1880, pp. 255—278.	II
1880	On the secular effects of tidal friction. Astr. Nachr., 96, 1880, col. 217—222.	omitted
1881	On the tidal friction of a planet attended by several satellites, and on the evolution of the solar system. Roy. Soc. Proc., 31, 1881, pp. 322—325; Phil. Trans., 172, 1881, pp. 491—535.	II
1881	On the stresses caused in the interior of the Earth by the weight of continents and mountains. Phil. Trans., 173, 1882, pp. 187—230; Amer. Journ. Sci., 24, 1882, pp. 256—269.	II
1881	(Together with Horace Darwin.) On an instrument for detecting and measuring small changes in the direction of the force of gravity. Brit. Assoc. Rep., 1881, pp. 93—126; Annal. Phys. Chem., Beibl. 6, 1882, pp. 59—62.	I
1882	On variations in the vertical due to elasticity of the Earth's surface. Brit. Assoc. Rep., 1882, pp. 106—119; Phil. Mag., 14, 1882, pp. 409—427.	I
1882	On the method of harmonic analysis used in deducing the numerical values of the tides of long period, and on a misprint in the Tidal Report for 1872. Brit. Assoc. Rep., 1882, pp. 319—327.	omitted
1882	A numerical estimate of the rigidity of the Earth. Brit. Assoc. Rep., 1882, pp. 472—474; § 848, Thomson and Tait's Nat. Phil. second edition.	I

YEAR	TITLE AND REFERENCE	Probable volume in collected papers
1883	Report on the Harmonic analysis of tidal observations. Brit. Assoc. Rep., 1883, pp. 49—117.	I
1883	On the figure of equilibrium of a planet of heterogeneous density. Roy. Soc. Proc., 36, pp. 158—166.	III
1883	On the horizontal thrust of a mass of sand. Instit. Civ. Engin. Proc., 71, 1883, pp. 350—378.	IV
1884	On the formation of ripple-mark in sand. Roy. Soc. Proc., 36, 1884, pp. 18—43.	IV
1884	Second Report of the Committee, consisting of Professors G. H. Darwin and J. C. Adams, for the harmonic analysis of tidal observations. Drawn up by Professor G. H. Darwin. Brit. Assoc. Rep., 1884, pp. 33—35.	omitted
1885	Note on a previous paper. Roy. Soc. Proc., 38, pp. 322—328.	II
1885	Results of the harmonic analysis of tidal observations. (Jointly with A. W. Baird.) Roy. Soc. Proc., 39, pp. 135—207.	omitted
1885	Third Report of the Committee, consisting of Professors G. H. Darwin and J. C. Adams, for the harmonic analysis of tidal observations. Drawn up by Professor G. H. Darwin. Brit. Assoc. Rep., 1885, pp. 35—60.	I
1886	Report of the Committee, consisting of Professor G. H. Darwin, Sir W. Thomson, and Major Baird, for preparing instructions for the practical work of tidal observation; and Fourth Report of the Committee, consisting of Professors G. H. Darwin and J. C. Adams, for the harmonic analysis of tidal observations. Drawn up by Professor G. H. Darwin. Brit. Assoc. Rep., 1886, pp. 40—58.	I
1886	Presidential Address. Section A, Mathematical and Physical Science. Brit. Assoc. Rep., 1886, pp. 511—518.	IV
1886	On the correction to the equilibrium theory of tides for the continents. I. By G. H. Darwin. II. By H. H. Turner. Roy. Soc. Proc., 40, pp. 303—315.	I
1886	On Jacobi's figure of equilibrium for a rotating mass of fluid. Roy. Soc. Proc., 41, pp. 319—336.	III
1886	On the dynamical theory of the tides of long period. Roy. Soc. Proc., 41, pp. 337—342.	I
1886	Article 'Tides.' (Admiralty) Manual of Scientific Inquiry.	I
1887	On figures of equilibrium of rotating masses of fluid. Roy. Soc. Proc., 42, pp. 359—362; Phil. Trans., 178A, pp. 379—428.	III
1887	Note on Mr Davison's Paper on the straining of the Earth's crust in cooling. Phil. Trans., 178A, pp. 242—249.	IV
1888	Article 'Tides.' Encyclopaedia Britannica.	Certain sections in I
1888	On the mechanical conditions of a swarm of meteorites, and on theories of cosmogony. Roy. Soc. Proc., 45, pp. 3—16; Phil. Trans., 180A, pp. 1—69.	IV

YEAR	TITLE AND REFERENCE	Probable volume in collected papers
1889	Second series of results of the harmonic analysis of tidal observations. Roy. Soc. Proc., 45, pp. 556—611.	omitted
1889	Meteorites and the history of Stellar systems. Roy. Inst. Rep., Friday, Jan. 25, 1889.	omitted
1890	On the harmonic analysis of tidal observations of high and low water. Roy. Soc. Proc., 48, pp. 278—340.	I
1891	On tidal prediction. Bakerian Lecture. Roy. Soc. Proc., 49, pp. 130—133; Phil. Trans., 182A, pp. 159—229.	I
1892	On an apparatus for facilitating the reduction of tidal observations. Roy. Soc. Proc., 52, pp. 345—389.	I
1896	On periodic orbits. Brit. Assoc. Rep., 1896, pp. 708, 709.	omitted
1897	Periodic orbits. Acta Mathematica, 21, pp. 101—242, also (with omission of certain tables of results) Mathem. Annalen, 51, pp. 523—583.	IV
	[by S. S. Hough. On certain discontinuities connected with periodic orbits. Acta Math., 24 (1901), pp. 257—288.]	IV
1899	The theory of the figure of the Earth carried to the second order of small quantities. Roy. Astron. Soc. Month. Not., 60, pp. 82—124.	III
1900	Address delivered by the President, Professor G. H. Darwin, on presenting the Gold Medal of the Society to M. H. Poincaré. Roy. Astron. Soc. Month. Not., 60, pp. 406—415.	IV
1901	Ellipsoidal harmonic analysis. Roy. Soc. Proc., 68, pp. 248—252; Phil. Trans., 197A, pp. 461—557.	III
1901	On the pear-shaped figure of equilibrium of a rotating mass of liquid. Roy. Soc. Proc., 69, pp. 147, 148; Phil. Trans., 198A, pp. 301—331.	III
1902	Article 'Tides.' Encyclopaedia Britannica, supplementary volumes.	Certain sections in I
1902	The stability of the pear-shaped figure of equilibrium of a rotating mass of liquid. Roy. Soc. Proc., 71, pp. 178—183; Phil. Trans., 200A, pp. 251—314.	III
1903	On the integrals of the squares of ellipsoidal surface harmonic functions. Roy. Soc. Proc., 72, p. 492; Phil. Trans., 203A, pp. 111—137.	III
1903	The approximate determination of the form of Maclaurin's spheroid. Trans. Amer. Math. Soc., 4, pp. 113—133.	III
1903	The Eulerian nutation of the Earth's axis. Bull. Acad. Roy. de Belgique (Sciences), pp. 147—161.	IV
1905	The analogy between Lesage's theory of gravitation and the repulsion of light. Roy. Soc. Proc., 76A, pp. 387—410.	IV
1905	Address by Professor G. H. Darwin, President. Brit. Assoc. Rep., 1905, pp. 3—32.	IV
1906	On the figure and stability of a liquid satellite. Roy. Soc. Proc., 77A, pp. 422—425; Phil. Trans., 206A, pp. 161—248.	III

YEAR	TITLE AND REFERENCE	Probable volume in collected papers
1908	Tidal observations of the 'Discovery.' National Antarctic Expedition 1901—4, Physical Observations, pp. 1—12.	I
1908	Discussion of the tidal observations of the 'Scotia.' National Antarctic Expedition 1901—4, Physical Observations, p. 16.	omitted
1908	Further consideration of the figure and stability of the pear-shaped figure of a rotating mass of liquid. Roy. Soc. Proc., 80A, pp. 166—7 ; Phil. Trans., 208A, pp. 1—19.	III
1908	Further note on Maclaurin's Ellipsoid. Trans. Amer. Math. Soc., 9, pp. 34—38.	III
1908	(Together with S. S. Hough.) Article 'Bewegung der Hydrosphäre' (The Tides). Encyklopädie der mathematischen Wissenschaften, VI. 1, 6. 83 pp.	IV
	Unpublished Article 'Tides.' Encyclopaedia Britannica, new edition to be published hereafter (by permission of the proprietors).	Certain sections in I

ERRATUM IN VOL. I.

p. 275, equation (26), line 9 from foot of page, should read

$$\vartheta_{2s} = 4\mathfrak{D} + 2p_s\kappa_m - \kappa_{2s}.$$

1.

ON THE BODILY TIDES OF VISCOUS AND SEMI-ELASTIC SPHEROIDS, AND ON THE OCEAN TIDES UPON A YIELDING NUCLEUS*.

[*Philosophical Transactions of the Royal Society*, Part I. Vol. 170 (1879), pp. 1—35.]

In a well-known investigation Sir William Thomson has discussed the problem of the bodily tides of a homogeneous elastic sphere, and has drawn therefrom very important conclusions as to the great rigidity of the earth†.

Now it appears improbable that the earth should be perfectly elastic; for the contortions of geological strata show that the matter constituting the earth is somewhat plastic, at least near the surface. We know also that even the most refractory metals can be made to flow under the action of sufficiently great forces.

Although Sir W. Thomson's investigation has gone far to overthrow the old idea of a semi-fluid interior to the earth, yet geologists are so strongly impressed by the fact that enormous masses of rock are being, and have been, poured out of volcanic vents in the earth's surface, that the belief is not yet extinct that we live on a thin shell over a sea of molten lava. Under these circumstances it appears to be of interest to investigate the consequences which would arise from the supposition that the matter constituting the earth is of a viscous or imperfectly elastic nature; for if the interior *is*

* [Since the date of this paper important contributions to the subject have been made by Professor Horace Lamb in his papers on "The Oscillations of a Viscous Spheroid," *Proc. Lond. Math. Soc.*, Vol. XIII. (1881–2), p. 51; "On the Vibrations of an Elastic Sphere," *ibid.*, p. 189, and "On the Vibrations of a Spherical Shell," *ibid.*, Vol. XIV. (1882–3), p. 50. See also a paper by T. J. Bromwich, *Proc. Lond. Math. Soc.*, Vol. XXX. (1898–9), p. 98.]

† Sir William states that M. Lamé had treated the subject at an earlier date, but in an entirely different manner. I am not aware, however, that M. Lamé had fully discussed the subject in its physical aspect.

constituted in this way, then the solid crust, unless very thick, cannot possess rigidity enough to repress the tidal surgings, and these hypotheses must give results fairly conformable to the reality. The hypothesis of imperfect elasticity will be principally interesting as showing how far Sir W. Thomson's results are modified by the supposition that the elasticity breaks down under continued stress.

In this paper, then, I follow out these hypotheses, and it will be seen that the results are fully as hostile to the idea of any great mobility of the interior of the earth as is that of Sir W. Thomson.

The only terrestrial evidence of the existence of a bodily tide in the earth would be that the ocean tides would be less in height than is indicated by theory. The subject of this paper is therefore intimately connected with the theory of the ocean tides.

In the first part the equilibrium tide-theory is applied to estimate the reduction and alteration of phase of ocean tides as due to bodily tides, but that theory is acknowledged on all hands to be quite fallacious in its explanation of tides of short period.

In the second part of this paper, therefore, I have considered the dynamical theory of tides in an equatorial canal running round a tidally-distorted nucleus, and the results are almost the same as those given by the equilibrium theory.

The first two sections of the paper are occupied with the adaptation of Sir W. Thomson's work* to the present hypotheses; as, of course, it was impossible to reproduce the whole of his argument, I fear that the investigation will only be intelligible to those who are either already acquainted with that work, or who are willing to accept my quotations therefrom as established.

As some readers may like to know the results of this inquiry without going into the mathematics by which they are established, I have given in Part III. a summary of the whole, and have as far as possible relegated to that part of the paper the comments and conclusions to be drawn. I have tried, however, to give so much explanation in the body of the paper as will make it clear whither the argument is tending.

The case of pure viscosity is considered first, because the analysis is somewhat simpler, and because the results will afterwards admit of an easy extension to the case of elastico-viscosity.

* His paper will be found in *Phil. Trans.*, 1863, p. 573, and §§ 733—737 and 834—846 of Thomson and Tait's *Natural Philosophy*.

I.

The Bodily Tides of Viscous and Elastico-viscous Spheroids.

1. *Analogy between the flow of a viscous body and the strain of an elastic one.*

The general equations of flow of a viscous fluid, *when the effects of inertia are neglected*, are

$$\left.\begin{aligned} -\frac{dp}{dx}+\upsilon\nabla^2\alpha+X=0 \\ -\frac{dp}{dy}+\upsilon\nabla^2\beta+Y=0 \\ -\frac{dp}{dz}+\upsilon\nabla^2\gamma+Z=0 \end{aligned}\right\}\text{..........................(1)}$$

where x, y, z are the rectangular coordinates of a point of the fluid; α, β, γ are the component velocities parallel to the axes; p is the mean of the three pressures across planes perpendicular to the three axes respectively; X, Y, Z are the component forces acting on the fluid, estimated per unit volume; υ is the coefficient of viscosity; and ∇^2 is the Laplacian operation

$$\frac{d^2}{dx^2}+\frac{d^2}{dy^2}+\frac{d^2}{dz^2}$$

Besides these we have the equation of continuity $\frac{d\alpha}{dx}+\frac{d\beta}{dy}+\frac{d\gamma}{dz}=0$.

Also if P, Q, R, S, T, U are the normal and tangential stresses estimated in the usual way across three planes perpendicular to the axes

$$\left.\begin{aligned} P=-p+2\upsilon\frac{d\alpha}{dx},\quad Q=-p+2\upsilon\frac{d\beta}{dy},\quad R=-p+2\upsilon\frac{d\gamma}{dz} \\ S=\upsilon\left(\frac{d\beta}{dz}+\frac{d\gamma}{dy}\right),\quad T=\upsilon\left(\frac{d\gamma}{dx}+\frac{d\alpha}{dz}\right),\quad U=\upsilon\left(\frac{d\alpha}{dy}+\frac{d\beta}{dx}\right) \end{aligned}\right\}\text{......(2)}$$

Now in an elastic solid, if α, β, γ be the displacements, $m-\frac{1}{3}n$ the coefficient of dilatation, and n that of rigidity, and if $\delta=\frac{d\alpha}{dx}+\frac{d\beta}{dy}+\frac{d\gamma}{dz}$; the equations of equilibrium are

$$\left.\begin{aligned} m\frac{d\delta}{dx}+n\nabla^2\alpha+X=0 \\ m\frac{d\delta}{dy}+n\nabla^2\beta+Y=0 \\ m\frac{d\delta}{dz}+n\nabla^2\gamma+Z=0 \end{aligned}\right\}\text{.........................(3)*}$$

* Thomson and Tait's *Natural Philosophy*, § 698, eq. (7) and (8).

Also

$$P=(m-n)\,\delta+2n\frac{d\alpha}{dx},\quad Q=(m-n)\,\delta+2n\frac{d\beta}{dy},\quad R=(m-n)\,\delta+2n\frac{d\gamma}{dz} \qquad \text{......(4)}$$

and S, T, U have the same forms as in (2), with n written instead of v.

Therefore if we put $-p=\frac{1}{3}(P+Q+R)$, we have $p=-(m-\frac{1}{3}n)\,\delta$, so that (3) may be written

$$-\frac{m}{m-\frac{1}{3}n}\frac{dp}{dx}+n\nabla^2\alpha+X=0, \text{ \&c., \&c.}$$

Also $$P=-\frac{m-n}{m-\frac{1}{3}n}p+2n\frac{d\alpha}{dx},\quad Q=\text{\&c.},\quad R=\text{\&c.}$$

Now if we suppose the elastic solid to be incompressible, so that m is infinitely large compared to n, then it is clear that the equations of equilibrium of the incompressible elastic solid assume exactly the same form as those of flow of the viscous fluid, n merely taking the place of v.

Thus every problem in the equilibrium of an incompressible elastic solid has its counterpart in a problem touching the state of flow of an incompressible viscous fluid, when the effects of inertia are neglected; and the solution of the one may be made applicable to the other by merely reading for "displacements" "velocities," and for the coefficient of "rigidity" that of "viscosity."

2. *A sphere under influence of bodily force.*

Sir W. Thomson has solved the following problem:

To find the displacement of every point of the substance of an elastic sphere exposed to no surface traction, but deformed infinitesimally by an equilibrating system of forces acting *bodily* through the interior.

If for "displacement" we read velocity, and for "elastic" viscous, we have the corresponding problem with respect to a viscous fluid, and *mutatis mutandis* the solution is the same.

But we cannot find the tides of a viscous sphere by merely making the equilibrating system of forces equal to the tide-generating influence of the sun or moon, because the substance of the sphere must be supposed to have the power of gravitation.

For suppose that at any time the equation to the free surface of the earth (as the viscous sphere may be called for brevity) is $r=a+\sum_{2}^{\infty}\sigma_i$, where σ_i is a surface harmonic. Then the matter, positive or negative, filling the space represented by $\Sigma\sigma_i$ exercises an attraction on every point of the interior; and this attraction, together with that of a homogeneous sphere of radius a, must be added to the tide-generating influence to form the whole force in the interior of the sphere. Also it is a spheroid, and no longer a true sphere with which we have to deal. If, however, we cut a true sphere of radius a out of the spheroid (leaving out $\Sigma\sigma_i$), then by a proper choice of surface actions, the tidal problem may be reduced to finding the state of flow in a true sphere under the action of (i) an external tide-generating influence, (ii) the attraction of the true sphere, and of the positive and negative matter filling the space $\Sigma\sigma_i$, but (iii) subject to certain surface forces.

Since (i) and (ii) together constitute a bodily force, the problem only differs from that of Sir W. Thomson in the fact that there are forces acting on the surface of the sphere.

Now as we are only going to consider small deviations from sphericity, these surface actions will be of small amount, and an approximation will be permissible.

It is clear that rigorously there is tangential action* between the layer of matter $\Sigma\sigma_i$ and the true sphere, but by far the larger part of the action is normal, and is simply the weight (either positive or negative) of the matter which lies above or below any point on the surface of the true sphere.

Thus, in order to reduce the earth to sphericity, the appropriate surface action is a normal traction equal to $-gw\Sigma\sigma_i$, where g is gravity at the surface, and w is the mass per unit volume of the matter constituting the earth.

In order to show what alteration this normal surface traction will make in Sir W. Thomson's solution, I must now give a short account of his method of attacking the problem.

He first shows that, where there is a potential function, the solution of the problem may be subdivided, and that the complete values of α, β, γ consist of the sums of two parts which are to be found in different ways. The first part consists of *any* values of α, β, γ, which satisfy the equations throughout the sphere, without reference to surface conditions. As far as regards the second part, the bodily force is deemed to be non-existent and is replaced by certain surface actions, so calculated as to counteract the surface actions which correspond to the values of α, β, γ found in the first part of the solution. Thus the first part satisfies the condition that there is a

* I shall consider some of the effects of this tangential action in a future paper, viz.: "Problems connected with the Tides of a Viscous Spheroid," read before the Royal Society on December 19th, 1878. [Paper 4.]

bodily force, and the second adds the condition that the surface forces are zero. The first part of the solution is easily found, and for the second part Sir W. Thomson discusses the case of an elastic sphere under the action of any surface tractions, but without any bodily force acting on it. The component surface tractions parallel to the three axes, in this problem, are supposed to be expanded in a series of surface harmonics; and the harmonic terms of any order are shown to have an effect on the displacements independent of those of every other order. Thus it is only necessary to consider the typical component surface tractions A_i, B_i, C_i of the order i.

He proves that (for an incompressible elastic solid for which m is infinite) this one surface traction A_i, B_i, C_i produces a displacement throughout the sphere given by

$$\alpha = \frac{1}{na^{i-1}}\left\{\frac{a^2-r^2}{2(2i^2+1)}\frac{d\Psi_{i-1}}{dx} + \frac{1}{i-1}\left[\frac{i+2}{(2i^2+1)(2i+1)} r^{2i+1}\frac{d}{dx}(\Psi_{i-1}r^{-2i+1}) + \frac{1}{2i(2i+1)}\frac{d\Phi_{i+1}}{dx} + A_i r^i\right]\right\} \quad\ldots\ldots\ldots\ldots\ldots\ldots(5)^*$$

with symmetrical expressions for β and γ; where Ψ and Φ are auxiliary functions defined by

$$\left.\begin{aligned}\Psi_{i-1} &= \frac{d}{dx}(A_i r^i) + \frac{d}{dy}(B_i r^i) + \frac{d}{dz}(C_i r^i)\\ \Phi_{i+1} &= r^{2i+3}\left\{\frac{d}{dx}(A_i r^{-i-1}) + \frac{d}{dy}(B_i r^{-i-1}) + \frac{d}{dz}(C_i r^{-i-1})\right\}\end{aligned}\right\}\quad\ldots\ldots\ldots(6)$$

In the case considered by Sir W. Thomson of an elastic sphere deformed by bodily stress and subject to no surface action, we have to substitute in (5) and (6) only those surface actions which are equal and opposite to the surface forces corresponding to the first part of the solution †; but in the case which we now wish to consider, we must add to these latter the components of the normal traction $-gw\Sigma\sigma_i$, and besides must include in the bodily force both the external disturbing force, and the attraction of the matter of the spheroid on itself.

Now from the forms of (5) and (6) it is obvious that the tractions which correspond to the first part of the solution, and the traction $-gw\Sigma\sigma_i$ produce quite independent effects, and therefore we need only add to the complete solution of Sir W. Thomson's problem of the elastic sphere, the terms which arise from the normal traction $-gw\Sigma\sigma_i$. Finally we must pass from the elastic problem to the viscous one, by reading v for n, and velocities for displacements.

* Thomson and Tait's *Natural Philosophy*, 1867, § 737, equation (52).

† Where the solid is incompressible, this surface traction is normal to the sphere at every point, provided that the potential of the bodily force is expressible in a series of solid harmonics.

I proceed then to find the state of internal flow in the viscous sphere, which results from a normal traction at every point of the surface of the sphere, given by the surface harmonic S_i.

In order to use the formulæ (5) and (6), it is first necessary to express the component tractions $\frac{x}{a}S_i$, $\frac{y}{a}S_i$, $\frac{z}{a}S_i$ as surface harmonics.

Now if V_i be a solid harmonic,

$$\frac{d}{dx}(r^{-2i-1}V_i) = -(2i+1)\,r^{-(2i+3)}\,xV_i + r^{-(2i+1)}\frac{dV_i}{dx}$$

So that $$xV_i = \frac{1}{2i+1}\left\{r^2\frac{dV_i}{dx} - r^{2i+3}\frac{d}{dx}(r^{-2i-1}V_i)\right\}$$

Therefore $$\frac{x}{a}S_i = \frac{1}{2i+1}\left\{\left[r^{-i+1}\frac{d}{dx}(r^iS_i)\right] - \left[r^{i+2}\frac{d}{dx}(r^{-i-1}S_i)\right]\right\}$$

The quantities within the brackets [ ] being independent of r, and being surface harmonics of orders $i-1$ and $i+1$ respectively, we have $\frac{x}{a}S_i$ expressed as the sum of two surface harmonics A_{i-1}, A_{i+1}, where

$$A_{i-1} = \frac{1}{2i+1}r^{-i+1}\frac{d}{dx}(r^iS_i), \quad A_{i+1} = -\frac{1}{2i+1}r^{i+2}\frac{d}{dx}(r^{-i-1}S_i)$$

Similarly $\frac{y}{a}S_i$, $\frac{z}{a}S_i$ may be expressed as $B_{i-1}+B_{i+1}$ and $C_{i-1}+C_{i+1}$, where the B's and C's only differ from the A's in having y, z written for x.

We have now to form the auxiliary functions Ψ_{i-2}, Φ_i corresponding to A_{i-1}, B_{i-1}, C_{i-1} and Ψ_i, Φ_{i+2} corresponding to A_{i+1}, B_{i+1}, C_{i+1}.

Then by the formulæ (6)

$$(2i+1)\Psi_{i-2} = \left(\frac{d^2}{dx^2}+\frac{d^2}{dy^2}+\frac{d^2}{dz^2}\right)r^iS_i = 0$$

$$\frac{2i+1}{r^{2i+1}}\Phi_i = \frac{d}{dx}\left[r^{-2i+1}\frac{d}{dx}(r^iS_i)\right] + \frac{d}{dy}\left[\quad\right] + \frac{d}{dz}\left[\quad\right] = -\frac{i(2i-1)}{r^{2i+1}}r^iS_i$$

$$-(2i+1)\Psi_i = \frac{d}{dx}\left[r^{2i+3}\frac{d}{dx}(r^{-i-1}S_i)\right] + \frac{d}{dy}\left[\quad\right] + \frac{d}{dz}\left[\quad\right] = -(i+1)(2i+3)r^iS_i$$

$$-\frac{2i+1}{r^{2i+5}}\Phi_{i+2} = \left(\frac{d^2}{dx^2}+\frac{d^2}{dy^2}+\frac{d^2}{dz^2}\right)r^{-i-1}S_i = 0$$

Thus

$$\Psi_{i-2} = 0,\ \Phi_i = -\frac{i(2i-1)}{2i+1}r^iS_i,\ \Psi_i = \frac{(i+1)(2i+3)}{2i+1}r^iS_i,\ \Phi_{i+2} = 0$$

Then by (5) we form α corresponding to A_{i-1}, B_{i-1}, C_{i-1}, and also to

A_{i+1}, B_{i+1}, C_{i+1}, and add them together. The final result is that a normal traction S_i gives

$$\alpha' = \frac{1}{va^i}\left[\left\{\frac{i(i+2)}{2(i-1)[2(i+1)^2+1]}a^2 - \frac{(i+1)(2i+3)}{2(2i+1)[2(i+1)^2+1]}r^2\right\}\frac{d}{dx}(r^i S_i) - \frac{i}{(2i+1)[2(i+1)^2+1]}r^{2i+3}\frac{d}{dx}(r^{-i-1}S_i)\right] \quad \text{........ (7)}$$

and symmetrical expressions for β' and γ'.

α', β', γ' are here written for α, β, γ to show that this is only a partial solution, and v is written for n to show that it corresponds to the viscous problem. If we now put $S_i = -gw\sigma_i$, we get the state of flow of the fluid due to the transmitted pressure of the deficiencies and excesses of matter below and above the true spherical surface. This constitutes the solution as far as it depends on (iii).

There remain the parts dependent on (i) and (ii), which may for the present be classified together; and for this part Sir W. Thomson's solution is directly applicable. The state of internal strain of an elastic sphere, subject to no surface action, but under the influence of a bodily force of which the potential is W_i, may be at once adapted to give the state of flow of a viscous sphere under like conditions. The solution is

$$\alpha'' = \frac{1}{v}\left[\left\{\frac{i(i+2)}{2(i-1)[2(i+1)^2+1]}a^2 - \frac{(i+1)(2i+3)}{2(2i+1)[2(i+1)^2+1]}r^2\right\}\frac{dW_i}{dx} - \frac{i}{(2i+1)[2(i+1)^2+1]}r^{2i+3}\frac{d}{dx}(r^{-2i-1}W_i)\right] \quad \text{............(8)*}$$

with symmetrical expressions for β'' and γ''.

I will first consider (ii); *i.e.*, the matter of the earth is now supposed to possess the power of gravitation.

The gravitation potential of the spheroid $r = a + \sigma_i$ (taking only a typical term of σ) at a point in the interior, estimated per unit volume, is

$$\frac{gw}{2a}(3a^2 - r^2) + \frac{3gw}{2i+1}\left(\frac{r}{a}\right)^i \sigma_i$$

according to the usual formula in the theory of the potential.

The first term, being symmetrical round the centre of the sphere, can clearly cause no flow in the incompressible viscous sphere. We are therefore left with $\frac{3gw}{2i+1}\left(\frac{r}{a}\right)^i \sigma_i$.

* *Natural Philosophy*, § 834, equation (8) when m is infinite compared with n, and $i-1$ written for i, and v replaces n.

Now if $\frac{3gw}{2i+1}\left(\frac{r}{a}\right)^i \sigma_i$ be substituted for W_i in (8), and if the resulting expression be compared with (7) when $-gw\sigma_i$ is written for S_i, it will be seen that $-\alpha'' = \frac{3}{2i+1}\alpha'$.

Thus $$\alpha' + \alpha'' = \alpha''\left(1 - \frac{2i+1}{3}\right)^* = -\tfrac{2}{3}(i-1)\,\alpha''$$

And if $$V_i = \frac{3gw}{2i+1}\left(\frac{r}{a}\right)^i \sigma_i$$

$$\alpha' + \alpha'' = -\frac{1}{v}\left[\left\{\frac{i(i+2)}{2(i-1)[2(i+1)^2+1]}a^2 - \frac{(i+1)(2i+3)}{2(2i+1)[2(i+1)^2+1]}r^2\right\}\frac{d}{dx}\{\tfrac{2}{3}(i-1)V_i\}\right.$$
$$\left. - \frac{i}{(2i+1)[2(i+1)^2+1]}r^{2i+3}\frac{d}{dx}\{r^{-2i-1}\tfrac{2}{3}(i-1)V_i\}\right]\text{........(9)}$$

with symmetrical expressions for $\beta' + \beta''$ and $\gamma' + \gamma''$.

Equation (9) then embodies the solution as far as it depends on (ii) and (iii). And since (9) is the same as (8) when $-\frac{2}{3}(i-1)V_i$ is written for W_i, we may include all the effects of mutual gravitation in producing a state of flow in the viscous sphere, by adopting Thomson's solution (8), and taking instead of the true potential of the layer of matter σ_i, $-\frac{2}{3}(i-1)$ times that potential, and by adding to it the external disturbing potential.

We have now learnt how to include the surface action in the potential; and if W_i be the potential of the external disturbing influence, the *effective* potential per unit volume at a point within the sphere, now free of surface action and of mutual gravitation, is $W_i - \frac{2gw(i-1)}{2i+1}\left(\frac{r}{a}\right)^i \sigma_i = r^i T_i$ suppose.

The complete solution of our problem is then found by writing $r^i T_i$ in place of W_i in Thomson's solution (8)†.

In order however to apply the solution to the case of the earth, it will be convenient to use polar coordinates. For this purpose, write $wr^i S_i$ for W_i, and let r be the radius vector; θ the colatitude; ϕ the longitude. Let ρ, ϖ, ν be the velocities radially, and along and perpendicular to the meridian respectively. Then the expressions for ρ, ϖ, ν will be precisely the same as those for α, β, γ in (8), save that for $\frac{d}{dx}$ we must put $\frac{d}{dr}$; for $\frac{d}{dy}, \frac{d}{r\sin\theta\, d\phi}$; and for $\frac{d}{dz}, \frac{d}{r\,d\theta}$.

* The case of § 815 in Thomson and Tait's *Natural Philosophy* is a special case of this.

† The introduction of the effects of gravitation may be also carried out synthetically, as is done by Sir W. Thomson (§ 840, *Natural Philosophy*); but the effects of the lagging of the tide-wave render this method somewhat artificial, and I prefer to exhibit the proof in the manner here given. Conversely, the elastic problem may be solved as in the text.

Then after some reductions we have

$$\left.\begin{aligned}\rho &= \frac{i^2(i+2)a^2 - i(i^2-1)r^2}{2(i-1)[2(i+1)^2+1]v} r^{i-1}\mathrm{T}_i \\ \varpi &= \frac{i(i+2)a^2-(i-1)(i+3)r^2}{2(i-1)[2(i+1)^2+1]v} r^{i-1}\frac{d\mathrm{T}_i}{d\theta} \\ \nu &= \frac{i(i+2)a^2-(i-1)(i+3)r^2}{2(i-1)[2(i+1)^2+1]v} \frac{r^{i-1}}{\sin\theta}\frac{d\mathrm{T}_i}{d\phi}\end{aligned}\right\} \quad\ldots\ldots\ldots\ldots(10)^*$$

where $\mathrm{T}_i = w\left(\mathrm{S}_i - 2g\dfrac{i-1}{2i+1}\dfrac{\sigma_i}{a^i}\right)$.

These equations for ρ, ϖ, ν give us the state of internal flow corresponding to the external disturbing potential $r^i\mathrm{S}_i$, including the effects of the mutual gravitation of the matter constituting the spheroid.

3. *The form of the free surface at any time.*

If ρ' be the surface value of ρ, then

$$\rho' = \frac{i(2i+1)}{2(i-1)[2(i+1)^2+1]}\frac{a^{i+1}}{v}\mathrm{T}_i$$

Hence after a short interval of time δt, the equation to the bounding surface of the spheroid becomes $r = a + \sigma_i + \rho'\delta t$; but during this same interval, σ_i has become $\dfrac{d\sigma_i}{dt}\delta t$, whence

$$\frac{d\sigma_i}{dt} = \rho' = \frac{i(2i+1)}{2(i-1)[2(i+1)^2+1]}\frac{wa^{i+1}}{v}\mathrm{S}_i - \frac{i}{2(i+1)^2+1}\frac{gwa}{v}\sigma_i$$

or

$$\frac{d\sigma_i}{dt} + \frac{i}{2(i+1)^2+1}\frac{gwa}{v}\sigma_i = \frac{i(2i+1)}{2(i-1)[2(i+1)^2+1]}\frac{wa^{i+1}}{v}\mathrm{S}_i \quad\ldots.(11)$$

This differential equation gives the manner in which the surface changes, under the influence of the external potential $r^i\mathrm{S}_i$.

If S_i be not a function of the time, and if s_i be the value of σ_i when $t=0$,

$$\sigma_i = \frac{2i+1}{2(i-1)}\frac{a^i\mathrm{S}_i}{g}\left[1 - \exp\left(\frac{-gwait}{[2(i+1)^2+1]v}\right)\right] + s_i\exp\left(\frac{-gwait}{[2(i+1)^2+1]v}\right) \quad (12)\dagger$$

When t is infinite

$$\sigma_i = \frac{2i+1}{2(i-1)}\frac{a^i\mathrm{S}_i}{g} \quad\ldots\ldots\ldots\ldots\ldots\ldots\ldots\ldots(13)$$

and there is no further state of flow, for the fluid has assumed the form

* There seems to be a misprint as to the signs of the 𝕲's in the second and third of equations (13) of § 834 of the *Natural Philosophy* (1867). When this is corrected μ and ν admit of reduction to tolerably simple forms. It appears to me also that the differentiation of ρ in (15) is incorrect; and this falsifies the argument in three following lines. The correction is not, however, in any way important.

† I write "exp" for "e to the power of."

which it would have done if it had not been viscous. This result is of course in accordance with the equilibrium theory of tides.

If S_i be zero, the equation shows how the inequalities on the surface of a viscous globe would gradually subside under the influence of simple gravity. We see how much more slowly the change takes place if i be large; that is to say, inequalities of small extent die out much more slowly than wide-spread inequalities. Is it not possible that this solution may throw some light on the laws of geological subsidence and upheaval?

4. *Digression on the adjustments of the earth to a form of equilibrium.*

In a former paper I had occasion to refer to some points touching the precession of a viscous spheroid, and to consider its rate of adjustment to a new form of equilibrium, when its axis of rotation had come to depart from its axis of symmetry*. I propose then to discuss the subject shortly, and to establish the law which was there assumed.

Suppose that the earth is rotating with an angular velocity ω about the axis of z, but that at the instant at which we commence our consideration the axis of symmetry is inclined to the axis of z at an angle α in the plane of xy, and that at that instant the equation to the free surface is

$$r = a\left\{1 + \tfrac{5}{4}m\left(\tfrac{1}{3} - [\cos\alpha\cos\theta + \sin\alpha\sin\theta\cos\phi]^2\right)\right\}$$

where m is the ratio of centrifugal force at the equator to pure gravity, and therefore equal to $\dfrac{\omega^2 a}{g}$.

Then putting $i = 2$ in (12), and dropping the suffixes of S, s, σ,

$$s = \tfrac{5}{4}ma\left(\tfrac{1}{3} - [\quad]^2\right)$$

We may conceive the earth to be at rest, if we apply a potential

$$wr^2 S = \tfrac{1}{2}\omega^2 wr^2\left(\tfrac{1}{3} - \cos^2\theta\right)$$

so that

$$S = \tfrac{1}{2}\omega^2\left(\tfrac{1}{3} - \cos^2\theta\right)$$

By (12) we have

$$\sigma = \frac{5a^2 S}{2g}\left[1 - \exp\left(-\frac{2wgat}{19v}\right)\right] + s\exp\left(-\frac{2wgat}{19v}\right)$$

Then, substituting for S and s, and putting $\kappa = \dfrac{2wga}{19v}$

$$\sigma = \tfrac{5}{4}ma\left\{\left(\tfrac{1}{3} - \cos^2\theta\right)[1 - \exp(-\kappa t)] + \left(\tfrac{1}{3} - [\cos\alpha\cos\theta + \sin\alpha\sin\theta\cos\phi]^2\right)\exp(-\kappa t)\right\}$$

* "On the Influence of Geological Changes on the Earth's Axis of Rotation," *Phil. Trans.*, Vol. 167, Part I., sec. 5. [To be included in Vol. III. of these Collected Papers.]

Now
$$[1-\exp(-\kappa t)]\cos^2\theta+\exp(-\kappa t)(\cos\alpha\cos\theta+\sin\alpha\sin\theta\cos\phi)^2$$
$$=\cos^2\theta\,[1-\sin^2\alpha\exp(-\kappa t)]+\sin^2\alpha\sin^2\theta\cos^2\phi\exp(-\kappa t)$$
$$+2\sin\alpha\cos\alpha\sin\theta\cos\theta\cos\phi\exp(-\kappa t)$$

Therefore the Cartesian equation to the spheroid at the time t is

$$\frac{x^2+y^2+z^2}{1+\frac{5}{6}m}=a^2-\tfrac{5}{2}m\,\{z^2(1-\sin^2\alpha\exp(-\kappa t))$$
$$+x^2\sin^2\alpha\exp(-\kappa t)+2xz\sin\alpha\cos\alpha\exp(-\kappa t)\}$$

or
$$x^2\{1+\tfrac{5}{2}m\sin^2\alpha\exp(-\kappa t)\}+y^2+z^2\{1+\tfrac{5}{2}m(1-\sin^2\alpha\exp(-\kappa t))\}$$
$$+5m\sin\alpha\cos\alpha\,xz\exp(-\kappa t)=a^2(1+\tfrac{5}{6}m)$$

Let α' be the inclination of the principal axis at this time to the axis of z, then

$$\tan 2\alpha'=\frac{\sin 2\alpha\exp(-\kappa t)}{1-2\sin^2\alpha\exp(-\kappa t)}$$

If α be small, as it was in the case I considered in my former paper, then

$$\alpha'=\alpha\exp(-\kappa t)\text{ and }\frac{d\alpha'}{dt}=-\kappa\alpha'$$

Therefore the velocity of approach of the principal axis to the axis of rotation varies as the angle between them, which is the law assumed.

Also $\kappa=\dfrac{2wga}{19v}$, so that κ (the ν of my former paper) varies inversely as the coefficient of viscosity,—as was also assumed.

5. *Bodily tides in a viscous earth*.*

The only case of interest in which S_i of equation (11) is a function of the time, is where it is a surface harmonic of the second order, and is periodic in time; for this will give the solution of the tidal problem. Since, moreover, we are only interested in the case where the motion has attained a permanently periodic character, the exponential terms in the solution of (11) may be set aside.

Let $S_2=S\cos(vt+\eta)$, and in accordance with Thomson's notation†, let $\dfrac{2g}{5a}=\mathfrak{g}$, and $\dfrac{19v}{5wa^2}=\mathfrak{r}$; and therefore $\dfrac{2gwa}{19v}=\dfrac{\mathfrak{g}}{\mathfrak{r}}$.

* In certain cases the forces do not form a rigorously equilibrating system, but there is a very small couple tending to turn the earth. The effects of this unbalanced couple, which varies as the square of $\dfrac{3}{2}\dfrac{m}{c^3}$, will be considered in a succeeding paper [Paper 3] on the "Precession of a Viscous Spheroid." (Read before the Royal Society, December 19th, 1878.)

† *Natural Philosophy*, § 840, eq. (27).

Then putting $i=2$ in (11), and omitting the suffix of σ for brevity, we have

$$\frac{d\sigma}{dt}+\frac{\mathfrak{g}}{\mathfrak{r}}\sigma=\frac{a}{\mathfrak{r}}\mathrm{S}\cos(vt+\eta)\text{.......................}(14)$$

It is evident that σ must be of the form $\mathrm{A}\cos(vt+\mathrm{B})$, and therefore

$$\mathrm{A}\{-v\mathfrak{r}\sin(vt+\mathrm{B})+\mathfrak{g}\cos(vt+\mathrm{B})\}=a\mathrm{S}\cos(vt+\eta)$$

or if we put $\tan\epsilon=\dfrac{v\mathfrak{r}}{\mathfrak{g}}$,

$$\mathrm{A}\mathfrak{g}\sec\epsilon\cos(vt+\mathrm{B}+\epsilon)=a\mathrm{S}\cos(vt+\eta)$$

Hence $\mathrm{A}=\dfrac{a}{\mathfrak{g}}\mathrm{S}\cos\epsilon$, and $\mathrm{B}=\eta-\epsilon$.

Therefore the solution of (14) is

$$\sigma=\frac{a}{\mathfrak{g}}\mathrm{S}\cos\epsilon\cos(vt+\eta-\epsilon)\text{.......................}(15)$$

where $\tan\epsilon=\dfrac{v\mathfrak{r}}{\mathfrak{g}}=\dfrac{19\upsilon v}{2ga w}$.

But if the globe were a perfect fluid, and if the equilibrium theory of tides were true, we should have by (13),

$$\sigma=\frac{5a}{2g}\cdot a\mathrm{S}\cos(vt+\eta)=\frac{\mathrm{S}a}{\mathfrak{g}}\cos(vt+\eta)$$

Thus we see that the tides of the viscous sphere are to the equilibrium tides of a fluid sphere as $\cos\epsilon:1$, and that there is a retardation in time of $\dfrac{\epsilon}{v}$.

A parallel investigation will be applicable to the general case where the disturbing potential is $wr^i\,\mathrm{S}_i\cos(vt+\eta)$; and the same solution will be found to hold save that we now have $\tan\epsilon=\dfrac{2(i+1)^2+1}{i}\cdot\dfrac{\upsilon v}{ga w}$, and that in place of $\mathfrak{g}$ we have $\dfrac{2(i-1)g}{(2i+1)a}$.

6. *Diminution of ocean tides on equilibrium theory.*

Suppose now that there is a shallow ocean on the viscous nucleus, and let us find the effects on the ocean tides of the motion of the nucleus according to the equilibrium theory, neglecting the gravitation of the water.

The potential at a point outside the nucleus is

$$g\frac{a^2}{r}+\tfrac{3}{5}g\left(\frac{a}{r}\right)^2\sigma+r^2\mathrm{S}\cos(vt+\eta)\text{.....................}(16)$$

and if this be put equal to a constant, we get the form which the ocean must assume. Let $r = a + u$ be the equation to the surface of the ocean. Then substituting for r in the potential, and neglecting u in the small terms, and equating the whole to a constant, we find

$$-gu + \tfrac{3}{5}g\sigma + a^2 S \cos(vt + \eta) = 0$$

or

$$u = \tfrac{3}{5}\sigma + \frac{a^2}{g} S \cos(vt + \eta)$$

But the rise and fall of the tide relative to the nucleus is given by $u - \sigma$, and

$$\begin{aligned} u - \sigma &= \frac{a^2 S}{g} \cos(vt + \eta) - \tfrac{2}{5}\sigma \\ &= \tfrac{2}{5}\frac{aS}{\mathfrak{g}}[\cos(vt + \eta) - \cos\epsilon \cos(vt + \eta - \epsilon)] \\ &= -\tfrac{2}{5}\frac{aS}{\mathfrak{g}} \sin\epsilon \sin(vt + \eta - \epsilon) \quad\text{.........................(17)} \end{aligned}$$

Now if the nucleus had been rigid, the rise and fall would have been given by

$$\tfrac{2}{5}\frac{aS}{\mathfrak{g}} \cos(vt + \eta) = H \cos(vt + \eta) \text{ suppose}$$

Therefore

$$u - \sigma = -H \sin\epsilon \sin(vt + \eta - \epsilon) \quad\text{.................(18)}$$

Hence the apparent tides on the yielding nucleus are equal to the tides on a rigid nucleus reduced in the proportion $\sin\epsilon : 1$; and since

$$-\sin(vt + \eta - \epsilon) = \cos(vt + \eta + \tfrac{1}{2}\pi - \epsilon)$$

they are retarded by $\frac{1}{v}(\epsilon - \frac{1}{2}\pi)$. As ϵ is necessarily less than $\frac{1}{2}\pi$, this is equivalent to an acceleration of the time of high water equal to $\frac{1}{v}(\frac{1}{2}\pi - \epsilon)$.

It is, however, worthy of notice that this is only an acceleration of phase relatively to the nucleus, and there is an absolute retardation of phase equal to arc-tan $\dfrac{3 \sin\epsilon \cos\epsilon}{3 + 2\cos^2\epsilon}$.

7. *Semidiurnal and fortnightly tides.*

Let the axis of z be the earth's axis of rotation, and let the plane of xz be fixed in the earth; let c be the moon's distance, and m its mass.

Suppose the moon to move in the equator with an angular velocity ω relatively to the earth, and let the moon's terrestrial longitude, measured from the plane of xz, at the time t be ωt.

Then at the time t, the gravitation potential of the tide-generating force, estimated per unit volume of the earth's mass, is

$$-\tfrac{3}{2}\frac{m}{c^3} wr^2 \{\tfrac{1}{3} - \sin^2\theta \cos^2(\phi - \omega t)\}$$

which is equal to

$$\tfrac{3}{4}\frac{m}{c^3}wr^2\left(\tfrac{1}{3}-\cos^2\theta\right)+\tfrac{3}{4}\frac{m}{c^3}wr^2\left\{\sin^2\theta\cos 2\phi\cos 2\omega t+\sin^2\theta\sin 2\phi\sin 2\omega t\right\}$$

The first term of this expression is independent of the time, and therefore produces an effect on the viscous earth, which will have died out when the motion has become steady; its only effect is slightly to increase the ellipticity of the earth's surface.

The two latter terms give rise to two tides, in one of which (according to previous notation)

$$S\cos(vt+\eta)=\tfrac{3}{4}\frac{m}{c^3}\sin^2\theta\cos 2\phi\cos 2\omega t$$

and in the second of which

$$S\cos(vt+\eta)=-\tfrac{3}{4}\frac{m}{c^3}\sin^2\theta\sin 2\phi\cos\left(2\omega t+\tfrac{1}{2}\pi\right)$$

Now ϵ, which depends on the frequency of the tide-generating potential, will clearly be the same for both these tides; and therefore they will each be equal to the corresponding tides of a fluid spheroid, reduced by the same amount and subject to the same retardation. They may therefore be recompounded into a single tide; and since v will here be equal to 2ω, it follows that the retardation of the bodily semidiurnal tide is $\frac{\epsilon}{2\omega}$, where $\tan\epsilon=\frac{2\omega\mathfrak{r}}{\mathfrak{g}}=\frac{19v\omega}{gaw}$. Also the height of the tide is less than the corresponding equilibrium tide of a fluid spheroid in the proportion of $\cos\epsilon$ to unity.

Similarly by section (6) the height of the ocean tide on the yielding nucleus is given by the corresponding tide on a rigid nucleus multiplied by $\sin\epsilon$, and there is an acceleration of relative high water equal to $\frac{\pi}{4\omega}-\frac{\epsilon}{2\omega}$.

The case of the fortnightly tide is somewhat simpler.

If Ω be the moon's orbital angular velocity, and I the inclination of the plane of the orbit to the earth's equator, then the part of the tide-generating potential, on which the fortnightly tide depends, is

$$\tfrac{9}{8}\frac{m}{c^3}wr^2\sin^2 I\left(\tfrac{1}{3}-\cos^2\theta\right)\cos 2\Omega t$$

and we see at once by sections (5) and (6) that $\tan\epsilon=\frac{19v\Omega}{gaw}$. The bodily tide is the tide of a fluid spheroid multiplied by $\cos\epsilon$; the reduction of ocean tide is given by $\sin\epsilon$; and there is a time-acceleration of relative high water of $\frac{\pi}{4\Omega}-\frac{\epsilon}{2\Omega}$ or $\frac{1}{2}-\frac{\epsilon}{\pi}$ of a week.

In order to make the meaning of the previous analytical results clearer, I have formed the following numerical tables, to show the effects of this hypothesis on the semidiurnal and fortnightly tides. The coefficient of viscosity is usually expressed in gravitation units of force so that the formula for ϵ becomes, $\tan\epsilon = \frac{19\,\upsilon\omega}{wa}$. In the tables υ is expressed in the centimetre-gramme-second system, and in gravitation units of force; a is taken as $6{\cdot}37 \times 10^8$, and w as 5·5, and the angular velocity ω of the moon relatively to the earth as ·00007025 radians per second.

With these data I find $\upsilon = 10^{12} \times 2{\cdot}625 \tan\epsilon$. As a standard of comparison with the coefficients of viscosity given in the tables, I may mention that, according to some rough experiments of my own, the viscosity of British pitch at near the freezing temperature (34° Fahr.), when it is hard and brittle, is about $10^8 \times 1{\cdot}3$ when measured in the same units.

Lunar Semidiurnal Tide

Coefficient of viscosity $\times 10^{-10}$ $(\upsilon \times 10^{-10})$	Retardation of bodily tide $\left(\frac{\epsilon}{2\omega}\right)$		Height of bodily tide is tide of fluid spheroid multiplied by $(\cos\epsilon)$	Height of ocean tide is tide on rigid nucleus multiplied by $(\sin\epsilon)$	High tide relatively to viscous nucleus accelerated by $\frac{1}{\omega}\left(\frac{\pi}{4} - \frac{\epsilon}{2}\right)$	
	Hrs.	min.			Hrs.	min.
Fluid 0	0	0	1·000	·000	3	6
46	0	21	·985	·174	2	46
96	0	41	·940	·342	2	25
152	1	2	·866	·500	2	4
220	1	23	·766	·643	1	44
313	1	44	·643	·766	1	23
455	2	4	·500	·866	1	2
721	2	25	·342	·940	0	41
1,488	2	46	·174	·985	0	21
Rigid ∞	3	6	·000	1·000	0	0

Fortnightly Tide

	Days	hrs.			Days	hrs.
Fluid 0	0	0	1·000	·000	3	10
1,200	0	9	·985	·174	3	1
2,500	0	18	·940	·342	2	16
4,000	1	3	·866	·500	2	6
5,800	1	12	·766	·643	1	21
8,300	1	21	·643	·766	1	12
12,000	2	6	·500	·866	1	3
19,000	2	16	·342	·940	0	18
39,300	3	1	·174	·985	0	9
Rigid ∞	3	10	·000	1·000	0	0

I now pass on to a case which is intermediate between the hypothesis of Sir W. Thomson and that just treated.

8. *The tides of an elastico-viscous spheroid.*

The term elastico-viscous is used to denote that the stresses requisite to maintain the body in a given strained configuration decrease the longer the body is thus constrained, and this is undoubtedly the case with many solids. In the particular case which is here treated, it is assumed that the stresses diminish in geometrical progression, as the time increases in arithmetical progression. If, for example, a cubical block of the substance be strained to a given amount by a shearing stress T, and maintained in that position, then after a time t, the shearing stress, is $\mathrm{T}\exp\left(-\frac{t}{\mathfrak{t}}\right)$. The time $\mathfrak{t}$ measures the rate at which the stress falls off, and is called (I believe by Professor Maxwell) "the modulus of the time of relaxation of rigidity"; it is the time in which the initial stress has been reduced to e^{-1} or ·3679 of its initial value. I do not suppose, however, that any solid conforms exactly to this law; but I conceive that it is often useful in physical problems to discuss mathematically an ideal case, which presents a sufficiently marked likeness to the reality, where we are unable to determine exactly what that reality may be.

Mr J. G. Butcher has found the equations of motion of such an ideal substance from the consideration that the elasticity of groups of molecules is continually breaking down, and that the groups rearrange themselves afterwards*. These considerations lead him to the following results for the stresses across rectangular planes at any point in the interior, viz. (with the notation of § 1):

$$\mathrm{P} = (m-n)\,\delta + 2n\left(\frac{1}{\mathfrak{t}} + \frac{d}{dt}\right)^{-1}\left(\frac{d\alpha}{dx} + \frac{1}{3\mathfrak{t}}\,\alpha\right), \qquad \mathrm{S} = n\left(\frac{1}{\mathfrak{t}} + \frac{d}{dt}\right)^{-1}\left(\frac{d\beta}{dz} + \frac{d\gamma}{dy}\right)$$

and similar expressions for Q, R, T, U; where $m - \frac{1}{3}n$ is the coefficient of dilatation, n that of rigidity, δ the dilatation, and α, β, γ, the components of flow.

These expressions are clearly in accordance with the above definition of elastico-viscosity, for $\frac{d\mathrm{S}}{dt} + \frac{\mathrm{S}}{\mathfrak{t}} = n\left(\frac{d\beta}{dz} + \frac{d\gamma}{dy}\right)$.

If the expressions for P, S, &c., be substituted in the equations of equilibrium of the elementary parallelopiped, it is found by aid of the equation of continuity $\frac{d\delta}{dt} = \frac{d\alpha}{dx} + \frac{d\beta}{dy} + \frac{d\gamma}{dz}$, that when inertia is neglected

$$\left(\frac{1}{\mathfrak{t}} + \frac{d}{dt}\right)^{-1}\left\{\left[(m - \tfrac{1}{3}n)\frac{1}{\mathfrak{t}} + m\frac{d}{dt}\right]\frac{d\delta}{dx} + n\nabla^2\alpha\right\} + \mathrm{X} = 0$$

and two similar equations.

* *Proc. Lond. Math. Soc.*, Dec. 14, 1876, pp. 107–9. It seems to me that the hypothesis ought to represent the elastico-viscosity of ice very closely.

By the same reasoning as in § 1, we may put, $\delta = \frac{-p}{m - \frac{1}{3}n}$, and the equations become

$$-\left(\frac{1}{\mathfrak{t}} + \frac{d}{dt}\right)^{-1} \left\{\left[\frac{1}{\mathfrak{t}} + \frac{m}{m - \frac{1}{3}n}\frac{d}{dt}\right]\frac{dp}{dx} - n\nabla^2\alpha\right\} + \mathrm{X} = 0$$

Then supposing the substance to be incompressible, so that m is infinitely large compared to n, and therefore $m \div m - \frac{1}{3}n$ is unity, the equations become

$$-\frac{dp}{dx} + n\left(\frac{1}{\mathfrak{t}} + \frac{d}{dt}\right)^{-1}\nabla^2\alpha + \mathrm{X} = 0$$

and two similar equations.

Now these equations have exactly the same form as those for the motion of a viscous fluid, save that the coefficient of viscosity υ is replaced by $n\left(\frac{1}{\mathfrak{t}} + \frac{d}{dt}\right)^{-1}$. We may therefore at once pass to the differential equation (11) which gives the form of the surface of the spheroid at any time.

Substituting, therefore $\frac{1}{n}\left(\frac{1}{\mathfrak{t}} + \frac{d}{dt}\right)$ in (11) for $\frac{1}{\upsilon}$, we get

$$\left[1 + \frac{i}{2(i+1)^2+1}\frac{gwa}{n}\right]\frac{d\sigma_i}{dt} + \frac{i}{2(i+1)^2+1}\frac{gwa}{n\mathfrak{t}}\sigma_i$$
$$= \frac{i(2i+1)}{2(i-1)[2(i+1)^2+1]}\frac{wa^{i+1}}{n}\left(\frac{1}{\mathfrak{t}} + \frac{d}{dt}\right)\mathrm{S}_i$$

This equation admits of solution just in the same way that equation (11) was solved; but I shall confine myself to the case of the tidal problem, where $i = 2$ and $\mathrm{S}_2 = \mathrm{S}\cos(vt + \eta)$. In this special case the equation becomes

$$\left(1 + \frac{2gwa}{19n}\right)\frac{d\sigma}{dt} + \frac{2gwa}{19n\mathfrak{t}}\sigma = \frac{5wa^3}{19n}\left[\frac{1}{\mathfrak{t}}\cos(vt+\eta) - v\sin(vt+\eta)\right]\mathrm{S}$$

And if we put $\frac{19n}{2gwa} + 1 = \frac{1}{k}$, $\tan\psi = v\mathfrak{t}$, and $\mathfrak{g} = \frac{2g}{5a}$, this may be written

$$\frac{d\sigma}{dt} + \frac{k}{\mathfrak{t}}\sigma = \frac{vak}{\mathfrak{g}\sin\psi}\mathrm{S}\cos(vt + \eta + \psi)$$

In the solution appropriate to the tidal problem, we may omit the exponential term, and assume $\sigma = \mathrm{A}\cos(vt + \mathrm{B})$. Then if we put $\tan\chi = \frac{v\mathfrak{t}}{k}$

$$\frac{d\sigma}{dt} + \frac{k}{\mathfrak{t}}\sigma = \frac{\mathrm{A}v}{\sin\chi}\cos(vt + \mathrm{B} + \chi)$$

Whence it follows that $\mathrm{B} = \eta + \psi - \chi$, and

$$\mathrm{A} = \frac{a}{\mathfrak{g}}k\frac{\sin\chi}{\sin\psi} = \frac{a}{\mathfrak{g}}\frac{\cos\chi}{\cos\psi}$$

so that
$$\sigma = \frac{a\mathrm{S}}{\mathfrak{g}}\frac{\cos\chi}{\cos\psi}\cos(vt + \eta + \psi - \chi)$$

Hence the bodily tide of the elastico-viscous spheroid is equal to the equilibrium tide of a fluid spheroid multiplied by $\dfrac{\cos\chi}{\cos\psi}$, and high tide is retarded by $\chi - \psi \div v$.

The formula for $\tan\chi$ may be expressed in a somewhat more convenient form; we have $\tan\psi = v\mathfrak{t}$, and therefore $\tan\chi = \tan\psi + \dfrac{19nv\mathfrak{t}}{2gwa}$.

But $n\mathfrak{t}$ is the coefficient of viscosity, and in treating the tides of the purely viscous spheroid we put $\tan\epsilon = \dfrac{19v}{2gwa} \times$ coefficient of viscosity; therefore adopting the same notation here, we have $\tan\chi = \tan\psi + \tan\epsilon$.

If the modulus of relaxation $\mathfrak{t}$ be zero, whilst the coefficient of rigidity n becomes infinite, but $n\mathfrak{t}$ finite, the substance is purely viscous, and we have $\psi = 0$ and $\chi = \epsilon$, so that the solution reduces to the case already considered. If $\mathfrak{t}$ be infinite, the substance is purely elastic, and we have $\psi = \frac{1}{2}\pi$, $\chi = \frac{1}{2}\pi$ and since $\dfrac{\cos\chi}{\cos\psi} = k\dfrac{\sin\chi}{\sin\psi}$, therefore

$$\sigma = \frac{ak}{\mathfrak{g}}\,\mathrm{S}\cos(vt+\eta)$$

But according to Thomson's notation* $\dfrac{10n}{2gwa} = \dfrac{\mathfrak{r}}{\mathfrak{g}}$, so that $\sigma = \dfrac{a}{\mathfrak{r}+\mathfrak{g}}\,\mathrm{S}\cos(vt+\eta)$, which is the solution of Thomson's problem of the purely elastic spheroid.

The present solution embraces, therefore, both the case considered by him, and that of the viscous spheroid.

9. *Ocean tides on an elastico-viscous nucleus.*

If $r = a + u$ be the equation to the ocean spheroid, we have, as in sec. (6), that the height of tide relatively to the nucleus is given by

$$u - \sigma = \frac{a^2}{g}\,\mathrm{S}\cos(vt+\eta) - \tfrac{3}{5}\sigma$$

and substituting the present value of σ,

$$u - \sigma = \tfrac{2}{5}\frac{a}{\mathfrak{g}}\,\mathrm{S}\left[\cos(vt+\eta) - \frac{\cos\chi}{\cos\psi}\cos(vt+\eta+\psi-\chi)\right]$$

$$= -\tfrac{2}{5}\frac{a}{\mathfrak{g}}\,\mathrm{S}\,\frac{\sin(\chi-\psi)}{\cos\psi}\sin(vt+\eta-\chi)$$

If the nucleus had been rigid the rise and fall would have been given by

* *Natural Philosophy*, § 840.

$H \cos(vt + \eta)$, where $H = \frac{2}{5}\frac{a}{\mathfrak{g}}S$; therefore on the yielding nucleus it is given by

$$u - \sigma = -H \frac{\sin(\chi - \psi)}{\cos \psi} \sin(vt + \eta - \chi)$$

$$= -H \cos \chi (\tan \chi - \tan \psi) \sin(vt + \eta - \chi)$$

$$= -H \cos \chi \tan \epsilon \sin(vt + \eta - \chi)$$

Hence the apparent tides on the yielding nucleus are equal to the corresponding tides on a rigid nucleus reduced in the proportion of $\cos \chi \tan \epsilon$ to unity, and there is an acceleration of the time of high water equal to $(\frac{1}{2}\pi - \chi)/v$.

As these analytical results present no clear meaning to the mind, I have compiled the following tables. In these tables I have taken the two cases considered by Sir W. Thomson, where the spheroid has the rigidity of glass, and that of iron, and have worked out the results for various times of relaxation of rigidity, for the semidiurnal and fortnightly tides. The last line in each division of each table is Thomson's result.

SPHEROID with Rigidity of Glass ($2{\cdot}44 \times 10^8$).

Modulus of relaxation of rigidity (t)	Coefficient of viscosity ($nt \times 10^{-10}$)	Ocean tide is tide on rigid nucleus multiplied by ($\cos \chi \tan \epsilon$)	High tide relatively to nucleus is accelerated by $(\frac{1}{2}\pi - \chi)\frac{1}{v}$
Lunar Semidiurnal Tide			
Hrs.			Hrs. min.
Fluid 0	0	·000	3 6
1	88	·256	1 44
2	176	·342	1 3
3	264	·370	0 45
4	351	·382	0 34
5	439	·388	0 28
Elastic ∞	∞	·398	0 0
Fortnightly Tide			
Days hrs.			Days hrs.
Fluid 0 0	0	·000	3 10
0 6	500	·099	2 21
0 12	1,100	·181	2 9
1 0	2,100	·285	1 16
2 0	4,200	·357	1 0
3 0	6,300	·379	0 16
Elastic ∞	∞	·398	0 0

SPHEROID with Rigidity of Iron ($7{\cdot}8 \times 10^8$).

Lunar Semidiurnal Tide

	Modulus of relaxation (Hrs. min.)	Viscosity	Reduction of ocean tide	Acceleration of high water (Hrs. min.)
Fluid	0 0	0	·000	3 6
	0 30	140	·420	1 47
	1 0	280	·573	1 7
	2 0	560	·647	0 36
	3 0	840	·665	0 25
Elastic	∞	∞	·679	0 0

Fortnightly Tide

	Modulus of relaxation (Days hrs.)	Viscosity	Reduction of ocean tide	Acceleration of high water (Days hrs.)
Fluid	0 0	0	·000	3 10
	0 6	1,700	·294	2 11
	0 12	3,400	·470	1 18
	1 0	6,700	·602	1 1
	2 0	13,500	·657	0 13
	3 0	20,200	·669	0 9
Elastic	∞	∞	·679	0 0

I may remind the reader that the modulus of relaxation of rigidity is the time in which the stress requisite to retain the body in its strained configuration falls to ·368 of its initial value.

10. *The influence of inertia.*

In establishing these results inertia has been neglected, and I will now show that this neglect is not such as to materially vitiate my results*.

Suppose that the spheroid is constrained to execute such a vibration as it would do if it were a perfect fluid, and if the equilibrium theory of tides were true. Then the effective forces which are the equivalent of inertia, according to D'Alembert's principle, are found by multiplying the acceleration of each particle by its mass.

Inertia may then be safely neglected if the effective force on that particle which has the greatest amplitude of vibration is small compared with the

* In a future paper (read on December 19th, 1878) [Paper 4] I shall give an approximate solution of the problem, inclusive of the effects of inertia.

tide-generating force on it. In the case of a viscous spheroid, the inertia will have considerably less effect than it would have in the supposed constrained oscillation.

Now suppose we have a tide-generating potential $wr^2 \mathrm{S} \cos(vt + \eta)$, then, according to the equilibrium theory of tides, the form of the surface is given by

$$\sigma = \frac{5a^2}{2g} \mathrm{S} \cos(vt + \eta)$$

and this function gives the proposed constrained oscillation. It is clear that it is the particles at the surface which have the widest amplitude of oscillation. The effective force on a unit element at the surface is

$$-w \frac{d^2\sigma}{dt^2} = \frac{5a^2}{2g} wv^2 \mathrm{S} \cos(vt + \eta)$$

But the normal disturbing force at the surface is $2wa \mathrm{S} \cos(vt + \eta)$. Therefore inertia may be neglected if $\frac{5a^2}{2g} wv^2$ is small compared with $2wa$, or if $\frac{5a}{4g} v^2$ is a small fraction. The tide of the shortest period with which we have to deal is that in which $v = 2\omega$, so that we must consider the magnitude of the fraction $4 \times \frac{5a\omega^2}{4g}$. If ω were the earth's true angular velocity, instead of its angular velocity relatively to the moon, then $\frac{5a\omega^2}{4g}$ would be the ellipticity of its surface if it were homogeneous. This ellipticity is, as is well known, $\frac{1}{232}$. Hence the fraction, which is the criterion of the negligeability of inertia, is about $\frac{1}{58}$.

If, then, it be considered that this way of looking at the subject certainly exaggerates the influence of inertia, it is clear that the neglect of inertia is not such as to materially vitiate the results given above.

II.

A Tidal Yielding of the Earth's Mass, and the Canal-theory of Tides.

In the first part of this paper the equilibrium theory has been used for the determination of the reduction of the height of tide, and the alteration of phase, due to bodily tides in the earth.

Sir W. Thomson remarks, with reference to a supposed elastic yielding of the earth's body: "Imperfect as the comparisons between theory and observation as to the actual height of the tides have been hitherto, it is scarcely possible to believe that the height is in reality only two-fifths of what it would be if, as has been universally assumed in tidal theories, the earth were perfectly rigid. It seems, therefore, nearly certain, with no other evidence than is afforded by the tides, that the tidal effective rigidity of the earth must be greater than that of glass*."

The equilibrium theory is quite fallacious in its explanation of the semidiurnal tide, but Sir W. Thomson is of opinion that it must give approximately correct results for tides of considerable period. It is therefore on the observed amount of the fortnightly tide that he places reliance in drawing the above conclusion. Under these circumstances, a dynamical investigation of the effects of a tidal yielding of the earth on a tide of short period, according to the canal theory, is likely to be interesting.

The following investigation will be applicable either to the case of the earth's mass yielding through elasticity, plasticity, or viscosity; it thus embraces Sir W. Thomson's hypothesis of elasticity, as well as mine of viscosity and elastico-viscosity.

11. *Semidiurnal tide in an equatorial canal on a yielding nucleus.*

I shall only consider the simple case of the moon moving uniformly in the equator, and raising tide waves in a narrow shallow equatorial canal of depth h.

The potential of the tide-generating force, as far as concerns the present inquiry, is, with the old notation, $\frac{1}{2}\frac{\tau r^2}{a^2}\sin^2\theta\cos 2(\phi-\omega t)$, where $\tau=\frac{3}{2}\frac{ma^2}{c^3}$. This force will raise a bodily tide in the earth, whether it be elastic, plastic, or viscous. Suppose, then, that the greatest range of the bodily tide at the equator is 2E, and that it is retarded after the passage of the moon over the meridian by an angle $\frac{1}{2}\epsilon$. Then the equation to the bounding surface of the solid earth, at the time t, is $r=a+\mathrm{E}\sin^2\theta\cos[2(\phi-\omega t)+\epsilon]$; or with former notation $\sigma=\mathrm{E}\sin^2\theta\cos[2(\phi-\omega t)+\epsilon]$.

The whole potential V, at a point outside the nucleus, is the sum of the potential of the earth's attraction, and of the potential of the tide-generating force. Therefore

$$\begin{aligned}\mathrm{V}&=g\frac{a^2}{r}+\tfrac{3}{5}g\frac{r^2}{a^2}\mathrm{E}\sin^2\theta\cos[2(\phi-\omega t)+\epsilon]+\tfrac{1}{2}\frac{\tau r^2}{a^2}\sin^2\theta\cos 2(\phi-\omega t)\\&=g\frac{a^2}{r}+\{\mathrm{F}\cos[2(\phi-\omega t)+\epsilon]+\mathrm{G}\sin[2(\phi-\omega t)+\epsilon]\}\frac{r^2}{a^2}\sin^2\theta\end{aligned}$$

where $\mathrm{F}=\frac{3}{5}g\mathrm{E}+\frac{1}{2}\tau\cos\epsilon$, $\mathrm{G}=\frac{1}{2}\tau\sin\epsilon$.

* *Natural Philosophy*, § 843.

Sir George Airy shows, in his article on "Tides and Waves" in the *Encyclopœdia Metropolitana*, that the motion of the tide-wave in a canal running round the earth is the same as though the canal were straight, and the earth at rest, whilst the disturbing body rotates round it. This simplification will be applicable here also.

As before stated, the canal is supposed to be equatorial and of depth h.

After the canal has been developed, take the origin of rectangular coordinates in the undisturbed surface of the water, and measure x along the canal in the direction of the moon's motion, and y vertically downwards.

We have now to transform the potential V, and the equation to the surface of the solid earth, so as to make them applicable to the supposed development. If v be the velocity of the tide-wave, then $\omega a = v$; also the wave length is half the circumference of the earth's equator, or πa; and let $m = 2/a$. Then we have the following transformations:

$$\theta = \tfrac{1}{2}\pi, \quad \phi = \tfrac{1}{2}mx, \quad r = a + h - y$$

Also in the small terms we may put $r = a$. Thus the potential becomes

$$V = \text{const.} + gy + F\cos[m(x - vt) + \epsilon] + G\sin[m(x - vt) + \epsilon]$$

Again, to find the equation to the bottom of the canal, we have to transform the equation

$$r = a + E\sin^2\theta\cos[2(\phi - \omega t) + \epsilon]$$

If y' be the ordinate of the bottom of the canal, corresponding to the abscissa x, this equation becomes after development

$$y' = h - E\cos[m(x - vt) + \epsilon]$$

We now have to find the forced waves in a horizontal shallow canal, under the action of a potential V, whilst the bottom executes a simple harmonic motion. As the canal is shallow, the motion may be treated in the same way as Professor Stokes has treated the long waves in a shallow canal, of which the bottom is stationary. In this method it appears that the particles of water, which are at any time in a vertical column, remain so throughout the whole motion.

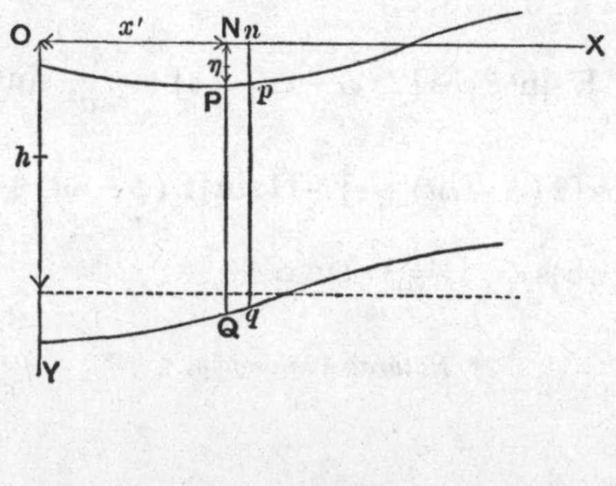

Suppose, then, that $x + \xi = x'$ is the abscissa of a vertical line of particles PQ, which, when undisturbed, had an abscissa x.

Let η be the ordinate of the surface corresponding to the abscissa x'.

Let pq be a neighbouring line of particles, which when undisturbed were distant from PQ by a small length k.

Conceive a slice of water cut off by planes through PQ, pq perpendicular to the length of the canal, of which the breadth is b. Then the volume of this slice is $b \times \text{PQ} \times \text{N}n$.

Now $$\text{PQ} = h - \text{E} \cos [m(x' - vt) + \epsilon] - \eta$$

and $$\text{N}n = k\left(1 + \frac{d\xi}{dx}\right)$$

Hence treating E and η as small compared with h, the volume of the slice is

$$bhk\left\{1 + \frac{d\xi}{dx} - \frac{\text{E}}{h}\cos[m(x' - vt) + \epsilon] - \frac{\eta}{h}\right\}$$

But this same slice, in its undisturbed condition, had a volume bhk. Therefore the equation of continuity is

$$\eta = h\frac{d\xi}{dx} - \text{E}\cos[m(x' - vt) + \epsilon]$$

Now the hydrodynamical equation of motion is approximately

$$\frac{dp}{dx'} = \frac{d\text{V}}{dx'} - \frac{d^2\xi}{dt^2}$$

The difference of the pressures on the two sides of the slice $PQqp$ at any depth is $\text{N}n \times \frac{dp}{dx'}$; and this only depends on the difference of the depressions of the wave surface below the axis of x on the two sides of the slice, viz. at P and p. Thus $\frac{dp}{dx'} = -g\frac{d\eta}{dx'}$.

Substituting then for η from the equation of continuity, and observing that $\frac{d^2\xi}{dx\,dx'}$ is very nearly the same as $\frac{d^2\xi}{dx^2}$, we have as the equation of wave motion,

$$gh\frac{d^2\xi}{dx^2} + mg\,\text{E}\sin[m(x' - vt) + \epsilon] = -\frac{d\text{V}}{dx'} + \frac{d^2\xi}{dt^2}$$

But $\frac{d\text{V}}{dx'} = -m\,\text{F}\sin[m(x' - vt) + \epsilon] + m\,\text{G}\cos[m(x' - vt) + \epsilon]$

So that

$$\frac{d^2\xi}{dt^2} = gh\frac{d^2\xi}{dx^2} + m\{\text{G}\cos[m(x' - vt) + \epsilon] - (\text{F} - \text{E}g)\sin[m(x' - vt) + \epsilon]\}$$

In obtaining the integral of this equation, we may omit the terms which are independent of G, F, E, because they only indicate free waves, which may be supposed not to exist.

The approximation will also be sufficiently close, if x be written for x' on the right hand side.

Assume, then, that

$$\xi = \mathrm{A}\cos[m(x-vt)+\epsilon] + \mathrm{B}\sin[m(x-vt)+\epsilon]$$

By substitution in the equation of motion and omitting $m(x-vt)+\epsilon$ for brevity, we find

$$-m^2(v^2-gh)\{\mathrm{A}\cos+\mathrm{B}\sin\} = m\{\mathrm{G}\cos-(\mathrm{F}-\mathrm{E}g)\sin\}$$

And as this must hold for all times and places,

$$\mathrm{A} = -\frac{\mathrm{G}}{m(v^2-gh)} = \frac{-\frac{1}{2}a\tau\sin\epsilon}{2(a^2\omega^2-gh)}$$

$$\mathrm{B} = \frac{\mathrm{F}-\mathrm{E}g}{m(v^2-gh)} = \frac{a(\frac{1}{2}\tau\cos\epsilon-\frac{2}{5}g\mathrm{E})}{2(a^2\omega^2-gh)}$$

In the case of such seas as exist in the earth, the tide-wave travels faster than the free-wave, so that $a^2\omega^2$ is greater than gh; and the denominators of A and B are positive.

We have then

$$\xi = \frac{a}{2(a^2\omega^2-gh)}\{(\tfrac{1}{2}\tau\cos\epsilon-\tfrac{2}{5}g\mathrm{E})\sin-\tfrac{1}{2}\tau\sin\epsilon\cos\}$$

But the present object is to find the motion of the wave-surface relatively to the bottom of the canal, for this will give the tide relatively to the dry land. Now the height of the wave relatively to the bottom is

$$\mathrm{PQ} = h - \mathrm{E}\cos[m(x-vt)+\epsilon]-\eta$$
$$= h - h\frac{d\xi}{dx}$$

And
$$\frac{d\xi}{dx} = \frac{1}{a^2\omega^2-gh}\{(\tfrac{1}{2}\tau\cos\epsilon-\tfrac{2}{5}g\mathrm{E})\cos+\tfrac{1}{2}\tau\sin\epsilon\sin\}$$

Hence reverting to the sphere, and putting a for $a+h$, we get as the equation to the relative spheroid of which the wave-surface in the equatorial canal forms part

$$r = a - \frac{h\sin^2\theta}{a^2\omega^2-gh}\{\tfrac{1}{2}\tau\cos 2(\phi-\omega t)-\tfrac{2}{5}g\mathrm{E}\cos[2(\phi-\omega t)+\epsilon]\}$$

But according to the equilibrium theory, if V has the same form as above, viz.

$$g\frac{a^2}{r}+\tfrac{3}{5}g\frac{r^2}{a^2}\mathrm{E}\sin^2\theta\cos[2(\phi-\omega t)+\epsilon]+\tfrac{1}{2}\tau\frac{r^2}{a^2}\sin^2\theta\cos 2(\phi-\omega t)$$

and if $r = a+u$ be the equation to the tidal spheroid, we have, as in Part I.,

$$u = \frac{\sin^2\theta}{g}\{\tfrac{1}{2}\tau\cos 2(\phi-\omega t)+\tfrac{3}{5}g\mathrm{E}\cos[2(\phi-\omega t)+\epsilon]\}$$

and the equation to the relative tidal spheroid is

$$r = a + u - \sigma$$

$$= a + \frac{\sin^2\theta}{g}\{\tfrac{1}{2}\tau\cos 2(\phi - \omega t) - \tfrac{2}{5}g\mathrm{E}\cos[2(\phi - \omega t) + \epsilon]\}$$

Now either in the case of the dynamical theory or of the equilibrium theory, if E be put equal to zero, we get the equations to the tidal spheroid on a rigid nucleus. A comparison, then, of the above equations shows at once that both the reduction of tide and the acceleration of phase are the same in one theory as in the other. But where the one gives high water, the other gives low water. The result is applicable to any kind of supposed yielding of the earth's mass; and in the special case of viscosity, the table of results for the fortnightly tide at the end of Part I. is applicable.

III.

Summary and Conclusions.

In § 1 an analogy is shown between problems about the state of strain of incompressible elastic solids, and the flow of incompressible viscous fluids, when inertia is neglected; so that the solutions of the one class of problems may be made applicable to the other. Sir W. Thomson's problem of the bodily tides of an elastic sphere is then adapted so as to give the bodily tides of a viscous spheroid. The adaptation is rendered somewhat complex by the necessity of introducing the effects of the mutual gravitation of the parts of the spheroid.

The solution is only applicable where the disturbing potential is capable of expansion as a series of solid harmonics, and it appears that each harmonic term in the potential then acts as though all the others did not exist; in consequence of this it is only necessary to consider a typical term in the potential.

In § 3 an equation is found which gives the form of the free surface of the spheroid at any time, under the action of any disturbing potential, which satisfies the condition of expansibility. By putting the disturbing potential equal to zero, the law is found which governs the subsidence of inequalities on the surface of the spheroid, under the influence of mutual gravitation alone. If the form of the surface be expressed as a series of surface harmonics, it appears that any harmonic diminishes in geometrical progression as the time increases in arithmetical progression, and harmonics of higher orders subside much more slowly than those of lower orders. Common sense, indeed, would tell us that wide-spread inequalities must subside much more quickly than wrinkles, but only analysis could give the law connecting the rapidity of the subsidence with the magnitude of the inequality*.

I hope at some future time to try whether it will not be possible to throw some light on the formation of parallel mountain chains and the direction of faults, by means of this equation. Probably the best way of doing this will be to transform the surface harmonics, which occur here, into Bessel's functions.

In § 4 the rate is considered at which a spheroid would adjust itself to a new form of equilibrium, when its axis of rotation had separated from that of figure; and the law is established which was assumed in a previous paper†.

In § 5 I pass to the case where the disturbing potential is a solid harmonic of the second degree, multiplied by a simple time harmonic. This is the case to be considered for the problem of a tidally distorted spheroid. A remarkably simple law is found connecting the viscosity, the height of tide, and the amount of lagging of tide; it is shown that if v be the speed of the tide, and if $\tan \epsilon$ varies jointly as the coefficient of viscosity and v, then the height of bodily tide is equal to that of the equilibrium tide of a perfectly fluid spheroid multiplied by $\cos \epsilon$, and the tide lags by a time equal to $\frac{\epsilon}{v}$.

It is then shown (§ 6) that in the equilibrium theory the *ocean* tides on the yielding nucleus will be equal in height to the ocean tides on a rigid nucleus multiplied by $\sin \epsilon$, and that there will be an acceleration of the time of high water equal to $\frac{\pi}{2v} - \frac{\epsilon}{v}$.

The tables in § 7 give the results of the application of the preceding theories to the lunar semidiurnal and fortnightly tides for various degrees of viscosity. A comparison of the numbers in the first columns with the viscosity of pitch at near the freezing temperature (viz., about $1{\cdot}3 \times 10^8$, as found by me), when it is hard, apparently solid and brittle, shows how enormously stiff

* On this Lord Rayleigh remarks, that if we consider the problem in two dimensions, and imagine a number of parallel ridges, the distance between which is λ, then inertia being neglected, the elements on which the time of subsidence depends are gw (force per unit mass due to weight), υ the coefficient of viscosity, and λ. Thus the time T must have the form

$$T = (gw)^x \upsilon^y \lambda^z$$

The dimensions of gw, υ, λ are respectively $ML^{-2}T^{-2}$, $ML^{-1}T^{-1}$, L; hence

$$x + y = 0$$
$$-2x - y + z = 0$$
$$-2x - y = 1$$

And $x = -1$, $y = 1$, $z = -1$, so that T varies as $\frac{\upsilon}{gw\lambda}$.

If we take the case on the sphere, then when i, the order of harmonics, is great, λ compares with $\frac{a}{i}$; so that T varies as $\frac{\upsilon i}{gwa}$.

† *Phil. Trans.*, Vol. 167, Part I., sec. 5 of my paper. [To be included in Vol. III. of these collected papers.]

the earth must be to resist the tidally deforming influence of the moon. For unless the viscosity were very much larger than that of pitch, the viscous sphere would comport itself sensibly like a perfect fluid, and the ocean tides would be quite insignificant. It follows, therefore, that no very considerable portion of the interior of the earth can even distantly approach the fluid state.

This does not, however, seem to be conclusive against the existence of bodily tides in the earth of the kind here considered; for although (as remarked by Sir W. Thomson) a very great hydrostatic pressure probably has a tendency to impart rigidity to a substance, yet the very high temperature which must exist in the earth at a small depth would tend to induce a sort of viscosity—at least if we judge by the behaviour of materials at the earth's surface.

In § 8 the theory of the tides of an imperfectly elastic spheroid is developed. The kind of imperfection of elasticity considered is where the forces requisite to maintain the body in any strained configuration diminish in geometrical progression as the time increases in arithmetical progression. There can be no doubt that all bodies *do* possess an imperfection in their elasticity of this general nature, but the exact law here assumed has not, as far as I am aware, any experimental justification; its adoption was rather due to mathematical necessities than to any other reason.

It would, of course, have been much more interesting if it had been possible to represent more exactly the mechanical properties of solid matter. One of the most important of these is that form of resistance to relative displacement, to which the term "plasticity" has been specially appropriated. This form of resistance is such that there is a change in the law of resistance to the relative motion of the parts, when the forces tending to cause flow have reached a certain definite intensity. This idea was founded, I believe, by MM. Tresca and St Vénant on a long course of experiments on the punching and squeezing of metals*; and they speak of a solid being reduced to the state of fluidity by stresses of a given magnitude. This theory introduces a discontinuity, since it has to be determined what parts of the body are reduced to the state of fluidity and what are not. But apart from this difficulty, there is another one which is almost insuperable, in the fact that the differential equations of flow are non-linear.

The hope of introducing this form of resistance must be abandoned, and the investigation must be confined to the inclusion of those two other continuous laws of resistance to relative displacement—elasticity and viscosity.

As above stated, the law of elastico-viscosity assumed in this paper has not got an experimental foundation. Indeed, Kohlrausch's experiments on

* "Sur l'écoulement des Corps Solides," *Mém. des Savants Étrangers*, Tom. XVIII. and Tom. XX., p. 75 and p. 137. See also *Comptes Rendus*, Tom. LXVI., LXVIII., and *Liouville's Journ.*, 2me série, XIII., p. 379, and XVI., p. 308, for papers on this subject.

glass* show that the elasticity degrades rapidly at first, and that it tends to attain a final condition, from which it does not seem to vary for an almost indefinite time. But glass is one of the most perfectly elastic substances known, and, by the light of Tresca's experiments, it seems probable that experiments with lead would have brought out very different results. It seems, moreover, hardly reasonable to suppose that the materials of the earth possess much mechanical similarity with glass. Notwithstanding all these objections, I think, for my part, that the results of this investigation of the tides of an ideal elastico-viscous sphere are worthy of attention.

There are two constants which determine the nature of this ideal solid: first, the coefficient of rigidity, at the instant immediately after the body has been placed in its strained configuration; and secondly, "the modulus of the time of relaxation of rigidity," which is the time in which the force requisite to retain the body in its strained configuration has fallen away to ·368 of its initial value.

In this section it is shown that the equations of flow of this incompressible elastico-viscous body have the same mathematical form as those for a purely viscous body; so that the solutions already attained are easily adapted to the new hypothesis.

The only case where the problem is completely worked out, is when the disturbing potential has the form appropriate to the tidal problem. The laws of reduction of bodily tide, of its lagging, of the reduction of ocean tide, and of its acceleration, are somewhat more complex than in the case of pure viscosity; and the reader is referred to § 8 for the statement of those laws. It is also shown that by appropriate choice of the values of the two constants, the solutions may be made either to give the results of the problem for a purely viscous sphere, or for a purely elastic one.

The tables give the results of this theory, for the semidiurnal and fortnightly tides, for spheroids which have the rigidity of glass or of iron—the two cases considered by Sir W. Thomson. As it is only possible to judge of the amount of bodily tide by the reduction of the ocean tide, I have not given the heights and retardations of the bodily tide.

It appears that if the time of relaxation of rigidity is about one-quarter of the tidal period, then the reduction of ocean tide does not differ much from what it would be if the spheroid were perfectly elastic. The amount of tidal acceleration still, however, remains considerable. A like observation may be made with respect to the acceleration of tide in the case of pure viscosity approaching rigidity: and this leads me to think that one of the most promising ways of detecting such tides in the earth would be by the

* *Poggendorff's Ann.*, Vol. 119, p. 337.

determination of the periods of maximum and minimum in a tide of long period, such as the fortnightly in a high latitude.

In § 10 it is shown that the effects of inertia, which had been neglected in finding the laws of the tidal movements, cannot be such as to materially affect the accuracy of the results.

[*The hypothesis of a viscous or imperfectly elastic nature for the matter of the earth would be rendered extremely improbable, if the ellipticity of an equatorial section of the earth were not very small. An ellipsoidal figure with three unequal axes, even if theoretically one of equilibrium, could not continue to subsist very long, because it is a form of greater potential energy than the oblate spheroidal form, which is also a figure of equilibrium.

Now, according to the results of geodesy, which until very recently have been generally accepted as the most accurate—namely, those of Colonel A. R. Clarke†—there is a difference of 6,378 feet between the major and minor equatorial radii, and the meridian of the major axis is 15° 34′ East of Greenwich.

The heterogeneity of the earth would have to be very great to permit so large a deviation from the oblate spheroidal shape to be either permanent, or to subside with extreme slowness. But since this paper was read, Colonel Clarke has published a revision of his results, founded on new data‡; and he now finds the difference between the equatorial radii to be only 1,524 feet, whilst the meridian of the greatest axis is 8° 15′ West. This exhibits a change of meridian of 24°, and a reduction of equatorial ellipticity to about one-quarter of the formerly-received value. Moreover, the new value of the polar axis is about 1,000 feet larger than the old one.

Colonel Clarke himself obviously regards the ellipsoidal form of the equator as doubtful. Thus there is at all events no proved result of geodesy opposed to the present hypothesis concerning the constitution of the earth. Sir W. Thomson remarks in a letter to me that "we may look to further geodetic observations and revisals of such calculations as those of Colonel Clarke for verification or disproof of your viscous theory."]

In the first part of the paper the equilibrium theory is used in discussing the question of ocean tides; in the second part I consider what would be the tides in a shallow equatorial canal running round the equator, if the nucleus yielded tidally at the same time. The reasons for undertaking this investigation are given at the beginning of that part. In § 11 it is shown that the height of tide relatively to the nucleus bears the same proportion to the

* The part within brackets [ ] was added in November, 1878, in consequence of a conversation with Sir W. Thomson.

† Quoted in Thomson and Tait, *Natural Philosophy*, § 797.

‡ *Phil. Mag.*, August, 1878.

height of tide on a rigid nucleus as in the equilibrium theory, and the alteration of phase is also the same; but where the one theory gives high water the other gives low water.

The chief practical result of this paper may be summed up by saying that it is strongly confirmatory of the view that the earth has a very great effective rigidity. But its chief value is that it forms a necessary first chapter to the investigation of the precession of imperfectly elastic spheroids, which will be considered in a future paper*. I shall there, as I believe, be able to show, by an entirely different argument, that the bodily tides in the earth are probably exceedingly small at the present time.

Appendix. (November 7, 1878.)

On the observed height and phase of the fortnightly oceanic tide.

[This contained an incomplete investigation and is replaced by Paper 9, Vol. I. p. 340.]

* Read before the Royal Society on December 19th, 1878. [Paper 3 in this volume.]

2.

NOTE ON THOMSON'S THEORY OF THE TIDES OF AN ELASTIC SPHERE*.

[*Messenger of Mathematics*, VIII. (1879), pp. 23—26.]

THE results of the theory of the elastic yielding of the earth would of course be more interesting, if it were possible fully to introduce the effects of the want of homogeneity of elasticity and density of the interior of the earth; but besides the mathematical difficulties of the case, the complete absence of data as to the nature of the deep-seated matter makes it impossible to do so. It is, however, possible to make a more or less probable estimate of the extent to which a given yielding of the *surface* will affect the ocean tide-wave, when the earth is treated as heterogeneous. And as we can only judge of the amount of the bodily tide in the earth by observations on the ocean tides, this estimate may be of some value.

The heterogeneity of the interior must of course be accompanied by heterogeneity of elasticity†, and under the influence of a given tide-generating force, this will affect the internal distribution of strain, and the form of the surface to an unknown extent. The diminution of ocean tide which arises from the yielding of the nucleus is entirely due to the alteration in the form of the level surfaces outside the nucleus. But it is by no means obvious how far the potential of the earth, when its surface is distorted to a given amount, may differ from that of the homogeneous spheroid considered by Sir W. Thomson; and in face of our ignorance of the law of internal elasticity, the problem does not admit of a precise solution.

I propose, however, to make an hypothesis, which seems as probable as any other, as to the law of the ellipticity of the internal strain ellipsoids, when the surface is strained to a given amount, and then to find the potential at an external point.

* [This subject has since been treated more fully by Dr G. Herglotz, *Zeitschr. für Math. und Physik*, Vol. LII. (1905), p. 273.]

† That is to say, if the earth is elastic at all.

Suppose that under the influence of a bodily harmonic potential of the second degree the earth's surface assumes the form $r=a+\sigma$, where σ is a surface harmonic of the second order. Then I propose to assume that the ellipticity of any internal strain ellipsoid is related to that of the surface by the same law as though the earth were homogeneous, elastic, and incompressible, and had its surface brought into the form $r=a+\sigma$ by a tide-generating potential of the second order. If μ be the coefficient of rigidity of an elastic incompressible sphere under the action of a bodily force, of which the potential is wr^2S_2, then Sir W. Thomson's solution* shows that the radial displacement at any point r is given by

$$\rho = \frac{8a^2 - 3r^2}{19\mu} rS_2$$

Putting $r=a$, we have $\sigma = \dfrac{5a^3}{19\mu} S_2$. And if $r=a'+\sigma'$ be the equation to a strain ellipsoid of mean radius a', we have by our hypothesis

$$\frac{\sigma'}{a'} \div \frac{\sigma}{a} = \frac{8a^2 - 3a'^2}{5a^2}$$

and
$$\sigma' = \frac{a'}{a}\left\{\tfrac{8}{5} - \tfrac{3}{5}\left(\frac{a'}{a}\right)^2\right\}\sigma = f(a')\,\sigma \text{ suppose}$$

Now the potential of a homogeneous spheroid $r=a'+\sigma f(a')$, of density q, at an external point is

$$\tfrac{4}{3}\pi q \frac{a'^3}{r} + \frac{4\pi q}{5r^2} a'^3 f(a')\,\sigma$$

and therefore the potential of a spheroidal shell of density q, whose inner and outer surfaces are given by $r=a'+f(a')\,\sigma$ and $r=a'+\delta a'+f(a'+\delta a')\,\sigma$, is

$$4\pi q \frac{a'^2}{r}\delta a' + \frac{4\pi q}{5r^2}\frac{d}{da'}\{a'^3 f(a')\}\,\sigma\delta a'$$

If then we integrate this expression from $a'=a$ to $a'=0$, and treat q as a function of a', we have the potential of the earth on the present hypothesis. The integral is

$$\frac{4\pi}{r}\int_0^a qa'^2 da' + \frac{4\pi\sigma}{5r^2}\int_0^a q\frac{d}{da'}\{a'^3 f(a')\}\,da'$$

The first of these two terms is clearly $g\dfrac{a^2}{r}$ (where g is gravity), and is the same as though the earth were homogeneous; and it only remains to evaluate the second. Now, according to the Laplacean law of internal density of the earth, if D be the mean density, and f the ratio of D to the surface density, and θ a certain angle which is about 144°,

$$q = \frac{D}{f}\frac{a}{\sin\theta}\cdot\frac{\sin a'\theta/a}{a'}$$

* Thomson and Tait's *Natural Philosophy*, § 834, equation (14).

Substituting this value for q, and for $f(a')$ its value, we have

$$q\frac{d}{da'}[a'^3 f(a')]\,da' = \tfrac{2}{5}\frac{D}{f\sin\theta}\frac{\sin a'\theta/a}{a'}\left(16a'^3 - \frac{9a'^5}{a^2}\right)da'$$

Then if we change the variable of integration by putting $x = \frac{a'\theta}{a}$, and put $4\pi D = \frac{3g}{a}$, we get for the second term of the earth's potential

$$\left\{\tfrac{3}{5}g\left(\frac{a}{r}\right)^2\sigma\right\}\frac{2}{5f\theta^3\sin\theta}\int_0^\theta \sin x\left(16x^2 - \frac{9x^4}{\theta^2}\right)dx$$

Now $\frac{3}{5}g\left(\frac{a}{r}\right)^2\sigma$ would be the potential of the surface layer given by $r = a + \sigma$, *if the earth were homogeneous and had a density* D, and the rest of the expression is a numerical factor (which may be called K), by which this potential must be reduced in order to get the potential of the heterogeneous earth on the present hypothesis.

If the integration be effected it will be found that

$$\tfrac{5}{2}fK = -\cot\theta\left[\frac{7}{\theta} + \frac{76}{\theta^3} - \frac{216}{\theta^5}\right] - \operatorname{cosec}\theta\left[\frac{32}{\theta^3} + \frac{216}{\theta^5}\right] - \frac{4}{\theta^2} + \frac{216}{\theta^4}$$

$$= 5\cdot1442, \text{ when } \theta = 144^\circ$$

Whence $$K = \frac{2\cdot0577}{f}$$

Also $$f = 3\left(\frac{1}{\theta^2} - \frac{\cot\theta}{\theta}\right) = 2\cdot1178 \text{ by Laplace's theory.}$$

Therefore $$K = \frac{2\cdot0577}{2\cdot1178} = \cdot972$$

Hence, on the present hypothesis, the potential of the earth at a point outside its mass is

$$g\frac{a^2}{r} + (\cdot972)\,\tfrac{3}{5}g\left(\frac{a}{r}\right)^2\sigma$$

This differs by very little from what it would be if the earth were homogeneous; for in that case ·972 would be merely replaced by unity.

Therefore, if at any future time it should be found that the fortnightly tide* is less than it would be theoretically on a rigid nucleus, it will then be probable that the surface of the earth rises and falls by about the same amount as would follow from the theory of the bodily tides of a *homogeneous* elastic sphere whose density is equal to the earth's mean density. This investigation being founded on conjecture, cannot claim anything better than a probability for its result; but without calculation, I, at least, could not form any sort of guess of what the result might be, and the question is of undoubted interest in the physics of the earth.

* Sir W. Thomson relies principally on observation of the fortnightly ocean tide for detecting bodily tides in the earth. [See Paper 9, Vol. I., and W. Schweydar, *Beiträgen zur Geophysik*, Vol. IX. (1907), p. 41. See also an important paper by Lord Rayleigh on the fortnightly tide in *Phil. Mag.*, Jan. 1903, p. 136.]

3.

ON THE PRECESSION OF A VISCOUS SPHEROID, AND ON THE REMOTE HISTORY OF THE EARTH.

[*Philosophical Transactions of the Royal Society*, Part II. Vol. 170 (1879), pp. 447—530.]

THE following paper contains the investigation of the mass-motion of viscous and imperfectly elastic spheroids, as modified by a relative motion of their parts, produced in them by the attraction of external disturbing bodies; it must be regarded as the continuation of my previous paper*, where the theory of the bodily tides of such spheroids was given.

The problem is one of theoretical dynamics, but the subject is so large and complex, that I thought it best, in the first instance, to guide the direction of the speculation by considerations of applicability to the case of the earth, as disturbed by the sun and moon.

In order to avoid an incessant use of the conditional mood, I speak simply of the earth, sun, and moon; the first being taken as the type of the rotating body, and the two latter as types of the disturbing or tide-raising bodies. This course will be justified, if these ideas should lead (as I believe they will) to important conclusions with respect to the history of the evolution of the solar system. This plan was the more necessary, because it seemed to me impossible to attain a full comprehension of the physical meaning of the long and complex formulæ which occur, without having recourse to numerical values; moreover, the differential equations to be integrated were so complex, that a laborious treatment, partly by analysis and partly by numerical quadratures, was the only method that I was able to devise. Accordingly, the earth, sun, and moon form the system from which the requisite numerical data are taken.

* "On the Bodily Tides of Viscous and Semi-elastic Spheroids," &c., *Phil. Trans.*, 1879, Part I. [Paper 1.]

It will of course be understood that I do not conceive the earth to be really a homogeneous viscous or elastico-viscous spheroid, but it does seem probable that the earth still possesses some plasticity, and if at one time it was a molten mass (which is highly probable), then it seems certain that some changes in the configuration of the three bodies must have taken place, closely analogous to those hereafter determined. And even if the earth has always been quite rigid, the greater part of the same effects would result from oceanic tidal friction, although probably they would have taken place with less rapidity.

As some persons may wish to obtain a general idea of the drift of the inquiry without reading a long mathematical argument, I have adhered to the plan adopted in my former paper, of giving at the end (in Part III.) a general view of the whole subject, with references back to such parts as it did not seem desirable to reproduce. In order not to interrupt the mathematical argument in the body of the paper, the discussion of the physical significance of the several results is given along with the summary; such discussions will moreover be far more satisfactory when thrown into a continuous form than when scattered in isolated paragraphs throughout the paper. I have tried, however, to prevent the mathematical part from being too bald of comments, and to place the reader in a position to comprehend the general line of investigation.

Before entering on analysis, it is necessary to give an explanation of how this inquiry joins itself on to that of my previous paper.

In that paper it was shown that, if the influence of the disturbing body be expressed in the form of a potential, and if that potential be expressed as a series of solid harmonic functions of points within the disturbed spheroid, each multiplied by a simple time-harmonic, then each such harmonic term raises a tide in the disturbed spheroid, which is the same as though all the other terms were non-existent. This is true, whether the spheroid be fluid, elastic, viscous, or elastico-viscous. Further, the free surface of the spheroid, as tidally distorted by any term, is expressible by a surface harmonic of the same type as that of the generating term; and where there is a frictional resistance to the tidal motion, the phase of the corresponding simple time harmonic is retarded. The height of each tide, and the retardation of phase (or the lag) are functions of the frequency of the tide, and of the constants expressive of the physical constitution of the spheroid.

Each such term in the expression for the form of the tidally distorted spheroid may be conveniently referred to as a simple tide.

Hence if we regard the whole tide-wave as a modification of the equilibrium tide-wave of a perfectly fluid spheroid, it may be said that the effect of the resistances to relative displacement is a disintegration of the whole wave into its constituent simple tides, each of which is reduced in height, and lags in time by its own special amount. In fact, the mathematical expansion in surface harmonics exactly corresponds to the physical breaking up of a single wave into a number of secondary waves.

It was remarked in the previous paper*, that when the tide-wave lags the attraction of the external tide-generating body gives rise to forces on the spheroid which are not rigorously equilibrating. Now it was a part of the assumptions, under which the theory of viscous and elastico-viscous tides was formed, that the whole forces which act on the spheroid *should* be equilibrating; but it was there stated that the couples arising from the non-equilibration of the attractions on the lagging tides were proportional to the square of the disturbing influence, and it was on this account that they were neglected in forming that theory of tides. The investigation of the effects which they produce in modifying the relative motion of the parts of the spheroid, that is to say in distorting the spheroid, must be reserved for a future occasion†.

The effect of these couples, in modifying the motion of the rotating spheroid as a whole, affords the subject of the present paper.

According to the ordinary theory, the tide-generating potential of the disturbing body is expressible as a series of Legendre's coefficients; the term of the first order is non-existent, and the one of the second order has the type $\frac{3}{2}\cos^2 - \frac{1}{2}$. Throughout this paper the potential is treated as though the term of the second order existed alone, but at the end it is shown that the term of the third order (of the type $\frac{5}{2}\cos^3 - \frac{3}{2}\cos$) will have an effect which is fairly negligeable compared with that of the first term.

In order to apply the theory of elastic, viscous, and elastico-viscous tides, the first task is to express the tide-generating potential in the form of a series of solid harmonics relatively to axes fixed in the spheroid, each harmonic being multiplied by a simple time-harmonic.

Afterwards it will be necessary to express that the wave surface of the distorted spheroid is the disintegration into simple lagging tides of the equilibrium tide-wave of a perfectly fluid spheroid.

The symbols expressive of the disintegration and lagging will be kept perfectly general, so that the theory will be applicable either to the assumptions of elasticity, viscosity, or elastico-viscosity, and probably to any other continuous law of resistance to relative displacement. It would not, however, be applicable to such a law as that which is *supposed* to govern the resistance to slipping of loose earth, nor to any law which assumes that there is no relative displacement of the parts of the solid, until the stresses have reached a definite magnitude.

* "Bodily Tides," &c. [Paper 1.] Sec. 5.

† See the next paper "On Problems connected with the Tides of a Viscous Spheroid." Part I. [Paper 4.]

After the form of the distorted spheroid has been found, the couples which arise from the attraction of the disturbing body on the wave surface will be found, and the rotation of the spheroid and the reaction on the disturbing body will be considered.

This preliminary explanation will, I think, make sufficiently clear the objects of the rather long introductory investigations which are necessary.

PART I.

§ 1. *The tide-generating potential.*

The disturbing body, or moon, is supposed to move in a circular orbit, with a uniform angular velocity $-\Omega$. The plane of the orbit is that of the ecliptic; for the investigation is sufficiently involved without complicating it by giving the true inclined eccentric orbit, with revolving nodes. I hope however in a future paper to consider the secular changes in the inclination and eccentricity of the orbit and the modifications to be made in the results of the present investigation.

Let m be the moon's mass, c her distance, and $\tau = \frac{3}{2}\frac{m}{c^3}$.

Let X, Y, Z (fig. 1) be rectangular axes fixed in space, XY being the ecliptic.

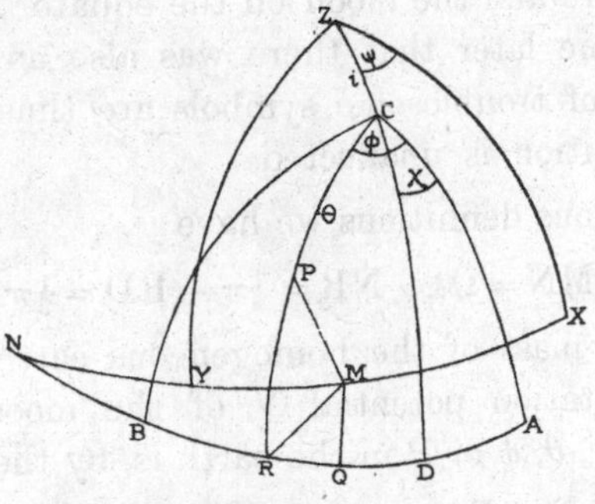

FIG. 1.

Let M be the moon in her orbit moving from Y towards X, with an angular velocity Ω*.

Let A, B, C be rectangular axes fixed in the earth, AB being the equator.

Let i, ψ be the coordinates of the pole C referred to X, Y, Z, so that i is the obliquity of the ecliptic, and $\frac{d\psi}{dt}$ the precession of the equinoxes.

* [The system of coordinates chosen is unfortunately what Lord Kelvin calls "perverted," but I do not think it worth while to go through the whole investigation and change the signs.]

Let r, θ, ϕ be the polar coordinates of any point P in the earth referred to A, B, C, as indicated in the figure.

Let $\omega_1, \omega_2, \omega_3$ be the component angular velocities of the earth about the instantaneous positions of A, B, C.

Then we have, as usual, the geometrical equations,

$$\left.\begin{aligned} \frac{di}{dt} \qquad\quad &= -\omega_1 \sin\chi + \omega_2 \cos\chi \\ \frac{d\psi}{dt}\sin i &= -\omega_1 \cos\chi - \omega_2 \sin\chi \\ -\frac{d\chi}{dt} + \frac{d\psi}{dt}\cos i &= \quad \omega_3 \end{aligned}\right\} \quad \ldots\ldots\ldots\ldots\ldots\ldots (1)$$

Let $\Pi \operatorname{cosec} i$ be the precession of the equinoxes, or $\frac{d\psi}{dt}$, so that $\frac{d\chi}{dt} = \Pi \cot i - \omega_3$*. Now the earth rotates with a negative angular velocity, that is from B to A; therefore if we put $\frac{d\chi}{dt} = n$, n is equal to the true angular velocity of the earth $+ \Pi \cot i$. But for purposes of numerical calculation n may be taken as the earth's angular velocity; and care need merely be taken that inequalities of very long period are not mistaken for secular changes.

Let the epoch be taken as the time when the colure ZC was in the plane of ZX, when χ was zero and the moon on the equator at Y. It will be convenient also to assume later that there was also an eclipse at the same instant. A number of troublesome symbols are thus got rid of, whilst the generality of the solution is unaffected.

Then by the previous definitions we have

$$\chi = nt, \quad MN = \Omega t, \quad NR = \tfrac{1}{2}\pi - RD = \tfrac{1}{2}\pi - (\phi - \chi)$$

Now if w be the mass of the homogeneous earth per unit volume, the tide-generating gravitation potential V of the moon, estimated per unit volume, at the point r, θ, ϕ or P in the earth is, by the well-known formula,

$$V = w\tau r^2 (\cos^2 PM - \tfrac{1}{3})$$

This is the function on which the tides depend, and as above explained, it must be expanded in a series of solid harmonics of r, θ, ϕ, each multiplied by a simple time harmonic, which will involve n and Ω.

For brevity of notation nt, Ωt are written simply n, Ω, but wherever these symbols occur in the argument of a trigonometrical term they must be understood to be multiplied by t the time.

* The limit of $\Pi \cot i$ is still small when i is zero. In considering the precession with one disturbing body only, $\Pi \operatorname{cosec} i$ is merely the precession due to that body; but afterwards when the effect of the sun is added it must be taken as the full precession.

We have $\cos \text{PM} = \sin\theta \cos \text{MR} + \cos\theta \sin \text{MR} \sin \text{MRQ}$

and $\cos \text{MR} = \cos \text{MN} \cos \text{NR} + \sin \text{MN} \sin \text{NR} \cos i$

$$= \cos\Omega \sin(\phi - n) + \sin\Omega \cos(\phi - n)\cos i$$

also $\sin \text{MR} \sin \text{MRQ} = \sin \text{MQ} = \sin\Omega \sin i$

Therefore

$$\cos \text{PM} = \sin\theta \sin(\phi - n)\cos\Omega + \sin\theta\cos(\phi - n)\sin\Omega\cos i + \cos\theta\sin\Omega\sin i$$
$$= \tfrac{1}{2}\sin\theta\,\{\sin[\phi - (n - \Omega)] + \sin[\phi - (n + \Omega)]\}$$
$$+ \tfrac{1}{2}\sin\theta\cos i\,\{\sin[\phi - (n - \Omega)] - \sin[\phi - (n + \Omega)]\} + \cos\theta\sin\Omega\sin i$$

Let $p = \cos\tfrac{1}{2}i, \quad q = \sin\tfrac{1}{2}i$

Then

$$\cos \text{PM} = p^2\sin\theta\sin[\phi - (n - \Omega)] + 2pq\cos\theta\sin\Omega + q^2\sin\theta\sin[\phi - (n + \Omega)] \quad \text{......(2)}$$

Therefore

$$\cos^2 \text{PM} = \tfrac{1}{2}p^4\sin^2\theta\,\{1 - \cos[2\phi - 2(n - \Omega)]\} + 2p^2q^2\cos^2\theta\,(1 - \cos 2\Omega)$$
$$+ \tfrac{1}{2}q^4\sin^2\theta\,\{1 - \cos[2\phi - 2(n + \Omega)]\} + 2p^3q\sin\theta\cos\theta\,\{\cos(\phi - n) - \cos[\phi - (n - 2\Omega)]\}$$
$$+ 2pq^3\sin\theta\cos\theta\,\{\cos[\phi - (n + 2\Omega)] - \cos(\phi - n)\}$$
$$+ p^2q^2\sin^2\theta\,\{\cos 2\Omega - \cos(2\phi - 2n)\}$$

Collecting terms, and noticing that

$$\tfrac{1}{2}(p^4 + q^4)\sin^2\theta + 2p^2q^2\cos^2\theta = \tfrac{1}{3} + \tfrac{1}{2}(1 - 6p^2q^2)(\tfrac{1}{3} - \cos^2\theta)$$

we have

$$\frac{V}{w\tau r^2} = \cos^2 \text{PM} - \tfrac{1}{3}$$
$$= -\tfrac{1}{2}\sin^2\theta\,\{p^4\cos[2\phi - 2(n - \Omega)] + 2p^2q^2\cos[2\phi - 2n] + q^4\cos[2\phi - 2(n + \Omega)]\}$$
$$- 2\sin\theta\cos\theta\,\{p^3q\cos[\phi - (n - 2\Omega)] - pq(p^2 - q^2)\cos(\phi - n) - pq^3\cos[\phi - (n + 2\Omega)]\}$$
$$+ (\tfrac{1}{3} - \cos^2\theta)\,\{3p^2q^2\cos 2\Omega + \tfrac{1}{2}(1 - 6p^2q^2)\} \quad \text{......(3)*}$$

Now if all the cosines involving ϕ be expanded, it is clear that we have V consisting of thirteen terms which have the desired form, and a fourteenth which is independent of the time.

It will now be convenient to introduce some auxiliary functions, which may be defined thus,

$$\left.\begin{aligned}
\Phi(2n) &= \tfrac{1}{2}p^4\cos 2(n - \Omega) + p^2q^2\cos 2n + \tfrac{1}{2}q^4\cos 2(n + \Omega)\\
\Psi(n) &= 2p^3q\cos(n - 2\Omega) - 2pq(p^2 - q^2)\cos n - 2pq^3\cos(n + 2\Omega)\\
X(2\Omega) &= 3p^2q^2\cos 2\Omega
\end{aligned}\right\} \quad (4)$$

* [This transformation is obtained by a neater process in the paper on "Harmonic Analysis of the Tides," p. 7, Vol. I.]

$\Phi(2n-\frac{1}{2}\pi)$, $\Psi(n-\frac{1}{2}\pi)$, $X(2\Omega-\frac{1}{2}\pi)$ are functions of the same form with sines replacing cosines. When the arguments of the functions are simply $2n$, n, 2Ω respectively, they will be omitted and the functions written simply Φ, Ψ, X; and when the arguments are simply $2n-\frac{1}{2}\pi$, $n-\frac{1}{2}\pi$, $2\Omega-\frac{1}{2}\pi$, they will be omitted and the functions written Φ', Ψ', X'. These functions may of course be expanded like sines and cosines, *e.g.*, $\Psi(n-\alpha)=\Psi\cos\alpha+\Psi'\sin\alpha$ and $\Psi'(n-\alpha)=\Psi'\cos\alpha-\Psi\sin\alpha$.

If now these functions are introduced into the expression for V, and if we replace the direction cosines $\sin\theta\cos\phi$, $\sin\theta\sin\phi$, $\cos\theta$ of the point P by ξ, η, ζ, we have

$$\frac{V}{w\tau r^2}=-(\xi^2-\eta^2)\Phi-2\xi\eta\Phi'-\xi\zeta\Psi-\eta\zeta\Psi'+\tfrac{1}{3}(\xi^2+\eta^2-2\zeta^2)[X+\tfrac{1}{2}(1-6p^2q^2)] \quad \text{......(5)}$$

$\xi^2-\eta^2$, $2\xi\eta$, $\xi\zeta$, $\eta\zeta$, $\frac{1}{3}(\xi^2+\eta^2-2\zeta^2)$ are surface harmonics of the second order, and the auxiliary functions involve only simple harmonic functions of the time. Hence we have obtained V in the desired form.

We shall require later certain functions of the direction cosines of the moon referred to A, B, C expressed in terms of the auxiliary functions. The formation of these functions may be most conveniently done before proceeding further.

Let x, y, z be these direction cosines, then

$$\cos PM = x\xi+y\eta+z\zeta$$

whence

$$\begin{aligned}\cos^2 PM-\tfrac{1}{3}&=(x\xi+y\eta+z\zeta)^2-\tfrac{1}{3}(\xi^2+\eta^2+\zeta^2)\\&=\xi^2(x^2-\tfrac{1}{3})+\eta^2(y^2-\tfrac{1}{3})+\zeta^2(z^2-\tfrac{1}{3})+2\eta\zeta yz+2\zeta\xi zx+2\xi\eta xy\end{aligned} \quad \text{......(6)}$$

But from (5) we have on rearranging the terms,

$$\begin{aligned}\cos^2 PM-\tfrac{1}{3}&=\xi^2\{-\Phi+\tfrac{1}{3}X+\tfrac{1}{6}(1-6p^2q^2)\}+\eta^2\{\Phi+\tfrac{1}{3}X+\tfrac{1}{6}(1-6p^2q^2)\}\\&\quad+\zeta^2\{-\tfrac{2}{3}X-\tfrac{1}{3}(1-6p^2q^2)\}-2\eta\zeta\,.\,\tfrac{1}{2}\Psi'-2\zeta\xi\,.\,\tfrac{1}{2}\Psi-2\xi\eta\Phi'\end{aligned} \quad \text{...(5')}$$

Equating coefficients in these two expressions (5') and (6)

$$\begin{aligned}x^2-\tfrac{1}{3}&=-\Phi+\tfrac{1}{3}X+\tfrac{1}{6}(1-6p^2q^2)\\y^2-\tfrac{1}{3}&=\Phi+\tfrac{1}{3}X+\tfrac{1}{6}(1-6p^2q^2)\\z^2-\tfrac{1}{3}&=-\tfrac{2}{3}X-\tfrac{1}{3}(1-6p^2q^2)\end{aligned}$$

Whence

$$\left.\begin{aligned}y^2-z^2&=\Phi+X+\tfrac{1}{2}(1-6p^2q^2)\\z^2-x^2&=\Phi-X-\tfrac{1}{2}(1-6p^2q^2)\\x^2-y^2&=-2\Phi\\ \text{also}\quad yz&=-\tfrac{1}{2}\Psi'\\zx&=-\tfrac{1}{2}\Psi\\xy&=-\Phi'\end{aligned}\right\}\quad \text{..................(7)}$$

These six equations (7) are the desired functions of x, y, z in terms of the auxiliary functions.

§ 2. *The form of the spheroid as tidally distorted.*

The tide-generating potential has thirteen terms, each consisting of a solid harmonic of the second degree multiplied by a simple harmonic function of the time, viz.: three in Φ, three in Φ', three in Ψ, three in Ψ', and one in X. The fourteenth term of V can raise no proper tide, because it is independent of the time, but it produces a permanent increment to the ellipticity of the mean spheroid.

Hence according to our hypothesis, explained in the introductory remarks, there will be thirteen distinct simple tides; the three tides corresponding to Φ' may however be compounded with the three in Φ, and similarly the Ψ' tides with the Ψ tides. Hence there are seven tides with speeds* $[2n - 2\Omega,\ 2n,\ 2n + 2\Omega]$, $[n - 2\Omega,\ n,\ n + 2\Omega]$, $[2\Omega]$, and each of these will be retarded by its own special amount.

The Φ tides have periods of nearly a half-day, and will be called the slow, sidereal, and fast semi-diurnal tides, the Ψ tides have periods of nearly a day, and will be called the slow, sidereal, and fast diurnal tides, and the X tide has a period of a fortnight, and is called the fortnightly tide.

The retardation of phase of each tide will be called the "lag," and the height of each tide will be expressed as a fraction of the corresponding equilibrium tide of a perfectly fluid spheroid. The following schedule gives the symbols to be introduced to express lag and reduction of tide:

	Semi-diurnal			Diurnal			Fortnightly
	Slow	Sidereal	Fast	Slow	Sidereal	Fast	
Tide	$(2n-2\Omega)$	$(2n)$	$(2n+2\Omega)$	$(n-2\Omega)$	(n)	$(n+2\Omega)$	(2Ω)
Height	E_1	E	E_2	E_1'	E'	E_2'	E''
Lag.	$2\epsilon_1$	2ϵ	$2\epsilon_2$	ϵ_1'	ϵ'	ϵ_2'	$2\epsilon''$

The E's are proper fractions, and the ϵ's are angles.

Let $r = a + \sigma$ be the equation to the surface of the spheroid as tidally distorted, a being the radius of the mean sphere,—for we may put out of account the permanent equatorial protuberance due to rotation, and to the non-periodic term of V.

It is a well-known result that, if $wr^2 \mathrm{S} \cos(vt + \eta)$ be a tide-generating potential, estimated per unit volume of a homogeneous perfectly fluid spheroid of density w, (S being of the second order of surface harmonics), the

* The useful term "speed" is due, I believe, to Sir William Thomson, and is much wanted to indicate the angular velocity of the radius of a circle, the inclination of which to a fixed radius gives the argument of a trigonometrical term. It will be used throughout this paper to indicate v, as it occurs in expressions of the type $\cos(vt + \eta)$.

equilibrium tide due to this potential is given by $\sigma = \frac{5a^2}{2g} \mathrm{S} \cos(vt + \eta)$. If we write $\mathfrak{g} = \frac{2g}{5a}$, this result may be written $\frac{\sigma}{a} = \frac{\mathrm{S}}{\mathfrak{g}} \cos(vt + \eta)$.

Now consider a typical term—say one part of the slow semi-diurnal term —of the tide-generating potential, as found in (3): it was

$$- wr^2\tau \tfrac{1}{2}p^4 \sin^2\theta \cos 2\phi \cos 2(n - \Omega)$$

The equilibrium value of the corresponding tide is found by putting $\frac{\sigma}{a}$ equal to this expression divided by $wr^2\mathfrak{g}$.

If we suppose that there is a frictional resistance to the tidal motion, the tide will lag and be reduced in height, and according to the preceding definitions the corresponding tide of our spheroid is expressed by

$$\frac{\sigma}{a} = -\frac{\tau}{\mathfrak{g}} E_1 \tfrac{1}{2} p^4 \sin^2\theta \cos 2\phi \cos[2(n - \Omega) - 2\epsilon_1]$$

All the other tides may be treated in the same way, by introducing the proper E's and ϵ's.

Thus if we write

$$\left.\begin{aligned} \Phi_\epsilon &= E_1 \tfrac{1}{2}p^4 \cos(2n - 2\Omega - 2\epsilon_1) + Ep^2q^2 \cos(2n - 2\epsilon) \\ &\qquad + E_2 \tfrac{1}{2}q^4 \cos(2n + 2\Omega - 2\epsilon_2) \\ \Psi_\epsilon &= E_1' \, 2p^3q \cos(n - 2\Omega - \epsilon_1') - E' \, 2pq(p^2 - q^2)\cos(n - \epsilon') \\ &\qquad - E_2' \, 2pq^3 \cos(n + 2\Omega - \epsilon_2') \\ \mathrm{X}_\epsilon &= E'' \, 3p^2q^2 \cos(2\Omega - 2\epsilon'') \end{aligned}\right\} \quad \ldots(8)$$

and if in the same symbols accented sines replace cosines, then, by comparison with (5), we see that

$$\frac{\mathfrak{g}}{\tau}\frac{\sigma}{a} = -(\xi^2 - \eta^2)\Phi_\epsilon - 2\xi\eta\Phi_\epsilon' - \xi\zeta\Psi_\epsilon - \eta\zeta\Psi_\epsilon' + \tfrac{1}{3}(\xi^2 + \eta^2 - 2\zeta^2)\mathrm{X}_\epsilon \ldots(9)$$

This is merely a symbolical way of writing down that every term in the tide-generating potential raises a lagging tide of its own type, but that tides of different speeds have different heights and lags.

This same expression may also be written

$$\frac{\mathfrak{g}}{\tau}\frac{\sigma}{a} = -\xi^2\{\Phi_\epsilon - \tfrac{1}{3}\mathrm{X}_\epsilon\} - \eta^2\{-\Phi_\epsilon - \tfrac{1}{3}\mathrm{X}_\epsilon\} - \zeta^2 \tfrac{2}{3}\mathrm{X}_\epsilon - 2\eta\zeta\tfrac{1}{2}\Psi_\epsilon' - 2\zeta\xi\tfrac{1}{2}\Psi_\epsilon - 2\xi\eta\Phi_\epsilon' \quad \ldots\ldots(9')$$

Then if we put

$$\left.\begin{aligned} \mathrm{c} - \mathrm{b} &= \Phi_\epsilon + \mathrm{X}_\epsilon \\ \mathrm{a} - \mathrm{c} &= \Phi_\epsilon - \mathrm{X}_\epsilon \\ \mathrm{b} - \mathrm{a} &= -2\Phi_\epsilon \\ \mathrm{c} &= \tfrac{2}{3}\mathrm{X}_\epsilon \\ \mathrm{d} &= -\tfrac{1}{2}\Psi_\epsilon' \\ \mathrm{e} &= -\tfrac{1}{2}\Psi_\epsilon \\ \mathrm{f} &= -\Phi_\epsilon' \end{aligned}\right\} \quad \ldots\ldots\ldots\ldots\ldots\ldots\ldots(10)$$

it is clear that

$$\frac{\mathfrak{g}}{\tau}\frac{\sigma}{a} = -\mathrm{a}\xi^2 - \mathrm{b}\eta^2 - \mathrm{c}\zeta^2 + 2\mathrm{d}\eta\zeta + 2\mathrm{e}\zeta\xi + 2\mathrm{f}\xi\eta \quad \ldots\ldots\ldots\ldots(11)$$

Whence

$$\left.\begin{aligned}
\frac{\mathfrak{g}}{2\tau}\left(\eta\frac{d}{d\zeta} - \zeta\frac{d}{d\eta}\right)\frac{\sigma}{a} &= -\{(\mathrm{c}-\mathrm{b})\,\eta\zeta - \mathrm{d}\,(\eta^2-\zeta^2) - \mathrm{e}\xi\eta + \mathrm{f}\zeta\xi\}\\
\frac{\mathfrak{g}}{2\tau}\left(\zeta\frac{d}{d\xi} - \xi\frac{d}{d\zeta}\right)\frac{\sigma}{a} &= -\{(\mathrm{a}-\mathrm{c})\,\zeta\xi - \mathrm{e}\,(\zeta^2-\xi^2) - \mathrm{f}\eta\zeta + \mathrm{d}\xi\eta\}\\
\frac{\mathfrak{g}}{2\tau}\left(\xi\frac{d}{d\eta} - \eta\frac{d}{d\xi}\right)\frac{\sigma}{a} &= -\{(\mathrm{b}-\mathrm{a})\,\xi\eta - \mathrm{f}\,(\xi^2-\eta^2) - \mathrm{d}\zeta\xi + \mathrm{e}\eta\zeta\}
\end{aligned}\right\} \ldots(12)$$

Of which expressions use will be made shortly.

§ 3. *The couples about the axes* A, B, C *caused by the moon's attraction.*

The earth is supposed to be a homogeneous spheroid of mean radius a, and mass w per unit volume, so that its mass $M = \frac{4}{3}\pi wa^3$. When undisturbed by tidal distortion it is a spheroid of revolution about the axis C, and its greatest and least principal moments of inertia are C, A. Upon this mean spheroid of revolution is superposed the tide-wave σ.

The attraction of the moon on the mean spheroid produces the ordinary precessional couples $2\tau\,(\mathrm{C}-\mathrm{A})\,yz$, $-2\tau\,(\mathrm{C}-\mathrm{A})\,zx$, 0 about the axes A, B, C respectively; besides these there are three couples, $\mathfrak{L}$, $\mathfrak{M}$, $\mathfrak{N}$ suppose, caused by the attraction on the wave surface σ.

As it is only desired to determine the corrections to the ordinary theory of precession, the former may be omitted from consideration, and attention confined to the determination of $\mathfrak{L}$, $\mathfrak{M}$, $\mathfrak{N}$.

The moon will be treated as an attractive particle of mass m.

Now σ as defined by (9) is a surface harmonic of the second order; hence by the ordinary formula in the theory of the potential, the gravitation potential of the tide-wave at a point whose coordinates referred to A, B, C are $r\xi$, $r\eta$, $r\zeta$ is $\frac{4}{5}\pi wa\left(\frac{a}{r}\right)^3\sigma$ or $\frac{3}{5}\frac{Ma}{r^3}\sigma$. Hence the moments about the axes A, B, C of the forces which act on a particle of mass m, situated at that point, are $\frac{3}{5}\frac{mMa}{r^3}\left(\eta\frac{d\sigma}{d\zeta} - \zeta\frac{d\sigma}{d\eta}\right)$, &c., &c. Then if this particle has the mass of the moon; if r be put equal to c, the moon's distance; and if ξ, η, ζ be replaced in σ by x, y, z (the moon's direction cosines) in the previous expressions, it is clear that $-\frac{2}{5}Ma\tau\left(y\frac{d\sigma}{dz} - z\frac{d\sigma}{dy}\right)$, &c., &c., are the couples on the earth caused by the moon's attraction.

These reactive couples are the required $\mathfrak{L}$, $\mathfrak{M}$, $\mathfrak{N}$.

Hence referring back to (12) and remarking that $\frac{2}{5}Ma^2 = C$, the earth's moment of inertia, we see at once that

$$\left.\begin{aligned} \frac{\mathfrak{L}}{C} &= \frac{2\tau^2}{\mathfrak{g}}[(\mathrm{c}-\mathrm{b})\,yz - \mathrm{d}\,(y^2-z^2) - \mathrm{e}xy + \mathrm{f}zx] \\ \frac{\mathfrak{M}}{C} &= \frac{2\tau^2}{\mathfrak{g}}[(\mathrm{a}-\mathrm{c})\,zx - \mathrm{e}\,(z^2-x^2) - \mathrm{f}yz + \mathrm{d}xy] \\ \frac{\mathfrak{N}}{C} &= \frac{2\tau^2}{\mathfrak{g}}[(\mathrm{b}-\mathrm{a})\,xy - \mathrm{f}\,(x^2-y^2) - \mathrm{d}zx + \mathrm{e}yz] \end{aligned}\right\} \quad \text{...........(13)}$$

where the quantities on the right-hand side are defined by the thirteen equations (7) and (10).

I shall confine my attention to determining the alteration in the uniform precession, the change in the obliquity of the ecliptic, and the tidal friction; because the nutations produced by the tidal motion will be so small as to possess no interest.

In developing $\mathfrak{L}$ and $\mathfrak{M}$ I shall only take into consideration the terms with argument n, and in $\mathfrak{N}$ only constant terms; for it will be seen, when we come to the equations of motion, that these are the only terms which can lead to the desired end.

§ 4. *Development of the couples $\mathfrak{L}$ and $\mathfrak{M}$.*

Now substitute from (7) and (10) in the first of (13), and we have

$$\frac{\mathfrak{L}}{C} \div \frac{2\tau^2}{\mathfrak{g}} = -\tfrac{1}{2}\{\Phi_\epsilon + X_\epsilon\}\Psi' + \tfrac{1}{2}\Psi_\epsilon'\{\Phi + X + \tfrac{1}{2}(1-6p^2q^2)\} - \tfrac{1}{2}\Psi_\epsilon\Phi' + \tfrac{1}{2}\Phi_\epsilon'\Psi \quad \text{......(14)}$$

A number of multiplications have now to be performed, and only those terms which contain the argument n to be retained.

The particular argument n can only arise in six ways, viz.: from products of terms with arguments

$2(n-\Omega),\ n-2\Omega;\ 2n,\ n;\ 2(n+\Omega),\ n+2\Omega;\ n-2\Omega,\ 2\Omega;\ n+2\Omega,\ 2\Omega$

and from terms of argument n multiplied by constant terms.

If Φ and Ψ, and Φ' and Ψ' be written underneath one another in the various combinations in which they occur in the above expression, it will be obvious that the desired argument can only arise from terms which stand one vertically over the other; this renders the multiplication easier. The Ψ, X products are comparatively easy.

Then we have

$$(\alpha)\quad -\tfrac{1}{2}\Phi_\epsilon\Psi' = -\tfrac{1}{4}[-E_1p^7q\sin(n-2\epsilon_1) + 2Ep^3q^3(p^2-q^2)\sin(n-2\epsilon) + E_2pq^7\sin(n-2\epsilon_2)]$$

$(\beta) \quad + \tfrac{1}{2}\Psi_\epsilon{}'\Phi = + \tfrac{1}{4}\left[- E_1{}'p^7q \sin(n + \epsilon_1{}') + 2E'p^3q^3(p^2 - q^2)\sin(n + \epsilon')\right.$
$\left. + E_2{}'pq^7 \sin(n + \epsilon_2{}')\right]$

$(\gamma) \quad - \tfrac{1}{2}\Psi_\epsilon\Phi' =$ same as (β)

$(\delta) \quad + \tfrac{1}{2}\Phi_\epsilon{}'\Psi =$ same as (α)

$(\epsilon) \quad - \tfrac{1}{2}X_\epsilon\Psi' = - \tfrac{1}{4}\left[E''6p^5q^3 \sin(n - 2\epsilon'') - E''6p^3q^5 \sin(n + 2\epsilon'')\right]$

$(\zeta) \quad + \tfrac{1}{2}\Psi_\epsilon{}'X = + \tfrac{1}{4}\left[E_1{}'6p^5q^3 \sin(n - \epsilon_1{}') - E_2{}'6p^3q^5 \sin(n - \epsilon_2{}')\right]$

$(\eta) \quad + \tfrac{1}{4}\Psi_\epsilon{}'(1 - 6p^2q^2) = - \tfrac{1}{4}E'2pq(p^2 - q^2)(1 - 6p^2q^2)\sin(n - \epsilon')$

Put $\dfrac{\mathfrak{L}}{C} = F \sin n + G \cos n$. Then if the expressions (α), (β) ... (ζ) be added up when $n = \tfrac{1}{2}\pi$, and the sum multiplied by $2\tau^2/\mathfrak{g}$, we shall get F; and if we perform the same addition and multiplication when $n = 0$, we shall get G.

In performing the first addition the terms (α), (δ) do not combine with any other, but the terms (β), (γ), (ζ), (η) combine.

Now

$$- \tfrac{1}{2}p^7q + \tfrac{3}{2}p^5q^3 = - \tfrac{1}{2}p^5q(p^2 - 3q^2)$$

$$p^3q^3(p^2 - q^2) - \tfrac{1}{2}pq(p^2 - q^2)(1 - 6p^2q^2) = - \tfrac{1}{2}pq(p^2 - q^2)(p^4 + q^4 - 6p^2q^2)$$

$$\tfrac{1}{2}pq^7 - \tfrac{3}{2}p^3q^5 = - \tfrac{1}{2}pq^5(3p^2 - q^2)$$

$$- \tfrac{3}{2}p^5q^3 + \tfrac{3}{2}p^3q^5 = - \tfrac{3}{2}p^3q^3(p^2 - q^2)$$

Hence

$$\begin{aligned} F \div \frac{2\tau^2}{\mathfrak{g}} = \; & \tfrac{1}{2}E_1p^7q \cos 2\epsilon_1 - Ep^3q^3(p^2 - q^2)\cos 2\epsilon - \tfrac{1}{2}E_2pq^7 \cos 2\epsilon_2 \\ & - \tfrac{1}{2}E_1{}'p^5q(p^2 - 3q^2)\cos\epsilon_1{}' - \tfrac{1}{2}E'pq(p^2 - q^2)(p^4 + q^4 - 6p^2q^2)\cos\epsilon' \\ & - \tfrac{1}{2}E_2{}'pq^5(3p^2 - q^2)\cos\epsilon_2{}' \\ & - \tfrac{3}{2}E''p^3q^3(p^2 - q^2)\cos 2\epsilon'' \quad \ldots\ldots\ldots\ldots\ldots\ldots (15) \end{aligned}$$

Again for the second addition when $n = 0$, we have

$$- \tfrac{1}{2}p^7q - \tfrac{3}{2}p^5q^3 = - \tfrac{1}{2}p^5q(p^2 + 3q^2)$$

$$p^3q^3(p^2 - q^2) + \tfrac{1}{2}pq(p^2 - q^2)(1 - 6p^2q^2) = \tfrac{1}{2}pq(p^2 - q^2)^3$$

$$\tfrac{1}{2}pq^7 + \tfrac{3}{2}p^3q^5 = \tfrac{1}{2}pq^5(3p^2 + q^2)$$

$$\tfrac{3}{2}p^5q^3 + \tfrac{3}{2}p^3q^5 = \tfrac{3}{2}p^3q^3$$

So that

$$\begin{aligned} G \div \frac{2\tau^2}{\mathfrak{g}} = & - \tfrac{1}{2}E_1p^7q \sin 2\epsilon_1 + Ep^3q^3(p^2 - q^2)\sin 2\epsilon + \tfrac{1}{2}E_2pq^7 \sin 2\epsilon_2 \\ & - \tfrac{1}{2}E_1{}'p^5q(p^2 + 3q^2)\sin\epsilon_1{}' + \tfrac{1}{2}E'pq(p^2 - q^2)^3 \sin\epsilon' \\ & + \tfrac{1}{2}E_2{}'pq^5(3p^2 + q^2)\sin\epsilon_2{}' \\ & + \tfrac{3}{2}E''p^3q^3 \sin 2\epsilon'' \quad \ldots\ldots\ldots\ldots\ldots\ldots (16) \end{aligned}$$

And
$$\frac{\mathfrak{L}}{\mathrm{C}} = \mathrm{F} \sin n + \mathrm{G} \cos n \qquad (17)$$

To find $\mathfrak{M}$ it is only necessary to substitute $n - \frac{1}{2}\pi$ for n, and we have

$$\frac{\mathfrak{M}}{\mathrm{C}} = -\mathrm{F} \cos n + \mathrm{G} \sin n \qquad (18)$$

There is a certain approximation which gives very nearly correct results and which simplifies these expressions very much. It has already been remarked that the three Φ-tides have periods of nearly a half-day and the three Ψ-tides of nearly a day, and this will continue to be true so long as Ω is small compared with n; hence it may be assumed with but slight error that the semi-diurnal tides are all retarded by the same amount and that their heights are proportional to the corresponding terms in the tide-generating potential. That is, we may put $\epsilon_1 = \epsilon_2 = \epsilon$ and $E_1 = E_2 = E$. The similar argument with respect to the diurnal tides permits us to put $\epsilon_1' = \epsilon_2' = \epsilon'$ and $E_1' = E_2' = E'$.

Introducing the quantities $P = p^2 - q^2 = \cos i$, $Q = 2pq = \sin i$ and observing that

$$\tfrac{1}{2}p^7q - p^3q^3(p^2 - q^2) - \tfrac{1}{2}pq^7$$
$$= \tfrac{1}{2}pq\left[(p^2 - q^2)(p^4 + p^2q^2 + q^4) - 2p^2q^2(p^2 - q^2)\right] = \tfrac{1}{4}PQ(1 - \tfrac{3}{4}Q^2)$$
$$\tfrac{1}{2}p^5q(p^2 - 3q^2) + \tfrac{1}{2}pq(p^2 - q^2)(p^4 + q^4 - 6p^2q^2) + \tfrac{1}{2}pq^5(3p^2 - q^2)$$
$$= pq(p^2 - q^2)(1 - 6p^2q^2) = \tfrac{1}{2}PQ(1 - \tfrac{3}{2}Q^2)$$
$$\tfrac{1}{2}p^5q(p^2 + 3q^2) - \tfrac{1}{2}pq(p^2 - q^2)^3 - \tfrac{1}{2}pq^5(3p^2 + q^2)$$
$$= \tfrac{1}{2}pq(p^2 - q^2)(1 + 2p^2q^2 - 1 + 4p^2q^2) = \tfrac{3}{8}PQ^3$$

we have

$$\left.\begin{aligned} \mathrm{F} \div \frac{\tau^2}{\mathfrak{g}} &= \tfrac{1}{2}EPQ(1 - \tfrac{3}{4}Q^2)\cos 2\epsilon - E'PQ(1 - \tfrac{3}{2}Q^2)\cos\epsilon' - \tfrac{3}{8}E''PQ^3\cos 2\epsilon'' \\ \mathrm{G} \div \frac{\tau^2}{\mathfrak{g}} &= -\tfrac{1}{2}EPQ(1 - \tfrac{3}{4}Q^2)\sin 2\epsilon - \tfrac{3}{4}E'PQ^3\sin\epsilon' + \tfrac{3}{8}E''Q^3\sin 2\epsilon'' \end{aligned}\right\} \qquad (19)$$

§ 5. *Development of the couple $\mathfrak{N}$.*

In the couple $\mathfrak{N}$ about the axis of rotation of the earth we only wish to retain non-periodic terms, and these can only arise from the products of terms with the same argument.

By substitution from (7) and (10) in the last of (13)

$$\frac{\mathfrak{N}}{\mathrm{C}} \div \frac{2\tau^2}{\mathfrak{g}} = 2\Phi_\epsilon\Phi' - 2\Phi_\epsilon'\Phi - \tfrac{1}{4}\Psi_\epsilon'\Psi + \tfrac{1}{4}\Psi_\epsilon\Psi' \qquad (20)$$

As far as we are now interested,

$$2\Phi_\epsilon\Phi' = -2\Phi_\epsilon'\Phi = E_1\tfrac{1}{4}p^8\sin 2\epsilon_1 \quad + Ep^4q^4\sin 2\epsilon \qquad\qquad + E_2\tfrac{1}{4}q^8\sin 2\epsilon_2$$

$$-\tfrac{1}{4}\Psi_\epsilon'\Psi = \tfrac{1}{4}\Psi_\epsilon\Psi' = E_1'\tfrac{1}{2}p^6q^2\sin\epsilon_1' + E'\tfrac{1}{2}p^2q^2(p^2-q^2)^2\sin\epsilon' + E_2'\tfrac{1}{2}p^2q^6\sin\epsilon_2'$$

Hence

$$\frac{\mathfrak{N}}{\text{C}} \div \frac{\tau^2}{\mathfrak{g}} = E_1p^8\sin 2\epsilon_1 \quad + E4p^4q^4\sin 2\epsilon \qquad\qquad + E_2q^8\sin 2\epsilon_2$$
$$+ E_1'2p^6q^2\sin\epsilon_1' + E'2p^2q^2(p^2-q^2)^2\sin\epsilon' + E_2'2p^2q^6\sin\epsilon_2' \quad (21)$$

If as in the last section we group the semi-diurnal and diurnal terms together and put $E_1 = E_2 = E$, &c., and observe that

$$p^8 + 4p^4q^4 + q^8 = (p^4+q^4)^2 + 2p^4q^4 = (1-\tfrac{1}{2}Q^2)^2 + \tfrac{1}{8}Q^4 = P^2 + \tfrac{3}{8}Q^4,$$

$$2p^6q^2 + 2p^2q^2(p^2-q^2)^2 + 2p^2q^6 = 4p^2q^2[p^4+q^4-p^2q^2] = Q^2(1-\tfrac{3}{4}Q^2),$$

whence $$\frac{\mathfrak{N}}{\text{C}} \div \frac{\tau^2}{\mathfrak{g}} = E(P^2+\tfrac{3}{8}Q^4)\sin 2\epsilon + E'Q^2(1-\tfrac{3}{4}Q^2)\sin\epsilon' \quad \text{.........(22)}$$

§ 6. *The equations of motion of the earth about its centre of inertia.*

In forming the equations of motion we are met by a difficulty, because the axes A, B, C are neither principal axes, nor can they rigorously be said to be fixed in the earth. But M. Liouville has given the equations of motion of a body which is changing its shape, using any set of rectangular axes which move in any way with reference to the body, except that the origin always remains at the centre of inertia.

If A, B, C, D, E, F be the moments and products of inertia of the body about these axes of reference at any time; H_1, H_2, H_3 the moments of momentum of the motion of all the parts of the body relative to the axes; ω_1, ω_2, ω_3 the component angular velocities of the axes about their instantaneous positions, the equations may be written

$$\frac{d}{dt}(\text{A}\omega_1 - \text{F}\omega_2 - \text{E}\omega_3 + \text{H}_1) + \text{D}(\omega_3^2 - \omega_2^2) + (\text{C}-\text{B})\,\omega_2\omega_3$$
$$+ \text{F}\omega_3\omega_1 - \text{E}\omega_2\omega_1 + \omega_2\text{H}_3 - \omega_3\text{H}_2 = \text{L} \quad \text{........(23)}$$

and two other equations found from this by cyclical changes of letters and suffixes*.

Now in the case to be considered here the axes A, B, C always occupy the average position of the same line of particles, and they move with very nearly an ordinary uniform precessional motion. Also the moments and products of inertia may be written A + a′, B + b′, C + c′, d′, e′, f′, where a′, b′, c′, d′, e′, f′ are small periodic functions of the time and a′ + b′ + c′ = 0, and where

* Routh's *Rigid Dynamics* (first edition only), p. 150, or my paper in the *Phil. Trans.*, 1877, Vol. 167, p. 272 [to be reproduced in Vol. III.]. The original is in *Liouville's Journal*, 2nd series, Vol. III., 1858, p. 1.

A, B, C are the principal moments of inertia of the undisturbed earth, so that B is equal to A.

The quantities a′, b′, &c., have in effect been already determined, as may be shown as follows: By the ordinary formula* the force function of the moon's action on the earth is $\frac{Mm}{c} + \frac{1}{3}\tau(A + B + C - 3I)$, where I is the moment of inertia of the earth about the line joining its centre to the moon, and is therefore

$$= Ax^2 + By^2 + Cz^2 + a'x^2 + b'y^2 + c'z^2 - 2d'yz - 2e'zx - 2f'xy$$

But the first three terms of I only give rise to the ordinary precessional couples, and a comparison of the last six with (11) and (13) shows that

$$\frac{a'}{a} = \frac{b'}{b} = \frac{c'}{c} = \frac{d'}{d} = \frac{e'}{e} = \frac{f'}{f} = \frac{\tau}{\mathfrak{g}} . C$$

Also in the small terms we may ascribe to ω_1, ω_2, ω_3 their uniform precessional values, viz.: $\omega_1 = -\Pi \cos n$, $\omega_2 = -\Pi \sin n$, $\omega_3 = -n$.

When these values are substituted in (23), we get some small terms of the form $a'\Pi^2 \sin n$, and others of the form $a'\Pi n \sin n$; both these are very small compared to the terms in $\mathfrak{L}$ and $\mathfrak{M}$—the fractions which express their relative magnitude being Π^2/τ and $\Pi n/\tau$.

There is also a term $-\Pi H_3 \sin n$, which I conceive may also be safely neglected, as also the similar terms in the second and third equations.

It is easy, moreover, to show that according to the theories of the tidal motion of a homogeneous viscous spheroid given in the previous paper, and according to Sir William Thomson's theory of elastic tides, H_1, H_2, H_3 are all zero. Those theories both neglect inertia but the actuality is not likely to differ materially therefrom.

Thus every term where ω_1 and ω_2 occur may be omitted and the equations reduced to

$$\left.\begin{aligned} A\frac{d\omega_1}{dt} + (C - B)\,\omega_2\omega_3 + n\frac{de'}{dt} + n^2 d' + \frac{dH_1}{dt} + nH_2 &= \mathfrak{L} \\ B\frac{d\omega_2}{dt} + (A - C)\,\omega_3\omega_1 + n\frac{dd'}{dt} - n^2 e' + \frac{dH_2}{dt} - nH_1 &= \mathfrak{M} \\ C\frac{d\omega_3}{dt} + (B - A)\,\omega_1\omega_2 + \frac{dH_3}{dt} &= \mathfrak{N} \end{aligned}\right\} \quad \ldots(24)$$

As before with the couples, so here, we are only interested in terms with the argument n in the small terms on the left-hand side of the first two of equations (24), and in non-periodic terms in the last of them.

* Routh's *Rigid Dynamics*, 1877, p. 495.

Now for each term in the moon's potential, as developed in Section 1, there is (by hypothesis) a corresponding co-periodic flux and reflux throughout the earth's mass, and therefore the H_1, H_2, H_3 must each have periodic terms corresponding to each term in the moon's potential. Hence the only term in the moon's potential to be considered is that with argument n, with respect to H_1 and H_2 in the first two equations; and H_3 may be omitted from the third as being periodic.

Suppose that H_1 was equal to $h \cos n + h' \sin n$, then precisely as we found $\mathfrak{M}$ from $\mathfrak{L}$ by writing $n - \frac{1}{2}\pi$ for n we have $H_2 = h \sin n - h' \cos n$. Thus $\frac{dH_1}{dt} + nH_2 = 0$, $\frac{dH_2}{dt} - nH_1 = 0$, and the H's disappear from the first two equations.

Next retaining only terms in argument n in d′ and e′, we have from (10)

$$e' = C \frac{\tau}{\mathfrak{g}} E' pq (p^2 - q^2) \cos (n - \epsilon'), \quad d' = C \frac{\tau}{\mathfrak{g}} E' pq (p^2 - q^2) \sin (n - \epsilon')$$

Therefore $\frac{de'}{dt} + nd' = 0$, $\frac{dd'}{dt} - ne' = 0$, and those terms also disappear.

Lastly, put $B = A$, and our equations reduce simply to those of Euler, viz.:

$$\left.\begin{aligned} A \frac{d\omega_1}{dt} + (C - A)\,\omega_2\omega_3 &= \mathfrak{L} \\ A \frac{d\omega_2}{dt} - (C - A)\,\omega_3\omega_1 &= \mathfrak{M} \\ C \frac{d\omega_3}{dt} \qquad\qquad &= \mathfrak{N} \end{aligned}\right\} \quad \text{.....................(25)}$$

Now $\mathfrak{N}$ is small, and therefore ω_3 remains approximately constant and equal to $-n$ for long periods, and as $C - A$ is small compared to A, we may put $\omega_3 = -n$ in the first two equations. But when $C - A$ is neglected compared to C, the integrals of these equations are the same as those of

$$\frac{d\omega_1}{dt} = \frac{\mathfrak{L}}{C}, \quad \frac{d\omega_2}{dt} = \frac{\mathfrak{M}}{C}, \quad \frac{d\omega_3}{dt} = \frac{\mathfrak{N}}{C} \quad \text{.................(26)}$$

apart from the complementary function, which may obviously be omitted. The two former of (26) give the change in the precession and the obliquity of the ecliptic, and the last gives the tidal friction.

§ 7. *Precession and change of obliquity.*

By (17), (18), and (26) the equations of motion are

$$\left.\begin{aligned}\frac{d\omega_1}{dt} &= \mathrm{F}\sin n + \mathrm{G}\cos n\\ \frac{d\omega_2}{dt} &= -\mathrm{F}\cos n + \mathrm{G}\sin n\end{aligned}\right\}\quad\ldots\ldots\ldots\ldots(27)$$

and by integration

$$\omega_1 = \frac{1}{n}[-\mathrm{F}\cos n + \mathrm{G}\sin n],\quad \omega_2 = \frac{1}{n}[-\mathrm{F}\sin n - \mathrm{G}\cos n]\ldots\ldots(28)$$

The geometrical equations (1) give

$$\frac{di}{dt} = -\omega_1\sin n + \omega_2\cos n$$

$$\frac{d\psi}{dt}\sin i = -\omega_1\cos n - \omega_2\sin n$$

Therefore, as far as concerns non-periodic terms,

$$\frac{di}{dt} = -\frac{\mathrm{G}}{n},\quad \frac{d\psi}{dt}\sin i = \frac{\mathrm{F}}{n}\ldots\ldots\ldots\ldots(29)$$

If we wish to keep all the seven tides distinct (as will have to be done later), we may write down the result for $\frac{di}{dt}$ and $\frac{d\psi}{dt}$ from (15) and (16).

But it is of more immediate interest to consider the case where the semi-diurnal tides are grouped together, as also the diurnal ones. In this case we have by (19)

$$\frac{di}{dt} = \frac{\tau^2}{\mathfrak{g}n}\left\{\tfrac{1}{2}PQ\left(1-\tfrac{3}{4}Q^2\right)E\sin 2\epsilon + \tfrac{3}{4}PQ^3E'\sin\epsilon' - \tfrac{3}{8}Q^3E''\sin 2\epsilon''\right\}\ldots\ldots(30)$$

and since $\sin i = Q$

$$\frac{d\psi}{dt} = \frac{\tau^2}{\mathfrak{g}n}\left\{\tfrac{1}{2}P\left(1-\tfrac{3}{4}Q^2\right)E\cos 2\epsilon - P\left(1-\tfrac{3}{2}Q^2\right)E'\cos\epsilon' - \tfrac{3}{8}PQ^2E''\cos 2\epsilon''\right\}\ (31)$$

In these equations P and Q stand for the cosine and sine of the obliquity of the ecliptic.

Several conclusions may be drawn from this result.

If ϵ, ϵ', ϵ'' are zero the obliquity remains constant.

Now if the spheroid be perfectly elastic, the tides do not lag, and therefore the obliquity remains unchanged; it would also be easy to find the correction to the precession to be applied in the case of elasticity.

It is possible that the investigation is not, strictly speaking, applicable to the case of a perfect fluid; I shall, however, show to what results it leads if we make the application to that case. Sir William Thomson has shown that the period of free vibration of a fluid sphere of the density of the earth would be about 1 hour 34 minutes*. And as this free period is pretty small compared to the forced period of the tidal oscillation, it follows that E, E', E'', will not differ much from unity. Putting them equal to unity, and putting $\epsilon, \epsilon', \epsilon''$ zero, since the tides do not lag, we find that the obliquity remains constant, and

$$\frac{d\psi}{dt} = -\frac{\tau^2}{\mathfrak{g}n}\tfrac{1}{2}P\left(1-\tfrac{3}{2}Q^2\right) = -\tfrac{1}{2}\frac{\tau^2}{\mathfrak{g}n}\cos i\left(1-\tfrac{3}{2}\sin^2 i\right) \text{......(32)}$$

This equation gives the correction to be applied to the precession as derived from the assumption that the rotating spheroid of fluid is rigid. This result is equally true if all the seven tides are kept distinct. Now if the spheroid were rigid its precession would be $\tau e \cos i/n$, where e is the ellipticity of the spheroid.

The ellipticity of a fluid spheroid rotating with an angular velocity n is $\frac{5}{4}n^2a/g$ or $\frac{1}{2}n^2/\mathfrak{g}$; but besides this, there is ellipticity due to the non-periodic part of the tide-generating potential.

By (3) § 1 the non-periodic part of V is $\frac{1}{2}w\tau r^2(\frac{1}{3}-\cos^2\theta)(1-6p^2q^2)$; such a disturbing potential will clearly produce an ellipticity $\frac{1}{2}\frac{\tau}{\mathfrak{g}}(1-6p^2q^2)$.

If therefore we put $e_0 = \frac{1}{2}\frac{n^2}{\mathfrak{g}}$, and remember that $6p^2q^2 = \frac{3}{2}\sin^2 i$, we have

$$e = e_0 + \tfrac{1}{2}\frac{\tau}{\mathfrak{g}}\left(1-\tfrac{3}{2}\sin^2 i\right)$$

Hence if the spheroid were rigid, and had its actual ellipticity, we should have

$$\frac{d\psi}{dt} = \frac{\tau e_0}{n}\cos i + \tfrac{1}{2}\frac{\tau^2}{\mathfrak{g}n}\cos i\left(1-\tfrac{3}{2}\sin^2 i\right) \text{...............(32')}$$

Adding (32′) to (32), the whole precession is

$$\frac{d\psi}{dt} = \frac{\tau e_0}{n}\cos i \text{..............................(32'')}$$

We thus see that the effect of the non-periodic part of the tide-generating potential, which may be conveniently called a permanent tide, is just such as to neutralise the effects of the tidal action. The result (32″) may be expressed as follows:

The precession of a fluid spheroid is the same as that of a rigid one which has an ellipticity equal to that due to the rotation of the spheroid.

* *Phil. Trans.*, 1863, p. 608.

From this it follows that the precession of a fluid spheroid will differ by little from that of a rigid one of the same ellipticity, if the additional ellipticity due to the non-periodic part of the tide-generating influence is small compared with the whole ellipticity.

Sir William Thomson has already expressed himself to somewhat the same effect in an address to the British Association at Glasgow*.

Since $e_0 = \frac{1}{2}\frac{n^2}{\mathfrak{g}}$, the criterion is the smallness of $\frac{\tau}{n^2}$.

It may be expressed in a different form; for τ/n^2 is small when $(\tau e/n) \div n$ is small compared with e, and $(\tau e/n) \div n$ is the reciprocal of the precessional period expressed in days. Hence the criterion may be stated thus: *The precession of a fluid spheroid differs by little from that of a rigid one of the same ellipticity, when the precessional period of the spheroid expressed in terms of its rotation is large compared with the reciprocal of its ellipticity.*

In his address, Sir William Thomson did not give a criterion for the case of a fluid spheroid without any confining shell, but for the case of a thin rigid spheroidal shell enclosing fluid he gave a statement which involves the above criterion, save that the ellipticity referred to is that of the shell itself; for he says, "The amount of this difference (in precession and nutation) bears the same proportion to the actual precession or nutation as the fraction measuring the periodic speed of the disturbance (in terms of the period of rotation as unity) bears to the fraction measuring the interior ellipticity of the shell."

This is, in fact, almost the same result as mine.

This subject is again referred to in Part III. of the succeeding paper.

§ 8. *The disturbing action of the sun.*

Now suppose that there is a second disturbing body, which may be conveniently called the sun†.

* See *Nature*, September 14, 1876, p. 429. [See G. H. Bryan, *Phil. Trans.*, Vol. 180, A (1889), p. 187.]

† It is not at first sight obvious how it is physically possible that the sun should exercise an influence on the moon-tide, and the moon on the sun-tide, so as to produce a secular change in the obliquity of the ecliptic and to cause tidal friction, for the periods of the sun and moon about the earth are different. It seems, therefore, interesting to give a physical meaning to the expansion of the tide-generating potential; it will then be seen that the interaction with which we are here dealing must occur.

The expansion of the potential given in Section 1 is equivalent to the following statement:—

The tide-generating potential of a moon of mass m, moving in a circular orbit of obliquity i at a distance c, is equal to the tide-generating potential of ten satellites at the same distance, whose orbits, masses, and angular velocities are as follows:—

$\Pi \operatorname{cosec} i$ must henceforth be taken as the full precession of the earth, and the time may be conveniently measured from an eclipse of the sun or moon.

---

1. A satellite of mass $m \cos^4 \frac{1}{2}i$, moving in the equator in the same direction and with the same angular velocity as the moon, and coincident with it at the nodes. This gives the slow semi-diurnal tide of speed $2(n - \Omega)$.

2. A satellite of mass $m \sin^4 \frac{1}{2}i$, moving in the equator in the opposite direction from that of the moon, but with the same angular velocity, and coincident with it at the nodes. This gives the fast semi-diurnal tide of speed $2(n + \Omega)$.

3. A satellite of mass $m \cdot 2 \sin^2 \frac{1}{2}i \cos^2 \frac{1}{2}i$, fixed at the moon's node. This gives the sidereal semi-diurnal tide of speed $2n$.

4. A repulsive satellite of mass $-m \cdot 2 \sin \frac{1}{2}i \cos^3 \frac{1}{2}i$, moving in N. declination 45° with twice the moon's angular velocity, in the same direction as the moon, and on the colure 90° in advance of the moon, when she is in her node.

5. A satellite of mass $m \sin i \cos^3 \frac{1}{2}i$, moving in the equator with twice the moon's angular velocity, and in the same direction, and always on the same meridian as the fourth satellite. (4) and (5) give the slow diurnal tide of speed $n - 2\Omega$.

6. A satellite of mass $m \sin^3 \frac{1}{2}i \cos \frac{1}{2}i$, moving in N. declination 45° with twice the moon's angular velocity, but in the opposite direction, and on the colure 90° in advance of the moon when she is in her node.

7. A repulsive satellite of mass $-m \cdot \frac{1}{2} \sin^3 \frac{1}{2}i \cos \frac{1}{2}i$, moving in the equator with twice the moon's angular velocity, but in the opposite direction, and always on the same meridian as the sixth satellite. (6) and (7) give the fast diurnal tide of speed $n + 2\Omega$.

8. A satellite of mass $m \sin i \cos i$ fixed in N. declination 45° on the colure.

9. A repulsive satellite of mass $-m \cdot \frac{1}{2} \sin i \cos i$, fixed in the equator on the same meridian as the eighth satellite. (8) and (9) give the sidereal diurnal tide of speed n.

10. A ring of matter of mass m, always passing through the moon and always parallel to the equator. This ring, of course, executes a simple harmonic motion in declination, and its mean position is the equator. This gives the fortnightly tide of speed 2Ω.

Now if we form the potentials of each of these satellites, and omit those parts which, being independent of the time, are incapable of raising tides, and add them altogether, we shall obtain the expansion for the moon's tide-generating potential used above; hence this system of satellites is mechanically equivalent to the action of the moon alone. The satellites 1, 2, 3, in fact, give the semi-diurnal or Φ terms; satellites 4, 5, 6, 7, 8, 9 give the diurnal or Ψ terms; and satellite 10 gives the fortnightly or X term.

This is analogous to "Gauss's way of stating the circumstances on which 'secular' variations in the elements of the solar system depend"; and the analysis was suggested to me by a passage in Thomson and Tait's *Natural Philosophy*, § 809, referring to the annular satellite 10.

It will appear in Section 22 that the 3rd, 8th, and 9th satellites, which are fixed in the heavens and which give the sidereal tides, are equivalent to a distribution of the moon's mass in the form of a uniform circular ring coincident with her orbit. And perhaps some other simpler plan might be given which would replace the other repulsive satellites.

These tides, here called "sidereal," are known, in the reports of the British Association on tides for 1872 and 1876, as the K tides [see Vol. I., Paper 1].

In a precisely similar way, it is clear that the sun's influence may be analysed into the influence of nine other satellites and one ring, or else to seven satellites and two rings. Then, with regard to the interaction of sun and moon, it is clear that those satellites of each system which are fixed in each system (viz.: 3, 8, and 9), or their equivalent rings, will not only exercise an influence on the tides raised by themselves, but each will necessarily exercise an influence on

Let $m_{,}$, $c_{,}$ be the sun's mass and distance; Ω, the earth's angular velocity in a circular orbit; and let $\tau_{,} = \frac{3}{2}\frac{m_{,}}{c_{,}^3}$.

It would be rigorously necessary to introduce a new set of quantities to give the heights and lagging of the seven solar tides: but of the three solar semi-diurnal tides, one has rigorously the same period as one of the three lunar semi-diurnal tides (viz.: the sidereal semi-diurnal with a speed $2n$), and the others have nearly the same period; a similar remark applies to the solar diurnal tides. Hence we may, without much error, treat E, ϵ, E', ϵ' as the same both for lunar and solar tides; but E''', ϵ''' must replace E'', ϵ'', because the semi-annual replaces the fortnightly tide.

If new auxiliary functions $\Phi_{,}$, $\Psi_{,}$, $X_{,}$ be introduced, the whole tide-generating potential V per unit volume of the earth at the point $r\xi$, $r\eta$, $r\zeta$ is given by

$$\frac{\mathrm{V}}{wr^2} = -(\tau\Phi + \tau_{,}\Phi_{,})(\xi^2 - \eta^2) \text{ \&c.}$$

Next if, as in (10), we put

$$\mathrm{c} - \mathrm{b} = \Phi_\epsilon + \mathrm{X}_\epsilon, \text{ \&c.}, \quad \mathrm{c}_{,} - \mathrm{b}_{,} = \Phi_{,\epsilon} + \mathrm{X}_{,\epsilon}, \text{ \&c.}$$

the equation to the tidally-distorted earth is $r = a + \sigma + \sigma_{,}$, where

$$\frac{\mathfrak{g}}{\tau}\frac{\sigma}{a} = -\mathrm{a}\xi^2 - \text{\&c.}, \quad \frac{\mathfrak{g}}{\tau_{,}}\frac{\sigma_{,}}{a} = -\mathrm{a}_{,}\xi^2 - \text{\&c.}$$

Also if x, y, z and $x_{,}, y_{,}, z_{,}$ be the moon's and sun's direction cosines, we have as in (7),

$$y^2 - z^2 = \Phi + \mathrm{X} + \tfrac{1}{2}(1 - 6p^2q^2), \text{ \&c.}, \quad y_{,}^2 - z_{,}^2 = \Phi_{,} + \mathrm{X}_{,} + \tfrac{1}{2}(1 - 6p^2q^2), \text{ \&c.}$$

Then using the same arguments as in Section 3, the couples about the three axes in the earth may be found, and we have

$$\frac{\mathfrak{L}}{\mathrm{C}} = -\left\{\tau\left(y\frac{d}{dz} - z\frac{d}{dy}\right)\left(\frac{\sigma}{a} + \frac{\sigma_{,}}{a}\right) + \tau_{,}\left(y_{,}\frac{d}{dz_{,}} - z_{,}\frac{d}{dy_{,}}\right)\left(\frac{\sigma}{a} + \frac{\sigma_{,}}{a}\right)\right\}$$

where in the first term x, y, z are written for ξ, η, ζ in $\sigma + \sigma_{,}$, and in the second term $x_{,}, y_{,}, z_{,}$ are similarly written for ξ, η, ζ.

Now let $\mathfrak{L}_{m^2}$, $\mathfrak{L}_{m_{,}^2}$, $\mathfrak{L}_{mm_{,}}$ indicate the parts of the couple $\mathfrak{L}$ which depend on the moon's action on the lunar tides, the sun's action on the solar tides,

---

the tides raised by the other, so as to produce tidal friction. All the other satellites will, of course, attract or repel the tides of all the other satellites of the other systems; but this interaction will necessarily be periodic, and will not cause any interaction in the way of tidal friction or change of obliquity, and as such periodic interaction is of no interest in the present investigation it may be omitted from consideration. In the analysis of the present section, this omission of all but the fixed satellites appears in the form of the omission of all terms involving the moon's or sun's angular velocity round the earth.

and the moon's and sun's action on the solar and lunar tides respectively, then

$$\frac{\mathfrak{L}}{\mathrm{C}} = \frac{\mathfrak{L}_{m^2}}{\mathrm{C}} + \frac{\mathfrak{L}_{m,^2}}{\mathrm{C}} - \left\{\tau\left(y\frac{d}{dz} - z\frac{d}{dy}\right)\frac{\sigma_,}{a} + \tau_,\left(y_,\frac{d}{dz_,} - z_,\frac{d}{dy_,}\right)\frac{\sigma}{a}\right\}$$

Obviously

$$\frac{\mathfrak{L}_{mm,}}{\mathrm{C}} \div \frac{2\tau\tau_,}{\mathfrak{g}} = (\mathrm{c} - \mathrm{b})\, y_, z_, + (\mathrm{c}_, - \mathrm{b}_,)\, yz + \&\mathrm{c}.$$

As before, we only want terms with argument n in $\mathfrak{L}_{mm,}$, $\mathfrak{M}_{mm,}$, and non-periodic terms in $\mathfrak{N}_{mm,}$.

The quantities a, b, &c., x, y, z with suffixes differ from those without in having $\Omega_,$ in place of Ω, and it is clear that no combination of terms which involve $\Omega_,$ and Ω can give the desired terms in the couples. Hence, as far as $\mathfrak{L}_{mm,}$, $\mathfrak{M}_{mm,}$, $\mathfrak{N}_{mm,}$ are concerned, the auxiliary functions may be abridged by the omission of all terms involving Ω or $\Omega_,$.

Therefore, from (4), we now simply have

$$\Phi = \Phi_, = p^2q^2 \cos 2n, \quad \Psi = \Psi_, = -2pq(p^2 - q^2)\cos n, \quad \mathrm{X} = \mathrm{X}_, = 0$$

But $\mathrm{c} - \mathrm{b}$ only differs from $\mathrm{c}_, - \mathrm{b}_,$ in that the latter involves $\Omega_,$ instead of Ω, and the same applies to yz and $y_, z_,$.

Hence, as far as we are now concerned,

$$(\mathrm{c} - \mathrm{b})\, y_, z_, = (\mathrm{c}_, - \mathrm{b}_,)\, yz$$

and similarly each pair of terms in $\mathfrak{L}_{mm,}$ are equal *inter se.*

Thus $$\frac{\mathfrak{L}_{mm,}}{\mathrm{C}} \div \frac{4\tau\tau_,}{\mathfrak{g}} = (\mathrm{c} - \mathrm{b})\, yz - \mathrm{d}(y^2 - z^2) - \mathrm{e}xy + \mathrm{f}zx$$

Comparing this with (14), when X is put equal to zero, we have

$$\frac{\mathfrak{L}_{mm,}}{\mathrm{C}} \div \frac{4\tau\tau_,}{\mathfrak{g}} = -\tfrac{1}{2}\Phi_\epsilon \Psi' + \tfrac{1}{2}\Psi_\epsilon'\{\Phi + \tfrac{1}{2}(1 - 6p^2q^2)\} - \tfrac{1}{2}\Psi_\epsilon \Phi' + \tfrac{1}{2}\Phi_\epsilon' \Psi$$

This quantity may be evaluated at once by reference to (15), (16), and (17), for it is clear that $\mathfrak{L}_{mm,}$ is what $\mathfrak{L}_{m^2}$ becomes when $E_1 = E_2 = 0$, $E_1' = E_2' = 0$, and when $2\tau\tau_,$ replaces τ^2.

If, therefore, we put $\dfrac{\mathfrak{L}_{mm,}}{\mathrm{C}} = \mathrm{F}_{mm,} \sin n + \mathrm{G}_{mm,} \cos n$, and remark that

$$4p^3q^3(p^2 - q^2) = \tfrac{1}{2}PQ^3, \quad 2pq(p^2 - q^2)(p^4 + q^4 - 6p^2q^2) = PQ(1 - 2Q^2)$$
$$2pq(p^2 - q^2)^3 = P^3Q$$

we have by selecting the terms in E, E' out of (15) and (16),

$$\left.\begin{aligned} \mathrm{F}_{mm,} \div \frac{\tau\tau_,}{\mathfrak{g}} &= -\tfrac{1}{2}EPQ^3 \cos 2\epsilon - E'PQ(1 - 2Q^2)\cos\epsilon' \\ \mathrm{G}_{mm,} \div \frac{\tau\tau_,}{\mathfrak{g}} &= \tfrac{1}{2}EPQ^3 \sin 2\epsilon + E'P^3Q \sin\epsilon' \end{aligned}\right\} \quad \ldots\ldots(33)$$

It may be shown in a precisely similar way by selecting terms out of (21) that

$$\frac{\mathfrak{N}_{mm_{\prime}}}{\mathrm{C}} \div \frac{\tau\tau_{\prime}}{\mathfrak{g}} = \tfrac{1}{2} EQ^4 \sin 2\epsilon + E'P^2Q^2 \sin \epsilon' \quad \text{...........(34)}$$

It is worthy of notice that (33) and (34) would be exactly the same, even if we did not put $E_1 = E_2 = E$; $E_1' = E_2' = E'$; $\epsilon_1 = \epsilon_2 = \epsilon$; $\epsilon_1' = \epsilon_2' = \epsilon'$, because these new terms depend entirely on the sidereal semi-diurnal and diurnal tides. The new expressions which ought rigorously to give the heights and lagging of the solar semi-diurnal and diurnal tides would only occur in $\mathfrak{L}_{m_{\prime}^2}$.

In the two following sections the results are collected with respect to the rate of change of obliquity and with respect to the tidal friction.

§ 9. *The rate of change of obliquity due to both sun and moon.*

The suffixes m^2, $m_{\prime}^2$, $mm_{\prime}$ to $\frac{di}{dt}$ will indicate the rate of change of obliquity due to the moon alone, to the sun alone, and to the sun and moon jointly.

Writing for P and Q their values, $\cos i$ and $\sin i$, we have by (19) and (29), or by (30),

$$\left.\begin{aligned}
\frac{n\mathfrak{g}}{\tau^2}\frac{di_{m^2}}{dt} &= \tfrac{1}{2}\sin i \cos i\,(1 - \tfrac{3}{4}\sin^2 i)\, E \sin 2\epsilon + \tfrac{3}{4}\sin^3 i \cos i\, E' \sin\epsilon' \\
&\qquad - \tfrac{3}{8}\sin^3 i\, E'' \sin 2\epsilon'' \\
\frac{n\mathfrak{g}}{\tau_{\prime}^2}\frac{di_{m_{\prime}^2}}{dt} &= \tfrac{1}{2}\sin i \cos i\,(1 - \tfrac{3}{4}\sin^2 i)\, E \sin 2\epsilon + \tfrac{3}{4}\sin^3 i \cos i\, E' \sin\epsilon' \\
&\qquad - \tfrac{3}{8}\sin^3 i\, E''' \sin 2\epsilon'''
\end{aligned}\right\} \quad \text{......(35)}$$

and by (33) and analogy with (19) and (29)

$$\frac{n\mathfrak{g}}{\tau\tau_{\prime}}\frac{di_{mm_{\prime}}}{dt} = -\tfrac{1}{2}\sin^3 i \cos i\, E \sin 2\epsilon - \sin i \cos^3 i\, E' \sin\epsilon' \quad \text{......(36)}$$

The sum of these three values of $\frac{di}{dt}$ gives the total rate of change of obliquity due both to sun and moon, on the assumption that the three semi-diurnal terms may be grouped together, as also the three diurnal ones.

It will be observed that the joint effect tends to counteract the separate effects; this arises from the fact that, as far as regards the joint effect, the two disturbing bodies may be replaced by rings of matter concentric with the earth but oblique to the equator, and such a ring of matter would cause the obliquity to diminish, as was shown by general considerations, in the abstract of this paper (*Proc. Roy. Soc.*, No. 191, 1878)*, must be the case.

* [Portion of this abstract is given in an Appendix to this paper.]

§ 10. *The rate of tidal friction due to both sun and moon.*

The equation which gives the rate of retardation of the earth's rotation is by (26) $\frac{d\omega_3}{dt} = \frac{\mathfrak{N}}{C}$; it will however be more convenient henceforward to replace ω_3 by $-n$ and to regard n as a variable, and to indicate by n_0 the value of n at the epoch from which the time is measured.

Generally the suffix 0 to any symbol will indicate its value at the epoch.

The equation of tidal friction may therefore be written

$$-\frac{d}{dt}\left(\frac{n}{n_0}\right) = \frac{\mathfrak{N}_{m^2}}{Cn_0} + \frac{\mathfrak{N}_{m_,^2}}{Cn_0} + \frac{\mathfrak{N}_{mm_,}}{Cn_0} \quad \text{..................(37)}$$

By (22) and (34), in which the semi-diurnal and diurnal terms are grouped together, we have

$$\left.\begin{aligned}\left(\frac{\mathfrak{g}n_0}{\tau^2}\right)\frac{\mathfrak{N}_{m^2}}{Cn_0} &= (\cos^2 i + \tfrac{3}{8}\sin^4 i)\,E\sin 2\epsilon + \sin^2 i\,(1 - \tfrac{3}{4}\sin^2 i)\,E'\sin\epsilon' \\ &= \left(\frac{\mathfrak{g}n_0}{\tau_,^2}\right)\frac{\mathfrak{N}_{m_,^2}}{Cn_0} \\ \left(\frac{\mathfrak{g}n_0}{\tau\tau_,}\right)\frac{\mathfrak{N}_{mm_,}}{Cn_0} &= \tfrac{1}{2}\sin^4 i E\sin 2\epsilon + \sin^2 i\cos^2 i E'\sin\epsilon'\end{aligned}\right\} \text{...(38)}$$

§ 11. *The rate of change of obliquity when the earth is viscous.*

In order to understand the physical meaning of the equations giving the rate of change of obliquity (viz., (35) and (36) if there be two disturbing bodies, or (29) if there be only one) it is necessary to use numbers. The subject will be illustrated in two cases: first, for the sun, moon, and earth with their present configurations; and secondly, for the case of a planet perturbed by a single satellite. For the first illustration I accordingly take the following data: $g = 32{\cdot}19$ (feet, seconds), the earth's mean radius $a = 20{\cdot}9 \times 10^6$ feet, the sidereal day ·9973 m. s. days, the sidereal year $= 365{\cdot}256$ m. s. days, the moon's sidereal period 27·3217 m. s. days, the ratio of the earth's mass to that of the moon $\nu = 82$, and the unit of time the tropical year 365·242 m. s. days.

With these values we have

$$n_0 = 2\pi \div {\cdot}9973 \text{ in radians per m. s. day}$$

$$\mathfrak{g} = \frac{2g}{5a}$$

$$\tau = \tfrac{3}{2} \times \tfrac{1}{83} \text{ of } 4\pi^2 \div (\text{month})^2$$

$$\tau_, = \tfrac{3}{2} \text{ of } 4\pi^2 \div (\text{sidereal year})^2$$

It will be found that these values give

$$\left.\begin{aligned}\frac{\tau^2}{\mathfrak{g}n_0} &= \cdot 6598 \text{ degrees per million tropical years}\\ \frac{\tau_,^2}{\mathfrak{g}n_0} &= \cdot 1423 \quad \text{,,} \qquad \text{,,} \qquad \text{,,}\\ \frac{\tau\tau_,}{\mathfrak{g}n_0} &= \cdot 3064 \quad \text{,,} \qquad \text{,,} \qquad \text{,,}\end{aligned}\right\}\ldots\ldots\ldots\ldots(39)$$

These three quantities will henceforth be written u^2, $u_,^2$, $uu_,$.

For the purpose of analysing the physical meaning of the *differential* equations for $\frac{di}{dt}$ and $\frac{d}{dt}\left(\frac{n}{n_0}\right)$, no distinction will be made between $\frac{\tau^2}{\mathfrak{g}n}$ and $\frac{\tau^2}{\mathfrak{g}n_0}$, &c., for it is here only sought to discover the *rates* of changes. But when we come to *integrate* and find the total changes in a given time, regard will have to be paid to the fact that both τ and n are variables.

For the immediate purpose of this section the numerical values of u^2, $u_,^2$, $uu_,$ given in (39), will be used.

I will now apply the foregoing results to the particular case where the earth is a viscous spheroid.

Let $\rho = \frac{2gaw}{19\upsilon}$, where υ is the coefficient of viscosity.

Then by the theory of bodily tides as developed in my last paper

$$\left.\begin{aligned}E=\cos 2\epsilon,\quad E'=\cos\epsilon',\quad E''=\cos 2\epsilon'',\quad E'''=\cos 2\epsilon'''\\ \tan 2\epsilon=\frac{2n}{\rho},\quad \tan\epsilon'=\frac{n}{\rho},\quad \tan 2\epsilon''=\frac{2\Omega}{\rho},\quad \tan 2\epsilon'''=\frac{2\Omega_,}{\rho}\end{aligned}\right\}\ldots(40)$$

Rigorously, we should add to these

$$\left.\begin{aligned}E_1=\cos 2\epsilon_1,\qquad E_2=\cos 2\epsilon_2,\qquad E_1'=\cos\epsilon_1',\qquad E_2'=\cos\epsilon_2'\\ \tan 2\epsilon_1=\frac{2(n-\Omega)}{\rho},\quad \tan 2\epsilon_2=\frac{2(n+\Omega)}{\rho},\quad \tan\epsilon_1'=\frac{n-2\Omega}{\rho},\quad \tan\epsilon_2'=\frac{n+2\Omega}{\rho}\end{aligned}\right\}\ldots\ldots\ldots(40')$$

But for the present we classify the three semi-diurnal tides together, as also the three diurnal ones.

Then we have

$$\begin{aligned}\frac{di}{dt} &= [\tfrac{1}{4}\sin i\cos i\,(1-\tfrac{3}{4}\sin^2 i)\sin 4\epsilon+\tfrac{3}{8}\sin^3 i\cos i\sin 2\epsilon']\,(u^2+u_,^2)\\ &\quad -\tfrac{3}{16}\sin^3 i\sin 4\epsilon'' u^2-\tfrac{3}{16}\sin^3 i\sin 4\epsilon''' u_,^2\\ &\quad -(\tfrac{1}{4}\sin^3 i\cos i\sin 4\epsilon+\tfrac{1}{2}\sin i\cos^3 i\sin 2\epsilon')\,uu_,\end{aligned}$$

Now

$$\tfrac{1}{4}\sin i\cos i\,(1-\tfrac{3}{4}\sin^2 i)=\tfrac{1}{64}\sin 2i\,(5+3\cos 2i)=\tfrac{1}{64}(5\sin 2i+\tfrac{3}{2}\sin 4i)$$

$$\tfrac{3}{8}\sin^3 i\cos i=\tfrac{3}{32}\sin 2i\,(1-\cos 2i)=\tfrac{3}{64}(2\sin 2i-\sin 4i)$$

$$\tfrac{3}{16}\sin^3 i=\tfrac{3}{64}(3\sin i-\sin 3i),\qquad \tfrac{1}{4}\sin^3 i\cos i=\tfrac{2}{64}(2\sin 2i-\sin 4i)$$

$$\tfrac{1}{2}\sin i\cos^3 i=\tfrac{1}{8}\sin 2i\,(1+\cos 2i)=\tfrac{4}{64}(2\sin 2i+\sin 4i)$$

If these transformations be introduced, the equation for $\dfrac{di}{dt}$ may be written

$$\left.\begin{aligned}64\frac{di}{dt}=&-9\,(u^2\sin 4\epsilon''+u_,^2\sin 4\epsilon''')\sin i+3\,(u^2\sin 4\epsilon''+u_,^2\sin 4\epsilon''')\sin 3i\\&+[(5\sin 4\epsilon+6\sin 2\epsilon')(u^2+u_,^2)-(4\sin 4\epsilon+8\sin 2\epsilon')\,uu_,]\sin 2i\\&+[(\tfrac{3}{2}\sin 4\epsilon-3\sin 2\epsilon')(u^2+u_,^2)+(2\sin 4\epsilon-4\sin 2\epsilon')\,uu_,]\sin 4i\end{aligned}\right\}\quad\text{.........(41)}$$

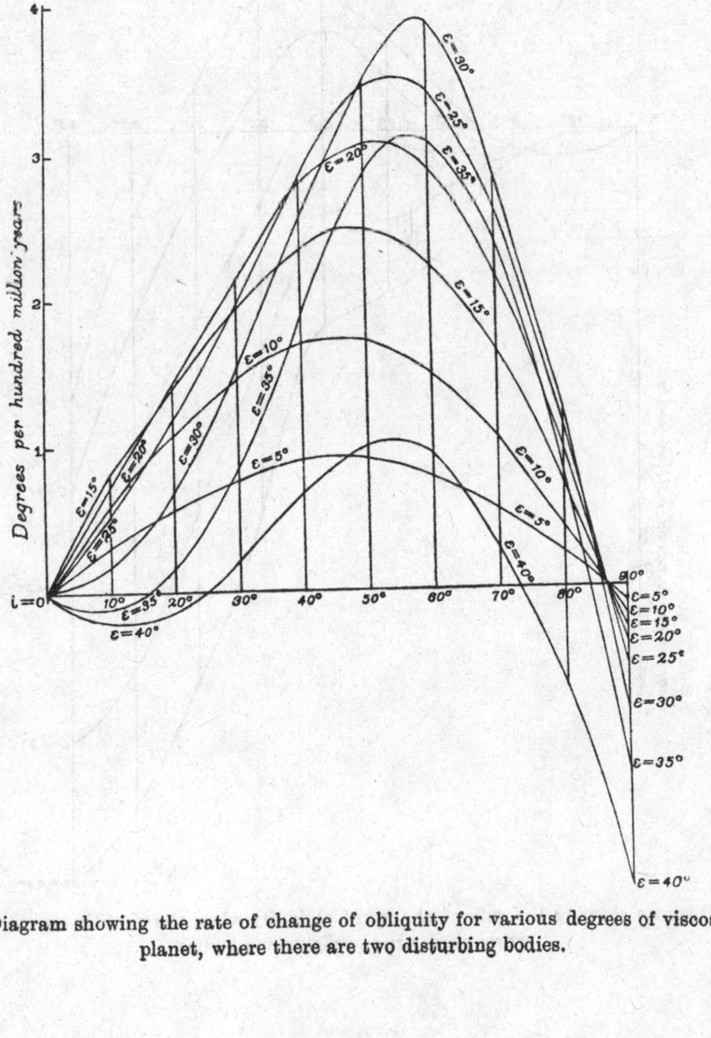

Fig. 2. Diagram showing the rate of change of obliquity for various degrees of viscosity of the planet, where there are two disturbing bodies.

Substituting for u and $u_,$ their numerical values (39), and omitting the term depending on the semi-annual tide as unimportant, I find

$$\left.\begin{aligned}64\frac{di}{dt} = &-5\cdot9378\sin 4\epsilon''\sin i + 1\cdot9793\sin 4\epsilon''\sin 3i\\ &+\{2\cdot7846\sin 4\epsilon + 2\cdot3611\sin 2\epsilon'\}\sin 2i\\ &+\{1\cdot8159\sin 4\epsilon - 3\cdot6317\sin 2\epsilon'\}\sin 4i\end{aligned}\right\}\quad\ldots\ldots(42)$$

The numbers are such that $\frac{di}{dt}$ is expressed in degrees per million years.

The various values which $\frac{di}{dt}$ is capable of assuming as the viscosity and obliquity vary are best shown graphically. In figs. 2 and 3, each curve corresponds to a given degree of viscosity, that is to say to a given value of ϵ,

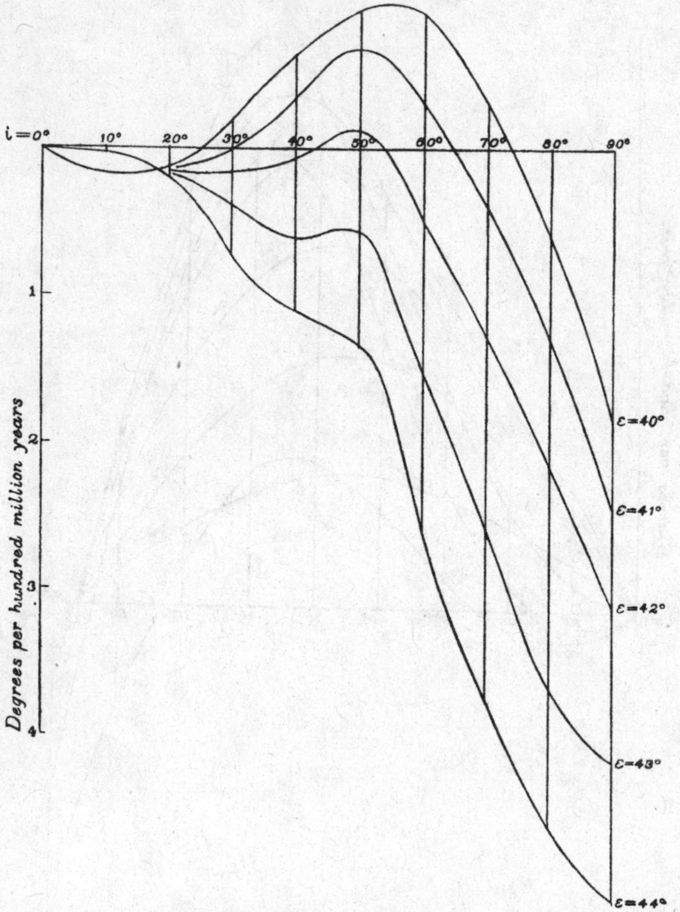

FIG. 3. Diagram showing the rate of change of obliquity when the viscosity is very great, and where there are two disturbing bodies.

and the ordinates give the values of $\frac{di}{dt}$ as the obliquity increases from 0° to 90°. The scale at the side of each figure is a scale of degrees per hundred million years—*e.g.*, if we had $\epsilon = 30°$ and i about 57°, the obliquity would be increasing at the rate of about 3° 45′ per hundred million years.

The behaviour of this family of curves is so very peculiar for high degrees of viscosity, that I have given a special figure (viz.: fig. 3) for the viscosities for which $\epsilon = 40°, 41°, 42°, 43°, 44°$.

The peculiarly rapid variation of the forms of the curves for these values of ϵ is due to the rising of the fortnightly tide into prominence for high degrees of viscosity. The matter of the spheroid is in fact so stiff that there is not time in 12 hours or a day to raise more than a very small tide, whilst in a fortnight a considerable lagging tide is raised.

For $\epsilon = 44°$ the fortnightly tide has risen to give its maximum effect (*i.e.*, $\sin 4\epsilon'' = 1$), whilst the effects of the other tides only remain evident in the hump in the middle of the curve. Between $\epsilon = 44°$ and 45° the ordinates of the curve diminish rapidly and the hump is smoothed down, so that when $\epsilon = 45°$ the curve is reduced to the horizontal axis.

By the theory of Paper 1, the values of ϵ when divided by 15 give the corresponding retardation of the bodily semi-diurnal tide—*e.g.*, when $\epsilon = 30°$ the tide is two hours late. Also the height of the tide is $\cos 2\epsilon$ of the height of the equilibrium tide of a perfectly fluid spheroid—*e.g.*, when $\epsilon = 30°$ the height of tide is reduced by one-half. In the tables given in Part I., Section 7, of Paper 1, will be found approximate values of the viscosity corresponding to each value of ϵ.

The numerical work necessary to draw these figures was done by means of Crelle's multiplication table, and as to fig. 2 in duplicate mechanically with a sector; the ordinates were thus only determined with sufficient accuracy to draw a fairly good figure. For the two figures I found 108 values of each of the seven terms of $\frac{di}{dt}$ (nine values of i and twelve of ϵ), and from the seven tables thus formed, the values corresponding to each ordinate of each member of the family were selected and added together.

From this figure several remarkable propositions may be deduced. When the ordinates are positive, it shows that the obliquity tends to increase, and when negative to diminish. Whenever, then, any curve cuts the horizontal axis there is a position of dynamical equilibrium; but when the curve passes from above to below, it is one of stability, and when from below to above, of instability. It follows from this that the positions of stability and instability must occur alternately. When $\epsilon = 0°$ or 45° (fluidity or rigidity) the curve reduces to the horizontal axis, and every position of the earth's axis is one of neutral equilibrium.

But in every other case the position of 90° of obliquity is not a position of equilibrium, but the obliquity tends to diminish. On the other hand, from $\epsilon = 0°$ to about 30° (infinitely small viscosity to tide-retardation of two hours), the position of zero obliquity is one of dynamical instability, whilst from then onwards to rigidity it becomes a position of stability.

For viscosities ranging from $\epsilon = 0°$ to about $42\frac{1}{4}°$ there is a position of stability which lies between about 50° to 87° of obliquity; and the obliquity of dynamical stability diminishes as the viscosity increases.

For viscosities ranging from $\epsilon = 30°$ nearly to about $42\frac{1}{4}°$, there is a second position of dynamical equilibrium, at an obliquity which increases from 0° to about 50°, as the viscosity increases from its lower to its higher value. But this position is one of instability.

From $\epsilon =$ about $42\frac{1}{4}°$ there is only one position of equilibrium, and that stable, viz.: when the obliquity is zero.

If the obliquity be supposed to increase past 90°, it is equivalent to supposing the earth's diurnal rotation reversed, whilst the orbital motion of the earth and moon remains the same as before; but it did not seem worth while to prolong the figure, as it would have no applicability to the planets of the solar system. And, indeed, the figure for all the larger obliquities would hardly be applicable, because any planet whose obliquity increased very much, must gradually make the plane of the orbit of its satellite become inclined to that of its own orbit, and thus the hypothesis that the satellite's orbit remains coincident with the ecliptic would be very inexact.

It follows from an inspection of the figure that for all obliquities there are two degrees of viscosity, one of which will make the rate of change of obliquity a maximum and the other minimum. A graphical construction showed that for obliquities of about 5° to 20°, the degree of viscosity for a maximum corresponds to about $\epsilon = 17\frac{1}{2}°$*, whilst that for a minimum to about $\epsilon = 40°$. In order, however, to check this conclusion, I determined the values of ϵ analytically when $i = 15°$, and when the fortnightly tide (which has very little effect for small obliquities) is neglected. I find that the values are given by the roots of the equation

$$x^3 + 10x^2 + 13{\cdot}660x - 20{\cdot}412 = 0, \text{ where } x = 3 \cos 4\epsilon$$

This equation has three real roots, of which one gives a hyperbolic cosine,

* I may here mention that I found when $\epsilon = 17\frac{1}{2}°$, that it would take about a thousand million years for the obliquity to increase from 5° to $23\frac{1}{2}°$, if regard was only paid to this equation of change of obliquity. The equations of tidal friction and tidal reaction will, however, entirely modify the aspects of the case.

and the other two give $\epsilon = 18° \, 15'$ and $\epsilon = 41° \, 37'$. This result therefore confirms the geometrical construction fairly well.

It is proper to mention that the expressions of dynamical stability and instability are only used in a modified sense, for it will be seen when the effects of tidal friction come to be included, that these positions are continually shifting, so that they may be rather described as positions of instantaneous stability and instability.

I will now illustrate the case where there is only one satellite to the planet, and in order to change the point of view, I will suppose that the periodic time of the satellite is so short that we cannot classify the semi-diurnal and diurnal terms together, but must keep them all separate.

Suppose that $n = 5\Omega$; then the speeds of the seven tides are proportional to the following numbers, 8, 10, 12 (semi-diurnal); 3, 5, 7 (diurnal); 2 (fortnightly).

These are all the data which are necessary to draw a family of curves similar to those in figs. 2 and 3, because the scale, to which the figure is drawn, is determined by the mass of the satellite, the mass and density of the planet, and the actual velocity of rotation of the planet.

By (16) and (29) we have

$$\frac{di}{dt} = \frac{\tau^2}{\mathfrak{g}n} [\tfrac{1}{2}p^7 q \sin 4\epsilon_1 - p^3 q^3 (p^2 - q^2) \sin 4\epsilon - \tfrac{1}{2}pq^7 \sin 4\epsilon_2 - \tfrac{3}{2}p^3 q^3 \sin 4\epsilon''$$
$$+ \tfrac{1}{2}p^5 q (p^2 + 3q^2) \sin 2\epsilon_1' - \tfrac{1}{2}pq (p^2 - q^2)^3 \sin 2\epsilon' - \tfrac{1}{2}pq^5 (3p^2 + q^2) \sin 2\epsilon_2']$$

where $p = \cos \tfrac{1}{2}i$ and $q = \sin \tfrac{1}{2}i$.

This equation may be easily reduced to the form

$$\frac{di}{dt} = \frac{\tau^2}{\mathfrak{g}n} \frac{1}{128} \sin i \, \{[10 \sin 4\epsilon_1 - 10 \sin 4\epsilon_2 + 16 \sin 2\epsilon_1' - 16 \sin 2\epsilon_2' - 12 \sin 4\epsilon'']$$
$$+ \cos i \, [15 \sin 4\epsilon_1 - 4 \sin 4\epsilon + 15 \sin 4\epsilon_2 + 18 \sin 2\epsilon_1' - 24 \sin 2\epsilon' + 18 \sin 2\epsilon_2']$$
$$+ \cos 2i \, [6 \sin 4\epsilon_1 - 6 \sin 4\epsilon_2 + 12 \sin 4\epsilon'']$$
$$+ \cos 3i \, [\sin 4\epsilon_1 + 4 \sin 4\epsilon + \sin 4\epsilon_2 - 2 \sin 2\epsilon_1' - 8 \sin 2\epsilon' - 2 \sin 2\epsilon_2']\}$$

which is convenient for the computation of the ordinates of the family of curves which illustrate the various values of $\frac{di}{dt}$ for various obliquities and viscosities.

In fig. 4, the lag (ϵ) of the sidereal semi-diurnal tide is taken as the standard of viscosity. The abscissæ represent the various obliquities of the planet's equator to the plane of the satellite's orbit; the ordinates represent the values of $\frac{di}{dt}$ (the actual scale depending on the value of $\frac{\tau^2}{\mathfrak{g}n}$); and each curve represents one degree of viscosity, viz.: when $\epsilon = 10°, 20°, 30°, 40°$ and $44°$.

The computation of the ordinates was done by Crelle's three-figure multiplication table, and thus the figure does not profess to be very rigorously exact.

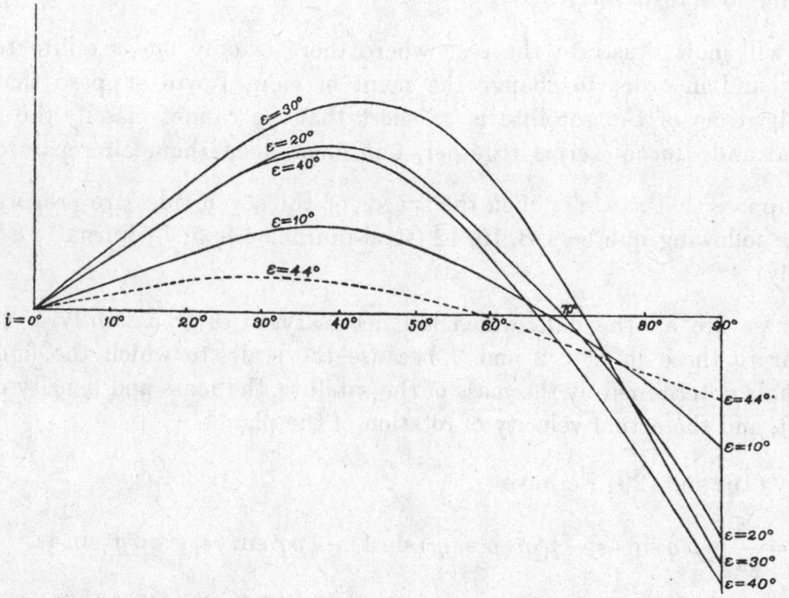

FIG. 4. Diagram showing the rate of change of obliquity for various degrees of viscosity of the planet, where there is one disturbing body.

This family of curves differs much from the preceding one. For moderate obliquities there is no degree of viscosity which tends to make the obliquity diminish, and thus there is no position of dynamically unstable equilibrium of the system except that of zero obliquity. Thus we see that the decrease of obliquity for small obliquities and large viscosities in the previous case was due to the attraction of the sun on the lunar tides and the moon on the solar tides.

In the present case the position of zero obliquity is never stable, as it was before. The dynamically stable position at a large obliquity still remains as before, but in consequence of the largeness of the ratio $\Omega \div n$ ($\frac{1}{5}$th instead of $\frac{1}{27}$th), this obliquity of dynamical stability is not nearly so great as in the previous case. As the ratio $\Omega \div n$ increases, the position of dynamical stability is one of smaller and smaller obliquity, until when $\Omega \div n$ is equal to a half, zero obliquity becomes stable,—as we shall see later on.

§ 12. *Rate of tidal friction when the earth is viscous.*

If in the same way the equations (37) and (38) be applied to the case where the earth is purely viscous, when the semi-diurnal and diurnal tides are grouped together, we have

$$\left.\begin{aligned}-\frac{d}{dt}\left(\frac{n}{n_0}\right) = (u^2 + u_{,}^2)\,[\tfrac{1}{2}(\cos^2 i + \tfrac{3}{8}\sin^4 i)\sin 4\epsilon + \tfrac{1}{2}\sin^2 i\,(1 - \tfrac{3}{4}\sin^2 i)\sin 2\epsilon'] \\ + uu_{,}\,[\tfrac{1}{4}\sin^4 i \sin 4\epsilon + \tfrac{1}{2}\sin^2 i \cos^2 i \sin 2\epsilon']\end{aligned}\right\} \quad \text{......(43)}$$

Fig. 5 exhibits the various values of $\frac{d}{dt}\left(\frac{n}{n_0}\right)$ for the various obliquities and degrees of viscosity, just as the previous figures exhibited $\frac{di}{dt}$. The calculations were done in the same way as before, after the various functions of the obliquity were expressed in terms of cos $2i$ and cos $4i$.

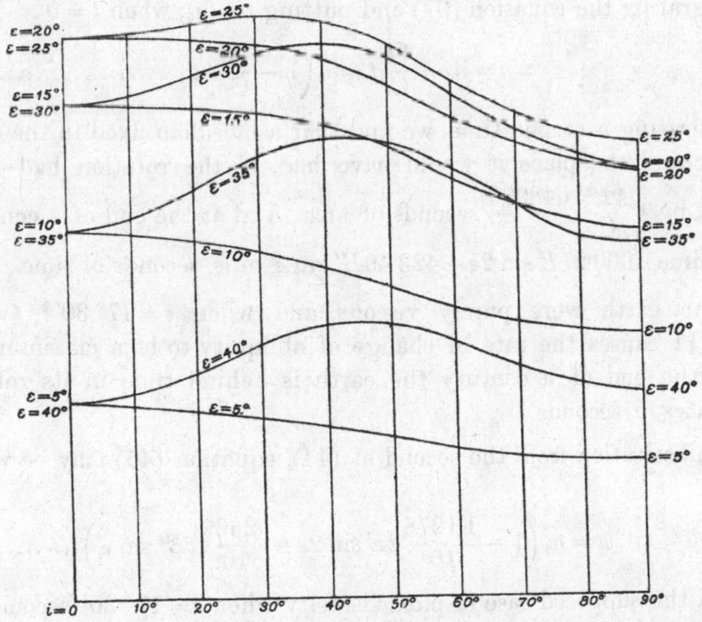

FIG. 5. Diagram showing the amount of tidal friction for various viscosities and obliquities, where there are two disturbing bodies.

The only remarkable point in these curves is that, for the higher degrees of viscosity, the tidal friction rises to a maximum for about 45° of obliquity. The tidal friction rises to its greatest value when $\epsilon = 22\tfrac{1}{2}°$ nearly; this is explained by the fact that by far the largest part of the friction arises from the semi-diurnal tide, which has its greatest effect when sin 4ϵ is unity.

§ 13. *Tidal friction and apparent secular acceleration of the moon.*

I now set aside again the hypothesis that the earth is purely viscous, and return to that of there being any kind of lagging tides.

I shall first find at what rate the earth is being retarded when it is moving with its present diurnal rotation, and when the moon is moving in her present orbit, and no distinction will be made between n and n_0; all the secular changes will be considered later.

The numerical data of Section 11 are here used, and the obliquity of the ecliptic $i = 23° \, 28'$; then u and $u_{,}$ being expressed in radians per tropical year, I find

$$\left.\begin{aligned}\frac{\mathfrak{N}}{C} &= \frac{2{\cdot}7563}{10^5} E \sin 2\epsilon + \frac{{\cdot}6143}{10^5} E' \sin \epsilon' \\ \frac{\mathfrak{N}}{Cn} &= \frac{1{\cdot}1978}{10^8} E \sin 2\epsilon + \frac{{\cdot}2669}{10^8} E' \sin \epsilon'\end{aligned}\right\} \quad \text{..............(44)}$$

Integrating the equation (37) and putting $n = n_0$, when $t = 0$

$$n = n_0 - \frac{\mathfrak{N}}{C} t = n_0 \left(1 - \frac{\mathfrak{N}}{Cn_0} t\right) \quad \text{.....................(45)}$$

Integrating a second time, we find that a meridian fixed in the earth has fallen behind the place it would have had, if the rotation had not been retarded, by $\frac{1}{2} \frac{\mathfrak{N} t^2}{C} \cdot \frac{648000}{\pi}$ seconds of arc. And at the end of a century it is behind time $1900{\cdot}27 E \sin 2\epsilon + 423{\cdot}49 E' \sin \epsilon'$ m. s. seconds of time.

If the earth were purely viscous, and when $\epsilon = 17° \, 30'$* (which by Section 11 causes the rate of change of obliquity to be a maximum), I find that at the end of a century the earth is behind time in its rotation by 17 minutes 5 seconds.

By substitution from the second of (44), equation (45) may be written in the form

$$n = n_0 \left(1 - \frac{1{\cdot}1978}{10^8} tE \sin 2\epsilon - \frac{{\cdot}2669}{10^8} tE' \sin \epsilon'\right) \text{...........(46)}$$

which in the supposed case of pure viscosity when $\epsilon = 17° \, 30'$ becomes

$$n = n_0 \left(1 - \frac{{\cdot}006460}{10^6} t\right) \quad \text{........................(47)}$$

All these results would, however, cease to be even approximately true after a few millions of years.

* This calculation was done before I perceived that I had not chosen that degree of viscosity which makes the tidal friction a maximum, but as all the other numerical calculations have been worked out for this degree of viscosity I adhere to it here also.

The effect of the failure of the earth to keep true time is to cause an apparent acceleration of the moon's motion; and if the moon's motion were really unaffected by the tides in the earth, there would be an apparent acceleration of the moon in a century of

$$1043''\cdot28E \sin 2\epsilon + 232''\cdot50E' \sin \epsilon' \quad \text{.................}(48)$$

for the moon moves over $0''\cdot5490$ of her orbit in one second of time.

This apparent acceleration would however be considerably diminished by the effects of tidal reaction on the moon, which will now be considered.

§ 14. *Tidal reaction on the moon.*

The action of the tides on the moon gives rise to a small force tangential to the orbit accelerating her linear motion. The spiral described by the moon about the earth will differ insensibly from a circle, and therefore we may assume throughout that the centrifugal force of the earth's and moon's orbital motion round their common centre of inertia is equal and opposite to the attraction between them.

We shall now find the tangential force on the moon in terms of the couples which we have already found acting on the earth. These couples consist of the sum of three parts, viz.: that due (i) to the moon alone, (ii) to the sun alone, and (iii) to the action of the sun on the lunar tides and of the moon on the solar tides, the latter two being equal *inter se*.

Now since action and reaction are equal and opposite, the only parts of these couples which correspond with the tangential force on the moon are those which arise from (i), and one-half those which arise from (iii).

We may thus leave the sun out of account if we suppose the earth only to be acted on by the couples $\mathfrak{L}_{m^2} + \frac{1}{2}\mathfrak{L}_{mm_{,}}$, $\mathfrak{M}_{m^2} + \frac{1}{2}\mathfrak{M}_{mm_{,}}$, $\mathfrak{N}_{m^2} + \frac{1}{2}\mathfrak{N}_{mm_{,}}$; these couples will be called $\mathfrak{L}'$, $\mathfrak{M}'$, $\mathfrak{N}'$, and the part of the change of obliquity which is due to $\mathfrak{L}'$, $\mathfrak{M}'$ will be called $\dfrac{di'}{dt}$.

Let r and $-\Omega$ be the moon's distance, and angular velocity at any time, and ν the ratio of the earth's mass to the moon's.

Let T be the force which acts on the moon perpendicular to her radius vector, in the direction of her motion.

From the equality of action and reaction, it follows that Tr must be equal to the couple which is produced by the moon's action on the tides in the earth, acting in the direction tending to retard the earth's diurnal rotation about the normal to the ecliptic. Referring to fig. 1, we see that the direction cosines of this normal are $-\sin i \cos n$, $-\sin i \sin n$, $\cos i$; hence

$$\text{Tr} = -\sin i\,(\mathfrak{L}' \cos n + \mathfrak{M}' \sin n) + \mathfrak{N}' \cos i$$

But by (17) and (18)

$$\frac{\mathfrak{L}'}{C} = (F_{m^2} + \tfrac{1}{2}F_{mm_,}) \sin n + (G_{m^2} + \tfrac{1}{2}G_{mm_,}) \cos n$$

$$\frac{\mathfrak{M}'}{C} = -(F_{m^2} + \tfrac{1}{2}F_{mm_,}) \cos n + (G_{m^2} + \tfrac{1}{2}G_{mm_,}) \sin n$$

Hence $$\frac{\mathfrak{L}'}{C} \cos n + \frac{\mathfrak{M}'}{C} \sin n = G_{m^2} + \tfrac{1}{2}G_{mm_,} = -n\frac{di'}{dt}$$

Thus $$Tr = C\left\{\frac{\mathfrak{N}'}{C} \cos i + n \sin i \frac{di'}{dt}\right\} \text{......................(49)}$$

In order to apply the ordinary formula for the motion of the moon, the earth must be reduced to rest, and therefore T must be augmented by the factor $(M+m) \div M$. Then if ϑ be the moon's longitude, the equation of motion of the moon is

$$m\frac{d}{dt}\left(r^2 \frac{d\vartheta}{dt}\right) = \frac{M+m}{M} Tr \text{.......................(50)}$$

But since the orbit is approximately circular $\frac{d\vartheta}{dt} = \Omega$.

Also $m = C \div \tfrac{2}{5}\nu a^2$, and $\frac{M+m}{M} = \frac{1+\nu}{\nu}$.

Therefore by (49) and (50)

$$\frac{d(\Omega r^2)}{dt} = \tfrac{2}{5}\nu a^2 \frac{1+\nu}{\nu}\left\{\frac{\mathfrak{N}'}{C} \cos i + n \sin i \frac{di'}{dt}\right\}$$

Now let $\xi = \left(\frac{\Omega_0}{\Omega}\right)^{\frac{1}{3}}$, whence $\Omega^2 = \Omega_0{}^2 \div \xi^6$.

The suffix 0 to Ω indicates the value of Ω when the time is zero, and no confusion will arise by this second use of the symbol ξ.

But since the centrifugal force is equal to the attraction between the two bodies, and the orbit is circular, $\Omega^2 r^3 = M + m$; thus $\Omega_0{}^2 r^3 = (M+m)\xi^6$.

Therefore

$$r^2 = (M+m)^{\frac{2}{3}} \xi^4 \Omega_0{}^{-\frac{4}{3}}, \text{ and } \Omega r^2 = (M+m)^{\frac{2}{3}} \Omega_0{}^{-\frac{1}{3}} \xi$$

and hence $$\frac{d}{dt}(\Omega r^2) = (M+m)^{\frac{2}{3}} \Omega_0{}^{-\frac{1}{3}} \frac{d\xi}{dt}$$

But $M + m = ga^2 \frac{1+\nu}{\nu}$, because M and m are here measured in astronomical units of mass.

Therefore our equation may be written

$$\left(ga^2 \frac{1+\nu}{\nu}\right)^{\frac{2}{3}} \Omega_0{}^{-\frac{1}{3}} \frac{d\xi}{dt} = \tfrac{2}{5}a^2(1+\nu)\left\{\frac{\mathfrak{N}}{C} \cos i + n \sin i \frac{di'}{dt}\right\}$$

Now let

$$s = \tfrac{2}{5}\left[\left(\frac{a}{g}\right)^2 \nu^2 (1+\nu)\right]^{\frac{1}{3}}, \text{ let } sn_0\Omega_0^{\frac{1}{3}} = \frac{1}{\mu}, \text{ and let } N = \frac{n}{n_0} \quad \text{......(51)}$$

and we have

$$\mu\frac{d\xi}{dt} = \frac{\mathfrak{N}'}{\text{C}n_0}\cos i + N\sin i\frac{di'}{dt} \quad \text{......................(52)}$$

It is not hard to show that the moment of momentum of the orbital motion of the two bodies is $\text{C} \div s\Omega^{\frac{1}{3}}$, and that of the earth's rotation is obviously Cn. Hence $sn\Omega^{\frac{1}{3}}$ is the ratio of the two momenta, and μ is the ratio of the two momenta at the fixed moment of time, which is the epoch.

In the similar equation expressive of the rate of change in the earth's orbital motion round the sun, it is obvious that the orbital moment of momentum is so very large compared with the earth's moment of momentum of rotation, that μ is very large and the earth's mean distance from the sun remains sensibly constant (see Section 19).

Then by (16) and (29), remembering that

$$p = \cos\tfrac{1}{2}i, \quad q = \sin\tfrac{1}{2}i, \quad \frac{di_{m^2}}{dt} = -\frac{\mathfrak{G}_{m^2}}{n}, \quad \text{and } N = \frac{n}{n_0}$$

we have

$$N\sin i\frac{di_{m^2}}{dt} = \frac{\tau^2}{\mathfrak{g}n_0}2pq\,[E_1p^7q\sin 2\epsilon_1 - E\,2p^3q^3(p^2-q^2)\sin 2\epsilon - E_2pq^7\sin 2\epsilon_2$$
$$+ E_1'p^5q(p^2+3q^2)\sin\epsilon_1' - E'pq(p^2-q^2)^3\sin\epsilon' - E_2'pq^5(3p^2+q^2)\sin\epsilon_2'$$
$$- E''3p^3q^3\sin 2\epsilon''] \quad \text{...(53)}$$

And by (21)

$$\cos i\frac{\mathfrak{N}_{m^2}}{\text{C}n_0} = \frac{\tau^2}{\mathfrak{g}n_0}(p^2-q^2)\,[E_1p^8\sin 2\epsilon_1 + E\,4p^4q^4\sin 2\epsilon + E_2q^8\sin 2\epsilon_2$$
$$+ E_1'2p^6q^2\sin\epsilon_1' + E'2p^2q^2(p^2-q^2)^2\sin\epsilon' + E_2'2p^2q^6\sin\epsilon_2'] \quad \text{...(54)}$$

By (33) and (34), and remembering to take the halves of $\mathfrak{G}_{mm_\prime}$ and $\mathfrak{N}_{mm_\prime}$, and that $\sin i = Q$, $\cos i = P$

$$N\sin i\left(\frac{1}{2}\frac{di_{mm_\prime}}{dt}\right) = -\frac{\tau\tau_\prime}{\mathfrak{g}n_0}Q\,[\tfrac{1}{4}EPQ^3\sin 2\epsilon + \tfrac{1}{2}E'P^3Q\sin\epsilon'] \quad \text{.....(55)}$$

$$\cos i\,\tfrac{1}{2}\frac{\mathfrak{N}_{mm_\prime}}{\text{C}n_0} = \frac{\tau\tau_\prime}{\mathfrak{g}n_0}P\,[\tfrac{1}{4}EQ^4\sin 2\epsilon + \tfrac{1}{2}E'P^2Q^2\sin\epsilon'] \quad \text{........(56)}$$

To obtain $\mu\frac{d\xi}{dt}$, we have to add the last four expressions together, and we observe that the last two cut one another out, so that the expression for $\frac{d\xi}{dt}$ is independent of the solar tides; also the terms in $\sin 2\epsilon$, $\sin\epsilon'$ cut one another out in the sum of the first two expressions, and hence it follows that $\frac{d\xi}{dt}$ is independent of the sidereal semi-diurnal and diurnal terms.

Thus we have

$$\mu \frac{d\xi}{dt} = \frac{\tau^2}{\mathfrak{g}n_0} [E_1 p^8 \sin 2\epsilon_1 - E_2 q^8 \sin 2\epsilon_2 + 4E_1' p^6 q^2 \sin \epsilon_1' - 4E_2' p^2 q^6 \sin \epsilon_2' - 6E'' p^4 q^4 \sin 2\epsilon''] \quad \text{......(57)}$$

This equation will be referred to hereafter as that of tidal reaction*. From its form we see that the tides of speeds $2(n+\Omega)$, $n+2\Omega$, and 2Ω tend to make the moon approach the earth, whilst the other tides tend to make it recede.

If, as in previous cases, we put $E_1 = E_2 = E$; $E_1' = E_2' = E'$; $\epsilon_1 = \epsilon_2 = \epsilon$; $\epsilon_1' = \epsilon_2' = \epsilon'$ (which is justifiable so long as the moon's orbital motion is slow compared with that of the earth's rotation), we have, after noticing that

$$p^8 - q^8 = (p^2 - q^2)(p^4 + q^4) = \cos i\,(1 - \tfrac{1}{2}\sin^2 i)$$

$$4p^6q^2 - 4p^2q^6 = 4p^2q^2(p^2 - q^2) = \sin^2 i \cos i$$

$$6p^4q^4 = \tfrac{3}{8}\sin^4 i$$

$$\mu \frac{d\xi}{dt} = \frac{\tau^2}{\mathfrak{g}n_0} [\cos i\,(1 - \tfrac{1}{2}\sin^2 i)\,E \sin 2\epsilon + \sin^2 i \cos i E' \sin \epsilon' - \tfrac{3}{8}\sin^4 i E'' \sin 2\epsilon''] \quad \text{........(58)}$$

If the present values of n, Ω, i be substituted in this equation (58) (*i.e.*, with the present day, month, and obliquity), and if the tropical year be the unit of time, it will be found that

$$10^{10} \frac{d\xi}{dt} = \frac{1}{\xi^{12}} (24{\cdot}27 E \sin 2\epsilon + 4{\cdot}18 E' \sin \epsilon' - {\cdot}271 E'' \sin 2\epsilon'')$$

ξ^{12} enters into this equation because τ varies as Ω^2 and therefore as ξ^{-6}.

But we may here put $\xi = 1$, because we only want the present instantaneous rate of increase of Ω.

Now $\dfrac{d\xi}{dt} = -\tfrac{1}{3}\Omega^{-\frac{4}{3}}\Omega_0^{\frac{1}{3}} \dfrac{d\Omega}{dt} = -\dfrac{1}{3\Omega_0}\dfrac{d\Omega}{dt}$ when $\Omega = \Omega_0$; hence multiplying the equation by $3\Omega_0$ we have at the present time

$$-10^{10}\frac{d\Omega}{dt} = 6115 E \sin 2\epsilon + 1053 E' \sin \epsilon' - 68{\cdot}28 E'' \sin 2\epsilon'' \quad \text{...(59)}$$

in radians per annum.

If for the moment we call the right-hand of this equation k, we have $\Omega = \Omega_0 - k\dfrac{t}{10^{10}}$. Integrating a second time, we find that the moon has fallen

* In a future paper [Paper 6] on the perturbations of a satellite revolving about a viscous primary, I shall obtain this equation by the method of the disturbing function.

behind her proper place in her orbit $\frac{1}{2}t^2 \frac{k}{10^{10}} \cdot \frac{648000}{\pi}$ seconds of arc in the time t. Put t equal to a century, and substitute for k, and it will then be found that the moon lags in a century

$$630{\cdot}7E \sin 2\epsilon + 108{\cdot}6E' \sin \epsilon' - 7{\cdot}042E'' \sin 2\epsilon'' \text{ seconds of arc} \quad \ldots(60)$$

But it was shown in Section 13 (48) that the moon, if unaffected by tidal reaction, would have been apparently accelerated in a century

$$1043{\cdot}3E \sin 2\epsilon + 232{\cdot}5E' \sin \epsilon' \text{ seconds of arc.}$$

Hence taking the difference between these two, we find that there is an apparent acceleration in a century of the moon's motion of

$$412{\cdot}6E \sin 2\epsilon + 123{\cdot}9E' \sin \epsilon' + 7{\cdot}042E'' \sin 2\epsilon'' \quad \ldots\ldots\ldots(61)$$

seconds of arc.

Now according to Adams and Delaunay, there is at the present time an unexplained acceleration of the moon's motion of about 4″ in a century. For the present I will assume that the whole of this 4″ is due to the bodily tidal friction and reaction, leaving nothing to be accounted for by oceanic tidal friction and reaction, to which the whole has hitherto been attributed. Then we must have

$$412{\cdot}6E \sin 2\epsilon + 123{\cdot}9E' \sin \epsilon' + 7{\cdot}042E'' \sin 2\epsilon'' = 4 \quad \ldots\ldots(62)$$

This equation gives a relation which must subsist between the heights E, E', E'', of the semi-diurnal, diurnal, and fortnightly bodily tides, and their retardations $\epsilon, \epsilon', \epsilon''$, in order that the observed amount of tidal friction may not be exceeded. But no further deduction can be made, without some assumption as to the nature of the matter constituting the earth.

I shall first assume then that the matter is purely viscous, so that $E = \cos 2\epsilon$, $E' = \cos \epsilon'$, $E'' = \cos 2\epsilon''$, and $\tan 2\epsilon = \frac{2n}{\rho}$, $\tan \epsilon' = \frac{n}{\rho}$, $\tan 2\epsilon'' = \frac{2\Omega}{\rho}$.

The equation then becomes

$$412{\cdot}6 \sin 4\epsilon + 123{\cdot}9 \sin 2\epsilon' + 7{\cdot}042 \sin 4\epsilon'' = 8 \quad \ldots\ldots\ldots\ldots(63)$$

If the values of $\epsilon, \epsilon', \epsilon''$ be substituted, we get an equation of the sixth degree for ρ, but it will not be necessary to form this equation, because the question may be more simply treated by the following approximation.

There are obviously two solutions of the equation, one of which represents that the earth is very nearly fluid, and the other that it is very nearly rigid.

In the first case, that of approximate fluidity, $\epsilon, \epsilon', \epsilon''$ are very small, and therefore

$$\sin 4\epsilon = 4\epsilon, \quad \sin 2\epsilon' = 2\epsilon' = 2\epsilon, \quad \sin 4\epsilon'' = 4\epsilon'' = 4\frac{\Omega}{n}\epsilon = \frac{4}{27{\cdot}32}\epsilon$$

Hence $$\left(1650 + 248 + \frac{4}{27 \cdot 32} \text{ of } 7 \cdot 04\right) \epsilon = 8$$

whence $$\epsilon = \tfrac{1}{237} = 14'$$

That is to say, the semi-diurnal tide only lags by the small angle 14′. But this is not the solution which is interesting in the case of the earth, for we know that the earth does not behave approximately as a fluid body*.

In the other solution, 2ϵ and ϵ' approach 90°, so that ρ is small; hence

$$\sin 4\epsilon = \frac{4n\rho}{\rho^2 + 4n^2} = \frac{\rho}{n}, \quad \sin 2\epsilon' = \frac{2n\rho}{\rho^2 + n^2} = \frac{2\rho}{n} \text{ very nearly,}$$

and $$\sin 4\epsilon'' = \frac{4\Omega\rho}{\rho^2 + 4\Omega^2}$$

Hence we have

$$412 \cdot 6 \left(\frac{\rho}{n}\right) + 123 \cdot 9 \left(\frac{2\rho}{n}\right) + 7 \cdot 042 \frac{4\Omega\rho}{\rho^2 + 4\Omega^2} = 8$$

Put $\frac{\rho}{2\Omega} = x$, so that $x = \cot 2\epsilon''$; then substituting for $\frac{\Omega}{n}$ its value $\frac{1}{27 \cdot 32}$, we have

$$\frac{1320 \cdot 7}{27 \cdot 32} x + 7 \cdot 042 \frac{2x}{x^2 + 1} = 8$$

whence $$x^3 - \cdot 1655x^2 + 1 \cdot 2921x - \cdot 1655 = 0$$

This equation has two imaginary roots, and one real one, viz.: ·12858. Hence the desired solution is given by $\cot 2\epsilon'' = \cdot 12858$; and $2\epsilon'' = \frac{1}{2}\pi - 7° \, 20'$, and the corresponding values of 2ϵ and ϵ' are $2\epsilon = \frac{1}{2}\pi - 16'$, and $\epsilon' = \frac{1}{2}\pi - 32'$. If these values for ϵ, ϵ', ϵ'' be used in the original equation (63), they will be found to satisfy it very closely; and it appears that there is a true retardation of the moon of 3″·1 in a century, whilst the lengthening of the day would make an apparent acceleration of 7″·1,—the difference between the two being the observed 4″.

With these values the semi-diurnal and diurnal ocean-tides are, according to the equilibrium theory of ocean-tides, sensibly the same as those on a rigid nucleus, whilst the fortnightly tide is reduced to $\sin 2\epsilon''$ or ·992 of its theoretical amount; and the time of high tide is accelerated by $\frac{\pi}{4\Omega} - \frac{\epsilon''}{\Omega}$, or $6\frac{1}{2}$ hours in advance of its theoretical time.

If these values be substituted in the equation giving the rate of variation of the obliquity, it will be found that the obliquity must be decreasing at the rate of 0°·00197 per million years, or 1° in 500 million years. Thus in 100 million years it would only decrease by 12′. So, also, it may be shown

[* If it is the oceanic tides which now make the principal contribution to the tidal retardation of the earth's rotation, the preceding solution must be approximately correct.]

that the moon's sidereal period is being increased by 2 hours 20 minutes in 100 million years.

Lastly, the earth considered as a clock is losing 13 seconds in a century*.

---

* [Appendix G, (*a*) to Thomson and Tait's *Natural Philosophy*, p. 503, by G. H. Darwin.]

The retardation of the earth's rotation, as deduced from the secular acceleration of the Moon's mean motion.

In my paper on the precession of a viscous spheroid [this present paper], all the data are given which are requisite for making the calculations for Professor Adams' result in § 830 [of Thomson and Tait's *Natural Philosophy*], viz.: that if there is an unexplained part in the coefficient of the secular acceleration of the moon's mean motion amounting to 6″, and if this be due to tidal friction, then in a century the earth gets 22 seconds behind time, when compared with an ideal clock, going perfectly for a century, and perfectly rated at the beginning of the century. In the paper referred to however the earth is treated as homogeneous, and the tides are supposed to consist in a bodily deformation of the mass. The numerical results there given require some modification on this account.

If E, E', E'' be the heights of the semi-diurnal, diurnal and fortnightly tides, expressed as fractions of the equilibrium tides of the same denominations; and if ϵ, ϵ', ϵ'' be the corresponding retardations of phase of these tides due to friction; it is shown on p. [69 of this volume] and in equation (48), that in consequence of lunar and solar tides, at the end of a century, the earth, as a time-keeper, is behind the time indicated by the ideal perfect clock

$$1900{\cdot}27\, E \sin 2\epsilon + 423{\cdot}49\, E' \sin \epsilon' \text{ seconds of time} \quad \ldots\ldots\ldots\ldots\ldots\ldots(a)$$

and that if the motion of the moon were unaffected by the tides, an observer, taking the earth as his clock, would note that at the end of the century the moon was in advance of her place in her orbit by

$$1043''{\cdot}28\, E \sin 2\epsilon + 232''{\cdot}50\, E' \sin \epsilon' \quad \ldots\ldots\ldots\ldots\ldots\ldots\ldots\ldots(b)$$

This is of course merely the expression of the same fact as (*a*), in a different form.

Lastly it is shown in equation (60) that from these causes the moon actually lags in a century behind her place

$$630''{\cdot}7\, E \sin 2\epsilon + 108''{\cdot}6\, E' \sin \epsilon' - 7''{\cdot}042\, E'' \sin 2\epsilon'' \quad \ldots\ldots\ldots\ldots\ldots\ldots(c)$$

In adapting these results to the hypothesis of oceanic tides on a heterogeneous earth, we observe in the first place that, if the fluid tides are inverted, that is to say if for example it is low water under the moon, then friction advances the fluid tides†, and therefore in that case the ϵ's are to be interpreted as advancements of phase; and secondly that the E's are to be multiplied by $\frac{2}{11}$, which is the ratio of the density of water to the mean density of the earth. Next the earth's moment of inertia (as we learn from col. vii. of the table in § 824) is about ·83 of its amount on the hypothesis of homogeneity, and therefore the results (*a*) and (*b*) have both to be multiplied by 1/·83 or 1·2; the result (*c*) remains unaffected except as to the factor $\frac{2}{11}$.

Thus subtracting (*c*) from (*b*) as amended, we find that to an observer, taking the earth as a true time-keeper, the moon is, at the end of the century, in advance of her place by

$$\tfrac{2}{11}\{(1{\cdot}2 \times 1043''{\cdot}28 - 630''{\cdot}7)\, E \sin 2\epsilon + (1{\cdot}2 \times 232''{\cdot}50 - 108''{\cdot}6)\, E' \sin \epsilon' + 7''{\cdot}042\, E'' \sin 2\epsilon''\}$$

which is equal to

$$\tfrac{2}{11}\{621''{\cdot}24\, E \sin 2\epsilon + 170''{\cdot}40\, E' \sin \epsilon' + 7''{\cdot}04\, E'' \sin 2\epsilon''\} \quad \ldots\ldots\ldots\ldots\ldots(d)$$

† That this is true may be seen from considerations of energy. If it were approximately low water under the moon, the earth's rotation would be accelerated by tidal friction, if the tides of short period lagged; and this would violate the principles of energy.

There is another supposition as to the physical constitution of the earth, which will lead to interesting results.

If the earth be elastico-viscous, then for the semi-diurnal and diurnal tides it might behave nearly as though it were perfectly elastic, whilst for the fortnightly tide it might behave nearly as though it were perfectly viscous. With the law of elastico-viscosity used in my previous paper*, it is not possible to satisfy these conditions very exactly. But there is no reason to suppose that that law represents anything but an ideal sort of matter; it is as likely that the degradation of elasticity immediately after straining is not so rapid as that law supposes. I shall therefore take a limiting case, and suppose that, for the semi-diurnal and diurnal tides, the earth is perfectly

---

and from (a) as amended that the earth, as a time-keeper, is behind the time indicated by the ideal clock, perfectly rated at the beginning of the century, by

$$\tfrac{2}{11}\{2280{\cdot}32\, E \sin 2\epsilon + 508{\cdot}19\, E' \sin \epsilon'\} \text{ seconds of time} \quad\ldots\ldots\ldots\ldots(e)$$

Now if we suppose that the tides have their equilibrium height, so that the E's are each unity; and that ϵ' is one-half of ϵ (which must roughly correspond to the state of the case), and that ϵ'' is insensible, and ϵ small, (d) becomes

$$\tfrac{4}{11}\{621''{\cdot}24 + \tfrac{1}{4} \times 170''{\cdot}40\}\,\epsilon \quad\ldots\ldots\ldots\ldots(f)$$

and (e) becomes

$$\tfrac{4}{11}\{2280{\cdot}32 + \tfrac{1}{4} \times 508{\cdot}19\}\,\epsilon \text{ seconds of time} \quad\ldots\ldots\ldots\ldots(g)$$

If (f) were equal to 1″, then (g) would clearly be

$$\frac{2280{\cdot}32 + \frac{1}{4} \times 508{\cdot}19}{621{\cdot}24 + \frac{1}{4} \times 170{\cdot}40} \text{ seconds of time} \quad\ldots\ldots\ldots\ldots(h)$$

The second term, both in the numerator and denominator of (h), depends on the diurnal tide, which only exists when the ecliptic is oblique. Now Adams' result was obtained on the hypothesis that the obliquity of the ecliptic was nil, therefore according to his assumption, 1″ in the coefficient of lunar acceleration means that the earth, as compared with a perfect clock rated at the beginning of the century, is behind time

$$\frac{2280{\cdot}32}{621{\cdot}24} = 3\tfrac{2}{3} \text{ seconds at the end of a century}$$

Accordingly 6″ in the coefficient gives 22 secs. at the end of a century, which is his result given in § 830. If however we include the obliquity of the ecliptic and the diurnal tide, we find that 1″ in the coefficient means that the earth, as compared with the perfect clock, is behind time

$$\frac{2407{\cdot}37}{663{\cdot}80} = 3{\cdot}6274 \text{ seconds at the end of a century}$$

Thus taking Hansen's 12″·56 with Delaunay's 6″·1, we have the earth behind

$$6{\cdot}46 \times 3{\cdot}6274 = 23{\cdot}4 \text{ sec.}$$

and taking Newcomb's 8″·4 with Delaunay's 6″·1, we have the earth behind $2{\cdot}3 \times 3{\cdot}6274 = 8{\cdot}3$ sec.

It is worthy of notice that this result would be only very slightly vitiated by the incorrectness of the hypothesis made above as to the values of the E's and ϵ's; for $E \sin 2\epsilon$ occurs in the important term both in the numerator and denominator of the result for the earth's defect as a time-keeper, and thus the hypothesis only enters in determining the part played by the diurnal tide. Hence the result is not sensibly affected by some inexactness in this hypothesis, nor by the fact that the oceans in reality only cover a portion of the earth's surface.

* Namely, that if the solid be strained, the stress required to maintain it in the strained configuration diminishes in geometrical progression as the time, measured from the epoch of straining, increases in arithmetical progression. See Section 8 of the paper on "Bodily Tides," &c., *Phil. Trans.*, Part I., 1879. [Paper 1.]

elastic, whilst for the fortnightly one it is perfectly viscous. This hypothesis, of course, will give results in excess of what is rigorously possible, at least without a discontinuity in the law of degradation of elasticity.

It is accordingly assumed that the semi-diurnal and diurnal bodily tides do not lag, and therefore $\epsilon = \epsilon' = 0$; whilst the fortnightly tide does lag, and $E'' = \cos 2\epsilon''$.

Thus by (38) there is no tidal friction, and by (60) there is a true acceleration of the moon's motion of $\frac{1}{2}$ of $7{\cdot}042 \sin 4\epsilon''$ seconds of arc in a century. Then if we take the most favourable case, namely, when $\epsilon'' = 22^\circ\, 30'$, there is a true secular acceleration of $3''{\cdot}521$ per century.

It follows, therefore, that the whole of the observed secular acceleration of the moon might be explained by this hypothesis as to the physical constitution of the earth. On this hypothesis the fortnightly ocean tide should amount to $\sin 22^\circ\, 30'$, or ·38 of its theoretical height on a rigid nucleus, and the time of high water should be accelerated by 1 day 17 hours. Again, by (35) $\frac{di}{dt} = -\frac{3}{16} u^2 \sin^3 i$, from whence it may be shown that the obliquity of the ecliptic would be decreasing at the rate of 1° in 128 million years.

The conclusion to be drawn from all these calculations is that, at the present time, the bodily tides in the earth, except perhaps the fortnightly tide, must be exceedingly small in amount; that it is utterly uncertain how much of the observed 4″ of acceleration of the moon's motion must be referred to the moon itself, and how much to the tidal friction, and accordingly that it is equally uncertain at what rate the day is at present being lengthened; lastly, that if there is at present any change in the obliquity of the ecliptic, it must be very slowly decreasing.

The result of this hypothesis of elastico-viscosity appears to me so curious that I shall proceed to show what might possibly have been the state of things a very long time ago, if the earth had been perfectly elastic for the tides of short period, but viscous for the fortnightly tide.

There will now be no tidal friction, and the length of day remains constant. The equation of tidal reaction reduces to

$$\mu \frac{d\xi}{dt} = -\frac{u^2}{\xi^{12}} \frac{3}{16} \sin^4 i \sin 4\epsilon''$$

Here u^2 is a constant, being the value of $\frac{\tau^2}{\mathfrak{g} n_0}$ at the epoch; and $u^2 \div \xi^{12}$ is the value of $\frac{\tau^2}{\mathfrak{g} n_0}$ at the time t.

The equation giving the rate of change of obliquity becomes

$$\frac{di}{dt} = -\frac{u^2}{\xi^{12}} \frac{3}{16} \sin^3 i \sin 4\epsilon''$$

Dividing the latter by the former, we have*

$$\sin i\,di = \mu\,d\xi$$

And by integration $\cos i = \cos i_0 - \mu(\xi - 1)$

If we look back long enough in time, we may find $\xi = 1{\cdot}01$, and μ being 4·007, we have

$$\cos i = \cos i_0 - {\cdot}04007$$

Taking $i_0 = 23° 28'$, we find $i = 28° 40'$.

This result is independent of the degree of viscosity. When, however, we wish to find how long a period is requisite for this amount of change, some supposition as to viscosity is necessary. The time cannot be less than if $\sin 4\epsilon'' = 1$, or $\epsilon'' = 22° 30'$, and we may find a rough estimate of the time by writing the equation of tidal reaction

$$\mu \frac{d\xi}{dt} = -\tfrac{3}{16}\frac{u^2}{\xi^{12}}\sin^4 \mathrm{I}$$

where I is constant and equal to 24°, suppose. Then integrating we have

$$\mu(\xi^{13} - 1) = -t\tfrac{39}{16}u^2\sin^4 \mathrm{I}$$

or

$$t = -\tfrac{16}{39}\frac{\mu}{u^2}\operatorname{cosec}^4 \mathrm{I}\,(\xi^{13} - 1)$$

When $\xi = 1{\cdot}01$, we find from this that $-t = 720$ million years, and that the length of the month is 28·15 m.s. days. Hence, if we look back 700 million years or more, we might find the obliquity 28° 40′, and the month 28·15 m.s. days, whilst the length of day might be nearly constant. It must, however, be reiterated, that on account of our assumptions the change of obliquity is greater than would be possible, whilst the time occupied by the change is too short. In any case, any change in this direction approaching this in magnitude seems excessively improbable.

PART II.

§ 15. *Integration of the differential equations for secular changes in the variables in the case of viscosity.*

It is now supposed that the earth is a purely viscous spheroid, and I shall proceed to find the changes which would occur in the obliquity of the ecliptic and the lengths of the day and month when very long periods of time are taken into consideration.

I have been unable to find even an approximate general analytical solution of the problem, and have therefore worked the problem by a laborious arithmetical method, when the earth is supposed to have a particular degree of viscosity.

* Concerning the legitimacy of this change of variable, see the following section.

The viscosity chosen is such that, with the present length of day, the semi-diurnal tide lags by 17° 30′. It was shown above that this viscosity makes the rate of change of obliquity nearly a maximum*. It does not follow that the whole series of changes will proceed with maximum velocity, yet this supposition will, I think, give a very good idea of the minimum time, and of the nature of the changes which may have occurred in the course of the development of the moon-earth system.

The three semi-diurnal tides will be supposed to lag by the same amount and to be reduced in the same proportion; as also will be the three diurnal tides.

There are three simultaneous differential equations to be treated, viz.: those giving (1) the rate of change of the obliquity of the ecliptic, (2) the rate of alteration of the earth's diurnal rotation, (3) the rate of tidal reaction on the moon. They will be referred to hereafter as the *equations of obliquity, of friction, and reaction* respectively.

To write these equations more conveniently a partly new notation is advantageous, as follows:

The suffix 0 to any symbol denotes the initial value of the quantity in question.

Let $u^2 = \dfrac{\tau_0^{\,2}}{\mathfrak{g}n_0}$, $u_{,}^{\,2} = \dfrac{\tau_{,}^{\,2}}{\mathfrak{g}n_0}$, $uu_{,} = \dfrac{\tau_0\tau_{,}}{\mathfrak{g}n_0}$; these three quantities are constant.

Since the tidal reaction on the sun is neglected, $\tau_{,}$ is a constant, and since τ varies as Ω^2 (and therefore as ξ^{-6}); hence

$$\frac{\tau^2}{\mathfrak{g}n} = \frac{n_0}{n}\frac{u^2}{\xi^{12}}, \quad \frac{\tau_{,}^{\,2}}{\mathfrak{g}n} = \frac{n_0}{n}u_{,}^{\,2}, \quad \frac{\tau\tau_{,}}{\mathfrak{g}n} = \frac{n_0}{n}\frac{uu_{,}}{\xi^6}$$

Let ρ be equal to $\dfrac{2gaw}{19v}$, where v is the coefficient of viscosity of the earth. Then according to the theory developed in my paper on tides†

$$\tan 2\epsilon = \frac{2n}{\rho}, \quad \tan\epsilon' = \frac{n}{\rho}, \quad \tan 2\epsilon'' = \frac{2\Omega}{\rho} \quad \text{..............(64)}$$

To simplify the work, terms involving the fourth power of the sine of the obliquity will be neglected.

Now let

$$\left.\begin{aligned} &P = \tfrac{1}{4}\log_{10} e, \quad Q = \tfrac{3}{8}\sin^2 i \log_{10} e, \quad R = \tfrac{3}{16}\frac{\sin^2 i}{\cos i}\log_{10} e = \tfrac{1}{2}Q\sec i \\ &U = \tfrac{1}{4}\sin^2 i \log_{10} e, \quad V = \frac{\tfrac{1}{2}\cos^2 i}{1 - \tfrac{3}{4}\sin^2 i}\log_{10} e \\ &W = \tfrac{1}{2}\cos^2 i, \quad X = \tfrac{1}{2}\sin^2 i\cos i, \quad Z = \tfrac{1}{2}\sin^2 i\cos^2 i \end{aligned}\right\} \text{...(65)}$$

* If I had to make the choice over again I should choose a slightly greater viscosity as being more interesting.

† *Phil. Trans.*, 1879, Part I. [Paper 1.]

Also let $sn_0\Omega_0^{\frac{1}{3}} = \frac{1}{\mu}$, $\frac{n}{n_0} = N$; and it may be called to mind that $\xi = \left(\frac{\Omega_0}{\Omega}\right)^{\frac{1}{3}}$, $s = \frac{2}{5}\left[\left(\frac{av}{g}\right)^2 (1+\nu)\right]^{\frac{1}{3}}$.

The terms depending on the semi-annual tide will be omitted throughout. With this notation the equation of obliquity (35) and (36) may be written

$$\log_{10} e \frac{di}{dt} = \sin i \cos i \,(1 - \tfrac{3}{4}\sin^2 i) \left[\left(\frac{u^2}{\xi^{12}} + u_{\prime}^2\right)(\mathrm{P}\sin 4\epsilon + \mathrm{Q}\sin 2\epsilon') - \frac{uu_{\prime}}{\xi^6}(\mathrm{U}\sin 4\epsilon + \mathrm{V}\sin 2\epsilon') - \frac{u^2}{\xi^{12}}\mathrm{R}\sin 4\epsilon''\right] \quad \text{...(66)}$$

The equation (43) of friction becomes

$$-\frac{dN}{dt} = \left(\frac{u^2}{\xi^{12}} + u_{\prime}^2\right)(\mathrm{W}\sin 4\epsilon + \mathrm{X}\sin 2\epsilon') + \frac{uu_{\prime}}{\xi^6}\mathrm{Z}\sin 2\epsilon' \quad \text{......(67)}$$

And by (58), Section 14, the equation of reaction becomes

$$\mu\frac{d\xi}{dt} = \frac{u^2}{\xi^{12}}(\mathrm{W}\sin 4\epsilon + \mathrm{X}\sin 2\epsilon') \quad \text{.................(68)}$$

This is the third of the simultaneous differential equations which have to be treated. The four variables involved are i, N, ξ, t, which give the obliquity, the earth's rotation, the square root of the moon's distance and the time. Besides where they are involved explicitly, they enter implicitly in Q, R, U, V, W, X, Z, $\sin 4\epsilon$, $\sin 2\epsilon'$, $\sin 4\epsilon''$.

Q, R, &c., are functions of the obliquity i only, but P is a constant. Also

$$\sin 4\epsilon = \frac{4n\rho}{4n^2+\rho^2} = \frac{4n_0\rho N}{4n_0^2N^2+\rho^2}, \quad \sin 2\epsilon' = \frac{2n_0\rho N}{n_0^2N^2+\rho^2}, \quad \sin 4\epsilon'' = \frac{4\Omega_0\rho\xi^3}{4\Omega_0^2+\rho^2\xi^6}$$

I made several attempts to solve these equations by retaining the time as independent variable, and substituting for ξ and N approximate values, but they were all unsatisfactory, because of the high powers of ξ which occur, and no security could be felt that after a considerable time the solutions obtained did not differ a good deal from the true one. The results, however, were confirmatory of those given hereafter.

The method finally adopted was to change the independent variable from t to ξ. A new equation was thus formed between N and ξ, which involved the obliquity i only in a subordinate degree, and which admitted of approximate integration. This equation is in fact that of conservation of moment of momentum, modified by the effects of the solar tidal friction. Afterwards the time and the obliquity were found by the method of quadratures. As, however, it was not safe to push this solution beyond a certain point, it was carried as far as seemed safe, and then a new set of equations were formed, in which the final values of the variables, as found from the previous integration, were used as the initial values. A similar operation was carried out a third and fourth time. The operations were thus divided into a series of periods, which will be referred to as periods of integration. As the error in the final values in any one period is carried on to the next period, the error tends to accumulate; on this account the integration in the first and second periods was carried out with greater accuracy than would in general be necessary for a speculative inquiry like the present one. The first step is to form the approximate equation of conservation of moment of momentum above referred to.

Let $A = W\sin 4\epsilon + X\sin 2\epsilon'$, $B = Z\sin 2\epsilon'$.

Then the equations of friction (67) and reaction (68) may be written

$$-n_0\mathfrak{g}\frac{dN}{dt} = \left(\frac{\tau_0^2}{\xi^{12}} + \tau_{,}^2\right)A + \frac{\tau_0\tau_{,}}{\xi^6}B \quad \text{.................(69)}$$

$$n_0\mathfrak{g}\mu\frac{d\xi}{dt} = \frac{\tau_0^2}{\xi^{12}}A \quad \text{.............................(70)}$$

We now have to consider the proposed change of variable from t to ξ.

The full expression for $\frac{dN}{dt}$ contains a number of periodic terms; $\frac{d\xi}{dt}$ also contains terms which are co-periodic with those in $\frac{dN}{dt}$. The object which is here in view is to determine the increase in the average value of N per unit increase of the average value of ξ. The proposed new independent variable is therefore not ξ, but it is the average value of ξ; but as no occasion will arise for the use of ξ as involving periodic terms, I shall retain the same symbol.

In order to justify the procedure to be adopted, it is necessary to show that, if $f(t)$ be a function of t, then the rate of increase of its average value estimated over a period T, of which the beginning is variable, is equal to the average rate of its increase estimated over the same period. The average value of $f(t)$ estimated over the period T, beginning at the time t is $\frac{1}{T}\int_t^{t+T} f(t)\,dt$, and therefore the rate of the increase of the average value is $\frac{d}{dt}\frac{1}{T}\int_t^{t+T} f(t)\,dt$, which is equal to $\frac{1}{T}\int_t^{t+T} f'(t)\,dt$; and this last expression is the average rate of increase of $f(t)$ estimated over the same period. This therefore proves the proposition in question.

Suppose we have $\frac{dN}{dt} = -M +$ periodic terms, where M varies very slowly; then $-M$ is the average value of the rate of increase of N estimated over a period which is the least common multiple of the periods of the several periodic terms. Hence by the above proposition $-M$ is also the rate of increase of the average value of N estimated over the like period.

Similarly if $\frac{d\xi}{dt} = X +$ periodic terms, X is the rate of increase of the average value of ξ estimated over a period, which will be the same as in the former case.

But the average value of N is the proposed new dependent variable, and the average value of ξ the new independent variable. Hence, from the present point of view, $\frac{dN}{d\xi} = -\frac{M}{X}$. This argument is, however, only strictly applicable, supposing there are not periodic terms in $\frac{dN}{dt}$ or $\frac{d\xi}{dt}$ of incommensurable periods, and supposing the periodic terms are rigorously circular functions, so that their amplitudes and frequencies are not functions of the time.

It is obvious, however, that if the incommensurable terms do not represent long inequalities, and if M and X vary slowly, the theorem remains very nearly true. With respect to the variability of amplitude and frequency, it is only necessary for the applicability of the argument to postulate that the so-called periodic terms are so nearly true circular functions that the integrals of them over any moderate multiple of their period are sensibly zero.

Suppose, for example, $\psi(t)\cos[vt + \chi(t)]$ were one of the periodic terms, we have only to suppose that $\psi(t)$ and $\chi(t)$ vary so slowly that they remain sensibly constant during a period $2\pi/v$ or any moderately small multiple of it, in order to be safe in assuming $\int_0^{\frac{2\pi}{v}} \psi(t)\cos(vt + \chi(t))\,dt$ as sensibly zero. Now in all the inequalities in N and ξ it is a question of days or weeks, whilst in the variations of the amplitudes and frequencies of the inequalities it is a question of millions of years. Hence the above method is safely applicable here.

It is worthy of remark that it has been nowhere assumed that the amplitudes of the periodic inequalities are small compared with the non-periodic parts of the expression.

A precisely similar argument will be applicable to every case where occasion will arise to change the independent variable. The change will accordingly be carried out without further comment, it being always understood that both dependent and independent variable are the average values of the quantities for which their symbols would in general stand*.

* In order to feel complete confidence in my view, I placed the question before Mr E. J. Routh, and with great kindness he sent me some remarks on the subject, in which he confirmed the correctness of my procedure, although he arrived at the conclusion from rather a different point of view.

Then dividing (69) by (70) we have

$$-\frac{dN}{\mu d\xi} = 1 + \left(\frac{\tau_{,}}{\tau_0}\right)^2 \xi^{12} + \frac{B}{A}\left(\frac{\tau_{,}}{\tau_0}\right)\xi^6 \quad\ldots\ldots\ldots\ldots(71)$$

Now $\dfrac{B}{A} = \dfrac{Z}{W\dfrac{\sin 4\epsilon}{\sin 2\epsilon'} + X} = \sin^2 i\dfrac{\sin 2\epsilon'}{\sin 4\epsilon}$ approximately. This approximation will be sufficiently accurate, because the last term is small and is diminishing. For the same reason, only a small error will be incurred by treating it as constant, provided the integration be not carried over too large a field—a condition satisfied by the proposed "periods of integration." Attribute then to i, ϵ, ϵ' average values, and put

$$\beta = \tfrac{1}{13}\left(\frac{\tau_{,}}{\tau_0}\right)^2, \quad \gamma = \tfrac{1}{7}\frac{\tau_{,}}{\tau_0}\sin^2 i\frac{\sin 2\epsilon'}{\sin 4\epsilon} \quad\ldots\ldots\ldots\ldots(72)$$

and integrate, and we have

$$N = 1 + \mu\{(1-\xi) + \beta(1-\xi^{13}) + \gamma(1-\xi^7)\} \quad\ldots\ldots\ldots\ldots(73)$$

This is the approximate form of the equation of conservation of moment of momentum, and it is very nearly accurate, provided ξ does not vary too widely.

By putting $\beta = 0$, $\gamma = 0$, we see that, if there be only two bodies, the earth and moon, the equation is independent of the obliquity, provided we neglect the fourth power of the sine of the obliquity.

The equation of reaction (68) may be written

$$\frac{dt}{d\xi} = \mu \div \frac{u^2}{\xi^{12}}(W\sin 4\epsilon + X\sin 2\epsilon') \quad\ldots\ldots\ldots\ldots(74)$$

Also, multiplying the equation of obliquity (66) by $\dfrac{dt}{d\xi}$, we have

$$\frac{\log_{10} e}{\sin i\cos i\,(1-\frac{3}{4}\sin^2 i)}\frac{di}{d\xi} = \frac{1}{N}\frac{dt}{d\xi}\left[\left(\frac{u^2}{\xi^{12}} + u_{,}^2\right)(P\sin 4\epsilon + Q\sin 2\epsilon') - \frac{uu_{,}}{\xi^6}(U\sin 4\epsilon + V\sin 2\epsilon') - \frac{u^2}{\xi^{12}}R\sin 4\epsilon''\right]$$

By far the most important term in $\dfrac{d\xi}{dt}$ is that in which W occurs, and therefore $\dfrac{1}{2W}\dfrac{d\xi}{dt}$ only depends on the obliquity in its smaller term. Then, since $2W = \cos^2 i$,

$$\frac{dt}{d\xi} = \frac{1}{\cos^2 i}\left(2W\frac{dt}{d\xi}\right)$$

Also
$$\frac{\cos^2 i}{\sin i\cos i\,(1-\frac{3}{4}\sin^2 i)}\,di = d\,.\log_e\frac{\sin i}{\sqrt{1-\frac{3}{4}\sin^2 i}}$$
$$= d\,.\log_e\tan i\,(1-\tfrac{1}{8}\sin^2 i)$$

when the fourth power of $\sin i$ is neglected.

Hence the equation may be written

$$\frac{d}{d\xi}\log_{10}\tan i\,(1-\tfrac{1}{8}\sin^2 i)=\frac{1}{N}\left(2\mathrm{W}\frac{dt}{d\xi}\right)\left[\left(\frac{u^2}{\xi^{12}}+u_{,}^2\right)(\mathrm{P}\sin 4\epsilon+\mathrm{Q}\sin 2\epsilon')\right.$$

$$\left.-\frac{u^2}{\xi^{12}}\mathrm{R}\sin 4\epsilon''-\frac{uu_{,}}{\xi^6}(\mathrm{U}\sin 4\epsilon+\mathrm{V}\sin 2\epsilon')\right]\text{.........}(75)$$

The term in P (which is a constant) is by far the most important of those within brackets [ ] on the right-hand side, and $2\mathrm{W}\frac{dt}{d\xi}$ has been shown only to involve i in its smaller term. Hence the whole of the right-hand side only involves the obliquity to a subordinate degree, and, in as far as it does so, an average value may be assigned to i without producing much error.

In the equation of tidal reaction (68) or (74) also, I attribute to i in W and X an average value, and treat them as constants. As the accumulation of the error of time from period to period is unimportant, this method of approximation will give quite good enough results.

We are now in a position to trace the changes in the obliquity, the day, and the month, and to find the time occupied by the changes by the method of quadratures.

First estimate an average value of i and compute Q, R ... Z, β, γ. Take seven values of ξ, viz.: 1, ·98, ·96 ... ·88, and calculate seven corresponding values of N; then calculate seven corresponding values of $\sin 4\epsilon$, $\sin 2\epsilon'$, $\sin 4\epsilon''$. Substitute these values in $\frac{d\xi}{dt}$, and take the reciprocals so as to get seven equidistant values of $\frac{dt}{d\xi}$.

Combine these seven values by Weddle's rule, viz.:

$$\int_0^{6h} u_x dx = \tfrac{3}{10}h\,[u_0+u_2+u_3+u_4+u_6+5\,(u_1+u_3+u_5)]$$

and so find the time corresponding to $\xi=·88$. It must be noted that the time is negative because $d\xi$ is negative.

In the course of the work the values of $\frac{dt}{d\xi}$ corresponding to $\xi=1$, ·96, ·92, ·88 have been obtained. Multiply them by 2W; these values, together with the four values of $\sin 4\epsilon$, $\sin 2\epsilon'$, $\sin 4\epsilon''$ and the four of N, enable us to compute four of $\frac{d}{d\xi}\log_{10}\tan i\,(1-\frac{1}{8}\sin^2 i)$, as given in (75).

Combine these four values by the rule

$$\int_0^{3h} u_x dx = \tfrac{3}{8}h\,[u_0+u_3+3\,(u_1+u_2)]$$

and we get $$\log_{10}\frac{\tan i\,(1-\frac{1}{8}\sin^2 i)}{\tan i_0\,(1-\frac{1}{8}\sin^2 i_0)}$$

from which the value of i corresponding to $\xi = \cdot 88$ may easily be found. It is here useless to calculate more than four values, because the function to be integrated does not vary rapidly.

We have now obtained final values of i, N, t corresponding to $\xi = \cdot 88$.

Since the earth is supposed to be viscous throughout the changes, its figure must always be one of equilibrium, and its ellipticity of figure $e = N^2 e_0$.

Also since $\xi = \left(\frac{\Omega_0}{\Omega}\right)^{\frac{1}{3}} = \sqrt{\frac{c}{c_0}}$, where c is the moon's distance from the earth, $\frac{c}{a} = \xi^2 \left(\frac{c_0}{a}\right)$, which gives the moon's distance in earth's mean radii.

The fifth and sixth columns of Table IV. were calculated from these formulæ.

The seventh column of Table IV. shows the distribution of moment of momentum in the system; it gives μ the ratio of the moment of momentum of the moon's and earth's motion round their common centre of inertia to that of the earth's rotation round its axis, at the beginning of each period of integration.

Table I. shows the values of ϵ, ϵ', ϵ'' the angles of lagging of the semi-diurnal, diurnal, and fortnightly tides at the beginning of each period.

Tables II. and III. show the relative importance of the contributions of each term to the values of $\frac{d\xi}{dt}$ and $\frac{d}{d\xi}\log_{10}\tan i\,(1 - \frac{1}{8}\sin^2 i)$ at the beginning of each period.

The several *lines* of the Tables II. and III. are not comparable with one another, because they are referred to different initial values of Ω and n in each line.

I will now give some details of the numerical results of each integration. The computation as originally carried out* was based on a method slightly different from that above explained, but I was able to adapt the old computation to the above method by the omission of certain terms and the application of certain correcting factors. For this reason the results in the first three tables are only given in round numbers. In the fourth table the length of day is given to the nearest five minutes, and the obliquity to the nearest five minutes of arc.

The integration begins when the length of the sidereal day is 23 hrs. 56 m., the moon's sidereal period 27·3217 m.s. days, the obliquity of the ecliptic 23° 28′, and the time zero.

* I have to thank Mr E. M. Langley, of Trinity College, for carrying out the laborious computations. The work was checked throughout by myself.

First period. Integration from $\xi = 1$ to $\cdot 88$; seven equidistant values computed for finding the time, and four for the obliquity.

For the obliquity the integration was not carried out exactly as above explained, in as much as that $\frac{d}{d\xi}\log_{10}\tan i$ was found instead of $\frac{d}{d\xi}\log_{10}\tan i\,(1 - \frac{1}{8}\sin^2 i)$, but the difference in method is quite unimportant. The result marked * in Table III. is $\frac{d}{d\xi}\log_{10}\tan i$.

The estimated average value of i was 22° 15′.

The final result is

$$N = 1\cdot 550, \quad i = 20^\circ\ 42', \quad -t = 46{,}301{,}000$$

Second period. Integration from $\xi = 1$ to $\cdot 76$; seven values computed for the time, and four for the obliquity.

The estimated average for i was 19°.

The final result is

$$N = 1\cdot 559, \quad i = 17^\circ\ 21', \quad -t = 10{,}275{,}000$$

Third period. Integration from $\xi = 1$ to $\cdot 76$; four values computed.

The estimated average for i was 16° 30′.

The final result is

$$N = 1\cdot 267, \quad i = 15^\circ\ 30', \quad -t = 326{,}000$$

Fourth period. Integration from $\xi = 1$ to $\cdot 76$; four values computed.

The estimated average for i was 15°. The small terms in β and γ were omitted in the equation of conservation of moment of momentum. All the solar and combined terms, except that in V in the equation of obliquity, were omitted.

The final result is

$$N = 1\cdot 160, \quad i = 14^\circ\ 25', \quad -t = 10{,}300$$

TABLE I. Showing the lagging of the several tides at the beginning of each period.

	Semi-diurnal (ϵ)	Diurnal (ϵ')	Fortnightly (ϵ'')
I.	17½°	19½°	0° 44′
II.	23½°	28½°	1° 5′
III.	29½°	40°	2° 27′
IV.	32½°	46½°	5° 30′

TABLE II. Showing the contribution of the several tidal effects to tidal reaction $\left(i.e., \text{to } \frac{d\xi}{dt}\right)$ at the beginning of each period. The numbers to be divided by 10^{10}.

	Semi-diurnal	Diurnal
I.	12·	1·2
II.	69·	6·3
III.	2200·	200·
IV.	70000·	6100·

TABLE III. Showing the contributions of the several tidal effects to the change of obliquity $\left(i.e., \text{to } \frac{d}{d\xi} \log_{10} \tan i \,(1 - \tfrac{1}{8}\sin^2 i)\right)$ at the beginning of each period.

	Lunar semi-diurnal	Lunar diurnal	Solar semi-diurnal	Solar diurnal	Combined semi-diurnal	Combined diurnal	Fortnightly	$\frac{d}{d\xi} \log \tan i\,(1 - \frac{1}{8}\sin^2 i)$
*I.	·82	·13	·18	03	− ·06	− ·48	− ·000	·60*
II.	·44	·06	·02	...	− ·01	− ·16	− ·003	·36
III.	·22	·03	...	...	...	− ·02	− ·003	·23
IV.	·13	·02	...	...	...	...	− ·004	·14

TABLE IV. Showing the physical meaning of the results of the integration.

	Time (− t)	Sidereal day in m. s. hours	Moon's sidereal period in m. s. days	Obliquity of ecliptic (i)	Reciprocal of ellipticity of figure	Moon's distance in earth's mean radii	Ratio of m. of m. of orbital motion to m. of m. of earth's rotation	Heat generated (see Section 16)
	Years	h. m.	d.					Degrees Fahr.
Initial state	0	23 56	27·32	23° 28′	232	60·4	4·01	0°
I.	46,300,000	15 30	18·62	20° 40′	96	46·8	2·28	225°
II.	56,600,000	9 55	8·17	17° 20′	40	27·0	1·11	760°
III.	56,800,000	7 50	3·59	15° 30′*	25	15·6	·67	1300°
IV.	56,810,000	6 45	1·58	14° 25′*	18	9·0	·44	1760°

The whole of these results are based on the supposition that the plane of the lunar orbit will remain very nearly coincident with the ecliptic throughout these changes. I now (July, 1879), however, see reason to believe that the secular changes in the plane of the lunar orbit will have an important influence on the obliquity of the ecliptic. Up to the end of the second period the change of obliquity as given in Table IV. will be approximately correct, but I find that during the third and fourth periods of integration there will be a phase of considerable nutation. The results in the column of obliquity marked (*) have not, therefore, very much value as far as regards the explanation of the obliquity of the ecliptic; they are, however, retained as being instructive from a dynamical point of view.

§ 16. *The loss of energy of the system.*

It is obvious that as there is tidal friction the moon-earth system must be losing energy, and I shall now examine how much of this lost energy turns into heat in the interior of the earth. The expressions potential and kinetic energy will be abbreviated by writing them P.E. and K.E.

The K.E. of the earth's rotation is $\frac{1}{5}Ma^2n^2$.

The K.E. of the earth's and moon's orbital motion round their common centre of inertia is

$$\tfrac{1}{2}M\left(\frac{m\text{r}}{m+M}\right)^2\Omega^2+\tfrac{1}{2}m\left(\frac{M\text{r}}{m+M}\right)^2\Omega^2=\tfrac{1}{2}M\text{r}^2\frac{\Omega^2}{1+\nu}$$

But since the moon's orbit is circular $\Omega^2\text{r}=g\left(\frac{a}{\text{r}}\right)^2\frac{1+\nu}{\nu}$, so that $\frac{\Omega^2\text{r}^2}{1+\nu}=\frac{ga^2}{\nu\text{r}}$. Hence the whole K.E. of the moon-earth system is

$$Ma^2\left(\tfrac{1}{5}n^2+\tfrac{1}{2}\frac{g}{\nu\text{r}}\right)$$

The P.E. of the system is $\quad -\frac{Mm}{\text{r}}=-\frac{M}{\nu}\frac{ga^2}{\text{r}}$

Therefore the whole energy E of the system is

$$Ma^2\left\{\tfrac{1}{5}n^2-\tfrac{1}{2}\frac{g}{\nu\text{r}}\right\}$$

and in gravitation units $\quad E=Ma\left\{\tfrac{1}{5}\frac{n^2a}{g}-\tfrac{1}{2}\frac{a}{\nu\text{r}}\right\}$

Since the earth is supposed to be plastic throughout all these changes, its ellipticity of figure

$$e=\tfrac{5}{4}\frac{n^2a}{g}$$

and $\quad E=Ma\left\{\tfrac{4}{25}e-\tfrac{1}{2}\frac{a}{\nu\text{r}}\right\}$

If e, $e+\Delta e$ and r, r + Δr be the ellipticity of figure, and the moon's distance at two epochs, if J be Joule's equivalent, and σ the specific heat of the matter constituting the earth; then the loss of energy of the system between these two epochs is sufficient to heat unit mass of the matter constituting the earth

$$-\frac{Ma}{J\sigma}\left\{\tfrac{4}{25}\Delta e - \frac{1}{2\nu}\Delta\frac{a}{r}\right\} \text{ degrees}$$

and is therefore enough to heat the whole mass of the earth

$$-\frac{a}{J\sigma}\left\{\tfrac{4}{25}\Delta e - \frac{1}{2\nu}\Delta\frac{a}{r}\right\} \text{ degrees}$$

It must be observed that in this formula the whole loss of K.E. of the earth's rotation, due both to solar and lunar tidal friction, is included, whilst only the gain of the moon's P.E. is included, and the effect of the solar tidal reaction in giving the earth greater potential energy relatively to the sun is neglected.

In the fifth and sixth columns of Table IV. of the last section the ellipticity of figure and the moon's distance in earth's radii are given; and these numbers were used in calculating the eighth column of the same table.

I used British units, so that 772 foot-pounds being required to heat 1 lb. of water 1° Fahr., J = 772; the specific heat of the earth was taken as $\frac{1}{9}$th, which is about that of iron, many of the other metals having a still smaller specific heat; the earth's radius was taken, as before, equal to 20·9 million feet. The last column states that energy enough has been turned into heat in the interior of the earth to warm its whole mass so many degrees Fahrenheit within the times given in the first column of the same table.

The consideration of the distribution of the generation of heat and the distortion of the interior of the earth must be postponed to a future occasion.

In the succeeding paper [Paper 4] I have considered the bearing of these results on the secular cooling of the earth, and in a subsequent paper [Paper 5] the general problem of tidal friction is considered by the aid of the principle of energy.

§ 17. *Integration in the case of small variable viscosity*.*

In the solution of the problem which has just been given, where the viscosity is constant, the obliquity of the ecliptic does not diminish as fast as it might do as we look backwards. The reason of this is that the ratio of the negative terms to the positive ones in the equation of obliquity is not as small as it might be; that ratio principally depends on the fraction $\frac{\sin 2\epsilon'}{\sin 4\epsilon}$, which has its smallest value when ϵ is very small.

I shall now, therefore, consider the case where the viscosity is small, and where it so varies that ϵ always remains small.

This kind of change of viscosity is in general accordance with what one may suppose to have been the case, if the earth was a cooling body, gradually freezing as it cooled.

The preceding solution is moreover somewhat unsatisfactory, inasmuch as the three semi-diurnal tides are throughout supposed to suffer the same retardation, as also are the three diurnal tides; and this approximation ceases to be sufficiently accurate towards the end of the integration.

In the present solution the retardations of all the lunar tides will be kept distinct.

By (40) and (40′), Section 11, for the lunar tides,

$$\tan 2\epsilon_1 = \frac{2(n-\Omega)}{\rho}, \quad \tan 2\epsilon = \frac{2n}{\rho}, \quad \tan 2\epsilon_2 = \frac{2(n+\Omega)}{\rho}, \quad \tan 2\epsilon'' = \frac{2\Omega}{\rho}$$

$$\tan \epsilon_1' = \frac{n-2\Omega}{\rho}, \qquad \tan \epsilon' = \frac{n}{\rho}, \qquad \tan \epsilon_2' = \frac{n+2\Omega}{\rho}$$

For the solar tides we may safely neglect $\Omega_{,}$ compared with n, and we have $\tan 2\epsilon = \frac{2n}{\rho}$, $\tan \epsilon' = \frac{n}{\rho}$ for the semi-diurnal and diurnal tides respectively. The semi-annual tide will be neglected.

If the viscosity so varies that all the ϵ's are always small, and if we put $\frac{\Omega}{n} = \lambda$, we have

$$\left.\begin{aligned} \frac{\sin 4\epsilon_1}{\sin 4\epsilon} &= 1-\lambda, & \frac{\sin 4\epsilon''}{\sin 4\epsilon} &= \lambda, & \frac{\sin 4\epsilon_2}{\sin 4\epsilon} &= 1+\lambda \\ \frac{\sin 2\epsilon_1'}{\sin 4\epsilon} &= \tfrac{1}{2}-\lambda, & \frac{\sin 2\epsilon'}{\sin 4\epsilon} &= \tfrac{1}{2}, & \frac{\sin 2\epsilon_2'}{\sin 4\epsilon} &= \tfrac{1}{2}+\lambda \end{aligned}\right\} \quad \ldots\ldots\ldots(76)$$

By means of these equations we may express all the sines of the ϵ's in terms of $\sin 4\epsilon$.

* This section has been partly rewritten and rearranged, and wholly recomputed since the paper was presented. The alterations are in the main dated December 19, 1878.

Remembering that the spheroid is viscous, and that therefore $E_1 = \cos 2\epsilon_1$, $E_1' = \cos \epsilon_1'$, &c., we have by Sections 4 and 7, equations (16) and (29),

$$\frac{di_{m^2}}{dt} = \frac{1}{N}\frac{\tau^2}{\mathfrak{g}n_0}\left[\tfrac{1}{2}p^7 q \sin 4\epsilon_1 - p^3q^3(p^2-q^2)\sin 4\epsilon - \tfrac{1}{2}pq^7\sin 4\epsilon_2 - \tfrac{3}{2}p^3q^3\sin 4\epsilon'' + \tfrac{1}{2}p^5q(p^2+3q^2)\sin 2\epsilon_1' - \tfrac{1}{2}pq(p^2-q^2)^3\sin 2\epsilon' - \tfrac{1}{2}pq^5(3p^2+q^2)\sin 2\epsilon_2'\right] \quad \text{.........(77)}$$

$$-\frac{dN_{m^2}}{dt} = \frac{\tau^2}{\mathfrak{g}n_0}\left[\tfrac{1}{2}p^8\sin 4\epsilon_1 + 2p^4q^4\sin 4\epsilon + \tfrac{1}{2}q^8\sin 4\epsilon_2 + p^6q^2\sin 2\epsilon_1' + p^2q^2(p^2-q^2)^2\sin 2\epsilon' + p^2q^6\sin 2\epsilon_2'\right] \quad \text{......(78)}$$

And by (57), Section 14,

$$\mu\frac{d\xi}{dt} = \frac{\tau^2}{\mathfrak{g}n_0}\left[\tfrac{1}{2}p^8\sin 4\epsilon_1 - \tfrac{1}{2}q^8\sin 4\epsilon_2 - 3p^4q^4\sin 4\epsilon'' + 2p^6q^2\sin 2\epsilon_1' - 2p^2q^6\sin 2\epsilon_2'\right] \quad \text{......(79)}$$

The first two of these equations only refer to the action of the moon on the lunar tides, but the last is the same whether there be solar tides or not.

If we substitute from (76) for all the ϵ's in terms of $\sin 4\epsilon$, and introduce $\cos i = P = p^2 - q^2$, $\sin i = Q = 2pq$, we find on reduction

$$\left.\begin{aligned} \frac{di_{m^2}}{dt} &= \frac{1}{N}\frac{\tau^2}{\mathfrak{g}n_0}\sin 4\epsilon\left[\tfrac{1}{4}PQ - \tfrac{1}{2}\lambda Q\right] \\ -\frac{dN_{m^2}}{dt} &= \tfrac{1}{2}\frac{\tau^2}{\mathfrak{g}n_0}\sin 4\epsilon\left[1 - \tfrac{1}{2}Q^2 - \lambda P\right] \\ \mu\frac{d\xi}{dt} &= \tfrac{1}{2}\frac{\tau^2}{\mathfrak{g}n_0}\sin 4\epsilon\left[P - \lambda\right] \end{aligned}\right\} \quad \text{..............(80)}$$

The parts of $\frac{di}{dt}$ and $\frac{dN}{dt}$ which arise from the attraction of the sun on the solar tides may be at once written down by symmetry, and $\lambda_{,} = \frac{\Omega}{n}$, may be considered as a small fraction to be neglected compared with unity. Thus we have

$$\left.\begin{aligned} \frac{di_{m_{,}^2}}{dt} &= \frac{1}{N}\frac{\tau_{,}^2}{\mathfrak{g}n_0}\sin 4\epsilon\,.\,\tfrac{1}{4}PQ \\ -\frac{dN_{m_{,}^2}}{dt} &= \tfrac{1}{2}\frac{\tau_{,}^2}{\mathfrak{g}n_0}\sin 4\epsilon\,(1 - \tfrac{1}{2}Q^2) \end{aligned}\right\} \quad \text{.....................(81)}$$

Lastly as to the terms due to the combined action of the two disturbing bodies, it was remarked that they only involved ϵ and ϵ', which are independent of the orbital motions.

Thus by (33) we have

$$\left.\begin{aligned} \frac{di_{mm_{,}}}{dt} &= -\frac{1}{N}\frac{\tau\tau_{,}}{\mathfrak{g}n_0}\sin 4\epsilon\,.\,\tfrac{1}{4}PQ \\ -\frac{dN_{mm_{,}}}{dt} &= \frac{\tau\tau_{,}}{\mathfrak{g}n_0}\sin 4\epsilon\,.\,\tfrac{1}{4}Q^2 \end{aligned}\right\} \quad \text{.....................(82)}$$

Collecting results from the last three sets of equations and substituting $\cos i$ and $\sin i$ for P and Q, and $\frac{\Omega}{n}$ for λ, we have

$$\left.\begin{aligned} \frac{di}{dt} &= \frac{1}{N}\frac{\sin 4\epsilon}{\mathfrak{g}n_0}\tfrac{1}{4}\sin i\cos i\left[\tau^2+\tau_{,}^2-\tau\tau_{,}-\frac{2\Omega}{n}\tau^2\sec i\right] \\ -\frac{dN}{dt} &= \tfrac{1}{2}\frac{\sin 4\epsilon}{\mathfrak{g}n_0}\left[(1-\tfrac{1}{2}\sin^2 i)(\tau^2+\tau_{,}^2)+\tfrac{1}{2}\tau\tau_{,}\sin^2 i-\frac{\Omega}{n}\tau^2\cos i\right] \\ \mu\frac{d\xi}{dt} &= \tfrac{1}{2}\frac{\sin 4\epsilon}{\mathfrak{g}n_0}\cos i\,\tau^2\left(1-\frac{\Omega}{n}\sec i\right) \end{aligned}\right\} \quad (83)$$

These are the simultaneous equations which are to be solved.

Subject to the special hypothesis regarding the relationship between the retardations of the several tides, and except for the neglect of a term $-\frac{2\Omega_{,}}{n}\tau_{,}^2\sec i$ in the first of them, and of $-\frac{\Omega_{,}}{n}\tau_{,}^2\cos i$ in the second, they are rigorously true.

We will first change the independent variable in the first two equations from t to ξ.

Dividing the first and second equations by the third, and observing that

$$\frac{2di}{\sin i} = d\log\tan^2\tfrac{1}{2}i$$

we have

$$\left.\begin{aligned} \frac{d}{\mu d\xi}\log\tan^2\tfrac{1}{2}i &= \frac{1+\left(\frac{\tau_{,}}{\tau}\right)^2-\left(\frac{\tau_{,}}{\tau}\right)-\frac{2\Omega}{n}\sec i}{N\left(1-\frac{\Omega}{n}\sec i\right)} \\ -\frac{dN}{\mu d\xi} &= \frac{\frac{1-\frac{1}{2}\sin^2 i}{\cos i}\left[1+\left(\frac{\tau_{,}}{\tau}\right)^2\right]+\frac{1}{2}\left(\frac{\tau_{,}}{\tau}\right)\sin i\tan i-\frac{\Omega}{n}}{1-\frac{\Omega}{n}\sec i} \end{aligned}\right\} \quad \text{.........(84)}$$

If there be only one disturbing body, which is an interesting case from a theoretical point of view, the equations may be found by putting $\tau_{,}=0$, and may then be written

$$\left.\begin{aligned} \frac{d}{\mu d\xi}\log\tan^2\tfrac{1}{2}i &= \frac{1}{N}\frac{\cos i-\frac{2\Omega}{n}}{\cos i-\frac{\Omega}{n}} \\ -\frac{dN}{\mu d\xi} &= \frac{1-\frac{1}{2}\sin^2 i-\frac{\Omega}{n}\cos i}{\cos i-\frac{\Omega}{n}} \\ \mu\frac{d\xi}{dt} &= \tfrac{1}{2}\sin 4\epsilon\,.\,\frac{\tau^2}{\mathfrak{g}n_0}\left(\cos i-\frac{\Omega}{n}\right) \end{aligned}\right\} \quad \text{..................(85)}$$

From these equations we see that so long as Ω is less than $n \cos i$, the satellite recedes from the planet as the time increases, and the planet's rotation diminishes, because the numerator of the second equation may be written $\cos i \left(\cos i - \frac{\Omega}{n}\right) + \frac{1}{2}\sin^2 i$, which is essentially positive so long as Ω is less than $n \cos i$. But the tidal friction vanishes whenever $\Omega = n \frac{1 + \cos^2 i}{2 \cos i}$. The fraction $\frac{1 + \cos^2 i}{2 \cos i}$ is however necessarily greater than unity, and therefore the tidal friction cannot vanish, unless the month be as short or shorter than the day. The obliquity increases if Ω be less than $\frac{1}{2}n \cos i$, but diminishes if it be greater than $\frac{1}{2}n \cos i$. Hence the equation $\Omega = \frac{1}{2}n \cos i$ gives the relationship which determines the position and configuration of the system for instantaneous dynamical stability with regard to the obliquity (compare the figures 2, 3, 4). From this it follows that the position of zero obliquity is one of dynamical stability for all values of n between Ω and 2Ω, but if n be greater than 2Ω, this position is unstable*.

* Added on September 25, 1879.—The result in the text applies to the case of evanescent viscosity. If the viscosity be infinitely large the sines of twice the angles of lagging will be inversely instead of directly proportional to the speeds of the corresponding tides (compare p. [74]). Thus we must here invert the right-hand sides of the six equations (76). If the obliquity be very small (77), (78), (79) become

$$\left.\begin{aligned} \frac{di}{dt} &= \frac{1}{N}\frac{\tau^2}{gn_0}\tfrac{1}{4}\sin i \sin 4\epsilon_1 \left[1 + \frac{2(1-\lambda)}{1-2\lambda} - 2(1-\lambda)\right] \\ &= \frac{1}{N}\frac{\tau^2}{gn_0}\tfrac{1}{4}\sin i \sin 4\epsilon_1 \left(\frac{1+2\lambda-4\lambda^2}{1-2\lambda}\right) \\ -\frac{dN}{dt} &= \mu\frac{d\xi}{dt} = \frac{\tau^2}{gn_0}\tfrac{1}{2}\sin 4\epsilon_1 \end{aligned}\right\} \quad \text{(85')}$$

When $2\lambda = 1$, $\frac{di}{dt}$ apparently becomes infinite; but in this case the viscosity must be infinitely large in order to make the tide of speed $n - 2\Omega$ lag at all, and if it be infinitely large $\sin 4\epsilon_1$ is infinitely small. If the viscosity be large but finite, then when $2\lambda = 1$, the slow diurnal tide of speed $n - 2\Omega$ is no longer a true tide, but is a permanent alteration of figure of the spheroid. Thus $\epsilon_1' = 0$ and $\frac{di}{dt}$ depends on $[\sin 4\epsilon_1 - \sin 2\epsilon']$ which is equal to $\sin 4\epsilon_1 [1 - 2(1-\lambda)]$ when the viscosity is large, and vanishes when $2\lambda = 1$. Thus when the viscosity is very large (not infinite) $\frac{di}{dt}$ vanishes when $2\Omega \div n = 1$, as it does when the viscosity is very small.

When $1 + 2\lambda - 4\lambda^2 = 0$, that is, when $\lambda = \frac{\sqrt{5}+1}{4} = 1 \div 1{\cdot}236$, $\frac{di}{dt}$ vanishes; and it is negative if λ be a little greater, and positive if a little less than $1 \div 1{\cdot}236$. And $1 - 2\lambda$ is negative if λ be greater than $\frac{1}{2}$.

Hence it follows that *for large viscosity of the planet, zero obliquity is dynamically unstable, if the satellite's period be less than* 1·236 *of the planet's period of rotation; is stable if the satellite's period be between* 1·236 *and* 2 *of the planet's period; and is unstable for longer periods of the satellite.*

If the viscosity be very large $\frac{N}{\mu}\frac{d}{d\xi}\log\tan^2\tfrac{1}{2}i = \frac{1+2\lambda-4\lambda^2}{1-2\lambda}$, but if the viscosity be very small

We will now return to the problem regarding the earth. We may here regard $\frac{\Omega}{n}$ as a small fraction, and i as sufficiently small to permit us to neglect $\frac{1}{8}\sin^4 i$; also $\left(\frac{\Omega}{n}\sec i\right)^2$, $\frac{\tau_{,}}{\tau}\frac{\Omega}{n}\sec i$, $\left(\frac{\tau_{,}}{\tau}\right)^2\frac{\Omega}{n}\sec i$ will be neglected.

Our equations thus become

$$\left.\begin{aligned} \frac{d}{\mu d\xi}\log_e \tan^2 \tfrac{1}{2}i &= \frac{1+\left(\frac{\tau_{,}}{\tau}\right)^2-\left(\frac{\tau_{,}}{\tau}\right)-\frac{2\Omega}{n}\sec i}{N\left(1-\frac{\Omega}{n}\sec i\right)} \\ -\frac{dN}{\mu d\xi} &= 1+\left(\frac{\tau_{,}}{\tau}\right)^2+\tfrac{1}{2}\frac{\tau_{,}}{\tau}\sin i\tan i+\frac{\Omega}{n}(\sec i-1) \end{aligned}\right\} \text{............(86)}$$

The experience of the preceding integration shows that i varies very slowly compared with the other variables N and ξ; hence in integrating these equations an average value will be attributed to i, as it occurs in small terms on the right-hand sides of these equations.

The second equation will be considered first.

We have $\tau = \frac{\tau_0}{\xi^6}$, so that if we put $\beta = \frac{1}{13}\left(\frac{\tau_{,}}{\tau_0}\right)^2$, $\gamma = \frac{1}{14}\frac{\tau_{,}}{\tau_0}\sin i\tan i$, and omit the last term, we get by integrating from 1 to N and from 1 to ξ

$$N = 1+\mu\{1-\xi+\beta(1-\xi^{13})+\gamma(1-\xi^7)\} \text{ ....(87)}$$

as a first approximation. This is the form which was used in the previous solution, for, by classifying the tides in three groups as regards retardation of phase, we virtually neglected Ω compared with n.

This equation will be sufficiently accurate so long as $\frac{\Omega}{n}$ is a moderately small fraction; but we may obtain a second approximation by taking account of the last term.

Now $$\frac{\Omega}{n}(\sec i-1) = \tfrac{1}{2}\sin^2 i\,\frac{\Omega_0}{n_0}\cdot\frac{1}{N\xi^3} \text{ very nearly}$$

$$= \tfrac{1}{2}\sin^2 i\,\frac{\Omega_0}{\mu n_0}\cdot\frac{1}{\xi^3\left[\frac{1+\mu}{\mu}-\xi\right]}$$

by substituting an approximate value for N.

---

the same expression $= \frac{1-2\lambda}{1-\lambda}$. For positive values of λ, less than 1 and greater than ·6910 or $1\div1{\cdot}447$, the former is less than the latter, and if λ be less than $1\div1{\cdot}447$ and greater than 0 the former is greater than the latter.

Hence if there be only a single satellite, as soon as the month is longer than two days, the obliquity of the planet's axis to the plane of the satellite's orbit will increase more, in the course of evolution, for large than for small viscosities. This result is reversed if there be two satellites, as we see by comparing figs. 2 and 4.

A more correct form for the equation of conservation of moment of momentum will be given by adding to the right-hand side of equation (87) the integral of this last expression from 1 to ξ and multiplying it by μ. And in effecting this integration i may be regarded as constant.

Let $k=\dfrac{1+\mu}{\mu}$. Then since

$$\frac{1}{\xi^3(k-\xi)}=\frac{1}{k\xi^3}+\frac{1}{k^2\xi^2}+\frac{1}{k^3\xi}+\frac{1}{k^3(k-\xi)}$$

therefore

$$\int_\xi^1 \frac{d\xi}{\xi^3(k-\xi)}=\frac{1}{2k}\left(\frac{1}{\xi^2}-1\right)+\frac{1}{k^2}\left(\frac{1}{\xi}-1\right)+\frac{1}{k^3}\log\frac{k-\xi}{\xi(k-1)}$$

$$=\frac{\mu}{2(1+\mu)}\left(\frac{1}{\xi}-1\right)\left(\frac{1}{\xi}+\frac{1+3\mu}{1+\mu}\right)+\left(\frac{\mu}{\mu+1}\right)^3\log\left[\frac{1+\mu(1-\xi)}{\xi}\right]$$

Hence the second approximation is

$$N=1+\mu\{(1-\xi)+\beta(1-\xi^{13})+\gamma(1-\xi^7)\}$$
$$+\tfrac{1}{4}\sin^2 i\,\frac{\Omega_0}{n_0}\,\frac{\mu}{1+\mu}\left(\frac{1}{\xi}-1\right)\left(\frac{1}{\xi}+\frac{1+3\mu}{1+\mu}\right)$$
$$+\tfrac{1}{2}\sin^2 i\,\frac{\Omega_0}{n_0}\left(\frac{\mu}{\mu+1}\right)^3\log\left[\frac{1+\mu(1-\xi)}{\xi}\right] \quad\ldots\ldots\ldots\ldots\ldots(88)$$

It would no doubt be possible to substitute this approximate value of N in terms of ξ, in the equation which gives the rate of change of obliquity, and then to find an approximate analytical integral of the first equation. But the integral would be very long and complicated, and I prefer to determine the amount of change of obliquity by the method of quadratures.

In the present case it is obviously useless to try to obtain the time occupied by the changes, without making some hypothesis with regard to the law governing the variations of viscosity; and even supposing the viscosity small but constant during the integration, the time would vary inversely as the coefficient of viscosity, and would thus be arbitrary. The only thing which can be asserted is that if the viscosity be small, the changes proceed more slowly than in the case which has been already solved numerically.

To return, then, to the proposed integration by quadratures: by means of the equation (88) we may compute four values of N (corresponding, say, to $\xi=1$, ·96, ·92, ·88); and since $\tau=\dfrac{\tau_0}{\xi^6}$, and $\dfrac{\Omega}{n}=\dfrac{\Omega_0}{n_0}\dfrac{1}{N\xi^3}$, we may compute four equidistant values of all the terms on the right-hand side of the first of equations (86), except in as far as i is involved. Now i being only involved in small terms, we may take as an approximate final value of i that which is given by the solution of Section 15, and take as the four corresponding values i_0, $i_0+\frac{1}{3}(i-i_0)$, $i_0+\frac{2}{3}(i-i_0)$, i.

Hence four equidistant values of the right-hand side may be computed, and combined by the rule $\int_0^{3h} u_x dx = \frac{3}{8}h\,[u_0 + u_3 + 3\,(u_1 + u_2)]$, which will give the integral of the right-hand side from ξ to 1; and this is equal to

$$\log \tan^2 \tfrac{1}{2}i - \log \tan^2 \tfrac{1}{2}i_0$$

The integration was divided into a number of periods, just as in the solution of Section 15. The following were the results:

First period. From $\xi = 1$ to $\cdot 88$; $\mu = 4{\cdot}0074$; $i = 20^\circ\ 28'$; $N = 1{\cdot}5478$. The term in $\dfrac{\Omega_0}{n_0}$ in the expression for N added $\cdot 0012$ to the value of N.

Second period. From $\xi = 1$ to $\cdot 76$; $\mu = 2{\cdot}2784$; $i = 17^\circ\ 4'$; $N = 1{\cdot}5590$. The term in $\dfrac{\Omega_0}{n_0}$ added $\cdot 0011$ to the value of N.

Third period. From $\xi = 1$ to $\cdot 76$; $\mu = 1{\cdot}1107$; $i = 15^\circ\ 22'$; $N = 1{\cdot}2677$. The term in $\dfrac{\Omega_0}{n_0}$ added $\cdot 0007$ to the value of N.

It may be observed that during the first period of integration Ω/n diminishes, and reaches its minimum about the end of the period. During the rest of the integration it increases. If we neglect the solar action and the obliquity, it is easy to find the minimum value of $\dfrac{\Omega}{n}$. For $\dfrac{\Omega}{n} = \dfrac{\Omega_0}{n_0}\dfrac{1}{N\xi^3}$ and reaches its minimum when $\dfrac{dN}{d\xi} = -\dfrac{3N}{\xi}$; but $\dfrac{dN}{d\xi} = -\mu$. Therefore $N = \frac{1}{3}\xi\mu$. Now $N = 1 + \mu\,(1 - \xi)$, and hence $\xi = \frac{3}{4}\dfrac{1+\mu}{\mu}$. If $\mu = 4$, $\xi = \frac{15}{16} = \cdot 9375$. This value of ξ is passed through at near the end of the first period of integration. At this period there are 19·2 mean solar hours in the day; 22½ mean solar days in the sidereal month; and $28\frac{1}{7}$ rotations of the earth in the sidereal month. This result $28\frac{1}{7}$ is, of course, only approximate, the true result being about 29*.

The physical meaning of these results is given in a table below.

At the end of the third period of integration the solar terms (those in $\tau_{,}/\tau$) have become small in all the equations, and as they are rapidly diminishing they may be safely neglected. To continue the integration from this point a slight variation of method will be convenient.

* The subject is referred to from a more general point of view in a paper on the "Secular Effects of Tidal Friction," see *Proc. Roy. Soc.*, No. 197, 1879. [Paper 5.]

Our equations may now be written approximately

$$N = 1 + \mu(1 - \xi)$$

$$-\frac{d}{dN}\log\tan^2\tfrac{1}{2}i = \frac{1}{N}\,\frac{1 - \frac{2\Omega}{n}\sec i}{1 - \frac{\Omega}{n}}$$

In order to find how large a diminution of obliquity is possible if the integration be continued, we require to stop at the point where $n\cos i = 2\Omega$.

The equation $N = 1 + \mu(1 - \xi)$ may be written

$$\frac{n}{n_0} = 1 + \mu\left(1 - \sqrt[3]{\frac{\Omega_0}{\Omega}}\right)$$

If therefore we put $x = \sqrt[3]{\Omega}$, we must stop the integration at the point where $n = 2x^3\sec i$, x being given by the equation

$$\frac{2x^3\sec i}{n_0} = 1 + \mu\left[1 - \frac{\sqrt[3]{\Omega_0}}{x}\right]$$

If we assume $i = 14^\circ$, since $\mu = 1 \div sn_0\Omega_0^{\frac{1}{3}}$, x is given by

$$x^4 - \tfrac{1}{2}n_0\cos 14^\circ\,(1 + \mu)\,x + \frac{1}{2s}\cos 14^\circ = 0$$

At the end of the third period of integration, which is the beginning of the new period, I found

$$\log n_0 = 3{\cdot}84753,\quad \log\mu = 9{\cdot}82338 - 10,\quad \text{and}\quad \log s = 5{\cdot}39378 - 10$$

the unit of time being the present tropical year.

Hence the equation is

$$x^4 - 5690x + 19586 = 0$$

The required root is nearly $\sqrt[3]{5690}$, and a second approximation gives $x = \Omega^{\frac{1}{3}} = 16{\cdot}703$ (16·51 would have been more accurate).

But $\Omega_0^{\frac{1}{3}} = 8{\cdot}616$. Hence we desire to stop the integration when

$$\xi = \left(\frac{\Omega_0}{\Omega}\right)^{\frac{1}{3}} = \frac{8{\cdot}616}{16{\cdot}703} = {\cdot}516$$

Now $\mu = {\cdot}6659$; hence when $\xi = {\cdot}516$, $N = 1{\cdot}322$.

In order to integrate the equation of obliquity by quadratures, I assume the four equidistant values,

$$N = 1{\cdot}000,\quad 1{\cdot}107,\quad 1{\cdot}214,\quad 1{\cdot}321$$

And by means of the equation $\xi = 1 - \dfrac{N-1}{{\cdot}6659} = 1 - (N-1)(1{\cdot}502)$ the corresponding values of ξ are found to be

$$1{\cdot}000,\quad {\cdot}8393,\quad {\cdot}6786,\quad {\cdot}5179$$

By means of the formula $\frac{\Omega}{n} = \frac{\Omega_0}{n_0} \frac{1}{N\xi^3}$, the corresponding values of $\frac{\Omega}{n}$ are found to be

$$\cdot 0909, \quad \cdot 1388, \quad \cdot 2395, \quad \cdot 4951$$

I assumed conjecturally four values of i lying between $i_0 = 15° \, 22'$ and $i = 14°$, which I knew would be very nearly the final value of i; and then computed four equidistant values of $-\frac{d}{dN} \log_{10} \tan \frac{1}{2} i$.

The values were

$$\cdot 19381, \quad \cdot 16230, \quad \cdot 11882, \quad - \cdot 00684$$

The fact that the last value is negative shows that the integration is carried a little beyond the point where $n \cos i = 2\Omega$, but this is unimportant.

Combining these values by the rules of the calculus of finite differences, I find $i = 13° \, 59'$.

This final value of ξ (viz.: ·5179) makes the moon's sidereal period 12 hours, and the value of N (viz.: 1·321) makes the day 5 hours 55 minutes.

These results complete the integration of the fifth period.

The physical meaning of the results for all five periods is given in the following table:—

Sidereal day in m.s. hours and minutes	Moon's sidereal period in m.s. days	Obliquity of ecliptic
h. m.		
Initial 23 56	27·32 days	23° 28′
15 28	18·62 "	20° 28′
9 55	8·17 "	17° 4′
7 49	3·59 "	15° 22′ *
Final 5 55	12 hours	14° 0′ *

It is worthy of notice that at the end of the first period there were 28·9 days of that time in the then sidereal month; whilst at the end of the second period there were only 19·7. It seems then that at the present time tidal friction has, in a sense, done more than half its work, and that the number of days in the month has passed its maximum on its way towards the state of things in which the day and month are of equal length—as investigated in the following section.

In the last column of the preceding table the last two results in the column giving the obliquity of the ecliptic (which are marked with asterisks) cannot safely be accepted, because, as I have reason to believe, the simultaneous changes of inclination of the lunar orbit will, after the end of the second period of integration, have begun to influence the results perceptibly.

For this same reason the integration, which has been carried to the critical point where $n \cos i = 2\Omega$, and where di/dt changes sign, will not be pursued any further. Nevertheless we shall be able to trace the moon's periodic time, and the length of day to their initial condition. It is obvious that as long as n is greater than Ω, there will be tidal friction, and n will continue to approach Ω, whilst both increase retrospectively in magnitude.

I shall now refer to a critical phase in the relationship between n and Ω, of a totally different character from the preceding one, and which must occur at a point a little more remote in time than that at which the above integration stops.

This critical phase occurs when the free nutation of the oblate spheroid has a frequency equal to that of the forced fortnightly nutation.

In the ordinary theory of the precession and nutation of a rigid oblate spheroid, the fortnightly nutation arises out of terms in the couples acting about a pair of axes fixed in the equator, which have speeds $n - 2\Omega$ and $n + 2\Omega$. If C and A be the greatest and least principal moments of inertia, on integration these terms are divided by $\frac{C-A}{A} n + n \mp 2\Omega$ and give rise to terms in $\frac{di}{dt}$ and $\frac{d\psi}{dt} \sin i$ of speed 2Ω. When 2Ω is neglected compared with n, we obtain the formula for the fortnightly nutation given in any work on physical astronomy.

It is obvious that if $\frac{C-A}{A} n + n = 2\Omega$, the former of these two terms becomes infinite. Since in our case the spheroid is homogeneous $\frac{C-A}{A} = e$, the ellipticity of the spheroid; and since the spheroid is viscous $e = \frac{1}{2}\frac{n^2}{\mathfrak{g}}$. Therefore the critical relationship is $\frac{1}{2}\frac{n^3}{\mathfrak{g}} + n = 2\Omega$.

When this condition is satisfied the ordinary solution is nugatory, and the true solution represents a nutation the amplitude of which increases with the time.

The critical point where the above integration stops is given by $\frac{2\Omega}{n} = \cos i$, and this critical point by $\frac{2\Omega}{n} = 1 + \frac{1}{2}\frac{n^2}{\mathfrak{g}}$; it follows therefore that $\frac{\Omega}{n}$ is little larger in the second case than in the first. Therefore this critical point has not been already reached where the integration stops, but will occur shortly afterwards.

It is obvious that the amplitude of the nutation cannot increase for an indefinite time, because the critical relationship is only exactly satisfied for a single instant. In fact, the problem is one of far greater complexity than that of ordinary disturbed rotation. The system is disturbed periodically, but the periodic time of the disturbance slowly increases, passing through a phase of equality to the free periodic time; the problem is to find the amplitude of the oscillations when they are at their maximum, and to find the mean configuration of the system some time before and some time after the maximum, when the oscillations are small. This problem does not seem to be soluble, unless we take into account the slow variation of the argument in the periodic disturbing term; and when the argument varies, the disturbing term is not strictly a simple time-harmonic.

In the case of the viscous spheroid, the question would be further complicated by the fact that when the nutation becomes large, a new series of bodily tides is set up by the effects of inertia.

I have been unable to make a satisfactory examination of this problem, but as far as I have gone it appeared to me probable that the mean obliquity of the axis of the spheroid would not be affected by the passage of the system through a phase of large nutation; and although I cannot pretend to say how large the nutation might be, yet I consider it probable that the amplitude would not have time to increase to a very wide extent*.

Throughout all the preceding investigations, the periodic inequalities have been neglected. Now a full development of the couples $\mathfrak{L}$, $\mathfrak{M}$, $\mathfrak{N}$, which are due to the tides, shows that there occur terms of speeds $n-2\Omega$, and $n-4\Omega$ in the first two, and of speeds 2Ω and 4Ω in the last. The terms in $n-2\Omega$ in $\mathfrak{L}$ and $\mathfrak{M}$ will clearly give rise to an increasing nutation at the critical point which we are considering, but they will be so very much smaller than those arising out of the attraction on the permanent equatorial protuberance that they may be neglected. The terms in $n-4\Omega$ are multiplied by very small quantities, and I think it may safely be assumed that the system would pass through the critical phase where $\frac{1}{2}n^3/\mathfrak{g}+n=4\Omega$ with sufficient rapidity to prevent the nutation becoming large.

If we were to go to higher orders of approximation in the disturbing forces, it is clear that we should meet with an infinite number of critical phases, but the coefficients representing the amplitudes of the resulting nutations would be multiplied by such small quantities that they may safely be neglected.

* I believe that I shall be able to show in an investigation, as yet incomplete, that when this critical phase is reached, the plane of the lunar orbit is nearly coincident with the equator of the earth. As the amplitude of this nutation depends on the sine of the obliquity of the equator to the lunar orbit, it seems probable that the nutation would not become considerable.—June 30, 1879.

§ 18. *The initial condition of the earth and moon**.

It is now supposed that, when the earth's rotation has been traced back to where it is equal to twice the moon's orbital motion, the obliquity to the plane of the lunar orbit has become zero. It is clear that, as long as there is any relative motion of the earth and moon, the tidal friction and reaction must continue to exist, and n and Ω must tend to an equality. The previous investigation shows also that for small viscosity, however nearly n approaches Ω, the position of zero obliquity is dynamically stable.

As n is approaching Ω, the changes must have taken place more and more slowly in time. For if the earth was a cooling spheroid, it is unreasonable to suppose that the process of becoming less stiff in consistency (which has hitherto been supposed to be taking place, as we go backwards in time) could ever have been reversed; and if it were not reversed, the lunar tides must have lagged by less and less, as more and more time was given by the slow relative motion of the two bodies for the moon's attraction to have its full effect. Hence the effects of the sun's attraction must again become sensible, after passing through a phase of insensibility—a phase perhaps short in time, but fertile in changes in the system. I shall not here make the attempt to trace the reappearance of these solar terms.

It is, however, possible to make a rough investigation of what must have been the initial state from which the earth and moon started the course of development, which has been traced back thus far. To do this, it is only necessary to consider the equation of conservation of moment of momentum.

When the obliquity is neglected, that equation may be written

$$\frac{n}{n_0} = 1 + \mu \left\{1 - \left(\frac{\Omega_0}{\Omega}\right)^{\frac{1}{3}}\right\}$$

and it is proposed to find what values of n would make n equal to Ω.

In the course of the above investigation four different starting points were taken, viz.: those at the beginning of each period of integration. There are objections to taking any one of these, to give the numerical values required for the solution of the above equation; for, on the one hand, the errors of each period accumulate on the next, and therefore it is advantageous to take one of the early periods; whilst, on the other hand, in the early periods the values of the quantities are affected by the sensibility of the solar terms, and by the obliquity of the ecliptic. The beginning of the fourth period was chosen, because by that time the solar terms had become insignificant. At that epoch I found $\log n_0 = 3{\cdot}84753$, when the present tropical year is the unit

* For further consideration of this subject, see a paper on the "Secular Effects of Tidal Friction," *Proc. Roy. Soc.*, No. 197, 1879. [Paper 5.] The arithmetic of this section has been recomputed since the paper was presented.

of time, and $\mu = \cdot 6659$, μ being the ratio of the orbital moment of momentum to the earth's moment of momentum; also $\log s = 5\cdot 39378 - 10$, s being a constant. Now put $x^3 = n = \Omega$, and we have

$$x^4 - (1 + \mu) n_0 x + \frac{1}{s} = 0$$

Substituting the numerical values,

$$x^4 - 11727x + 40385 = 0$$

This equation has two real roots, one of which is nearly equal to $\sqrt[3]{11727}$, and the other to $40385 \div 11727$. By Horner's method these roots are found to be $21\cdot 4320$ and $3\cdot 4559$ respectively. These are the two values of the cube root of the earth's rotation, for which the earth and moon move round as a rigid body.

The first gives a day of 5 hours 36 minutes, and the second a day of about $55\frac{1}{2}$ m. s. days.

The latter is the state to which the earth and moon tend, under the influence of tidal friction (whether of oceanic or bodily tides) in the far distant future. For this case Thomson and Tait give a day of 48 of our present days*; the discrepancy between my value and theirs is explicable by the fact that they are considering a heterogeneous earth, whilst I treat a homogeneous one. Since on the hypothesis of heterogeneity the earth's moment of inertia is about $\frac{1}{3}Ma^2$, whilst on that of homogeneity it is $\frac{2}{5}Ma^2$, and since the $\frac{2}{5}$ which occurs in the quantity s enters by means of the expression for the earth's moment of inertia, it follows that in my solution μ has been taken too small in the proportion $5:6$. Hence if we wish to consider the case of heterogeneity, we must solve the equation $x^4 - 12664x + 48462 = 0$. The two roots of this equation are such that they give as the corresponding lengths of the day, 5 hours 16 minutes and $40\cdot 4$ days respectively. The remaining discrepancy (between 40 and 48) is doubtless due in part to the crude method of amending the solution, but also to the fact that they partly include the obliquity in one way, whilst I partly include it in another way, and I include a large part of the solar tidal friction whilst they neglect it. It is interesting to note that the larger root, which gives the shorter length of day, is but little affected by the consideration of the earth's heterogeneity.

With respect to the second solution (56 days), it must be remarked that the sun's tidal friction will go on lengthening the day even beyond this point, but then the lunar tides will again come into existence, and the lunar tidal friction will tend in part to counteract the solar. The tidal reaction will also be reversed, so that the moon will again approach the earth. Thus the effect of the sun is to make this a state of dynamical instability.

* *Natural Philosophy*, § 276. They say:—"It is probable that the moon, in ancient times liquid or viscous in its outer layer or throughout, was thus brought to turn always the same face to the earth." In the new edition (1879) the ultimate effects of tidal friction are considered.

The first solution, where both the day and month are 5 hours 36 minutes in length, is the one which is of interest in the present inquiry, for this is the initial state towards which the integration has been running back.

This state of things is one of dynamical instability, as may be shown as follows:—

First consider the case where the sun does not exist. Suppose the earth to be rotating in about $5\frac{1}{2}$ hours, and the moon moving orbitally around it in a little less than that time. Then the motion of the moon relatively to the earth is consentaneous with the earth's rotation, and therefore the tidal friction, small though it be, tends to accelerate the earth's rotation; the tidal reaction is such as to tend to retard the moon's linear velocity, and therefore increase her orbital angular velocity, and reduce her distance from the earth. The end will be that the moon falls into the earth.

This subject is graphically illustrated in a paper on the "Secular Effects of Tidal Friction," read before the Royal Society on June 19, 1879*.

Secondly, take the case where the sun also exists, and suppose the system started in the same way as before. Now the motion of the earth relatively to the sun is rapid, and such that the solar tidal friction retards the earth's rotation; whilst the lunar tidal friction is, as before, such as to accelerate the rotation.

Hence if the viscosity be very large the earth's rotation may be accelerated, but if it be not very large it will be retarded. The tidal reaction, which depends on the lunar tides alone, continues negative, and the moon approaches the earth as before. Thus after a short time the motion of the moon relatively to the earth is more rapid than in the previous case, whatever be the ratio between solar and lunar tidal friction. Hence in this case the moon will fall into the earth more rapidly than if the sun did not exist, and the dynamical instability is more marked.

If, however, the day were shorter than the month, the moon must continually recede from the earth, until it reaches the outer limit of a day of 56 m. s. days.

There is one circumstance which might perhaps decide that this should be the direction in which the equilibrium would break down; for the earth was a cooling body, and therefore probably a contracting one, and therefore its rotation would tend to increase. Of course this increase of rotation is partly counteracted by the solar tidal friction, but on the present theory, the mere existence of the moon seems to show that it was not more than counteracted, for if it had been so the moon must have been drawn into and confounded with the earth.

This month of 5 hours 36 minutes corresponds to a lunar distance of 2·52 earth's mean radii, or about 10,000 miles; the month of 5 hours 16 minutes

* Paper 5 in this volume.

corresponds to 2·39 earth's mean radii; so that in the case of the earth's homogeneity only 6,000 miles intervene between the moon's centre and the earth's surface, and even this distance would be reduced if we treated the earth as heterogeneous. This small distance seems to me to point to a break-up of the earth-moon mass into two bodies at a time when they were rotating in about 5 hours; for of course the precise figures given above cannot claim any great exactitude (see also Section 23).

It is a material circumstance in the conditions of the breaking-up of the earth into two bodies to consider what would have been the ellipticity of the earth's figure when rotating in $5\frac{1}{2}$ hours. Now the reciprocal of the ellipticity of a homogeneous fluid or viscous spheroid varies as the square of the period of rotation of the spheroid. The reciprocal of the ellipticity for a rotation in 24 hours is 232, and therefore the reciprocal of the ellipticity for a rotation in $5\frac{1}{2}$ hours is $\left(\frac{5\frac{1}{2}}{24}\right)^2$ of $232 = \frac{121}{2304} \times 232 = 12{\cdot}2$.

Hence the ellipticity of the earth when rotating in $5\frac{1}{2}$ hours is $\frac{1}{12}$th.

The conditions of stability of a rotating mass of fluid are as yet unknown, but when we look at the planets Jupiter and Saturn, it is not easy to believe that an ellipticity of $\frac{1}{12}$th is sufficiently great to cause the break-up of the spheroid.

A homogeneous fluid spheroid of the same density as the earth has its greatest ellipticity compatible with equilibrium when rotating in 2 hours 24 minutes*.

The maximum ellipticity of all fluid spheroids of the same density is the same, and their periods of rotation multiplied by the square root of their densities is a function of the ellipticity only. Hence a spheroid, which rotates in 4 hours 48 minutes, will be in limiting equilibrium if its density is $\left(\frac{2{\cdot}4}{4{\cdot}8}\right)^2$ or $\frac{1}{4}$ of that of the earth. If this latter spheroid had the same mass as the earth, its radius would be $\sqrt[3]{4}$ or 1·59 of that of the earth. If therefore the earth had a radius of 6,360 miles, and rotated in 4 hours 48 minutes, it would just have the maximum ellipticity compatible with equilibrium. It is, however, by no means certain that instability would not have set in long before this limiting ellipticity was reached.

In Part III. I shall refer to another possible cause of instability, which may perhaps be the cause of the break-up of the earth into two bodies.

It is easy to find the minimum time in which the system can have passed from this initial configuration, where the day and month are both $5\frac{1}{2}$ hours, down to the present condition. If we neglect the obliquity of the ecliptic,

* Pratt's *Figure of the Earth*, 2nd edition, Arts. 68 and 70. [The figure is however unstable for this rapid rotation.]

the equation (57) of tidal reaction, when adapted to the case of a viscous spheroid, becomes

$$\mu \frac{d\xi}{dt} = \frac{1}{2} \frac{\tau^2}{\mathfrak{g} n_0} \sin 4\epsilon_1$$

It is clear that the rate of tidal reaction can never be greater than when $\sin 4\epsilon_1 = 1$, when the lunar semi-diurnal tide lags by $22\frac{1}{2}°$. Then since $\tau = \tau_0/\xi^6$, we shall obtain the minimum time by integrating the equation

$$\frac{dt}{d\xi} = 2\mu \frac{\mathfrak{g} n_0}{\tau_0^2} \xi^{12}$$

Whence
$$-t = \frac{2\mu}{13} \frac{\mathfrak{g} n_0}{\tau_0^2} (1 - \xi^{13})$$

Now $\xi = \left(\frac{\Omega_0}{\Omega}\right)^{\frac{1}{3}}$, and we have found by the solution of the biquadratic that the initial condition is given by $\Omega^{\frac{1}{3}} = 21{\cdot}4320$; also with the present value of the month $\Omega_0^{\frac{1}{3}} = 4{\cdot}38$, the present year being in both cases the unit of time. Hence it follows that ξ is very nearly $\cdot 2$, and ξ^{13} may be neglected compared with unity. Thus $-t = \frac{2\mu}{13} \frac{\mathfrak{g} n_0}{\tau_0^2}$.

Now $\mu = 4{\cdot}007$ and $\frac{\mathfrak{g} n_0}{\tau_0^2}$ is 86,844,000 years.

Hence $-t = 53{,}540{,}000$ years.

Thus we see that tidal reaction is competent to reduce the system from the initial state to the present state in something over 54 million years.

The rest of the paper is occupied with the consideration of a number of miscellaneous points, which it was not convenient to discuss earlier.

§ 19. *The change in the length of year.*

The effects of tidal reaction on the earth's orbit round the sun have been neglected; I shall now justify that neglect, and show by how much the length of the year may have been altered.

It is easy to show that the moment of momentum of the orbital motion of the moon and earth round their common centre of inertia is $\frac{C}{s\Omega^{\frac{1}{3}}}$, where C is the earth's moment of inertia, and $s = \frac{2}{5}\left[\left(\frac{av}{g}\right)^2 (1+\nu)\right]^{\frac{1}{3}}$.

The moment of nomentum of the earth's rotation is obviously Cn. The normal to the lunar orbit is inclined to the earth's axis at an angle i. Hence the resultant moment of momentum of the moon and earth is

$$C\left\{n^2 + \frac{1}{(s\Omega^{\frac{1}{3}})^2} + \frac{2n}{s\Omega^{\frac{1}{3}}} \cos i\right\}^{\frac{1}{2}}$$

The change in this quantity from one epoch to another is the amount of moment of momentum of the moon-earth system which has been destroyed by solar tidal friction. This destroyed moment of momentum reappears in the form of moment of momentum of the moon and earth in their orbital motion round the sun.

At the beginning of the integration of Section 17, that is to say at the present time, I find that when the present year is taken as the unit of time, the resultant moment of momentum of the moon and earth is 11369 C.

At the end of the third period of integration (after which the solar terms were neglected), and when the obliquity has become 15° 22′, I find the same quantity to be 11625 C.

Hence the loss of moment of momentum is 256 C, or 102·4 Ma^2.

At the present time the moment of momentum of the moon and earth in their orbit is $(M+m)\Omega_{,}c_{,}^2 = Ma^2 \frac{1+\nu}{\nu}\left(\frac{c_{,}}{a}\right)^2 \Omega_{,}$; $\frac{a}{c_{,}}$ is clearly the sun's parallax, and with the present unit of time $\Omega_{,}$ is 2π.

Hence the loss of moment of momentum is equal to the present moment of momentum of orbital motion multiplied by $\frac{102{\cdot}4}{2\pi}\frac{\nu}{1+\nu}$ (sun's parallax)^2.

But the moment of momentum of the earth's and moon's orbital motion round the sun varies as $\Omega_{,}^{-\frac{1}{3}}$; hence the loss of moment of momentum corresponding to a change of $\Omega_{,}$ to $\Omega_{,}+\delta\Omega_{,}$ is the present moment of momentum multiplied by $\frac{1}{3}\frac{\delta\Omega_{,}}{\Omega_{,}}$, whence it is clear that

$$\frac{\delta\Omega_{,}}{\Omega_{,}} = 3\,\frac{102{\cdot}4}{2\pi}\,\frac{\nu}{1+\nu}\,(\text{sun's parallax})^2$$

But the shortening of the year is $\frac{\delta\Omega_{,}}{\Omega_{,}}$ of a year; taking therefore the sun's parallax as 8″·8, we find that at the end of the third period of integration the year was shorter than at present by

$$3 \times \frac{102{\cdot}4}{2\pi} \times \frac{82}{83} \times \left(\frac{8{\cdot}8\pi}{648{,}000}\right)^2 \times 365{\cdot}25 \times 86{,}400 \text{ seconds}$$

which will be found equal to 2·77 seconds.

Thus the solar tidal reaction had only the effect of lengthening the year by $2\frac{3}{4}$ seconds, since the epoch specified as the end of the third period of integration. The whole change in the length of year since the initial condition to which we traced back the moon would probably be very small indeed, but it is impossible to make this assertion positively, because, as observed above, the solar effects must have again become sensible, after passing through a period of insensibility.

§ 20. *Terms of the second order in the tide-generating potential.*

The whole of the previous investigation has been conducted on the hypothesis that the tide-generating potential, estimated per unit volume of the earth's mass, is $w\tau r^2 (\cos^2 \text{PM} - \frac{1}{3})$*, but in fact this expression is only the first term of an infinite series. I shall now show what quantities have been neglected by this treatment. According to the ordinary theory, the next term of the tide-generating potential is

$$V_2 = w\frac{m}{c}\left(\frac{r}{c}\right)^3 (\tfrac{5}{2}\cos^3 \text{PM} - \tfrac{3}{2}\cos \text{PM})$$

Although for my own satisfaction I have completely developed the influence of this term in a similar way to that exhibited at the beginning of this paper, yet it does not seem worth while to give so long a piece of algebra; and I shall here confine myself to the consideration of the terms which will arise in the tidal friction from this term in the potential, when the obliquity is neglected. A comparison of the result with the value of the tidal friction, as already obtained, will afford the requisite information as to what has been neglected.

When the obliquity is put equal to zero (see fig. 1),

$$\cos \text{PM} = \sin\theta \sin(\phi - \omega)$$

where ω is written for $n - \Omega$ for brevity. Then

$$\cos^3 \text{PM} = \tfrac{3}{4}\sin^3\theta \sin(\phi - \omega) - \tfrac{1}{4}\sin^3\theta \sin 3(\phi - \omega)$$

and

$$\cos^3 \text{PM} - \tfrac{3}{5}\cos \text{PM} = \tfrac{3}{20}\sin\theta(1 - 5\cos^2\theta)\sin(\phi - \omega) - \tfrac{1}{4}\sin^3\theta \sin 3(\phi - \omega)$$

Since

$$w\frac{m}{c}\left(\frac{r}{c}\right)^3 \tfrac{5}{2} = w\tau\frac{r^3}{c}\tfrac{5}{3}$$

we have

$$V_2 \div w\frac{\tau}{c}r^3 = -\tfrac{5}{12}\sin^3\theta \sin 3(\phi - \omega) + \tfrac{1}{4}\sin\theta(1 - 5\cos^2\theta)\sin(\phi - \omega)$$

If $\sin 3(\phi - \omega)$ and $\sin(\phi - \omega)$ be expanded, we have V_2 in the desired form, viz.: a series of solid harmonics of the third degree, each multiplied by a simple time-harmonic. If $wr^3 S_3 \cos(vt + \eta)$ be a tide-generating potential, estimated per unit volume of a homogeneous perfectly fluid spheroid of density w, S_3 being a surface harmonic of the third order, the equilibrium tide due to this potential is given by $\sigma = \dfrac{7a^3}{4g} S_3 \cos(vt + \eta)$, or $\dfrac{\sigma}{a} = \dfrac{7a}{10\mathfrak{g}} S_3 \cos(vt + \eta)$. Hence just as in Section 2, the tide-generating

* See Section 1.

potential of the third order of harmonics due to the moon will raise tides in the earth, when there is a frictional resistance to the internal motion, given by

$$\frac{\sigma}{a} = \tfrac{7}{10}\frac{\tau}{\mathfrak{g}}\frac{a}{c}\left[-\tfrac{5}{12}F\sin^3\theta\sin 3(\phi-\omega+f) + \tfrac{1}{4}F'\sin\theta(1-5\cos^2\theta)\sin(\phi-\omega+f')\right]$$

Now σ is a surface harmonic of the third order, and therefore the potential of this layer of matter, at an external point whose coordinates are r, θ, ϕ, is

$$\tfrac{4}{7}\pi a w\left(\frac{a}{r}\right)^4\sigma = \tfrac{3}{7}\frac{Ma^2}{r^4}\sigma$$

Hence the moment about the earth's axis of the forces which the attraction of the distorted spheroid exercises on a particle of mass m, situated at r, θ, ϕ, is $\tfrac{3}{7}\frac{Mma}{r^4}\frac{d\sigma}{d\phi}$. If this mass be equal to that of the moon, and $r=c$, then $\tfrac{3}{7}\frac{Mma^2}{r^4} = \tfrac{2}{7}\frac{\tau}{c}Ma^2 = \tfrac{5}{7}\frac{\tau}{c}\mathrm{C}$, where, as before, C is the moment of inertia of the earth.

Hence the couple $\mathfrak{N}_2$, which the moon's attraction exercises on the earth, is given by $\mathfrak{N}_2 = -\tfrac{5}{7}\frac{\tau}{c}\mathrm{C}\frac{d\sigma}{d\phi}$, where after differentiation we put $\theta=\frac{1}{2}\pi$ and $\phi=\frac{1}{2}\pi+\omega$.

Now

$$-\frac{d\sigma}{d\phi} = \tfrac{7}{10}\frac{\tau}{\mathfrak{g}}\frac{a^2}{c}\left[\tfrac{5}{4}F\sin^3\theta\cos 3(\phi-\omega+f) - \tfrac{1}{4}F'\sin\theta(1-5\cos^2\theta)\cos(\phi-\omega+f')\right]$$

Hence $$\frac{\mathfrak{N}_2}{\mathrm{C}} \div \tfrac{1}{2}\frac{\tau^2}{\mathfrak{g}}\left(\frac{a}{c}\right)^2 = \tfrac{5}{4}F\cos(\tfrac{3}{2}\pi+3f) - \tfrac{1}{4}F'\cos(\tfrac{1}{2}\pi+f')$$
$$= \tfrac{5}{4}F\sin 3f + \tfrac{1}{4}F'\sin f'$$

In the case of viscosity

$$F=\cos 3f, \quad F'=\cos f'$$

Therefore $$\frac{\mathfrak{N}_2}{\mathrm{C}} = \left(\frac{a}{c}\right)^2\frac{\tau^2}{\mathfrak{g}}\left(\tfrac{5}{16}\sin 6f + \tfrac{1}{16}\sin 2f'\right)$$

If the obliquity had been neglected, the tidal friction $\mathfrak{N}_1$, due to the term of the first order in the tide-generating potential, would have been given by $\frac{\mathfrak{N}_1}{\mathrm{C}} = \frac{\tau^2}{\mathfrak{g}}\tfrac{1}{2}\sin 4\epsilon_1$.

Hence $$\frac{\mathfrak{N}_2}{\mathfrak{N}_1} = \tfrac{1}{8}\left(\frac{a}{c}\right)^2\left(\frac{5\sin 6f+\sin 2f'}{\sin 4\epsilon_1}\right)$$

That is to say, this is the ratio of the terms neglected previously to those included.

According to the theory of viscous tides*,

$$\tan 3f = \frac{2 \cdot 4^2 + 1}{3} \frac{(3\omega)}{gwa} \upsilon = \tfrac{22}{19} (3\omega) \left(\frac{19\upsilon}{2gwa}\right)$$

where υ is the coefficient of viscosity.

But throughout the previous work we have written $\rho = \dfrac{2gwa}{19\upsilon}$.

Hence $\tan 3f = \tfrac{22}{19} \dfrac{3\omega}{\rho}$, and similarly $\tan f = \tfrac{22}{19} \dfrac{\omega}{\rho}$.

Also $\tan 2\epsilon_1 = \dfrac{2\omega}{\rho}$.

I will now consider two cases:—

1st. Suppose the viscosity to be small, then f, f', ϵ_1 are all small, and

$$\frac{\sin 6f}{\sin 4\epsilon_1} = \frac{\tan 3f}{\tan 2\epsilon_1} = \tfrac{22}{19} \times \tfrac{3}{2}, \qquad \frac{\sin 2f'}{\sin 4\epsilon_1} = \frac{\tan f'}{\tan 2\epsilon_1} = \tfrac{22}{19} \times \tfrac{1}{2}$$

Therefore

$$\frac{\mathfrak{P}_2}{\mathfrak{P}_1} = \tfrac{22}{19} \left(\frac{a}{c}\right)^2$$

2nd. Suppose the viscosity very great, then $3f, f', 2\epsilon_1$ are very nearly equal to $\frac{1}{2}\pi$, and

$$\tan(\tfrac{1}{2}\pi - 3f) = \tfrac{19}{22} \frac{\rho}{3\omega}, \qquad \tan(\tfrac{1}{2}\pi - f') = \tfrac{19}{22} \frac{\rho}{\omega}, \qquad \tan(\tfrac{1}{2}\pi - 2\epsilon_1) = \frac{\rho}{2\omega},$$

so that we have approximately

$$\frac{\sin 6f}{\sin 4\epsilon_1} = \frac{\sin(\pi - 6f)}{\sin(\pi - 4\epsilon_1)} = \tfrac{19}{22} \times \tfrac{2}{3}$$

and similarly

$$\frac{\sin 2f'}{\sin 4\epsilon_1} = \tfrac{19}{22} \times 2$$

So that

$$\frac{\mathfrak{P}_2}{\mathfrak{P}_1} = \left(\frac{a}{c}\right)^2 \tfrac{1}{8} \times \tfrac{19}{22} (\tfrac{10}{3} + 2) = \tfrac{19}{33} \left(\frac{a}{c}\right)^2$$

Hence it follows that the terms of the second order may bear a ratio to those of the first order lying between $\tfrac{22}{19} \left(\dfrac{a}{c}\right)^2$, or $1{\cdot}16 \left(\dfrac{a}{c}\right)^2$, and $\tfrac{19}{33} \left(\dfrac{a}{c}\right)^2$, or $\cdot 576 \left(\dfrac{a}{c}\right)^2$.

At the end of the fourth period of integration in the solution of Section 15, c/a or the moon's distance in earth's mean radii was 9; hence the terms of the second order in the equation of tidal friction must at that epoch lie in magnitude between $\frac{1}{70}$th and $\frac{1}{141}$st of those of the first order. It follows

* "Bodily Tides," &c., *Phil. Trans.*, 1879, Part I., Section 5. [Paper 1.]

that even at that stage, when the moon is comparatively near the earth, the effect of the tides of the second order (the third degree of harmonics) is insignificant, and the neglect of them is justified.

In the case of those terms of this order, which affect the obliquity, a very similar relationship to the terms of the lower order would be found to hold good.

§ 21. *On certain other small terms.*

It will be well to advert to certain terms, the neglect of which might be suspected of vitiating my results.

According to the hypothesis of the plastic nature of the earth's mass, that body must always have been a figure of equilibrium throughout the series of changes which are to be followed out. In consequence of tidal friction the earth's rotation is diminishing, and therefore its ellipticity (which by the ordinary theory is $\frac{5}{4}n^2a/g$) is also diminishing; this change of figure might be supposed to exercise a material influence on the results, but I will now show that in one respect at least its effects are unimportant.

In a previous paper* I showed that, neglecting (C − A)/A compared with unity, when the earth's figure changed symmetrically with respect to the axis of rotation,

$$\frac{di}{dt} = -\frac{\tau + \tau_{\prime}}{Cn^2} \sin i \cos i \frac{d}{dt}(C - A)$$

But if e be the ellipticity of figure,

$$C - A = \tfrac{2}{3} Ma^2 e$$

So that
$$\frac{1}{C}\frac{d}{dt}(C - A) = \frac{de}{dt} = \tfrac{5}{2}\frac{na}{g}\frac{dn}{dt} = -\frac{n}{\mathfrak{g}}\frac{\mathfrak{N}}{C}$$

and therefore
$$\frac{di}{dt} = \frac{\tau + \tau_{\prime}}{\mathfrak{g}n} \sin i \cos i \frac{\mathfrak{N}}{C}$$

Numerical calculation shows that at present $\frac{\tau + \tau_{\prime}}{\mathfrak{g}} = \frac{3{\cdot}04}{10^7}$, and since $\frac{\mathfrak{N}}{Cn} \sin i \cos i$ is of the same order of magnitude as $\frac{\mathfrak{L}}{Cn}, \frac{\mathfrak{M}}{Cn}$ (on which the changes of obliquity have been shown to depend), it follows that this term is fairly negligeable compared with those already included in the equations.

* "On the Influence of Geological Changes," &c., *Phil. Trans.*, Vol. 167, Part I., page 272, Section 8. [To be reproduced in Vol. III.] The notation is changed, and the equation presented in a form suitable for the present purpose.

As far as it goes, however, this term tends in the direction of increasing the obliquity with the time*.

It will however appear, I believe, that this secular change of ellipticity of the earth's figure will exercise an important influence on the plane of the lunar orbit and thereby will affect the secular change in the obliquity of the ecliptic. The investigation of this point is however as yet incomplete.

The other small term which I shall consider arises out of the ordinary precession, together with the fact that the tide-generating force diminishes with the time on account of the tidal reaction on the moon.

The differential equations which give the ordinary precession are in effect (compare equations (26))

$$\frac{d\omega_1}{dt} = \tau\frac{C-A}{C}\sin i\cos i\sin n$$

$$\frac{d\omega_2}{dt} = -\tau\frac{C-A}{C}\sin i\cos i\cos n$$

and they give rise to no change of obliquity if τ be constant, but

$$\tau = \frac{\tau_0}{\xi^6} = \tau_0\left\{1 - 6\left(\frac{d\xi}{dt}\right)t\right\}$$

when t is small.

Also $\dfrac{C-A}{C} = e = \dfrac{5n^2a}{4g} = \frac{1}{2}\dfrac{n^2}{\mathfrak{g}}$. Hence as far as regards the change of obliquity the equations may be written

$$\frac{d\omega_1}{dt} = -\frac{3\tau_0 n^2}{\mathfrak{g}}\left(\frac{d\xi}{dt}\right)\sin i\cos i\, t\sin n$$

$$\frac{d\omega_2}{dt} = \frac{3\tau_0 n^2}{\mathfrak{g}}\left(\frac{d\xi}{dt}\right)\sin i\cos i\, t\cos n$$

If we regard all the quantities, except t, on the right-hand sides of these equations as constants and integrate, we have

$$\omega_1 = \frac{3\tau_0}{\mathfrak{g}}\left(\frac{d\xi}{dt}\right)\sin i\cos i\,\{nt\cos n - \sin n\}$$

$$\omega_2 = \frac{3\tau_0}{\mathfrak{g}}\left(\frac{d\xi}{dt}\right)\sin i\cos i\,\{nt\sin n + \cos n\}$$

* In a paper in the *Phil. Mag.*, March, 1877, I suggested that the obliquity might possibly be due to the contraction of the terrestrial nebula in cooling; I there neglected tidal friction and assumed the conservation of moment of momentum to hold good for the earth by itself, so that the ellipticity was continually increasing with the time. I did not at that time perceive that this increase of ellipticity was antagonistic to the effects of contraction. Though the work of that paper is correct, as I believe, yet the fundamental assumption is incorrect, and therefore the results are not worthy of attention. [This paper will be reproduced in Vol. III.]

And if these be substituted in the geometrical equations (1) we have

$$\frac{di}{dt} = \frac{3\tau_0}{\mathfrak{g}} \sin i \cos i \frac{d\xi}{dt}$$

On comparing this with the small term due to the secular change of figure of the earth, we see that it is fairly negligeable, being of the same order of magnitude as that term. As far as it goes, however, it tends to increase the obliquity of the ecliptic.

§ 22. *The change of obliquity and tidal friction due to an annular satellite.*

Conceive the ring to be rotating round the planet with an angular velocity Ω, let its radius be c, and its mass per unit length of its arc $m/2\pi c$, so that its mass is m. Let cl be the length of the arc measured from some point fixed in the ring up to the element $c\delta l$; and let Ωt be the longitude of the fixed point in the ring at the time t. Let δV be the tide-generating potential due to the element $\frac{m}{2\pi}\delta l$. Then we have by (5)

$$\delta V \div wr^2 \frac{3}{2c^3}\left(\frac{m}{2\pi}\delta l\right) = -(\xi^2 - \eta^2)\Phi_l - 2\xi\eta\Phi_l' - \&c.$$

where the suffixes to the functions indicate that $\Omega + l$ is to be written for Ω. Integrating all round the ring from $l = 0$ to $l = 2\pi$ it is clear that

$$\frac{V}{w\tau r^2} = -p^2q^2 \sin^2\theta \cos 2(\phi - n) + 2pq(p^2 - q^2)\sin\theta\cos\theta\cos(\phi - n) + (\tfrac{1}{3} - \cos^2\theta)\tfrac{1}{2}(1 - 6p^2q^2)$$

which is the tide-generating potential of the ring.

Hence, as in Section 2, the form of the tidally-distorted spheroid is given by (9), save that E_1, E_2, E_1', E_2', E'' are all zero. Also, as in that section, the moments of the forces which the tidally-distorted spheroid exerts on the element of ring are $\frac{3}{5}\left(\frac{m}{2\pi}\delta l\right)\frac{Ma}{r^3}\left(\eta\frac{d\sigma}{d\zeta} - \zeta\frac{d\sigma}{d\eta}\right)$, &c., &c., where ξr, ηr, ζr are put equal to the rectangular coordinates of the element of ring, whose annular coordinate is l.

If x, y, z are the direction cosines of the element, equations (7) are simply modified by Ω being written $\Omega + l$. Hence the couples due to one element of ring may be found just as the whole couples were found before, and the integrals of the elementary couples from $l = 0$ to 2π are the desired couples due to the whole ring. A little consideration shows that the results of this integration may be written down at once by putting E_1, E_2, E_1', E_2', E'' zero in (15), (16), and (21). Thus in order to determine the change of obliquity and the tidal friction due to an annular satellite, we have simply the expressions (33) and (34), save that $\tau\tau_{,}$ must be replaced by $\frac{1}{2}\tau^2$.

It thus appears that an annular satellite causes tidal friction in its planet, and that the obliquity of the planet's axis to the ring tends to diminish, but both these effects are evanescent with the obliquity. Since this ring only raises the tides which are called sidereal semi-diurnal and sidereal diurnal, and since we see by (57), Section 14, that tidal reaction is independent of those tides, it follows that there is no tangential force on the ring tending to accelerate its linear motion. If, however, the arc of the ring be not of uniform density, there is a slight tendency for the lighter parts to gain on the heavier, and for the heavier parts to become more remote from the planet than the lighter.

§ 23. *Double tidal reaction.*

Throughout the whole of this investigation the moon has been supposed to be merely an attractive particle, but there can be no doubt but that if the earth was plastic, the moon was so also. To take a simple case, I shall now suppose that both the earth and moon are homogeneous viscous spheres revolving round their common centre of inertia, and that the moon is rotating on her own axis with an angular velocity ω, and that their axes are parallel and perpendicular to the plane of their orbit. Then the whole of the argument with respect to the earth as disturbed by the moon, may be transferred to the case of the moon as disturbed by the earth.

All symbols which apply to the moon will be distinguished from those which apply to the earth by an accent.

From (21) or (43) we have

$$\frac{\mathfrak{N}'}{C'} = \frac{1}{2}\frac{\tau'^2}{\mathfrak{g}'}\sin 4\epsilon_1'$$

and the equation which gives the lunar tidal friction is

$$\frac{d\omega}{dt} = -\frac{1}{2}\frac{\tau'^2}{\mathfrak{g}'}\sin 4\epsilon_1' \text{ ....(89)}$$

Now
$$\tau' = \frac{3}{2}\frac{M}{c^3} = \nu\tau = \frac{wa^3}{w'a'^3}\tau$$

and
$$\mathfrak{g}' = \frac{2}{5}\frac{g'}{a'} = \frac{2g}{5a}\frac{w'}{w} = \frac{w'}{w}\mathfrak{g}$$

So that
$$\frac{\tau'^2}{\mathfrak{g}'} = \left(\frac{wa^2}{w'a'^2}\right)^3\frac{\tau^2}{\mathfrak{g}} \text{ ....(90)}$$

Also
$$\frac{C'}{C} = \frac{w'a'^5}{wa^5}$$

and therefore
$$\frac{\mathfrak{N}'}{C} = \frac{1}{2}\frac{\tau^2}{\mathfrak{g}}\frac{w^2a}{w'^2a'}\sin 4\epsilon_1'$$

The force on the moon tangential to her orbit, results from a double tidal reaction. By the method employed in Section 14, the tangential force due to the earth's tides is

$$T = \frac{\mathfrak{N}}{r} = \frac{C}{2r}\frac{\tau^2}{\mathfrak{g}}\sin 4\epsilon_1$$

and similarly the tangential force due to the moon's tides is

$$T' = \frac{\mathfrak{N}'}{r} = \frac{C}{2r}\frac{\tau^2}{\mathfrak{g}}\frac{w^2 a}{w'^2 a'}\sin 4\epsilon_1'$$

and the whole tangential force is $(T+T')$.

Hence following the argument of that section, the equation of tidal reaction becomes

$$\mu\frac{d\xi}{dt} = \tfrac{1}{2}\frac{\tau^2}{\mathfrak{g}n_0}\left[\sin 4\epsilon_1 + \frac{w^2 a}{w'^2 a'}\sin 4\epsilon_1'\right]$$

Taking the moon's apparent radius as 16′, and the ratio of the earth's mass to that of the moon as 82, we have $\frac{a}{a'} = 3{\cdot}567$ and $\frac{w}{w'} = 1{\cdot}806$ (so that taking w as $5\frac{1}{2}$, the specific gravity of the moon is 3), and hence $\frac{w^2 a}{w'^2 a'} = 11{\cdot}64$.

At first sight it would appear from this that the effect of the tides in the moon was nearly twelve times as important as the effect of those in the earth, as far as concerns the influence on the moon's orbit, and hence it would seem that a grave oversight has been made in treating the moon as a simple attractive particle; a little consideration will show, however, that this is by no means the case.

Supposing that υ', υ are the coefficients of viscosity of the moon and earth respectively, the only tides which exist in each body being those of which the speeds are $2(\omega - \Omega)$, $2(n - \Omega)$ in the moon and earth respectively, we have

$$\tan 2\epsilon_1' = \frac{19\upsilon'(\omega - \Omega)}{g'a'w'} \quad \text{and} \quad \tan 2\epsilon_1 = \frac{19\upsilon(n - \Omega)}{gaw}$$

But

$$g'a'w' = gaw\left(\frac{w'a'}{wa}\right)^2$$

and hence

$$\tan 2\epsilon_1' = \frac{\omega - \Omega}{n - \Omega}\frac{\upsilon'}{\upsilon}\left(\frac{wa}{w'a'}\right)^2 \tan 2\epsilon_1$$

It will be found that $\left(\frac{wa}{w'a'}\right)^2 = 41{\cdot}10$. It is also almost certain that υ' must for a long time be greater than υ, because the moon being a smaller body must have stiffened quicker than the earth. Hence unless $\omega - \Omega$ is very much less than $n - \Omega$, ϵ_1' must be larger than ϵ_1. Therefore if in the early stages of development the earth had a small viscosity, it is probable that the effects of the moon's tides on her own orbit must have had a much more important influence than had the tides in the earth.

I shall now show, however, that this state of things must probably have had so short a duration as not to seriously affect the investigation of this paper. By (89) and (90) we have, as the equation which determines the rate of tidal friction reducing the moon's rotation round her axis,

$$\frac{d\omega}{dt} = -\tfrac{1}{2}\frac{\tau^2}{\mathfrak{g}}\left(\frac{wa^2}{w'a'^2}\right)^3 \sin 4\epsilon_1'$$

Now $\left(\frac{wa^2}{w'a'^2}\right)^3 = 12{,}148$; and hence, for the same values of ϵ_1' and ϵ_1, the moon's rotation round her axis is reduced 12,000 times as rapidly as that of the earth round its axis, and therefore in a very short period the moon's rotation round her axis must have been reduced to a sensible identity with the orbital motion. As ω becomes very nearly equal to Ω, $\sin 4\epsilon_1'$ becomes very small. Hence the term in the equation of tidal reaction dependent on the moon's own tides must have become rapidly evanescent. While this shows that the main body of our investigation is unaffected by the lunar tide, there is one slight modification to which it leads.

In Section 18 we traced back the moon to the initial condition, when her centre was 10,000 miles from the earth's centre. If lunar tidal friction had been included, this distance would have been increased; for the coefficient of x in the biquadratic (viz.: 11,727) would have to be diminished by $\frac{w'a'^5}{wa^5}(\omega - \omega_0)$. Now $\frac{w'a'^5}{wa^5}$ is very nearly $\frac{1}{1000}$th, and the unit of time being the year, it follows that we should have to suppose an enormously rapid primitive rotation of the moon round her axis, to make any sensible difference in the configuration of the two bodies when her centre of inertia moved as though rigidly connected with the earth's surface.

The supposition of two viscous globes moving orbitally round their common centre of inertia, and one having a congruent and the other an incongruent axial rotation, would lead to some very curious results.

§ 24. *Secular contraction of the earth**.

If the earth be contracting as it cools, it follows, from the principle of conservation of moment of momentum, that the angular velocity of rotation is being increased. Sir William Thomson has, however, shown that the contraction (which probably now only takes place in the superficial strata) cannot be sufficiently rapid to perceptibly counteract the influence of tidal friction at the present time.

* Rewritten in July, 1879.

The enormous height of the lunar mountains compared to those in the earth seems, however, to give some indications that a cooling celestial orb must contract by a perceptible fraction of its radius after it has consolidated*. Perhaps some of the contraction might be due to chemical combinations in the interior, when the heat had departed, so that the contraction might be deep-seated as well as superficial.

It will be well, therefore, to point out how this contraction will influence the initial condition to which we have traced back the earth and moon, when they were found rotating as parts of a rigid body in a little more than 5 hours.

Let C, C_0 be the moment of inertia of the earth at any time, and initially. Then the equation of conservation of moment of momentum becomes

$$\frac{Cn}{C_0 n_0} = 1 + \mu \left(1 - \left(\frac{\Omega_0}{\Omega}\right)^{\frac{1}{3}}\right)$$

And the biquadratic of Section 18 which gives the initial configuration becomes

$$x^4 - (1+\mu)\frac{C_0 n_0}{C} x + \frac{C_0}{Cs} = 0$$

The required root of this equation is very nearly equal to $\left[(1+\mu)\frac{C_0 n_0}{C}\right]^{\frac{1}{3}}$. Now $x^3 = \Omega$; hence Ω is nearly equal to $(1+\mu)\frac{C_0 n_0}{C}$. But in Section 18, when C was equal to C_0, it was nearly equal to $(1+\mu) n_0$. Therefore on the present hypothesis, the value of Ω as given in that section must be multiplied

* Suppose a sphere of radius a to contract until its radius is $a + \delta a$, but that, its surface being incompressible, in doing so it throws up n conical mountains, the radius of whose bases is b, and their height h, and let b be large compared with h. The surface of such a cone is $\pi b \sqrt{h^2 + b^2} = \pi (b^2 + \frac{1}{2}h^2)$. Hence the excess of the surface of the cone above the area of the base is $\frac{1}{2}\pi h^2$, and $4\pi a^2 = 4\pi (a + \delta a)^2 + \frac{1}{2} n \pi h^2$. Therefore $-\frac{\delta a}{a} = \frac{n}{16}\left(\frac{h}{a}\right)^2$.

Suppose we have a second sphere of primitive radius a', which contracts and throws up the same number of mountains; then similarly $-\frac{\delta a'}{a'} = \frac{n}{16}\left(\frac{h'}{a'}\right)^2$ and $\frac{\delta a'}{a'} \div \frac{\delta a}{a} = \left(\frac{h'a}{ha'}\right)^2$. Now let these two spheres be the earth and moon. The height of the highest lunar mountain is 23,000 feet (Grant's *Physical Astron.*, p. 229), and the height of the highest terrestrial mountain is 29,000 feet; therefore we may take $\frac{h'}{h} = \frac{23}{29}$. Also $\frac{a'}{a} = {\cdot}2729$ (Herschel's *Astron.*, Section 404). Therefore $\frac{ha'}{h'a} = \frac{29}{23}$ of $\cdot 2729 = \cdot 344$, and $\left(\frac{ha'}{h'a}\right)^2 = \cdot 1183$ or $\left(\frac{h'a}{ha'}\right)^2 = 8{\cdot}45$. Hence $\frac{\delta a'}{a'} \div \frac{\delta a}{a} = 8\frac{1}{2}$; whence it appears that, if both lunar and terrestrial mountains are due to the crumpling of the surfaces of those globes in contraction, the moon's radius has been diminished by about eight times as large a fraction as the earth's.

This is, no doubt, a very crude way of looking at the subject, because it entirely omits volcanic action from consideration, but it seems to justify the assertion that the moon has contracted much more than the earth, since both bodies solidified.

by $\frac{C_0}{C}$; and the periodic time must be multiplied by $\frac{C}{C_0}$. But in this initial state C is greater than C_0; hence the periodic time when the two bodies move round as a rigid body is longer, and the moon is more distant from the earth, if the earth has sensibly contracted since this initial configuration.

If, then, the theory here developed of the history of the moon is the true one, as I believe it is, it follows that the earth cannot have contracted since this initial state by so much as to considerably diminish the effects of tidal friction, and it follows that Sir William Thomson's result as to the present unimportance of the contraction must have always been true.

If the moon once formed a part of the earth we should expect to trace the changes back until the two bodies were in actual contact. But it is obvious that the data at our disposal are not of sufficient accuracy, and the equations to be solved are so complicated, that it is not to be expected that we should find a closer accordance, than has been found, between the results of computation and the result to be expected, if the moon was really once a part of the earth.

It appears to me, therefore, that the present considerations only negative the hypothesis of any large contraction of the earth since the moon has existed.

PART III.

Summary and discussion of results.

The general object of the earlier or preparatory part of the paper is sufficiently explained in the introductory remarks.

The earth is treated as a homogeneous spheroid, and in what follows, except where otherwise expressly stated, the matter of which it is formed is supposed to be purely viscous. The word "earth" is thus an abbreviation of the expression "a homogeneous rotating viscous spheroid"; also wherever numerical values are given they are taken from the radius, mean density, mass, &c., of the earth.

The case is considered first of the action of one tide-raising body, namely, the moon. To simplify the problem the moon is supposed to move in a circular orbit in the ecliptic*—that plane being the average position of the

* The effect of neglecting the eccentricity of the moon's orbit is, that we underestimate the efficiency of the tidal effects. Those effects vary as the inverse sixth power of r the radius vector, and if T be the periodic time of the moon, the average value of $\frac{1}{r^6}$ is $\frac{1}{T}\int_0^T \frac{dt}{r^6}$. If c be the mean distance and e the eccentricity of the orbit, this integral will be found equal to $\frac{1}{c^6}\frac{1+3e^2+\frac{3}{8}e^4}{(1-e^2)^{\frac{9}{2}}}$. If the eccentricity be small the average value of $\frac{1}{r^6}$ is $\frac{1}{c^6}\left(1+\frac{15}{2}e^2\right)$; if e is $\frac{1}{20}$ this is $\frac{54}{53}$ of $\frac{1}{c^6}$. There are obviously forces tending to modify the eccentricity of the moon's orbit.

lunar orbit with respect to the earth's axis. The case becomes enormously more complex if we suppose the moon to move in an inclined eccentric orbit with revolving nodes. The consideration of the secular changes in the inclination of the lunar orbit and of the eccentricity will form the subject of another investigation [in Paper 6].

The expression for the moon's tide-generating potential is shown to consist of 13 simple tide-generating terms, and the physical meaning of this expansion is given in the note to Section 8. The physical causes represented by these 13 terms raise 13 simple tides in the earth, the heights and retardations of which depend on their speeds and on the coefficient of viscosity.

The 13 simple tides may be more easily represented both physically and analytically as seven tides, of which three are approximately semi-diurnal, three approximately diurnal, and one has a period equal to a half of the sidereal month, and is therefore called the fortnightly tide.

By an approximation which is sufficiently exact for a great part of the investigation, the semi-diurnal tides may be grouped together, and the diurnal ones also. Hence the earth may be regarded as distorted by two complex tides, namely, the semi-diurnal and diurnal, and one simple tide, namely, the fortnightly. The absolute heights and retardations of these three tides are expressed by six functions of their speeds and of the coefficient of viscosity (Sections 1 and 2).

When the form of the distorted spheroid is thus given, the couples about three axes fixed in the earth due to the attraction of the moon on the tidal protuberances are found. It must here be remarked that this attraction must in reality cause a tangential stress between the tidal protuberances and the true surface of the mean oblate spheroid. This tangential stress must cause a certain very small tangential flow*, and hence must ensue a very small diminution of the couples. The diminution of couple is here neglected, and the tidal spheroid is regarded as being instantaneously rigidly connected with the rotating spheroid. The full expressions for the couples on the earth are long and complex, but since the nutations to which they give rise are exceedingly minute, they may be much abridged by the omission of all terms except such as can give rise to secular changes in the precession, the obliquity of the ecliptic, and the diurnal rotation. The terms retained represent that there are three couples independent of the time, the first of which tends to make the earth rotate about an axis in the equator which is always 90° from the nodes of the moon's orbit: this couple affects the obliquity of the ecliptic; second, there is a couple about an axis in the equator which is always coincident with the nodes: this affects the precession; third, there is a couple about the earth's axis of rotation, and this affects the length of the day

* See Part I. of Paper 5.

(Sections 3, 4, and 5). All these couples vary as the fourth power of the moon's orbital angular velocity, or as the inverse sixth power of her distance.

These three couples give the alteration in the precession due to the tidal movement, the rate of increase of obliquity, and the rate at which the diurnal rotation is being diminished, or in other words the tidal friction. The change of obliquity is in reality due to tidal friction, but it is convenient to retain the term specially for the change of rotation alone.

It appears that if the bodily tides do not lag, which would be the case if the earth were perfectly fluid or perfectly elastic, there is no alteration in the obliquity, nor any tidal friction (Section 7). The alteration in the precession is a very small fraction of the precession due to the earth considered as a rigid oblate spheroid. I have some doubts as to whether this result is properly applicable to the case of a perfectly fluid spheroid.* At any rate, Sir William Thomson has stated, in agreement with this result, that a perfectly fluid spheroid has a precession scarcely differing from that of a perfectly rigid one. Moreover, the criterion which he gives of the negligeability of the additional terms in the precession in a closely analogous problem appears to be almost identical with that found by me (Section 7). I am not aware that the investigation on which his statement is founded has ever been published. The alteration in the precession being insignificant, no more reference will be made to it. This concludes the analytical investigation as far as concerns the effects on the disturbed spheroid, where there is only one disturbing body.

The sun is now (Section 8) introduced as a second disturbing body. Its independent effect on the earth may be determined at once by analogy with the effect of the moon. But the sun attracts the tides raised by the moon, and *vice versâ*. Notwithstanding that the periods of the sun and moon about the earth have no common multiple, yet the interaction is such as to produce a secular alteration in the position of the earth's axis and in the angular velocity of its diurnal rotation. A physical explanation of this curious result is given in the note to Section 8. I have distinguished this from the separate effect of each disturbing body, as a combined effect.

The combined effects are represented by two terms in the tide-generating potential, one of which goes through its period in 12 sidereal hours, and the other in a sidereal day*; the latter being much more important than the former for moderate obliquities of the ecliptic. Both these terms vanish when the earth's axis is perpendicular to the plane of the orbit.

As far as concerns the combined effects, the disturbing bodies may be

* These combined effects depend on the tides which are designated as K_1 and K_2 in the British Association's Report on Tides for 1872 and 1876, and which I have called the sidereal semi-diurnal and diurnal tides. [See Paper 1, Vol. I.] For a general explanation of this result see the abstract of this paper in the *Proceedings of the Royal Society*, No. 191, 1878. [See the Appendix to this paper.]

conceived to be replaced by two circular rings of matter coincident with their orbits and equal in mass to them respectively. The tidal friction due to these rings is insignificant compared with that arising separately from the sun and moon. But the diurnal combined effect has an important influence in affecting the rate of change of obliquity. The combined effects are such as to cause the obliquity of the ecliptic to diminish, whereas the separate effects on the whole make it increase—at least in general (see Section 22).

The relative importance of all the effects may be seen from an inspection of Table III., Section 15.

Section 11 contains a graphical analysis of the physical meaning of the equations, giving the rate of change of obliquity for various degrees of viscosity and obliquity.

Figures 2 and 3 refer to the case where the disturbed planet is the earth, and the disturbing bodies the sun and moon.

This analysis gives some remarkable results as to the dynamical stability or instability of the system.

It will be here sufficient to state that, for moderate degrees of viscosity, the position of zero obliquity is unstable, but that there is a position of stability at a high obliquity. For large viscosities the position of zero obliquity becomes stable, and (except for a very close approximation to rigidity) there is an unstable position at a larger obliquity, and again a stable one at a still larger one*.

These positions of dynamical equilibrium do not strictly deserve the name, since they are slowly shifting in consequence of the effects of tidal friction; they are rather positions in which the rate of change of obliquity becomes of a higher order of small quantities.

It appears that the degree of viscosity of the earth which at the present time would cause the obliquity of the ecliptic to increase most rapidly is such that the bodily semi-diurnal tide would be retarded by about 1 hour and 10 minutes; and the viscosity which would cause the obliquity to decrease most rapidly is such that the bodily semi-diurnal tide would be retarded by about $2\frac{3}{4}$ hours.

The former of these two viscosities was the one which I chose for subsequent numerical application, and for the consideration of secular changes in the system.

Figure 4 (Section 11) shows a similar analysis of the case where there is only one disturbing satellite, which moves orbitally with one-fifth of the velocity of rotation of the planet. This case differs from the preceding one in the fact that the position of zero obliquity is now unstable for all

* For a general explanation of some part of these results, see the abstract of this paper in the *Proceedings of the Royal Society*, No. 191, 1878. [See the Appendix below.]

viscosities, and that there is always one other, and only one other position of equilibrium, and that is a stable one.

This shows that the fact that the earth's obliquity would diminish for large viscosity is due to the attraction of the sun on the lunar tides, and of the moon on the solar tides.

It is not shown by these figures, but it is the fact that if the motion of the satellite relatively to the planet be slow enough (viz.: the month less than twice the day), the obliquity will diminish.

This result, taken in conjunction with results given later with regard to the evolution of satellites, shows that the obliquity of a planet perturbed by a single satellite must rise from zero to a maximum and then decrease again to zero. If we regard the earth as a satellite of the moon, we see that this must have been the case with the moon.

Figure 5 (Section 12) contains a similar graphical analysis of the various values which may be assumed by the tidal friction. As might be expected, the tidal friction always tends to stop the planet's rotation, unless indeed the satellite's period is less than the planet's day, when the friction is reversed.

This completes the consideration of the effect on the earth, at any instant, of the attraction of the sun and moon on their tides; the next subject is to consider the reaction on the disturbing bodies.

Since the moon is tending to retard the earth's diurnal rotation, it is obvious that the earth must exercise a force on the moon tending to accelerate her linear velocity. The effect of this force is to cause her to recede from the earth and to decrease her orbital angular velocity. Hence tidal reaction causes a secular retardation of the moon's mean motion.

The tidal reaction on the sun is shown to have a comparatively small influence on the earth's orbit and is neglected (Sections 14 and 19).

The influence of tidal reaction on the lunar orbit is determined by finding the disturbing force on the moon tangential to her orbit, in terms of the couples which have been already found as perturbing the earth's rotation; and hence the tangential force is found in terms of the rate of tidal friction and of the rate of change of obliquity.

It appears that the non-periodic part of the force, on which the secular change in the moon's distance depends, involves the lunar tides alone.

By the consideration of the effects of the perturbing force on the moon's motion, an equation is found which gives the rate of increase of the square root of the moon's distance, in terms of the heights and retardations of the several lunar tides (Section 14).

Besides the interaction of the two bodies which affects the moon's mean motion, there is another part which affects the plane of the lunar orbit; but this latter effect is less important than the former, and in the present paper is neglected, since the moon is throughout supposed to remain in the ecliptic. The investigation of the subject will, however, lead to interesting results, since a complete solution of the problem of the obliquity of the ecliptic cannot be attained without a simultaneous tracing of the secular changes in the plane of the lunar orbit.

It appears that the influence of the tides, here called slow semi-diurnal and slow diurnal, is to increase the moon's distance from the earth, whilst the influence of the fast semi-diurnal, fast diurnal, and fortnightly tide tends to diminish the moon's distance; also the sidereal semi-diurnal and diurnal tides exercise no effects in this respect. The two tides which tend to increase the moon's distance are much larger than the others, so that the moon in general tends to recede from the earth. The increase of distance is, of course, accompanied by an increase of the moon's periodic time, and hence there is in general a true secular retardation of the moon's motion. But this change is accompanied by a retardation of the earth's diurnal rotation, and a terrestrial observer, taking the earth as his clock, would conceive that the angular velocity of an ideal moon, which was undisturbed by tidal reaction, was undergoing a secular acceleration. The apparent acceleration of the ideal undisturbed moon must considerably exceed the true retardation of the real disturbed moon, and the difference between these two will give an apparent acceleration.

It is thus possible to give an equation connecting the apparent acceleration of the moon's motion and the heights and retardations of the several bodily tides in the earth.

There is at the present time an unexplained secular acceleration of the moon of about 4″ per century, and therefore if we attribute the whole of this to the action of the bodily tides in the earth, instead of to the action of ocean tides, as was done by Adams and Delaunay, we get a numerical relation which must govern the actual heights and retardations of the bodily tides in the earth at the present time.

This equation involves the six constants expressive of the heights and retardations of the three bodily tides, and which are determined by the physical constitution of the earth. No further advance can therefore be made without some theory of the earth's nature. Two theories are considered.

First, that the earth is purely viscous. The result shows that the earth is either nearly fluid—which we know it is not—or exceedingly nearly rigid. The only traces which we should ever be likely to find of such a high degree of viscosity would be in the fortnightly ocean tide; and even here the influence would be scarcely perceptible, for its height would be ·992 of its theoretical amount according to the equilibrium theory, whilst the time of high water would be only accelerated by six hours and a half.

It is interesting to note that the indications of a fortnightly ocean tide, as deduced from tidal observations, are exceedingly uncertain, as is shown in a preceding paper*, where I have made a comparison of the heights and phases of such small fortnightly tides as have hitherto been observed. And now (July, 1879) Sir William Thomson has informed me that he thinks it very possible that the effects of the earth's rotation may be such as to prevent our trusting to the equilibrium theory to give even approximately the height of the fortnightly tide. He has recently read a paper on this subject before the Royal Society of Edinburgh†.

With the degree of viscosity of the earth, which gives the observed amount of secular acceleration to the moon, it appears that the moon is subject to such a true secular retardation that at the end of a century she is 3″·1 behind the place in her orbit which she would have occupied if it were not for the tidal reaction, whilst the earth, considered as a clock, is losing 13 seconds in the same time. This rate of retardation of the earth is such that an observer taking the earth as his clock would conceive a moon, which was undisturbed by tidal reaction, to be 7″·1 in advance of her place at the end of a century. But the actual moon is 3″·1 behind her true place, and thus our observer would suppose the moon to be in advance 7·1 − 3·1 or 4″ at the end of the century. Lastly, the obliquity of the ecliptic is diminishing at the rate of 1° in 500 million years.

The other hypothesis considered is that the earth is very nearly perfectly elastic. In this case the semi-diurnal and diurnal tides do not lag perceptibly, and the whole of the reaction is thrown on to the fortnightly tide, and moreover there is no perceptible tidal frictional couple about the earth's axis of rotation. From this follows the remarkable conclusion that the moon may be undergoing a true secular acceleration of motion of something less than 3″·5 per century, whilst the length of day may remain almost unaffected. Under these circumstances the obliquity of the ecliptic must be diminishing at the rate of 1° in something like 130 million years.

This supposition leads to such curious results, that I investigated what state of things we should arrive at if we look back for a very long period, and I found that 700 million years ago the obliquity might have been 5° greater than at present, whilst the month would only be a little less than a day longer. The suppositions on which these results are based are such that they *necessarily* give results more striking than would be physically possible.

The enormous lapse of time which has to be postulated renders it in the

* See the Appendix to my paper on the "Bodily Tides," &c. [This Appendix is however omitted from the present volume on account of its incompleteness, and is replaced by Paper 9, p. 340, Vol. I.]

† [See Paper 11, Vol. I. An investigation by Dr W. Schweydar (*Beiträgen zur Geophysik*, Vol. IX. p. 41) seems to indicate that the equilibrium theory is nearly fulfilled by the fortnightly tide, and it is explained by Lord Rayleigh (*Phil. Mag.* 1903) how this may be the case for an ocean interrupted by barriers of land.]

highest degree improbable that more than a very small change in this direction has been taking place, and moreover the action of the ocean tides has been entirely omitted from consideration.

The results of these two hypotheses show what fundamentally different interpretations may be put to the phenomenon of the secular acceleration of the moon.

Sir William Thomson also has drawn attention to another disturbing cause in the fall of meteoric dust on to the earth*.

Under these circumstances, I cannot think that any estimate having any pretension to accuracy can be made as to the present rate of tidal friction.

Since the obliquity of the ecliptic, the diurnal rotation of the earth, and the moon's distance change, the whole system is in a state of flux; and the next question to be considered is to determine the state of things which existed a very long time ago (Part II.). This involved the integration of three simultaneous differential equations; the mathematical difficulties were, however, so great, that it was found impracticable to obtain a general analytical solution. I therefore had to confine myself to a numerical solution adapted to the case of the earth, sun, and moon, for one particular degree of viscosity of the earth. The particular viscosity was such that, with the present values of the day and month, the time of the lunar semi-diurnal tide was retarded by 1 hour and 10 minutes; the greatest possible lagging of this tide is 3 hours, and therefore this must be regarded as a very moderate degree of viscosity. It was chosen because initially it makes the rate of change of obliquity a maximum, and although it is not that degree of viscosity which will make all the changes proceed with the greatest possible rapidity, yet it is sufficiently near that value to enable us to estimate very well the smallest time which can possibly have elapsed in the history of the earth, if changes of the kind found really have taken place. This estimate of time is confirmed by a second method, which will be referred to later.

The changes were traced backwards in time from the present epoch, and for convenience of diction I shall also reverse the form of speech—*e.g.*, a true loss of energy as the time increases will be spoken of as a gain of energy as we look backwards.

I shall not enter at all into the mathematical difficulties of the problem, but shall proceed at once to comment on the series of tables at the end of Section 15, which give the results of the solution.

The whole process, as traced backwards, exhibits a gain of kinetic energy to the system (of which more presently), accompanied by a transference of moment of momentum from that of orbital motion of the moon and earth to that of rotation of the earth. The last column but one of Table IV. exhibits the fall of the ratio of the two moments of momentum from 4·01

* *Proceedings of Glasgow Geological Society*, Vol. III. Address "On Geological Time."

down to ·44. The whole moment of momentum of the moon-earth system rises slightly, because of solar tidal friction. The change is investigated in Section 19.

Looked at in detail, we see the day, month, and obliquity all diminishing, and the changes proceeding at a rapidly increasing rate, so that an amount of change which at the beginning required many millions of years, at the end only requires as many thousands. The reason of this is that the moon's distance diminishes with great rapidity; and as the effects vary as the square of the tide-generating force, they vary as the inverse sixth power of the moon's distance, or, in physical language, the height of the tides increases with great rapidity, and so also does the moon's attraction. But there is a counteracting principle, which to some extent makes the changes proceed slower. It is obvious that a disturbing body will not have time to raise such high tides in a rapidly rotating spheroid as in one which rotates slowly. As the earth's rotation increases, the lagging of the tides increases. The first column of Table I. shows the angle by which the crest of the lunar semi-diurnal tide precedes the moon; we see that the angle is almost doubled at the end of the series of changes, as traced backwards. It is not quite so easy to give a physical meaning to the other columns, although it might be done. In fact, as the rotation increases, the effect of each tide rises to a maximum, and then dies away; the tides of longer period reach their maximum effect much more slowly than the ones of short period. At the point where I have found it convenient to stop the solution (see Table IV.), the semi-diurnal effect has passed its maximum, the diurnal tide has just come to give its maximum effect, whilst the fortnightly tide has not nearly risen to that point.

As the lunar effects increase in importance (when we look backwards), the relative value of the solar effects decreases rapidly, because the solar tidal reaction leaves the earth's orbit sensibly unaffected (see Section 19), and thus the solar effects remain nearly constant, whilst the lunar effects have largely increased. The relative value of the several tidal effects is exhibited in Tables II. and III.

Table IV. exhibits the length of day decreasing to a little more than a quarter of its present value, whilst the obliquity diminishes through 9°. But the length of the month is the element which changes to the most startling extent, for it actually falls to $\frac{1}{17}$th of its primitive value.

It is particularly important to notice that all the changes might have taken place in 57 million years; and this is far within the time which physicists admit that the earth and moon may have existed. It is easy to find a great many *veræ causæ* for changes in the planetary system; but it is in general correspondingly hard to show that they are competent to produce any marked effects, without exorbitant demands on the efficiency of the causes and on lapse of time.

It is a question of great interest to geologists to determine whether any part of these changes could have taken place during geological history. It seems to me that this question must be decided by whether or not a globe, such as has been considered, could have afforded a solid surface for animal life, and whether it might present a superficial appearance such as we know it. These questions must, I think, be answered in the affirmative, for the following reasons.

The coefficient of viscosity of the spheroid with which the previous solution deals is given by the formula $\frac{wa}{19n}\tan 35^\circ$ (see Section 11, (40)), when gravitation units of force are used. This, when turned into numbers, shows that $2{\cdot}055 \times 10^7$ grammes weight are required to impart unit shear to a cubic centimetre block of the substance in 24 hours, or 2,055 kilogs. per square centimetre acting tangentially on the upper face of a slab one centimetre thick for 24 hours, would displace the upper surface through a millimetre relatively to the lower, which is held fixed. In British units this becomes,—$13\frac{1}{2}$ tons to the square inch, acting for 24 hours on a slab an inch thick, displaces the upper surface relatively to the lower through one-tenth of an inch. It is obvious that such a substance as this would be called a solid in ordinary parlance, and in the tidal problem this must be regarded as a rather small viscosity.

It seems to me, then, that we have only got to postulate that the upper and cool surface of the earth presents such a difference from the interior that it yields with extreme slowness, if at all, to the weight of continents and mountains, to admit the possibility that the globe on which we live may be like that here treated of. If, therefore, astronomical facts should confirm the argument that the world has really gone through changes of the kind here investigated, I can see no adequate reason for assuming that the whole process was pre-geological. Under these circumstances it must be admitted that the obliquity to the ecliptic is now probably slowly decreasing; that a long time ago it was perhaps a degree greater than at present, and that it was then nearly stationary for another long time, and that in still earlier times it was considerably less*.

The violent changes which some geologists seem to require in geologically recent times would still, I think, not follow from the theory of the earth's viscosity.

According to the present hypothesis (and for the moment looking forward

* In my paper "On the Effects of Geological Changes on the Earth's Axis," *Phil. Trans.*, 1877, p. 271 [Vol. III.], I arrived at the conclusion that the obliquity had been unchanged throughout geological history. That result was obtained on the hypothesis of the earth's rigidity, except as regards geological upheavals. The result at which I now arrive affords a warning that every conclusion must always be read along with the postulates on which it is based.

in time), the moon-earth system is, from a dynamical point of view, continually losing energy from the internal tidal friction. One part of this energy turns into potential energy of the moon's position relatively to the earth, and the rest developes heat in the interior of the earth. Section 16 contains the investigation of the amount which has turned to heat between any two epochs. The heat is estimated by the number of degrees Fahrenheit, which the lost energy would be sufficient to raise the temperature of the whole earth's mass, if it were all applied at once, and if the earth had the specific heat of iron.

The last column of Table IV., Section 15, gives the numerical results, and it appears therefrom that, during the 57 million years embraced by the solution, the energy lost suffices to heat the whole earth's mass 1760° Fahr.

It would appear at first sight that this large amount of heat, generated internally, must seriously interfere with the accuracy of Sir William Thomson's investigation of the secular cooling of the earth*; but a further consideration of the subject in the next paper will show that this cannot be the case.

There are other consequences of interest to geologists which flow from the present hypothesis. As we look at the whole series of changes from the remote past, the ellipticity of figure of the earth must have been continually diminishing, and thus the polar regions must have been ever rising and the equatorial ones falling; but, as the ocean always followed these changes, they might quite well have left no geological traces.

The tides must have been very much more frequent and larger, and accordingly the rate of oceanic denudation much accelerated.

The more rapid alternations of day and night† would probably lead to more sudden and violent storms, and the increased rotation of the earth would augment the violence of the trade winds, which in their turn would affect oceanic currents.

Thus there would result an acceleration of geological action.

The problem, of which the solution has just been discussed, deals with a spheroid of constant viscosity; but there is every reason to believe that the earth is a cooling body, and has stiffened as it cooled. We therefore have to deal with a spheroid whose viscosity diminishes as we look backwards.

A second solution is accordingly given (Section 17) where the viscosity is variable; no definite law of diminution of viscosity is assumed, however, but it is merely supposed that the viscosity always remains small from a tidal point of view. This solution gives no indication of the time which may have elapsed, and differs chiefly from the preceding one in the fact that the change in the obliquity is rather greater for a given amount of change in the moon's distance.

* *Nat. Phil.*, Appendix.

† At the point where the solution stops there are just 1,300 of the sidereal days of that time in the year, instead of 366 as at present.

There is not much to say about it here, because the two solutions follow closely parallel lines as far as the place where the former one left off.

The first solution was not carried further, because as the month approximates in length to the day, the three semi-diurnal tides cease to be of nearly equal frequencies, and so likewise do the three diurnal tides; hence the assumption on which the solution was founded, as to their approximately equal speeds, ceases to be sufficiently accurate.

In this second solution all the seven tides are throughout distinguished from one another. At about the stage where the previous solution stops the solar terms have become relatively unimportant, and are dropped out. It appears that (still looking backwards in time) the obliquity will only continue to diminish a little more beyond the point it had reached when the previous method had become inapplicable. For when the month has become equal to twice the day, there is no change of obliquity; and for yet smaller values of the month the change is the other way.

This shows that for small viscosity of the planet the position of zero obliquity is dynamically stable for values of the month which are less than twice the day, while for greater values it is unstable; and the same appears to be true for very large viscosity of the planet (see the foot-note on p. 93).

If the integration be carried back as far as the critical point of relationship between the day and month, it appears that the whole change of obliquity since the beginning is $9\frac{1}{2}°$.

The interesting question then arises—Does the hypothesis of the earth's viscosity afford a complete explanation of the obliquity of the ecliptic? It does not seem at present possible to give any very conclusive answer to this question; for the problem which has been solved differs in many respects from the true problem of the earth.

The most important difference from the truth is in the neglect of the secular changes of the plane of the lunar orbit; and I now (September, 1879) see reason to believe that that neglect will make a material difference in the results given for the obliquity at the end of the third and fourth periods of integration in both solutions. It will not, therefore, be possible to discuss this point adequately at present; but it will be well to refer to some other points in which our hypothesis must differ from reality.

I do not see that the heterogeneity of density and viscosity would make any very material difference in the solution, because both the change of obliquity and the tidal friction would be affected *pari passû*, and therefore the change of obliquity for a given amount of change in the day would not be much altered.

Although the effects of the contraction of the earth in cooling would be certainly such as to render the changes more rapid in time, yet as the tidal friction would be somewhat counteracted, the critical point where the month is equal to twice the day would be reached when the moon was further from the earth than in my problem. I think, however, that there is reason to believe that the whole amount of contraction of the earth, since the moon has existed, has not been large (Section 24).

There is one thing which might exercise a considerable influence favourable to change of obliquity. We are in almost complete ignorance of the behaviour of semi-solids under very great pressures, such as must exist in the earth, and there is no reason to suppose that the amount of relative displacement is simply proportional to the stress and the time of its action. Suppose that the displacement varied as some other function of the time, then clearly the relative importance of the several tides might be much altered.

Now, the great obstacle to a large change of obliquity is the diurnal combined effect (see Table IV., Section 15); and so any change in the law of viscosity which allowed a relatively greater influence to the semi-diurnal tides would cause a greater change of obliquity, and this without much affecting the tidal friction and reaction. Such a law seems quite within the bounds of possibility. The special hypothesis, however, of elastico-viscosity, used in the previous paper, makes the other way, and allows greater influence to the tides of long period than to those of short. This was exemplified where it was shown that the tidal reaction might depend principally on the fortnightly tide.

The whole investigation is based on a theory of tides in which the effects of inertia are neglected. Now it will be shown in Part III. of the next paper that the effect of inertia will be to make the crest of the tidal spheroid lag more for a given height of tide than results from the theory founded on the neglect of inertia. An analysis of the effect produced on the present results, by the modification of the theory of tides introduced by inertia, is given in the next paper [Paper 4].

On the whole, we can only say at present that it seems probable that a part of the obliquity of the ecliptic may be referred to the causes here considered; but a complete discussion of the subject must be deferred to a future occasion, when the secular changes in the plane of the lunar orbit will be treated.

The question of the obliquity is now set on one side, and it is supposed that when the moon has reached the critical point (where the month is twice the day) the obliquity of the plane of the lunar orbit was zero. In the more remote past the obliquity had no tendency to alter, except under the influence of certain nutations, which are referred to at the end of Section 17.

The manner in which the moon's periodic time approximates to the day is an inducement to speculate as to the limiting or initial condition from which the earth and moon started their course of development.

So long as there is any relative motion of the two bodies there must be tidal friction, and therefore the moon's period must continue to approach the day. It would be a problem of extreme complication to trace the changes in detail to their end, and fortunately it is not necessary to do so.

The principle of conservation of moment of momentum, which has been used throughout in tracing the parallel changes in the moon and earth, affords the means of leaping at once to the conclusion (Section 18). The equation expressive of that principle involves the moon's orbital angular velocity and the earth's diurnal rotation as its two variables; and it is only necessary to equate one to the other to obtain an equation, which will give the desired information.

As we are now supposed to be transported back to the initial state, I shall henceforth speak of time in the ordinary way; there is no longer any convenience in speaking of the past as the future, and *vice versâ*.

The equation above referred to has two solutions, one of which indicates that tidal friction has done its work, and the other that it is just about to begin. Of the first I shall here say no more, but refer the reader to Section 18.

The second solution indicates that the moon (considered as an attractive particle) moves round the earth as though it were rigidly fixed thereto in 5 hours 36 minutes. This is a state of dynamical instability; for if the month is a little shorter than the day, the moon will approach the earth, and ultimately fall into it; but if the day is a little shorter than the month, the moon will continually recede from the earth, and pass through the series of changes which were traced backwards.

Since the earth is a cooling and contracting body, it is likely that its rotation would increase, and therefore the dynamical equilibrium would be more likely to break down in the latter than in the former way.

The continuous solution of the problem is taken up at the point where the moon has receded from the earth so far that her period is twice that of the earth's rotation.

I have calculated that the heat generated in the interior of the earth in the course of the lengthening of the day from 5 hours 36 minutes to 23 hours 56 minutes would be sufficient, if applied all at once, to heat the whole earth's mass about 3000° Fahr., supposing the earth to have the specific heat of iron (see Section 16).

A rough calculation shows that the minimum time in which the moon can have passed from the state where it had a period of 5 hours 36 minutes to the present state, is 54 million years, and this confirms the previous estimates of time.

This periodic time of the moon corresponds to an interval of only 6,000 miles between the earth's surface and the moon's centre. If the earth had been treated as heterogeneous, this distance, and with it the common periodic time both of moon and earth, would be still further diminished.

These results point strongly to the conclusion that, if the moon and earth were ever molten viscous masses, they once formed parts of a common mass.

We are thus led at once to the inquiry as to how and why the planet broke up. The conditions of stability of rotating masses of fluid are unfortunately unknown *, and it is therefore impossible to do more than speculate on the subject.

The most obvious explanation is similar to that given in Laplace's nebular hypothesis, namely, that the planet being partly or wholly fluid, contracted, and thus rotated faster and faster until the ellipticity became so great that the equilibrium was unstable, and then an equatorial ring separated itself, and the ring finally conglomerated into a satellite. This theory, however, presents an important difference from the nebular hypothesis, in as far as that the ring was not left behind 240,000 miles away from the earth, when the planet was a rare gas, but that it was shed only 4,000 or 5,000 miles from the present surface of the earth, when the planet was perhaps partly solid and partly fluid.

This view is to some extent confirmed by the ring of Saturn, which would thus be a satellite in the course of formation.

It appears to me, however, that there is a good deal of difficulty in the acceptance of this view, when it is considered along with the numerical results of the previous investigation.

At the moment when the ring separated from the planet it must have had the same linear velocity as the surface of the planet; and it appears from Section 22 that such a ring would not tend to expand from tidal reaction, unless its density varied in different parts. Thus we should hardly expect the distance from the earth of the chain of meteorites to have increased much, until it had agglomerated to a considerable extent. It follows, therefore, that we ought to be able to trace back the moon's path, until she was nearly in contact with the earth's surface, and was always opposite the same face of the earth. Now this is exactly what has been done in the previous investigation. But there is one more condition to be satisfied, namely, that the common speed of rotation of the two bodies should be so great that the equilibrium of the rotating spheroid should be unstable. Although we do not know what is the limiting angular velocity of a rotating spheroid consistent with stability, yet it seems improbable that a rotation in a little over 5 hours, with an ellipticity of one-twelfth would render the system unstable.

Now notwithstanding that the data of the problem to be solved are to some extent uncertain, and notwithstanding the imperfection of the solution of the problem here given, yet it hardly seems likely that better data and a

* [This statement is now no longer correct.]

more perfect solution would largely affect the result, so as to make the common period of revolution of the two bodies in the initial configuration very much less than 5 hours*. Moreover we obtain no help from the hypothesis that the earth has contracted considerably since the shedding of the satellite, but rather the reverse; for it appears from Section 24 that if the earth has contracted, then the common period of revolution of the two bodies in the initial configuration must have been slower, and the moon more distant from the earth. This slower revolution would correspond with a smaller ellipticity, and thus the system would probably be less nearly unstable.

The following appears to me at least a possible cause of instability of the spheroid when rotating in about 5 hours. Sir William Thomson has shown that a fluid spheroid of the same mean density as the earth would perform a complete gravitational oscillation in 1 hour 34 minutes. The speed of oscillation varies as the square root of the density, hence it follows that a less dense spheroid would oscillate more slowly, and therefore a spheroid of the same mean density as the earth, but consisting of a denser nucleus and a rarer surface, would probably oscillate in a longer time than 1 hour 34 minutes. It seems to be quite possible that two complete gravitational oscillations of the earth in its primitive state might occupy 4 or 5 hours. But if this were the case, the solar semi-diurnal tide would have very nearly the same period as the free oscillation of the spheroid, and accordingly the solar tides would be of enormous height.

Does it not then seem possible that, if the rotation were fast enough to bring the spheroid into anything near the unstable condition, the large solar tides might rupture the body into two or more parts? In this case one would conjecture that it would not be a ring which would detach itself†.

It seems highly probable that the moon once did rotate more rapidly round her own axis than in her orbit, and if she was formed out of the fusion together of a ring of meteorites, this rotation would necessarily result.

In Section 23 it is shown that the tidal friction due to the earth's action on the moon must have been enormous, and it must necessarily have soon brought her to present the same face constantly to the earth. This explanation was, I believe, first given by Helmholtz‡. In the process, the inclination of her axis to the plane of her orbit must have rapidly increased, and then, as she rotated more and more slowly, must have slowly diminished again. Her present aspect is thus in strict accordance with the results of the purely theoretical investigation.

* This is illustrated by my paper on "The Secular Effects of Tidal Friction," [Paper 5], where it appears that the "line of momentum" does not cut the "curve of rigidity" at a very small angle, so that a small error in the data would not make a very large one in the solution.

† [On this subject see Professor A. E. H. Love, "On the oscillations of a rotating liquid Spheroid and the genesis of the Moon," *Phil. Mag.*, March, 1889.]

‡ [Both Kant and Laplace gave the same explanation many years before.]

It would perhaps be premature to undertake a complete review of the planetary system, so as to see how far the ideas here developed accord with it. Although many facts which could be adduced seem favourable to their acceptance, I will only refer to two. The satellites of Mars appear to me to afford a remarkable confirmation of these views. Their extreme minuteness has prevented them from being subject to any perceptible tidal reaction, just as the minuteness of the earth compared with the sun has prevented the earth's orbit from being perceptibly influenced (see Section 19); they thus remain as a standing memorial of the primitive periodic time of Mars round his axis. Mars, on the other hand, has been subjected to solar tidal friction. This case, however, deserves to be submitted to numerical calculation.

The other case is that of Uranus, and this appears to be somewhat unfavourable to the theory; for on account of the supposed adverse revolution of the satellites, and of the high inclinations of their orbits, it is not easy to believe that they could have arisen from a planet which ever rotated about an axis at all nearly perpendicular to the ecliptic.

The system of planets revolving round the sun presents so strong a resemblance to the systems of satellites revolving round the planets, that we are almost compelled to believe that their modes of development have been somewhat alike. But in applying the present theory to explain the orbits of the planets, we are met by the great difficulty that the tidal reaction due to solar tides in the planet is exceedingly slow in its influence; and not much help is got by supposing the tides in the sun to react on the planet. Thus enormous periods of time would have to be postulated for the evolution.

If, however, this theory should be found to explain the greater part of the configurations of the satellites round the planets, it would hardly be logical to refuse it some amount of applicability to the planets. We should then have to suppose that before the birth of the satellites the planets occupied very much larger volumes, and possessed much more moment of momentum than they do now. If they did so, we should not expect to trace back the positions of the axes of the planets to the state when they were perpendicular to the ecliptic, as ought to be the case if the action of the satellites, and of the sun after their birth, is alone concerned.

Whatever may be thought of the theory of the viscosity of the earth, and of the large speculations to which it has given rise, the fact remains that nearly all the effects which have been attributed to the action of bodily tides would also follow, though probably at a somewhat less rapid rate, from the influence of oceanic tides on a rigid nucleus. The effect of oceanic tidal friction on the obliquity of the ecliptic has already been considered by Mr Stone, in the only paper on the subject which I have yet seen*. His argument is based on what I conceive to be an incorrect assumption as to the nature of

* *Ast. Soc. Monthly Notices*, March 8, 1867.

the tidal frictional couple, and he neglects tidal reaction; he finds that the effects would be quite insignificant. This result would, I think, be modified by a more satisfactory assumption.

APPENDIX.

An extract from the abstract of the foregoing paper, Proc. Roy. Soc., Vol. XXVIII. (1879), pp. 184—194.

. . . .

I will now show, from geometrical considerations, how some of the results previously stated come to be true. It will not, however, be possible to obtain a quantitative estimate in this way.

The three following propositions do not properly belong to an abstract, since they are not given in the paper itself; they merely partially replace the analytical method pursued therein. The results of the analysis were so wholly unexpected in their variety, that I have thought it well to show that the more important of them are conformable to common sense. These general explanations might doubtless be multiplied by some ingenuity, but it would not have been easy to discover the results, unless the way had been first shown by analysis.

PROP. I. *If the viscosity be small the earth's obliquity increases, the rotation is retarded, and the moon's distance and periodic time increase.*

The figure represents the earth as seen from above the South Pole, so that S is the Pole, and the outer circle the Equator. The earth's rotation is in the direction of the curved arrow at S. The half of the inner circle which is drawn with a full line is a semi-small-circle of S. lat., and the dotted semi-circle is a semi-small-circle in the same N. lat.

Generally dotted lines indicate parts of the figure which are below the plane of the paper.

It will make the explanation somewhat simpler, if we suppose the tides to be raised by a moon and anti-moon diametrically opposite to one another. Accordingly let M and M′ be the projections of the moon and anti-moon on to the terrestrial sphere.

If the substance of the earth were a perfect fluid or perfectly elastic, the apices of the tidal spheroid would be at M and M′. If, however, there is internal friction due to any sort of viscosity, the tides will lag, and we may suppose the tidal apices to be at T and T′.

Suppose the tidal protuberances to be replaced by two equal heavy particles at T and T′, which are instantaneously rigidly connected with the earth. Then the attraction of the moon on T is greater than on T′; and of the anti-moon on T′ is greater than on T. The resultant of these forces is clearly a pair of forces acting on the earth in the direction of TM, T′M′.

The effect on the obliquity will be considered first.

These forces TM, T′M′, clearly cause a couple about the axis in the equator, which lies in the same meridian as the moon and anti-moon. The direction of the couple is shown by the curved arrows at L, L′.

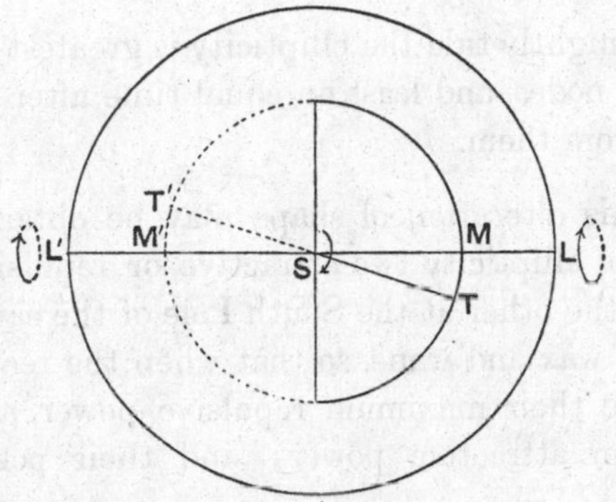

FIG. 6.

If the effects of this couple be compounded with the existing rotation of the earth, according to the principle of the gyroscope, it will be seen that the South Pole S tends to approach M, and the North Pole to approach M′. Hence supposing the moon to move in the ecliptic, the inclination of the earth's axis to the ecliptic diminishes, or the obliquity increases.

Next, the forces TM, T′M′, clearly produce a couple about the earth's polar axis, which tends to retard the diurnal rotation.

Lastly, since action and reaction are equal and opposite, and since the moon and anti-moon cause the forces TM, T′M′, on the earth, therefore the earth must cause forces on those two bodies (or on their equivalent single moon) in the directions MT and M′T′. These forces are in the direction of the moon's orbital motion, and therefore her linear velocity is augmented. Since the centrifugal force of her orbital motion must remain constant, her distance increases, and with the increase of distance comes an increase of periodic time round the earth.

This general explanation remains a fair representation of the state of the case so long as the different harmonic constituents of the aggregate tide-wave do not suffer very different amounts of retardation; and this is the case so long as the viscosity is not great.

PROP. II. *The attraction of the moon on a lagging fortnightly tide causes the earth's obliquity to diminish, but does not affect the diurnal rotation; the reaction on the moon causes a diminution of her distance and periodic time.*

The fortnightly tide of a perfectly fluid earth is a periodic increase and diminution of the ellipticity of figure; the increment of ellipticity varies as the square of the sine of the obliquity of the equator to the ecliptic, and as the cosine of twice the moon's longitude from her node. Thus the ellipticity is greatest when the moon is in her nodes, and least when she is 90° removed from them.

In a lagging fortnightly tide the ellipticity is greatest some time after the moon has passed the nodes, and least an equal time after she has passed the point 90° removed from them.

The effects of this alteration of shape may be obtained by substituting for these variations of ellipticity two attractive or repulsive particles, one at the North Pole and the other at the South Pole of the earth. These particles must be supposed to wax and wane, so that when the real ellipticity of figure is greatest they have their maximum repulsive power, and when least they have their maximum attractive power; and their positive and negative maxima are equal.

We will now take the extreme case when the obliquity is 90°; this makes the fortnightly tide as large as possible.

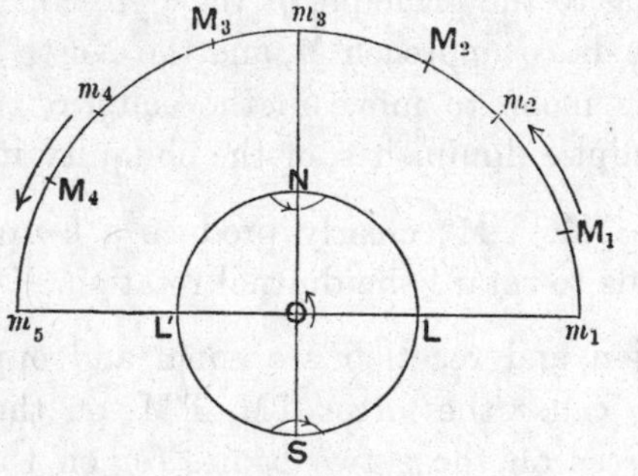

FIG. 7.

Let the plane of the paper be that of the ecliptic, and let the outer semicircle be the moon's orbit, which she describes in the direction of the arrows. Let NS be the earth's axis, which by hypothesis lies in the ecliptic, and let L, L′ be the nodes of the orbit. Let N be the North Pole; that is to say, if the earth were turned about the line LL′, so that N rises above the plane of the paper, the earth's rotation would be in the same direction as the moon's orbital motion.

First consider the case where the earth is perfectly fluid, so that the tides do not lag.

Let m_2, m_4 be points in the orbit whose longitudes are 45° and 135°; and suppose that couples acting on the earth about an axis at O perpendicular to the plane of the paper are called positive when they are in the direction of the curved arrow at O. When the moon is at m_1 the particles at N and S have their maximum repulsion. But at this instant the moon is equidistant from both, and there is no couple about O. As, however, the moon passes to m_2 there is a positive couple, which vanishes when the moon is at m_2, because the particles have waned to zero. From m_2 to m_3 the couple is negative; from m_3 to m_4 positive; and from m_4 to m_5 negative. Now, the couple goes through just the same changes of magnitude, as the moon passes from m_1 to m_2, as it does while the moon passes from m_4 to m_5, but in the reverse order; the like may be said of the arcs m_2m_3 and m_3m_4. Hence it follows that the average effect, as the moon passes through half its course, is *nil*, and therefore there can be no secular change in the position of the earth's axis.

But now consider the case when the tide lags. When the moon is at m_1 the couple is zero, because she is equally distant from both particles. The particles have not, however, reached their maximum of repulsiveness; this they do when the moon has reached M_1, and they do not cease to be repulsive until the moon has reached M_2. Hence, during the description of the arc m_1M_2, the couple round O is positive.

Throughout the arc M_2m_3 the couple is negative, but it vanishes when the moon is at m_3, because the moon and the two particles are in a straight line. The particles reach their maximum of attractiveness when the moon is at M_3, and the couple continues to be positive until the moon is at M_4.

Lastly, during the description of the arc M_4m_5 the couple is negative.

But now there is no longer a balance between the arcs m_1M_2 and M_4m_5, nor between M_2m_3 and m_3M_4. The arcs during which the couples are positive are longer and the couples are more intense than in the rest of the semi-orbit. Hence the average effect of the couples is a positive couple, that is to say, in the direction of the curved arrow round O.

It may be remarked that if the arcs m_1M_1, m_2M_2, m_3M_3, m_4M_4 had been 45°, there would have been no negative couples at all, and the positive couples would merely have varied in intensity.

A couple round O in the direction of the arrow, when combined with the earth's rotation, would, according to the principle of the gyroscope, cause the pole N to rise above the plane of the paper, that is to say, the obliquity of the ecliptic would diminish. The same thing would happen, but to a less extent, if the obliquity had been less than 90°; it would not, however, be nearly so easy to show this from general considerations.

Since the forces which act on the earth always pass through N and S, there can be no moment about the axis NS, and the rotation about that axis remains unaffected. This can hardly be said to amount to strict proof that the diurnal rotation is unaffected by the fortnightly tide, because it has not been rigorously shown that the two particles at N and S are a complete equivalent to the varying ellipticity of figure.

Lastly, the reaction on the moon must obviously be in the opposite direction to that of the curved arrow at O; therefore there is a force retarding her linear motion, the effect of which is a diminution of her distance and of her periodic time.

The fortnightly tidal effect must be far more efficient for very great viscosities than for small ones, for, unless the viscosity is very great, the substance of the spheroid has time to behave sensibly like a perfect fluid, and the tide hardly lags at all.

PROP. III. *An annular satellite not parallel to the planet's equator attracts the lagging tides raised by it, so as to diminish the inclination of the planet's equator to the plane of the ring, and to diminish the planet's rotation. The effects of the joint action of sun and moon may be explained from this.*

Suppose the figure to represent the planet as seen from vertically over the South Pole S; let L, L′ be the nodes of the ring, and LRL′ the projection of half the ring on to the planetary sphere.

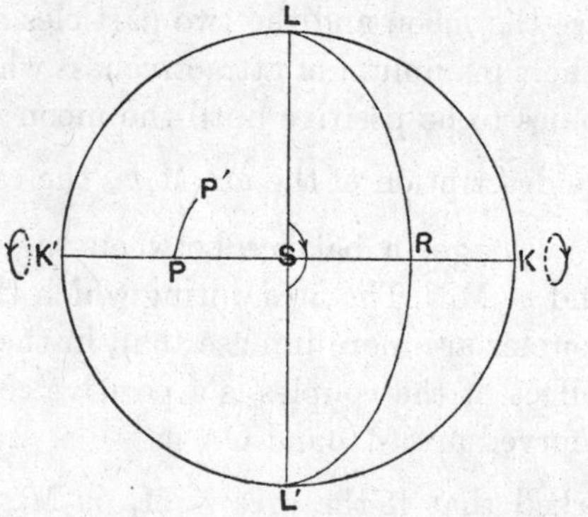

FIG. 8.

If the planet were perfectly fluid the attraction of the ring would produce a ridge of elevation all along the neighbourhood of the arc LRL′, together with a compression in the direction of an axis perpendicular to the plane of the ring. This tidal spheroid may be conceived to be replaced by a repulsive particle placed at P, the pole of the ring, and an equal repulsive particle at its antipodes, which is not shown in the figure.

Suppose that the spheroid is viscous, and that the tide lags; then since the planet rotates in the direction of the curved arrow at S, the repulsive particle is carried past its place, P, to P′. The angle PSP′ is a measure of the lagging of the tide.

We have to consider the effect of the repulsion of the ring on a particle which is instantaneously and rigidly connected with the planet at P′.

Since P′ is nearer to the half L of the ring, than to the half L′, the general effect of the repulsion must be a force somewhere in the direction P′P.

Now this force P′P must cause a couple in the direction of the curved arrows K, K′ about an axis, KK′, perpendicular to LL′, the nodes of the ring. The effects of this couple, when compounded with the planet's rotation, is to cause the pole S to recede from the ring LRL′. Hence the inclination of the planet's equator to the ring diminishes.

Secondly, the force P′P produces a couple about S, adverse to the planet's rotation about its axis S. If the obliquity of the ring be small, this couple will be small, because P′ will lie close to S.

Lastly, it may be shown analytically that the tangential force on the ring in the direction of the planet's rotation, corresponding with the tidal friction, is exactly counterbalanced by a tangential force in the opposite direction, corresponding with the change of the obliquity. Thus the diameter of the ring remains constant. It would not be very easy to prove this from general considerations.

It may be shown that, as far as concerns their joint action, the sun and moon may be conceived to be replaced by a pair of rings, and these rings may be replaced by a single one; hence the above proposition is also applicable to the explanation of the joint action of the two bodies on the earth, and numerical calculation shows that these joint effects exercise a very important influence on the rate of variation of obliquity.

4.

PROBLEMS CONNECTED WITH THE TIDES OF A VISCOUS SPHEROID.

[*Philosophical Transactions of the Royal Society,* Part II. Vol. 170 (1879), pp. 539—593.]

CONTENTS.

		PAGE
I.	Secular distortion of the spheroid, and certain tides of the second order	141
II.	Distribution of heat generated by internal friction, and secular cooling	155
III.	The effects of inertia in the forced oscillations of viscous, fluid, and elastic spheroids	167
IV.	Discussion of the applicability of the results to the history of the earth	187

In the following paper several problems are considered, which were alluded to in my two previous papers on this subject*.

The paper is divided into sections which deal with the problems referred to in the table of contents. It was found advantageous to throw the several investigations together, because their separation would have entailed a good deal of repetition, and one system of notation now serves throughout.

It has, of course, been impossible to render the mathematical parts entirely independent of the previous papers, to which I shall accordingly have occasion to make a good many references.

As the whole inquiry is directed by considerations of applicability to the earth, I shall retain the convenient phraseology afforded by speaking of the tidally distorted spheroid as the earth, and of the disturbing body as the moon.

It is probable that but few readers will care to go through the somewhat complex arguments and analysis by which the conclusions are supported, and

* [Papers 1 and 3 above. They will be referred to hereafter as "Tides" and "Precession" respectively.]

therefore in the fourth part a summary of results is given, together with some discussion of their physical applicability to the case of the earth.

I. *Secular distortion of the spheroid, and certain tides of the second order.*

In considering the tides of a viscous spheroid, it was supposed that the tidal protuberances might be considered as the excess and deficiency of matter above and below the mean sphere—or more strictly the mean spheroid of revolution which represents the average shape of the earth. The spheroid was endued with the power of gravitation, and it was shown that the action of the spheroid on its own tides might be found approximately by considering the state of flow in the mean sphere caused by the attraction of the protuberances, and also by supposing the action of the protuberances on the sphere to be normal thereto, and to consist, in fact, merely of the weight (either positive or negative) of the protuberances.

Thus if a be the mean radius of the sphere, w its density, g mean gravity at the surface, and $r = a + \sigma_i$ the equation to the tidal protuberance, where σ_i is a surface harmonic of order i, the potential per unit volume of the protuberance in the interior of the sphere is $\frac{3gw}{2i+1}\left(\frac{r}{a}\right)^i \sigma_i$, and the sphere is subjected to a normal traction per unit area of surface equal to $-gw\sigma_i$.

It was also shown that these two actions might be compounded by considering the interior of the sphere (now free of gravitation) to be under the action of a potential $-\frac{2(i-1)}{2i+1} gw \left(\frac{r}{a}\right)^i \sigma_i$.

This expression therefore gave the effective potential when the sphere was treated as devoid of gravitational power.

It was remarked* that, strictly speaking, there is tangential action between the protuberance and the surface of the sphere. And later† it was stated that the action of an external tide-generating body on the lagging tides was not such as to form a rigorously equilibrating system of forces. The effects of this non-equilibration, in as far as it modifies the rotation of the spheroid as a whole, were considered in the paper on "Precession."

It is easy to see from general considerations that these previously neglected tangential stresses on the surface of the sphere, together with the effects of inertia due to the secular retardation of the earth's rotation (produced by the non-equilibrating forces), must cause a secular distortion of the spheroid.

This distortion I now propose to investigate.

* "Tides," Section 2. † "Tides," Section 5.

In order to avoid unnecessary complication, the tides will be supposed to be raised by a single disturbing body or moon moving in the plane of the earth's equator.

Let $r = a + \sigma$ be the equation to the bounding surface of the tidally-distorted earth, where σ is a surface harmonic of the second order.

I shall now consider how the equilibrium is maintained of the layer of matter σ, as acted on by the attraction of the spheroid and under the influence of an external disturbing potential V, which is a solid harmonic of the second degree of the coordinates of points within the sphere*. The object to be attained is the evaluation of the stresses tangential to the surface of the sphere, which are exercised by the layer σ on the sphere.

Let θ, ϕ be the colatitude and longitude of a point in the layer. Then consider a prismatic element bounded by the two cones θ, $\theta + \delta\theta$, and by the two planes ϕ, $\phi + \delta\phi$.

The radial faces of this prism are acted on by the pressures and tangential stresses communicated by the four contiguous prisms. But the tangential stresses on these faces only arise from the fact that contiguous prisms are solicited by slightly different forces, and therefore the action of the four prisms, surrounding the prism in question, must be principally pressure. I therefore propose to consider that the prism resists the tendency of the impressed forces to move tangentially along the surface of the sphere, by means of hydrostatic pressures on its four radial faces, and by a tangential stress across its base.

This approximation by which the whole of the tangential stress is thrown on to the base, is clearly such as slightly to accentuate, as it were, the distribution of the tangential stresses on the surface of the sphere, by which the equilibrium of the layer σ is maintained. For consider the following special case:—Suppose σ to be a surface of revolution, and V to be such that only a single small circle of latitude is solicited by a tangential force everywhere perpendicular to the meridian. Then it is obvious that, strictly speaking, the elements lying a short way north and south of the small circle would tend to be carried with it, and the tangential stress on the sphere would be a maximum along the small circle, and would gradually die away to the north and south. In the approximate method, however, which it is proposed to use, such an application of external force would be deemed to cause no tangential stress to the surface of the sphere to the north and south of the small circle acted on. This special case is clearly a great exaggeration of what holds in our problem, because it postulates a finite difference of disturbing force between elements infinitely near to one another.

* A parallel investigation would be applicable, when σ and V are of any orders.

We will first find what are the hydrostatic pressures transmitted by the four prisms contiguous to the one we are considering.

Let p be the hydrostatic pressure at the point r, θ, ϕ of the layer σ. If we neglect the variations of gravity due to the layer σ and to V, p is entirely due to the attraction of the mean sphere of radius a.

The mean pressure on the radial faces at the point in question is $\frac{1}{2}gw\sigma$; where σ is negative the pressures are of course tractions.

We will first resolve along the meridian.

The excess of the pressure acting on the face $\theta + \delta\theta$ over that on the face θ (whose area is $\sigma a \sin\theta\,\delta\phi$) is

$$\frac{d}{d\theta}\left[\tfrac{1}{2}gw\sigma\,.\,\sigma a\sin\theta\,\delta\phi\right]\delta\theta, \quad \text{or} \quad \tfrac{1}{2}gwa\frac{d}{d\theta}(\sigma^2\sin\theta)\,\delta\theta\,\delta\phi$$

and it acts towards the pole.

The resolved part of the pressures on the faces $\phi + \delta\phi$ and ϕ (whose area is $\sigma a\delta\theta$) along the meridian is

$$(\tfrac{1}{2}gw\sigma)(\sigma a\delta\theta)(\cos\theta\,\delta\phi) \quad \text{or} \quad \tfrac{1}{2}gwa\sigma^2\cos\theta\,\delta\theta\,\delta\phi$$

and it acts towards the equator.

Hence the whole force due to pressure on the element resolved along the meridian towards the equator is

$$\tfrac{1}{2}gwa\,\delta\theta\,\delta\phi\left[\sigma^2\cos\theta - \frac{d}{d\theta}(\sigma^2\sin\theta)\right], \quad \text{or} \quad -gwa\,\delta\theta\,\delta\phi\sin\theta\,\sigma\frac{d\sigma}{d\theta}$$

But the mass of the elementary prism $\delta m = wa^2\sin\theta\,\delta\theta\,\delta\phi\,.\,\sigma$.

Hence the meridional force due to pressure is $-\dfrac{g}{a}\,\delta m\dfrac{d\sigma}{d\theta}$.

We will next resolve the pressures perpendicular to the meridian.

The excess of pressure on the face $\phi + \delta\phi$ over that on the face ϕ (whose area is $\sigma a\delta\theta$), measured in the direction of ϕ increasing, is

$$-\frac{d}{d\phi}\left[\tfrac{1}{2}gw\sigma\,.\,\sigma a\delta\theta\right]\delta\phi = -gwa\sigma\frac{d\sigma}{d\phi}\,\delta\theta\,\delta\phi = -\frac{g}{a}\,\delta m\frac{1}{\sin\theta}\frac{d\sigma}{d\phi}$$

Hence the force due to pressure perpendicular to the meridian is

$$-\frac{g}{a}\,\delta m\frac{1}{\sin\theta}\frac{d\sigma}{d\phi}$$

We have now to consider the impressed forces on the element.

Since σ is a surface harmonic of the second degree, the potential of the layer of matter σ at an external point is $\frac{3}{5}g\sigma\left(\dfrac{a}{r}\right)^3$. Therefore the forces along and perpendicular to the meridian on a particle of mass δm, just outside the layer σ but infinitely near the prismatic element, are $\frac{3}{5}\frac{g}{a}\,\delta m\,\frac{d\sigma}{d\theta}$ and $\frac{3}{5}\frac{g}{a}\,\delta m\,\frac{1}{\sin\theta}\frac{d\sigma}{d\phi}$, and these are also the forces acting on the element δm due to the attraction of the rest of the layer σ.

Lastly, the forces due to the external potential V are clearly $\delta m\,\frac{1}{a}\frac{dV}{d\theta}$ and $\delta m\,\frac{1}{a\sin\theta}\frac{dV}{d\phi}$.

Collecting results we get for the forces due both to pressure and attraction, along the meridian towards the equator

$$\delta m\left[-\frac{g}{a}\frac{d\sigma}{d\theta}+\frac{3}{5}\frac{g}{a}\frac{d\sigma}{d\theta}+\frac{dV}{ad\theta}\right]=\delta m\,\frac{d}{ad\theta}(V-\tfrac{2}{5}g\sigma)$$

and perpendicular to the meridian, in the direction of ϕ increasing,

$$\delta m\left[-\frac{g}{a\sin\theta}\frac{d\sigma}{d\phi}+\frac{3g}{5a\sin\theta}\frac{d\sigma}{d\phi}+\frac{dV}{a\sin\theta\,d\phi}\right]=\delta m\,\frac{1}{a\sin\theta}\frac{d}{d\phi}(V-\tfrac{2}{5}g\sigma)$$

Henceforward $\frac{2g}{5a}$ will be written $\mathfrak{g}$, as in the previous papers.

These are the forces on the element which must be balanced by the tangential stresses across the base of the prismatic element.

It follows from the above formulæ that the tangential stresses communicated by the layer σ to the surface of the sphere are those due to a potential $V-\mathfrak{g}a\sigma$ acting on the layer σ.

If $\sigma = V/\mathfrak{g}a$ there is no tangential stress. But this is the condition that σ should be the equilibrium tidal spheroid due to V, so that the result fulfils the condition that if σ be the equilibrium tidal spheroid of V there is no tendency to distort the spheroid further; this obviously ought to be the case.

In the problem before us, however, σ does not fulfil this condition, and therefore there is tangential stress across the base of each prismatic element tending to distort the sphere.

Suppose $V=r^2S$ where S is a surface harmonic.

Then at the surface $V=a^2S$. If δm be the mass of a prism cut out of the layer σ, which stands on unit area as base, $\delta m = w\sigma$.

Therefore the tangential stresses per unit area communicated to the sphere are

$$\left.\begin{aligned}&wa^2\frac{\sigma}{a}\frac{d}{d\theta}\left(S-\mathfrak{g}\frac{\sigma}{a}\right)\text{ along the meridian}\\ \text{and}\quad &wa^2\frac{\sigma}{a}\frac{1}{\sin\theta}\frac{d}{d\phi}\left(S-\mathfrak{g}\frac{\sigma}{a}\right)\text{ perpendicular to the meridian}\end{aligned}\right\}\ldots\ldots(1)$$

Besides these tangential stresses there is a small radial stress over and above the radial traction $-gw\sigma$, which was taken into account in forming the tidal theory. But we remark that the part of this stress, which is periodic in time, will cause a very small tide of the second order, and the part which is non-periodic will cause a very small permanent modification of the figure of the sphere. These effects are, however, so minute as not to be worth investigating.

We will now apply these results to the tidal problem.

Let X, Y, Z (fig. 1) be rectangular axes fixed in the earth, Z being the axis of rotation and XZ the plane from which longitudes are measured.

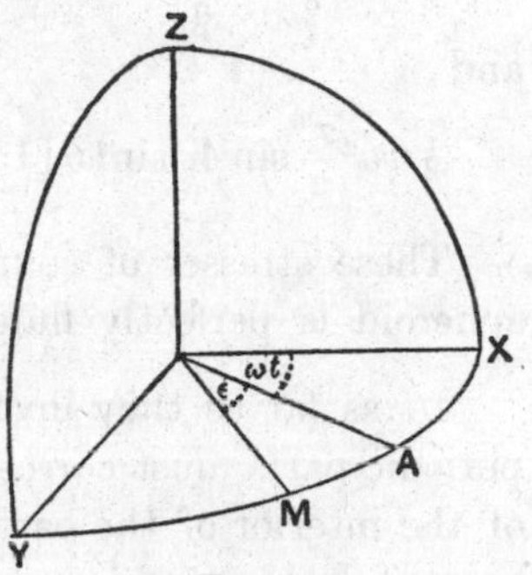

FIG. 1.

Let M be the projection of the moon on the equator, and let ω be the earth's angular velocity of rotation relatively to the moon.

Let A be the major axis of the tidal ellipsoid.

Let $AX = \omega t$, where t is the time, and let $MA = \epsilon$.

Let m be the moon's mass measured astronomically, and c her distance, and $\tau = \frac{3}{2}\frac{m}{c^3}$.

According to the usual formula, the moon's tide-generating potential is

$$\tau r^2 [\sin^2\theta \cos^2(\phi - \omega t - \epsilon) - \tfrac{1}{3}]$$

which may be written

$$\tfrac{1}{2}\tau r^2 (\tfrac{1}{3} - \cos^2\theta) + \tfrac{1}{2}\tau r^2 \sin^2\theta \cos 2(\phi - \omega t - \epsilon)$$

The former of these terms is not a function of the time, and its effect is to cause a permanent small increase of ellipticity of figure of the earth, which may be neglected. We are thus left with

$$\tfrac{1}{2}\tau r^2 \sin^2\theta \cos 2(\phi - \omega t - \epsilon)$$

as the true tide-generating potential.

If $\tan 2\epsilon = \dfrac{19 v\omega}{\mathfrak{g} a w}$, where v is the coefficient of viscosity of the spheroid, by the theory of the paper on "Tides," such a potential will raise a tide expressed by

$$\frac{\sigma}{a} = \tfrac{1}{2}\frac{\tau}{\mathfrak{g}} \cos 2\epsilon \sin^2\theta \cos 2(\phi - \omega t)^* \quad \text{.................}(2)$$

If we put $$S = \tfrac{1}{2}\tau \sin^2\theta \cos 2(\phi - \omega t - \epsilon) \quad \text{....................}(3)$$

$$S - \mathfrak{g}\frac{\sigma}{a} = \tfrac{1}{2}\tau \sin 2\epsilon \sin^2\theta \sin 2(\phi - \omega t) \quad \text{..............}(4)$$

* "Tides," Section 5.

and $$\frac{d}{d\theta}\left(\mathrm{S} - \mathfrak{g}\frac{\sigma}{a}\right) = \tau \sin 2\epsilon \sin\theta \cos\theta \sin 2(\phi - \omega t)$$

$$\frac{1}{\sin\theta}\frac{d}{d\phi}\left(\mathrm{S} - \mathfrak{g}\frac{\sigma}{a}\right) = \tau \sin 2\epsilon \sin\theta \cos 2(\phi - \omega t)$$

Multiplying these by $wa^2\frac{\sigma}{a}$, we find from (1) the tangential stresses communicated by the layer σ to the sphere.

They are

$$\tfrac{1}{8}wa^2\frac{\tau^2}{\mathfrak{g}}\sin 4\epsilon \sin^3\theta \cos\theta \sin 4(\phi - \omega t) \text{ along the meridian}$$

and

$$\tfrac{1}{8}wa^2\frac{\tau^2}{\mathfrak{g}}\sin 4\epsilon \sin^3\theta\,[1 + \cos 4(\phi - \omega t)] \text{ perpendicular to the meridian}$$

These stresses of course vanish when ϵ is zero, that is to say when the spheroid is perfectly fluid.

In as far as they involve $\phi - \omega t$ these expressions are periodic, and the periodic parts must correspond with periodic inequalities in the state of flow of the interior of the earth. These small tides of the second order have no present interest and may be neglected.

We are left, therefore, with a non-periodic tangential stress per unit area of the surface of the sphere perpendicular to the meridian from east to west equal to $\frac{1}{8}wa^2\frac{\tau^2}{\mathfrak{g}}\sin 4\epsilon \sin^3\theta$.

The sum of the moments of these stresses about the axis Z constitutes the tidal frictional couple $\mathfrak{N}$, which retards the earth's rotation.

Therefore

$$\mathfrak{N} = \tfrac{1}{8}wa^2\frac{\tau^2}{\mathfrak{g}}\sin 4\epsilon \iint \sin^3\theta \,.\, a\sin\theta \,.\, a^2\sin\theta\, d\theta\, d\phi$$

integrated all over the surface of the sphere.

On effecting the integration we have $\mathfrak{N} = \frac{4}{15}\pi wa^5 \,.\, \frac{\tau^2}{\mathfrak{g}}\sin 4\epsilon$.

But if C be the earth's moment of inertia, $\mathrm{C} = \frac{8}{15}\pi wa^5$.

Therefore $$\frac{\mathfrak{N}}{\mathrm{C}} = \tfrac{1}{2}\frac{\tau^2}{\mathfrak{g}}\sin 4\epsilon \qquad (5)$$

This expression agrees with that found by a different method in the paper on "Precession*."

We may now write the tangential stress on the surface of the sphere as

* "Precession," Section 5 (22), when $i=0$.

$\frac{1}{4}wa^3 \frac{\mathfrak{N}}{C} \sin^3\theta$; and the components of this stress parallel to the axes X, Y, Z are

$$-\tfrac{1}{4}wa^2 \frac{\mathfrak{N}}{C} \sin^3\theta \sin\phi, \quad +\tfrac{1}{4}wa^2 \frac{\mathfrak{N}}{C} \sin^3\theta \cos\phi, \quad 0 \quad \text{........(6)}$$

We next have to consider those effects of inertia which equilibrate this system of surface forces.

The couple $\mathfrak{N}$ retards the earth's rotation very nearly as though it were a rigid body. Hence the effective force due to inertia on a unit of volume of the interior of the earth at a point r, θ, ϕ is $wr\sin\theta \frac{\mathfrak{N}}{C}$, and it acts in a small circle of latitude from west to east. The sum of the moments of these forces about the axis of Z is of course equal to $\mathfrak{N}$, and therefore this bodily force would equilibrate the surface forces found in (6), if the earth were rigid.

The components of the bodily force parallel to the axes are in rectangular coordinates

$$wy\frac{\mathfrak{N}}{C}, \quad -wx\frac{\mathfrak{N}}{C}, \quad 0 \quad \text{........................(7)}$$

The problem is therefore reduced to that of finding the state of flow in the interior of a viscous sphere, which is subject to a bodily force of which the components are (7) and to the surface stresses of which the components are (6).

Let α, β, γ be the component velocities of flow at the point x, y, z, and υ the coefficient of viscosity. Neglecting inertia because the motion is very slow, the equations of motion are

$$\left.\begin{aligned} -\frac{dp}{dx} + \upsilon\nabla^2\alpha + w\frac{\mathfrak{N}}{C}y &= 0 \\ -\frac{dp}{dy} + \upsilon\nabla^2\beta - w\frac{\mathfrak{N}}{C}x &= 0 \\ -\frac{dp}{dz} + \upsilon\nabla^2\gamma \qquad &= 0 \\ \frac{d\alpha}{dx} + \frac{d\beta}{dy} + \frac{d\gamma}{dz} \qquad &= 0 \end{aligned}\right\} \quad \text{......................(8)}$$

We have to find a solution of these equations, subject to the condition above stated, as to surface stress.

Let $\alpha', \beta', \gamma', p'$ be functions which satisfy the equations (8) throughout the sphere. If we put $\alpha = \alpha' + \alpha_{,}$, $\beta = \beta' + \beta_{,}$, $\gamma = \gamma' + \gamma_{,}$, $p = p' + p_{,}$, we see that to complete the solution we have to find $\alpha_{,}, \beta_{,}, \gamma_{,}, p_{,}$, as determined by the equations

$$\frac{dp_{,}}{dx} + \upsilon\nabla^2\alpha_{,} = 0, \quad \frac{dp_{,}}{dy}\ \&\text{c.} = 0, \quad \frac{dp_{,}}{dz}\ \&\text{c.} = 0, \quad \frac{d\alpha_{,}}{dx} + \frac{d\beta_{,}}{dy} + \frac{d\gamma_{,}}{dz} = 0 \text{...(9)}$$

which they are to satisfy throughout the sphere. They must also satisfy certain equations to be found by subtracting from the given surface stresses (6), components of surface stress to be calculated from $\alpha', \beta', \gamma', p'$*.

We have first to find $\alpha', \beta', \gamma', p'$.

Conceive the symbols in equations (8) to be accented, and differentiate the first three by x, y, z respectively and add them; then bearing in mind the fourth equation, we have $\nabla^2 p' = 0$, of which $p' = 0$ is a solution.

Thus the equations to be satisfied become

$$\nabla^2\alpha' = -\frac{w}{v}\frac{\mathfrak{N}}{C}y, \quad \nabla^2\beta' = \frac{w}{v}\frac{\mathfrak{N}}{C}x, \quad \nabla^2\gamma' = 0$$

Solutions of these are obviously

$$\left.\begin{aligned} \alpha' &= -\tfrac{1}{10}\frac{w}{v}\frac{\mathfrak{N}}{C}r^2 y, & \beta' &= \tfrac{1}{10}\frac{w}{v}\frac{\mathfrak{N}}{C}r^2 x, & \gamma' &= 0 \\ &= -\tfrac{1}{10}\frac{w}{v}\frac{\mathfrak{N}}{C}r^3\sin\theta\sin\phi & &= \tfrac{1}{10}\frac{w}{v}\frac{\mathfrak{N}}{C}r^3\sin\theta\cos\phi & & \end{aligned}\right\} \quad \ldots(10)$$

These values satisfy the last of (8), viz.: the equation of continuity, and therefore together with $p' = 0$, they form the required values of $\alpha', \beta', \gamma', p'$.

We have next to compute the surface stresses corresponding to these values.

Let P, Q, R, S, T, U be the normal and tangential stresses (estimated as is usual in the theory of elastic solids) across three planes at right angles at the point x, y, z.

Then
$$P = -p' + 2v\frac{d\alpha'}{dx}, \quad S = v\left(\frac{d\beta'}{dz} + \frac{d\gamma'}{dy}\right) \quad \ldots\ldots\ldots\ldots(11)$$

Q, R, T, U being found by cyclical changes of symbols.

If F, G, H denote the component stresses across a plane perpendicular to the radius vector r at the point x, y, z

$$\left.\begin{aligned} Fr &= Px + Uy + Tz \\ Gr &= Ux + Qy + Sz \\ Hr &= Tx + Sy + Rz \end{aligned}\right\} \ldots\ldots\ldots\ldots\ldots\ldots\ldots\ldots(12)$$

Substitute in (12) for P, Q, &c., from (11), and put $\zeta' = \alpha' x + \beta' y + \gamma' z$, and $r\dfrac{d}{dr}$ for $x\dfrac{d}{dx} + y\dfrac{d}{dy} + z\dfrac{d}{dz}$. Then

$$Fr = -p'x + v\left\{\left(r\frac{d}{dr} - 1\right)\alpha' + \frac{d\zeta'}{dx}\right\}, \quad Gr = \&c., \quad Hr = \&c. \ldots(13)$$

These formulæ give the stresses across any of the concentric spherical surfaces.

* This statement of method is taken from Thomson and Tait's *Natural Philosophy*, § 733.

In the particular case in hand $p'=0$, $\gamma'=0$, $\zeta'=0$, and α', β' are homogeneous functions of the third degree; hence

$$F=-\tfrac{1}{5}w\frac{\mathfrak{N}}{C}r^2\sin\theta\sin\phi,\quad G=\tfrac{1}{5}w\frac{\mathfrak{N}}{C}r^2\sin\theta\cos\phi,\quad H=0 \ldots (14)$$

and at the surface of the sphere $r=a$.

According to the principles above explained, we have to find $\alpha_{,}$, $\beta_{,}$, $\gamma_{,}$ so that they may satisfy

$$-\frac{dp_{,}}{dx}+\upsilon\nabla^2\alpha_{,}=0, \text{ \&c., \&c.}$$

throughout a sphere, which is subject to surface stresses given by subtracting from (6) the surface values of F, G, H in (14). Hence the surface stresses to be satisfied by $\alpha_{,}$, $\beta_{,}$, $\gamma_{,}$, have components

$$A_3=\tfrac{1}{4}w\frac{\mathfrak{N}}{C}a^2\left(\tfrac{4}{5}-\sin^2\theta\right)\sin\theta\sin\phi$$

$$B_3=-\tfrac{1}{4}w\frac{\mathfrak{N}}{C}a^2\left(\tfrac{4}{5}-\sin^2\theta\right)\sin\theta\cos\phi,\quad C_3=0$$

These are surface harmonics of the third order as they stand.

The solution of Thomson's problem of the state of strain of an incompressible elastic sphere, subject only to surface stress, is applicable to an incompressible viscous sphere, *mutatis mutandis*. His solution* shows that a surface stress, of which the components are A_i, B_i, C_i (surface harmonics of the ith order), gives rise to a state of flow expressed by

$$\alpha=\frac{1}{\upsilon a^{i-1}}\left\{\frac{(a^2-r^2)}{2(2i^2+1)}\frac{d\Psi_{i-1}}{dx}+\frac{1}{i-1}\left[\frac{(i+2)r^{2i+1}}{(2i^2+1)(2i+1)}\frac{d}{dx}\left(\Psi_{i-1}r^{-2i+1}\right)+\frac{1}{2i(2i+1)}\frac{\Phi_{i+1}}{dx}+A_ir^i\right]\right\} \ldots\ldots (15)$$

and symmetrical expressions for β, γ, where Ψ and Φ are auxiliary functions defined by

$$\left.\begin{aligned}\Psi_{i-1}&=\frac{d}{dx}(A_ir^i)+\frac{d}{dy}(B_ir^i)+\frac{d}{dz}(C_ir^i)\\ \Phi_{i+1}&=r^{2i+3}\left\{\frac{d}{dx}(A_ir^{-i-1})+\frac{d}{dy}(B_ir^{-i-1})+\frac{d}{dz}(C_ir^{-i-1})\right\}\end{aligned}\right\} \ldots\ldots (16)$$

In our case $i=3$, and it is easily shown that the auxiliary functions are both zero, so that the required solution is

$$\alpha_{,}=\frac{w}{8\upsilon}\frac{\mathfrak{N}}{C}\left(\tfrac{4}{5}-\sin^2\theta\right)r^3\sin\theta\sin\phi$$

$$\beta_{,}=-\frac{w}{8\upsilon}\frac{\mathfrak{N}}{C}\left(\tfrac{4}{5}-\sin^2\theta\right)\sin\theta\cos\phi,\quad \gamma_{,}=0$$

* Thomson and Tait's *Natural Philosophy*, § 737.

If we add to these the values of α', β', γ' from (10), we have as the complete solution of the problem,

$$\alpha = -\frac{w}{8v}\frac{\mathfrak{P}}{C} r^3 \sin^3\theta \sin\phi, \quad \beta = \frac{w}{8v}\frac{\mathfrak{P}}{C} r^3 \sin^3\theta\cos\phi, \quad \gamma = 0 \text{...(17)}$$

These values show that the motion is simply cylindrical round the earth's axis, each point moving in a small circle of latitude from east to west with a linear velocity $\frac{w}{8v}\frac{\mathfrak{P}}{C} r^3 \sin^3\theta$, or with an angular velocity about the axis equal to $\frac{w}{8v}\frac{\mathfrak{P}}{C} r^2 \sin^2\theta$*.

In this statement a meridian at the pole is the curve of reference, but it is more intelligible to state that each particle moves from west to east with an angular velocity about the axis equal to $\frac{w}{8v}\frac{\mathfrak{P}}{C}(a^2 - r^2\sin^2\theta)$, with reference to a point on the surface at the equator.

The easterly rate of change of the longitude L of any point on the surface in colatitude θ is therefore $\frac{wa^2}{8v}\frac{\mathfrak{P}}{C}\cos^2\theta$.

Since $\frac{\mathfrak{P}}{C} = \frac{\tau^2}{\mathfrak{g}}\sin 2\epsilon\cos 2\epsilon$, and $\tan 2\epsilon = \frac{2}{5}\cdot\frac{19v\omega}{\mathfrak{g}wa^2}$,

$$\frac{dL}{dt} = \frac{19}{20}\left(\frac{\tau}{\mathfrak{g}}\cos 2\epsilon\right)^2 \omega\cos^2\theta \text{ ....(17')}$$

This equation gives the rate of change of longitude. The solution is not applicable to the case of perfect fluidity, because the terms introduced by inertia in the equations of motion have been neglected; and if the viscosity be infinitely small, the inertia terms are no longer small compared with those introduced by viscosity.

In order to find the total change of longitude in a given period, it will be more convenient to proceed from a different formula.

Let n, Ω be the earth's rotation, and the moon's orbital motion at any time; and let the suffix 0 to any symbol denote its initial value, also let $\xi = \left(\frac{\Omega_0}{\Omega}\right)^{\frac{1}{3}}$.

It was shown in the paper on "Precession" that the equation of conservation of moment of momentum of the moon-earth system is

$$\frac{n}{n_0} = 1 + \mu(1-\xi)\dagger \text{ ....(18)}$$

* The problem might probably be solved more shortly without using the general solution, but the general solution will be required in Part III.

† "Precession," equation (73), when $i = 0$ and $\tau' = 0$.

where μ is a certain constant, which in the case of the homogeneous earth with the present lengths of day and month, is almost exactly equal to 4.

By differentiation of (18)

$$\frac{dn}{dt} = -\mu n_0 \frac{d\xi}{dt} \quad \text{......................(19)}$$

But the equation of tidal friction is $\frac{dn}{dt} = -\frac{\mathfrak{N}}{C}$.

Therefore $$\frac{d\xi}{dt} = \frac{1}{\mu} \frac{\mathfrak{N}}{Cn_0}$$

Now $$\frac{dL}{dt} = \frac{wa^2}{8v} \frac{\mathfrak{N}}{C} \cos^2\theta$$

Therefore $$\frac{dL}{d\xi} = \mu n_0 \frac{wa^2}{8v} \cos^2\theta \quad \text{......................(19')}$$

All the quantities on the right-hand side of this equation are constant, and therefore by integration we have for the change of longitude

$$\Delta L = \mu n_0 \frac{wa^2}{8v} (\xi - 1) \cos^2\theta$$

But since $\omega_0 = n_0 - \Omega_0$, and $\tan 2\epsilon_0 = \frac{2}{5} \cdot \frac{19v\omega_0}{\mathfrak{g}wa^2}$, therefore in degrees of arc,

$$\Delta L = \frac{180}{\pi} \mu n_0 \tfrac{19}{20} \frac{n_0 - \Omega_0}{\mathfrak{g}} \cot 2\epsilon_0 (\xi - 1) \cos^2\theta$$

In order to make the numerical results comparable with those in the paper on "Precession," I will apply this to the particular case which was the subject of the first method of integration of that paper*. It was there supposed that $\epsilon_0 = 17° 30'$, and it was shown that looking back about 46 million years ξ had fallen from unity to ·88. Substituting for the various quantities their numerical values, I find that

$$-\Delta L = 0°{\cdot}31 \cos^2\theta = 19' \cos^2\theta$$

Hence looking back 46 million years, we find the longitude of a point in latitude 30°, further west by $4\tfrac{3}{4}'$ than at present, and a point in latitude 60°, further west by $14\tfrac{1}{4}'$—both being referred to a point on the equator.

Such a shift is obviously quite insignificant, but in order to see whether this screwing motion of the earth's mass could have had any influence on the crushing of the surface strata, it will be well to estimate the amount by which a cubic foot of the earth's mass at the surface would have been distorted.

The motion being referred to the pole, it appears from (17) that a point distant ρ from the axis shifts through $\frac{w}{8v} \frac{\mathfrak{N}}{C} \rho^3 \delta t$ in the time δt. There would

* "Precession," Section 15.

be no shearing if a point distant $\rho + \delta\rho$ shifted through $\frac{w}{8\upsilon}\frac{\mathfrak{N}}{C}\rho^2(\rho + \delta\rho)\,\delta t$; but this second point does shift through $\frac{w}{8\upsilon}\frac{\mathfrak{N}}{C}(\rho + \delta\rho)^3\,\delta t$.

Hence the amount of shear in unit time is

$$\frac{1}{\delta\rho} \times \frac{w}{8\upsilon}\frac{\mathfrak{N}}{C}\left[(\rho + \delta\rho)^3 - (\rho + \delta\rho)\,\rho^2\right] = \frac{w}{4\upsilon}\frac{\mathfrak{N}}{C}\rho^2$$

Therefore at the equator, at the surface where the shear is greatest, the shear per unit time is

$$\frac{wa^2}{4\upsilon}\frac{\mathfrak{N}}{C} = \tfrac{19}{10}\left(\frac{\tau}{\mathfrak{g}}\right)^2 \cos^2 2\epsilon\,.\,\omega$$

With the present values of τ and ω, $\tfrac{19}{10}\left(\frac{\tau}{\mathfrak{g}}\right)^2 \omega$ is a shear of $\frac{1{\cdot}84}{10^{10}}$ per annum.

Hence at the equator a slab one foot thick would have one face displaced with reference to the other at the rate of $\tfrac{1}{500}\cos^2 2\epsilon$ of an inch in a million years.

The bearing of these results on the history of the earth will be considered in Part IV.

The next point which will be considered is certain tides of the second order.

We have hitherto supposed that the tides are superposed upon a sphere; it is, however, clear that besides the tidal protuberance there is a permanent equatorial protuberance. Now this permanent protuberance is by hypothesis not rigidly connected with the mean sphere; and, as the attraction of the moon on the equatorial regions produces the uniform precession and the fortnightly nutation, it might be (and indeed has been) supposed that there would arise a shifting of the surface with reference to the interior, and that this change in configuration would cause the earth to rotate round a new axis, and so there would follow a geographical shifting of the poles. I will now show, however, that the only consequence of the non-rigid attachment of the equatorial protuberance to the mean sphere is a series of tides of the second order in magnitude, and of higher orders of harmonics than the second.

For a complete solution of the problem the task before us would be to determine what are the additional tangential and normal stresses existing between the protuberant parts and the mean sphere, and then to find the tides and secular distortion (if any) to which they give rise.

The first part of these operations may be done by the same process which has just been carried out with reference to the secular distortion due to tidal friction.

The additional normal stress (in excess of $-gw\sigma$, the mean weight of an element of the protuberance) can have no part in the precessional and nutational couples, and the remark may be repeated that, that part of it which is non-periodic will only cause a minute change in the mean figure of the spheroid which is negligeable, and the part which is periodic will cause small tides of about the same magnitude as those caused by the tangential stresses. With respect to the tangential stresses, it is *à priori* possible that they may cause a continued distortion of the spheroid, and they will cause certain small tides, whose relative importance we have to estimate.

The expressions for the tangential stresses, which we have found above in (1), are not linear, and therefore we must consider the phenomenon in its entirety, and must not seek to consider the precessional and nutational effects apart from the tidal effects.

The whole bodily potential which acts on the earth is that due to the moon (of which the full expression is given in equation (3) of "Precession"), together with that due to the earth's diurnal rotation (being $\frac{1}{2}n^2r^2(\frac{1}{3} - \cos^2\theta)$); the whole may be called r^2S. The form of the surface σ is that due to the tides and to the non-periodic part of the moon's potential, together with that due to rotation—being $\frac{1}{2}\frac{n^2a}{\mathfrak{g}}(\frac{1}{3} - \cos^2\theta)$.

If we form the effective potential $a^2\left(S - \mathfrak{g}\frac{\sigma}{a}\right)$, which determines the tangential stresses between σ and the mean sphere, we shall find that all except periodic terms disappear. This is so whether we suppose the earth's axis to be oblique or not to the lunar orbit, and also if the sun be supposed to act.

If we differentiate these and form the expressions

$$wa^2\frac{\sigma}{a}\frac{d}{d\theta}\left(S - \mathfrak{g}\frac{\sigma}{a}\right), \quad wa^2\frac{\sigma}{a}\frac{d}{\sin\theta\, d\phi}\left(S - \mathfrak{g}\frac{\sigma}{a}\right)$$

we shall find that there are no non-periodic terms in the expression giving the tangential stress along the meridian; and that the only non-periodic terms which exist in the expression giving the tangential stress perpendicular to the meridian are precisely those whose effects have been already considered as causing secular distortion, and which have their maximum effect when the obliquity is zero.

Hence the whole result must be—

(1) A very minute change in the permanent or average figure of the globe;

(2) The secular distortion already investigated;

(3) Small tides of the second order.

The only question which is of interest is—Can these small tides be of any importance?

The sum of the moments of all the tangential stresses which result from the above expressions, about a pair of axes in the equator, one 90° removed from the moon's meridian and the other in the moon's meridian, together give rise to the precessional and nutational couples.

Hence it follows that part of the tangential stresses form a non-equilibrating system of forces acting on the sphere's surface. In order to find the distorting effects on the globe, we should, therefore, have to equilibrate the system by bodily forces arising from the effects of the inertia due to the uniform precession and the fortnightly nutation—just as was done above with the tidal friction. This would be an exceedingly laborious process; and although it seems certain that the tides thus raised would be very small, yet we are fortunately able to satisfy ourselves of the fact more rigorously. Certain parts of the tangential stresses *do* form an equilibrating system of forces, and these are precisely those parts of the stresses which are the most important, because they do not involve the sine of the obliquity.

I shall therefore evaluate the tangential stresses when the obliquity is zero.

The complete potential due both to the moon and to the diurnal rotation is

$$r^2 S = \tfrac{1}{2} r^2 (n^2 + \tau)(\tfrac{1}{3} - \cos^2\theta) + \tfrac{1}{2} r^2 \tau \sin^2\theta \cos 2(\phi - \omega t - \epsilon)$$

and the complete expression for the surface of the spheroid is given by

$$\mathfrak{g}\frac{\sigma}{a} = \tfrac{1}{2}(n^2 + \tau)(\tfrac{1}{3} - \cos^2\theta) + \tfrac{1}{2}\tau \cos 2\epsilon \sin^2\theta \cos 2(\phi - \omega t)$$

Hence
$$S - \mathfrak{g}\frac{\sigma}{a} = \tfrac{1}{2}\tau \sin 2\epsilon \sin^2\theta \sin 2(\phi - \omega t)$$

Neglecting τ^2 compared with τn^2, and omitting the terms which were previously considered as giving rise to secular distortion, we find

$$wa^2 \frac{\sigma}{a}\frac{d}{d\theta}\left(S - \mathfrak{g}\frac{\sigma}{a}\right) = wa^2 \tau \,\tfrac{1}{2}\frac{n^2}{\mathfrak{g}} \sin 2\epsilon \sin\theta \cos\theta (\tfrac{1}{3} - \cos^2\theta) \sin 2(\phi - \omega t)$$

$$wa^2 \frac{\sigma}{a}\frac{d}{\sin\theta\, d\phi}\left(S - \mathfrak{g}\frac{\sigma}{a}\right) = wa^2 \tau \,\tfrac{1}{2}\frac{n^2}{\mathfrak{g}} \sin 2\epsilon \sin\theta (\tfrac{1}{3} - \cos^2\theta) \cos 2(\phi - \omega t)$$

The former gives the tangential stress along, and the latter perpendicular to, the meridian.

If we put $e = \tfrac{1}{2} n^2/\mathfrak{g}$, the ellipticity of the spheroid, we see that the intensity of the tangential stresses is estimated by the quantity $wa^2 . \tau e \sin 2\epsilon$. But we must now find a standard of comparison, in order to see what height of tide such stresses would be competent to produce.

It appears from a comparison of equations (7) and (8) of Section 2 of the paper on "Tides," that a surface traction S_i (a surface harmonic) everywhere normal to the sphere produces the same state of flow as that caused by a bodily force, whose potential per unit volume is $\left(\frac{r}{a}\right)^i S_i$; and conversely a potential W_i is mechanically equivalent to a surface traction $\left(\frac{a}{r}\right)^i W_i$.

Now the tides of the first order are those due to an effective potential $wr^2\left(S - \mathfrak{g}\frac{\sigma}{a}\right)$, and hence the surface normal traction which is competent to produce the tides of the first order is $wa^2\left(S - \mathfrak{g}\frac{\sigma}{a}\right)$, which is equal to $\frac{1}{2}wa^2\tau \sin 2\epsilon \sin^2\theta \sin 2(\phi - \omega t)$. Hence the intensity of this normal traction is estimated by the quantity $\frac{1}{2}wa^2\tau \sin 2\epsilon$, and this affords a standard of comparison with the quantity $wa^2\tau e \sin 2\epsilon$, which was the estimate of the intensity of the secondary tides. The ratio of the two is $2e$, and since the ellipticity of the mean spheroid is small, the secondary tides must be small compared with the primary ones. It cannot be asserted that the ratio of the heights of the two tides will be $2e$, because the secondary tides are of a higher order of harmonics than the primary, and because the tangential stresses have not been reduced to harmonics and the problem completely worked out. I think it probable that the height of the secondary tides would be considerably less than is expressed by the quantity $2e$, but all that we are concerned to know is that they will be negligeable, and this is established by the preceding calculations.

It follows, then, that the precessional and nutational forces will cause no secular shifting of the surface with reference to the interior, and therefore cannot cause any such geographical deplacement of the poles, as has been sometimes supposed to have taken place.

II. *The distribution of heat generated by internal friction, and secular cooling.*

In the paper on "Precession" (Section 16) the total amount of heat was found, which was generated in the interior of the earth, in the course of its retardation by tidal friction. The investigation was founded on the principle that the energy, both kinetic and potential, of the moon-earth system, which was lost during any period, must reappear as heat in the interior of the earth. This method could of course give no indication of the manner and distribution of the generation of heat in the interior. Now the distribution of heat must have a very important influence on the way it will affect the secular cooling of the earth's mass, and I therefore propose to investigate the subject from a different point of view.

It will be sufficient for the present purpose if we suppose the obliquity of the ecliptic to be zero, and the earth to be tidally distorted by the moon alone.

It has already been explained in the first section how we may neglect the mutual gravitation of a spheroid tidally distorted by an external disturbing potential $wr^2\mathrm{S}$, if we suppose the disturbing potential to be $wr^2\left(\mathrm{S} - \mathfrak{g}\dfrac{\sigma}{a}\right)$, where $r = a + \sigma$ is the equation to the tidal protuberance.

It is shown in (4) that

$$\mathrm{S} - \mathfrak{g}\frac{\sigma}{a} = \tfrac{1}{2}\tau \sin 2\epsilon \sin^2\theta \sin 2(\phi - \omega t)$$

If we refer the motion to rectangular axes rotating so that the axis of x is the major axis of the tidal spheroid, and that of z is the earth's axis of rotation, and if W be the effective disturbing potential estimated per unit volume, we have

$$\mathrm{W} = wr^2\left(\mathrm{S} - \mathfrak{g}\frac{\sigma}{a}\right) = w\tau \sin 2\epsilon \,.\, xy \quad \text{..................(20)}$$

It was also shown in Paper 1 that the solution of Thomson's problem of the state of internal strain of an elastic sphere, devoid of gravitation, as distorted by a bodily force, of which the potential is expressible as a solid harmonic function of the second degree, is identical in form with the solution of the parallel problem for a viscous spheroid.

That solution is as follows:

$$\alpha = \frac{1}{19v}\left[(4a^2 - \tfrac{21}{10}r^2)\frac{d\mathrm{W}}{dx} - \tfrac{2}{5}r^7\frac{d}{dx}\left(\frac{\mathrm{W}}{r^5}\right)\right]^*$$

with symmetrical expressions for β and γ.

Since $\dfrac{d}{dx}\left(\dfrac{\mathrm{W}}{r^5}\right) = \dfrac{1}{r^5}\dfrac{d\mathrm{W}}{dx} - \dfrac{5x}{r^7}\mathrm{W}$, the solution may be written

$$\alpha = \frac{1}{38v}\left[(8a^2 - 5r^2)\frac{d\mathrm{W}}{dx} + 4x\mathrm{W}\right], \quad \beta = \&c., \quad \gamma = \&c.$$

Substituting for W from (20) we have

$$\left.\begin{aligned} \alpha &= \frac{w\tau}{38v}\sin 2\epsilon\,[(8a^2 - 5r^2)\,y + 4x^2y] \\ \beta &= \frac{w\tau}{38v}\sin 2\epsilon\,[(8a^2 - 5r^2)\,x + 4xy^2] \\ \gamma &= \frac{w\tau}{38v}\sin 2\epsilon\, 4xyz \end{aligned}\right\} \quad \text{..............(21)}$$

Putting $\mathrm{K} = \dfrac{w\tau}{19v}\sin 2\epsilon$, we have

$$\left.\begin{aligned} &\frac{d\alpha}{dx} = -\mathrm{K}xy, && \frac{d\alpha}{dy} = \tfrac{1}{2}\mathrm{K}\,[8a^2 - (x^2 + 15y^2 + 5z^2)], && \frac{d\alpha}{dz} = -5\mathrm{K}yz \\ &\frac{d\beta}{dx} = \tfrac{1}{2}\mathrm{K}\,[8a^2 - (15x^2 + y^2 + 5z^2)], && \frac{d\beta}{dy} = -\mathrm{K}xy, && \frac{d\beta}{dz} = -5\mathrm{K}xz \\ &\frac{d\gamma}{dx} = 2\mathrm{K}yz, && \frac{d\gamma}{dy} = 2\mathrm{K}zx, && \frac{d\gamma}{dz} = 2\mathrm{K}xy \end{aligned}\right\} \quad (22)$$

* See Thomson and Tait's *Natural Philosophy*, § 834, or "Tides," Section 3.

And

$$\frac{d\beta}{dz}+\frac{d\gamma}{dy}=-3\mathrm{K}zx,\quad \frac{d\gamma}{dx}+\frac{d\alpha}{dz}=-3\mathrm{K}yz,\quad \frac{d\alpha}{dy}+\frac{d\beta}{dx}=\mathrm{K}[8(a^2-x^2-y^2)-5z^2]$$

$$\text{.........(23)}$$

If P, Q, R, S, T, U be the stresses across three mutually rectangular planes at x, y, z, estimated in the usual way, the work done per unit time on a unit of volume situated at x, y, z is

$$\mathrm{P}\frac{d\alpha}{dx}+\mathrm{Q}\frac{d\beta}{dy}+\mathrm{R}\frac{d\gamma}{dz}+\mathrm{S}\left(\frac{d\beta}{dz}+\frac{d\gamma}{dy}\right)+\mathrm{T}\left(\frac{d\gamma}{dx}+\frac{d\alpha}{dz}\right)+\mathrm{U}\left(\frac{d\alpha}{dy}+\frac{d\beta}{dx}\right)^*$$

But $\mathrm{P}=-p+2v\frac{d\alpha}{dx}$, $\mathrm{S}=v\left(\frac{d\beta}{dz}+\frac{d\gamma}{dy}\right)$, and Q, R, T, U have symmetrical forms. Therefore, substituting in the expression for the work $\left(\text{which will be called } \frac{d\mathrm{E}}{dt}\right)$, and remembering that

$$\frac{d\alpha}{dx}+\frac{d\beta}{dy}+\frac{d\gamma}{dz}=0$$

we have

$$\frac{1}{v}\frac{d\mathrm{E}}{dt}=2\left\{\left(\frac{d\alpha}{dx}\right)^2+\left(\frac{d\beta}{dy}\right)^2+\left(\frac{d\gamma}{dz}\right)^2\right\}+\left(\frac{d\beta}{dz}+\frac{d\gamma}{dy}\right)^2+\left(\frac{d\gamma}{dx}+\frac{d\alpha}{dz}\right)^2+\left(\frac{d\alpha}{dy}+\frac{d\beta}{dx}\right)^2$$

Now from (22)

$$\frac{2}{\mathrm{K}^2}\left[\left(\frac{d\alpha}{dx}\right)^2+\left(\frac{d\beta}{dy}\right)^2+\left(\frac{d\gamma}{dz}\right)^2\right]=12x^2y^2=\tfrac{3}{2}r^4\sin^4\theta\,[1-\cos 4(\phi-\omega t)]\text{...(24)}$$

and from (23)

$$\frac{1}{\mathrm{K}^2}\left[\left(\frac{d\beta}{dz}+\frac{d\gamma}{dy}\right)^2+\left(\frac{d\gamma}{dx}+\frac{d\alpha}{dz}\right)^2+\left(\frac{d\alpha}{dy}+\frac{d\beta}{dx}\right)^2\right]$$

$$=9z^2(x^2+y^2)+[8(a^2-x^2-y^2)-5z^2]^2$$

$$=9r^4\sin^2\theta\cos^2\theta+(8a^2-5r^2-3r^2\sin^2\theta)^2\text{ ......(25)}$$

Adding (24) and (25) and rearranging the terms

$$\frac{1}{\mathrm{K}^2 v}\frac{d\mathrm{E}}{dt}=-\tfrac{3}{2}r^4\sin^4\theta\cos 4(\phi-\omega t)+(8a^2-5r^2)^2$$

$$-\tfrac{3}{2}r^2\sin^2\theta\,[32a^2-r^2(26+\sin^2\theta)]$$

The first of these terms is periodic, going through its cycle of changes in six lunar hours, and therefore the average rate of work, or the average rate of heat generation, is given by

$$\frac{d\mathrm{E}}{dt}=\frac{1}{v}\left(\frac{w\tau}{19}\sin 2\epsilon\right)^2[(8a^2-5r^2)^2-\tfrac{3}{2}r^2\sin^2\theta\,\{32a^2-r^2(26+\sin^2\theta)\}]\text{...(26)}$$

It will now be well to show that this formula leads to the same results as those given in the paper on "Precession."

* Thomson and Tait, *Natural Philosophy*, § 670.

In order to find the whole heat generated per unit time throughout the sphere, we must find the integral $\iiint \frac{dE}{dt} r^2 \sin\theta\, dr\, d\theta\, d\phi$, from $r = a$ to 0, $\theta = \pi$ to 0, $\phi = 2\pi$ to 0.

In a later investigation we shall require a transformation of the expression for $\frac{dE}{dt}$, and as it will here facilitate the integration, it will be more convenient to effect the transformation now.

If Q_2, Q_4 be the zonal harmonics of the second and fourth orders,

$$\cos^2\theta = \tfrac{2}{3}Q_2 + \tfrac{1}{3}$$

$$\cos^4\theta = \tfrac{8}{35}Q_4 + \tfrac{4}{7}Q_2 + \tfrac{1}{5}\text{*}$$

Now

$$(8a^2 - 5r^2)^2 - \tfrac{3}{2}r^2 \sin^2\theta\,[32a^2 - (26 + \sin^2\theta)\,r^2]$$

$$= (8a^2 - 5r^2)^2 - r^2\,[48a^2 - \tfrac{81}{2}r^2 - \tfrac{3}{2}(32a^2 - 28r^2)\cos^2\theta - \tfrac{3}{2}r^2\cos^4\theta]$$

$$= \tfrac{1}{5}\{320a^4 - 560a^2r^2 + 259r^4\} - \tfrac{2}{7}(112a^2 - 95r^2)\,r^2Q_2 + \tfrac{12}{35}r^4Q_4 \ldots\ldots(27)$$

the last transformation being found by substituting for $\cos^2\theta$ and $\cos^4\theta$ in terms of Q_2 and Q_4, and rearranging the terms.

The integrals of Q_2 and Q_4 vanish when taken all round the sphere, and

$$\iiint \tfrac{1}{5}(320a^4 - 560a^2r^2 + 259r^4)\,r^2 \sin\theta\, dr\, d\theta\, d\phi = \tfrac{4}{5}\pi a^7 \{\tfrac{320}{3} - \tfrac{560}{5} + \tfrac{259}{7}\}$$

$$= \frac{Ca^2}{w} \times \tfrac{5}{2} \times 19$$

where C is the earth's moment of inertia, and therefore equal to $\tfrac{8}{15}\pi w a^5$.

Hence we have

$$\iiint \frac{dE}{dt} r^2 \sin\theta\, dr\, d\theta\, d\phi = \frac{w}{v}\left(\frac{\tau}{19}\sin 2\epsilon\right)^2 Ca^2 . \tfrac{5}{2} \times 19 = \frac{5wa^2}{38v}(\tau \sin 2\epsilon)^2\, C$$

But $\tan 2\epsilon = \frac{19v\omega}{gaw} = 2 . \frac{19v\omega}{5\mathfrak{g}wa^2}$, so that $\frac{5wa^2}{38v} = \frac{\omega}{\mathfrak{g}} \cot 2\epsilon$.

Therefore the whole work done on the sphere per unit time is $\tfrac{1}{2}\frac{\tau^2}{\mathfrak{g}} \sin 4\epsilon . C\omega$.

Now, it was shown in equation (5) of Part I. of the present paper that, if $\mathfrak{N}$ be the tidal frictional couple, $\frac{\mathfrak{N}}{C} = \tfrac{1}{2}\frac{\tau^2}{\mathfrak{g}} \sin 4\epsilon$.

Therefore the work done on the sphere per unit time is $\mathfrak{N}\omega$.

It is worth mentioning, in passing, that if the integral be taken from $\tfrac{1}{2}a$ to 0, we find that ·32 of the whole heat is generated within the central eighth of the volume; and by taking the integral from $\tfrac{7}{8}a$ to a, we find that one-tenth of the whole heat is generated within 500 miles of the surface.

* Todhunter's *Functions of Laplace, &c.* p. 13; or any other work on the subject.

It remains to show the identity of this remarkably simple result, for the whole work done on the sphere, with that used in the paper on "Precession." It was there shown (Section 16) that if n be the earth's rotation, r the moon's distance at any time, ν the ratio of the earth's mass to the moon's, then the whole energy both potential and kinetic of the moon-earth system is

$$\tfrac{1}{2}\mathrm{C}\left(n^2 - \frac{5g}{2\nu \mathrm{r}}\right)$$

Now c being the moon's distance initially, since the lunar orbit is supposed to be circular,

$$\Omega_0^2 c^3 = g a^2 \frac{1+\nu}{\nu}$$

Also
$$\frac{c}{\mathrm{r}} = \left(\frac{\Omega}{\Omega_0}\right)^{\frac{2}{3}} = \frac{1}{\xi^2}$$

Therefore
$$\tfrac{2}{5}\frac{c\nu}{g} = \tfrac{2}{5}\left\{\left(\frac{a}{g}\right)^2 \nu^2 (1+\nu)\right\}^{\frac{1}{3}} \Omega_0^{-\frac{2}{3}} = s\Omega_0^{-\frac{2}{3}}$$

according to the notation of the paper on "Precession."

In that paper I also put $\frac{1}{\mu} = sn_0\Omega_0^{\frac{1}{3}}$.

Therefore $\frac{5g}{2\nu \mathrm{r}} = \frac{\mu n_0 \Omega_0}{\xi^2}$.

And the whole energy of the system is $\tfrac{1}{2}\mathrm{C}\left(n^2 - \frac{\mu n_0\Omega_0}{\xi^2}\right)$.

Therefore the rate of loss of energy is $-\mathrm{C}\left(n\frac{dn}{dt} + \frac{\mu n_0}{\xi^3}\Omega_0\frac{d\xi}{dt}\right)$.

But $\frac{dn}{dt} = -\frac{\mathfrak{N}}{\mathrm{C}}$, and as shown in the first part (19), $\mu n_0 \frac{d\xi}{dt} = \frac{\mathfrak{N}}{\mathrm{C}}$, also $\frac{\Omega_0}{\xi^3} = \Omega$.

Therefore the rate of loss of energy is $\mathfrak{N}(n-\Omega)$ or $\mathfrak{N}\omega$, which expression agrees with that obtained above. The two methods therefore lead to the same result.

I will now return to the investigation in hand.

The average throughout the earth of the rate of loss of energy is $\mathfrak{N}\omega \div \tfrac{4}{3}\pi a^3$, which quantity will be called H. Then

$$\mathrm{H} = \frac{3}{4\pi a^3}\mathfrak{N}\omega = \frac{w}{\mathrm{M}} \cdot \tfrac{2}{5}\mathrm{M}a^2 \cdot \tfrac{1}{2}\frac{\tau^2}{\mathfrak{g}}\sin 4\epsilon \cdot \omega = \tfrac{1}{5}wa^2 \cdot \frac{\tau^2}{\mathfrak{g}}\sin 4\epsilon \cdot \omega$$

Now

$$\frac{1}{v}(\tfrac{1}{19}w\tau\sin 2\epsilon)^2 a^4 = \frac{2\omega}{5\mathfrak{g}}\cot 2\epsilon \cdot \tfrac{1}{19}\tau^2\sin^2 2\epsilon \cdot wa^2 = \tfrac{1}{95}wa^2\omega\frac{\tau^2}{\mathfrak{g}}\sin 4\epsilon = \tfrac{1}{19}\mathrm{H}$$

Hence (26) may be written

$$\frac{d\mathrm{E}}{dt} = \tfrac{1}{19}\mathrm{H}\left[\left\{8 - 5\left(\frac{r}{a}\right)^2\right\}^2 - \tfrac{3}{2}\left(\frac{r}{a}\right)^2 \sin^2\theta\left\{32 - (26+\sin^2\theta)\left(\frac{r}{a}\right)^2\right\}\right] \quad \text{...(28)}$$

This expression gives the rate of generation of heat at any point in terms of the average rate, and if we equate it to a constant we get the equation to the family of surfaces of equal heat-generation.

We may observe that the heat generated at the centre is $3\frac{7}{19}$ times the average, at the pole $1/2\frac{1}{9}$ of the average, and at the equator $1/12\frac{2}{3}$ of the average.

The accompanying figure exhibits the curves of equal heat-generation; the dotted line shows that of $\frac{1}{4}$ of the average, and the others those of $\frac{1}{2}$, 1, $1\frac{1}{2}$, 2, $2\frac{1}{2}$, and 3 times the average. It is thus obvious from inspection of the figure that by far the largest part of the heat is generated in the central regions.

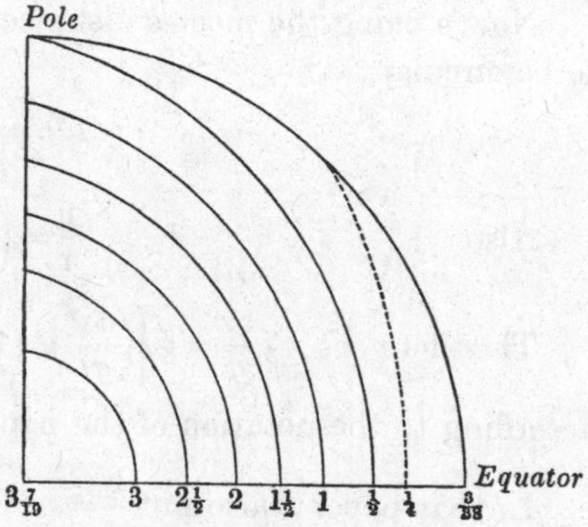

FIG. 2.

The next point to consider is the effect which the generation of heat will have on underground temperature, and how far it may modify the investigation of the secular cooling of the earth.

It has already been shown* that the total amount of heat which might be generated is very large, and my impression was that it might, to a great extent, explain the increase of temperature underground, until a conversation with Sir William Thomson led me to undertake the following calculations:—

We will first calculate in what length of time the earth is losing by cooling an amount of energy equal to its present kinetic energy of rotation.

The earth's conductivity may be taken as about ·004 according to the results given in Everett's illustrations of the centimetre-gramme-second system of units, and the temperature gradient at the surface as 1° C. in $27\frac{1}{2}$ metres, which is the same as 1° Fahr. in 50 feet—the rate used by Sir W. Thomson in his paper on the cooling of the earth†.

This temperature gradient is $\frac{4}{11 \times 10^3}$ degrees C. per centimetre, and since there are 31,557,000 seconds in a year, therefore in centimetre-gramme-second units,

the heat lost by earth per annum

$$= \text{earth's surface in square centimetres} \times \frac{4}{10^3} \times \frac{4}{11 \times 10^3} \times 3{\cdot}1557 \times 10^7$$

$$= \text{earth's surface} \times 45{\cdot}9 \text{ (centimetre-gramme-second heat units)}$$

* "Precession," Section 15, Table IV., and Section 16.

† Thomson and Tait's *Natural Philosophy*, Appendix D.

If J be Joule's equivalent

the earth's kinetic energy of rotation in heat units

$$= \tfrac{1}{2}\frac{Cn_0^2}{gJ} = \frac{Ma}{J}\left(\tfrac{2}{5}\right)^2\left(\tfrac{5}{4}\frac{n_0^2 a}{g}\right), \text{ where } C = \tfrac{2}{5}Ma^2$$

$$= \text{earth's surface} \times \frac{wa^2}{3J}\left(\tfrac{2}{5}\right)^2 e_0, \text{ where } e_0 = \tfrac{5}{4}\frac{n_0^2 a}{g} = \frac{1}{232}$$

$$= \text{earth's surface} \times \frac{(5\cdot5)\times(6\cdot37)^2\times10^{16}\times(\cdot4)^2}{3\times4\cdot34\times10^4\times232}, \quad \text{for } a = 6\cdot37\times10^8 \text{ centimetres}, \; J = 4\cdot34\times10^4 \text{ gram centim. and } w = 5\tfrac{1}{2}$$

$$= \text{earth's surface} \times 1\cdot2\times10^{10} \text{ nearly}$$

Therefore at the present rate of loss the earth is losing energy by cooling equivalent to its kinetic energy of rotation in $\frac{1\cdot2\times10^{10}}{45\cdot9} = 262$ million years.

If we had taken the earth as heterogeneous and $C = \tfrac{1}{3}Ma^2$ we should have found 218 million years.

We will next find how much energy is lost to the moon-earth system in the series of changes investigated in the paper on "Precession."

In that paper (Section 16) it was shown that the whole energy of the system is $\tfrac{1}{5}Ma^2\left(n^2 - \frac{5g}{2\nu r}\right)$, where ν is earth ÷ moon, r moon's distance, n earth's diurnal rotation.

Hence the loss of energy $= \tfrac{1}{5}Ma^2 n_0^2\left[\left(\frac{n}{n_0}\right)^2 - 1 - \frac{5g}{2\nu n_0^2}\left(\frac{1}{r} - \frac{1}{r_0}\right)\right]$, while n passes from n to n_0, and r from r to r_0.

Taking $\nu = 82$, and $\frac{4g}{5n_0^2 a} = 232$,

$$\frac{5g}{2\nu n_0^2} = \frac{25}{8\nu}\left(\frac{4g}{5n_0^2 a}\right)a = \frac{100\times232}{32\times82}a = 8\cdot84a$$

If D be the length of the day, $\frac{n}{n_0} = \frac{D_0}{D}$; and if Π be the moon's distance in earth's radii,

$$\text{the loss of energy} = \left[\left(\frac{D_0}{D}\right)^2 - 1 - 8\cdot84\left(\frac{1}{\Pi} - \frac{1}{\Pi_0}\right)\right] \times \text{earth's present K.E. of rotation}$$

In the paper on "Precession" we showed the system passing from a day of 5 hours 40 minutes*, and a lunar distance of 2·547 earth's radii, to a day of 24 hours, and a lunar distance of 60·4 earth's radii.

Now $24 \div 5\tfrac{2}{3} = 4\cdot23$, and $(2\cdot547)^{-1} - (60\cdot4)^{-1} = \cdot376$.

* A recalculation in the paper on "Precession" gave 5 hours 36 minutes, but I have not thought it worth while to alter this calculation.

Therefore the loss of energy

$$= [(4{\cdot}23)^2 - 1 - {\cdot}376 \times 8{\cdot}84] \times \text{earth's present K.E.}$$

$$= 13{\cdot}57 \times \text{earth's present K.E. of rotation}$$

Hence the whole heat, generated in the earth from first to last, gives a supply of heat, at the present rate of loss, for $13{\cdot}6 \times 262$ million years, or 3,560 million years.

This amount of heat is certainly prodigious, and I found it hard to believe that it should not largely affect the underground temperature. But Sir William Thomson pointed out to me that the distribution of its generation would probably be such as not materially to affect the temperature gradient at the earth's surface; this remarkable prevision on his part has been confirmed by the results of the following problem, which I thought might be taken roughly to represent the state of the case.

Conceive an infinite slab of rock of thickness $2a$ (or 8,000 miles) being part of an infinite mass of rock; suppose that in a unit of volume, distant x from the medial plane, there is generated, per unit time, a quantity of heat equal to $\mathfrak{h}\,[320a^4 - 560a^2x^2 + 259x^4]$; suppose that initially the slab and the whole mass of rock have a uniform temperature V; let the heat begin to be generated according to the above law, and suppose that the two faces of the slab are for ever maintained at the constant temperature V; it is required to find the distribution of temperature within the slab after any time.

This problem roughly represents the true problem to be considered, because if we replace x by the radius vector r, we have the average distribution of internal heat-generation due to friction; also the maintenance of the faces of the slab at a constant temperature represents the rapid cooling of the earth's surface, as explained by Sir William Thomson in his investigation.

If ϑ be temperature, γ thermal capacity, k conductivity, the equation of heat-flow is

$$\gamma \frac{d\vartheta}{dt} = k \frac{d^2\vartheta}{dx^2} + \mathfrak{h}\,[320a^4 - 560a^2x^2 + 259x^4]$$

Let $320\frac{\mathfrak{h}}{k} = 2\text{L}$, $560\frac{\mathfrak{h}}{k} = 12\text{M}$, $259\frac{\mathfrak{h}}{k} = 30\text{N}$, and let the thermometric conductivity $\kappa = \frac{k}{\gamma}$. Then

$$\frac{d\vartheta}{dt} = \kappa \frac{d^2}{dx^2}[\vartheta + \text{L}a^4x^2 - \text{M}a^2x^4 + \text{N}x^6 - \text{R}]$$

Let the constant $\text{R} = (\text{L} - \text{M} + \text{N})\,a^6$, and put

$$\psi = \vartheta + \text{L}a^4x^2 - \text{M}a^2x^4 + \text{N}x^6 - \text{R}$$

$$= \vartheta - \text{L}a^4(a^2 - x^2) + \text{M}a^2(a^4 - x^4) - \text{N}(a^6 - x^6)$$

When $x = \pm a$, $\psi = \vartheta$.

Since L, M, N, R are constants as regards the time,

$$\frac{d\psi}{dt} = \kappa \frac{d^2\psi}{dx^2}$$

and $\psi = \mathrm{V} - \Sigma \mathrm{P} e^{-\kappa q^2 t} \cos qx$ is obviously a solution of this equation.

Now we wish to make $\vartheta = \mathrm{V}$, when $x = \pm a$, for all values of t; since $\psi = \vartheta$ when $x = \pm a$, this condition is clearly satisfied by making

$$q = (2i+1)\frac{\pi}{2a}$$

Hence the solution may be written,

$$\vartheta = \mathrm{V} - [\mathrm{L}a^4x^2 - \mathrm{M}a^2x^4 + \mathrm{N}x^6 - \mathrm{R}] - \sum_0^\infty \mathrm{P}_{2i+1} e^{-\kappa t\left[\frac{(2i+1)\pi}{2a}\right]^2} \cos(2i+1)\frac{\pi x}{2a} \quad (29)$$

and it satisfies all the conditions except that, initially, when $t = 0$, the temperature everywhere should be V. This last condition is satisfied if

$$\sum_0^\infty \mathrm{P}_{2i+1} \cos(2i+1)\frac{\pi x}{2a} = \mathrm{R} - \mathrm{L}a^4x^2 + \mathrm{M}a^2x^4 - \mathrm{N}a^6$$

for all values between $x = \pm a$.

The expression on the right must therefore be expanded by Fourier's Theorem; but we need only consider the range from $x = a$ to 0, because the rest, from $x = 0$ to $-a$, will follow of its own accord.

Let $\chi = \frac{\pi x}{2a}$; let ϖ be written for $\frac{1}{2}\pi$; let $\mathrm{M}' = \frac{\mathrm{M}}{\varpi^2}$, $\mathrm{N}' = \frac{\mathrm{N}}{\varpi^4}$ and $\mathrm{R}' = \mathrm{R}\frac{\varpi^2}{a^6}$.

Then

$$\mathrm{R} - \mathrm{L}a^4x^2 + \mathrm{M}a^2x^4 - \mathrm{N}x^6 = \frac{a^6}{\varpi^2}[\mathrm{R}' - \mathrm{L}\chi^2 + \mathrm{M}'\chi^4 - \mathrm{N}'\chi^6]$$

and this has to be equal to $\sum_0^\infty \mathrm{P}_{2i+1} \cos(2i+1)\chi$ from $\chi = \frac{1}{2}\pi$ to 0.

Since $$\int_0^{\frac{1}{2}\pi} \cos(2i+1)\chi \cos(2j+1)\chi \, d\chi = 0 \text{ unless } j = i$$

and $$\int_0^{\frac{1}{2}\pi} \cos^2(2i+1)\chi \, d\chi = \tfrac{1}{4}\pi = \tfrac{1}{2}\varpi$$

therefore

$$\tfrac{1}{2}\varpi \mathrm{P}_{2i+1} = \frac{a^6}{\varpi^2}\int_0^{\frac{1}{2}\pi} [\mathrm{R}' - \mathrm{L}\chi^2 + \mathrm{M}'\chi^4 - \mathrm{N}'\chi^6] \cos(2i+1)\chi \, d\chi$$

Now

$$\int_0^{\frac{1}{2}\pi} \chi^{2j} \cos(2i+1)\chi \, d\chi = \frac{1}{2i+1}\left[\chi^{2j}\sin(2i+1)\chi + \frac{\frac{d\chi^{2j}}{d\chi}}{2i+1}\cos(2i+1)\chi \right.$$
$$\left. - \frac{\frac{d^2\chi^{2j}}{d\chi^2}}{(2i+1)^2}\sin(2i+1)\chi - \frac{\frac{d^3\chi^{2j}}{d\chi^3}}{(2i+1)^3}\cos(2i+1)\chi + \&c.\right]_0^{\frac{1}{2}\pi}$$
$$= \frac{(-)^i}{2i+1}\left[1 - \frac{\frac{d^2}{d\varpi^2}}{(2i+1)^2} + \frac{\frac{d^4}{d\varpi^4}}{(2i+1)^4} - \&c.\right]\varpi^{2j}$$

If therefore $f(\chi)$ be a function of χ involving only even powers of χ,

$$\int_0^{\frac{1}{2}\pi} f(\chi)\cos(2i+1)\chi d\chi = \frac{(-)^i}{(2i+1)}\left[1+\left(\frac{1}{2i+1}\frac{d}{d\varpi}\right)^2\right]^{-1} f(\varpi)$$

This theorem will make the calculation of the coefficients very easy, for we have at once

$$\begin{aligned}\frac{\varpi^3}{2a^6}\mathrm{P}_{2i+1} = \frac{(-)^i}{2i+1}\Big\{&\mathrm{R}' - \mathrm{L}\varpi^2 + \mathrm{M}'\varpi^4 - \mathrm{N}'\varpi^6 \\ &- \frac{1}{(2i+1)^2}[-2\mathrm{L} + 4.3\mathrm{M}'\varpi^2 - 6.5\mathrm{N}'\varpi^4] \\ &+ \frac{1}{(2i+1)^4}[4.3.2.1\,\mathrm{M}' - 6.5.4.3\mathrm{N}'\varpi^2] \\ &- \frac{1}{(2i+1)^6}[-6.5.4.3.2.1\mathrm{N}']\Big\}\end{aligned}$$

Substituting for R′, L, M′, N′ their values in terms of $\mathfrak{h}/k$ we find

$$\mathrm{P}_{2i+1} = \frac{(-)^i.2a^6}{(2i+1)^3\varpi^3}\frac{\mathfrak{h}}{k}\left[19 - \frac{1988}{(2i+1)^2\varpi^2} + \frac{6216}{(2i+1)^4\varpi^4}\right]$$

Putting for ϖ its value, viz.: $\frac{1}{2}$ of 3·14159, and i successively equal to 0, 1, 2, it will be found that

$$\mathrm{P}_1 = \frac{\mathfrak{h}a^6}{k}(120{\cdot}907), \quad \mathrm{P}_3 = \frac{\mathfrak{h}a^6}{k}(1{\cdot}107), \quad \mathrm{P}_5 = -\frac{\mathfrak{h}a^6}{k}({\cdot}048)$$

Thus the Fourier expansion is

$$120{\cdot}907\cos\frac{\pi x}{2a} + 1{\cdot}107\cos\frac{3\pi x}{2a} - {\cdot}048\cos\frac{5\pi x}{2a}$$

which will be found to differ by not so much as one per cent. from the function

$$\tfrac{320}{2}\left(1-\frac{x^2}{a^2}\right) - \tfrac{560}{12}\left(1-\frac{x^4}{a^4}\right) + \tfrac{259}{30}\left(1-\frac{x^6}{a^6}\right)$$

to which it should be equal.

By substitution in (29) we have, therefore, as the complete solution of the problem satisfying all the conditions

$$\begin{aligned}\vartheta = \mathrm{V} + \frac{\mathfrak{h}a^6}{k}\Big\{&\left(1 - e^{-\kappa\left(\frac{\pi}{2a}\right)^2 t}\right)120{\cdot}907\cos\frac{\pi x}{2a} \\ &+ \left(1 - e^{-\kappa\left(\frac{3\pi}{2a}\right)^2 t}\right)1{\cdot}107\cos\frac{3\pi x}{2a} - \left(1 - e^{-\kappa\left(\frac{5\pi}{2a}\right)^2 t}\right){\cdot}048\cos\frac{5\pi x}{2a}\Big\}\end{aligned}$$

The only quantity, which it is of interest to determine, is the temperature gradient at the surface, which is equal to $-\dfrac{d\vartheta}{dx}$ when $x = \pm a$.

When $x = \pm a$,

$$\frac{d\vartheta}{dx} = \frac{\mathfrak{h}a^5}{k}\frac{\pi}{2}\left\{120{\cdot}907\left(1 - e^{-\kappa\left(\frac{\pi}{2a}\right)^2 t}\right) - 3{\cdot}321\left(1 - e^{-\kappa\left(\frac{3\pi}{2a}\right)^2 t}\right) - {\cdot}240\left(1 - e^{-\kappa\left(\frac{5\pi}{2a}\right)^2 t}\right)\right\}$$

If t be not so large but that $\kappa(\frac{5}{2}\pi/a)^2 t$ is a small fraction, we have approximately

$$-\frac{d\vartheta}{dx} = \frac{\mathfrak{h}a^5}{k}\frac{\pi}{2}\kappa\left(\frac{\pi}{2a}\right)^2 t\,\{120{\cdot}907 - 9 \times 3{\cdot}321 - 25 \times {\cdot}240\}$$

and since $\dfrac{\kappa}{k} = \dfrac{1}{\gamma}$

$$-\frac{d\vartheta}{dx} = (\tfrac{1}{2}\pi a)^3 \frac{\mathfrak{h}}{\gamma} t \times (85)$$

This formula will give the temperature gradient at the surface when a proper value is assigned to $\mathfrak{h}$, and if t be not taken too large.

With respect to the value of t, Sir W. Thomson took $\kappa = 400$ in British units, the year being the unit of time; and $a = 21 \times 10^6$ feet.

Hence $$\kappa\left(\frac{\pi}{2a}\right)^2 = 4 \times 10^2\left(\frac{1{\cdot}5}{2{\cdot}1 \times 10^7}\right)^2 = \frac{2}{10^{12}} \text{ nearly}$$

and $\kappa(\frac{5}{2}\pi/a)^2 = 5 \times 10^{-11}$; if therefore t be 10^9 years, this fraction is $\frac{1}{20}$. Therefore the solution given above will hold provided the time t does not exceed 1,000 million years.

We next have to consider what is the proper value to assign to $\mathfrak{h}$.

By (27) and (28) it appears that $\mathfrak{h}a^4$ is $1/(5 \times 19)$ of the average heat generated throughout the whole earth, which we called H. Suppose that p times the present kinetic energy of the earth's rotation is destroyed by friction in a time T, and suppose the generation of heat to be uniform in time, then the average heat generated throughout the whole earth per unit time is

$$\frac{p}{g\mathrm{JT}} \cdot \tfrac{1}{5}\mathrm{M}a^2 n_0^2 \div \text{earth's volume}$$

Therefore $$\mathrm{H} = \frac{p}{5\mathrm{JT}} \cdot \frac{wa^2 n_0^2}{g} = \frac{4}{25}\frac{p}{\mathrm{JT}} wae_0$$

where e_0 is the ellipticity of figure of the homogeneous earth and is equal to $\frac{5}{4}n_0^2 a/g$, which I take as equal to $\frac{1}{232}$.

Hence $$\mathfrak{h}a^4 = \frac{16}{9500}\frac{p}{\mathrm{JT}} wae_0$$

and $$-\frac{d\vartheta}{dx} = \frac{16 \times 85}{9500}(\tfrac{1}{2}\pi)^3 \frac{w}{\gamma}\frac{pe_0}{\mathrm{J}}\frac{t}{\mathrm{T}}$$

But $\gamma = sw$, where s is specific heat.

Therefore $$-\frac{d\vartheta}{dx} = \frac{170\pi^3}{9500} \frac{pe_0}{s} \frac{1}{J} \frac{t}{T}$$

The dimensions of J are those of work (in gravitation units) per mass and per scale of temperature, that is to say, length per scale of temperature; p, e_0, and s have no dimensions, and therefore this expression is of proper dimensions.

Suppose the solution to run for the whole time embraced by the changes considered in "Precession," then $t = T$, and as we have shown $p = 13{\cdot}57$. Suppose the specific heat to be that of iron, viz.: $\frac{1}{9}$. Taking $J = 772$, so that the result will be given in degrees Fahrenheit per foot, we have

$$-\frac{d\vartheta}{dx} = \frac{17\pi^3}{950} \times \frac{13{\cdot}57 \times 9}{232 \times 772}$$
$$= \frac{1}{2650}$$

That is to say, at the end of the changes the temperature gradient would be 1° Fahr. per 2,650 feet, provided the whole operation did not take more than 1,000 million years.

It might, however, be thought that if the tidal friction were to operate very slowly, so that the whole series of changes from the day of 5 hours 36 minutes to that of 24 hours occupied much more than 1,000 million years, then the large amount of heat which is generated deep down would have time to leak out, so that finally the temperature gradient would be steeper than that just found. But this is not the case.

Consider only the first, and by far the most important, term of the expression for the temperature gradient. It has the form $\mathfrak{h}(1 - e^{-pT})$, when $t = T$ at the end of the series of changes. Now $\mathfrak{h}$ varies as T^{-1}, and $(1 - e^{-pT})/pT$ has its maximum value unity when $T = 0$. Hence, however slowly the tidal friction operates, the temperature gradient can never be greater than if the heat were all generated instantaneously; but the temperature gradient at the end of the changes is not sensibly less than it would be if all the heat were generated instantaneously, provided the series of changes do not occupy more than 1,000 million years*.

* [The conclusion reached in this section might be different if the earth were to consist of a rigid nucleus covered by a thick or thin stratum of viscous material.]

III. *The forced oscillations of viscous, fluid, and elastic spheroids.*

In investigating the tides of a viscous spheroid, the effects of inertia were neglected, and it was shown that the neglect could not have an important influence on the results*. I shall here obtain an approximate solution of the problem including the effects of inertia; that solution will easily lead to a parallel one for the case of an elastic sphere, and a comparison with the forced oscillations of a fluid spheroid will prove instructive as to the nature of the approximation.

If W be the potential of the impressed forces, estimated per unit volume of the viscous body, the equations of flow, with the same notation as before, are

$$\left.\begin{aligned} &-\frac{dp}{dx}+\upsilon\nabla^2\alpha+\frac{dW}{dx}-w\left(\frac{d\alpha}{dt}+\alpha\frac{d\alpha}{dx}+\beta\frac{d\alpha}{dy}+\gamma\frac{d\alpha}{dz}\right)=0 \\ &-\frac{dp}{dy}+\&c.=0, \qquad -\frac{dp}{dz}+\&c.=0 \\ &\frac{d\alpha}{dx}+\frac{d\beta}{dy}+\frac{d\gamma}{dz}=0 \end{aligned}\right\}\text{......(30)}$$

The terms $-w\left(\frac{d\alpha}{dt}+\&c.\right)$ are those due to inertia, which were neglected in the paper on "Tides."

It will be supposed that the tidal motion is steady, and that W consists of a series of solid harmonics each multiplied by a simple time-harmonic, also that W includes not only the potential of the external tide-generating body, but also the effective potential due to gravitation, as explained in the first part of this paper.

The tidal disturbance is supposed to be sufficiently slow to enable us to obtain a first approximation by the neglect of the terms due to inertia.

In proceeding to the second approximation, the inertia terms depending on the squares and products of the velocities, that is to say,

$$w\left(\alpha\frac{d\alpha}{dx}+\beta\frac{d\alpha}{dy}+\gamma\frac{d\alpha}{dz}\right)$$

may be neglected compared with $w\dfrac{d\alpha}{dt}$. A typical case will be considered in which $W=Y\cos(vt+\epsilon)$, where Y is a solid harmonic of the ith degree, and the ϵ will be omitted throughout the analysis for brevity.

If we write $I=2(i+1)^2+1$, the first approximation, when the inertia terms are neglected, is

$$\alpha=\frac{1}{I\upsilon}\left\{\left[\frac{i(i+2)}{2(i-1)}a^2-\frac{(i+1)(2i+3)}{2(2i+1)}r^2\right]\frac{dY}{dx}-\frac{i}{2i+1}r^{2i+3}\frac{d}{dx}(r^{-2i-1}Y)\right\}\cos vt\dagger \quad\text{........(31)}$$

* "Tides," Section 10.

† "Tides," Section 3, equation (8), or Thomson and Tait, *Natural Philosophy*, § 834 (8).

Hence for the second approximation we must put

$$-w\frac{d\alpha}{dt}=\frac{wv}{\mathrm{I}v}\left\{\ldots\right\}\sin vt$$

And the equations to be solved are

$$\left.\begin{aligned}-\frac{dp}{dx}+v\nabla^2\alpha+\frac{d\mathrm{Y}'}{dx}\cos vt+\frac{wv}{\mathrm{I}v}\left\{\left[\frac{i\,(i+2)}{2\,(i-1)}a^2-\frac{(i+1)\,(2i+3)}{2\,(2i+1)}r^2\right]\frac{d\mathrm{Y}}{dx}\right.&\\ \left.-\frac{i}{2i+1}r^{2i+3}\frac{d}{dx}(r^{-2i-1}\,\mathrm{Y})\right\}\sin vt=0&\\ -\frac{dp}{dy}+\&\text{c.}=0,\qquad -\frac{dp}{dz}+\&\text{c.}=0&\end{aligned}\right\}\quad\text{........(32)}$$

These equations are to be satisfied throughout a sphere subject to no surface stress. It will be observed that in the term due directly to the impressed forces, we write Y′ instead of Y; this is because the effective potential due to gravitation will be different in the second approximation from what it was in the first, on account of the different form which must now be attributed to the tidal protuberance.

The problem is now reduced to one strictly analogous to that solved in the paper on "Tides"; for we may suppose that the terms introduced by $w\dfrac{d\alpha}{dt}$, &c., are components of bodily force acting on the viscous spheroid, and that inertia is neglected.

The equations being linear, we consider the effects of the several terms separately, and indicate the partial values of α, β, γ, p by suffixes and accents.

First, then, we have

$$-\frac{dp_0}{dx}+v\nabla^2\alpha_0+\frac{d\mathrm{Y}'}{dx}\cos vt=0,\ \&\text{c.},\ \&\text{c.}$$

The solution of this has the same form as in the first approximation, viz.: equation (31), with α_0 written for α, and Y′ for Y.

We shall have occasion hereafter to use the velocity of flow resolved along the radius vector, which may be called ρ. Then

$$\rho_0=\alpha_0\frac{x}{r}+\beta_0\frac{y}{r}+\gamma_0\frac{z}{r}$$

Hence
$$\rho_0=\frac{1}{\mathrm{I}v}\left\{\frac{i^2\,(i+2)\,a^2-i\,(i^2-1)\,r^2}{2\,(i-1)}\right\}\frac{\mathrm{Y}'}{r}\cos vt\quad\text{.............(33)}$$

Observing that $\mathrm{Y}'\div r^i$ is independent of r, we have as the surface value

$$\rho_0=\frac{a^{i+1}}{\mathrm{I}v}\,\frac{i\,(2i+1)}{2\,(i-1)}\,\frac{\mathrm{Y}'}{r^i}\cos vt\quad\text{.....................(34)}$$

Secondly,

$$-\frac{dp_0'}{dx}+\nu\nabla^2\alpha_0'+\frac{wva^2}{I\upsilon}\frac{i(i+2)}{2(i-1)}\frac{dY}{dx}\sin vt=0, \text{ \&c., \&c.} \quad \ldots\ldots(35)$$

This, again, may clearly be solved in the same way, and we have

$$\alpha_0'=\frac{wva^2}{I^2\upsilon^2}\cdot\frac{i(i+2)}{2(i-1)}\left\{\left[\frac{i(i+2)}{2(i-1)}a^2-\frac{(i+1)(2i+3)}{2(2i+1)}r^2\right]\frac{dY}{dx}\right.$$
$$\left.-\frac{i}{2i+1}r^{2i+3}\frac{d}{dx}(Yr^{-2i-1})\right\}\sin vt \quad \ldots\ldots\ldots(36)$$

and $$\rho_0'=\frac{wva^2}{I^2\upsilon^2}\cdot\frac{i(i+2)}{2(i-1)}\left\{\frac{i^2(i+2)a^2-i(i^2-1)r^2}{2(i-1)}\right\}\frac{Y}{r}\sin vt \quad \ldots\ldots(37)$$

Its surface value is

$$\rho_0'=wva^{i+3}.\frac{i^2(i+2)(2i+1)}{[2I\upsilon(i-1)]^2}\frac{Y}{r^i}\sin vt \quad \ldots\ldots\ldots\ldots\ldots(38)$$

Thirdly, let $$U=\frac{wv}{I\upsilon}\frac{Y}{2(2i+1)}\sin vt \quad \ldots\ldots\ldots\ldots\ldots\ldots\ldots(39)$$

So that U is a solid harmonic of the *i*th degree multiplied by a simple time-harmonic. Then the rest of the terms to be satisfied are given in the following equations:

$$\left.\begin{aligned}&-\frac{dp}{dx}+\nu\nabla^2\alpha-\left[(i+1)(2i+3)r^2\frac{dU}{dx}+2ir^{2i+3}\frac{d}{dx}(Ur^{-2i-1})\right]=0\\&-\frac{dp}{dy}+\text{\&c.}=0, \qquad -\frac{dp}{dz}+\text{\&c.}=0\end{aligned}\right\}\ldots(40)$$

These equations have to be satisfied throughout a sphere subject to no surface stresses. The procedure will be exactly that explained in Part I., viz., put $\alpha=\alpha'+\alpha_{,}$, $\beta=\beta'+\beta_{,}$, $\gamma=\gamma'+\gamma_{,}$, $p=p'+p_{,}$, and find α', β', γ', p' any functions which satisfy the equations (40) throughout the sphere.

Differentiate the three equations (40) as to x, y, z respectively and add them together, and notice that

$$(i+1)(2i+3)\left\{\frac{d}{dx}\left(r^2\frac{dU}{dx}\right)+\frac{d}{dy}\left(\ \right)+\frac{d}{dz}\left(\ \right)\right\}$$
$$+2i\left\{\frac{d}{dx}\left(r^{2i+3}\frac{d}{dx}(Ur^{-2i-1})\right)+\frac{d}{dy}\left(\ \right)+\frac{d}{dz}\left(\ \right)\right\}=0$$

and that $$\frac{d\alpha'}{dx}+\frac{d\beta'}{dy}+\frac{d\gamma'}{dz}=0$$

Then we have $\nabla^2p'=0$, of which $p'=0$ is a solution.

If V_n be a solid harmonic of degree n,

$$\nabla^2r^mV_n=m(2n+m+1)r^{m-2}V_n$$

Hence $$\left.\begin{aligned}r^2\frac{dU}{dx}&=\nabla^2\frac{r^4}{4(2i+3)}\frac{dU}{dx}\\r^{2i+3}\frac{d}{dx}(Ur^{-2i-1})&=\nabla^2\frac{r^{2i+5}}{2(2i+5)}\frac{d}{dx}(Ur^{-2i-1})\end{aligned}\right\}\ldots\ldots\ldots\ldots(41)$$

Substituting from (41) in the equations of motion (40), and putting $p'=0$, our equations become

$$\left.\begin{aligned}&\nabla^2\left\{\upsilon\alpha'-\frac{(i+1)}{4}r^4\frac{dU}{dx}-\frac{i}{2i+5}r^{2i+5}\frac{d}{dx}(Ur^{-2i-1})\right\}=0\\&\nabla^2\{\upsilon\beta'-\&c.\}=0,\quad\nabla^2\{\upsilon\gamma'-\&c.\}=0\end{aligned}\right\}\text{......(42)}$$

of which a solution is obviously

$$\left.\begin{aligned}&\alpha'=\frac{1}{\upsilon}\left\{\frac{i+1}{4}r^4\frac{dU}{dx}+\frac{i}{2i+5}r^{2i+5}\frac{d}{dx}(Ur^{-2i-1})\right\}\\&\beta'=\&c.,\quad\gamma'=\&c.\end{aligned}\right\}\text{.........(43)}$$

It may easily be shown that these values satisfy the equation of continuity, and thus together with $p'=0$ they are the required values of α', β', γ', p', which satisfy the equations throughout the sphere.

The next step is to find the surface stresses to which these values give rise. The formulæ (13) of Part I. are applicable

$$\begin{aligned}\upsilon\zeta'&=\upsilon(\alpha'x+\beta'y+\gamma'z)\\&=\frac{i(i+1)}{4}r^4U-\frac{i(i+1)}{2i+5}r^4U=\frac{i(i+1)(2i+1)}{4(2i+5)}r^4U\end{aligned}$$

Remembering that

$$xU=\frac{1}{2i+1}\left\{r^2\frac{dU}{dx}-r^{2i+3}\frac{d}{dx}(r^{-2i-1}U)\right\}$$

we have

$$\begin{aligned}\upsilon\frac{d\zeta'}{dx}&=\frac{i(i+1)(2i+1)}{4(2i+5)}\left\{r^4\frac{dU}{dx}+\frac{4r^2}{2i+1}r^2\frac{dU}{dx}-\frac{4r^2}{2i+1}r^{2i+3}\frac{d}{dx}(r^{-2i-1}U)\right\}\\&=\frac{i(i+1)}{4(2i+5)}\left\{(2i+5)r^4\frac{dU}{dx}-4r^{2i+5}\frac{d}{dx}(r^{-2i-1}U)\right\}\text{..................(44)}\end{aligned}$$

Again, by the properties of homogeneous functions,

$$\begin{aligned}\upsilon\left(r\frac{d}{dr}-1\right)\alpha'&=\upsilon\left(x\frac{d}{dx}+y\frac{d}{dy}+z\frac{d}{dz}\right)\alpha'-\upsilon\alpha'\\&=\tfrac{1}{4}(i+1)(i+2)r^4\frac{dU}{dx}+\frac{i(i+2)}{2i+5}r^{2i+5}\frac{d}{dx}(r^{-2i-1}U)\text{...(45)}\end{aligned}$$

Also $p'=0$.

Adding (44) and (45) together, we have for the component of stress parallel to the axis of x across any of the concentric spherical surfaces,

$$\begin{aligned}Fr&=-p'x+\upsilon\left[\left(r\frac{d}{dr}-1\right)\alpha'+\frac{d\zeta'}{dx}\right]\text{ by (13), Part I.}\\&=\frac{(i+1)^2}{2}r^4\frac{dU}{dx}+\frac{i}{2i+5}r^{2i+5}\frac{d}{dx}(r^{-2i-1}U)\text{ by (44) and (45)}\end{aligned}$$

And at the surface of the sphere, where $r=a$,

$$F=\frac{(i+1)^2}{2}a^{i+2}\left[r^{-i+1}\frac{dU}{dx}\right]+\frac{i}{2i+5}a^{i+2}\left[r^{i+2}\frac{d}{dx}(r^{-2i-1}U)\right]\text{...(46)}$$

The quantities in square brackets are independent of r, and are surface harmonics of orders $i-1$ and $i+1$ respectively.

Let
$$\left.\begin{aligned} F &= -A_{i-1} - A_{i+1} \\ \text{where}\quad A_{i-1} &= -\frac{(i+1)^2}{2}a^{i+2}\left[r^{-i+1}\frac{dU}{dx}\right] \\ A_{i+1} &= -\frac{i}{2i+5}a^{i+2}\left[r^{i+2}\frac{d}{dx}(r^{-2i-1}U)\right] \end{aligned}\right\} \quad\text{.........(47)}$$

Also let the other two components G and H of the surface stress due to α', β', γ', p' be given by

$$G = -B_{i-1} - B_{i+1}, \qquad H = -C_{i-1} - C_{i+1} \quad\text{..............(47)}$$

By symmetry it is clear that the B's and C's only differ from the A's in having y and z in place of x.

We now have got in (43) values of α', β', γ', which satisfy the equations (40) throughout the sphere, together with the surface stresses in (47) to which they correspond. Thus (43) would be the solution of the problem, if the surface of the sphere were subject to the surface stresses (47). It only remains to find $\alpha_{,}$, $\beta_{,}$, $\gamma_{,}$, to satisfy the equations

$$-\frac{dp_{,}}{dx} + v\nabla^2\alpha_{,} = 0, \quad -\frac{dp_{,}}{dy} + \&c. = 0, \quad -\frac{dp_{,}}{dz} + \&c. = 0 \quad\text{......(48)}$$

throughout the sphere, which is not under the influence of bodily force, but is subject to surface stresses of which $A_{i-1} + A_{i+1}$, $B_{i-1} + B_{i+1}$, $C_{i-1} + C_{i+1}$ are the components.

The sum of the solution of these equations and of the solutions (43) will clearly be the complete solution; for (43) satisfies the condition as to the bodily force in (40), and the two sets of surface actions will annul one another, leaving no surface action.

For the required solutions of (48), Thomson's solution given in (15) and (16) of Part I. is at once applicable.

We have first to find the auxiliary functions Ψ_{i-2}, Φ_i corresponding to A_{i-1}, B_{i-1}, C_{i-1}, and Ψ_i, Φ_{i+2} corresponding to A_{i+1}, B_{i+1}, C_{i+1}. It is easy to show that

$$\Psi_{i-2} = 0, \quad \Phi_{i+2} = 0$$

and
$$\Psi_i = -a^{i+2}\frac{i}{2i+5}\left[\frac{d}{dx}\left\{r^{2i+3}\frac{d}{dx}(r^{-2i-1}U)\right\} + \frac{d}{dy}\left\{\ \right\} + \frac{d}{dz}\left\{\ \right\}\right]$$
$$= a^{i+2}\frac{i(i+1)(2i+3)}{2i+5}U$$
$$\Phi_i = -\tfrac{1}{2}r^{2i+1}a^{i+2}(i+1)^2\left[\frac{d}{dx}\left(r^{-2i+1}\frac{dU}{dx}\right) + \frac{d}{dy}\left(\ \right) + \frac{d}{dz}\left(\ \right)\right]$$
$$= \tfrac{1}{2}a^{i+2}i(i+1)^2(2i-1)U$$

We have next to substitute these values of the auxiliary functions in Thomson's solution (15), Part I. It will be simpler to perform the substitutions piece-meal, and to indicate the various parts which go to make up the complete value of $\alpha_{,}$ by accents to that symbol.

First. For the terms in $\alpha_{,}$ depending on A_{i-1}, Ψ_{i-2}, Φ_i, we have

$$\alpha_{,}' = \frac{1}{va^{i-2}}\left\{\frac{1}{2(i-2)(i-1)(2i-1)}\frac{d\Phi_i}{dx} + \frac{1}{i-2}A_{i-1}r^{i-1}\right\}$$

$$= \frac{a^4}{v}\left\{\frac{i(i+1)^2}{4(i-1)(i-2)}\frac{dU}{dx} - \frac{(i+1)^2}{2(i-2)}\frac{dU}{dx}\right\}$$

$$= -\frac{a^4}{v}\frac{(i+1)^2}{4(i-1)}\frac{dU}{dx} \quad\text{......(49)}$$

(Note that $i-2$ divides out, so that the solution is still applicable when $i=2$.)

Second. In finding the terms dependent on A_{i+1}, Ψ_i, Φ_{i+2} it will be better to subdivide the process further.

$$\text{(i)}\quad \alpha_{,}'' = \frac{1}{va^i}\frac{1}{2I}(a^2-r^2)\frac{d\Psi_i}{dx}$$

$$= \frac{a^2}{v}\frac{i(i+1)(2i+3)}{2I(2i+5)}(a^2-r^2)\frac{dU}{dx} \quad\text{......(50)}$$

$$\text{(ii)}\quad \alpha_{,}''' = \frac{1}{va^i}\left\{\frac{i+3}{Ii(2i+3)}r^{2i+3}\frac{d}{dx}(r^{-2i-1}\Psi_i) + \frac{1}{i}A_{i+1}r^{i+1}\right\}$$

$$= \frac{a^2}{v}\left\{\frac{(i+1)(i+3)}{I(2i+5)}r^{2i+3}\frac{d}{dx}(r^{-2i-1}U) - \frac{1}{2i+5}r^{2i+3}\frac{d}{dx}(r^{-2i-1}U)\right\}$$

Since $\qquad (i+3)(i+1) - I = i^2 + 4i + 3 - 2i^2 - 4i - 3 = -i^2$

therefore

$$\alpha_{,}''' = -\frac{a^2}{v}\frac{i^2}{I(2i+5)}r^{2i+3}\frac{d}{dx}(r^{-2i-1}U) \quad\text{......(51)}$$

This completes the solution for $\alpha_{,}$.

Collecting results from (49), (50), and (51), we have

$$\alpha_{,} = \alpha_{,}' + \alpha_{,}'' + \alpha_{,}''' = -\frac{a^2}{v}\left\{\frac{(i+1)^2}{4(i-1)}a^2\frac{dU}{dx} - \frac{i(i+1)(2i+3)}{2I(2i+5)}(a^2-r^2)\frac{dU}{dx}\right.$$

$$\left. + \frac{i^2}{I(2i+5)}r^{2i+3}\frac{d}{dx}(r^{-2i-1}U)\right\} \quad\text{...(52)}$$

Further collecting the several results, the complete value of α as the solution of the second approximation is

$$\alpha = \alpha_0 + \alpha_0' + \alpha' + \alpha_{,}$$

so that it is only necessary to collect the results of equations (31), (with Y' written for Y), (36), (43), and (52), and to substitute for U its value from (39) in order to obtain the solution required. The values of β and γ may then at once be written down by symmetry. The expressions are naturally very long, and I shall not write them down in the general case.

The radial velocity ρ is however an important expression, because it alone is necessary to enable us to obtain the second approximation to the form of the spheroid, and accordingly I will give it.

It may be collected from (33), (37), and by forming ρ' and $\rho_,$ from (43) and (52).

I find then after some rather tedious analysis, which I did in order to verify my solution, that as far as concerns the inertia terms alone

$$\rho = \frac{wv}{v^2}\frac{\mathrm{Y}}{r}\sin vt\,\{\mathfrak{A}r^4 - \mathfrak{B}a^2r^2 + \mathfrak{C}a^4\}$$

where

$$\mathfrak{A} = \frac{i(i+1)}{2\,.\,4(2i+5)\,\mathrm{I}},\quad \mathfrak{B} = \frac{i^2(i+1)(2i^2+10i+9)}{4(i-1)(2i+5)\,\mathrm{I}^2}$$

and
$$\mathfrak{C} = \left(\frac{i}{2\mathrm{I}}\right)^2\left[i\left(\frac{i+2}{i-1}\right)^2 + \frac{(i+1)(2i+3)}{(2i+1)(2i+5)}\right] - \frac{i(i+1)^2}{2\,.\,4(i-1)(2i+1)\,\mathrm{I}}$$

If $\mathfrak{C}$ be reduced to the form of a single fraction, I think it probable that the numerator would be divisible by $2i+1$, but I do not think that the quotient would divide into factors, and therefore I leave it as it stands.

In the case where $i=2$ this formula becomes

$$\rho = \frac{wv}{v^2}\frac{\mathrm{Y}}{r}\sin vt\,\frac{1}{2^2\,.\,3\,.\,19^2}\{19r^4 - 148a^2r^2 + 287a^4\}$$

which agrees (as will appear presently) with the same result obtained in a different way.

I shall now go on to the special case where $i=2$, which will be required in the tidal problem.

From (39) we have $\qquad \mathrm{U} = \dfrac{wv}{v}\,.\,\dfrac{1}{2\,.\,5\,.\,19}\,\mathrm{Y}\sin vt$

From (36)

$$\alpha_0' = \frac{wva^2}{v^2}\,.\,\frac{4}{19^2}\left[\left(4a^2 - \frac{3\,.\,7}{2\,.\,5}r^2\right)\frac{d\mathrm{Y}}{dx} - \frac{2}{5}r^7\frac{d}{dx}(\mathrm{Y}r^{-5})\right]\sin vt$$

From (43)

$$\alpha' = \frac{wv}{v^2}\,.\,\frac{1}{2^3\,.\,3\,.\,5\,.\,19}\left[9r^4\frac{d\mathrm{Y}}{dx} + \frac{2\,.\,4}{3}r^9\frac{d}{dx}(\mathrm{Y}r^{-5})\right]\sin vt$$

From (52)

$$\alpha_, = -\frac{wva^2}{v^2}\,.\,\frac{1}{2^3\,.\,3\,.\,5\,.\,19^2}\left[(5\,.\,97a^2 + 7\,.\,4r^2)\frac{d\mathrm{Y}}{dx} - \frac{4\,.\,4}{3}r^7\frac{d}{dx}(r^{-5}\mathrm{Y})\right]$$

Adding these expressions together, and adding α_0, we get

$$\alpha = \alpha_0 + \frac{wv}{v^2} \cdot \frac{1}{2^3 . 3 . 5 . 19^2} \Big[(5 . 287a^4 - 37 . 4 . 7a^2r^2 + 9 . 19r^4) \frac{dY}{dx} - \tfrac{8}{3} (2 . 37a^2 - 19r^2) r^7 \frac{d}{dx} (Yr^{-5}) \Big] \sin vt \quad \text{...(53)}$$

and symmetrical expressions for β and γ.

In order to obtain the radial flow we multiply α by $\frac{x}{r}$, β by $\frac{y}{r}$, γ by $\frac{z}{r}$, and add, and find

$$\rho = \rho_0 + \frac{wv}{v^2} \cdot \frac{1}{2^2 . 3 . 19^2} (287a^4 - 4 . 37a^2r^2 + 19r^4) \frac{Y}{r} \sin (vt + \epsilon) \quad \text{...(54)}$$

the ϵ which was omitted in the trigonometrical term being now replaced.

The surface value of ρ when $r = a$ is

$$\rho = \rho_0 + \frac{wva^5}{v^2} \frac{79}{2 . 3 . 19^2} \frac{Y}{r^2} \sin (vt + \epsilon) \quad \text{..................(55)}$$

where ρ_0 is given by (34).

If we write $-\frac{1}{2}\pi - \epsilon$ for ϵ we see that a term $Y \sin (vt - \epsilon)$ in the effective disturbing potential will give us

$$\rho = \rho_0 - \frac{wva^5}{v^2} \frac{79}{2 . 3 . 19^2} \frac{Y}{r^2} \cos (vt - \epsilon) \quad \text{..................(56)}$$

Suppose $wr^2 S \cos vt$ to be an external disturbing potential per unit volume of the earth, not including the effective potential due to gravitation, and let $r = a + \sigma_{\prime}$ be the first approximation to the form of the tidal spheroid. Then by the theory of tides as previously developed (see equation (15), Section 5, "Tides")

$$\frac{\sigma_{\prime}}{a} = \frac{S}{\mathfrak{g}} \cos \epsilon \cos (vt - \epsilon), \text{ where } \tan \epsilon = \frac{19 \upsilon v}{2 g a w}$$

When the sphere is deemed free of gravitation the effective disturbing potential is $wr^2 \left(S \cos vt - \mathfrak{g} \frac{\sigma_{\prime}}{a} \right)$; this is equal to $- wr^2 \sin \epsilon \, S \sin (vt - \epsilon)$.

In proceeding to a second approximation we must put in equation (56) $Y = - wr^2 \sin \epsilon \, S$.

Thus we get from (56), at the surface where $r = a$,

$$\rho = \rho_0 + \frac{w^2 va^5}{v^2} \cdot \frac{79}{2 . 3 . 19^2} \sin \epsilon \, S \cos (vt - \epsilon) \quad \text{..............(57)}$$

To find ρ_0 we must put $r = a + \sigma$ as the equation to the second approximation.

ρ_0 is the surface radial velocity due directly to the external disturbing potential $wr^2 S \cos vt$ and to the effective gravitation potential. The sum of these two gives an effective potential $wr^2\left(\mathrm{S}\cos vt - \mathfrak{g}\frac{\sigma}{a}\right)$, which is the $\mathrm{Y}'\cos vt$ of (34).

ρ_0 is found by writing this expression in place of $\mathrm{Y}'\cos vt$ in equation (34), and we have

$$\rho_0 = \frac{5wa^3}{19\upsilon}\left(\mathrm{S}\cos vt - \mathfrak{g}\frac{\sigma}{a}\right)$$

Substituting in (57) we have

$$\rho = \frac{5wa^3}{19\upsilon}\left(\mathrm{S}\cos vt - \mathfrak{g}\frac{\sigma}{a} + \frac{5wva^2}{19\upsilon}\frac{79}{2\,.\,3\,.\,5^2}\sin\epsilon\,\mathrm{S}\cos(vt-\epsilon)\right) \quad \ldots(58)$$

Since $\tan\epsilon = \dfrac{19\upsilon v}{2gaw}$, therefore $\dfrac{5wa^2}{19\upsilon} = \dfrac{v}{\mathfrak{g}}\cot\epsilon$, and (58) becomes

$$\rho = a\frac{v}{\mathfrak{g}}\cot\epsilon\left(\mathrm{S}\cos vt - \mathfrak{g}\frac{\sigma}{a} + \tfrac{79}{150}\frac{v^2}{\mathfrak{g}}\cos\epsilon\,\mathrm{S}\cos(vt-\epsilon)\right)$$

But the radial surface velocity is equal to $\dfrac{d\sigma}{dt}$, and therefore $\dfrac{d\sigma}{dt} = \rho$, so that

$$\frac{d\sigma}{dt} + v\cot\epsilon\,.\,\sigma = a\frac{v}{\mathfrak{g}}\cot\epsilon\left(\mathrm{S}\cos vt + \tfrac{79}{150}\frac{v^2}{\mathfrak{g}}\cos\epsilon\,\mathrm{S}\cos(vt-\epsilon)\right) \ldots(59)$$

If we divide σ into two parts, σ', σ'', to satisfy the two terms on the right respectively, we have

$$\frac{\sigma'}{a} = \cos\epsilon\,.\,\frac{\mathrm{S}}{\mathfrak{g}}\cos(vt-\epsilon)$$

which is the first approximation over again, and

$$\frac{\sigma''}{a} = \cos\epsilon\,.\,\frac{\mathrm{S}}{\mathfrak{g}}\,.\,\tfrac{79}{150}\frac{v^2}{\mathfrak{g}}\cos\epsilon\cos(vt-2\epsilon)$$

Therefore

$$\frac{\sigma}{a} = \cos\epsilon\,.\,\frac{\mathrm{S}}{\mathfrak{g}}\left\{\cos(vt-\epsilon) + \tfrac{79}{150}\frac{v^2}{\mathfrak{g}}\cos\epsilon\cos(vt-2\epsilon)\right\} \quad \ldots\ldots(60)$$

This gives the second approximation to the form of the tidal spheroid. We see that the inertia generates a second small tide which lags twice as much as the primary one.

Although this expression is more nearly correct than subsequent ones, it will be well to group both these tides together and to obtain a single expression for σ.

Let
$$\tan\chi = \frac{\tfrac{79}{150}\dfrac{v^2}{\mathfrak{g}}\sin\epsilon\cos\epsilon}{1 + \tfrac{79}{150}\dfrac{v^2}{\mathfrak{g}}\cos^2\epsilon}$$

Then
$$\frac{\sigma}{a} = \frac{\mathrm{S}}{\mathfrak{g}}\frac{\cos\epsilon}{\cos\chi}\left(1 + \tfrac{79}{150}\frac{v^2}{\mathfrak{g}}\cos^2\epsilon\right)\cos(vt-\epsilon-\chi) \quad \ldots\ldots\ldots\ldots(61)$$

This shows that the tide lags by $(\epsilon + \chi)$, and is in height

$$\frac{\cos \epsilon}{\cos \chi}\left(1 + \tfrac{79}{150}\frac{v^2}{\mathfrak{g}}\cos^2 \epsilon\right)$$

of the equilibrium tide of a perfectly fluid spheroid.

By the method employed it is postulated that $\tfrac{79}{150}\frac{v^2}{\mathfrak{g}}$ is a small fraction, because the effects of inertia are supposed to be small. Hence χ must be a small angle, and there will not be much error in putting

$$\chi = \tfrac{79}{150}\frac{v^9}{\mathfrak{g}}\sin\epsilon\cos\epsilon, \quad \text{and} \quad \sec\chi = 1$$

We have for the lag of the tide $\left(\epsilon + \tfrac{79}{150}\frac{v^2}{\mathfrak{g}}\sin\epsilon\cos\epsilon\right)$, and for its height $\cos\epsilon\left(1 + \tfrac{79}{150}\frac{v^2}{\mathfrak{g}}\cos^2\epsilon\right)$.

Let η be the lag, so that

$$\eta = \epsilon + \tfrac{79}{150}\frac{v^2}{\mathfrak{g}}\sin\epsilon\cos\epsilon$$

whence $$\epsilon = \eta - \tfrac{79}{150}\frac{v^2}{\mathfrak{g}}\sin\eta\cos\eta \text{ very nearly}$$

Also $$\cos\epsilon = \cos\eta\left(1 + \tfrac{79}{150}\frac{v^2}{\mathfrak{g}}\sin^2\eta\right)$$

and $$\cos\epsilon\left(1 + \tfrac{79}{150}\frac{v^2}{\mathfrak{g}}\cos^2\epsilon\right) = \cos\eta\left(1 + \tfrac{79}{150}\frac{v^2}{\mathfrak{g}}\right)$$

Hence (61) becomes

$$\left.\begin{aligned} &\frac{\sigma}{a} = \frac{\mathrm{S}}{\mathfrak{g}}\cos\eta\left(1 + \tfrac{79}{150}\frac{v^2}{\mathfrak{g}}\right)\cos(vt - \eta) \\ \text{where}\quad &\eta - \tfrac{79}{150}\frac{v^2}{\mathfrak{g}}\sin\eta\cos\eta = \arctan\left(\frac{19\upsilon v}{2gaw}\right) \end{aligned}\right\} \quad \text{(62)}$$

This is probably the simplest form in which the result of the second approximation may be stated.

From it we see that with a given lag, the height of tide is a little greater than in the theory used in the two previous papers; and that for a given frequency of tide the lag is a little greater than was supposed.

The whole investigation of the precession of the viscous spheroid was based on the approximate theory of tides, when inertia is neglected. It will be well, therefore, to examine how far the present results will modify the conclusions there arrived at. It would, however, occupy too much space to recapitulate the methods employed, and therefore the following discussion will only be intelligible, when read in conjunction with that paper.

The couples on the earth, caused by the attraction of the disturbing bodies on the tidal protuberance, were found to be expressible by the sum of a number of terms, each of which corresponded to one of the constituent simple harmonic tides. Each such term involved two factors, one of which was the height of the tide, and the other the sine of the lag. Now if ϵ be the lag and v the speed of the tide, it was found in the first approximation that $\tan\epsilon = 19vv \div 2gaw$, and that the height of tide was proportional to $\cos\epsilon$; hence each term had a factor $\sin 2\epsilon$.

But from the present investigation it appears that, with the same value of ϵ, the height of tide is really proportional to $\cos\epsilon\left(1 + \frac{79}{150}\frac{v^2}{\mathfrak{g}}\cos^2\epsilon\right)$; whilst the lag is $\epsilon + \frac{79}{150}\frac{v^2}{\mathfrak{g}}\sin\epsilon\cos\epsilon$, so that its sine is $\left(1 + \frac{79}{150}\frac{v^2}{\mathfrak{g}}\cos^2\epsilon\right)\sin\epsilon$.

Hence in place of $\sin 2\epsilon$, we ought to have put $\sin 2\epsilon\left(1 + \frac{79}{150}\frac{v^2}{\mathfrak{g}}\cos^2\epsilon\right)^2$, or $\sin 2\epsilon\left(1 + \frac{79}{75}\frac{v^2}{\mathfrak{g}}\cos^2\epsilon\right)$.

Thus every term in the expressions for $\frac{di}{dt}, \frac{dN}{dt}, \frac{d\xi}{dt}$ should be augmented, each in a proportion depending on the speed and lag of the tide from which it takes its origin.

In the paper on "Precession," two numerical integrations were given of the differential equations for the secular changes in the variables; in the first of these, in Section 15, the viscosity was not supposed to be small, and was constant, in the second, in Section 17, it was merely supposed that the alteration of phase of each tide was small, and the viscosity was left indeterminate. It is not proposed to determine directly the correction to the first solution.

The correcting factor for the expression $\sin 2\epsilon$ is greatest when ϵ is small, because $\cos^2\epsilon$ may then be replaced in it by unity; hence the correction in the second integration will necessarily be larger than in the first, and a superior limit to the correction to the first integration may be found.

We have tides of the seven speeds $2(n-\Omega)$, $2n$, $2(n+\Omega)$, $n-2\Omega$, n, $n+2\Omega$, 2Ω; hence if the viscosity be small, the correcting factor for the expressions $\sin 4\epsilon_1$, $\sin 4\epsilon$, $\sin 4\epsilon_2$, $\sin 2\epsilon_1'$, $\sin 2\epsilon'$, $\sin 2\epsilon_2'$, $\sin 4\epsilon''$ is $1 + \frac{79}{75}\frac{(\text{speed})^2}{\mathfrak{g}}$, where the speed is one of the seven specified.

If $\lambda = \frac{\Omega}{n}$, the seven factors may be written

$$\left.\begin{aligned} &1 + \tfrac{316}{75}n^2[(1-\lambda)^2, \text{ or } 1, \text{ or } (1+\lambda)^2], \text{ for semi-diurnal terms}\\ &1 + \tfrac{79}{75}n^2[(1-2\lambda)^2, \text{ or } 1, \text{ or } (1+2\lambda)^2], \text{ for diurnal terms}\\ \text{and}\quad &1 + \tfrac{316}{75}\frac{n^2}{\mathfrak{g}}\lambda^2, \text{ for the fortnightly term}\end{aligned}\right\}\ldots(63)$$

Also we have the equations

$$\left.\begin{aligned}&\frac{\sin 4\epsilon_1}{\sin 4\epsilon}=1-\lambda, \quad \frac{\sin 4\epsilon}{\sin 4\epsilon}=1, \quad \frac{\sin 4\epsilon_2}{\sin 4\epsilon}=1+\lambda\\ &\frac{\sin 2\epsilon_1'}{\sin 4\epsilon}=\tfrac{1}{2}(1-2\lambda), \quad \frac{\sin 2\epsilon'}{\sin 4\epsilon}=\tfrac{1}{2}, \quad \frac{\sin 2\epsilon_2'}{\sin 4\epsilon}=\tfrac{1}{2}(1+2\lambda), \quad \frac{\sin 4\epsilon''}{\sin 4\epsilon}=\lambda\end{aligned}\right\} \quad (64)$$

We shall obtain a sufficiently accurate result, if the corrections be only applied to those terms in the differential equations which do not involve powers of q (or $\sin \frac{1}{2}i$), higher than the first. For the purpose of correction the differential equations to be corrected are by (77), (78), and (79) of Section 17 of "Precession," viz.:

$$\left.\begin{aligned}\frac{di_{m^2}}{dt}&=\frac{1}{N}\frac{\tau^2}{\mathfrak{g}n_0}\left[\tfrac{1}{2}p^7q\sin 4\epsilon_1+\tfrac{1}{2}p^5q(p^2+3q^2)\sin 2\epsilon_1'-\tfrac{1}{2}pq(p^2-q^2)^3\sin 2\epsilon'\right]\\ -\frac{dN_{m^2}}{dt}&=\tfrac{1}{2}\frac{\tau^2}{\mathfrak{g}n_0}p^8\sin 4\epsilon_1=\mu\frac{d\xi}{dt}\end{aligned}\right\} \quad \text{......(65)}$$

As we are treating the obliquity as small, we may put

$$\tfrac{1}{2}p^7q=\tfrac{1}{2}p^5q(p^2+3q^2)=\tfrac{1}{2}pq(p^2-q^2)^3=\tfrac{1}{4}PQ \quad \text{and} \quad p^8=P$$

where $P=\cos i$, $Q=\sin i$.

For the purpose of correction, the terms depending on the moon's influence become

$$\left.\begin{aligned}\frac{di_{m^2}}{dt}&=\frac{1}{N}\frac{\tau^2}{\mathfrak{g}n_0}\tfrac{1}{4}PQ\{\sin 4\epsilon_1+\sin 2\epsilon_1'-\sin 2\epsilon'\}\\ -\frac{dN_{m^2}}{dt}&=\tfrac{1}{2}\frac{\tau^2}{\mathfrak{g}n_0}P\sin 4\epsilon_1=\mu\frac{d\xi}{dt}\end{aligned}\right\} \quad \text{..........(66)}$$

And by symmetry (or by (81) "Precession") we have for the solar terms

$$\frac{di_{m_,^2}}{dt}=\frac{1}{N}\frac{\tau_,^2}{\mathfrak{g}n_0}\tfrac{1}{4}PQ\sin 4\epsilon, \quad -\frac{dN_{m_,^2}}{dt}=\tfrac{1}{2}\frac{\tau_,^2}{\mathfrak{g}n_0}P\sin 4\epsilon \quad \text{........(67)}$$

For the terms depending on the joint action of the sun and moon we have, by (82) and (33) "Precession," when the obliquity is treated as small,

$$\frac{di_{mm_,}}{dt}=-\frac{1}{N}\frac{\tau\tau_,}{\mathfrak{g}n_0}\tfrac{1}{2}PQ\sin 2\epsilon', \quad \frac{dN_{mm_,}}{dt}=0 \quad \text{..............(68)}$$

If we multiply each of the sines by its appropriate factor given in (63), and substitute from (64) for each of them in terms of $\sin 4\epsilon$, and collect the results from (66), (67), and (68), and express by the symbol δ the corrections to be introduced for the effects of inertia, we have

$$\begin{aligned}\delta\frac{di}{dt}&=\frac{1}{N}\frac{\sin 4\epsilon}{\mathfrak{g}n_0}\tfrac{1}{4}PQ\cdot\tfrac{79}{75}\frac{n^2}{\mathfrak{g}}\left[\{4(1-\lambda)^3+\tfrac{1}{2}(1-2\lambda)^3-\tfrac{1}{2}\}\tau^2+4\tau_,^2-\tau\tau_,\right]\\ -\delta\frac{dN}{dt}&=\tfrac{1}{2}\frac{\sin 4\epsilon}{\mathfrak{g}n_0}P\cdot\tfrac{316}{75}\frac{n^2}{\mathfrak{g}}\left[\tau^2(1-\lambda)^3+\tau_,^2\right]\\ \mu\delta\frac{d\xi}{dt}&=\tfrac{1}{2}\frac{\sin 4\epsilon}{\mathfrak{g}n_0}P\cdot\tfrac{316}{75}\frac{n^2}{\mathfrak{g}}\tau^2(1-\lambda)^3\end{aligned}$$

Now $4(1-\lambda)^3+\frac{1}{2}(1-2\lambda)^3-\frac{1}{2}=(1-2\lambda)(4-7\lambda+4\lambda^2)$. Therefore if we add these corrections to the full expressions for $\frac{di}{dt}$, $\frac{dN}{dt}$ (in which I put $1-\frac{1}{2}Q^2=P$) and $\mu\frac{d\xi}{dt}$, given in (83) "Precession," and write $K=\frac{316}{75}\frac{n^2}{\mathfrak{g}}$ for brevity, we have

$$\left.\begin{aligned}\frac{di}{dt}&=\frac{1}{N}\frac{\sin 4\epsilon}{\mathfrak{g}n_0}\tfrac{1}{4}PQ\left\{\tau^2\left(1-\frac{2\lambda}{P}\right)+\tau_{\prime}^2-\tau\tau_{\prime}\right.\\&\qquad\left.+K\left[(1-2\lambda)(1-\tfrac{7}{4}\lambda+\lambda^2)\tau^2+\tau_{\prime}^2-\tfrac{1}{4}\tau\tau_{\prime}\right]\right\}\\-\frac{dN}{dt}&=\tfrac{1}{2}\frac{\sin 4\epsilon}{\mathfrak{g}n_0}P\left\{\tau^2(1-\lambda)+\tau_{\prime}^2+\tfrac{1}{2}\tau\tau_{\prime}\frac{Q^2}{P}+K\left[(1-\lambda)^3\tau^2+\tau_{\prime}^2\right]\right\}\\\mu\frac{d\xi}{dt}&=\tfrac{1}{2}\frac{\tau^2}{\mathfrak{g}n_0}\sin 4\epsilon P\left[1-\frac{\lambda}{P}+K(1-\lambda)^3\right]\end{aligned}\right\}\quad\text{......(69)}$$

The last of these equations may be written approximately

$$\frac{dt}{\mu d\xi}=\left[\tfrac{1}{2}\frac{\tau^2}{\mathfrak{g}n_0}\sin 4\epsilon P\left(1-\frac{\lambda}{P}\right)\right]^{-1}\left[1-K(1-\lambda)^2\right]\quad\text{........(70)}$$

If we multiply the two former of equations (69) by (70), and notice that, when P is taken as unity,

$$(1-2\lambda)(1-\tfrac{7}{4}\lambda+\lambda^2)-\left(1-\frac{2\lambda}{P}\right)(1-\lambda)^2=\tfrac{1}{4}\lambda(1-2\lambda)$$

and that

$$1-(1-\lambda)^2=\lambda(2-\lambda)\ \text{ and }\ -\tfrac{1}{4}+(1-\lambda)^2=\tfrac{1}{4}(1-2\lambda)(3-2\lambda)$$

we have

$$\left.\begin{aligned}&\frac{d}{\mu d\xi}\log\tan^2\tfrac{1}{2}i\\&=\frac{1-\dfrac{2\lambda}{P}+\left(\dfrac{\tau_{\prime}}{\tau}\right)^2-\left(\dfrac{\tau_{\prime}}{\tau}\right)+K\left[\tfrac{1}{4}\lambda(1-2\lambda)+\lambda(2-\lambda)\left(\dfrac{\tau_{\prime}}{\tau}\right)^2+\tfrac{1}{4}(1-2\lambda)(3-2\lambda)\left(\dfrac{\tau_{\prime}}{\tau}\right)\right]}{N\left(1-\dfrac{\lambda}{P}\right)}\\&\qquad-\frac{dN}{\mu d\xi}=\frac{1-\lambda+\left(\dfrac{\tau_{\prime}}{\tau}\right)^2+\tfrac{1}{2}\dfrac{Q^2}{P}\left(\dfrac{\tau_{\prime}}{\tau}\right)+K\lambda(2-\lambda)\left(\dfrac{\tau_{\prime}}{\tau}\right)^2}{1-\dfrac{\lambda}{P}}\end{aligned}\right\}\quad\text{........(71)}$$

If K be put equal to zero, we have the equations (84) which were the subject of integration in Section 17 "Precession."

Since K, λ, and $\tau_{\prime}^2\div\tau^2$ are all small, the correction to the second equation is obviously insignificant, and we may take the term in K in the numerator of the first equation as being equal to $\frac{1}{4}K(1-2\lambda)(3-2\lambda)(\tau_{\prime}/\tau)$. This correction is small although not insensible. This shows that the amount of change of obliquity has been slightly under-estimated. It does not, however, seem worth while to compute the corrected value for the change of obliquity in the integrations of the preceding paper.

The equation of conservation of moment of momentum, which is derived from the integration of the second of (71), clearly remains sensibly unaffected.

We see also from (70) that the time required for the changes has been over-estimated. If K_0, λ_0; K, λ be the initial and final values of K and λ at the beginning and end of one of the periods of integration, it is obvious that our estimate of time should have been multiplied by some fraction lying between $1 - K_0(1-\lambda_0)^2$ and $1 - K(1-\lambda)^2$.

At the beginning of the first period $K_0 = \cdot 0364$ and $\lambda_0 = \cdot 0365$, and at the end $K = \cdot 0865$ and $\lambda = \cdot 0346$.

Whence $K_0(1-\lambda_0)^2 = \cdot 034$, $K(1-\lambda)^2 = \cdot 080$.

Hence it follows that the time, in the first period of the integration of Section 15, may have been over-estimated by some percentage less than some number lying between 3 and 8.

In fact, I have corrected the first period of that integration by a rather more tedious process than that here exhibited, and I found that the time was over-estimated by a little less than 3 per cent. And it was found that we ought to subtract from the 46,300,000 years comprised within the first period about 1,300,000 years. I also found that the error in the final value of the obliquity could hardly amount to more than 1′ or 2′.

In the later periods of integration the error in the time would no doubt be a little larger fraction of the time comprised within each period, but as it is not interesting to find the time in anything but round numbers, it is not worth while to find the corrections.

There is another point worth noticing. It might be suspected that when we approach the critical point where $n \cos i = 2\Omega$, where the rate of change of obliquity was found to vanish, the tidal movements might have become so rapid as seriously to affect the correctness of the tidal theory used; and accordingly it might be thought that the critical point was not reached even approximately when $n \cos i = 2\Omega$.

The preceding analysis will show at once that this is not the case. Near the critical point the solar terms have become negligeable; if we put $\tau_{,} = 0$ in the first of equations (69) we have

$$\frac{di}{dt} = \frac{1}{N} \cdot \frac{\tau^2}{\mathfrak{g} n_0} \sin 4\epsilon \cdot \tfrac{1}{4} PQ \left[1 - 2\lambda \sec i + K(1-2\lambda)(1 - \tfrac{7}{4}\lambda + \lambda^2)\right] \ldots (72)$$

The condition for the critical point in the first approximation was $2\lambda \sec i = 1$; if then i is so small that we may take $\sec i = 1$ in the inertia term, this condition also causes the inertia term to vanish.

Hence the corrected theory of tides makes no sensible difference in the critical point where $\frac{di}{dt}$ changes sign.

Having now disposed of these special points connected with previous results, I shall return to questions of general dynamics connected with the approximate solution of the forced vibrations of viscous spheroids; that is to say, I shall compare the results with those of—

*The forced oscillations of a liquid sphere**.

The same notation as before will serve again, and the equations of motion are

$$\left.\begin{array}{l} -\frac{dp}{dx}+\frac{dW}{dx}-w\frac{d\alpha}{dt}=0 \\ \text{two similar equations} \\ \text{and } \frac{d\alpha}{dx}+\frac{d\beta}{dy}+\frac{d\gamma}{dz}=0 \end{array}\right\} \quad \ldots\ldots\ldots\ldots\ldots\ldots(73)$$

If the external tide-generating forces be those due to a potential per unit volume equal to $wr^i S_i$, and $r=a+\sigma_i$ be the equation to the tidal spheroid, where S_i, σ_i are surface harmonics of the ith order, we must put

$$W=w\left[r^i S_i+\frac{3g}{2i+1}\left(\frac{r}{a}\right)^i\sigma_i+(3a^2-r^2)\frac{g}{2a}\right]$$

the second term being the potential of the tidal protuberance, and the last of the mean sphere.

Differentiate the three equations of motion by x, y, z and add them, and we have

$$\nabla^2\left[p-w(3a^2-r^2)\frac{g}{2a}\right]=0$$

Hence $\quad p=w(3a^2-r^2)\frac{g}{2a}+\text{solid harmonics}+\text{a constant}$

When $r=a$, at the mean surface of the sphere, $p=gw\sigma_i$, therefore

$$p=w(a^2-r^2)\frac{g}{2a}+gw\sigma_i\left(\frac{r}{a}\right)^i$$

Substituting this value of p in the equations of motion (73),

$$\frac{d\alpha}{dt}=\frac{d}{dx}\left[r^i S_i+\frac{3g}{2i+1}\sigma_i\left(\frac{r}{a}\right)^i+(3a^2-r^2)\frac{g}{2a}-(a^2-r^2)\frac{g}{2a}-g\sigma_i\left(\frac{r}{a}\right)^i\right]$$

whence

$$\left.\begin{array}{l} \frac{d\alpha}{dt}=\frac{d}{dx}\left[r^i S_i-\frac{2(i-1)}{2i+1}g\left(\frac{r}{a}\right)^i\sigma_i\right] \\ \text{and two similar equations} \end{array}\right\} \quad \ldots\ldots\ldots\ldots\ldots\ldots(74)$$

* This is a slight modification of Sir W. Thomson's investigation of the free oscillations of fluid spheres, *Phil. Trans.*, 1863, p. 608.

The expression within brackets [ ] on the right is the effective disturbing potential, inclusive of the effects of mutual gravitation, and thus this process is exactly parallel to that adopted above in order to include the effects of mutual gravitation in the disturbing potential in the case of the viscous spheroid.

Now ρ, the radial velocity of flow, is equal to $\alpha\frac{x}{r}+\beta\frac{y}{r}+\gamma\frac{z}{r}$.

Therefore multiplying the equations (74) by $\frac{x}{r}, \frac{y}{r}, \frac{z}{r}$ and adding them, we have, by the properties of homogeneous functions,

$$\frac{d\rho}{dt}=i\left[r^{i-1}S_i-\frac{2(i-1)}{2i+1}g\frac{r^{i-1}}{a^i}\sigma_i\right]$$

But when $r=a$, $\rho=\frac{d\sigma_i}{dt}$.

Therefore $$\frac{d^2\sigma_i}{dt^2}=ia^{i-1}S_i-\frac{2i(i-1)}{2i+1}\frac{g}{a}\sigma_i \quad\text{.....................(75)}$$

Suppose $S_i=Q_i\cos vt$, and that the tidal motion is steady, so that σ_i must be of the form $XQ_i\cos vt$; then substituting in (75) this form of σ_i, we find

$$X\left[-v^2+\frac{2i(i-1)}{2i+1}\frac{g}{a}\right]=ia^{i-1}$$

Whence $$\sigma_i=\frac{ia^{i-1}}{\frac{2i(i-1)}{2i+1}\frac{g}{a}-v^2}Q_i\cos vt \quad\text{.....................(76)}$$

This gives the equation to the tidal spheroid.

Since the equilibrium tide, due to the disturbing potential, would be given by

$$\sigma_i=\frac{a^{i-1}}{\frac{2(i-1)}{2i+1}\frac{g}{a}}Q_i\cos vt$$

it follows that inertia augments the height of tide in the proportion

$$1 : 1-\frac{(2i+1)}{2i(i-1)}\frac{a}{g}v^2$$

In the case where $i=2$, the augmentation is in the proportion $1 : 1-\frac{1}{2}\frac{v^2}{\mathfrak{g}}$.

We will now consider the nature of the motion by which each particle assumes its successive positions.

With the value of σ_i given in (76)

$$S_i-\frac{2(i-1)}{2i+1}\frac{g}{a^i}\sigma_i=\frac{-Q_iv^2}{\frac{2i(i-1)}{(2i+1)}\frac{g}{a}-v^2}\cos vt$$

Substituting in (74)

$$\left.\begin{array}{c}\dfrac{d\alpha}{dt} = -\dfrac{d}{dx}\dfrac{v^2\cos vt\, Q_i r^i}{\dfrac{2i(i-1)}{2i+1}\dfrac{g}{a} - v^2} \\ \text{and two similar equations}\end{array}\right\} \quad\ldots\ldots\ldots\ldots\ldots\ldots\ldots(77)$$

Integrating with regard to t

$$\left.\begin{array}{c}\alpha = -\dfrac{d}{dx}\dfrac{Q_i r^i v\sin vt}{\dfrac{2i(i-1)}{2i+1}\dfrac{g}{a} - v^2} \\ \text{and two similar equations}\end{array}\right\} \quad\ldots\ldots\ldots\ldots\ldots\ldots\ldots(78)$$

There might be a term introduced by integration, independent of the time, but this term must be zero, because if there were no disturbing force there would be no flow. Hence it is clear that there is a velocity potential ϑ, and that

$$\vartheta = \frac{1}{\dfrac{2i(i-1)}{2i+1}\dfrac{g}{a} - v^2}\frac{d}{dt}(r^i S_i) \quad\ldots\ldots\ldots\ldots\ldots\ldots\ldots(79)$$

Now however slowly the motion takes place, there will always be a velocity potential, and if it be slow enough we may omit v^2 in the denominator of (79). In other words, if inertia be neglected the velocity potential is

$$\vartheta = \frac{2i+1}{2i(i-1)}\frac{a}{g}\frac{d}{dt}(r^i S_i)$$

For the sake of comparison with the approximate solution for the tides of a viscous spheroid, a precisely parallel process will now be carried out with regard to the liquid sphere.

We obtain a first approximation for $\dfrac{d\alpha}{dt}$, when inertia is neglected, by omitting v^2 in the denominator of (77); whence

$$\frac{d\alpha}{dt} = -\frac{d}{dx}\left(\frac{2i+1}{2i(i-1)}\frac{g}{a}v^2\cos vt\, r^i Q_i\right)$$

Substituting this approximate value in the equations of motion (73) we have

$$\left.\begin{array}{c}-\dfrac{dp}{dx} + \dfrac{d}{dx}\left(W + w\dfrac{2i+1}{2i(i-1)}\dfrac{g}{a}v^2\cos vt\, r^i Q_i\right) = 0 \\ \text{and two similar equations}\end{array}\right\} \quad\ldots\ldots\ldots(80)$$

From these equations it is obvious that the second approximation to the form of the tidal spheroid is found by augmenting the equilibrium tide due to the tide-generating potential $r^i Q_i \cos vt$ in the proportion $1 + \dfrac{2i+1}{2i(i-1)}\dfrac{g}{a}v^2$ to unity.

When $i=2$ the augmenting factor is $1+\frac{1}{2}\frac{v^2}{\mathfrak{g}}$.

This is of course only an approximate result; the accurate value of the factor is $1\div\left(1-\frac{1}{2}\frac{v^2}{\mathfrak{g}}\right)$, and we see that the two agree if the squares and higher powers of $\frac{1}{2}\frac{v^2}{\mathfrak{g}}$ are negligeable.

Now in the case of the viscous tides we found the augmenting factor to be $1+\frac{79}{150}\frac{v^2}{\mathfrak{g}}\cos^2\epsilon$. When $\epsilon=0$, which corresponds to the case of fluidity, the expressions are closely alike, but we should expect that the 79 ought really to be 75.

The explanation which lies at the bottom of this curious discrepancy will be most easily obtained by considering the special case of a lunar semi-diurnal tide.

We found in Part II., equation (21), the following values for α, β, γ,

$$\left.\begin{aligned} \alpha &= \frac{w\tau}{38v}\sin 2\epsilon\,[(8a^2-5r^2)\,y+4x^2y] \\ \beta &= \frac{w\tau}{38}\sin 2\epsilon\,[(8a^2-5r^2)\,x+4xy^2] \\ \gamma &= \frac{w\tau}{38}\sin 2\epsilon\,.\,4xyz \end{aligned}\right\} \quad\ldots\ldots\ldots\ldots\ldots(81)$$

where

$$\left.\begin{aligned} x &= r\sin\theta\cos(\phi-\omega t) \\ y &= r\sin\theta\sin(\phi-\omega t) \\ z &= r\cos\theta \end{aligned}\right\}$$

Consider the case when the viscosity is infinitely small: here ϵ is small, and $\sin 2\epsilon=\tan 2\epsilon=\dfrac{38v\omega}{5\mathfrak{g}wa^2}$.

Hence $\dfrac{\omega\tau}{38v}\sin 2\epsilon=\dfrac{\omega\tau}{5\mathfrak{g}a^2}$, which is independent of the viscosity.

By substituting this value in (81), we see that however small the viscosity, the nature of the motion, by which each particle assumes its successive positions, always preserves the same character; and the motion always involves molecular rotation.

But it has been already proved that, however slow the tidal motion of a liquid sphere may be, yet the fluid motion is always irrotational.

Hence in the two methods of attacking the same problem, different first approximations have been used, whence follows the discrepancy of 79 instead of 75.

The fact is that in using the equations of flow of a viscous fluid, and neglecting inertia to obtain a first approximation, we postulate that $w\frac{d\alpha}{dt}$, $w\frac{d\beta}{dt}$, $w\frac{d\gamma}{dt}$, are less important than $\upsilon\nabla^2\alpha$, $\upsilon\nabla^2\beta$, $\upsilon\nabla^2\gamma$; and this is no longer the case if υ be very small.

It does not follow therefore that, in approaching the problem of fluidity from the side of viscosity, we must necessarily obtain even an approximate result.

But the comparison which has just been made, shows that as regards the form of the tidal spheroid the two methods lead to closely similar results.

It follows therefore that, in questions regarding merely the form of the spheroid, and not the mode of internal motion, we only incur a very small error by using the limiting case when $\upsilon = 0$ to give the solution for pure fluidity.

In the paper on "Precession" (Section 7), some doubt was expressed as to the applicability of the analysis, which gave the effects of tides on the precession of a rotating spheroid, to the limiting case of fluidity; but the present results seem to justify the conclusions there drawn.

The next point to be considered is the effects of inertia in—

*The forced oscillations of an elastic sphere**.

Sir William Thomson has found the form into which a homogeneous elastic sphere becomes distorted under the influence of a potential expressible as a solid harmonic of the points within the sphere. He afterwards supposed the sphere to possess the power of gravitation, and considered the effects by a synthetical method. The result is the equilibrium theory of the tides of an elastic sphere. When, however, the disturbing potential is periodic in time this theory is no longer accurate.

It has already been remarked that the approximate solution of the problem of determining the state of internal flow of a viscous spheroid when inertia is neglected, is identical in form with that which gives the state of internal strain of an elastic sphere; the velocities α, β, γ have merely to be read as displacements, and the coefficient of viscosity υ as that of rigidity.

The effects of mutual gravitation may also be introduced in both problems by the same artifice; for in both cases we may take, instead of the external disturbing potential $wr^2 \text{S} \cos vt$, an effective potential $wr^2\left(\text{S}\cos vt - \mathfrak{g}\frac{\sigma}{a}\right)$, and then deem the sphere free of gravitational power.

* [Professor Horace Lamb has treated the problem of the "Vibrations of an Elastic Sphere" in *Proc. London Math. Soc.*, Vol. XIII. (1882), p. 189. At p. 51 of the same volume he has also solved the problem of the "Oscillations of a Viscous Spheroid."]

Now Sir William Thomson's solution shows that the surface radial displacement (which is of course equal to σ) is equal to

$$\frac{5wa^3}{19\upsilon}\left(\mathrm{S}\cos vt - \mathfrak{g}\frac{\sigma}{a}\right) \text{.........................(82)}$$

If therefore we put (with Sir William Thomson) $\mathfrak{r} = \dfrac{19\upsilon}{5wa^2}$, we have

$$\frac{\sigma_{\prime}}{a} = \frac{\mathrm{S}}{\mathfrak{r}+\mathfrak{g}}\cos vt$$

This expression gives the equilibrium elastic tide, the suffix being added to the σ to indicate that it is only a first approximation.

Before going further we may remark that

$$\mathrm{S}\cos vt - \mathfrak{g}\frac{\sigma_{\prime}}{a} = \frac{\mathfrak{r}}{\mathfrak{r}+\mathfrak{g}}\mathrm{S}\cos vt \text{ ....(83)}$$

When we wish to proceed to a second approximation, including the effects of inertia, it must be noticed that the equations of motion in the two problems only differ in the fact that in that relating to viscosity the terms introduced by inertia are $-w\dfrac{d\alpha}{dt}, -w\dfrac{d\beta}{dt}, -w\dfrac{d\gamma}{dt}$, whilst in the case of elasticity they are $-w\dfrac{d^2\alpha}{dt^2}, -w\dfrac{d^2\beta}{dt^2}, -w\dfrac{d^2\gamma}{dt^2}$. Hence a very slight alteration will make the whole of the above investigation applicable to the case of elasticity; we have, in fact, merely to differentiate the approximate values for α, β, γ twice with regard to the time instead of once.

Just as before, we find the surface radial displacement, as far as it is due to inertia, to be (compare (55))

$$\frac{wv^2a^5}{\upsilon^2}\,\frac{79}{2\,.\,3\,.\,19^2}\,\frac{\mathrm{Y}}{r^2}\cos vt$$

and $\dfrac{\mathrm{Y}}{wr^2}\cos vt$ must be put equal to (the first approximation) $\mathrm{S}\cos vt - \mathfrak{g}\dfrac{\sigma_{\prime}}{a}$. Hence by (57) and (83) the surface radial displacement due to inertia is

$$\frac{w^2v^2a^5}{\upsilon^2}\,\frac{79}{2\,.\,3\,.\,19^2}\,\frac{\mathfrak{r}}{\mathfrak{r}+\mathfrak{g}}\mathrm{S}\cos vt$$

To this we must add the displacement due directly to the effective disturbing potential $wr^2\left(\mathrm{S}\cos vt - \mathfrak{g}\dfrac{\sigma}{a}\right)$, where σ is now the second approximation. This we know from (82) is equal to

$$\frac{5wa^3}{19\upsilon}\left(\mathrm{S}\cos vt - \mathfrak{g}\frac{\sigma}{a}\right)$$

Hence the total radial displacement is

$$\frac{5wa^3}{19\upsilon}\left(\mathrm{S}\cos vt - \mathfrak{g}\frac{\sigma}{a} + \frac{5wa^2}{19\upsilon}\cdot\frac{79v^2}{150}\,\frac{\mathfrak{r}}{\mathfrak{r}+\mathfrak{g}}\mathrm{S}\cos vt\right)$$

But the total radial displacement is itself equal to σ.

Therefore
$$\mathfrak{r}\frac{\sigma}{a} = \mathrm{S}\cos vt - \mathfrak{g}\frac{\sigma}{a} + \frac{79v^2}{150(\mathfrak{r}+\mathfrak{g})}\mathrm{S}\cos vt$$

and
$$\frac{\sigma}{a} = \frac{\mathrm{S}}{\mathfrak{r}+\mathfrak{g}}\cos vt\left(1 + \frac{79v^2}{150(\mathfrak{r}+\mathfrak{g})}\right)$$

This is the second approximation to the form of the tidal spheroid, and from it we see that inertia has the effect of increasing the ellipticity of the spheroid in the proportion $1 + \frac{79v^2}{150(\mathfrak{r}+\mathfrak{g})}$.

Analogy with (76) would lead one to believe that the period of the gravest vibration of an elastic sphere is $2\pi\left(\frac{79}{150\mathfrak{r}}\right)^{\frac{1}{2}}$; this result might be tested experimentally*.

If $\mathfrak{g}$ be put equal to zero, the sphere is devoid of gravitation, and if $\mathfrak{r}$ be put equal to zero the sphere becomes perfectly fluid; but the solution is then open to objections similar to those considered, when viscosity graduates into fluidity.

It is obvious that the whole of this present part might be easily adapted to that hypothesis of elastico-viscosity which was considered in the paper on "Tides," but it does not at present seem worth while to do so.

By substituting these second approximations in the equations of motion again, we might proceed to a third approximation, and so on; but the analytical labour of the process would become very great.

IV. *Discussion of the applicability of the results to the history of the earth.*

The first paper of this series was devoted to the consideration of inequalities of short period, in the state of flow of the interior, and in the form of surface, produced in a rotating viscous sphere by the attraction of an external disturbing body: this was the theory of tides. The investigation was admitted to be approximate from two causes—(i) the neglect of the inertia of the relative motion of the parts of the spheroid; (ii) the neglect of tangential action between the surface of the mean sphere and the tidal protuberances.

* [At p. 211 of the paper referred to above, Professor Lamb finds the period of this vibration to be $2\left(\frac{wa^2}{v}\right)^{\frac{1}{2}} \div \cdot 842$, the notation being changed so as to agree with mine. My result may be written $2\left(\frac{wa^2}{v}\right)^{\frac{1}{2}} \div \frac{1}{\pi}\sqrt{\frac{570}{79}}$. Now $\frac{1}{\pi}\sqrt{\frac{570}{79}}$ is equal to $\cdot 855$, so that there is a close agreement between my result and the rigorous solution.]

In the second paper the inertia was still neglected, but the effects of these tangential actions were considered, in as far as they modified the rotation of the spheroid as a whole. In that paper the sphere was treated as though it were rigid, but had rigidly attached to its surface certain inequalities, which varied in distribution from instant to instant according to the tidal theory.

In order to justify this assumption, it is now necessary to examine whether the tidal protuberances may be regarded as instantaneously and rigidly connected with the rotating sphere. If there is a secular distortion of the spheroid in excess of the regular tidal flux and reflux, the assumption is not rigorously exact; but if the distortion be very slow, the departure from exactness may be regarded as insensible.

The first problem in the present paper is the investigation of the amount of secular distortion, and it is treated only in the simple case of a single disturbing body, or moon, moving in the equator of the tidally-distorted spheroid or earth.

It is found, then, that the form of the lagging tide in the earth is not such that the pull, exercised by the moon on it, can retard the earth's rotation exactly as though the earth were a rigid body. In other words, there is an unequal distribution of the tidal frictional couple in various latitudes.

We may see in a general way that the tidal protuberance is principally equatorial, and that accordingly the moon tends to retard the diurnal rotation of the equatorial portions of the sphere more rapidly than that of the polar regions. Hence the polar regions tend to outstrip the equator, and there is a slow motion from west to east relatively to the equator.

When, however, we come to examine numerically the amount of this screwing motion of the earth's mass, it appears that the distortion is exceedingly slow, and accordingly the assumption of the instantaneous rigid connexion of the tidal protuberance with the mean sphere is sufficiently accurate to allow all the results of the paper on "Precession" to hold good.

In the special case, which was the subject of numerical solution in that paper, we were dealing with a viscous mass which in ordinary parlance would be called a solid, and it was maintained that the results might possibly be applicable to the earth within the limits of geological history.

Now the present investigation shows that if we look back 45,000,000 years from the present state of things, we might find a point in lat. 30° further west with reference to a point on the equator, by $4\frac{3}{4}'$ than at present, and a point in lat. 60° further west by $14\frac{1}{4}'$. The amount of distortion of the surface strata is also shown to be exceedingly minute.

From these results we may conclude that this cause has had little or nothing to do with the observed crumpling of strata, at least within recent geological times.

If, however, the views maintained in the paper on "Precession" as to the remote history of the earth are correct, it would not follow, from what has been stated above, that this cause has never played an important part; for the rate of the screwing of the earth's mass varies inversely as the sixth power of the moon's distance, multiplied by the angular velocity of the earth relatively to the moon. And according to that theory, in very early times the moon was very near the earth, whilst the relative angular velocity was comparatively great. Hence the screwing action may have been once sensible*.

Now this sort of motion, acting on a mass which is not perfectly homogeneous, would raise wrinkles on the surface which would run in directions perpendicular to the axis of greatest pressure.

In the case of the earth the wrinkles would run north and south at the equator, and would bear away to the eastward in northerly and southerly latitudes; so that at the north pole the trend would be north-east, and at the south pole north-west. Also the intensity of the wrinkling force varies as the square of the cosine of the latitude, and is thus greatest at the equator, and zero at the poles. Any wrinkle when once formed would have a tendency to

* This result is not strictly applicable to the case of infinitely small viscosity, because it gives a finite though very small circulation, if the coefficient of viscosity be put equal to zero.

By putting $\epsilon=0$ in (17′), Part I., we find a superior limit to the rate of distortion. With the present angular velocities of the earth and moon, $\frac{dL}{dt}$ must be less than $5\times10^{-9}\cos^2\theta$ in degrees per annum.

It is easy to find when $\frac{dL}{dt}$ would be a maximum in the course of development considered in "Precession"; for, neglecting the solar effects, it will be greatest when $\tau^2(n-\Omega)$ is greatest.

Now $\tau^2(n-\Omega)$ varies as $[1+\mu-\mu\xi-\frac{\Omega_0}{n_0}.\xi^{-3}]\xi^{-12}$, and this function is a maximum when

$$\xi^{-4}-\tfrac{12}{15}(1+\mu)\frac{n_0}{\Omega_0}\xi^{-1}+\tfrac{11}{15}\mu\frac{n_0}{\Omega_0}=0$$

Taking $\mu=4{\cdot}0074$, and $\frac{n_0}{\Omega_0}=27{\cdot}32$, we have $\xi^{-4}-109{\cdot}45\,\xi^{-1}+80{\cdot}293=0$.

The solution of this is $\xi={\cdot}2218$.

With this solution dL/dt will be found to be 56 million times as great as at present, being equal to $18'\cos^2\theta$ per annum. With this value of ξ, the length of the day is 5 hours 50 minutes, and of the month 7 hours 10 minutes.

This gives a superior limit to the greatest rate of distortion that can ever have occurred.

By (19′), however, we see that the rate of distortion per unit increment of the moon's distance may be made as large as we please by taking the coefficient of viscosity small enough.

These considerations seem to show that there is no reason why this screwing action of the earth should not once have had considerable effects. (Added October 15, 1879.)

turn slightly, so as to become more nearly east and west, than it was when first made.

The general configuration of the continents (the large wrinkles) on the earth's surface appears to me remarkable when viewed in connexion with these results.

There can be little doubt that, on the whole, the highest mountains are equatorial, and that the general trend of the great continents is north and south in those regions. The theoretical directions of coast line are not so well marked in parts removed from the equator.

The great line of coast running from North Africa by Spain to Norway has a decidedly north-easterly bearing, and the long Chinese coast exhibits a similar tendency. The same may be observed in the line from Greenland down to the Gulf of Mexico, but here we meet with a very unfavourable case in Panama, Mexico, and the long Californian coast line.

From the paucity of land in the southern hemisphere the indications are not so good, nor are they very favourable to these views. The great line of elevation which runs from Borneo through Queensland to New Zealand might perhaps be taken as an example of north-westerly trend. The Cordilleras run very nearly north and south, but exhibit a clear north-westerly twist in Tierra del Fuego, and there is another slight bend of the same character in Bolivia.

But if this cause was that which principally determined the direction of terrestrial inequalities, the view must be held that the general position of the continents has always been somewhat as at present, and that, after the wrinkles were formed, the surface attained a considerable rigidity, so that the inequalities could not entirely subside during the continuous adjustment to the form of equilibrium of the earth, adapted at each period to the lengthening day. With respect to this point, it is worthy of remark that many geologists are of opinion that the great continents have always been more or less in their present positions.

An inspection of Professor Schiapparelli's map of Mars*, I think, will prove that the north and south trend of continents is not something peculiar to the earth. In the equatorial regions we there observe a great many very large islands, separated by about twenty narrow channels running approximately north and south. The northern hemisphere is not given beyond lat. 40°, but the coast lines of the southern hemisphere exhibit a strongly marked north-westerly tendency. It must be confessed, however, that the case of Mars is almost too favourable, because we have to suppose, according

* *Appendice alle Memorie della Società degli Spettroscopisti Italiani*, 1878, Vol. VII., for a copy of which I have to thank M. Schiapparelli.

to the theory, that its distortion is due to the sun, from which the planet must always have been distant. The very short period of the inner satellite shows, however, that the Martian rotation must have been (according to the theory) largely retarded; and where there has been retardation, there must have been internal distortion.

The second problem which is considered in the first part of the present paper is concerned with certain secondary tides. My attention was called to these tides by some remarks of Dr Jules Carret*, who says:

"Les actions perturbatrices du soleil et de la lune, qui produisent les mouvements coniques de la précession des équinoxes et de la nutation, n'agissent que sur cette portion de l'ellipsoïde terrestre qui excède la sphère tangente aux deux pôles, c'est-à-dire, en admettant l'état pâteux de l'intérieur, à peu près uniquement sur ce que l'on est convenu d'appeler la croûte terrestre, et presque sur toute la croûte terrestre. La croûte glisse sur l'intérieur plastique. Elle parvient à entrainer l'intérieur, car, sinon, l'axe de la rotation du globe demeurerait parallèle à lui-même dans l'espace, ou n'éprouverait que des variations insignifiantes, et le phenomène de la précession des equinoxes n'existerait pas. Ainsi la croûte et l'intérieur se meuvent de quantités inégales, d'où le déplacement géographique du pôlo sur la sphère.

"Cette idée a été émise, je crois, pour la première fois, par M. Evans; depuis par M. J. Péroche."

Now with respect to this view, it appears to me to be sufficient to remark that, as the axes of the precessional and nutational couples are fixed relatively to the moon, whilst the earth rotates, therefore the tendency of any particular part of the crust to slide over the interior is reversed in direction every twelve lunar hours, and therefore the result is not a secular displacement of the crust, but a small tidal distortion.

As, however, it was just possible that this general method of regarding the subject overlooked some residual tendency to secular distortion, I have given the subject a more careful consideration. From this it appears that there is no other tendency to distortion besides that arising out of tidal friction, which has just been discussed. It is also found that the secondary tides must be very small compared with the primary ones; with the present angular velocity of diurnal rotation, probably not so much in height as one-hundredth of the primary lunar semi-diurnal bodily tide.

It seems out of the question that any heterogeneity of viscosity could alter this result, and therefore it may, I think, be safely asserted that any

* *Société Savoisienne d'Histoire et d'Archéologie*, May 23, 1878. He is also author of a work, *Le Déplacement Polaire*. I think Dr Carret has misunderstood Mr [now Sir John] Evans.

sliding of the crust over the interior is impossible—at least as arising from this set of causes.

The second part of the paper is an investigation of the amount of work done in the interior of the viscous sphere by the bodily tidal distortion.

According to the principles of energy, the work done on any element makes itself manifest in the form of heat. The whole work which is done on the system in a given time is equal to the whole energy lost to the system in the same time. From this consideration an estimate was given, in the paper on "Precession," of the whole amount of heat generated in the earth in a given time. In the present paper the case is taken of a moon moving round the earth in the plane of the equator, and the work done on each element of the interior is found. The work done on the whole earth is found by summing up the work on each element, and it appears that the work per unit time is equal to the tidal frictional couple multiplied by the relative angular velocity of the two bodies. This remarkably simple law results from a complex law of internal distribution of work, and its identity with the law found in "Precession," from simple considerations of energy, affords a valuable confirmation of the complete consistency of the theory of tides with itself.

Fig. 2 gives a graphical illustration of the distribution in the interior of the work done, or of the heat generated, which amounts to the same thing. The reader is referred to Part II. for an explanation of the figure. Mere inspection of the figure shows that by far the larger part of the heat is generated in the central parts, and calculation shows that about one-third of the whole heat is generated within the central one-eighth of the volume, whilst in a spheroid of the size of the earth only one-tenth is generated within 500 miles of the surface.

In the paper on "Precession" the changes in the system of the sun, moon, and earth were traced backwards from the present lengths of day and month back to a common length of day and month of 5 hours 36 minutes, and it was found that in such a change heat enough must have been generated within the earth to raise its whole mass 3000° Fahr. if applied all at once, supposing the earth to have the specific heat of iron. It appeared to me at that time that, unless these changes took place at a time very long antecedent to geological history, this enormous amount of internal heat generated would serve in part to explain the increase of temperature in mines and borings. Sir William Thomson, however, pointed out to me that the distribution of heat-generation would probably be such as to prevent the realisation of my expectations. I accordingly made the further calculations, connected with the secular cooling of the earth, comprised in the latter portion of Part II.

It is first shown that, taking certain average values for the increase of underground temperature and for the conductivity of the earth, the earth (considered homogeneous) must be losing by conduction outwards an amount of energy equal to its present kinetic energy of rotation in about 262 million years.

It is next shown that in the passage of the system from a day of 5 hours 40 minutes to one of 24 hours, there is lost to the system an amount of energy equal to 13½ times the present kinetic energy of rotation of the earth. Thus it appears that, at the present rate of loss, the internal friction gives a supply of heat for 3,560 million years. So far it would seem that internal friction might be a powerful factor in the secular cooling of the earth, and the next investigation is directly concerned with that question.

In the case of the tidally-distorted sphere the distribution of heat-generation depends on latitude as well as depth from the surface, but the average law of heat-generation, as dependent on depth alone, may easily be found. Suppose, then, that we imagine an infinite slab of rock 8,000 miles thick, and that we liken the medial plane to the earth's centre and suppose the heat to be generated uniformly in time, according to the average law above referred to. Conceive the two faces of the slab to be always kept at the same constant temperature, and that initially, when the heat-generation begins, the whole slab is at this same temperature. The problem then is, to find the rate of increase of temperature going inwards from either face of the slab after any time.

This problem is solved, and by certain considerations (for which the reader is referred back) is made to give results which must agree pretty closely with the temperature gradient at the surface of an earth in which 13½ times the present kinetic energy of earth's rotation, estimated as heat, is uniformly generated in time, with the average space distribution referred to. It appears that at the end of the heat-generation the temperature gradient at the surface is sensibly the same, at whatever rate the heat is generated, provided it is all generated within 1,000 million years; but the temperature gradient can never be quite so steep as if the whole heat were generated instantaneously. The gradient, if the changes take place within 1,000 million years, is found to be about 1° Fahr. in 2,600 feet. Now the actually observed increase of underground temperature is something like 1° Fahr. in 50 feet; it therefore appears that perhaps one-fiftieth of the present increase of underground temperature may possibly be referred to the effects of long past internal friction. It follows that Sir William Thomson's investigation of the secular cooling of the earth is not sensibly affected by these considerations*.

If at any time in the future we should attain to an accurate knowledge of the increase of underground temperature, it is just within the bounds of

* [The conclusion might be different if the earth were to consist of a rigid nucleus covered by a thick or thin stratum of viscous material.]

possibility that a smaller rate of increase of temperature may be observed in the equatorial regions than elsewhere, because the curve of equal heat generation, which at the equator is nearly 500 miles below the surface, actually reaches the surface at the pole.

The last problem here treated is concerned with the effects of inertia on the tides of a viscous spheroid. As this part will be only valuable to those who are interested in the actual theory of tides, it may here be dismissed in a few words. The theory used in the two former papers, and in the first two parts of the present one, was founded on the neglect of inertia; and although it was shown in the paper on "Tides" that the error in the results could not be important, in the case of a sphere disturbed by tides of a frequency equal to the present lunar and solar tides, yet this neglect left a defect in the theory which it was desirable to supply. Moreover it was possible that, when the frequency of the tides was much more rapid than at present (as was found to have been the case in the paper on "Precession"), the theory used might be seriously at fault.

It is here shown (see (62)) that for a given lag of tide the height of tide is a little greater, and that for a given frequency of tide the lag is a little greater than the approximate theory supposed.

A rough correction is then applied to the numerical results given in the paper on "Precession" for the secular changes in the configuration of the system; it appears that the time occupied by the changes in the first solution (Section 15) is overstated by about one-fortieth part, but that all the other results, both in this solution and the other, are left practically unaffected. To the general reader, therefore, the value of this part of the paper simply lies in its confirmation of previous work.

From a mathematical point of view, a comparison of the methods employed with those for finding the forced oscillations of liquid spheres is instructive.

Lastly, the analytical investigation of the effects of inertia on the forced oscillations of a viscous sphere is found to be applicable, almost verbatim, to the same problem concerning an elastic sphere. The results are complementary to those of Sir William Thomson's statical theory of the tides of an elastic sphere.

5.

THE DETERMINATION OF THE SECULAR EFFECTS OF TIDAL FRICTION BY A GRAPHICAL METHOD.

[*Proceedings of the Royal Society of London*, XXIX. (1879), pp. 168—181.]

SUPPOSE an attractive particle or satellite of mass m to be moving in a circular orbit, with an angular velocity Ω, round a planet of mass M, and suppose the planet to be rotating about an axis perpendicular to the plane of the orbit, with an angular velocity n; suppose, also, the mass of the planet to be partially or wholly imperfectly elastic or viscous, or that there are oceans on the surface of the planet; then the attraction of the satellite must produce a relative motion in the parts of the planet, and that motion must be subject to friction, or, in other words, there must be frictional tides of some sort or other. The system must accordingly be losing energy by friction, and its configuration must change in such a way that its whole energy diminishes.

Such a system does not differ much from those of actual planets and satellites, and, therefore, the results deduced in this hypothetical case must agree pretty closely with the actual course of evolution, provided that time enough has been and will be given for such changes.

Let C be the moment of inertia of the planet about its axis of rotation;

r the distance of the satellite from the centre of the planet;

h the resultant moment of momentum of the whole system;

e the whole energy, both kinetic and potential, of the system.

It will be supposed that the figure of the planet and the distribution of its internal density are such that the attraction of the satellite causes no couple about any axis perpendicular to that of rotation.

The two bodies revolve in circles about their common centre of inertia with an angular velocity Ω, and, therefore, the moment of momentum of orbital motion is

$$M\left(\frac{mr}{M+m}\right)^2 \Omega + m\left(\frac{Mr}{M+m}\right)^2 \Omega = \frac{Mm}{M+m} r^2\Omega$$

If μ be attraction between unit masses at unit distance, by the law of periodic times in a circular orbit,

$$\Omega^2 r^3 = \mu(M+m)$$

whence

$$\Omega r^2 = \mu^{\frac{2}{3}} (M+m)^{-\frac{4}{3}} \Omega^{-\frac{1}{3}}$$

And the moment of momentum of orbital motion $= \mu^{\frac{2}{3}} Mm (M+m)^{-\frac{1}{3}} \Omega^{-\frac{1}{3}}$

The moment of momentum of the planet's rotation is Cn, and therefore

$$h = C\left\{n + \mu^{\frac{2}{3}} \frac{Mm}{C} (M+m)^{-\frac{1}{3}} \Omega^{-\frac{1}{3}}\right\} \quad \text{..................(1)}$$

Again, the kinetic energy of orbital motion is

$$\tfrac{1}{2}M\left(\frac{mr}{M+m}\right)^2 \Omega^2 + \tfrac{1}{2}m\left(\frac{Mr}{M+m}\right)^2 \Omega^2 = \tfrac{1}{2}\frac{Mm}{M+m} r^2\Omega^2 = \tfrac{1}{2}\mu^{\frac{2}{3}} Mm (M+m)^{-\frac{1}{3}} \Omega^{\frac{2}{3}}$$

The kinetic energy of the planet's rotation is $\tfrac{1}{2}Cn^2$.

The potential energy of the system is

$$-\mu\frac{Mm}{r} = -\mu^{\frac{2}{3}} Mm (M+m)^{-\frac{1}{3}} \Omega^{\frac{2}{3}}$$

Adding the three energies together

$$2e = C\left\{n^2 - \mu^{\frac{2}{3}} \frac{Mm}{C} (M+m)^{-\frac{1}{3}} \Omega^{\frac{2}{3}}\right\} \quad \text{..................(2)}$$

Now, suppose that by a proper choice of the unit of time,

$$\mu^{\frac{2}{3}} \frac{Mm}{C} (M+m)^{-\frac{1}{3}}$$

is unity, and that by a proper choice of units of length or of mass C is unity*, and let

$$x = \Omega^{-\frac{1}{3}}, \quad y = n, \quad Y = 2e$$

* If g be the mean gravity at the surface of the planet, a its mean radius, and $\nu = M/m$,

$$\mu(M+m) = ga^2 \frac{1+\nu}{\nu}$$

and

$$\mu^{\frac{2}{3}} Mm (M+m)^{-\frac{1}{3}} = \left[ga^2 \frac{1+\nu}{\nu}\right]^{\frac{2}{3}} \frac{M}{1+\nu} = Ma^2 \div \left\{\left(\frac{a\nu}{g}\right)^2 (1+\nu)\right\}^{\frac{1}{3}}$$

If the planet be homogeneous, and differ infinitesimally from a sphere $C = \tfrac{2}{5}Ma^2$, and

$$\mu^{\frac{2}{3}} \frac{Mm}{C} (M+m)^{-\frac{1}{3}} = 1 \div \tfrac{2}{5} \left\{\left(\frac{a\nu}{g}\right)^2 (1+\nu)\right\}^{\frac{1}{3}} = \frac{1}{s}, \text{ suppose}$$

in the case of the earth, considered as heterogeneous, the $\tfrac{2}{5}$ would be replaced by about $\tfrac{1}{3}$.

It may be well to notice that x is proportional to the square root of the satellite's distance from the planet.

Then the equations (1) and (2) become

$$h = y + x \quad \text{.............................(3)}$$

$$Y = y^2 - \frac{1}{x^2} = (h - x)^2 - \frac{1}{x^2} \quad \text{.......................(4)}$$

(3) is the equation of conservation of moment of momentum, or shortly, the equation of momentum; (4) is the equation of energy.

Now, consider a system started with given positive (or say clockwise) moment of momentum h; we have all sorts of ways in which it may be started. If the two rotations be of opposite kinds, it is clear that we may start the system with any amount of energy however great, but the true maxima and minima of energy compatible with the given moment of momentum are given by $\frac{dY}{dx} = 0$, or

$$x - h + \frac{1}{x^3} = 0$$

or

$$x^4 - hx^3 + 1 = 0 \quad \text{..............................(5)}$$

We shall presently see that this biquadratic has either two real roots and two imaginary, or all imaginary roots.

This biquadratic may be derived from quite a different consideration, viz., by finding the condition under which the satellite may move round the planet, so that the planet shall always show the same face to the satellite, in fact, so that they move as parts of one rigid body.

The condition is simply that the satellite's orbital angular velocity $\Omega = n$ the planet's angular velocity round its axis; or since $n = y$ and $\Omega^{-\frac{1}{3}} = x$, therefore $y = 1/x^3$.

It is clear that $s^{\frac{3}{4}}$ is a time; and in the case of the earth and moon (with $\nu = 82$),

$s^{\frac{3}{4}} = 3$ hrs. $4\frac{1}{2}$ mins., if the earth be homogeneous

and $s^{\frac{3}{4}} = 2$ hrs. 41 mins., if the earth be heterogeneous

For the units of length and mass we have only to choose them so that $\frac{2}{5}Ma^2$, or $\frac{1}{3}Ma^2$, may be unity.

With these units it will be found that for the present length of day $n = \cdot 8056$ (homog.) or $\cdot 7026$ (heterog.), and that

$$h = \cdot 8056\,[1 + 4\cdot 01] = 4\cdot 03 \text{ (homog.)}$$

$$\text{or } h = \cdot 7026\,[1 + 4\cdot 38] = 3\cdot 78 \text{ (heterog.)}$$

For the value 4·38 see Thomson and Tait's *Natural Philosophy*, § 276, where tidal friction is considered.

By substituting this value of y in the equation of momentum (3), we get as before

$$x^4 - hx^3 + 1 = 0 \quad \text{.............................(5)}$$

In my paper on the "Precession of a Viscous Spheroid" [Paper 3] I obtained the biquadratic equation from this last point of view only, and considered analytically and numerically its bearings on the history of the earth.

Sir William Thomson, having read the paper, told me that he thought that much light might be thrown on the general physical meaning of the equation, by a comparison of the equation of conservation of moment of momentum with the energy of the system for various configurations, and he suggested the appropriateness of geometrical illustration for the purpose of this comparison. The method which is worked out below is the result of the suggestions given me by him in conversation.

The simplicity with which complicated mechanical interactions may be thus traced out geometrically to their results appears truly remarkable.

At present we have only obtained one result, viz.: that if with given moment of momentum it is possible to set the satellite and planet moving as a rigid body, then it is possible to do so in two ways, and one of these ways requires a maximum amount of energy and the other a minimum; from which it is clear that one must be a rapid rotation with the satellite near the planet, and the other a slow one with the satellite remote from the planet.

Now, consider the three equations,

$$h = y + x \quad \text{.....................................(6)}$$

$$Y = (h - x)^2 - \frac{1}{x^2} \quad \text{.............................(7)}$$

$$x^3 y = 1 \quad \text{.....................................(8)}$$

(6) is the equation of momentum; (7), that of energy; and (8) we may call the equation of rigidity, since it indicates that the two bodies move as though parts of one rigid body.

If we wish to illustrate these equations graphically, we may take as abscissa x, which is the moment of momentum of orbital motion; so that the axis of x may be called the axis of orbital momentum. Also, for equations (6) and (8) we may take as ordinate y, which is the moment of momentum of the planet's rotation; so that the axis of y may be called the axis of rotational momentum. For (7) we may take as ordinate Y, which is twice the energy of the system; so that the axis of Y may be called the axis of energy. As it will be convenient to exhibit all three curves in the same figure, with a parallel axis of x, we must have the axis of energy identical with that of rotational momentum.

It will not be necessary to consider the case where the resultant moment of momentum h is negative, because this would only be equivalent to reversing all the rotations; thus h is to be taken as essentially positive.

The line of momentum, whose equation is (6), is a straight line at 45° to either axis, having positive intercepts on both axes.

The curve of rigidity, whose equation is (8), is clearly of the same nature as a rectangular hyperbola, but having a much more rapid rate of approach to the axis of orbital momentum than to that of rotational momentum.

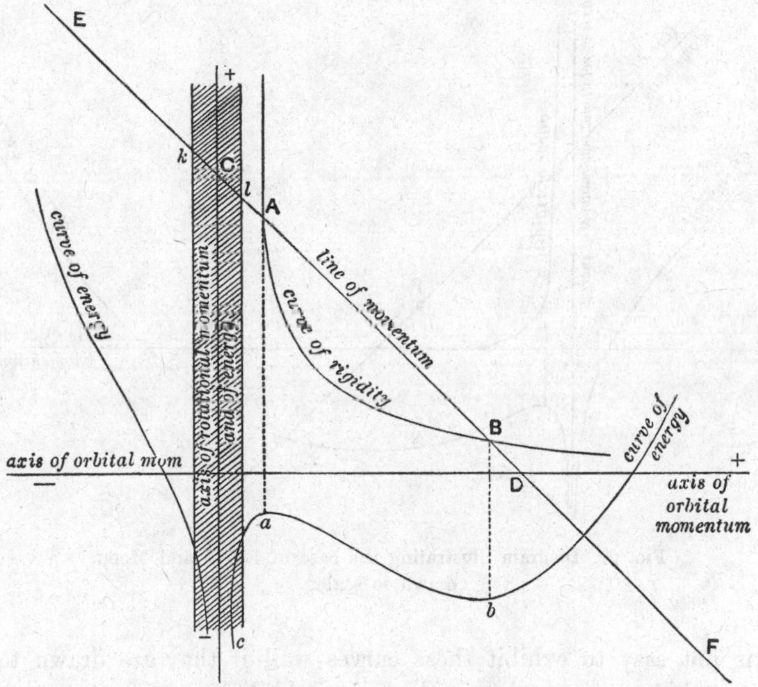

Fig. 1. Graphical illustration of the equations specifying the system.

The intersections (if any) of the curve of rigidity with the line of momentum have abscissæ which are the two roots of the biquadratic $x^4 - hx^3 + 1 = 0$. The biquadratic has, therefore, two real roots or all imaginary roots. Since $x = \Omega^{-\frac{1}{3}}$, it varies as $\sqrt{r}$, and, therefore, the intersection which is more remote from the origin, indicates a configuration where the satellite is remote from the planet; the other gives the configuration where the satellite is closer to the planet. We have already learnt that these two correspond respectively to minimum and maximum energy.

When x is very large, the equation to the curve of energy is $Y = (h - x)^2$, which is the equation to a parabola, with a vertical axis parallel to Y and distant h from the origin, so that the axis of the parabola passes through the intersection of the line of momentum with the axis of orbital momentum.

When x is very small the equation becomes $Y = -1/x^2$.

Hence, the axis of Y is asymptotic on both sides to the curve of energy.

If the line of momentum intersects the curve of rigidity, the curve of energy has a maximum vertically underneath the point of intersection nearer the origin, and a minimum underneath the point more remote. But if there are no intersections, it has no maximum or minimum.

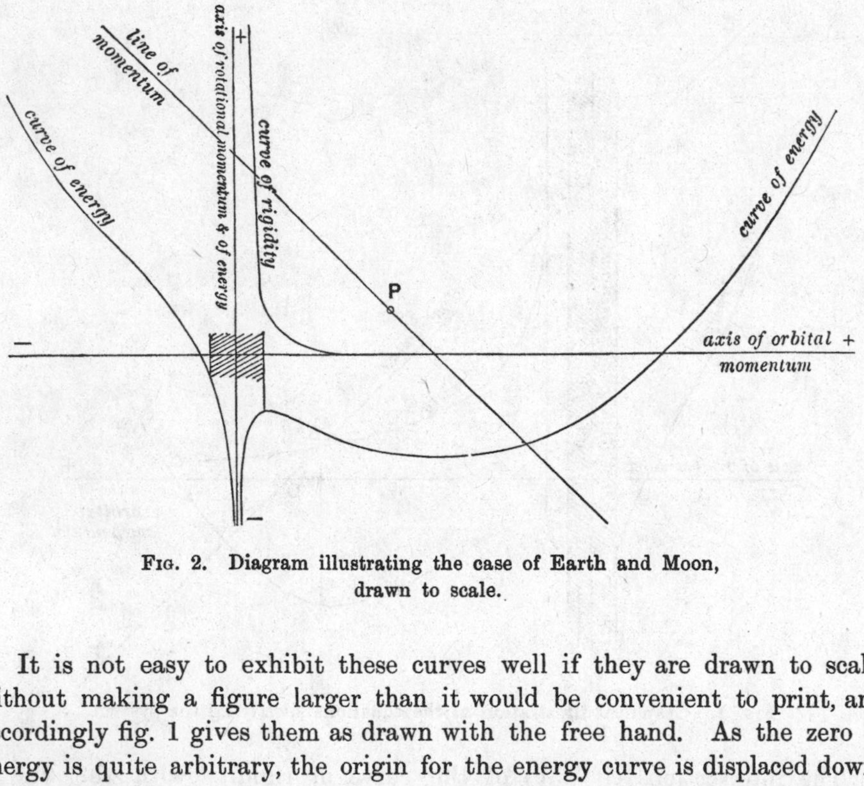

Fig. 2. Diagram illustrating the case of Earth and Moon, drawn to scale.

It is not easy to exhibit these curves well if they are drawn to scale, without making a figure larger than it would be convenient to print, and accordingly fig. 1 gives them as drawn with the free hand. As the zero of energy is quite arbitrary, the origin for the energy curve is displaced downwards, and this prevents the two curves from crossing one another in a confusing manner. The same remark applies also to figs. 2 and 3.

Fig. 1 is erroneous principally in that the curve of rigidity ought to approach its horizontal asymptote much more rapidly, so that it would be difficult in a drawing to scale to distinguish the points of intersection B and D.

Fig. 2 exhibits the same curves, but drawn to scale, and designed to be applicable to the case of the earth and moon, that is to say, when $h = 4$ nearly.

Fig. 3 shows the curves when $h=1$, and when the line of momentum does not intersect the curve of rigidity; and here there is no maximum or minimum in the curve of energy.

These figures exhibit all the possible methods in which the bodies may move with given moment of momentum, and they differ in the fact that in figs. 1 and 2 the biquadratic (5) has real roots, but in the case of fig. 3 this is not so. Every point of the line of momentum gives by its abscissa and ordinate the square root of the satellite's distance and the rotation of the planet, and the ordinate of the energy curve gives the energy corresponding to each distance of the satellite.

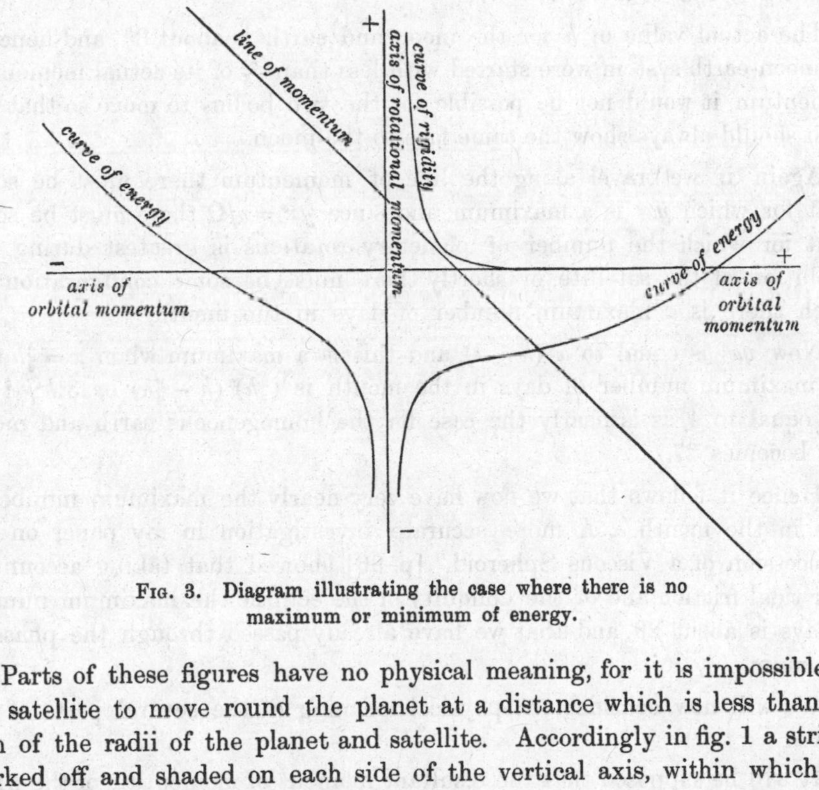

FIG. 3. Diagram illustrating the case where there is no maximum or minimum of energy.

Parts of these figures have no physical meaning, for it is impossible for the satellite to move round the planet at a distance which is less than the sum of the radii of the planet and satellite. Accordingly in fig. 1 a strip is marked off and shaded on each side of the vertical axis, within which the figure has no physical meaning.

Since the moon's diameter is about 2,200 miles, and the earth's about 8,000, therefore the moon's distance cannot be less than 5,100 miles; and in fig. 2, which is intended to apply to the earth and moon and is drawn to scale, the base only of the strip is shaded, so as not to render the figure confused. The strip has been accidentally drawn a very little too broad.

The point P in fig. 2 indicates the present configuration of the earth and moon.

The curve of rigidity $x^3y = 1$ is the same for all values of h, and by moving the line of momentum parallel to itself nearer or further from the origin, we may represent all possible moments of momentum of the whole system.

The smallest amount of moment of momentum with which it is possible to set the system moving as a rigid body, is when the line of momentum touches the curve of rigidity. The condition for this is clearly that the equation $x^4 - hx^3 + 1 = 0$ should have equal roots. If it has equal roots each root must be $\frac{3}{4}h$, and therefore

$$(\tfrac{3}{4}h)^4 - h\,(\tfrac{3}{4}h)^3 + 1 = 0$$

whence $h^4 = 4^4/3^3$ or $h = 4/3^{\frac{3}{4}} = 1{\cdot}75$.

The actual value of h for the moon and earth is about $3\frac{3}{4}$, and hence if the moon-earth system were started with less than $\frac{7}{15}$ of its actual moment of momentum, it would not be possible for the two bodies to move so that the earth should always show the same face to the moon.

Again if we travel along the line of momentum there must be some point for which yx^3 is a maximum, and since $yx^3 = n/\Omega$ there must be some point for which the number of planetary rotations is greatest during one revolution of the satellite, or shortly there must be some configuration for which there is a maximum number of days in the month.

Now yx^3 is equal to $x^3(h - x)$, and this is a maximum when $x = \frac{3}{4}h$ and the maximum number of days in the month is $(\frac{3}{4}h)^3(h - \frac{3}{4}h)$ or $3^3h^4/4^4$; if h is equal to 4, as is nearly the case for the homogeneous earth and moon, this becomes 27.

Hence it follows that we now have very nearly the maximum number of days in the month. A more accurate investigation in my paper on the "Precession of a Viscous Spheroid," [p. 96] showed that taking account of solar tidal friction and of the obliquity of the ecliptic the maximum number of days is about 29, and that we have already passed through the phase of maximum.

We will now consider the physical meaning of the several parts of the figures.

It will be supposed that the resultant moment of momentum of the whole system corresponds to a clockwise rotation.

Imagine two points with the same abscissa, one on the momentum line and the other on the energy curve, and suppose the one on the energy curve to guide that on the momentum line.

Since we are supposing frictional tides to be raised on the planet, the energy must degrade, and however the two points are set initially, the point on the energy curve must always slide down a slope carrying with it the other point.

Now looking at fig. 1 or 2, we see that there are four slopes in the energy curve, two running down to the planet, and two others which run down to the minimum. In fig. 3 on the other hand there are only two slopes, both of which run down to the planet.

In the first case there are four ways in which the system may degrade, according to the way it was started; in the second only two ways.

i. In fig. 1, for all points of the line of momentum from C through E to infinity, x is negative and y is positive; therefore this indicates an anti-clockwise revolution of the satellite, and a clockwise rotation of the planet, but the moment of momentum of planetary rotation is greater than that of the orbital motion. The corresponding part of the curve of energy slopes uniformly down, hence however the system be started, for this part of the line of momentum, the satellite must approach the planet, and will fall into it when its distance is given by the point k.

ii. For all points of the line of momentum from D through F to infinity, x is positive and y is negative; therefore the motion of the satellite is clockwise, and that of the planetary rotation anti-clockwise, but the moment of momentum of the orbital motion is greater than that of the planetary rotation. The corresponding part of the energy curve slopes down to the minimum b. Hence the satellite must approach the planet until it reaches a certain distance where the two will move round as a rigid body. It will be noticed that as the system passes through the configuration corresponding to D, the planetary rotation is zero, and from D to B the rotation of the planet becomes clockwise.

If the total moment of momentum had been as shown in fig. 3, the satellite would have fallen into the planet, because the energy curve would have no minimum.

From i. and ii. we learn that if the planet and satellite are set in motion with opposite rotations, the satellite will fall into the planet, if the moment of momentum of orbital motion be less than or equal to or only greater by a certain critical amount*, than the moment of momentum of planetary rotation, but if it be greater by more than a certain critical amount the satellite will approach the planet, the rotation of the planet will stop and reverse, and finally the system will come to equilibrium when the two bodies move round as a rigid body, with a long periodic time.

iii. We now come to the part of the figure between C and D. For the parts AC and BD of the line AB in fig. 1, the planetary rotation is slower than that of the satellite's revolution, or the month is shorter than the day, as in one of the satellites of Mars. In fig. 3 these parts together embrace the

* With the units which are here used the excess must be more than $4 \div 3^{\frac{3}{4}}$; see p. 202.

whole. In all cases the satellite approaches the planet. In the case of fig. 3, the satellite must ultimately fall into the planet; in the case of figs. 1 and 2 the satellite will fall in if its distance from the planet is small, or move round along with the planet as a rigid body if its distance be large.

For the part of the line of momentum AB, the month is longer than the day, and this is the case of all known satellites except the nearer one of Mars. As this part of the line is non-existent in fig. 3, we see that the case of all existing satellites (except the Martian one) is comprised within this part of figs. 1 and 2. If a satellite be placed in the condition A, that is to say, moving rapidly round a planet, which always shows the same face to the satellite, the condition is clearly dynamically unstable, for the least disturbance will determine whether the system shall degrade down the slopes *ac* or *ab*, that is to say, whether it falls into or recedes from the planet. If the equilibrium breaks down by the satellite receding, the recession will go on until the system has reached the state corresponding to B.

The point P, in fig. 2, shows approximately the present state of the earth and moon, viz., when $x = 3{\cdot}2$, $y = {\cdot}8$.

It is clear that, if the point *l*, which indicates that the satellite is just touching the planet, be identical with the point A, then the two bodies are in effect part of a single body in an unstable configuration. If, therefore, the moon was originally part of the earth, we should expect to find A and *l* identical. The figure 2, which is drawn to represent the earth and moon, shows that there is so close an approach between the edge of the shaded band and the intersection of the line of momentum and curve of rigidity, that it would be scarcely possible to distinguish them on the figure. Hence, there seems a considerable probability that the two bodies once formed parts of a single one, which broke up in consequence of some kind of instability. This view is confirmed by the more detailed consideration of the case in the paper on the "Precession of a Viscous Spheroid" [Paper 3].

Hitherto the satellite has been treated as an attractive particle, but the graphical method may be extended to the case where both the satellite and planet are spheroids rotating about axes perpendicular to the plane of the orbit.

Suppose, then, that k is the ratio of the moment of inertia of the satellite to that of the planet, and that z is equal to the angular velocity of the satellite round its axis, then kz is the moment of momentum of the satellite's rotation, and we have

$$h = x + y + kz \text{ for the equation to the plane of momentum}$$

$$2e = y^2 + kz^2 - \frac{1}{x^2} \text{ for the equation of energy}$$

and $x^3y = 1$, $x^3z = 1$ for the equation to the line of rigidity.

The most convenient form in which to put the equation to the surface of energy is

$$E = y^2 + kz^2 - \frac{1}{(h - y - kz)^2}$$

where E, y, z are the three ordinates.

The best way of understanding the surface is to draw the contour-lines of energy parallel to the plane of yz, as shown in fig. 4.

The case which I have considered may be called a double-star system, where the planet and satellite are equal and $k = 1$. Any other case may be easily conceived by stretching or contracting the surface parallel to z.

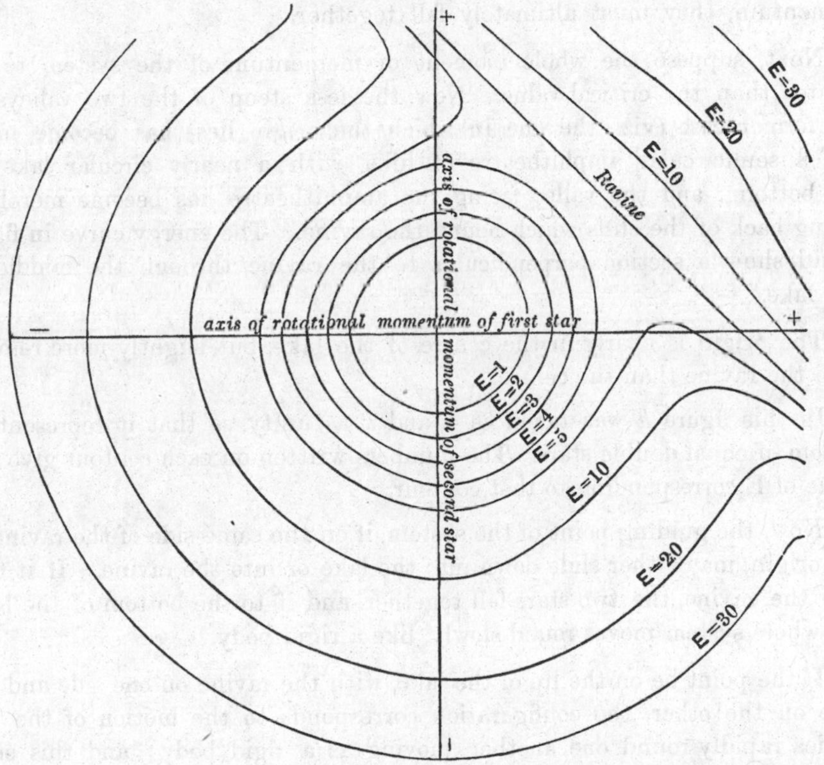

FIG. 4. Contour lines of energy surface for two equal stars, revolving about one another.

It will be found that, if the whole moment of momentum h has less than a certain critical value (found by the consideration that $x^4 - hx^3 + 2 = 0$ has equal roots), the surface may be conceived as an infinitely narrow and deep ravine, opening out at one part of its course into rounded valleys on each side of the ravine. In this case the contours would resemble those of fig. 4, supposing the round closed curves to be absent. The course of the ravine is at 45° to the axes of y and z, and the origin is situated in one of the valleys, which is less steep than the valley facing it on the opposite side of the ravine. The form of a section perpendicular to the ravine is such as the curve of energy in fig. 3, so that everywhere there is a slope towards the ravine.

Every point on the surface corresponds to one configuration of the system, and, if the system be guided by a point on the energy surface, that point must always slide down hill. It does not, however, necessarily follow that it will always slide down the steepest path. The fall of the guiding point into the ravine indicates the falling together of the two stars.

Thus, if the two bodies be started with less than a certain moment of momentum, they must ultimately fall together.

Next, suppose the whole moment of momentum of the system to be greater than the critical value. Now the less steep of the two valleys of the former case (viz., the one in which the origin lies) has become more like a semicircular amphitheatre of hills, with a nearly circular lake at the bottom; and the valley facing the amphitheatre has become merely a falling back of the cliffs which bound the ravine. The energy curve in fig. 2 would show a section perpendicular to the ravine through the middle of the lake.

The origin is nearly in the centre of the lake, but slightly more remote from the ravine than the centre.

In this figure h was taken as 4, and k as unity, so that it represents a system of equal double stars. The numbers written on each contour give the value of E corresponding to that contour.

Now, the guiding point of the system, if on the same side of the ravine as the origin, may either slide down into the lake or into the ravine. If it falls into the ravine, the two stars fall together, and if to the bottom of the lake, the whole system moves round slowly, like a rigid body.

If the point be on the lip of the lake, with the ravine on one side and the lake on the other, the configuration corresponds to the motion of the two bodies rapidly round one another, moving as a rigid body; and this state is clearly dynamically unstable.

If the point be on the other side of the ravine, it must fall into it, and the two stars fall together.

It has been remarked that the guiding point does not necessarily slide down the steepest gradient, and of such a mode of descent illustrations will be given hereafter.

Hence it is possible that, if the guiding point be started somewhere on the amphitheatre of hills, it may slide down until it comes to the lip of the lake. As far as one can see, however, such a descent would require a peculiar relationship of the viscosities of the two stars, probably varying from time to time. It is therefore possible, though improbable, that the unstable condition where the two bodies move rapidly round one another, always showing the same faces to one another, may be a degradation of a previous condition. If this state corresponds with a distance between the stars less than the sum of the radii of their masses, it clearly cannot be the result of such a degradation.

If, therefore, we can trace back a planet and satellite to this state, we have most probably found the state where the satellite first had a separate existence.

The conditions of stability of a rotating mass of fluid are very obscure, but it seems probable that, if the stability broke down and the mass gradually separated into two parts, the condition immediately after separation might be something like the unstable configuration described above.

In conclusion, I will add a few words to show that the guiding point on an energy surface need not necessarily move down the steepest path, but may even depart from the bottom of a furrow or move along a ridge. Of this two cases will be given.

The satellite will now be again supposed to be merely an attractive particle.

First, with given moment of momentum, the energy is greater when the axis of the planet is oblique to the orbit. Hence, if we draw an energy surface in which one of the co-ordinate axes corresponds to obliquity, there must be a furrow in the surface corresponding to zero obliquity. To conclude that the obliquity of the ecliptic must diminish in consequence of tidal friction would be erroneous. In fact, it appears, in my paper on the "Precession of a Viscous Spheroid" [Paper 3], that for a planet of small viscosity the position of zero obliquity is dynamically unstable, if the period of the satellite is greater than twice that of the planet's rotation. Thus the guiding point, though always descending on the energy surface, will depart from the bottom of the furrow.

Secondly. For given moment of momentum the energy is less if the orbit be eccentric, and an energy surface may be constructed in which zero eccentricity corresponds to a ridge. Now, I shall show in [Paper 6] that for small viscosity of the planet the circular orbit is dynamically stable if eighteen periods of the satellite be less than eleven periods of the planet's rotation. This will afford a case of the guiding point sliding down a ridge; when, however, the critical point is passed, the guiding point will depart from the ridge and the orbit become eccentric.

6.

ON THE SECULAR CHANGES IN THE ELEMENTS OF THE ORBIT OF A SATELLITE REVOLVING ABOUT A TIDALLY DISTORTED PLANET.

[*Philosophical Transactions of the Royal Society,* Vol. 171 (1880), pp. 713—891.]

TABLE OF CONTENTS.

	PAGE
Introduction	210
I. The Theory of the Disturbing Function.	
§ 1. Preliminary considerations	212
§ 2. Notation.—Equation of variation of elements	213
§ 3. To find spherical harmonic functions of Diana's coordinates with reference to axes fixed in the earth	218
§ 4. The disturbing function	222
II. Secular Changes in the Inclination of the Orbit of a Satellite.	
§ 5. The perturbed satellite moves in a circular orbit inclined to a fixed plane.—Subdivision of the problem	224
§ 6. Secular change of inclination of the orbit of a satellite, where there is a second disturbing body, and where the nodes revolve with sensible uniformity on the fixed plane of reference	229
§ 7. Application to the case where the planet is viscous	234
§ 8. Secular change in the mean distance of a satellite, where there is a second disturbing body, and where the nodes revolve with sensible uniformity on the fixed plane of reference	237
§ 9. Application to the case where the planet is viscous	237
§ 10. Secular change in the inclination of the orbit of a single satellite to the invariable plane, where there is no other disturbing body than the planet	238
§ 11. Secular change of mean distance under similar conditions.—Comparison with result of previous paper	241
§ 12. The method of the disturbing function applied to the motion of the planet	242

PAGE

III. THE PROPER PLANES OF THE SATELLITE, AND OF THE PLANET, AND THEIR SECULAR CHANGES.
§ 13. On the motion of a satellite moving about a rigid oblate spheroidal planet, and perturbed by another satellite . 250
§ 14. On the small terms in the equations of motion due directly to tidal friction 262
§ 15. On the secular changes of the constants of integration . 275
§ 16. Evaluation of a', a', &c., in the case of the earth's viscosity . 293
§ 17. Change of independent variable, and formation of equations for integration 296

IV. INTEGRATION OF THE DIFFERENTIAL EQUATIONS FOR CHANGES IN THE INCLINATION OF THE ORBIT AND THE OBLIQUITY OF THE ECLIPTIC.
§ 18. Integration in the case of small viscosity, where the nodes revolve uniformly 298
§ 19. Secular changes in the proper planes of the earth and moon when the viscosity is small 305
§ 20. Secular changes in the proper planes of the earth and moon when the viscosity is large 313
§ 21. Graphical illustration of the preceding integrations . 319
§ 22. The effects of solar tidal friction on the primitive condition of the earth and moon 322

V. SECULAR CHANGES IN THE ECCENTRICITY OF THE ORBIT.
§ 23. Formation of the disturbing function . . . 324
§ 24. Secular changes in eccentricity and mean distance . 336
§ 25. Application to the case where the planet is viscous . 338
§ 26. Secular change in the obliquity and diurnal rotation of the planet, when the satellite moves in an eccentric orbit . 342
§ 27. Verification of analysis, and effect of evectional tides . 345

VI. INTEGRATION FOR CHANGES IN THE ECCENTRICITY OF THE ORBIT.
§ 28. Integration in the case of small viscosity . . 346
§ 29. The change of eccentricity when the viscosity is large . 350

VII. SUMMARY AND DISCUSSION OF RESULTS.
§ 30. Explanation of problem.—Summary of Parts I. and II. 350
§ 31. Summary of Part III. 354
§ 32. Summary of Part IV. 358
§ 33. On the initial condition of the earth and moon . . . 363
§ 34. Summary of Parts V. and VI. 364

VIII. REVIEW OF THE TIDAL THEORY OF EVOLUTION AS APPLIED TO THE EARTH AND THE OTHER MEMBERS OF THE SOLAR SYSTEM . 367

APPENDIX.—A graphical illustration of the effects of tidal friction when the orbit of the satellite is eccentric 374

[APPENDIX A.—Extract from the abstract of the foregoing paper. *Proc. Roy. Soc.* Vol. xxx. (1880), pp. 1—10.] 380

Introduction.

THE following paper treats of the effects of frictional tides in a planet on the orbit of its satellite. It is the sequel to three previous papers on a similar subject*.

The investigation has proved to be one of unexpected complexity, and this must be my apology for the great length of the present paper. This was in part due to the fact that it was found impossible to consider adequately the changes in the orbit of the satellite, without a reconsideration of the parallel changes in the planet. Thus some of the ground covered in the previous paper on "Precession" had to be retraversed; but as the methods here employed are quite different from those used before, this repetition has not been without some advantage.

It will probably conduce to the intelligibility of what follows, if an explanatory outline of the contents of the paper is placed before the reader. Such an outline must of course contain references to future procedure, and cannot therefore be made entirely intelligible, yet it appears to me that some sort of preliminary notions of the nature of the subject will be advantageous, because it is sometimes difficult for a reader to retain the thread of the argument amidst the mass of details of a long investigation, which is leading him in some unknown direction.

Part VIII. contains a general review of the subject in its application to the evolution of the planets of the solar system. This is probably the only part of the paper which will have any interest to the general reader.

The mathematical reader, who merely wishes to obtain a general idea of the results, is recommended to glance through the present introduction, and then to turn to Part VII., which contains a summary, with references to such parts of the paper as it was not desirable to reproduce. This summary does not contain any analysis, and deals more especially with the physical aspects of the problem, and with the question of the applicability of the investigation to the history of the earth and moon, but of course it must not be understood

* "On the Bodily Tides of Viscous and Semi-elastic Spheroids, and on the Ocean Tides upon a Yielding Nucleus," *Phil. Trans.*, Part I., 1879. [Paper 1.]

"On the Precession of a Viscous Spheroid, and on the remote History of the Earth," *Phil. Trans.*, Part II., 1879. [Paper 3.]

"On Problems connected with the Tides of a Viscous Spheroid," *Phil. Trans.*, Part II., 1879. [Paper 4.]

These papers are hereafter referred to as "Tides," "Precession," and "Problems" respectively.

There is also a fourth paper, treating the subject from a different point of view, viz.: "The Determination of the Secular Effects of Tidal Friction by a Graphical Method," *Proc. Roy. Soc.*, No. 197, 1879. [Paper 5.] And lastly a fifth paper, "On the Analytical Expressions which give the history of a Fluid Planet of Small Viscosity, attended by a Single Satellite," *Proc. Roy. Soc.*, No. 202, 1880. [Paper 7.]

to contain references to every point which seems to be worthy of notice. I think also that a study of Part VII. will facilitate the comprehension of the analytical parts of the paper.

Part I. contains an explanation of the peculiarities of the method of the disturbing function as applied to the tidal problem. At the beginning there is a summary of the meaning to be attached to the principal symbols employed. The problem is divided into several heads, and the disturbing function is partially developed in such a way that it may be applicable either to finding the perturbations of the satellite, or of the planet itself.

In Part II. the satellite is supposed to move in a circular orbit, inclined to the fixed plane of reference. It here appears that the problem may be advantageously subdivided into the following cases: 1st, where the permanent oblateness of the planet is small, and where the satellite is directly perturbed by the action of a second large and distant satellite such as the sun; 2nd, where the planet and satellite are the only two bodies in existence; 3rd, where the permanent oblateness is considerable, and the action of the second satellite is not so important as in the first case. The first and second of these cases afford the subject for the rest of this part, and the laws are found which govern the secular changes in the inclination and mean distance of the satellite, and the obliquity and diurnal rotation of the planet.

Part III. is devoted to the third of the above cases. It was found necessary first to investigate the motion of a satellite revolving about a rigid oblate spheroidal planet, and perturbed by a second satellite. Here I had to introduce the conception of a pair of planes, to which the motions of the satellite and planet may be referred. The problem of the third case is then shown to resolve itself into a tracing of the secular changes in the positions of these two "proper" planes, under the influence of tidal friction. After a long analytical investigation differential equations are found for the rate of these changes.

Part IV. contains the numerical integration of the differential equations of Parts II. and III., in application to the case of the earth, moon, and sun, the earth being supposed to be viscous.

Part V. contains the investigation of the secular changes of the eccentricity of the orbit of a satellite, together with the corresponding changes in the planet's mode of motion.

Part VI. contains a numerical integration of the equations of Part V. in the case of the earth and moon. The objects of Parts VII. and VIII. have been already explained.

In the abstract of this paper in the *Proceedings of the Royal Society**, certain general considerations are adduced which throw light on the nature

* No. 200, 1879. [See Appendix A below.]

of the results here found. Such general reasoning could not lead to definite results, and it was only used in the Abstract as a substitute for analysis; [it is however given in an Appendix].

I.

THE THEORY OF THE DISTURBING FUNCTION.

§ 1. *Preliminary considerations.*

In the theory of disturbed elliptic motion the six elements of the orbit may be divided into two groups of three.

One set of three gives a description of the nature of the orbit which is being described at any epoch, and the second set is required to determine the position of the body at any instant of time. In a speculative inquiry like the present one, where we are only concerned with very small inequalities which would have no interest unless their effects could be cumulative from age to age, so that the orbit might become materially changed, it is obvious that the secular changes in the second set of elements need not be considered.

The three elements whose variations are not here found are the longitudes of the perigee, the node, and the epoch; but the subsequent investigation will afford the materials for finding their variations if it be desirable to do so.

The first set of elements whose secular changes are to be traced are, according to the ordinary system, the mean distance, the eccentricity, and the inclination of the orbit. We shall, however, substitute for the two former elements, viz.: mean distance and eccentricity, two other functions which define the orbit equally well; the first of these is a quantity proportional to the square root of the mean distance, and the second is the ellipticity of the orbit. The inclination will be retained as the third element.

The principal problem to be solved is as follows:

A planet is attended by one or more satellites which raise frictional tides (either bodily or oceanic) in their planet; it is required to find the secular changes in the orbits of the satellites due to tidal reaction.

This problem is however intimately related to a consideration of the parallel changes in the inclination of the planet's axis to a fixed plane, and in its diurnal rotation.

It will therefore be necessary to traverse again, to some extent, the ground covered by my previous paper "On the Precession of a Viscous Spheroid." [Paper 3.]

In the following investigation the tides are supposed to be a bodily deformation of the planet, but a slight modification of the analytical results would make the whole applicable to the case of oceanic tides on a rigid nucleus*. The analysis will be such that the results may be applied to any theory of tides, but particular application will be made to the case where the planet is a homogeneous viscous spheroid, and the present paper is thus a continuation of my previous ones on the tides and rotation of such a spheroid.

The general problem above stated may be conveniently divided into two:

First, to find the secular changes in mean distance and inclination of the orbit of a satellite moving in a circular orbit about its planet.

Second, to find the secular change in mean distance, and eccentricity of the orbit of a satellite moving in an elliptic orbit, but always remaining in a fixed plane.

As stated in the introductory remarks, it will also be necessary to investigate the secular changes in the diurnal rotation and in the obliquity of the planet's equator to the plane of reference.

The tidally distorted planet will be spoken of as the earth, and the satellites as the moon and sun.

This not only affords a useful vocabulary, but permits an easy transition from questions of abstract dynamics to speculations concerning the remote history of the earth and moon.

§ 2. *Notation.—Equation of variation of elements.*

The present section, and the two which follow it, are of general applicability to the whole investigation.

For reasons which will appear later it will be necessary to conceive the earth to have two satellites, which may conveniently be called Diana and the moon. The following are the definitions of the symbols employed.

The time is t, and the suffix 0 to any symbol indicates the value of the corresponding quantity initially, when $t=0$. The attraction of unit masses at unit distance is μ.

For the earth, let—

M = mass in ordinary units; a = mean radius; w = density, or mass per unit volume, the earth being treated as homogeneous; g = mean gravity; $\mathfrak{g} = \frac{2}{5}g/a$; C, A = the greatest and least moments of inertia of the earth; if we neglect the ellipticity they will be equal to $\frac{2}{5}Ma^2$; n = angular velocity of diurnal rotation; ψ = longitude of vernal equinox measured along the ecliptic from a fixed point in the ecliptic—the ecliptic being here a name for a plane fixed in space; i = obliquity of ecliptic; χ the angle between a point fixed on the equator and the vernal equinox; ρ the radius vector of any point measured from the earth's centre.

* Or, as to Part III., on a nucleus which is sufficiently plastic to adjust itself to a form of equilibrium.

For Diana, let—

c = mean distance; $\xi = (c/c_0)^{\frac{1}{2}}$; Ω = mean motion; e = eccentricity of orbit; η = ellipticity of orbit; ϖ = longitude of perigee; j = inclination of orbit to ecliptic; N = longitude of node; ϵ = longitude of epoch; m = mass; ν = ratio of earth's mass to Diana's or M/m; l = true longitude measured from the node; θ = true longitude measured from the vernal equinox; $\tau = \frac{3}{2}\mu m/c^3$, so that $\tau = 3\Omega^2/2(1+\nu)$, also $\tau = \tau_0/\xi^6$; r the radius vector measured from earth's centre.

Also $\lambda = \Omega/n$; $\mathfrak{m}$ the ratio of the earth's moment of momentum of rotation to that of the orbital motion of Diana (or the moon) and the earth round their common centre of inertia.

For the moon let all the same symbols apply when accents are added to them.

Where occasion arises to refer merely to the elements of a satellite in general, the unaccented symbols will be employed.

Let R be the disturbing function as ordinarily defined in works on physical astronomy.

Other symbols will be defined as the necessity for them arises.

Then the following are the well-known equations for the variation of the mean distance, eccentricity, inclination, and longitude of the node:

$$\frac{dc}{dt} = \frac{2\Omega c^2}{\mu(M+m)}\frac{dR}{d\epsilon} \quad\text{.........(1)}$$

$$\frac{de}{dt} = \frac{\Omega c}{\mu(M+m)}\left[\frac{1-e^2}{e}\frac{dR}{d\epsilon} - \frac{\sqrt{1-e^2}}{e}\left(\frac{dR}{d\epsilon}+\frac{dR}{d\varpi}\right)\right] \quad\text{.........(2)}$$

$$-\frac{dj}{dt} = \frac{\Omega c}{\mu(M+m)}\frac{1}{\sqrt{1-e^2}}\left[\frac{1}{\sin j}\frac{dR}{dN} + \tan\tfrac{1}{2}j\left(\frac{dR}{d\epsilon}+\frac{dR}{d\varpi}\right)\right] \quad\text{.....(3)}$$

$$\sin j\frac{dN}{dt} = \frac{\Omega c}{\mu(M+m)}\frac{1}{\sqrt{1-e^2}}\frac{dR}{dj} \quad\text{.........(4)}$$

The last of these equations will only be required in Part III.

Let R = WC$(M+m)/Mm$; then if we substitute this value for R in each of the equations (1—4), it is clear that the right hand side of each will involve a factor $\Omega c\mathrm{C}/\mu Mm$.

Let
$$k = \frac{\mathrm{C}}{\mu Mm}\Omega_0 c_0 \quad\text{.........(5)}$$

(For a homogeneous earth $\dfrac{\mathrm{C}}{\mu Mm} = \dfrac{2\nu}{5g}$, and $\Omega_0 c_0 = \left[(ga^2)\dfrac{1+\nu}{\nu}\right]^{\frac{1}{3}}\Omega_0^{\frac{1}{3}}$. Thus if we put

$$s = \tfrac{2}{5}\left[\left(\frac{a\nu}{g}\right)^2(1+\nu)\right]^{\frac{1}{3}} \quad\text{.........(6)}$$

$$k = s\Omega_0^{\frac{1}{3}} \quad\text{.........(7)}$$

$s^{\frac{3}{2}}$ is a time, being about 3 hrs. $4\frac{1}{2}$ mins. for the homogeneous earth. k is also a time, being about 57 minutes, with the present orbital angular velocity of the moon, and the earth being homogeneous.)

Since $\Omega = \Omega_0 \xi^{-3}$, $c = c_0 \xi^2$, therefore

$$\frac{\mathrm{C}}{\mu Mm} \Omega c = \frac{k}{\xi} \quad \text{.................................(8)}$$

Again, $(c/c_0)^{\frac{1}{2}} = \xi$, and therefore

$$\frac{1}{c}\frac{dc}{dt} = \frac{2}{\xi}\frac{d\xi}{dt} \quad \text{.................................(9)}$$

and since $\eta = 1 - \sqrt{1 - \mathrm{e}^2}$

$$\frac{d\eta}{dt} = \frac{\mathrm{e}}{\sqrt{1-\mathrm{e}^2}}\frac{d\mathrm{e}}{dt} \quad \text{.................................(10)}$$

Substituting for R in terms of W in the four equations (1—4), and using the transformations (8—10), we get

$$\frac{d\xi}{dt} = k\frac{d\mathrm{W}}{d\epsilon} \quad \text{.................................(11)}$$

$$\frac{d\eta}{dt} = -\frac{k}{\xi}\left(\eta \frac{d\mathrm{W}}{d\epsilon} + \frac{d\mathrm{W}}{d\varpi}\right) \quad \text{.................................(12)}$$

and if the orbit be circular, so that $\mathrm{e} = 0$, $d\mathrm{W}/d\varpi = 0$,

$$-\frac{dj}{dt} = \frac{k}{\xi}\left(\frac{1}{\sin j}\frac{d\mathrm{W}}{dN} + \tan \tfrac{1}{2}j \frac{d\mathrm{W}}{d\epsilon}\right) \quad \text{.................(13)}$$

$$\sin j \frac{dN}{dt} = \frac{k}{\xi}\frac{d\mathrm{W}}{dj} \quad \text{.................................(14)}$$

These are the equations of variation of elements which will be used below. The last two (13) and (14) will only be required in the case where the orbit is circular.

The function W only differs from the ordinary disturbing function by a constant factor, and so W will be referred to as the disturbing function.

I will now explain why it has been convenient to depart from ordinary usage, and will show how the same disturbing function W may be used for giving the perturbations of the rotation of the planet.

In the present problem all the perturbations, both of satellites and planet, arise from tides raised in the planet.

The only case treated will be where the tidal wave is expressible as a surface spherical harmonic of the second order.

Suppose then that $\rho = a + \sigma$ is the equation to the wave surface, superposed on the sphere of mean radius a.

The potential V of the wave σ, at an external point ρ, must be given by

$$V = \tfrac{4}{5}\pi\mu w a \left(\frac{a}{\rho}\right)^3 \sigma \quad \ldots\ldots\ldots\ldots\ldots\ldots\ldots\ldots(15)$$

Here w is the density of the matter forming the wave; in our case of a homogeneous earth, distorted by bodily tides, w is the mean density of the earth. (If we contemplate oceanic tides, the subsequent results for the disturbing function must be reduced by the factor $\frac{2}{11}$, this being the ratio of the density of water to the mean density of the earth.)

Now suppose the external point ρ to be at a satellite whose mass, radius vector, and mean distance are m, r, c. If we put $\tau = \frac{3}{2}\mu m/c^3$, and observe that $C = \frac{8}{15}\pi w a^5$, we have

$$V = \frac{C}{m}\tau\left(\frac{c}{r}\right)^3 \frac{\sigma}{a} \quad \ldots\ldots\ldots\ldots\ldots\ldots\ldots\ldots(16)$$

where σ is the height of tide, at the point where the wave surface is pierced by the satellite's radius vector.

But the ordinary disturbing function R for this satellite is this potential V augmented by the factor $(M+m)/M$, because the planet must be reduced to rest. Hence our disturbing function

$$W = \tau\left(\frac{c}{r}\right)^3 \frac{\sigma}{a} \quad \ldots\ldots\ldots\ldots\ldots\ldots\ldots\ldots(17)$$

where σ is the height of tide at the place where the wave surface is pierced by r.

Now let us turn to the case of the planet as perturbed by the attraction of the same satellite on the same wave surface. The whole force function of the action of the satellite on the planet is, by (16), clearly equal to

$$m\left[\frac{M}{c} + \frac{C}{m}\tau\left(\frac{c}{r}\right)^3 \frac{\sigma}{a}\right]$$

The latter term of this expression will give the perturbing couples; it is equal to CW.

In the accompanying fig. 1 let X, Y, Z be axes fixed in space, and (adopting the phraseology for the case of the earth) let XY be the ecliptic; let A, B, C be axes fixed in the planet; let χ be the angle AN or BCD; i the obliquity of the ecliptic; ψ the longitude of the vernal equinox from the fixed point X in the ecliptic.

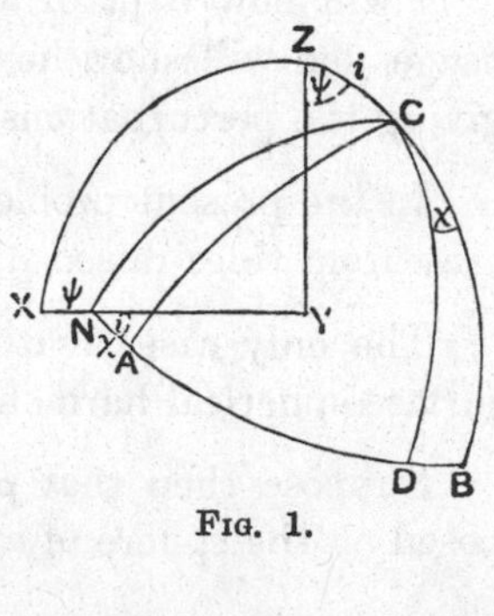

FIG. 1.

If W be expressed in terms of χ, i, ψ, the perturbing couples, which act on the planet, are

$$C\frac{dW}{di} \text{ about N, tending to increase } i,$$

$$C\frac{dW}{d\psi} \text{ about Z, tending to increase } \psi,$$

$$C\frac{dW}{d\chi} \text{ about C, tending to increase } \chi.$$

Let $\mathfrak{L}$, $\mathfrak{M}$, $\mathfrak{N}$ be the perturbing couples acting about A, B, C respectively. Then must

$$C\frac{dW}{d\psi} = -\mathfrak{L}\sin i\sin\chi - \mathfrak{M}\sin i\cos\chi + \mathfrak{N}\cos i$$

$$C\frac{dW}{di} = -\mathfrak{L}\cos\chi + \mathfrak{M}\sin\chi$$

$$C\frac{dW}{d\chi} = \mathfrak{N}$$

Whence

$$\frac{\mathfrak{L}}{C} = \frac{1}{\sin i}\left(\cos i\frac{dW}{d\chi} - \frac{dW}{d\psi}\right)\sin\chi - \frac{dW}{di}\cos\chi$$

$$\frac{\mathfrak{M}}{C} = \frac{1}{\sin i}\left(\cos i\frac{dW}{d\chi} - \frac{dW}{d\psi}\right)\cos\chi + \frac{dW}{di}\sin\chi$$

$$\frac{\mathfrak{N}}{C} = \frac{dW}{d\chi}$$

But if ω_1, ω_2, ω_3 be the component angular velocities of the planet about A, B, C respectively, and if we may neglect $(C-A)/A$ compared with unity, the equations of motion may be written

$$\frac{d\omega_1}{dt} = \frac{\mathfrak{L}}{C}, \quad \frac{d\omega_2}{dt} = \frac{\mathfrak{M}}{C}, \quad \frac{d\omega_3}{dt} = \frac{\mathfrak{N}}{C}$$

as was shown in section (6) [p. 51] of my previous paper on "Precession."

Since $\chi = nt$, we have by integration,

$$\omega_1 = -\frac{1}{n\sin i}\left(\cos i\frac{dW}{d\chi} - \frac{dW}{d\psi}\right)\cos\chi - \frac{1}{n}\frac{dW}{di}\sin\chi$$

$$\omega_2 = \frac{1}{n\sin i}\left(\cos i\frac{dW}{d\chi} - \frac{dW}{d\psi}\right)\sin\chi - \frac{1}{n}\frac{dW}{di}\cos\chi$$

These are to be substituted in the geometrical equations,

$$\frac{di}{dt} = -\omega_1\cos\chi + \omega_2\sin\chi$$

$$\sin i\frac{d\psi}{dt} = -\omega_1\sin\chi - \omega_2\cos\chi$$

Hence finally,

$$\left.\begin{aligned} n\sin i\frac{di}{dt} &= \cos i\frac{dW}{d\chi} - \frac{dW}{d\psi} \\ n\sin i\frac{d\psi}{dt} &= \frac{dW}{di} \\ \frac{dn}{dt} &= \frac{dW}{d\chi} \end{aligned}\right\} \text{..................}(18)^*$$

These are the equations which will be used for determining the perturbations of the planet's rotation.

We now see that the same disturbing function W will serve for finding both sets of perturbations.

It is clear that it is not necessary in the above investigation that σ should actually be a tide wave; it may just as well refer to the permanent oblateness of the planet. Thus the ordinary precession and nutations may be determined from these formulæ.

§ 3. *To find spherical harmonic functions of Diana's coordinates with reference to axes fixed in the earth.*

Let A, B, C be rectangular axes fixed in the earth, C being the pole and AB the equator.

Let X, Y, Z be a second set of rectangular axes, XY being the plane of Diana's orbit.

Let M be the projection of Diana in her orbit.

Let $i_{,} =$ ZC, the obliquity of the equator to the plane of Diana's orbit.

$\chi_{,} =$ AX = BCY.

$l_{,} =$ MX, Diana's longitude from the node X.

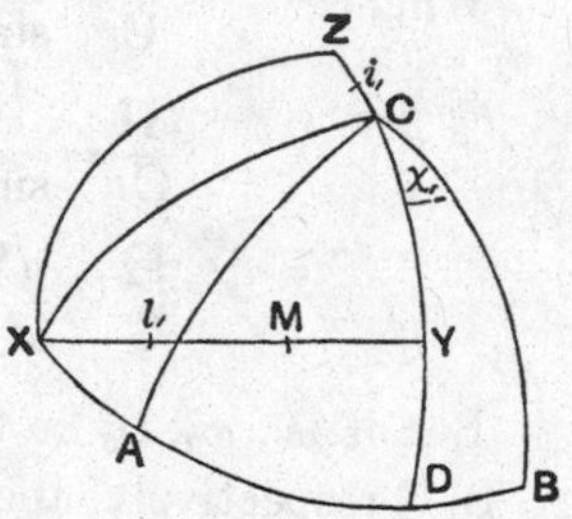

Fig. 2.

$$\left.\begin{aligned} \text{Let } M_1 &= \cos MA \\ M_2 &= \cos MB \\ M_3 &= \cos MC \end{aligned}\right\} \text{Diana's direction-cosines referred to A, B, C.}$$

Then

$$\left.\begin{aligned} M_1 &= \cos l_{,} \cos \chi_{,} + \sin l_{,} \sin \chi_{,} \cos i_{,} \\ M_2 &= -\cos l_{,} \sin \chi_{,} + \sin l_{,} \cos \chi_{,} \cos i_{,} \\ M_3 &= \sin l_{,} \sin i_{,} \end{aligned}\right\} \text{..............}(19)$$

We may observe that M_2 is derivable from M_1 by writing $\chi_{,} + \frac{1}{2}\pi$ in place of $\chi_{,}$.

These expressions refer to the plane of Diana's orbit, but we must now refer to the ecliptic.

* [Although established by approximate methods, these equations are rigorously true, provided that i, χ, ψ receive appropriate definitions.]

In fig. 3, let A be the vernal equinox, B the ascending node of the orbit, C the intersection of the orbit with the equator, being the X of fig. 2, and let D be a point fixed in the equator, being the A of fig. 2.

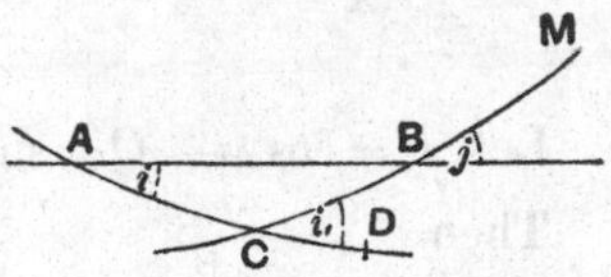

FIG. 3.

If we refer to the sides and angles of the spherical triangle ABC by the letters a, b, c, A, B, C as is usual in works on spherical trigonometry, we have

$A = i$, the obliquity of the ecliptic.

$B = j$, the inclination of the orbit.

$\pi - C = i_{,} = ZC$ of fig. 2.

$c = N$, the longitude of the node measured from A, for at present we may suppose $\psi = 0$, without loss of generality.

Let $\chi = DA$, and we have

$$\chi - b = DC = \chi_{,}$$

Again, if M be Diana in her orbit, $MB = l$, and since $MC = l_{,}$, therefore

$$l + a = l_{,}$$

Whence

$$\cos \chi_{,} = \cos \chi \cos b + \sin \chi \sin b$$
$$\sin \chi_{,} = \sin \chi \cos b - \cos \chi \sin b$$
$$\cos l_{,} = \cos l \cos a - \sin l \sin a$$
$$\sin l_{,} = \sin l \cos a + \cos l \sin a$$

Substituting these values in the first of (19) we have

$$\begin{aligned} M_1 = {} & \cos \chi \cos l (\cos a \cos b - \sin a \sin b \cos i_{,}) \\ & + \sin \chi \cos l (\cos a \sin b + \sin a \cos b \cos i_{,}) \\ & - \cos \chi \sin l (\sin a \cos b + \cos a \sin b \cos i_{,}) \\ & - \sin \chi \sin l (\sin a \sin b - \cos a \cos b \cos i_{,}) \end{aligned}$$

Now $\cos i_{,} = -\cos C$, and

$$\cos a \cos b + \sin a \sin b \cos C = \cos c = \cos N$$

$$\cos a \sin b - \sin a \cos b \cos C = \sin a [\cot a \sin b - \cos b \cos C] = \sin a \cot A \sin C$$
$$= \cos i \sin N$$

$$\sin a \cos b - \cos a \sin b \cos C = \sin b [\cot b \sin a - \cos a \cos C] = \sin b \cot B \sin C$$
$$= \cos j \sin N$$

$$\begin{aligned} \sin a \sin b + \cos a \cos b \cos C & \\ & = \sin a \sin b + \cos c \cos C - \sin a \sin b \cos^2 C \\ & = \sin a \sin b \sin^2 C + \cos c (-\cos A \cos B + \sin A \sin B \cos c) \\ & = \sin A \sin B \sin^2 c + \sin A \sin B \cos^2 c - \cos A \cos B \cos c \\ & = \sin i \sin j - \cos i \cos j \cos N \end{aligned}$$

Substituting in the expression for M_1,

$$M_1 = \cos\chi\cos l\cos N + \sin\chi\cos l\sin N\cos i - \cos\chi\sin l\sin N\cos j$$
$$- \sin\chi\sin l(\sin i\sin j - \cos i\cos j\cos N)$$

Let $P = \cos\frac{1}{2}i$, $Q = \sin\frac{1}{2}i$, $p = \cos\frac{1}{2}j$, $q = \sin\frac{1}{2}j$.

Then

$$M_1 = (P^2+Q^2)(p^2+q^2)\cos\chi\cos l\cos N + (P^2-Q^2)(p^2+q^2)\sin\chi\cos l\sin N$$
$$-(P^2+Q^2)(p^2-q^2)\cos\chi\sin l\sin N + (P^2-Q^2)(p^2-q^2)\sin\chi\sin l\cos N$$
$$-4PQpq\sin\chi\sin l$$
$$= P^2p^2\cos(\chi-l-N) + P^2q^2\cos(\chi+l-N) + Q^2p^2\cos(\chi+l+N)$$
$$+ Q^2q^2\cos(\chi-l+N) + 2PQpq[\cos(\chi+l) - \cos(\chi-l)] \quad \text{...(20)}$$

Since M_2 is derivable from M_1 by writing $\chi_, + \frac{1}{2}\pi$ for $\chi_,$, therefore it is also derivable by writing $\chi + \frac{1}{2}\pi$ for χ. Hence $-M_2$ is the same as M_1, save that sines replace cosines.

Again $\quad M_3 = \sin l_, \sin i_, = \sin l\cos \mathrm{a}\sin i_, + \cos l\sin \mathrm{a}\sin i_,$

But $\quad \sin \mathrm{a}\sin i_, = \sin i\sin N = 2PQ\sin N$

And $\quad \cos\mathrm{a}\sin i_, = \sin i\cot\mathrm{a}\sin\mathrm{c} = \sin i(\cot\mathrm{A}\sin\mathrm{B} + \cos\mathrm{c}\cos\mathrm{B})$

$$= \cos i\sin j + \sin i\cos j\cos N$$
$$= 2pq(P^2-Q^2) + 2PQ(p^2-q^2)\cos N$$

Therefore

$$M_3 = 2PQ[p^2\sin(l+N) - q^2\sin(l-N)] + 2pq(P^2-Q^2)\sin l \quad \text{...(21)}$$

For the sake of future developments it will be more convenient to replace the sines and cosines in the expressions for the M's by exponentials, and for brevity the $\sqrt{-1}$ will be omitted in the indices.

We have therefore

$$2M_1 = e^{\chi-l-N}[Pp - Qqe^N]^2 + e^{\chi+l+N}[Qp + Pqe^{-N}]^2$$

+ the same with the signs of the indices of the exponentials changed,

$-2M_2\sqrt{-1}$ = the same with sign of second line changed,

$$M_3\sqrt{-1} = e^{l+N}[Pp - Qqe^{-N}][Qp + Pqe^{-N}]$$

− same with signs of the indices of the exponentials changed.

Let

$$\left.\begin{aligned} \varpi &= Pp - Qqe^N, & \kappa &= Qp + Pqe^N \\ \underline{\varpi} &= Pp - Qqe^{-N}, & \underline{\kappa} &= Qp + Pqe^{-N} \end{aligned}\right\} \quad \text{...............(22)}$$

From these definitions it appears that ϖ and κ are two imaginary functions, which oscillate between the real values $\cos\frac{1}{2}(i+j)$ and $\cos\frac{1}{2}(i-j)$, and $\sin\frac{1}{2}(i+j)$ and $\sin\frac{1}{2}(i-j)$ as the node of the orbit moves round.

Also let $\theta = l + N$, the true longitude of Diana measured from the vernal equinox. Strictly speaking, when longitudes are measured from a fixed point in the ecliptic $\theta = l + N - \psi$, but in the present investigation nothing is lost by regarding ψ as zero; in § 12, and in Part III., we shall have however to introduce ψ.

Then

$$\left.\begin{aligned} 2\mathrm{M}_1 &= \varpi^2 e^{\chi-\theta} + \underline{\kappa}^2 e^{\chi+\theta} + \underline{\varpi}^2 e^{-\chi+\theta} + \kappa^2 e^{-\chi-\theta} \\ 2\mathrm{M}_2\sqrt{-1} &= -\varpi^2 e^{\chi-\theta} - \underline{\kappa}^2 e^{\chi+\theta} + \underline{\varpi}^2 e^{-\chi+\theta} + \kappa^2 e^{-\chi-\theta} \\ \mathrm{M}_3\sqrt{-1} &= \underline{\varpi}\underline{\kappa} e^{\theta} - \varpi\kappa e^{-\theta} \end{aligned}\right\} \quad \text{.........(23)}$$

The object of the present investigation is to find the following spherical harmonic functions of the second degree of M_1, M_2, M_3, viz.:

$$\mathrm{M}_1^2 - \mathrm{M}_2^2, \quad 2\mathrm{M}_1\mathrm{M}_2, \quad 2\mathrm{M}_2\mathrm{M}_3, \quad 2\mathrm{M}_1\mathrm{M}_3, \quad \tfrac{1}{3} - \mathrm{M}_3^2$$

By adding the squares of the first and second of (23), we have

$$\begin{aligned} 2(\mathrm{M}_1^2 - \mathrm{M}_2^2) = {} & \varpi^4 e^{2(\chi-\theta)} + 2\varpi^2\underline{\kappa}^2 e^{2\chi} + \underline{\kappa}^4 e^{2(\chi+\theta)} \\ & + \underline{\varpi}^4 e^{-2(\chi-\theta)} + 2\underline{\varpi}^2\kappa^2 e^{-2\chi} + \kappa^4 e^{-2(\chi+\theta)} \quad \text{.........(24)} \end{aligned}$$

From (20) we know that M_1 has the form $\Sigma \mathrm{A} \cos(\chi + \mathrm{B})$, and $-\mathrm{M}_2$ the form $\Sigma \mathrm{A} \sin(\chi + \mathrm{B})$; therefore $(\mathrm{M}_1 + \mathrm{M}_2)\, 2^{-\frac{1}{2}}$ has the form $\Sigma \mathrm{A} \cos(\chi + \frac{1}{4}\pi + \mathrm{B})$, and $(\mathrm{M}_1 - \mathrm{M}_2)\, 2^{-\frac{1}{2}}$ the form $\Sigma \mathrm{A} \sin(\chi + \frac{1}{4}\pi + \mathrm{B})$. Hence if we write $\chi - \frac{1}{4}\pi$ for χ in $\mathrm{M}_1^2 - \mathrm{M}_2^2$, we obtain $-2\mathrm{M}_1\mathrm{M}_2$. Therefore from (24) we obtain

$$\begin{aligned} -4\mathrm{M}_1\mathrm{M}_2\sqrt{-1} = {} & \varpi^4 e^{2(\chi-\theta)} + 2\varpi^2\underline{\kappa}^2 e^{2\chi} + \underline{\kappa}^4 e^{2(\chi+\theta)} \\ & - \underline{\varpi}^4 e^{-2(\chi-\theta)} - 2\underline{\varpi}^2\kappa^2 e^{-2\chi} - \kappa^4 e^{-2(\chi+\theta)} \quad \text{......(25)} \end{aligned}$$

The $\sqrt{-1}$ appears on the left hand side because $e^{\frac{1}{2}\pi} = -(-1)^{-\frac{1}{2}}$, $e^{-\frac{1}{2}\pi} = (-1)^{-\frac{1}{2}}$.

It is also easy to show that

$$\begin{aligned} 2\mathrm{M}_2\mathrm{M}_3 = {} & -\varpi^3\kappa e^{\chi-2\theta} + \varpi\underline{\kappa}(\varpi\underline{\varpi} - \kappa\underline{\kappa}) e^{\chi} + \underline{\varpi}\underline{\kappa}^3 e^{\chi+2\theta} \\ & - \underline{\varpi}^3\underline{\kappa} e^{-(\chi-2\theta)} + \underline{\varpi}\kappa(\underline{\varpi}\varpi - \underline{\kappa}\kappa) e^{-\chi} + \varpi\kappa^3 e^{-(\chi+2\theta)} \quad \text{...(26)} \end{aligned}$$

$$\begin{aligned} 2\mathrm{M}_1\mathrm{M}_3\sqrt{-1} = {} & -\varpi^3\kappa e^{\chi-2\theta} + \varpi\underline{\kappa}(\varpi\underline{\varpi} - \kappa\underline{\kappa}) e^{\chi} + \underline{\varpi}\underline{\kappa}^3 e^{\chi+2\theta} \\ & + \underline{\varpi}^3\kappa e^{-(\chi-2\theta)} - \underline{\varpi}\kappa(\underline{\varpi}\varpi - \underline{\kappa}\kappa) e^{-\chi} - \varpi\kappa^3 e^{-(\chi+2\theta)} \quad \text{...(27)} \end{aligned}$$

$$\tfrac{1}{3} - \mathrm{M}_3^2 = \tfrac{1}{3} - 2\varpi\underline{\varpi}\kappa\underline{\kappa} + \underline{\varpi}^2\underline{\kappa}^2 e^{2\theta} + \varpi^2\kappa^2 e^{-2\theta} \quad \text{........................(28)}$$

It may be here noted that $\varpi\underline{\varpi} + \kappa\underline{\kappa} = 1$, so that

$$\tfrac{1}{3} - 2\varpi\underline{\varpi}\kappa\underline{\kappa} = \tfrac{1}{3}(\varpi^2\underline{\varpi}^2 - 4\varpi\underline{\varpi}\kappa\underline{\kappa} + \kappa^2\underline{\kappa}^2)$$

These five formulæ (24) to (28) are clearly equivalent to the expansion of the harmonic functions as a series of sines and cosines of angles of the form $\alpha\chi + \beta l + \gamma N$. It remains to explain the uses to be made of these expressions.

§ 4. *The disturbing function.*

In the theory of the disturbing function the differentiation with respect to the elements of the orbit of the disturbed body is an artifice to avoid the determination of the three component disturbing forces, by means of differentiation with regard to the radius vector, longitude and latitude. In the present problem we have to determine the perturbation of a satellite under the influence of the tides raised by itself and by another satellite. Where the tides are raised by the satellite itself, the elements of that satellite's orbit of course enter in the disturbing function in expressing the state of tidal distortion of the planet, but they also enter as expressing the position of the satellite. It is clear that, in effecting the differentiations above referred to, we must only regard the elements of the orbit as entering in the disturbing function in the latter sense. Hence it follows that even although there may be only one satellite, yet in the evaluation of the disturbing function we must suppose that there are two satellites, viz.: one a tide-raising satellite and another a disturbed satellite.

In this place, where the planet is called the earth, the tide-raising satellite may be conveniently called Diana, and the satellite whose motion is disturbed may be called the moon. After the formation of the differential equations Diana may be made identical with the moon or with the sun at will, or the analysis may be made applicable to a planet with any number of satellites.

As above stated, unaccented symbols will be taken to apply to Diana, and accented symbols to the moon.

The first step, then, is to find the tidal distortion due to Diana.

Let M be the projection of Diana on the celestial sphere concentric with the earth, and P the projection of any point in the earth.

Let $\rho\xi$, $\rho\eta$, $\rho\zeta$ be the rectangular coordinates of P and rM_1, rM_2, rM_3 the rectangular coordinates of Diana referred to axes A, B, C fixed in the earth.

Since ρ, r are radii vectores, ξ, η, ζ and M_1, M_2, M_3 are direction-cosines.

The tide-generating potential V (of the second degree of harmonics, which will be alone considered) at P is given by

$$V = \tfrac{3}{2}\frac{\mu m}{r^3}\rho^2(\cos^2 PM - \tfrac{1}{3})$$

according to the usual theory.

Now
$$\cos PM = \xi M_1 + \eta M_2 + \zeta M_3$$
and
$$\cos^2 PM - \tfrac{1}{3} = 2\xi\eta M_1M_2 + 2\,\frac{\xi^2-\eta^2}{2}\,\frac{M_1^2-M_2^2}{2} + 2\eta\zeta M_2M_3 + 2\xi\zeta M_1M_3$$
$$+\tfrac{3}{2}\,\frac{\xi^2+\eta^2-2\zeta^2}{3}\,\frac{M_1^2+M_2^2-2M_3^2}{3}$$

Also by previous definition, $\tau = \frac{3}{2}\mu m/c^3$; so that

$$\frac{3}{2}\frac{\mu m}{r^3} = \frac{\tau}{(1-e^2)^3}\left[\frac{c(1-e^2)}{r}\right]^3$$

Let

$$X = \left[\frac{c(1-e^2)}{r}\right]^{\frac{3}{2}} M_1, \quad Y = \left[\frac{c(1-e^2)}{r}\right]^{\frac{3}{2}} M_2, \quad Z = \left[\frac{c(1-e^2)}{r}\right]^{\frac{3}{2}} M_3 \ldots(29)$$

Then clearly

$$V \div \frac{\tau}{(1-e^2)^3}\rho^2 = 2\xi\eta XY + 2\frac{\xi^2-\eta^2}{2}\frac{X^2-Y^2}{2} + 2\eta\zeta YZ + 2\xi\zeta XZ + \frac{3}{2}\frac{\xi^2+\eta^2-2\zeta^2}{3}\frac{X^2+Y^2-2Z^2}{3}$$

Now assume that the five functions $2XY$, X^2-Y^2, YZ, XZ, $X^2+Y^2-2Z^2$ are each expressed as a series of simple time-harmonics; it will appear below that this may always be done. We have therefore V expressed as the sum of five solid harmonics $\rho^2\xi\eta$, $\rho^2(\xi^2-\eta^2)$, &c., each multiplied by a simple time-harmonic. According to any tidal theory each such term must raise a tide expressible by a surface harmonic of the same type, and multiplied by a simple time-harmonic of the same speed; moreover, each such tide must have a height which is some fraction of the corresponding equilibrium tide of a perfectly fluid spheroid, but the simple time-harmonic will in general be altered in phase.

If $r = a + \sigma$ be the equation to the wave-surface, corresponding to a generating potential $V = [\tau/(1-e^2)^3]\rho^2 2\xi\eta XY$, then when the spheroid is perfectly fluid, $\sigma/a = [\tau/\mathfrak{g}(1-e^2)^3] 2\xi\eta XY$, where $\mathfrak{g} = \frac{2}{5}g/a$, according to the ordinary equilibrium theory of tides. (It will now be assumed that we are dealing with bodily tides of the spheroid; if the tides were oceanic a slight modification would have to be introduced.)

In a frictional fluid, the tide σ will be reduced in height and altered in phase.

Let $\mathfrak{X}\mathfrak{Y}$ represent a function of the same form as XY, save that each simple time-harmonic term of XY is multiplied by some fraction expressive of reduction of height of tide, and that the argument of each such simple harmonic term is altered in phase; the constants so introduced will be functions of the constitution of the spheroid, and of the speeds of the harmonic terms. Also extend the same notation to the other functions of X, Y, Z which occur in V.

Then it is clear that, if $r = a + \sigma$ be the equation to the complete wave surface corresponding to the potential V,

$$(1-e^2)^3\frac{\mathfrak{g}}{\tau}\frac{\sigma}{a} = 2\xi\eta\mathfrak{X}\mathfrak{Y} + 2\frac{\xi^2-\eta^2}{2}\frac{\mathfrak{X}^2-\mathfrak{Y}^2}{2} + 2\eta\zeta\mathfrak{Y}\mathfrak{Z} + 2\xi\zeta\mathfrak{X}\mathfrak{Z} + \frac{3}{2}\frac{\xi^2+\eta^2-2\zeta^2}{3}\frac{\mathfrak{X}^2+\mathfrak{Y}^2-2\mathfrak{Z}^2}{3} \ldots(30)$$

This expression shows that σ is a surface harmonic of the second order.

By (17) we have for the disturbing function for the moon, due to Diana's tides,

$$W = \tau' \left(\frac{c'}{r'}\right)^3 \left(\frac{\sigma}{a}\right)$$

where σ is the height of tide, at the point where the moon's radius vector pierces the wave surface.

Hence in the expression (30) for σ, we must put

$$\xi = M_1', \quad \eta = M_2', \quad \zeta = M_3'$$

By analogy with (29), let

$$X' = \left[\frac{c'(1-e'^2)}{r'}\right]^{\frac{3}{2}} M_1', \quad Y' = \left[\frac{c'(1-e'^2)}{r'}\right]^{\frac{3}{2}} M_2', \quad Z' = \left[\frac{c'(1-e'^2)}{r'}\right]^{\frac{3}{2}} M_3'$$

and we have

$$W = \frac{\tau\tau'}{\mathfrak{g}} \frac{1}{(1-e^2)^3(1-e'^2)^3} \left[2X'Y'\mathfrak{X}\mathfrak{Y} + 2\frac{X'^2 - Y'^2}{2}\frac{\mathfrak{X}^2 - \mathfrak{Y}^2}{2} + 2Y'Z'\mathfrak{Y}\mathfrak{Z} + 2X'Z'\mathfrak{X}\mathfrak{Z} + \tfrac{3}{2}\frac{X'^2 + Y'^2 - 2Z'^2}{3}\frac{\mathfrak{X}^2 + \mathfrak{Y}^2 - 2\mathfrak{Z}^2}{3}\right] \quad \ldots(31)$$

This is the required expression for the disturbing function on the moon, due to Diana's tides.

So far the investigation is general, but we now have to develop this function so as to make it applicable to the several problems to be considered.

II.

Secular Changes in the Inclination of the Orbit of a Satellite.

§ 5. *The perturbed satellite moves in a circular orbit inclined to a fixed plane. Subdivision of the problem.*

In this case $e = 0$, $e' = 0$, $r = c$, $r' = c'$, so that the functions X, Y, Z and X′, Y′, Z′ are simply the direction cosines of Diana and the moon, referred to the axes A, B, C fixed in the earth. Hence $X = M_1$, $Y = M_2$, $Z = M_3$, and the five formulæ (24—28) give the functions $X^2 - Y^2$, $2XY$, $2YZ$, $2ZX$, $\frac{1}{3} - Z^2$. In order to form the functions in gothic letters we must express these functions as simple time-harmonics.

The formulæ (24) to (28) are equivalent to the expression of the five functions as a series of terms of the type $A\cos(\alpha\chi + \beta\theta + \gamma N + \delta)$. Now χ is the angle between a point fixed on the equator and the vernal equinox, and therefore (neglecting alterations in the diurnal rotation and the precessional motion) increases uniformly with the time, being equal to nt + a constant, which constant may be treated as zero by a proper choice of axes A, B, C.

θ is the true longitude measured from the vernal equinox, and is equal to $\Omega t + \epsilon - \psi$, since the orbit is circular; also ψ may for the present be put equal to zero, without any loss of generality.

Then if in forming the expressions for the state of tidal distortion of the earth we neglect the motion of the node, the five functions are expressed as a series of simple time-harmonics of the type $A\cos(\alpha nt + \beta\Omega t + \zeta)$.

The corresponding term in the corresponding gothic-letter function will be $KA\cos(\alpha nt + \beta\Omega t + \zeta - \mathrm{k})$, where K is the fraction by which the tide is reduced and k is the alteration of phase.

It appears, from the inspection of the five formulæ (24–8), that there are tides of seven speeds, viz.: $2(n-\Omega)$, $2n$, $2(n+\Omega)$, $n-2\Omega$, n, $n+2\Omega$, 2Ω.

The following schedule gives the symbols to be introduced for reduction of tide and alteration of phase or lag.

	Semi-diurnal			Diurnal			Fortnightly
	Slow	Sidereal	Fast	Slow	Sidereal	Fast	
Speed	$2(n-\Omega)$	$2n$	$2(n+\Omega)$	$n-2\Omega$	n	$n+2\Omega$	2Ω
Fraction of equilibrium tide	F_1	F	F_2	G_1	G	G_2	H
Retardation of phase or lag	$2f_1$	2f	$2f_2$	g_1	g	g_2	2h

The gothic-letter functions may now at once be written down from (24–8).

Thus,

$$2(\mathfrak{X}^2 - \mathfrak{Y}^2) = F_1\varpi^4 e^{2(\chi-\theta)-2f_1} + F\,2\varpi^2\underline{\kappa}^2 e^{2\chi-2f} + F_2\underline{\kappa}^4 e^{2(\chi+\theta)-2f_2}$$
$$+ F_1\underline{\varpi}^4 e^{-2(\chi-\theta)+2f_1} + F\,2\underline{\varpi}^2\kappa^2 e^{-2\chi+2f} + F_2\kappa^4 e^{-2(\chi+\theta)+2f_2} \quad \ldots(32)$$

$$-4\mathfrak{X}\mathfrak{Y}\sqrt{-1} = \text{the same, with second line of opposite sign} \quad \ldots\ldots(33)$$

$$2\mathfrak{Y}\mathfrak{Z} = -G_1\varpi^3\kappa e^{\chi-2\theta-g_1} + G\varpi\underline{\kappa}(\varpi\underline{\varpi} - \kappa\underline{\kappa})e^{\chi-g} + G_2\underline{\varpi}\underline{\kappa}^3 e^{\chi+2\theta-g_2}$$
$$- G_1\underline{\varpi}^3\underline{\kappa}e^{-(\chi-2\theta)+g_1} + G\underline{\varpi}\kappa(\underline{\varpi}\varpi - \underline{\kappa}\kappa)e^{-\chi+g} + G_2\varpi\kappa^3 e^{-(\chi+2\theta)+g_2} \quad \ldots(34)$$

$$2\mathfrak{X}\mathfrak{Z}\sqrt{-1} = \text{the same, with second line of opposite sign} \quad \ldots\ldots\ldots\ldots\ldots\ldots(35)$$

$$\tfrac{1}{3} - \mathfrak{Z}^2 = \tfrac{1}{3} - 2\underline{\varpi}\varpi\underline{\kappa}\kappa + H\underline{\varpi}^2\underline{\kappa}^2 e^{2\theta-2h} + H\varpi^2\kappa^2 e^{-2\theta+2h} \quad \ldots\ldots\ldots(36)$$

The fact that there is no factor of the same kind as H in the first pair of (36) results from the assumption that the tides due to the motion of the nodes of the orbit are the equilibrium tides unaltered in phase.

The formulæ for $2(X'^2 - Y'^2)$, $-4X'Y'\sqrt{-1}$, $2Y'Z'$, $2X'Z'\sqrt{-1}$, $\tfrac{1}{3} - Z'^2$ are found by symmetry, by merely accenting all the symbols in the five formulæ (24–8) for the M functions. In the use made of these formulæ this accentuation will be deemed to be done.

At present we shall not regard χ as being accented, but in § 12 and in Part III. we shall have to regard χ as also accented.

We now have to develop the several products of the X′ functions multiplied by the $\mathfrak{X}$ functions.

Before making these multiplications, it must be considered what are the terms which are required for finding secular changes in the elements, since all others are superfluous for the problem in hand.

Such terms are clearly those in which θ and θ' are wanting, and also those where $\theta' - \theta$ occurs, for these will be wanting in θ when Diana is made identical with the moon. It follows that we need only multiply together terms of the like speeds. In the following developments all superfluous terms are omitted.

Semi-diurnal terms.

These are $2X'Y'\mathfrak{X}\mathfrak{Y} + 2\dfrac{X'^2 - Y'^2}{2}\dfrac{\mathfrak{X}^2 - \mathfrak{Y}^2}{2}$.

If we multiply (24) (with accented symbols) by (32), and (25) (with accented symbols) by (33), and subtract the latter from the former, we see that χ disappears from the expression, and that,

$$8X'Y'\mathfrak{X}\mathfrak{Y} + 2(X'^2 - Y'^2)(\mathfrak{X}^2 - \mathfrak{Y}^2) = \text{First line of (24)} \times \text{second of (32)}$$
$$+ \text{Second of (25)} \times \text{first of (33)}$$

Thus as far as we are concerned

$$2X'Y'\mathfrak{X}\mathfrak{Y} + 2\frac{X'^2 - Y'^2}{2}\frac{\mathfrak{X}^2 - \mathfrak{Y}^2}{2}$$
$$= \tfrac{1}{4}\left[F_1\varpi^4\underline{\varpi}'^4 e^{2(\theta'-\theta)-2f_1} + 4F\varpi^2\underline{\kappa}^2\underline{\varpi}'^2\kappa'^2 e^{-2f} + F_2\underline{\kappa}^4\kappa'^4 e^{-2(\theta'-\theta)-2f_2}\right]$$
$$+ \tfrac{1}{4}\left[F_1\underline{\varpi}^4\varpi'^4 e^{-2(\theta'-\theta)+2f_1} + 4F\underline{\varpi}^2\kappa^2\varpi'^2\underline{\kappa}'^2 e^{2f} + F_2\kappa^4\underline{\kappa}'^4 e^{2(\theta'-\theta)+2f_2}\right] \quad \ldots(37)$$

If χ had been accented in the X′ functions, we should have had $2(\chi - \chi')$ in all the indices of exponentials of the first line, and $-2(\chi - \chi')$ in all the indices of the second line. These three pairs of terms will be called W_{I}, W_{II}, W_{III}.

Diurnal terms.

These are $2Y'Z'\mathfrak{Y}\mathfrak{Z} + 2X'Z'\mathfrak{X}\mathfrak{Z}$.

If the multiplications be performed as in the previous case, it will be found that χ disappears in the sum of the two products, and, as far as concerns terms in $\theta' - \theta$ and those independent of θ and θ', we have

$$2Y'Z'\mathfrak{Y}\mathfrak{Z} + 2X'Z'\mathfrak{X}\mathfrak{Z}$$
$$= G_1\varpi^3\kappa\underline{\varpi}'^3\underline{\kappa}' e^{2(\theta'-\theta)-g_1} + G\varpi\underline{\kappa}(\varpi\underline{\varpi} - \kappa\underline{\kappa})\underline{\varpi}'\kappa'(\underline{\varpi}'\varpi' - \underline{\kappa}'\kappa')e^{-g}$$
$$+ G_2\underline{\varpi}\underline{\kappa}^3\varpi'\kappa'^3 e^{-2(\theta'-\theta)-g_2}$$
$$+ G_1\underline{\varpi}^3\underline{\kappa}\varpi'^3\kappa' e^{-2(\theta'-\theta)+g_1} + G\underline{\varpi}\kappa(\varpi\underline{\varpi} - \kappa\underline{\kappa})\varpi'\underline{\kappa}'(\underline{\varpi}'\varpi' - \underline{\kappa}'\kappa')e^{g}$$
$$+ G_2\varpi\kappa^3\underline{\varpi}'\underline{\kappa}'^3 e^{2(\theta'-\theta)+g_2} \quad \ldots\ldots(38)$$

If χ had been accented in the X′ functions we should have had $\chi - \chi'$ in all the indices of the exponentials of the first line, and $-(\chi - \chi')$ in all the indices of the second line. These three pairs of terms will be called W_1, W_2, W_3.

Fortnightly term.

This is $\frac{3}{2}(\frac{1}{3} - Z'^2)(\frac{1}{3} - \mathfrak{Z}^2)$.

Multiplying (36) by (28) when the symbols are accented, and only retaining desired terms,

$$\frac{3}{2}(\tfrac{1}{3} - Z'^2)(\tfrac{1}{3} - \mathfrak{Z}^2) = \frac{3}{2}(\tfrac{1}{3} - 2\varpi\underline{\varpi}\kappa\underline{\kappa})(\tfrac{1}{3} - 2\varpi'\underline{\varpi}'\kappa'\underline{\kappa}') + \tfrac{3}{2}\mathrm{H}\underline{\varpi}^2\underline{\kappa}^2\varpi'^2\kappa'^2 e^{-2(\theta'-\theta)-2\mathrm{h}} + \tfrac{3}{2}\mathrm{H}\varpi^2\kappa^2\underline{\varpi}'^2\underline{\kappa}'^2 e^{2(\theta'-\theta)+2\mathrm{h}} \ldots (39)$$

Even if χ had been accented in the X′ functions, neither χ nor χ' would have entered in this expression. These terms will be called W_0.

The sum of the three expressions (37), (38), and (39), when multiplied by $\tau\tau'/\mathfrak{g}$, is equal to W, the disturbing function.

If Diana be a different body from the moon the terms in $\theta' - \theta$ are periodic, and the only parts of W, from which secular changes in the moon's mean distance and inclination can arise, are the sidereal semi-diurnal and diurnal terms, viz.: those in F and G, and also the term independent of H in (39). These terms being independent of θ' are independent of ϵ', the moon's epoch. Hence it follows that, as far as concerns the influence of Diana's tides upon the moon, $dW/d\epsilon'$ is zero, and we conclude that—*the tides raised by any one satellite can produce directly no secular change in the mean distance of any other satellite**.

But Diana being still distinct from the moon, the F and G terms and part of the fortnightly term, which are independent of θ, do involve N and N'; for W contains terms of the forms $e^{\pm\alpha N}$, $e^{\pm\alpha N'}$, $e^{\pm(\alpha N + \beta N')}$, also it has terms independent of N, N'. Hence dW/dN' will contain terms of the forms $e^{\pm\alpha N'}$, $e^{\pm(\alpha N + \beta N')}$, or their equivalent sines or cosines.

Now by hypothesis there are two disturbing bodies, and we know by lunar theory that the direct influence of Diana on the moon is such as to tend to make the nodes of the moon's orbit revolve on the ecliptic; on the other hand, there is a direct influence of the permanent oblateness of the earth on the nodes of the moon's orbit.

If the oblateness of the earth be large, the result of the joint influence of these two causes may be such as either to make the nodes of the moon's orbit rotate with a very unequal angular velocity, or perform oscillations (possibly

* If there be a rigorous relationship between the mean motions of a pair of satellites this may not be true. This appears to be (at least very nearly) the case between two pairs of satellites of the planet Saturn.

large ones) about a mean position. If this be the case the mean value of dW/dN' may differ considerably from zero. This case is considered in detail in Part III. of this paper.

If on the other hand the oblateness be small the nodes of the orbit revolve with a sensibly uniform angular velocity on the ecliptic. This is the case at present with the earth and moon. Here then dW/dN', as far as concerns the influence of Diana's tides on the moon, is sensibly periodic according to simple harmonic functions of the time. From this we conclude that:

If the nodes of the satellites' orbits revolve uniformly on the plane of reference, the tides raised by any one satellite can produce no secular change in the inclination of the orbit of any other satellite.

There are thus two cases in which the problem is simplified by our being permitted to consider only the case of identity between Diana and the moon:

1st. Where there are two or more satellites, but where the nodes of the perturbed satellite's orbit revolve with sensible uniformity on the plane of reference.

2nd. Where the planet and satellite are the only bodies in existence.

In these two cases, after differentiation of the disturbing function with respect to the accented elements, we shall be able to drop the accents.

There is also a third case in which Diana's tides will produce a secular effect on the inclination of the moon's orbit, and this is where the nodes of the moon's orbit either revolve irregularly or oscillate. This case is enormously more complicated than the others, and forms the subject of Part III. of this paper; I have only attempted to solve it on the supposition of the smallness both of the inclination of the orbit, and of the obliquity of the ecliptic.

The first of these three cases is that which actually represents the moon and earth, together with solar perturbation of the moon at the present time.

In tracing the configuration of the lunar orbit backwards from the present state, we shall start with the first case; this will graduate into the third, and from this it will pass to a state represented to a very close degree of approximation by the second.

We are not at present concerned to know what are the conditions under which there may be approximate uniformity in the motion of the nodes; this will be investigated below.

We will begin with the first of the three cases, and will find also the rate of change of the diurnal rotation and of the obliquity of the planet.

The second case will then be taken, and afterwards the third case will have to be discussed almost *ab initio* in Part III.

§ 6. *Secular change of inclination of the orbit of a satellite, where there is a second disturbing body, and where the nodes revolve with sensible uniformity on the fixed plane of reference.*

By (13) the equation giving the change of inclination is

$$-\frac{\xi'}{k'}\frac{dj'}{dt} = \frac{1}{\sin j'}\frac{dW}{dN'} + \tan\tfrac{1}{2}j'\frac{dW}{d\epsilon'}$$

As shown above, however, we need here only deal with a single satellite, so that Diana and the moon may be considered as identical and the accents may be dropped to all the symbols, except in the differential coefficients of W. Also we need only maintain the distinction between Diana and the moon as regards N, N' and ϵ, ϵ'; and after the differentiations of W these distinctions must also be dropped. Hence ϖ only differs from ϖ', κ from κ', $\underline{\varpi}$ from $\underline{\varpi}'$, and $\underline{\kappa}$ from $\underline{\kappa}'$ in the accentuation of N.

Also since $\theta = \Omega t + \epsilon$, $\theta' = \Omega' t + \epsilon'$, we may replace $\theta' - \theta$ in the three expressions (37—39) by $\epsilon' - \epsilon$.

If we put $\sin j = 2pq$, $\tan\frac{1}{2}j = q/p$, and write $\phi(N, \epsilon)$ for the operation $\frac{1}{2pq}\frac{d}{dN'} + \frac{q}{p}\frac{d}{d\epsilon'}$, putting $N = N'$, $\epsilon = \epsilon'$ after differentiation; then from (13) we have

$$-\frac{\xi}{k}\frac{dj}{dt} = \phi(N, \epsilon)\,W$$

Also for brevity, let $\phi(N) = \frac{1}{2pq}\frac{d}{dN'}$, $\phi(\epsilon) = \frac{q}{p}\frac{d}{d\epsilon'}$; so that

$$\phi(N, \epsilon) = \phi(N) + \phi(\epsilon)$$

The terms corresponding to the tides of the seven speeds will now be taken separately, the coefficients in ϖ, κ will be developed, and the terms involving $N' - N$ selected, the operation $\phi(N, \epsilon)$ performed, and then N' put equal to N, and ϵ to ϵ'. For the sake of brevity the coefficient $\tau^2/\mathfrak{g}$ will be dropped and will be added in the final result. The component parts of W taken from the equations (37—39) will be indicated as W_I, W_{II}, W_{III} for the slow, sidereal, and fast semi-diurnal parts; as W_1, W_2, W_3 for the slow, sidereal, and fast diurnal parts; and as W_0 for the fortnightly part.

Slow semi-diurnal terms $(2n - 2\Omega)$.

$$W_I = \tfrac{1}{4}F_1\left[\varpi^4\underline{\varpi}'^4 e^{2(\epsilon'-\epsilon)-2\mathfrak{f}_1} + \underline{\varpi}^4\varpi'^4 e^{-2(\epsilon'-\epsilon)+2\mathfrak{f}_1}\right] \text{..............} (40)$$

Let
$$w_I = \tfrac{1}{4}\varpi^4\underline{\varpi}'^4 e^{2(\epsilon'-\epsilon)-2\mathfrak{f}_1}$$

Since
$$\varpi = Pp - qQe^N$$

Therefore

$$\varpi^4 = P^4p^4 - 4P^3Qp^3qe^N + 6P^2Q^2p^2q^2e^{2N} - 4PQ^3pq^3e^{3N} + Q^4q^4e^{4N}$$

$\underline{\varpi}'^4 =$ the same with $-N'$ in place of N

Therefore

$$w_I = \tfrac{1}{4}\{P^8p^8 + 16P^6Q^2p^6q^2e^{N-N'} + 36P^4Q^4p^4q^4e^{2(N-N')} + 16P^2Q^6p^2q^6e^{3(N-N')} + Q^8q^8e^{4(N-N')}\}\,e^{2(\epsilon'-\epsilon)-2\mathfrak{f}_1}$$

Therefore $\quad w_{\mathrm{I}} = \Sigma A_n P^{8-2n} Q^{2n} p^{8-2n} q^{2n} e^{n(N-N')+2(\epsilon'-\epsilon)-2f_1}$

where $n = 0, 1, 2, 3, 4$.

Then
$$\phi(N)\, w_{\mathrm{I}} = \Sigma \frac{n}{2\sqrt{-1}} A_n P^{8-2n} Q^{2n} p^{7-2n} q^{2n-1} e^{-2f_1}$$

$$\phi(\epsilon)\, w_{\mathrm{I}} = -\Sigma \frac{4}{2\sqrt{-1}} A_n P^{8-2n} Q^{2n} p^{7-2n} q^{2n+1} e^{-2f_1}$$

Therefore by addition
$$\phi(N, \epsilon)\, w_{\mathrm{I}} = \Sigma [n(p^2+q^2) - 4q^2] P^{8-2n} Q^{2n} p^{7-2n} q^{2n-1} \frac{e^{-2f_1}}{2\sqrt{-1}}$$

Now when
$$\begin{array}{lll} n=0, & A_n = \tfrac{1}{4}, & n(p^2+q^2) - 4q^2 = -4q^2 \\ \;\;=1, & \;\;\;\;= 4, & = p^2 - 3q^2 \\ \;\;=2, & \;\;\;\;= 9, & = 2(p^2-q^2) \\ \;\;=3, & \;\;\;\;= 4, & = 3p^2 - q^2 \\ \;\;=4, & \;\;\;\;= \tfrac{1}{4}, & = 4p^2 \end{array}$$

If we had taken the second term of W_{I} we should have had the same coefficients but multiplied by $-e^{2f_1}/2\sqrt{-1}$ instead of by $e^{-2f_1}/2\sqrt{-1}$. Therefore, since $(e^{2f_1} - e^{-2f_1})/2\sqrt{-1} = \sin 2f_1$,
$$\phi(N, \epsilon)\, W_{\mathrm{I}} = -F_1 \sin 2f_1 [-P^8 p^7 q + 4P^6 Q^2 p^5 q (p^2 - 3q^2) + 18 P^4 Q^4 p^3 q^3 (p^2 - q^2) \\ + 4P^2 Q^6 p q^5 (3p^2 - q^2) + Q^8 p q^7]$$

Then let
$$\mathfrak{F}_1 = \tfrac{1}{4} [P^8 p^6 - 4P^6 Q^2 p^4 (p^2 - 3q^2) - 18 P^4 Q^4 p^2 q^2 (p^2 - q^2) \\ - 4P^2 Q^6 q^4 (3p^2 - q^2) - Q^8 q^6] \quad \text{......(41)}$$

and remembering that $2pq = \sin j$, we have
$$\phi(N, \epsilon)\, W_{\mathrm{I}} = 2\mathfrak{F}_1 F_1 \sin 2f_1 \sin j \quad \text{......................(42)}$$

Sidereal semi-diurnal terms $(2n)$.
$$W_{\mathrm{II}} = F[\varpi^2 \underline{\kappa}^2 \underline{\varpi}'^2 \kappa'^2 e^{-2f} + \underline{\varpi}^2 \kappa^2 \varpi'^2 \underline{\kappa}'^2 e^{2f}] \quad \text{..............(43)}$$

Here the epoch is wanting, so that $\phi(N, \epsilon) = \phi(N)$.

Let
$$w_{\mathrm{II}} = (\varpi \underline{\kappa} \underline{\varpi}' \kappa')^2$$
$$\varpi = Pp - Qqe^N, \quad \underline{\kappa} = Qp + Pqe^{-N}$$
$$\varpi \underline{\kappa} = PQ(p^2 - q^2) + pq(P^2 e^{-N} - Q^2 e^N)$$
$$\underline{\varpi}' \kappa' = PQ(p^2 - q^2) + pq(P^2 e^{N'} - Q^2 e^{-N'})$$
$$\sqrt{w_{\mathrm{II}}} = P^2 Q^2 (p^2 - q^2)^2 + PQpq(p^2 - q^2)[P^2(e^{N'} + e^{-N}) - Q^2(e^{-N'} + e^N)] \\ + p^2 q^2 [P^4 e^{-(N-N')} + Q^4 e^{(N-N')} - P^2 Q^2 (e^{N+N'} + e^{-(N+N')})]$$
$$w_{\mathrm{II}} = P^4 Q^4 [(p^2 - q^2)^4 - 4p^2 q^2 (p^2 - q^2)^2 + 4p^4 q^4] + 4P^6 Q^2 p^2 q^2 (p^2 - q^2)^2 e^{-(N-N')} \\ + 4P^2 Q^6 p^2 q^2 (p^2 - q^2)^2 e^{N-N'} + p^4 q^4 [P^8 e^{-2(N-N')} + Q^8 e^{2(N-N')}]$$
$$\phi(N)\, w_{\mathrm{II}} = -\frac{1}{2\sqrt{-1}} [4pq(p^2 - q^2)^2 P^2 Q^2 (P^4 - Q^4) + 2p^3 q^3 (P^8 - Q^8)]$$

If we had operated on the other term of W_{II} we should have got the same with the opposite sign, and e^{2f} in place of e^{-2f}.

Then let

$$\mathfrak{F} = \tfrac{1}{2}(P^2 - Q^2)\{2(p^2 - q^2)^2 P^2 Q^2 + p^2 q^2 (P^4 + Q^4)\} \quad \text{.........(44)}$$

and we have

$$\phi(N, \epsilon)\, W_{II} = 2\mathfrak{F}\mathrm{F} \sin 2\mathrm{f} \sin j \quad \text{.....................(45)}$$

Fast semi-diurnal terms $(2n + 2\Omega)$.

$$W_{III} = \tfrac{1}{4}\mathrm{F}_2 [\underline{\kappa}^4 \kappa'^4 e^{-2(\epsilon'-\epsilon)-2\mathrm{f}_2} + \kappa^4 \underline{\kappa}'^4 e^{2(\epsilon'-\epsilon)+2\mathrm{f}_2}] \quad \text{..............(46)}$$

Since κ is obtained from ϖ by writing Q for P, and $-P$ for Q, therefore by writing $-2\mathrm{f}_2$ for $2\mathrm{f}_1$, and interchanging Q's and P's we may write down the result by symmetry with the slow semi-diurnal terms. Then let

$$\mathfrak{F}_2 = \tfrac{1}{4}[Q^8 p^6 - 4P^2 Q^6 p^4 (p^2 - 3q^2) - 18P^4 Q^4 p^2 q^2 (p^2 - q^2)$$
$$- 4P^6 Q^2 q^4 (3p^2 - q^2) - P^8 q^6] \quad \text{......(47)}$$

and

$$\phi(N, \epsilon)\, W_{III} = -2\mathfrak{F}_2 \mathrm{F}_2 \sin 2\mathrm{f}_2 \sin j \quad \text{..................(48)}$$

Slow diurnal terms.

$$W_1 = G_1 [\varpi^3 \kappa \underline{\varpi}'^3 \underline{\kappa}' e^{2(\epsilon'-\epsilon)-\mathrm{g}_1} + \underline{\varpi}^3 \underline{\kappa} \varpi'^3 \kappa' e^{-2(\epsilon'-\epsilon)+\mathrm{g}_1}] \quad \text{.........(49)}$$

Let $w_1 = \varpi^3 \kappa \underline{\varpi}'^3 \underline{\kappa}' e^{2(\epsilon'-\epsilon)-\mathrm{g}_1}$.

For the moment let $I = \tfrac{1}{2}i$, then since $\varpi = Pp - Qqe^N$, and since $P = \cos I$, $Q = \sin I$, therefore $d\varpi/dI = -\kappa$, and therefore $d\varpi^4/dI = -4\varpi^3\kappa$.

Hence from the slow semi-diurnal term we find

$$\varpi^3\kappa = P^3 Q p^4 + P^2(P^2 - 3Q^2) p^3 q e^N - 3PQ(P^2 - Q^2) p^2 q^2 e^{2N}$$
$$+ Q^2(3P^2 - Q^2) pq^3 e^{3N} - PQ^3 q^4 e^{4N}$$

$\underline{\varpi}'^3 \underline{\kappa}' =$ same with $-N'$ for N

Hence

$$w_1 = [P^6 Q^2 p^8 + P^4 (P^2 - 3Q^2)^2 p^6 q^2 e^{N-N'} + 9P^2 Q^2 (P^2 - Q^2)^2 p^4 q^4 e^{2(N-N')}$$
$$+ Q^4 (3P^2 - Q^2)^2 p^2 q^6 e^{3(N-N')} + P^2 Q^6 q^8 e^{4(N-N')}]\, e^{2(\epsilon'-\epsilon)-\mathrm{g}_1}$$

$$\phi(N)\, w_1 = \frac{e^{-\mathrm{g}_1}}{2\sqrt{-1}} \Big[P^4 (P^2 - 3Q^2)^2 p^5 q + 18P^2 Q^2 (P^2 - Q^2)^2 p^3 q^3$$
$$+ 3Q^4 (3P^2 - Q^2)^2 pq^5 + 4P^2 Q^6 \frac{q^7}{p} \Big]$$

$$\phi(\epsilon)\, w_1 = -\frac{e^{-\mathrm{g}_1}}{2\sqrt{-1}} \Big[4P^6 Q^2 p^7 q + 4P^4 (P^2 - 3Q^2)^2 p^5 q^3 + 36P^2 Q^2 (P^2 - Q^2)^2 p^3 q^5$$
$$+ 4Q^4 (3P^2 - Q^2)^2 pq^7 + 4P^2 Q^6 \frac{q^9}{p} \Big]$$

Adding

$$\phi(N, \epsilon)\, w_1 = -\frac{e^{-\mathrm{g}_1}}{2\sqrt{-1}} [4P^6 Q^2 p^7 q - P^4 (P^2 - 3Q^2)^2 p^5 q (p^2 - 3q^2)$$
$$- 18P^2 Q^2 (P^2 - Q^2)^2 p^3 q^3 (p^2 - q^2) - Q^4 (3P^2 - Q^2)^2 pq^5 (3p^2 - q^2) - 4P^2 Q^6 pq^7]$$

Let

$$\mathfrak{G}_1 = \tfrac{1}{4}[4P^6Q^2p^6 - P^4(P^2-3Q^2)^2p^4(p^2-3q^2) - 18P^2Q^2(P^2-Q^2)^2p^2q^2(p^2-q^2)$$
$$- Q^4(3P^2-Q^2)^2 q^4(3p^2-q^2) - 4P^2Q^6q^6] \ldots(50)$$

and we have $$\phi(N,\epsilon)\,\mathrm{W}_1 = 2\mathfrak{G}_1\mathrm{G}_1 \sin \mathrm{g}_1 \sin j \ldots\ldots\ldots\ldots(51)$$

Sidereal diurnal terms (n).

$$\mathrm{W}_2 = \mathrm{G}\,[\varpi\underline{\kappa}(\varpi\underline{\varpi} - \kappa\underline{\kappa})\underline{\varpi}'\kappa'(\varpi'\underline{\varpi}' - \kappa'\underline{\kappa}')e^{-\mathrm{g}} + \underline{\varpi}\kappa(\varpi\underline{\varpi} - \kappa\underline{\kappa})\varpi'\underline{\kappa}'(\varpi'\underline{\varpi}' - \underline{\kappa}'\kappa')e^{\mathrm{g}}]$$
$$\ldots\ldots(52)$$

Here the epoch is wanting, so that $\phi(N,\epsilon) = \phi(N)$.

Let

$$\mathrm{w}_2 = \varpi\underline{\kappa}(\varpi\underline{\varpi} - \kappa\underline{\kappa})\underline{\varpi}'\kappa'(\varpi'\underline{\varpi}' - \kappa'\underline{\kappa}')$$

$$\varpi\underline{\kappa} = PQ(p^2-q^2) + pq(P^2e^{-N} - Q^2e^{N})$$

$$\varpi\underline{\varpi} - \kappa\underline{\kappa} = (P^2-Q^2)(p^2-q^2) - 2PQpq(e^N + e^{-N})$$

$$\varpi\underline{\kappa}(\varpi\underline{\varpi} - \kappa\underline{\kappa}) = PQ(P^2-Q^2)[(p^2-q^2)^2 - 2p^2q^2] + P^2(P^2-3Q^2)pq(p^2-q^2)e^{-N}$$
$$- Q^2(3P^2-Q^2)pq(p^2-q^2)e^{N} - 2PQp^2q^2(P^2e^{-2N} - Q^2e^{2N})$$

$\underline{\varpi}'\kappa'(\varpi'\underline{\varpi}' - \kappa'\underline{\kappa}')$ = the same with $-N'$ instead of N

$$\mathrm{w}_2 = P^2Q^2(P^2-Q^2)^2[(p^2-q^2)^2 - 2p^2q^2]^2 + P^4(P^2-3Q^2)^2p^2q^2(p^2-q^2)^2e^{-(N-N')}$$
$$+ Q^4(3P^2-Q^2)^2p^2q^2(p^2-q^2)^2e^{N-N'} + 4P^2Q^2p^4q^4(P^4e^{-2(N-N')} + Q^4e^{2(N-N')})$$

$$\phi(N)\,\mathrm{w}_2 = -\frac{1}{2\sqrt{-1}}\{pq(p^2-q^2)^2[P^4(P^2-3Q^2)^2 - Q^4(3P^2-Q^2)^2]$$
$$+ 8P^2Q^2(P^4-Q^4)p^3q^3\}$$

Now

$$P^4(P^2-3Q^2)^2 - Q^4(3P^2-Q^2)^2 = (P^2-Q^2)(P^4+Q^4-6P^2Q^2)$$

Put therefore

$$\mathfrak{G} = \tfrac{1}{4}(P^2-Q^2)\{(p^2-q^2)^2(P^4+Q^4-6P^2Q^2) + 8P^2Q^2p^2q^2\} \ldots\ldots(53)$$

and we have $$\phi(N,\epsilon)\,\mathrm{W}_2 = 2\mathfrak{G}\mathrm{G}\sin\mathrm{g}\sin j \ldots\ldots\ldots\ldots(54)$$

Fast diurnal terms $(n + 2\Omega)$.

$$\mathrm{W}_3 = \mathrm{G}_2[\underline{\varpi}\underline{\kappa}^3\varpi'\kappa'^3e^{-2(\epsilon'-\epsilon)-\mathrm{g}_2} + \varpi\kappa^3\underline{\varpi}'\underline{\kappa}'^3e^{2(\epsilon'-\epsilon)+\mathrm{g}_2}] \ldots\ldots\ldots\ldots(55)$$

By an analogy similar to that by which the fast semi-diurnal was derived from the slow, we have

$$\mathfrak{G}_2 = \tfrac{1}{4}[4P^2Q^6p^6 - Q^4(3P^2-Q^2)^2p^4(p^2-3q^2) - 18P^2Q^2(P^2-Q^2)^2p^2q^2(p^2-q^2)$$
$$- P^4(P^2-3Q^2)^2q^4(3p^2-q^2) - 4P^6Q^2q^6] \ldots\ldots(56)$$

and $$\phi(N,\epsilon)\,\mathrm{W}_3 = -2\mathfrak{G}_2\mathrm{G}_2\sin\mathrm{g}_2\sin j \ldots\ldots\ldots\ldots(57)$$

Fortnightly terms (2Ω).

$$\mathrm{W}_0 = \tfrac{3}{2}[(\tfrac{1}{3} - 2\varpi\underline{\varpi}\kappa\underline{\kappa})(\tfrac{1}{3} - 2\varpi'\underline{\varpi}'\kappa'\underline{\kappa}') + \mathrm{H}\underline{\varpi}^2\underline{\kappa}^2\varpi'^2\kappa'^2e^{-2(\epsilon'-\epsilon)-2\mathrm{h}}$$
$$+ \mathrm{H}\varpi^2\kappa^2\underline{\varpi}'^2\underline{\kappa}'^2e^{2(\epsilon'-\epsilon)+2\mathrm{h}}] \ldots\ldots(58)$$

It will be found that $\phi(N)$ performed on the first term is zero, as it ought to be according to the general principles of energy—for the system is a conservative one as far as regards these terms.

Let

$$\mathrm{w}_0 = (\varpi\kappa\underline{\varpi}'\underline{\kappa}')^2\, e^{2(\epsilon'-\epsilon)+2\mathrm{h}}$$

$$\varpi\kappa = PQp^2 + pq\,(P^2 - Q^2)\,e^N - PQq^2e^{2N}$$

$$\varpi^2\kappa^2 = P^2Q^2p^4 + 2PQ\,(P^2 - Q^2)\,p^3qe^N + [(P^2 - Q^2)^2 - 2P^2Q^2]\,p^2q^2e^{2N} - 2PQ\,(P^2 - Q^2)\,pq^3e^{3N} + P^2Q^2q^4e^{4N}$$

$\underline{\varpi}'^2\underline{\kappa}'^2 =$ the same with $-N'$ for N

$$\mathrm{w}_0 = \{P^4Q^4p^8 + 4P^2Q^2\,(P^2 - Q^2)^2p^6q^2e^{N-N'} + [(P^2 - Q^2)^2 - 2P^2Q^2]\,p^4q^4e^{2(N-N')} + 4P^2Q^2\,(P^2 - Q^2)^2p^2q^6e^{3(N-N')} + P^4Q^4q^8e^{4(N-N')}\}\, e^{2(\epsilon'-\epsilon)+2\mathrm{h}}$$

$$\phi(N)\,\mathrm{w}_0 = \frac{e^{2\mathrm{h}}}{2\sqrt{-1}}\Big[4P^2Q^2\,(P^2 - Q^2)^2p^5q + 2\,[(P^2 - Q^2)^2 - 2P^2Q^2]^2\,p^3q^3 + 12P^2Q^2\,(P^2 - Q^2)^2\,pq^5 + 4P^4Q^4\frac{q^7}{p}\Big]$$

$$\phi(\epsilon)\,\mathrm{w}_0 = -\frac{e^{2\mathrm{h}}}{2\sqrt{-1}}\Big[4P^4Q^4p^7q + 16P^2Q^2\,(P^2 - Q^2)^2p^5q^3 + 4\,[(P^2 - Q^2)^2 - 2P^2Q^2]^2p^3q^5 + 16P^2Q^2\,(P^2 - Q^2)^2pq^7 + 4P^4Q^4\frac{q^9}{p}\Big]$$

Adding and arranging the terms

$$\phi(N, \epsilon)\,\mathrm{w}_0 = -\frac{e^{2\mathrm{h}}}{2\sqrt{-1}}\,pq\,\{4P^4Q^4\,(p^6 - q^6) - 4P^2Q^2\,(P^2 - Q^2)^2\,(p^2 - q^2)^3 - 2p^2q^2\,(p^2 - q^2)\,[(P^2 - Q^2)^2 - 2P^2Q^2]^2\}$$

Then let

$$\mathfrak{H} = \tfrac{3}{4}\,\{2P^4Q^4\,(p^6 - q^6) - 2P^2Q^2\,(P^2 - Q^2)^2\,(p^2 - q^2)^3 - p^2q^2\,(p^2 - q^2)\,[(P^2 - Q^2)^2 - 2P^2Q^2]^2\} \quad \text{...(59)}$$

and we have $\phi(N, \epsilon)\,\mathrm{W}_0 = -2\mathfrak{H}\mathrm{H}\sin 2\mathrm{h}\sin j$ ....(60)

This is the last of the seven sets of terms.

Collecting results from (42–5–8, 51–4–7, 60), we have

$$\frac{1}{\sin j}\frac{dj}{dt} = -\frac{\tau^2}{\mathfrak{g}}\frac{k}{\xi}\{2\mathfrak{F}_1\mathrm{F}_1\sin 2\mathrm{f}_1 + 2\mathfrak{F}\mathrm{F}\sin 2\mathrm{f} - 2\mathfrak{F}_2\mathrm{F}_2\sin 2\mathrm{f}_2 + 2\mathfrak{G}_1\mathrm{G}_1\sin \mathrm{g}_1 + 2\mathfrak{G}\mathrm{G}\sin \mathrm{g} - 2\mathfrak{G}_2\mathrm{G}_2\sin \mathrm{g}_2 - 2\mathfrak{H}\mathrm{H}\sin 2\mathrm{h}\} \quad \text{...(61)}$$

The seven gothic-letter functions defined by (41–4–7, 50–3–6–9) are functions of the sines and cosines of half the obliquity and of half the inclination, but they are reducible to forms which may be expressed in the following manner:

$$\left.\begin{aligned}
\mathfrak{F}_1 + \mathfrak{F}_2 &= \tfrac{1}{4}\cos j\,[1 - \tfrac{1}{4}\sin^2 j - 2\sin^2 i\,(1 - \tfrac{5}{8}\sin^2 j) + \tfrac{5}{8}\sin^4 i\,(1 - \tfrac{7}{4}\sin^2 j)]\\
\mathfrak{F}_1 - \mathfrak{F}_2 &= \tfrac{1}{4}\cos i\,[1 - \tfrac{3}{4}\sin^2 j - \tfrac{3}{2}\sin^2 i\,(1 - \tfrac{5}{4}\sin^2 j)]\\
\mathfrak{G}_1 + \mathfrak{G}_2 &= -\tfrac{1}{4}\cos j\,[1 - \sin^2 j - \tfrac{7}{2}\sin^2 i\,(1 - \tfrac{10}{7}\sin^2 j) + \tfrac{5}{2}\sin^4 i\,(1 - \tfrac{7}{4}\sin^2 j)]\\
\mathfrak{G}_1 - \mathfrak{G}_2 &= -\tfrac{1}{4}\cos i\,[1 - \tfrac{3}{2}\sin^2 j - 3\sin^2 i\,(1 - \tfrac{5}{4}\sin^2 j)]\\
\mathfrak{F} &= \tfrac{1}{4}\cos i\,[\tfrac{1}{2}\sin^2 j + \sin^2 i - \tfrac{5}{4}\sin^2 i\sin^2 j]\\
\mathfrak{G} &= \tfrac{1}{4}\cos i\,[1 - \sin^2 j - 2\sin^2 i + \tfrac{5}{2}\sin^2 i\sin^2 j]\\
\mathfrak{H} &= -\tfrac{1}{4}\cos j\,[\tfrac{3}{4}\sin^2 j + \tfrac{3}{2}\sin^2 i\,(1 - \tfrac{5}{2}\sin^2 j) - \tfrac{15}{8}\sin^4 i\,(1 - \tfrac{7}{4}\sin^2 j)]
\end{aligned}\right\} \quad \text{......(62)}$$

These coefficients will be applicable whatever theory of tides be used, and no approximation, as regards either the obliquity or inclination, has been used in obtaining them.

§ 7. *Application to the case where the planet is viscous.*

If the planet or earth be viscous with a coefficient of viscosity v, then according to the theory of viscous tides, when inertia is neglected, the tangent of the phase-retardation or lag of any tide is equal to $19v/2gaw$ multiplied by the speed of that tide; and the height of tide is equal to the equilibrium tide of a perfectly fluid spheroid multiplied by the cosine of the lag. If therefore we put $\frac{2gaw}{19v} = \mathfrak{p}$, we have

$$\tan 2\mathrm{f}_1 = \frac{2(n-\Omega)}{\mathfrak{p}}, \quad \tan 2\mathrm{f} = \frac{2n}{\mathfrak{p}}, \quad \tan 2\mathrm{f}_2 = \frac{2(n+\Omega)}{\mathfrak{p}}$$

$$\tan \mathrm{g}_1 = \frac{n-2\Omega}{\mathfrak{p}}, \quad \tan \mathrm{g} = \frac{n}{\mathfrak{p}}, \quad \tan 2\mathrm{g}_2 = \frac{n+2\Omega}{\mathfrak{p}}, \quad \tan 2\mathrm{h} = \frac{2\Omega}{\mathfrak{p}}$$

$$\mathrm{F}_1 = \cos 2\mathrm{f}_1, \quad \mathrm{F} = \cos 2\mathrm{f}, \quad \mathrm{F}_2 = \cos 2\mathrm{f}_2, \quad \mathrm{G}_1 = \cos \mathrm{g}_1, \quad \mathrm{G} = \cos \mathrm{g}, \quad \mathrm{G}_2 = \cos \mathrm{g}_2$$

and $\mathrm{H} = \cos 2\mathrm{h}$.

Therefore

$$-\frac{\xi}{k \sin j}\frac{dj}{dt} = \frac{\tau^2}{\mathfrak{g}}\{\mathfrak{F}_1 \sin 4\mathrm{f}_1 + \mathfrak{F} \sin 4\mathrm{f} - \mathfrak{F}_2 \sin 4\mathrm{f}_2 + \mathfrak{G}_1 \sin 2\mathrm{g}_1$$
$$+ \mathfrak{G} \sin 2\mathrm{g} - \mathfrak{G}_2 \sin 2\mathrm{g}_2 - \mathfrak{H} \sin 4\mathrm{h}\} \quad \text{...(63)}$$

This equation involves such complex functions of i and j, that it does not present to the mind any physical meaning. It will accordingly be illustrated graphically.

For this purpose the case is taken when the planet rotates fifteen times as fast as the satellite revolves. Then the speeds of the seven tides are proportional to the following numbers: 28, 30, 32 (semi-diurnal); 13, 15, 17 (diurnal); and 2 (fortnightly).

It would require a whole series of figures to illustrate the equation for all values of i and j, and for all viscosities. The case is therefore taken where the inclination j of the orbit to the ecliptic is so small that we may neglect squares and higher powers of $\sin j$. Then the formulæ (62) become

$$\mathfrak{F}_1 + \mathfrak{F}_2 = \tfrac{1}{4}(1 - 2\sin^2 i + \tfrac{5}{8}\sin^4 i)$$
$$\mathfrak{F}_1 - \mathfrak{F}_2 = \tfrac{1}{4}\cos i(1 - \tfrac{3}{2}\sin^2 i)$$
$$\mathfrak{G}_1 + \mathfrak{G}_2 = -\tfrac{1}{4}(1 - \tfrac{7}{2}\sin^2 i + \tfrac{5}{2}\sin^4 i)$$
$$\mathfrak{G}_1 - \mathfrak{G}_2 = -\tfrac{1}{4}\cos i(1 - 3\sin^2 i)$$
$$\mathfrak{F} = \tfrac{1}{4}\cos i \sin^2 i, \quad \mathfrak{G} = \tfrac{1}{4}\cos i(1 - 2\sin^2 i)$$
$$\mathfrak{H} = -\tfrac{3}{8}\sin^2 i(1 - \tfrac{5}{4}\sin^2 i)$$

From these we may compute a series of values corresponding to $i = 0°$, $15°$, $30°$, $45°$, $60°$, $75°$, $90°$. (I actually did compute them from the P, Q formulæ.)

I then took as five several standards of the viscosity of the planet, such viscosities as would make the lag $\mathfrak{f}_1$ of the slow semi-diurnal tide (of speed $2n - 2\Omega$) equal to $10°$, $20°$, $30°$, $40°$, $44°$. It is easy to compute tables giving the five corresponding values of each of the following, viz.: $\sin 4\mathfrak{f}_1$, $\sin 4\mathfrak{f}$, $\sin 4\mathfrak{f}_2$, $\sin 2\mathfrak{g}_1$, $\sin 2\mathfrak{g}$, $\sin 2\mathfrak{g}_2$, $\sin 4\mathfrak{h}$.

The numerical values were appropriately multiplied (with Crelle's three-figure table) by the sets of values before found for the $\mathfrak{F}$'s, $\mathfrak{G}$'s, &c.

From the sets of tables formed, the proper sets were selected and added up. The result was to have a series of numbers which were proportional to $dj/\sin j\, dt$.

The series corresponding to each degree of viscosity were set off in a curve, as shown in fig. 4.

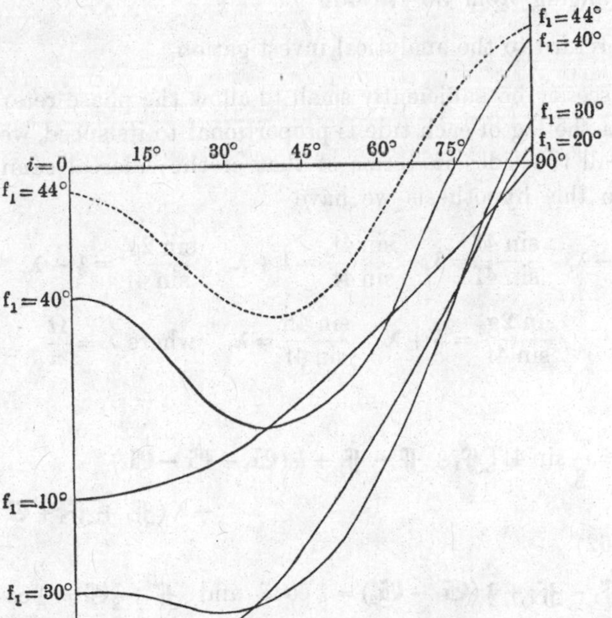

FIG. 4. Diagram illustrating the rate of change of the inclination of a satellite's orbit to a fixed plane on which its nodes revolve, for various obliquities and viscosities of the planet $\left(\frac{1}{\sin j}\frac{dj}{dt}\text{ when } j \text{ is small}\right)$.

The ordinates, which are generally negative, represent $dj/\sin j\,dt$, and the abscissæ correspond to i, the obliquity of the planet's equator to the ecliptic.

This figure shows that the inclination j of the orbit will diminish, unless the obliquity be very large.

It appears from the results of previous papers, that the satellite's distance will increase as the time increases, unless the obliquity be very large, and if the obliquity be very large the mean distance decreases more rapidly for large than for small viscosity. This statement, taken in conjunction with our present figure, shows that in general the inclination will decrease as long as the mean distance increases, and *vice versâ*. This is not, however, necessarily true for all speeds of rotation of the planet and revolution of the satellite.

The most remarkable feature in these curves is that they show that, for moderate degrees of viscosity (f_1 less than 20°), the inclination j decreases most rapidly when i the obliquity is zero; whilst for larger viscosities (f_1 between 20° and 45°), there is a very marked maximum rate of decrease for obliquities ranging from 30° to 40°.

We now return to the analytical investigation.

If the viscosity be sufficiently small to allow the phase retardations to be small, so that the lag of each tide is proportional to its speed, we may express the lags of all the tides in terms of that of the sidereal semi-diurnal tide, viz.: 2f. On this hypothesis we have

$$\frac{\sin 4f_1}{\sin 4f} = 1 - \lambda, \quad \frac{\sin 4f}{\sin 4f} = 1, \quad \frac{\sin 4f_2}{\sin 4f} = 1 + \lambda, \quad \frac{\sin 2g_1}{\sin 4f} = \tfrac{1}{2} - \lambda, \quad \frac{\sin 2g}{\sin 4f} = \tfrac{1}{2}$$

$$\frac{\sin 2g_2}{\sin 4f} = \tfrac{1}{2} + \lambda, \quad \frac{\sin 4h}{\sin 4f} = \lambda, \quad \text{where } \lambda = \frac{\Omega}{n}$$

And

$$-\frac{\xi}{k \sin j}\frac{dj}{dt} = \frac{\tau^2}{\mathfrak{g}} \sin 4f \left[\mathfrak{F}_1 + \mathfrak{F} - \mathfrak{F}_2 + \tfrac{1}{2}(\mathfrak{G}_1 + \mathfrak{G} - \mathfrak{G}_2) - \lambda(\mathfrak{F}_1 + \mathfrak{F}_2 + \mathfrak{G}_1 + \mathfrak{G}_2 + \mathfrak{H})\right]$$

But by (62)

$$\mathfrak{F}_1 - \mathfrak{F}_2 + \tfrac{1}{2}(\mathfrak{G}_1 - \mathfrak{G}_2) = \tfrac{1}{8}\cos i \quad \text{and} \quad \mathfrak{F} + \tfrac{1}{2}\mathfrak{G} = \tfrac{1}{8}\cos i$$

and

$$\mathfrak{F}_1 + \mathfrak{F}_2 + \mathfrak{G}_1 + \mathfrak{G}_2 + \mathfrak{H} = 0$$

These results may of course be also obtained when the functions are expressed in terms of P, Q, p, q.

Whence on this hypothesis

$$-\frac{\xi}{k \sin j}\frac{dj}{dt} = \frac{\tau^2}{\mathfrak{g}} \sin 4f \,.\, \tfrac{1}{4}\cos i \qquad \text{......................(64)}$$

§ 8. *Secular change in the mean distance of a satellite, where there is a second disturbing body, and where the nodes revolve with sensible uniformity on the fixed plane of reference.*

By (11) the equation giving the rate of change of ξ' is

$$\frac{1}{k'}\frac{d\xi'}{dt}=\frac{d\mathrm{W}}{d\epsilon'}$$

As before, we may drop the accents, except as regards ϵ'.

In § 6 we wrote $\phi(\epsilon)$ for the operation $\tan\frac{1}{2}j\frac{d}{d\epsilon'}$; hence $\frac{d\mathrm{W}}{d\epsilon'}=\frac{p}{q}\phi(\epsilon)\mathrm{W}$, and by reference to that section the result may be at once written down. We have

$$\frac{1}{k}\frac{d\xi}{dt}=\frac{\tau^2}{\mathfrak{g}}\{2\Phi_1\mathrm{F}_1\sin 2\mathrm{f}_1-2\Phi_2\mathrm{F}_2\sin 2\mathrm{f}_2+2\Gamma_1\mathrm{G}_1\sin\mathrm{g}_1-2\Gamma_2\mathrm{G}_2\sin\mathrm{g}_2 - 2\Lambda\mathrm{H}\sin 2\mathrm{h}\}\text{......}(65)$$

Where

$$\left.\begin{aligned}
\Phi_1&=\tfrac{1}{2}[P^8p^8+Q^8q^8+16P^2p^2Q^2q^2(P^4p^4+Q^4q^4)+36P^4Q^4p^4q^4]\\
\Phi_2&=\text{the same with } Q \text{ and } P \text{ interchanged}\\
\Gamma_1&=2[P^2Q^2(P^4p^8+Q^4q^8)+P^4(P^2-3Q^2)^2p^6q^2+Q^4(3P^2-Q^2)^2p^2q^6\\
&\qquad+9P^2Q^2(P^2-Q^2)^2p^4q^4]\\
\Gamma_2&=\text{the same with } Q \text{ and } P \text{ interchanged}\\
\Lambda&=3\{P^4Q^4(p^8+q^8)+4P^2Q^2(P^2-Q^2)^2p^2q^2(p^4+q^4)\\
&\qquad+p^4q^4[(P^2-Q^2)^2-2P^2Q^2]^2\}
\end{aligned}\right\}\text{(66)}$$

These functions are reducible to the following forms

$$\left.\begin{aligned}
2(\Phi_1+\Phi_2)&=1-\sin^2 j+\tfrac{1}{8}\sin^4 j-\sin^2 i(1-2\sin^2 j+\tfrac{5}{8}\sin^4 j)\\
&\qquad+\tfrac{1}{8}\sin^4 i(1-5\sin^2 j+\tfrac{35}{8}\sin^4 j)\\
2(\Phi_1-\Phi_2)&=\cos i\cos j[1-\tfrac{1}{2}\sin^2 j-\tfrac{1}{2}\sin^2 i(1-\tfrac{5}{2}\sin^2 j)]\\
2(\Gamma_1+\Gamma_2)&=\sin^2 j-\tfrac{1}{2}\sin^4 j+\sin^2 i(1-\tfrac{7}{2}\sin^2 j+\tfrac{5}{2}\sin^4 j)\\
&\qquad-\tfrac{1}{2}\sin^4 i(1-5\sin^2 j+\tfrac{35}{8}\sin^4 j)\\
2(\Gamma_1-\Gamma_2)&=\cos i\cos j[\sin^2 j+\sin^2 i(1-\tfrac{5}{2}\sin^2 j)]\\
2\Lambda\quad&=\tfrac{3}{8}\sin^4 j+\sin^2 i(\tfrac{3}{2}\sin^2 j-\tfrac{15}{8}\sin^4 j)\\
&\qquad+\tfrac{3}{8}\sin^4 i(1-5\sin^2 j+\tfrac{35}{8}\sin^4 j)
\end{aligned}\right\}\text{...(67)}$$

§ 9. *Application to the case where the planet is viscous.*

As in § 7

$$\frac{1}{k}\frac{d\xi}{dt}=\frac{\tau^2}{\mathfrak{g}}\{\Phi_1\sin 4\mathrm{f}_1-\Phi_2\sin 4\mathrm{f}_2+\Gamma_1\sin 2\mathrm{g}_1-\Gamma_2\sin 2\mathrm{g}_2-\Lambda\sin 4\mathrm{h}\}\text{...(68)}$$

If j be put equal to zero this equation will be found to be the same as that used as the equation of tidal reaction in the previous paper on "Precession."

If the viscosity be small, with the same notation as before

$$\frac{1}{k}\frac{d\xi}{dt} = \frac{\tau^2}{\mathfrak{g}} \sin 4f\,[\Phi_1 - \Phi_2 + \tfrac{1}{2}(\Gamma_1 - \Gamma_2) - \lambda(\Phi_1 + \Phi_2 + \Gamma_1 + \Gamma_2 + \Lambda)] \ldots (69)$$

Now $$\Phi_1 - \Phi_2 + \tfrac{1}{2}(\Gamma_1 - \Gamma_2) = \tfrac{1}{2}\cos i \cos j$$

and $$\Phi_1 + \Phi_2 + \Gamma_1 + \Gamma_2 + \Lambda = \tfrac{1}{2}$$

Therefore $$\frac{1}{k}\frac{d\xi}{dt} = \tfrac{1}{2}\frac{\tau^2}{\mathfrak{g}} \sin 4f\,[\cos i \cos j - \lambda] \ldots\ldots\ldots\ldots (70)$$

We see that the rate of tidal reaction diminishes as the inclination of the orbit increases.

§ 10. *Secular change in the inclination of the orbit of a single satellite to the invariable plane, where there is no other disturbing body than the planet.*

This is the second of the two cases into which the problem subdivides itself.

If there be only two bodies, the fixed plane of reference, which was called the ecliptic, may be taken as the invariable plane of the system. It follows from the principle of the composition of moments of momentum that the planet's axis of rotation, the normal to the satellite's orbit and the normal to the invariable plane, necessarily lie in one plane. Whence it follows that the orbit and the equator necessarily intersect in the invariable plane. From this principle it would of course be possible either to determine the motion of the node from the precession of the planet or *vice versâ*, and the change of obliquity of the planet's axis (if any) from the change in the plane of the orbit or *vice versâ*; this principle will be applied later.

We have found it convenient to measure longitudes from a line in the fixed plane, which is instantaneously coincident with the descending node of the equator on the fixed plane. Hence it follows that where there are only two bodies we shall after differentiation have to put $N = N' = 0$.

Then since $\varpi' = Pp - Qqe^{N'}$ therefore $\dfrac{d\varpi'}{dN'} = \dfrac{1}{\sqrt{-1}}Qq$, and similarly

$$\frac{d\underline{\varpi}'}{dN'} = -\frac{Qq}{\sqrt{-1}},\quad \frac{d\kappa'}{dN'} = -\frac{Pq}{\sqrt{-1}},\quad \frac{d\underline{\kappa}'}{dN'} = \frac{Pq}{\sqrt{-1}},\ \text{when } N' = 0$$

Also after differentiation when $N = 0$,

$$\varpi = \underline{\varpi} = \cos\tfrac{1}{2}(i + j),\quad \kappa = \underline{\kappa} = \sin\tfrac{1}{2}(i + j)$$

In order to find dj/dt we must, as before, perform $\phi(N, \epsilon)$ on W, and we take the same notation as before for the W's and w's with suffixes.

Slow semi-diurnal term.

$$\frac{d}{dN'}\left(\tfrac{1}{4}\varpi^4\underline{\varpi}'^4\right) = \varpi^7\frac{d\underline{\varpi}'}{dN'} = -\frac{\varpi^7 Qq}{\sqrt{-1}}$$

and

$$\phi(N)\left(\tfrac{1}{4}\varpi^4\underline{\varpi}'^4\right) = -\frac{1}{2\sqrt{-1}}\cdot\varpi^7\cdot\frac{Q}{p},\quad \text{also } \phi(\epsilon)\,e^{2(\epsilon'-\epsilon)-2\mathrm{f}_1} = -\frac{1}{2\sqrt{-1}}\cdot\frac{4q}{p}\,e^{-2\mathrm{f}_1}$$

Hence $$\phi(N,\epsilon)\,\mathrm{w}_{\mathrm{I}} = \frac{e^{-2\mathrm{f}_1}}{2\sqrt{-1}}\left[-\varpi^7\frac{Q}{p} - \varpi^8\cdot\frac{q}{p}\right] = -\varpi^7\kappa\frac{e^{-2\mathrm{f}_1}}{2\sqrt{-1}}$$

and $$\phi(N,\epsilon)\,\mathrm{W}_{\mathrm{I}} = \varpi^7\kappa\mathrm{F}_1\sin 2\mathrm{f}_1$$

Sidereal semi-diurnal term.

$$\frac{d\mathrm{w}_{\mathrm{II}}}{dN'} = 2\varpi^3\kappa^3\left(\varpi\frac{d\kappa'}{dN'} + \kappa\frac{d\underline{\varpi}'}{dN'}\right) = -\frac{2\varpi^3\kappa^3}{\sqrt{-1}}(\varpi Pq + \kappa Qq) = -\frac{2\varpi^3\kappa^3}{\sqrt{-1}}\,.\,pq$$

and since $\phi(\epsilon)\,\mathrm{W}_{\mathrm{II}} = 0$, therefore

$$\phi(N,\epsilon)\,\mathrm{W}_{\mathrm{II}} = 2\varpi^3\kappa^3\mathrm{F}\sin 2\mathrm{f}$$

Fast semi-diurnal term.

By symmetry $$\phi(N,\epsilon)\,\mathrm{W}_{\mathrm{III}} = \varpi\kappa^7\mathrm{F}_2\sin 2\mathrm{f}_2$$

Slow diurnal term.

$$\frac{d}{dN'}\underline{\varpi}'^3\underline{\kappa}' = 3\varpi^2\kappa\frac{d\underline{\varpi}'}{dN'} + \varpi^3\frac{d\underline{\kappa}'}{dN'} = -\frac{\varpi^2}{\sqrt{-1}}(3\kappa Qq - \varpi Pq)$$

$$\phi(N,\epsilon)\,\mathrm{w}_1 = -\frac{e^{-\mathrm{g}_1}}{2\sqrt{-1}}\frac{\varpi^5\kappa}{p}(3Q\kappa - P\varpi + 4q\varpi\kappa) = -\frac{e^{-\mathrm{g}_1}}{2\sqrt{-1}}\varpi^5\kappa(\varpi^2 - 3\kappa^2)$$

and $$\phi(N,\epsilon)\,\mathrm{W}_1 = -\varpi^5\kappa(\varpi^2 - 3\kappa^2)\,\mathrm{G}_1\sin\mathrm{g}_1$$

Sidereal diurnal term.

$$\frac{d}{dN'}\underline{\varpi}'\kappa' = -\frac{q}{\sqrt{-1}}(Q\kappa + P\varpi) = -\frac{pq}{\sqrt{-1}} \text{ and } \frac{d}{dN'}(\varpi'\underline{\varpi}' - \kappa'\underline{\kappa}') = 0$$

Therefore $$\phi(N,\epsilon)\,\mathrm{w}_2 = -\frac{e^{-\mathrm{g}}}{\sqrt{-1}}\varpi\kappa(\varpi^2 - \kappa^2)^2$$

and $$\phi(N,\epsilon)\,\mathrm{W}_2 = \varpi\kappa(\varpi^2 - \kappa^2)^2\,\mathrm{G}\sin\mathrm{g}$$

Fast diurnal term.

By symmetry $$\phi(N,\epsilon)\,\mathrm{W}_3 = \varpi\kappa^5(3\varpi^2 - \kappa^2)\,\mathrm{G}_2\sin\mathrm{g}_2$$

Fortnightly term.

$$-\frac{d}{dN'}(\underline{\varpi}'\underline{\kappa}')^2 = -\frac{2\varpi\kappa}{\sqrt{-1}}q(\kappa Q - P\varpi)$$

and

$$\phi(N,\epsilon)\,\mathrm{w}_0 = -\frac{e^{2\mathrm{h}}}{2\sqrt{-1}}\varpi^2\kappa^2\left[2\varpi\kappa(Q\kappa - P\varpi) + 4\varpi^2\kappa^2 q\right] = \frac{e^{2\mathrm{h}}}{2\sqrt{-1}}2\varpi\kappa(\varpi^2 - \kappa^2)$$

Whence $\phi(N, \epsilon)\, W_0 = 3\varpi^3\kappa^3(\varpi^2 - \kappa^2)\, H \sin 2h$

Collecting terms we have, on applying the result to the case of viscosity,

$$-\frac{\xi}{k}\frac{dj}{dt} = \frac{\tau^2}{\mathfrak{g}}\left[\tfrac{1}{2}\varpi^7\kappa \sin 4f_1 + \varpi^3\kappa^3 \sin 4f + \tfrac{1}{2}\varpi\kappa^7 \sin 4f_2 + \tfrac{3}{2}\varpi^3\kappa^3(\varpi^2 - \kappa^2)\sin 4h \right.$$
$$\left. - \tfrac{1}{2}\varpi^5\kappa(\varpi^2 - 3\kappa^2)\sin 2g_1 + \tfrac{1}{2}\varpi\kappa(\varpi^2 - \kappa^2)^2 \sin 2g + \tfrac{1}{2}\varpi\kappa^5(3\varpi^2 - \kappa^2)\sin 2g_2\right] \quad \text{......(71)}$$

In the particular case where the viscosity is small, this becomes

$$-\frac{\xi}{k}\frac{dj}{dt} = \tfrac{1}{2}\frac{\tau^2}{\mathfrak{g}} \sin 4f\varpi\kappa = \tfrac{1}{4}\frac{\tau^2}{\mathfrak{g}} \sin 4f \sin(i + j) \quad \text{...........(72)}$$

The right hand side is necessarily positive, and therefore the inclination of the orbit to the invariable plane will always diminish with the time.

The general equation (71) for any degree of viscosity is so complex as to present no idea to the mind, and it will accordingly be graphically illustrated.

The case taken is where $n/\Omega = 15$, which is the same relation as in the previous graphical illustration of § 7.

The general method of illustration is sufficiently explained in that section.

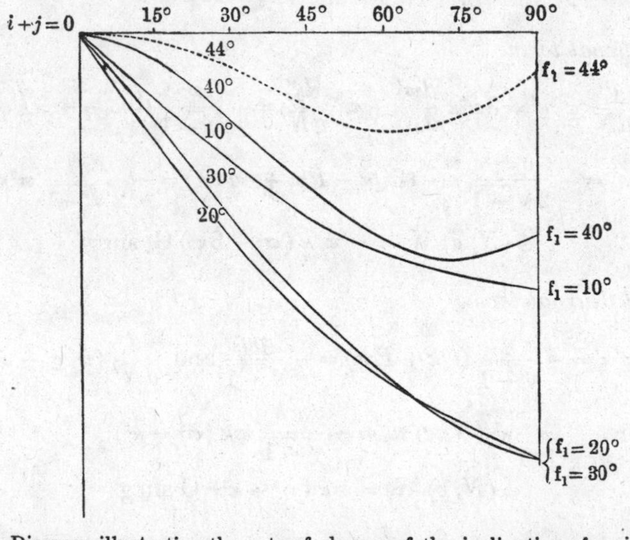

FIG. 5. Diagram illustrating the rate of change of the inclination of a single satellite's orbit to the invariable plane, for various viscosities of the planet, and various inclinations of the orbit to the planet's equator $\left(\frac{dj}{dt}\right)$.

Fig. 5 illustrates the various values which dj/dt (the rate of increase of inclination to the invariable plane) is capable of assuming for various viscosities of the planet, and for various inclinations of the satellite's orbit to the planet's equator. Each curve corresponds to one degree of viscosity, the viscosity being determined by the lag of the slow semi-diurnal tide of speed

$2n-2\Omega$. The ordinates give dj/dt (not as before $dj/\sin j\, dt$) and the abscissæ give $i+j$, the inclination of the orbit to the equator.

We see from this figure that the inclination to the invariable plane will always decrease as the time increases, and the only noticeable point is the maximum rate of decrease for large viscosities, for inclinations of the orbit and equator ranging from 60° to 70°. If n/Ω had been taken considerably smaller than 15, the inclination would have been found to increase with the time for large viscosity of the planet.

§ 11. *Secular change in the mean distance of the satellite, where there is no other disturbing body than the planet. Comparison with result of previous paper.*

To find the variation of ξ we have to differentiate with respect to ϵ', and the following result may be at once written down

$$\frac{d\xi}{kdt}=\tfrac{1}{2}\frac{\tau^2}{\mathfrak{g}}[\varpi^8\sin 4\mathfrak{f}_1-\kappa^8\sin 4\mathfrak{f}_2+4\varpi^6\kappa^2\sin 2\mathfrak{g}_1-4\varpi^2\kappa^6\sin 2\mathfrak{g}_2-6\varpi^4\kappa^4\sin 4\mathfrak{h}] \quad \text{......(73)}$$

This agrees with the result of a previous paper (viz.: (57) or (79) of "Precession"), obtained by a different method; but in that case the inclination of the orbit was zero, so that ϖ and κ were the cosine and sine of half the obliquity, instead of the cosine and sine of $\frac{1}{2}(i+j)$.

In the case where the viscosity is small this becomes

$$\frac{d\xi}{kdt}=\tfrac{1}{2}\frac{\tau^2}{\mathfrak{g}}\sin 4\mathfrak{f}\,[\cos(i+j)-\lambda] \quad \text{....................(74)}$$

It will now be shown that the preceding result (71) for dj/dt may be obtained by means of the principle of conservation of moment of momentum, and by the use of the results of a previous paper.

It is easily shown that the moment of momentum of orbital motion of the moon and earth round their common centre of inertia is $C\xi/k$, and the moment of momentum of the earth's rotation is clearly Cn. Also j and i are the inclinations of the two axes of moment of momentum to the axis of resultant moment of momentum of the system. Hence

$$\frac{\xi}{k}\sin j=n\sin i$$

By differentiation of this equation we have

$$\begin{aligned}\frac{\xi}{k}\frac{dj}{dt}\cos j&=\frac{dn}{dt}\sin i+n\cos i\frac{di}{dt}-\frac{d\xi}{kdt}\sin j\\ &=\left[\frac{dn}{dt}\sin(i+j)+n\cos(i+j)\frac{di}{dt}\right]\cos j\\ &\quad-\left[\frac{dn}{dt}\cos(i+j)-n\sin(i+j)\frac{di}{dt}+\frac{d\xi}{kdt}\right]\sin j\end{aligned}$$

Now from equation (52) of the paper on "Precession," the second term on the right hand side is zero, and therefore

$$\frac{\xi}{k}\frac{dj}{dt} = \frac{dn}{dt}\sin(i+j) + n\cos(i+j)\frac{di}{dt}$$

But by equations (21) and (16) and (29) of the paper on "Precession" (when ϖ and κ are written for the p, q of that paper)

$$\frac{dn}{dt} = -\frac{\tau^2}{\mathfrak{g}}[\tfrac{1}{2}\varpi^8 \sin 4\mathrm{f}_1 + 2\varpi^4\kappa^4 \sin 4\mathrm{f} + \tfrac{1}{2}\kappa^8 \sin 4\mathrm{f}_2 + \varpi^6\kappa^2 \sin 2\mathrm{g}_1$$
$$+ \varpi^2\kappa^2(\varpi^2 - \kappa^2)^2 \sin 2\mathrm{g} + \varpi^2\kappa^6 \sin 2\mathrm{g}_2]$$

$$n\frac{di}{dt} = \frac{\tau^2}{\mathfrak{g}}[\tfrac{1}{2}\varpi^7\kappa \sin 4\mathrm{f}_1 - \varpi^3\kappa^3(\varpi^2 - \kappa^2)\sin 4\mathrm{f} - \tfrac{1}{2}\varpi\kappa^7 \sin 4\mathrm{f}_2 + \tfrac{1}{2}\varpi^5\kappa(\varpi^2 + 3\kappa^2)\sin 2\mathrm{g}_1$$
$$- \tfrac{1}{2}\varpi\kappa(\varpi^2 - \kappa^2)^3 \sin 2\mathrm{g} - \tfrac{1}{2}\varpi\kappa^5(3\varpi^2 + \kappa^2)\sin 2\mathrm{g}_2 - \tfrac{3}{2}\varpi^3\kappa^3 \sin 4\mathrm{h}]$$

If we multiply the former of these by $\sin(i+j)$ or $2\varpi\kappa$, and the latter by $\cos(i+j)$ or $\varpi^2 - \kappa^2$, and add, we get the equation (71), which has already been established by the method of the disturbing function.

It seemed well to give this method, because it confirms the accuracy of the two long analytical investigations in the paper on "Precession" and in the present one.

§ 12. *The method of the disturbing function applied to the motion of the planet.*

In the case where there are only two bodies, viz.: the planet and the satellite, the problem is already solved in the paper on "Precession," and it is only necessary to remember that the p and q of that paper are really $\cos\frac{1}{2}(i+j)$, $\sin\frac{1}{2}(i+j)$, instead of $\cos\frac{1}{2}i$, $\sin\frac{1}{2}i$. This will not be reinvestigated, but we will now consider the case of two satellites, the nodes of whose orbits revolve with uniform angular velocity on the ecliptic. The results may be easily extended to the hypothesis of any number of satellites.

In (18) we have the equations of variation of i, ψ, χ in terms of W. But as the correction to the precession has not much interest, we will only take the two equations

$$\left.\begin{aligned} n\sin i\frac{di}{dt} &= \cos i\frac{d\mathrm{W}}{d\chi'} - \frac{d\mathrm{W}}{d\psi'} \\ \frac{dn}{dt} &= \frac{d\mathrm{W}}{d\chi'} \end{aligned}\right\} \quad\text{.....................(75)}$$

which give the rate of change of obliquity and the tidal friction.

In the development of W in § 5, it was assumed that ψ, ψ' were zero, and χ, χ' did not appear, because χ was left unaccented in the X′-Y′-Z′ functions.

Longitudes were there measured from the vernal equinox, but here we must conceive the N, N' of previous developments replaced by $N-\psi$, $N'-\psi'$; also $\Omega t+\epsilon$, $\Omega' t+\epsilon'$ must be replaced by $\Omega t+\epsilon-\psi$, $\Omega' t+\epsilon'-\psi'$.

It will not be necessary to redevelop W for the following reasons.

$\Omega' t+\epsilon'-\psi'$ occurs only in the exponentials, and $N'-\psi'$ does not occur there; and $N'-\psi'$ only occurs in the functions of ϖ and κ, and $\Omega' t+\epsilon'-\psi'$ does not occur there. Hence

$$-\frac{d\mathrm{W}}{d\psi'}=\frac{d\mathrm{W}}{d\epsilon'}+\frac{d\mathrm{W}}{dN'} \quad \text{.........................(76)}$$

Again, it will be seen by referring to the remarks made as to χ, χ' in the development of W in § 5, that we have the following identities:

$$\left.\begin{array}{l}
\text{For semi-diurnal terms,}\\
\dfrac{d\mathrm{W}_{\mathrm{I}}}{d\chi'}=-\dfrac{d\mathrm{W}_{\mathrm{I}}}{d\epsilon'},\quad \dfrac{d\mathrm{W}_{\mathrm{II}}}{d\chi'}=\dfrac{d\mathrm{W}_{\mathrm{II}}}{d\mathrm{f}},\quad \dfrac{d\mathrm{W}_{\mathrm{III}}}{d\chi'}=\dfrac{d\mathrm{W}_{\mathrm{III}}}{d\epsilon'}\\
\text{For diurnal terms,}\\
\dfrac{d\mathrm{W}_1}{d\chi'}=-\tfrac{1}{2}\dfrac{d\mathrm{W}_1}{d\epsilon'},\quad \dfrac{d\mathrm{W}_2}{d\chi'}=\dfrac{d\mathrm{W}_2}{dg},\quad \dfrac{d\mathrm{W}_3}{d\chi'}=\tfrac{1}{2}\dfrac{d\mathrm{W}_3}{d\epsilon'}\\
\text{For the fortnightly term,}\\
\qquad \dfrac{d\mathrm{W}_0}{d\chi'}=0\\
\text{Also}\qquad \dfrac{d\mathrm{W}_{\mathrm{II}}}{d\epsilon'}=0,\quad \dfrac{d\mathrm{W}_2}{d\epsilon'}=0
\end{array}\right\} \quad \text{......(77)}$$

Making use of (76) and (77), and remembering that $\cos i=P^2-Q^2$, $\sin i=2PQ$, we may write equations (75), thus

$$(2PQ)\,n\frac{di}{dt}=\frac{d}{d\epsilon'}\left[2Q^2\mathrm{W}_{\mathrm{I}}+2P^2\mathrm{W}_{\mathrm{III}}+\tfrac{1}{2}(P^2+3Q^2)\,\mathrm{W}_1+\tfrac{1}{2}(3P^2+Q^2)\,\mathrm{W}_3+\mathrm{W}_0\right]$$
$$+(P^2-Q^2)\left[\frac{d\mathrm{W}_{\mathrm{II}}}{d\mathrm{f}}+\frac{d\mathrm{W}_2}{dg}\right]+\frac{d}{dN'}(\Sigma\mathrm{W}) \quad \text{...(78)}$$

$$\frac{dn}{dt}=-\frac{d}{d\epsilon'}\left[\mathrm{W}_{\mathrm{I}}-\mathrm{W}_{\mathrm{III}}+\tfrac{1}{2}\mathrm{W}_1-\tfrac{1}{2}\mathrm{W}_3\right]+\frac{d\mathrm{W}_{\mathrm{II}}}{d\mathrm{f}}+\frac{d\mathrm{W}_2}{dg} \quad \text{......(79)}$$

It is clear that by using these transformations we may put $\psi=\psi'=0$, $\chi=\chi'$ before differentiation, so that ψ and χ again disappear, and we may use the old development of W.

The case where Diana and the moon are distinct bodies will be taken first, and it will now be convenient to make Diana identical with the sun.

In this case after the differentiations are made we are not to put $N=N'$, and $\epsilon=\epsilon'$.

The only terms, out of which secular changes in i and n can arise, are those depending on the sidereal semi-diurnal and diurnal tides, for all others are periodic with the longitudes of the two disturbing bodies. Hence the disturbing function is reduced to W_{II} and W_2. Also dW_{II}/dN' and dW_2/dN' can only contribute periodic terms, because $N-N'$ is not zero, and by hypothesis the nodes revolve uniformly on the ecliptic.

If we consider that here p' is not equal to p, nor q' to q, we see that, as far as is of present interest,

$$W_{II} = 2F \cos 2f\, P^4 Q^4 [(p^2 - q^2)^2 - 2p^2q^2][(p'^2 - q'^2)^2 - 2p'^2q'^2]$$

$$W_2 = 2G \cos g\, P^2 Q^2 (P^2 - Q^2)^2 [(p^2 - q^2)^2 - 2p^2q^2][(p'^2 - q'^2)^2 - 2p'^2q'^2]$$

Also the equations of variation of i and n are simply

$$(2PQ)\, n \frac{di}{dt} = (P^2 - Q^2)\left[\frac{dW_{II}}{df} + \frac{dW_2}{dg}\right]$$

$$\frac{dn}{dt} = \frac{dW_{II}}{df} + \frac{dW_2}{dg}$$

Thus if we put

$$\left.\begin{aligned} \phi &= 2P^4Q^4 [(p^2 - q^2)^2 - 2p^2q^2][(p'^2 - q'^2)^2 - 2p'^2q'^2] \\ &= \tfrac{1}{8}\sin^4 i\,(1 - \tfrac{3}{2}\sin^2 j)(1 - \tfrac{3}{2}\sin^2 j') \\ \tfrac{1}{2}\gamma &= P^2Q^2(P^2 - Q^2)^2[(p^2 - q^2)^2 - 2p^2q^2][(p'^2 - q'^2)^2 - 2p'^2q'^2] \\ &= \tfrac{1}{4}\sin^2 i\cos^2 i\,(1 - \tfrac{3}{2}\sin^2 j)(1 - \tfrac{3}{2}\sin^2 j') \end{aligned}\right\} \quad (80)$$

we have

$$\left.\begin{aligned} -\frac{dn}{dt} &= \frac{2\tau\tau'}{\mathfrak{g}}[2\phi F \sin 2f + \gamma G \sin g] \\ n\frac{di}{dt} &= -\frac{2\tau\tau'}{\mathfrak{g}}[2\phi F \sin 2f + \gamma G \sin g]\cot i \end{aligned}\right\} \quad \ldots\ldots\ldots\ldots(81)$$

It will be noticed that in (81) $2\tau\tau'$ has been introduced in the equations instead of $\tau\tau'$; this is because in the complete solution of the problem these terms are repeated twice, once for the attraction of the moon on the solar tides, and again for that of the sun on the lunar tides.

The case where Diana is identical with the moon must now be considered. This will enable us to find the effects of the moon's attraction on her own tides, and then by symmetry those of the sun's attraction on his tides.

We will begin with the *tidal friction.*

By comparison with (65)

$$\frac{d}{d\epsilon'}[W_I - W_{III} + \tfrac{1}{2}W_1 - \tfrac{1}{2}W_3] = 2\Phi_1 F_1 \sin 2f_1 + 2\Phi_2 F_2 \sin 2f_2 + \Gamma_1 G_1 \sin g_1 + \Gamma_2 G_2 \sin g_2 \quad \ldots\ldots(82)$$

When we put $N = N'$ (see (43) and (52))

$$W_{II} = 2F\cos 2f\,.\,w_{II} \quad\text{and}\quad \frac{dW_{II}}{df} = -4F\sin 2f\,.\,w_{II}$$

Also $\quad W_2 = 2G\cos g \,.\, w_2$ and $\dfrac{dW_2}{dg} = -2G\sin g \,.\, w_2$

Let

$$\Phi = 2w_{II} = 2P^4Q^4[(p^2-q^2)^2 - 2p^2q^2]^2 + 8P^2Q^2(P^4+Q^4)p^2q^2(p^2-q^2)^2 + 2p^4q^4(P^8+Q^8)$$

$$= 2P^4Q^4(p^2-q^2)^4 + 8p^2q^2(p^2-q^2)^2P^2Q^2(P^4+Q^4-P^2Q^2) + 2p^4q^4(P^8+4P^4Q^4+Q^8) \quad \text{...(83)}$$

and let

$$\tfrac{1}{2}\Gamma = w_2 = P^2Q^2(P^2-Q^2)^2[(p^2-q^2)^2-2p^2q^2]^2 + [P^4(P^2-3Q^2)^2 + Q^4(3P^2-Q^2)^2]p^2q^2(p^2-q^2)^2 + 4P^2Q^2(P^4+Q^4)p^4q^4$$

$$= P^2Q^2(P^2-Q^2)^2(p^2-q^2)^4 + [(P^2-Q^2)^4 - 6P^2Q^2(P^2-Q^2)^2 + 8P^4Q^4]p^2q^2(p^2-q^2)^2 + 8P^2Q^2(P^4+Q^4-P^2Q^2)p^4q^4 \quad \text{...(84)}$$

And we have

$$-\frac{dn}{dt} = \frac{\tau^2}{\mathfrak{g}}[2\Phi_1F_1\sin 2f_1 + 2\Phi F\sin 2f + 2\Phi_2F_2\sin 2f_2 + \Gamma_1G_1\sin g_1 + \Gamma G\sin g + \Gamma_2G_2\sin g_2] \quad \text{...(85)}$$

This is only a partial solution, since it only refers to the action of the moon on her own tides.

If the second satellite, say the sun, be introduced, the action of the sun on the solar tides may be written down by symmetry, and the elements of the solar (or terrestrial) orbit may be indicated by the same symbols as before, but with accents.

From (85) and (81) the complete solution may be collected.

In the case of viscosity, and where the viscosity is small, it will be found that the solution becomes

$$-\frac{dn}{dt} = \frac{1}{2}\frac{\sin 4f}{\mathfrak{g}}\Big\{(1-\tfrac{1}{2}\sin^2 i)(\tau^2+\tau'^2) - \tfrac{1}{2}(1-\tfrac{3}{2}\sin^2 i)(\tau^2\sin^2 j + \tau'^2\sin^2 j') - \tau^2\frac{\Omega}{n}\cos i\cos j - \tau'^2\frac{\Omega'}{n}\cos i\cos j' + \tfrac{1}{2}\tau\tau'\sin^2 i(1-\tfrac{3}{2}\sin^2 j)(1-\tfrac{3}{2}\sin^2 j')\Big\} \quad \text{......(86)}$$

If j and j' be put equal to zero and Ω'/n neglected, this result will be found to agree with that given in the paper on "Precession," § 17, (83).

We will next consider *the change of obliquity*.

The combined effect has already been determined in (81), but the separate effects of the two bodies remain to be found. The terms of different speeds must now be taken one by one.

Slow semi-diurnal term.

$$n\frac{di}{dt}\div\frac{\tau^2}{\mathfrak{g}}=\frac{Q}{P}\frac{d\mathrm{W_I}}{d\epsilon'}+\frac{1}{2PQ}\frac{d\mathrm{W_I}}{dN'}$$

We had before $$-\frac{\xi}{k}\frac{dj}{dt}\div\frac{\tau^2}{\mathfrak{g}}=\frac{q}{p}\frac{d\mathrm{W_I}}{d\epsilon'}+\frac{1}{2pq}\frac{d\mathrm{W_I}}{dN'}$$

Now $\mathrm{W_I}$ is symmetrical with regard to P and p, Q and q, and so are its differentials with regard to ϵ' and N'. Therefore the solution may be written down by symmetry with the "slow semi-diurnal" of § 6, by writing P for p and Q for q and *vice versâ.*

Let

$$\mathsf{F}_1=\tfrac{1}{4}\{P^6p^8-4P^4(P^2-3Q^2)p^6q^2-18P^2Q^2(P^2-Q^2)p^4q^4 -4Q^4(3P^2-Q^2)p^2q^6-Q^6q^8\} \quad \text{...(87)}$$

and $$n\frac{di}{dt}\div\frac{\tau^2}{\mathfrak{g}}=2\mathsf{F}_1\mathrm{F}_1\sin 2\mathrm{f}_1\sin i \quad \text{.....................(88)}$$

Sidereal semi-diurnal term.

$$n\frac{di}{dt}\div\frac{\tau^2}{\mathfrak{g}}=\frac{1}{2PQ}\left[(P^2-Q^2)\frac{d\mathrm{W_{II}}}{d\mathrm{f}}+\frac{d\mathrm{W_{II}}}{dN'}\right]$$

$$\frac{d\mathrm{W_{II}}}{d\mathrm{f}}=-2\Phi\mathrm{F}\sin 2\mathrm{f} \text{ and } \frac{d\mathrm{W_{II}}}{dN'}=4p^2q^2\,.\,2\mathfrak{F}\mathrm{F}\sin 2\mathrm{f}$$

Therefore

$$n\frac{di}{dt}\div\frac{\tau^2}{\mathfrak{g}}=2\mathrm{F}\sin 2\mathrm{f}\left[-\frac{P^2-Q^2}{2PQ}\Phi+\frac{2p^2q^2}{PQ}\mathfrak{F}\right]$$

On substitution from (44) and (83) for Φ and $\mathfrak{F}$ and simplification, we find that if

$$\mathsf{F}=\tfrac{1}{2}\{P^2Q^2(P^2-Q^2)[(p^2-q^2)^2-2p^2q^2]^2+2p^2q^2(p^2-q^2)^2(P^2-Q^2)^3 -2p^4q^4(P^6-Q^6)\} \quad \text{...(89)}$$

then $$n\frac{di}{dt}\div\frac{\tau^2}{\mathfrak{g}}=-2\mathsf{F}\mathrm{F}\sin 2\mathrm{f}\sin i \quad \text{.....................(90)}$$

Fast semi-diurnal term.

$$n\frac{di}{dt}\div\frac{\tau^2}{\mathfrak{g}}=\frac{P}{Q}\frac{d\mathrm{W_{III}}}{d\epsilon'}+\frac{1}{2PQ}\frac{d\mathrm{W_{III}}}{dN'}$$

Since $\mathrm{W_{III}}$ is found from $\mathrm{W_I}$ by writing Q for P, and $-P$ for Q, and $-2\mathrm{f}_2$ for $2\mathrm{f}_1$, therefore in this case $n\,di/dt$ is found from its value in the slow semi-diurnal term by the like changes, and if

$$\mathsf{F}_2=\tfrac{1}{4}\{Q^6p^8+4Q^4(3P^2-Q^2)p^6q^2+18P^2Q^2(P^2-Q^2)p^4q^4 +4P^4(P^2-3Q^2)p^2q^6-P^6q^8\} \quad \text{...(91)}$$

then $$n\frac{di}{dt}\div\frac{\tau^2}{\mathfrak{g}}=-2\mathsf{F}_2\mathrm{F}_2\sin 2\mathrm{f}_2\sin i \quad \text{.....................(92)}$$

Slow diurnal term.

$$n\frac{di}{dt}\div\frac{\tau^2}{\mathfrak{g}}=\frac{1}{2PQ}\left[\frac{P^2+3Q^2}{2}\frac{dW_1}{d\epsilon'}+\frac{dW_1}{dN'}\right]$$

$$\frac{dW_1}{dN'}=-2G_1\sin g_1\left[P^4(P^2-3Q^2)^2p^6q^2+18P^2Q^2(P^2-Q^2)^2p^4q^4+3Q^4(3P^2-Q^2)^2p^2q^6+4P^2Q^6q^8\right]$$

$$\frac{dW_1}{d\epsilon'}=\quad 2G_1\sin g_1\,.\,\Gamma_1$$

Substituting these values and simplifying, it will be found that if

$$\mathsf{G}_1=\tfrac{1}{4}\{P^4(P^2+3Q^2)p^8+2P^2(P^2-3Q^2)^2p^6q^2-9(P^2-Q^2)^3p^4q^4-2Q^2(3P^2-Q^2)^2p^2q^6-Q^4(3P^2+Q^2)q^8\}\;\ldots(93)$$

then
$$n\frac{di}{dt}\div\frac{\tau^2}{\mathfrak{g}}=2\mathsf{G}_1G_1\sin g_1\sin i\;\ldots\ldots\ldots\ldots\ldots\ldots(94)$$

Sidereal diurnal term.

$$n\frac{di}{dt}\div\frac{\tau^2}{\mathfrak{g}}=\frac{1}{2PQ}\left[(P^2-Q^2)\frac{dW_2}{dg}+\frac{dW_2}{dN'}\right]$$

$$\frac{dW_2}{dg}=-2G\sin g\,(\tfrac{1}{2}\Gamma)\quad\text{and}\quad\frac{dW_2}{dN'}=4p^2q^2\,2\mathfrak{G}G\sin g$$

Therefore
$$n\frac{di}{dt}\div\frac{\tau^2}{\mathfrak{g}}=2G\sin g\left[-\frac{P^2-Q^2}{4PQ}\Gamma+\frac{2p^2q^2}{PQ}\mathfrak{G}\right]$$

On substitution from (53) and (84) for Γ and $\mathfrak{G}$ and simplification, we find that if

$$\mathsf{G}=\tfrac{1}{4}\{(P^2-Q^2)^3[(p^2-q^2)^2-2p^2q^2]^2-2(P^2-Q^2)[(P^2-Q^2)^2-12P^2Q^2]p^2q^2(p^2-q^2)^2-4(P^2-Q^2)(P^4+4P^2Q^2+Q^4)p^4q^4\}\;\ldots(95)$$

then
$$n\frac{di}{dt}\div\frac{\tau^2}{\mathfrak{g}}=-2\mathsf{G}G\sin g\sin i\ldots\ldots\ldots\ldots\ldots\ldots(96)$$

Fast diurnal term.

$$\frac{di}{dt}\div\frac{\tau^2}{\mathfrak{g}}=\frac{1}{2PQ}\left[\frac{3P^2+Q^2}{2}\frac{dW_3}{d\epsilon'}+\frac{dW_3}{dN'}\right]$$

As the fast semi-diurnal is derived from the slow, so here also; and if

$$\mathsf{G}_2=\tfrac{1}{4}\{Q^4(3P^2+Q^2)p^8+2Q^2(3P^2-Q^2)^2p^6q^2+9(P^2-Q^2)^3p^4q^4-2P^2(P^2-3Q^2)^2p^2q^6-P^4(P^2+3Q^2)q^8\}\;\ldots(97)$$

then
$$n\frac{di}{dt}\div\frac{\tau^2}{\mathfrak{g}}=-2\mathsf{G}_2G_2\sin g_2\sin i\;\ldots\ldots\ldots\ldots\ldots\ldots(98)$$

Fortnightly term.

$$\frac{di}{dt}\div\frac{\tau^2}{\mathfrak{g}}=\frac{1}{2PQ}\left(\frac{dW_0}{d\epsilon'}+\frac{dW_0}{dN'}\right)$$

If we take the term in W_0 which has 2h positive in the exponential, we have

$$\frac{dW_0}{dN'} = \tfrac{3}{2}\,\frac{e^{2h}}{2\sqrt{-1}}\,[8P^2Q^2(P^2-Q^2)^2p^6q^2 + 4[(P^2-Q^2)^2 - 2P^2Q^2]^2p^4q^4 + 24P^2Q^2(P^2-Q^2)^2p^2q^6 + 8P^4Q^4q^8]$$

$$\frac{dW_0}{d\epsilon'} = -\tfrac{3}{2}\,\frac{e^{2h}}{2\sqrt{-1}}\,[4P^4Q^4p^8 + 16P^2Q^2(P^2-Q^2)^2p^6q^2 + 4[(P^2-Q^2)^2 - 2P^2Q^2]^2p^4q^4 + 16P^2Q^2(P^2-Q^2)^2p^2q^6 + 4P^4Q^4q^8]$$

If these be added and simplified, it will be found that if

$$\mathsf{H} = \tfrac{3}{4}(p^4-q^4)[(p^4+q^4)P^2Q^2 + 2p^2q^2(P^2-Q^2)^2] \quad \text{.........(99)}$$

then

$$n\frac{di}{dt} \div \frac{\tau^2}{\mathfrak{g}} = -2\mathsf{H}\text{H}\sin 2\text{h}\sin i \quad \text{..................(100)}$$

Collecting results from the seven equations (88, 90–2–4–6–8, 100),

$$n\frac{di}{dt} = \frac{\tau^2}{\mathfrak{g}}\sin i\,\{2\mathsf{F}_1\text{F}_1\sin 2\text{f}_1 - 2\mathsf{F}\text{F}\sin 2\text{f} - 2\mathsf{F}_2\text{F}_2\sin 2\text{f}_2 + 2\mathsf{G}_1\text{G}_1\sin\text{g}_1 - 2\mathsf{G}\text{G}\sin\text{g} - 2\mathsf{G}_2\text{G}_2\sin\text{g}_2 - 2\mathsf{H}\text{H}\sin 2\text{h}\} \quad \text{...(101)}$$

This is only a partial solution, and refers only to the action of the moon on her own tides; the part depending on the sun alone may be written down by symmetry.

The various functions of i and j here introduced admit of reduction to the following forms:

$$\left.\begin{aligned}\Phi &= \tfrac{1}{4}\{\tfrac{1}{2}\sin^4 i + \tfrac{1}{2}\sin^2 j(4\sin^2 i - 5\sin^4 i) + \tfrac{1}{2}\sin^4 j(1 - 5\sin^2 i + \tfrac{35}{8}\sin^4 i)\}\\ \tfrac{1}{2}\Gamma &= \tfrac{1}{4}\{\sin^2 i - \sin^4 i + \sin^2 j(1 - \tfrac{11}{2}\sin^2 i + 5\sin^4 i)\\ &\qquad - \sin^4 j(1 - 5\sin^2 i + \tfrac{35}{8}\sin^4 i)\}\end{aligned}\right\} \quad \text{......(102)}$$

$$\left.\begin{aligned}\mathsf{F}_1 + \mathsf{F}_2 &= \tfrac{1}{4}\cos j\,\{1 - \tfrac{3}{4}\sin^2 i - \tfrac{3}{2}\sin^2 j(1 - \tfrac{5}{4}\sin^2 i)\}\\ \mathsf{F}_1 - \mathsf{F}_2 &= \tfrac{1}{4}\cos i\,\{1 - \tfrac{1}{4}\sin^2 i - 2\sin^2 j(1 - \tfrac{5}{8}\sin^2 i) + \tfrac{5}{8}\sin^4 j(1 - \tfrac{7}{4}\sin^2 i)\}\\ \mathsf{G}_1 + \mathsf{G}_2 &= \tfrac{1}{4}\cos j\\ \mathsf{G}_1 - \mathsf{G}_2 &= \tfrac{1}{4}\cos i\,\{1 + \tfrac{1}{2}\sin^2 i - \tfrac{1}{2}\sin^2 j(1 + 5\sin^2 i) - \tfrac{5}{4}\sin^4 j(1 - \tfrac{7}{4}\sin^2 i)\}\\ \mathsf{F} &= \tfrac{1}{4}\cos i\,\{\tfrac{1}{2}\sin^2 i + \sin^2 j(1 - \tfrac{5}{2}\sin^2 i) - \tfrac{5}{4}\sin^4 j(1 - \tfrac{7}{4}\sin^2 i)\}\\ \mathsf{G} &= \tfrac{1}{4}\cos i\,\{1 - \sin^2 i - \tfrac{7}{2}\sin^2 j(1 - \tfrac{10}{7}\sin^2 i) + \tfrac{5}{2}\sin^4 j(1 - \tfrac{7}{4}\sin^2 i)\}\\ \mathsf{H} &= \tfrac{1}{4}\cos j\,\{\tfrac{3}{4}\sin^2 i + \tfrac{3}{2}\sin^2 j(1 - \tfrac{5}{4}\sin^2 i)\}\end{aligned}\right\} \quad \text{......(103)}$$

Φ_1, Φ_2, Γ_1, Γ_2 are given in equations (67), and ϕ and γ in equations (80).

The expressions for F_1 and F_2 are found by symmetry with those for $\mathfrak{F}_1$ and $\mathfrak{F}_2$, by interchanging i and j; the first of equations (62) then corresponds with the second of (103), and *vice versâ*.

From (103) it follows that

$$\mathsf{F}_1 - \mathsf{F}_2 + \tfrac{1}{2}(\mathsf{G}_1 - \mathsf{G}_2) = \tfrac{3}{8}\cos i\,(1 - \tfrac{3}{2}\sin^2 j)$$

and

$$\mathsf{F} + \tfrac{1}{2}\mathsf{G} = \tfrac{1}{8}\cos i\,(1 - \tfrac{3}{2}\sin^2 j)$$

Also

$$\mathsf{F}_1 + \mathsf{F}_2 + \mathsf{G}_1 + \mathsf{G}_2 + \mathsf{H} = \tfrac{1}{2}\cos j$$

The complete solution of the problem may be collected from the equations (101) and (81).

In the case of the viscosity of the earth, and when the viscosity is small, we easily find the complete solution to be

$$n\frac{di}{dt} = \frac{\sin 4\mathfrak{f}}{\mathfrak{g}} \cdot \tfrac{1}{4}\sin i \cos i \left\{\tau^2(1 - \tfrac{3}{2}\sin^2 j) + \tau'^2(1 - \tfrac{3}{2}\sin^2 j') - \frac{2\Omega}{n}\tau^2 \sec i \cos j \right.$$
$$\left. - \frac{2\Omega'}{n}\tau'^2 \sec i \cos j' - \tau\tau'(1 - \tfrac{3}{2}\sin^2 j)(1 - \tfrac{3}{2}\sin^2 j')\right\} \quad \text{...(104)}$$

This result agrees with that given in (83) of "Precession," when the squares of j and j' are neglected, and when Ω'/n is also neglected.

The preceding method of finding the tidal friction and change of obliquity is no doubt somewhat artificial, but as the principal object of the present paper is to discuss the secular changes in the elements of the satellite's orbit, it did not seem worth while to develop the disturbing function in such a form as would make it applicable both to the satellite and the planet; it seemed preferable to develop it for the satellite and to adapt it for the case of the perturbation of the planet.

In long analytical investigations it is difficult to avoid mistakes; it may therefore give the reader confidence in the correctness of the results and process if I state that I have worked out the preceding values of di/dt and dn/dt independently, by means of the determination of the disturbing couples $\mathfrak{L}$, $\mathfrak{M}$, $\mathfrak{N}$. That investigation separated itself from the present one at the point where the products of the X′-Y′-Z′ functions and $\mathfrak{X}$-$\mathfrak{Y}$-$\mathfrak{Z}$ functions are formed, for products of the form $\mathrm{Y'Z'} \times \mathfrak{X}\mathfrak{Y}$ had there to be found. From this early stage the two processes are quite independent, and the identity of the results is confirmatory of both. Moreover, the investigation here presented reposes on the values found for dj/dt and $d\xi/dt$, hence the correctness of the result of the first problem here treated was also confirmed.

III.

The Proper Planes of the Satellite, and of the Planet, and their Secular Changes.

§ 13. *On the motion of a satellite moving about a rigid oblate spheroidal planet, and perturbed by another satellite.*

The present problem is to determine the joint effects of the perturbing influence of the sun, and of the earth's oblateness upon the motion of the moon's nodes, and upon the inclination of the orbit to the ecliptic; and also to determine the effects on the obliquity of the ecliptic and on the earth's precession. In the present configuration of the three bodies the problem presents but little difficulty, because the influence of oblateness on the moon's motion is very small compared with the perturbation due to the sun; on the other hand, in the case of Jupiter, the influence of oblateness is more important than that of solar perturbation. In each of these special cases there is an appropriate approximation which leads to the result. In the present problem we have, however, to obtain a solution, which shall be applicable to the preponderance of either perturbing cause, because we shall have to trace, in retrospect, the evanescence of the solar influence, and the increase of the influence of oblateness.

The lunar orbit will be taken as circular, and the earth or planet as homogeneous and of ellipticity $\mathfrak{e}$, so that the equation to its surface is

$$\rho = a\{1 + \mathfrak{e}(\tfrac{1}{3} - \cos^2\theta)\}$$

The problem will be treated by the method of the disturbing function, and the method will be applied so as to give the perturbations both of the moon and earth.

First consider only the influence of oblateness.

Let ρ, θ be the coordinates of the moon, so that $\rho = c$ and $\cos\theta = \mathrm{M}_3$. In the formula (17) § 2, $r = c$ and $\dfrac{\sigma}{a} = \mathfrak{e}(\tfrac{1}{3} - \mathrm{M}_3^2)$, so that the disturbing function

$$\mathrm{W} = \tau\mathfrak{e}(\tfrac{1}{3} - \mathrm{M}_3^2)$$

This function, when suitably developed, will give the perturbation of the moon's motion due to oblateness, and the lunar precession and nutation of the earth.

By (21) we have

$$M_3 = \sin i\,[p^2 \sin(l+N) - q^2 \sin(l-N)] + \sin j \cos i \sin l$$

where l is the moon's longitude measured from the node, and N is the longitude of the ascending node of the lunar orbit measured from the descending node of the equator.

As we are only going to find secular inequalities, we may, in developing the disturbing function, drop out terms involving l; also we must write $N-\psi$ for N, because we cannot now take the vernal equinox as fixed.

Omitting all terms which involve l,

$$M_3^2 = \sin^2 i\,[\tfrac{1}{2}(p^4+q^4) - p^2q^2 \cos 2(N-\psi)] + \tfrac{1}{2}\sin^2 j \cos^2 i$$
$$+ \sin j \sin i \cos i\,[p^2 - q^2] \cos(N-\psi)$$

Since $p = \cos\tfrac{1}{2}j$, $q = \sin\tfrac{1}{2}j$, we have

$$p^4 + q^4 = 1 - \tfrac{1}{2}\sin^2 j, \quad p^2q^2 = \tfrac{1}{4}\sin^2 j, \quad p^2 - q^2 = \cos j$$

and

$$M_3^2 = \tfrac{1}{2}\sin^2 i\,(1 - \tfrac{1}{2}\sin^2 j) + \tfrac{1}{2}\sin^2 j\,(1 - \sin^2 i)$$
$$+ \tfrac{1}{4}\sin 2i \sin 2j \cos(N-\psi) - \tfrac{1}{4}\sin^2 i \sin^2 j \cos 2(N-\psi)$$

Now $\tfrac{1}{2}(\sin^2 i + \sin^2 j) - \tfrac{3}{4}\sin^2 i \sin^2 j - \tfrac{1}{3} = -\tfrac{1}{3}(1 - \tfrac{3}{2}\sin^2 i)(1 - \tfrac{3}{2}\sin^2 j)$

Wherefore

$$W = \tau\mathfrak{c}\,\{\tfrac{1}{3}(1 - \tfrac{3}{2}\sin^2 i)(1 - \tfrac{3}{2}\sin^2 j) - \tfrac{1}{4}\sin 2i \sin 2j \cos(N-\psi)$$
$$+ \tfrac{1}{4}\sin^2 i \sin^2 j \cos 2(N-\psi)\} \quad \ldots(105)$$

This is the disturbing function.

Before applying it, we will assume that i and j are sufficiently small to permit us to neglect $\sin^2 i \sin^2 j$ compared with unity.

Then

$$\tfrac{1}{3}(1 - \tfrac{3}{2}\sin^2 i)(1 - \tfrac{3}{2}\sin^2 j) = \tfrac{1}{12} + \tfrac{1}{4} - \tfrac{1}{2}\sin^2 i - \tfrac{1}{2}\sin^2 j + \sin^2 i \sin^2 j - \tfrac{1}{4}\sin^2 i \sin^2 j$$
$$= \tfrac{1}{12} + \tfrac{1}{4}\cos 2i \cos 2j - \tfrac{1}{4}\sin^2 i \sin^2 j$$

Hence, when we neglect the terms in $\sin^2 i \sin^2 j$,

$$W = \tfrac{1}{4}\tau\mathfrak{c}\,\{\tfrac{1}{3} + \cos 2i \cos 2j - \sin 2i \sin 2j \cos(N-\psi)\} \quad \ldots\ldots(106)$$

Since this disturbing function does not involve the epoch or χ, we have by (13), (14), and (18)

$$-\frac{\xi}{k}\sin j \frac{dj}{dt} = \frac{dW}{dN}, \quad \frac{\xi}{k}\sin j \frac{dN}{dt} = \frac{dW}{dj}$$

$$-n \sin i \frac{di}{dt} = \frac{dW}{d\psi}, \quad n \sin i \frac{d\psi}{dt} = \frac{dW}{di}$$

Thus as far as concerns the influence of the oblateness on the moon, and the reaction of the moon on the earth,

$$\left.\begin{aligned}
\frac{\xi}{k}\sin j\,\frac{dj}{dt} &= -\tfrac{1}{4}\tau\mathfrak{e}\sin 2i\sin 2j\sin(N-\psi)\\
\frac{\xi}{k}\sin j\,\frac{dN}{dt} &= -\tfrac{1}{2}\tau\mathfrak{e}\{\cos 2i\sin 2j+\sin 2i\cos 2j\cos(N-\psi)\}\\
n\sin i\,\frac{di}{dt} &= \tfrac{1}{4}\tau\mathfrak{e}\sin 2i\sin 2j\sin(N-\psi)\\
n\sin i\,\frac{d\psi}{dt} &= -\tfrac{1}{2}\tau\mathfrak{e}\{\sin 2i\cos 2j+\cos 2i\sin 2j\cos(N-\psi)\}
\end{aligned}\right\}\quad\ldots(107)$$

If there be no other disturbing body, and if we refer the motion to the invariable plane of the system, we must always have $N=\psi$.

In this case the first and third of (107) become

$$\frac{dj}{dt}=\frac{di}{dt}=0$$

and the second and fourth become

$$\frac{\xi}{k}\sin j\,\frac{dN}{dt}=n\sin i\,\frac{d\psi}{dt}=-\tfrac{1}{2}\tau\mathfrak{e}\sin 2(i+j)$$

But ξ/k is proportional to the moment of momentum of the orbital motion, and n is proportional to the moment of momentum of the earth's rotation, and so by the definition of the invariable plane

$$\frac{\xi}{k}\sin j=n\sin i\quad\ldots\ldots\ldots\ldots\ldots\ldots(108)$$

Wherefore $\dfrac{dN}{dt}=\dfrac{d\psi}{dt}$, and it follows that the two nodes remain coincident. This result is obviously correct.

In the present case, however, there is another disturbing body, and we must now consider—

The perturbing influence of the sun.

Accented symbols will here refer to the elements of the solar orbit.

We might of course form the disturbing function, but it is simpler to accept the known results of lunar theory; these are that the inclination of the lunar orbit to the ecliptic remains constant, whilst the nodes regrede with an angular velocity $\frac{3}{4}\left(\frac{\Omega'}{\Omega}\right)^2\left[1-\frac{3}{8}\frac{\Omega'}{\Omega}\right]\Omega\cos j$.

Now $\frac{3}{4}\left(\frac{\Omega'}{\Omega}\right)^2\Omega=\frac{1}{2}\left(\frac{3}{2}\Omega'^2\right)\times\frac{1}{\Omega}=\frac{1}{2}\frac{\tau'}{\Omega}$ in our notation. Hence I shall write $\frac{1}{2}\frac{\tau'}{\Omega}$ for $\frac{3}{4}\left(\frac{\Omega'}{\Omega}\right)^2\left[1-\frac{3}{8}\frac{\Omega'}{\Omega}\right]\Omega$, although if necessary (in Part IV.) I shall use the more accurate formula for numerical calculation.

For the solar precession and nutation we may obtain the results from (107) by putting $j=0$, and τ' for τ.

Thus for the solar effects we have

$$\left.\begin{aligned}\frac{dj}{dt}&=0\\ \frac{dN}{dt}&=-\tfrac{1}{2}\frac{\tau'}{\Omega}\cos j\\ \frac{di}{dt}&=0\\ n\sin i\frac{d\psi}{dt}&=-\tfrac{1}{2}\tau'\mathfrak{e}\sin 2i\end{aligned}\right\}\quad\text{.................(109)*}$$

When the system is perturbed both by the oblateness of the earth and by the sun, we have from (107) and (109),

$$\left.\begin{aligned}\frac{\xi}{k}\sin j\frac{dj}{dt}&=-\tfrac{1}{4}\tau\mathfrak{e}\sin 2i\sin 2j\sin(N-\psi)\\ \frac{\xi}{k}\sin j\frac{dN}{dt}&=-\tfrac{1}{2}\tau\mathfrak{e}\{\cos 2i\sin 2j+\sin 2i\cos 2j\cos(N-\psi)\}-\tfrac{1}{4}\frac{\tau'}{\Omega}\frac{\xi}{k}\sin 2j\\ n\sin i\frac{di}{dt}&=\tfrac{1}{4}\tau\mathfrak{e}\sin 2i\sin 2j\sin(N-\psi)\\ n\sin i\frac{d\psi}{dt}&=-\tfrac{1}{2}\tau\mathfrak{e}\{\sin 2i\cos 2j+\cos 2i\sin 2j\cos(N-\psi)\}-\tfrac{1}{2}\tau'\mathfrak{e}\sin 2i\end{aligned}\right\}$$

.........(110)

The second pair of equations is derivable from the first by writing i for j and j for i; N for ψ and ψ for N; n for ξ/k; n for Ω; and $\tfrac{1}{2}\mathfrak{e}$ for $\tfrac{1}{4}$ in the term in τ'.

The first pair of equations may be put into the form

$$\cos 2j\frac{d(2j)}{dt}=-\frac{k}{\xi}\tau\mathfrak{e}\sin 2i\cos j\cos 2j\sin(N-\psi)$$

$$\sin 2j\frac{dN}{dt}=-\frac{k}{\xi}\tau\mathfrak{e}\{\cos 2i\cos j\sin 2j+\sin 2i\cos j\cos 2j\cos(N-\psi)\}$$
$$-\tfrac{1}{2}\frac{\tau'}{\Omega}\sin 2j\cos j$$

* The following seems worthy of remark. By the last of (109) we have $d\psi/dt=-\tau'\mathfrak{e}\cos i/n$. In this formula $\mathfrak{e}$ is the precessional constant, because the earth is treated as homogeneous. The full expression for the precessional constant is (2C − A − B)/2C, where A, B, C are the three principal moments of inertia.

Now if we regard the earth and moon as being two particles rotating with an angular velocity Ω about their common centre of inertia, the three principal moments of inertia of the system are $Mmc^2/(M+m)$, $Mmc^2/(M+m)$, 0, and therefore the precessional constant of the system is $\tfrac{1}{2}$. Thus the formula for dN/dt is precisely analogous to that for $d\psi/dt$, each of them being equal to τ' × prec. const. × cos inclin. ÷ rotation.

Let
$$\left.\begin{aligned} y &= \tfrac{1}{2}\sin 2j \sin N, \quad & \eta &= \tfrac{1}{2}\sin 2i \sin\psi \\ z &= \tfrac{1}{2}\sin 2j \cos N, \quad & \zeta &= \tfrac{1}{2}\sin 2i \cos\psi \end{aligned}\right\} \quad \text{...........(111)}$$

Therefore

$$2\frac{dz}{dt} = \cos N \cos 2j \frac{d(2j)}{dt} - \sin N \sin 2j \frac{dN}{dt}$$

$$= \frac{k}{\xi}\tau\mathfrak{e}\,[\cos j \cos 2j \,.\, 2\eta + \cos 2i \cos j \,.\, 2y] + \tfrac{1}{2}\frac{\tau'}{\Omega}\cos j \,.\, 2y$$

or
$$\frac{dz}{dt} = \left(\frac{k\tau\mathfrak{e}}{\xi}\cos 2i \cos j + \tfrac{1}{2}\frac{\tau'}{\Omega}\cos j\right) y + \frac{k\tau\mathfrak{e}}{\xi}\cos j \cos 2j \,.\, \eta$$

Again

$$2\frac{dy}{dt} = \sin N \cos 2j \frac{d(2j)}{dt} + \cos N \sin 2j \frac{dN}{dt}$$

$$= -\frac{k\tau\mathfrak{e}}{\xi}[\cos j \cos 2j \,.\, 2\zeta + \cos 2i \cos j \,.\, 2z] - \tfrac{1}{2}\frac{\tau'}{\Omega}\cos j \,.\, 2z$$

or
$$\frac{dy}{dt} = -\left(\frac{k\tau\mathfrak{e}}{\xi}\cos 2i \cos j + \tfrac{1}{2}\frac{\tau'}{\Omega}\cos j\right) z - \frac{k\tau\mathfrak{e}}{\xi}\cos j \cos 2j \,.\, \zeta$$

Let
$$\left.\begin{aligned} a_1 &= \frac{k\tau\mathfrak{e}}{\xi}, \quad & a_2 &= \tfrac{1}{2}\frac{\tau'}{\Omega} \\ b_1 &= \frac{\tau\mathfrak{e}}{n}, \quad & b_2 &= \frac{\tau'\mathfrak{e}}{n} \end{aligned}\right\} \quad \text{.....................(112)}$$

and we have

$$\left.\begin{aligned} \frac{dz}{dt} &= (a_1 \cos 2i \cos j + a_2 \cos j)\, y + a_1 \cos j \cos 2j \,.\, \eta \\ \frac{dy}{dt} &= -(a_1 \cos 2i \cos j + a_2 \cos j)\, z - a_1 \cos j \cos 2j \,.\, \zeta \end{aligned}\right\} \text{......(113)}$$

and by symmetry from the two latter of (110)

$$\left.\begin{aligned} \frac{d\zeta}{dt} &= (b_1 \cos 2j \cos i + b_2 \cos i)\, \eta + b_1 \cos i \cos 2i \,.\, y \\ \frac{d\eta}{dt} &= -(b_1 \cos 2j \cos i + b_2 \cos i)\, \zeta - b_1 \cos i \cos 2i \,.\, z \end{aligned}\right\} \text{......(114)}$$

These four simultaneous differential equations have to be solved.

The a's and b's are constant, and if it were not for the cosines on the right the equations would be linear and easily soluble.

It has already been assumed that i and j are not very large, hence it would require large variations of i and j to make considerable variations in the coefficients, I shall therefore substitute for i and j, as they occur explicitly, mean values i_0 and j_0; and this procedure will be justifiable unless it be found subsequently that i and j vary largely.

Let

$$\left.\begin{array}{ll}\alpha = a_1 \cos 2i_0 \cos j_0 + a_2 \cos j_0, & \beta = b_1 \cos 2j_0 \cos i_0 + b_2 \cos i_0 \\ \mathrm{a} = a_1 \cos j_0 \cos 2j_0, & \mathrm{b} = b_1 \cos i_0 \cos 2i_0\end{array}\right\} \ldots (115)$$

(Hereafter i and j will be treated as small and the cosines as unity.)

Then

$$\left.\begin{array}{l}\dfrac{dz}{dt} = \alpha y + \mathrm{a}\eta \\ \dfrac{dy}{dt} = -\alpha z - \mathrm{a}\zeta \\ \dfrac{d\zeta}{dt} = \beta\eta + \mathrm{b}y \\ \dfrac{d\eta}{dt} = -\beta\zeta - \mathrm{b}z\end{array}\right\} \ldots\ldots\ldots\ldots\ldots\ldots (116)$$

These equations suggest the solutions

$$z = \Sigma L \cos(\kappa t + \mathrm{m}), \qquad \zeta = \Sigma L' \cos(\kappa t + \mathrm{m})$$

$$y = \Sigma L \sin(\kappa t + \mathrm{m}), \qquad \eta = \Sigma L' \sin(\kappa t + \mathrm{m})$$

Substituting in (116), we must have

$$-L\kappa = \alpha L + \mathrm{a}L'; \quad -L'\kappa = \beta L' + \mathrm{b}L$$

Wherefore

$$\frac{L'}{L} = -\frac{\kappa + \alpha}{\mathrm{a}} = -\frac{\mathrm{b}}{\kappa + \beta}$$

and $\quad (\kappa + \alpha)(\kappa + \beta) - \mathrm{ab} = 0 \quad$ or $\quad \kappa^2 + \kappa(\alpha + \beta) + \alpha\beta - \mathrm{ab} = 0$

This quadratic equation has two real roots (κ_1 and κ_2 suppose), because $(\alpha + \beta)^2 - 4(\alpha\beta - \mathrm{ab}) = (\alpha - \beta)^2 + 4\mathrm{ab}$ is essentially positive.

Let

$$\left.\begin{array}{l}\kappa_1 + \kappa_2 = -(\alpha + \beta) \\ \kappa_1 - \kappa_2 = -\{(\alpha - \beta)^2 + 4\mathrm{ab}\}^{\frac{1}{2}}\end{array}\right\} \ldots\ldots\ldots\ldots\ldots (117)$$

and the solution is

$$\left.\begin{array}{l}\tfrac{1}{2} \sin 2j \cos N = z = L_1 \cos(\kappa_1 t + \mathrm{m}_1) + L_2 \cos(\kappa_2 t + \mathrm{m}_2) \\ \tfrac{1}{2} \sin 2j \sin N = y = L_1 \sin(\kappa_1 t + \mathrm{m}_1) + L_2 \sin(\kappa_2 t + \mathrm{m}_2) \\ \tfrac{1}{2} \sin 2i \cos \psi = \zeta = L_1' \cos(\kappa_1 t + \mathrm{m}_1) + L_2' \cos(\kappa_2 t + \mathrm{m}_2) \\ \tfrac{1}{2} \sin 2i \sin \psi = \eta = L_1' \sin(\kappa_1 t + \mathrm{m}_1) + L_2' \sin(\kappa_2 t + \mathrm{m}_2) \\ \text{where} \quad \dfrac{L_1'}{L_1} = -\dfrac{\kappa_1 + \alpha}{\mathrm{a}} = -\dfrac{\mathrm{b}}{\kappa_1 + \beta}; \quad \dfrac{L_2'}{L_2} = -\dfrac{\kappa_2 + \alpha}{\mathrm{a}} = -\dfrac{\mathrm{b}}{\kappa_2 + \beta}\end{array}\right\} \ldots (118)$$

From these equations we have

$$\tfrac{1}{4} \sin^2 2j = L_1^2 + L_2^2 + 2L_1 L_2 \cos[(\kappa_1 - \kappa_2)t + \mathrm{m}_1 - \mathrm{m}_2]$$

$$\tfrac{1}{4} \sin^2 2i = L_1' + L_2' + 2L_1' L_2' \cos[(\kappa_1 - \kappa_2)t + \mathrm{m}_1 - \mathrm{m}_2]$$

From this we see that $\sin 2j$ oscillates between $2(L_1 + L_2)$ and $2(L_1 \sim L_2)$, and $\sin 2i$ between $2(L_1' + L_2')$ and $2(L_1' \sim L_2')$.

Let us change the constants introduced by integration, and write

$$L_1 = \tfrac{1}{2}\sin 2j_0, \quad L_2' = \tfrac{1}{2}\sin 2i_0$$

Then our solution is

$$\left.\begin{aligned}
\sin 2j \cos N &= \sin 2j_0 \cos(\kappa_1 t + \mathrm{m}_1) - \frac{\mathrm{a}}{\kappa_2 + \alpha}\sin 2i_0 \cos(\kappa_2 t + \mathrm{m}_2)\\
\sin 2j \sin N &= \sin 2j_0 \sin(\kappa_1 t + \mathrm{m}_1) - \frac{\mathrm{a}}{\kappa_2 + \alpha}\sin 2i_0 \sin(\kappa_2 t + \mathrm{m}_2)\\
\sin 2i \cos \psi &= -\frac{\kappa_1 + \alpha}{\mathrm{a}}\sin 2j_0 \cos(\kappa_1 t + \mathrm{m}_1) + \sin 2i_0 \cos(\kappa_2 t + \mathrm{m}_2)\\
\sin 2i \sin \psi &= -\frac{\kappa_1 + \alpha}{\mathrm{a}}\sin 2j_0 \sin(\kappa_1 t + \mathrm{m}_1) + \sin 2i_0 \sin(\kappa_2 t + \mathrm{m}_2)
\end{aligned}\right\} \quad \ldots(119)$$

From this it follows that

$$\sin 2i \sin 2j \cos(N - \psi) = -\frac{\kappa_1 + \alpha}{\mathrm{a}}\sin^2 2j_0 - \frac{\mathrm{a}}{\kappa_2 + \alpha}\sin^2 2i_0$$
$$+ \left(1 + \frac{\kappa_1 + \alpha}{\kappa_2 + \alpha}\right)\sin 2i_0 \sin 2j_0 \cos[(\kappa_1 - \kappa_2)t + \mathrm{m}_1 - \mathrm{m}_2]$$

$$\sin 2i \sin 2j \sin(N - \psi) = \left(1 - \frac{\kappa_1 + \alpha}{\kappa_2 + \alpha}\right)\sin 2i_0 \sin 2j_0 \sin[(\kappa_1 - \kappa_2)t + \mathrm{m}_1 - \mathrm{m}_2]$$

Now $$(\kappa_1 + \alpha)(\kappa_2 + \alpha) = -(\kappa_1 + \alpha)(\kappa_1 + \beta) = -\mathrm{ab}$$
$$\kappa_1 + \kappa_2 + 2\alpha = \alpha - \beta$$

Therefore

$$\left.\begin{aligned}
\sin 2i \sin 2j \cos(N - \psi) &= -\frac{1}{\kappa_2 + \alpha}\{\mathrm{a}\sin^2 2i_0 - \mathrm{b}\sin^2 2j_0\\
&\qquad - (\alpha - \beta)\sin 2i_0 \sin 2j_0 \cos[(\kappa_1 - \kappa_2)t + \mathrm{m}_1 - \mathrm{m}_2]\}\\
\sin 2i \sin 2j \sin(N - \psi) &= -\frac{\kappa_1 - \kappa_2}{\kappa_2 + \alpha}\sin 2i_0 \sin 2j_0 \sin[(\kappa_1 - \kappa_2)t + \mathrm{m}_1 - \mathrm{m}_2]
\end{aligned}\right\} \quad \ldots\ldots(120)$$

From (120) it is clear that the nodes of the lunar orbit will oscillate about the equinoctial line, if

$$\mathrm{a}\sin^2 2i_0 \sim \mathrm{b}\sin^2 2j_0 \text{ be greater than } (\alpha - \beta)\sin 2i_0 \sin 2j_0$$

but will rotate (although not uniformly) if the former be less than the latter.

With the present configuration of the earth and moon

$$\mathrm{a}\sin^2 2i_0 \sim \mathrm{b}\sin^2 2j_0 \text{ is very small compared with } (\alpha - \beta)\sin 2i_0 \sin 2j_0$$

and the nodes of the lunar orbit revolve very nearly uniformly on the ecliptic; also the inclination of the orbit varies very slightly, as the nodes revolve.

In the investigation in Part II. the secular rate of change in the inclination of the lunar orbit has been found, on the assumption that the nodes of the lunar orbit rotate uniformly.

It is intended to trace the effects of tidal friction on the earth and moon retrospectively. In the course of the solution the importance of the solar perturbation of the moon, relatively to the influence of the earth's oblateness, will wane; the nodes will cease to revolve uniformly, and the inclination of the lunar orbit and of the equator to the ecliptic will be subject to nutation. The differential equations of Part II. will then cease to be applicable, and new ones will have to be found.

The problem is one of such complication, that I have thought it advisable only to attempt to obtain a solution on the hypothesis of the smallness both of the obliquity and of the inclination of the orbit to the plane of reference or the ecliptic. It seems best however to give the preceding investigation, although it is more accurate than the solution subsequently used*.

The first step towards this further consideration is to obtain a clear idea of the nature of the motions represented by the analytical solutions (118) or (119) of the present problem.

Assuming then i and j to be small, we have from (112) and (115)

$$\alpha = a_1 + a_2, \quad \mathrm{a} = a_1, \quad \beta = b_1 + b_2, \quad \mathrm{b} = b_1 \quad \text{..............(121)}$$

$$\left.\begin{aligned} j\cos N &= L_1 \cos(\kappa_1 t + \mathrm{m}_1) + L_2 \cos(\kappa_2 t + \mathrm{m}_2) \\ j\sin N &= L_1 \sin(\kappa_1 t + \mathrm{m}_1) + L_2 \sin(\kappa_2 t + \mathrm{m}_2) \\ i\cos\psi &= L_1' \cos(\kappa_1 t + \mathrm{m}_1) + L_2' \cos(\kappa_2 t + \mathrm{m}_2) \\ i\sin\psi &= L_1' \sin(\kappa_1 t + \mathrm{m}_1) + L_2' \sin(\kappa_2 t + \mathrm{m}_2) \end{aligned}\right\} \text{.........(122)}$$

Take a set of rectangular axes; let the axis of x' pass through the fixed point in the ecliptic from which longitudes are measured, let the axis of z' be drawn perpendicular to the ecliptic northwards, and let the rotation from x' to y' be positive, and therefore consentaneous with the moon's orbital motion.

N is the longitude of the ascending node of the lunar orbit, and therefore the direction cosines of the normal to the lunar orbit drawn northwards are,

$$\sin j \cos(N - \tfrac{1}{2}\pi), \quad \sin j \sin(N - \tfrac{1}{2}\pi), \quad \cos j$$

or since j is small, $\quad j\sin N, \quad -j\cos N, \quad 1$

And ψ is the longitude of the descending node of the equator, and therefore the direction cosines of the earth's axis, drawn northwards are,

$$\sin i \cos(\psi + \tfrac{1}{2}\pi), \quad \sin i \sin(\psi + \tfrac{1}{2}\pi), \quad \cos i$$

or since i is small, $\quad -i\sin\psi, \quad i\cos\psi, \quad 1$

Draw a sphere of unit radius, with the origin as centre; draw a tangent plane to it at the point where the axis of z' meets the sphere, and project on this plane the poles of the lunar orbit and of the earth. We here in fact

* See the foot-note to § 18 for a comparison of these results with those ordinarily given.

map the motion of the two poles on a tangent plane to the celestial sphere. Let x', y' be a pair of axes in this plane parallel to our previous x', y'; and let x', y' be the coordinates of the pole of the lunar orbit, and ξ', η' be the coordinates of the earth's pole, so that

$$x' = j \sin N, \quad y' = -j \cos N; \quad \xi' = -i \sin \psi, \quad \eta' = i \cos \psi \;\ldots(123)$$

Let x, y, ξ, η be the coordinates of these same points referred to another pair of rectangular axes in this plane, inclined at an angle ϕ to the axes x', y'.

Then
$$\begin{aligned} x &= x' \cos\phi + y' \sin\phi, & \xi &= \xi' \cos\phi + \eta' \sin\phi \\ y &= -x' \sin\phi + y' \cos\phi, & \eta &= -\xi' \sin\phi + \eta' \cos\phi \end{aligned}$$

From (123) and (118) we have therefore

$$\left.\begin{aligned} x &= L_1 \sin(\kappa_1 t + \mathrm{m}_1 - \phi) + L_2 \sin(\kappa_2 t + \mathrm{m}_2 - \phi) \\ y &= -L_1 \cos(\kappa_1 t + \mathrm{m}_1 - \phi) - L_2 \cos(\kappa_2 t + \mathrm{m}_2 - \phi) \\ \xi &= -L_1' \sin(\kappa_1 t + \mathrm{m}_1 - \phi) - L_2' \sin(\kappa_2 t + \mathrm{m}_2 - \phi) \\ \eta &= L_1' \cos(\kappa_1 t + \mathrm{m}_1 - \phi) + L_2' \cos(\kappa_2 t + \mathrm{m}_2 - \phi) \end{aligned}\right\}$$

Now suppose the new axes to rotate with an angular velocity κ_2, and that $\phi = \kappa_2 t + \mathrm{m}_2$.

Then
$$\left.\begin{aligned} x &= L_1 \sin[(\kappa_1 - \kappa_2) t + \mathrm{m}_1 - \mathrm{m}_2] \\ y + L_2 &= -L_1 \cos[(\kappa_1 - \kappa_2) t + \mathrm{m}_1 - \mathrm{m}_2] \\ \xi &= -L_1' \sin[(\kappa_1 - \kappa_2) t + \mathrm{m}_1 - \mathrm{m}_2] \\ \eta - L_2' &= L_1' \cos[(\kappa_1 - \kappa_2) t + \mathrm{m}_1 - \mathrm{m}_2] \end{aligned}\right\} \ldots\ldots\ldots\ldots\ldots(124)$$

These four equations represent that each pole describes a circle, relatively to the rotating axes, with a negative angular velocity (because $\kappa_1 - \kappa_2$ is negative). The centres of the circles are on the axis of y. The ratio

$$\frac{\text{distance of centre of terrestrial circle}}{\text{distance of centre of lunar circle}} = \frac{L_2'}{-L_2} = \frac{\kappa_2 + \alpha}{\mathrm{a}} = \frac{\mathrm{b}}{\kappa_2 + \beta} \;\ldots(125)$$

the distances being measured from the pole of the ecliptic. And the ratio

$$\frac{\text{radius of terrestrial circle}}{\text{radius of lunar circle}} = \frac{L_1'}{L_1} = -\frac{\kappa_1 + \alpha}{\mathrm{a}} = -\frac{\mathrm{b}}{\kappa_1 + \beta} \;\ldots\ldots(126)$$

According to the definitions adopted in (117) of κ_1 and κ_2, $(\kappa_1 + \alpha)/\mathrm{a}$ is negative and $(\kappa_2 + \alpha)/\mathrm{a}$ is positive; hence L_1 has the same sign as L_1', and L_2 has the opposite sign from L_2'. When $t = -(\mathrm{m}_1 - \mathrm{m}_2)/(\kappa_1 - \kappa_2)$, we have

$$x = 0, \quad y = (-L_2) - L_1, \quad \xi = 0, \quad \eta = L_2' + L_1'$$

In fig. 6 let Ox, Oy be the rotating axes, which revolve with a rotation equal to κ_2, which is negative. Let M be the centre of the lunar circle, and Q of the terrestrial circle. Then we see that L and P must be simultaneous positions of the two poles, which revolve round their respective circles with an angular velocity $\kappa_2 - \kappa_1$, in the direction of the arrows.

M and Q are the poles of two planes, which may be appropriately called the proper planes of the moon and the earth. These proper planes are inclined at a constant angle to one another and to the ecliptic, and have a common node on the ecliptic, and a uniform slow negative precession relatively to the ecliptic.

The lunar orbit and the equator are inclined at constant angles to the lunar and terrestrial proper planes respectively, and the nodes of the orbit, and of the equator regrede uniformly on the respective proper planes.

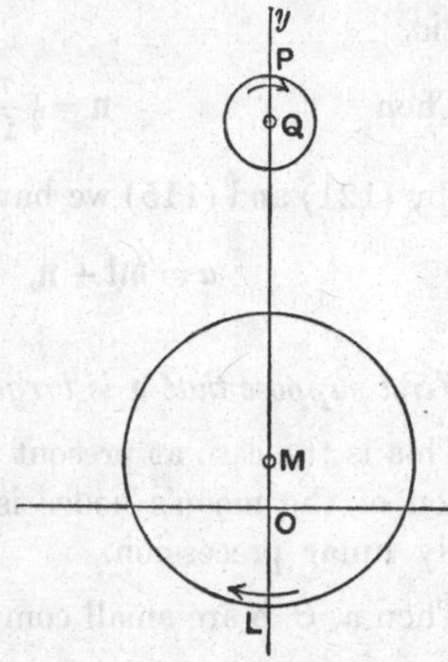

FIG. 6.

In the *Mécanique Céleste* (livre vii., chap. 2, sec. 20) Laplace refers to the proper plane of the lunar orbit, but the corresponding inequality of the earth is ordinarily referred to as the 19-yearly nutation. It will be proved later, that the above results are identical with those ordinarily given.

Suppose that

$$\left.\begin{array}{l} \mathrm{I} = \text{the inclination of the earth's proper plane to the ecliptic} \\ \mathrm{J} = \text{the inclination of the lunar orbit to its proper plane} \\ \mathrm{I}_, = \text{the inclination of the equator to the earth's proper plane} \\ \mathrm{J}_, = \text{the inclination of the moon's proper plane to the ecliptic} \\ \text{Then} \\ \mathrm{J} = L_1, \quad \mathrm{I} = L_2', \quad \mathrm{I}_, = L_1', \quad \mathrm{J}_, = -L_2 \\ \text{and by (125–6)} \\ \mathrm{I}_, = -\dfrac{\kappa_1+\alpha}{\mathrm{a}}\mathrm{J} = -\dfrac{\mathrm{b}}{\kappa_1+\beta}\mathrm{J}; \quad \mathrm{J}_, = -\dfrac{\mathrm{a}}{\kappa_2+\alpha}\mathrm{I} = \dfrac{\kappa_2+\beta}{\mathrm{b}}\mathrm{I} \end{array}\right\} \quad (127)$$

Thus I and J are the two constants introduced in the integration of the simultaneous differential equations (116).

It is interesting to examine the physical meaning of these results, and to show how the solution degrades into the two limiting cases, viz.: where the planet is spherical, and where the sun's influence is evanescent.

Let $\mathfrak{n}$ be the speed of motion of the nodes, when the ellipticity of the planet is zero.

Let $\mathfrak{l}$ be the purely lunar precession, or the precession when the solar influence is nil.

Let $\mathfrak{m}$ be the ratio of the moment of momentum of the earth's rotation to that of the orbital motion of the two bodies round their common centre of inertia.

Then $$\mathfrak{n} = \tfrac{1}{2}\frac{\tau'}{\Omega}, \quad \mathfrak{l} = \frac{\tau\mathfrak{k}}{n}, \quad \mathfrak{m} = \frac{kn}{\xi}$$

and by (121) and (115) we have

$$\alpha = \mathfrak{m}\mathfrak{l} + \mathfrak{n}, \quad \mathrm{a} = \mathfrak{m}\mathfrak{l}, \quad \beta = \mathfrak{l} + \frac{\tau'\mathfrak{k}}{n}, \quad \mathrm{b} = \mathfrak{l}$$

First suppose that $\mathfrak{n}$ *is large compared with* $\mathfrak{l}$.

This is the case at present with the earth and moon, because the speed of motion of the moon's nodes is very great compared with the speed of the purely lunar precession.

Then a, β, b are small compared with α.

Therefore by (117)

$$\kappa_1 - \kappa_2 = -\alpha + \beta, \quad \kappa_1 + \kappa_2 = -\alpha - \beta$$

and $$\kappa_1 = -\alpha, \quad \kappa_2 = -\beta$$

Therefore

$$-\frac{\mathrm{b}}{\kappa_1 + \beta} = \frac{\mathrm{b}}{\alpha - \beta} = \frac{\mathfrak{l}}{\mathfrak{n} - (1 - \mathfrak{m})\,\mathfrak{l} - \dfrac{\tau'\mathfrak{k}}{n}} = \frac{\mathfrak{l}}{\mathfrak{n}} \text{ approximately}$$

$$\frac{\mathrm{a}}{\kappa_2 + \alpha} = \frac{\mathrm{a}}{\alpha - \beta} = \mathfrak{m}\,\frac{\mathfrak{l}}{\mathfrak{n}} \text{ approximately}$$

$$\kappa_2 = -\frac{\tau + \tau'}{n}\mathfrak{k}, \quad \kappa_2 - \kappa_1 = \mathfrak{n} \text{ approximately}$$

And by (127) $$\mathrm{I}_{,} = \frac{\mathfrak{l}}{\mathfrak{n}}\mathrm{J}, \quad \mathrm{J}_{,} = \mathfrak{m}\frac{\mathfrak{l}}{\mathfrak{n}}\mathrm{I}$$

We have shown above that $-\kappa_2$ is the common angular velocity of the pair of proper planes, and the above results show that it is in fact the luni-solar precession.

$\kappa_2 - \kappa_1$ is the angular velocity of the two nodes on their proper planes, and it is nearly equal to $\mathfrak{n}$.

The ratio of the amplitude of the 19-yearly nutation to the inclination of the lunar orbit is $\mathfrak{l}/\mathfrak{n}$.

The ratio of the inclination of the lunar proper plane to the obliquity of the ecliptic is $\mathfrak{m}\mathfrak{l}/\mathfrak{n}$.

In this case, therefore, the lunar proper plane is inclined at a small angle to the ecliptic, and if the earth were spherical would be identical with the ecliptic.

Secondly, suppose that $\mathfrak{n}$ *is small compared with* $\mathfrak{l}$.

Then *à fortiori* $\frac{\tau'\mathfrak{k}}{n}$ is small compared with $\mathfrak{l}$. Hence we may put $\beta = \mathrm{b}$.

Therefore

$$\kappa_2 - \kappa_1 = \sqrt{(\alpha - \beta)^2 + 4\mathrm{ab}} = \mathrm{a} + \mathrm{b} + \frac{\mathrm{a} - \mathrm{b}}{\mathrm{a} + \mathrm{b}}\mathfrak{n}, \text{ nearly}$$

$$= (\mathfrak{m} + 1)\mathfrak{l} + \frac{\mathfrak{m} - 1}{\mathfrak{m} + 1}\mathfrak{n}$$

$$\kappa_2 + \kappa_1 = -(\mathfrak{m} + 1)\mathfrak{l} - \mathfrak{n}$$

$$\kappa_2 = -\frac{\mathfrak{n}}{\mathfrak{m} + 1}, \quad \kappa_1 = -(\mathfrak{m} + 1)\mathfrak{l} - \frac{\mathfrak{m}}{\mathfrak{m} + 1}\mathfrak{n}$$

$$\frac{\kappa_2 + \beta}{\mathrm{b}} = 1 - \frac{1}{\mathfrak{m} + 1}\frac{\mathfrak{n}}{\mathfrak{l}}; \quad -\frac{\kappa_1 + \alpha}{\mathrm{a}} = \frac{1}{\mathfrak{m}}\left(1 - \frac{1}{\mathfrak{m} + 1}\frac{\mathfrak{n}}{\mathfrak{l}}\right)$$

Therefore

$$\mathrm{I}_{,} = \frac{1}{\mathfrak{m}}\left(1 - \frac{1}{\mathfrak{m} + 1}\frac{\mathfrak{n}}{\mathfrak{l}}\right)J, \quad J_{,} = \left(1 - \frac{1}{\mathfrak{m} + 1}\frac{\mathfrak{n}}{\mathfrak{l}}\right)\mathrm{I}$$

From the last of these,

$$\mathrm{I} - J_{,} = \frac{1}{\mathfrak{m} + 1}\frac{\mathfrak{n}}{\mathfrak{l}}\mathrm{I}$$

$-\kappa_2$ is the precession of the system of proper planes, and the above results show that the solar precession of the planet and satellite together, considered as one system, is one $(\mathfrak{m} + 1)$th of the angular velocity which the nodes of the satellite would have, if the planet were spherical.

$\kappa_2 - \kappa_1$ is the lunar precession of the earth which goes on within the system, and it is approximately the same as though the sun did not exist. (Compare the second and fourth of (107) with $N = \psi$, and use (108).)

It also appears that the lunar proper plane is inclined to the planet's proper plane at a small angle the ratio of which to the inclination of the earth's proper plane to the ecliptic is equal to one $(\mathfrak{m} + 1)$th part of $\mathfrak{n}/\mathfrak{l}$.

If $\mathfrak{n}$ and $\mathfrak{l}$ are of approximately equal speeds the proper plane of the moon will neither be very near the ecliptic, nor very near the earth's proper plane. The results do not then appear to be reducible to very simple forms; nor are the angular velocities κ_2 and $\kappa_2 - \kappa_1$ so easily intelligible, each of them being a sort of compound precession.

If the solar influence were to wane, M and Q, the poles of the proper planes, would approach one another, and ultimately become identical. The two planes would have then become the invariable plane of the system; and the two circles would be concentric and their radii would be inversely proportional to the two moments of momentum (whose ratio is $\mathfrak{m}$).

Now in the problem which is to be considered here the solar influence will in effect wane, because the effect of tidal friction is, in retrospect, to bring the moon nearer and nearer to the earth, and to increase the ellipticity of the earth's figure; hence the relative importance of the solar influence diminishes.

We now see that the problem to be solved is to trace these proper planes, from their present condition when one is nearly identical with the ecliptic and the other is the mean equator, backwards until they are both sensibly coincident with the equator.

We also see that the present angular velocity of the moon's nodes on the ecliptic is analogous to and continuous with the purely lunar precession on the invariable plane of the moon-earth system; and that the present luni-solar precession is analogous to and continuous with a slow precessional motion of the same invariable plane.

Analytically the problem is to trace the secular changes in the constants of integration, when α, a, β, b, instead of being constant, are slowly variable under the influence of tidal friction, and when certain other small terms, also due to tides, are added to the differential equations of motion.

§ 14. *On the small terms in the equations of motion due directly to tidal friction.*

The first step is the formation of the disturbing function.

As we shall want to apply the function both to the case of the earth and to that of the moon, it will be necessary to measure longitudes from a fixed point in the ecliptic; also we must distinguish between the longitude of the equinox and the angle χ, as they enter in the two capacities (viz.: in the X′Y′ and $\mathfrak{X}\mathfrak{Y}$ functions); thus the N and N' of previous developments must become $N-\psi$, $N'-\psi'$; ϵ, ϵ' must become $\epsilon-\psi$, $\epsilon'-\psi'$; and $2(\chi-\chi')$ must be introduced in the arguments of the trigonometrical terms in the semi-diurnal terms, and $\chi-\chi'$ in the diurnal ones.

The disturbing function must be developed so that it may be applicable to the cases either where Diana, the tide-raiser, is or is not identical with the moon; but as we are only going to consider secular inequalities, all those terms which depend on the longitudes of Diana or the moon may be dropped.

In the previous development of Part II. we had terms whose arguments involved $\epsilon-\epsilon'$; in the present case this ought to be written

$$(\Omega t+\epsilon-\psi)-(\Omega' t+\epsilon'-\psi')$$

for which it is, in fact, only an abbreviation.

A term involving this expression can only give rise to secular inequalities, in the case where Diana is identical with the moon; and as we shall never want to differentiate the disturbing function with regard to Ω', we may in the present development drop the Ωt and $\Omega' t$.

Having made these preliminary explanations, we shall be able to use previous results for the development of the disturbing function. The work will be much abridged by the treatment of i, j, i', j' as small.

Unaccented symbols refer to the elements of the orbit of the tide-raiser Diana, or (in the case of i, χ, ψ) to the earth as a tidally distorted body; accented symbols refer to the elements of the orbit of the perturbed satellite, or to the earth as a body whose rotation is perturbed.

Since i, i' and j, j' are to be treated as small, (22) becomes

$$\left.\begin{array}{l} \left.\begin{array}{l}\varpi\\ \underline{\varpi}\end{array}\right\} = Pp - Qqe^{\pm(N-\psi)} = 1 - \frac{1}{8}i^2 - \frac{1}{8}j^2 - \frac{1}{4}ije^{\pm(N-\psi)} \\ \left.\begin{array}{l}\kappa\\ \underline{\kappa}\end{array}\right\} = Qp + Pqe^{\pm(N-\psi)} = \frac{1}{2}i + \frac{1}{2}je^{\pm(N-\psi)} \end{array}\right\} \quad \ldots\ldots\ldots(128)$$

The same quantities when accented are equal to the same quantities when i, j, N, ψ are accented.

Referring to the development in § 5 of the disturbing function, we see that, for the same reasons as before, we need only consider products of terms of the same kind in the sets of products of the type $X'Y' \times \mathfrak{X}\mathfrak{Y}$. Hence the disturbing function W is the sum of the three expressions (37–9) multiplied by $\tau\tau'/\mathfrak{g}$. Now since we only wish to develop the expression as far as the squares of i and j, we may at once drop out all those terms in these expressions, in which κ occurs raised to a higher power than the second. This at once relieves us of the sidereal and fast semi-diurnal terms, the fast diurnal and the true fortnightly term. We are, however, left with one part of $\frac{3}{2}(\frac{1}{3} - Z'^2)(\frac{1}{3} - \mathfrak{Z}^2)$, which is independent of the moon's longitude and of the earth's rotation; this part represents the permanent increase of ellipticity of the earth, due to Diana's attraction, and to that part of the tidal action which depends on the longitude of the nodes, in which the tides are assumed to have their equilibrium value. I shall refer to it as the permanent tide.

As before, it will be convenient to consider the constituent parts of the disturbing function separately, and to indicate the several parts of W by suffixes as in § 5 and elsewhere; as above explained, we need only consider W_I, W_1, W_2, and W_0.

Semi-diurnal term.

From (37) we have

$$W_I \Big/ \frac{\tau\tau'}{\mathfrak{g}} = \tfrac{1}{4}\left[F_1 \varpi^4 \underline{\varpi}'^4 e^{2(\theta'-\theta)-2f_1} + F_1 \underline{\varpi}^4 \varpi'^4 e^{-2(\theta'-\theta)+2f_1}\right]$$

To the indices of these exponentials we must add $\pm 2(\chi - \chi')$, and for θ write $\epsilon - \psi$, and for θ', $\epsilon' - \psi'$.

By (128)
$$\varpi^4 = 1 - \tfrac{1}{2}i^2 - \tfrac{1}{2}j^2 - ije^{(N-\psi)}$$
$$\underline{\varpi}'^4 = 1 - \tfrac{1}{2}i'^2 - \tfrac{1}{2}j'^2 - i'j'e^{-(N'-\psi')}$$

Hence

$$W_1 \Big/ \frac{\tau\tau'}{\mathfrak{g}} = \tfrac{1}{2}F_1 \{(1 - \tfrac{1}{2}i^2 - \tfrac{1}{2}j^2 - \tfrac{1}{2}i'^2 - \tfrac{1}{2}j'^2)\cos[2(\chi-\chi') + 2(\epsilon'-\epsilon) - 2(\psi'-\psi) - 2f_1]$$
$$- ij\cos[2(\chi-\chi') + 2(\epsilon'-\epsilon) - 2(\psi'-\psi) + (N-\psi) - 2f_1]$$
$$- i'j'\cos[2(\chi-\chi') + 2(\epsilon'-\epsilon) - 2(\psi'-\psi) - (N'-\psi') - 2f_1]\} \ldots(129)$$

Slow diurnal term.

From (38) we have

$$W_1 \Big/ \frac{\tau\tau'}{\mathfrak{g}} = G_1 [\varpi^3\kappa\underline{\varpi}'^3\underline{\kappa}'e^{2(\theta'-\theta)-g_1} + \underline{\varpi}^3\underline{\kappa}\varpi'^3\kappa'e^{-2(\theta'-\theta)+g_1}]$$

To the indices of the exponentials we must add $\pm(\chi-\chi')$; ϖ^3, ϖ'^3 may be obviously put equal to unity, and by (128)

$$\kappa\underline{\kappa}' = \tfrac{1}{4}[ii' + i'je^{(N-\psi)} + ij'e^{-(N'-\psi')} + jj'e^{(N-N')-(\psi-\psi')}]$$

Hence

$$W_1 \Big/ \frac{\tau\tau'}{\mathfrak{g}} = \tfrac{1}{2}G_1 \{ii'\cos[(\chi-\chi') + 2(\epsilon'-\epsilon) - 2(\psi'-\psi) - g_1]$$
$$+ i'j\cos[(\chi-\chi') + 2(\epsilon'-\epsilon) - 2(\psi'-\psi) + (N-\psi) - g_1]$$
$$+ ij'\cos[(\chi-\chi') + 2(\epsilon'-\epsilon) - 2(\psi'-\psi) - (N'-\psi') - g_1]$$
$$+ jj'\cos[(\chi-\chi') + 2(\epsilon'-\epsilon) - 2(\psi'-\psi) + (N-N') - (\psi-\psi') - g_1]\} \ldots\ldots(130)$$

Sidereal diurnal term.

From (38) we have

$$W_2 \Big/ \frac{\tau\tau'}{\mathfrak{g}} = G[\varpi\underline{\kappa}(\varpi\underline{\varpi} - \kappa\underline{\kappa})\underline{\varpi}'\kappa'(\varpi'\underline{\varpi}' - \kappa'\underline{\kappa}')e^{-g} + \underline{\varpi}\kappa(\varpi\underline{\varpi} - \kappa\underline{\kappa})\varpi'\underline{\kappa}'(\varpi'\underline{\varpi}' - \kappa'\underline{\kappa}')e^{g}]$$

To the indices of the exponentials must be added $\pm(\chi-\chi')$. ϖ, ϖ' may be treated as unity. Hence the expression becomes

$$G[\underline{\kappa}\kappa'e^{\chi-\chi'-g} + \kappa\underline{\kappa}'e^{-(\chi-\chi')+g}]$$

and
$$W_2 \Big/ \frac{\tau\tau'}{\mathfrak{g}} = \tfrac{1}{2}G\{ii'\cos(\chi-\chi'-g)$$
$$+ i'j\cos[(\chi-\chi') - (N-\psi) - g]$$
$$+ ij'\cos[(\chi-\chi') + (N'-\psi') - g]$$
$$+ jj'\cos[(\chi-\chi') - (N-N') + (\psi-\psi') - g]\} \ldots(131)$$

Permanent term.

From (39) we have

$$W_0 \Big/ \frac{\tau\tau'}{\mathfrak{g}} = \tfrac{3}{2}\left(\tfrac{1}{3} - 2\varpi\underline{\varpi}\kappa\underline{\kappa}\right)\left(\tfrac{1}{3} - 2\varpi'\underline{\varpi}'\kappa'\underline{\kappa}'\right)$$

$$= \tfrac{1}{6} - \kappa\underline{\kappa} - \kappa'\underline{\kappa}' \text{ to our degree of approximation}$$

Now

$$\kappa\underline{\kappa} = \tfrac{1}{4}\left[i^2 + j^2 + ij\left(e^{N-\psi} + e^{-(N-\psi)}\right)\right] = \tfrac{1}{4}\left[i^2 + j^2 + 2ij\cos(N-\psi)\right]$$

Hence

$$W_0 \Big/ \frac{\tau\tau'}{\mathfrak{g}} = \tfrac{1}{6} - \tfrac{1}{4}\left[i^2 + j^2 + 2ij\cos(N-\psi)\right] - \tfrac{1}{4}\left[i'^2 + j'^2 + 2i'j'\cos(N'-\psi')\right] \quad \text{.........(132)}$$

W_2 and W_0 are the only terms in W which can contribute anything to the secular inequalities, unless Diana and the satellite are identical; for all the other terms involve $\epsilon - \epsilon'$, and will therefore be periodic however differentiated, unless $\epsilon = \epsilon'$.

We now have to differentiate W with respect to i', χ', ψ', j', ϵ, N'. The results will then have to be applied in the following cases.

For the moon:

(i) When the tide-raiser is the moon.

(ii) When the tide-raiser is the sun.

For the earth:

(iii) When the tide-raiser is the moon, and the disturber the moon.

(iv) When the tide-raiser is the sun, and the disturber the sun.

(v) When the tide-raiser is the moon, and the disturber the sun.

(vi) When the tide-raiser is the sun, and the disturber the moon.

The sum of the values derived from the differentiations, according to these several hypotheses, will be the complete values to be used in the differential equations (13), (14) and (18) for dj/dt, dN/dt, di/dt, $d\psi/dt$.

A little preliminary consideration will show that the labour of making these differentiations may be considerably abridged.

In the present case i and j are small, and the equations (110) which give the position of the two proper planes, and the inclinations of the orbit and equator thereto, become

$$\left.\begin{aligned}
\frac{\xi}{k}\frac{dj}{dt} &= -\tau\mathfrak{e}i\sin(N-\psi)\\
\frac{\xi}{k}\sin j\,\frac{dN}{dt} &= -\left(\tau\mathfrak{e} + \tfrac{1}{2}\frac{\tau'}{\Omega}\frac{\xi}{k}\right)j - \tau\mathfrak{e}i\cos(N-\psi)\\
n\frac{di}{dt} &= \tau\mathfrak{e}j\sin(N-\psi)\\
n\sin i\,\frac{d\psi}{dt} &= -(\tau\mathfrak{e} + \tau'\mathfrak{e})\,i - \tau\mathfrak{e}j\cos(N-\psi)
\end{aligned}\right\} \quad \text{......(133)}$$

We are going to find certain additional terms, depending on frictional tides, to be added to these four equations. These terms will all involve τ^2, τ'^2, or $\tau\tau'$ in their coefficients, and will therefore be small compared with those in (133). If these small terms are of the same types as the terms in (133), they may be dropped; because the only effect of them will be to produce a very small and negligeable alteration in the position of the two proper planes*.

In consequence of this principle, we may entirely drop W_0 from our disturbing function, for W_0 only gives rise to a small permanent alteration of oblateness, and therefore can only slightly modify the positions of the proper planes.

Analytically the same result may be obtained, by observing that W_0 in (132) has the same form as W in (105), when i and j are treated as small.

In each case, after differentiation, the transition will be made to the case of viscosity of the planet, and the proper terms will be dropped out, without further comment.

First take the perturbations of the moon.

For this purpose we have to find dW/dj' and

$$\frac{dW}{\sin j'\, dN'} + \tan \tfrac{1}{2} j' \frac{dW}{d\epsilon'} \quad \text{or} \quad \frac{dW}{j' dN'} + \tfrac{1}{2} j' \frac{dW}{d\epsilon'}$$

By the above principle, in finding dW/dj' we may drop terms involving j and $i \cos (N - \psi)$, and in finding $dW/j'dN' + \tfrac{1}{2} j' dW/d\epsilon'$, we may drop terms involving $i \sin (N - \psi)$.

We may now suppose $\chi = \chi'$, $\psi = \psi'$.

Take the case (i), where the tide-raiser is the moon. As the perturbed body is also the moon, after differentiation we may drop the accents to all the symbols.

From (129)

$$\frac{dW_I}{dj'} \Big/ \frac{\tau^2}{\mathfrak{g}} = \tfrac{1}{2} F_1 \{-j \cos 2f_1 - i \cos (N - \psi + 2f_1)\}$$

$$= \tfrac{1}{4} i \sin (N - \psi) \sin 4f_1 \text{.............................} (134)$$

* For example, we should find the following terms in $\frac{\xi}{k} \sin j \frac{dN}{dt}$, viz.:

$$-\tfrac{1}{2} j \frac{\tau\tau'}{\mathfrak{g}} - \tfrac{1}{2} i \cos (N - \psi) \sin^2 g \frac{\tau\tau'}{\mathfrak{g}} + \tfrac{1}{2} [j + i \cos (N - \psi)] [\sin^2 2f_1 - \sin^2 g_1 - \sin^2 g] \frac{\tau^2}{\mathfrak{g}}$$

which may be all coupled up with those in the second of (133).

If the viscosity be small, so that the angles of lagging are small, it will be found that all the terms of this kind vanish in all four equations, excepting the first of those just written down, viz.: $-\tfrac{1}{2} j\tau\tau'/\mathfrak{g}$.

From (130)

$$\frac{d\mathrm{W}_1}{dj'}\Big/\frac{\tau^2}{\mathfrak{g}} = \tfrac{1}{2}\mathrm{G}_1\{i\cos(N-\psi+\mathrm{g}_1)+j\cos\mathrm{g}_1\}$$

$$= -\tfrac{1}{4}i\sin(N-\psi)\sin 2\mathrm{g}_1 \quad\text{.......................(135)}$$

From (131) and symmetry with (135)

$$\frac{d\mathrm{W}_2}{dj'}\Big/\frac{\tau^2}{\mathfrak{g}} = \tfrac{1}{4}i\sin(N-\psi)\sin 2\mathrm{g} \quad\text{.................(136)}$$

Adding these three (134–6) together, we have for the whole effect of the lunar tides on the moon

$$\frac{d\mathrm{W}}{dj'}\Big/\frac{\tau^2}{\mathfrak{g}} = \tfrac{1}{4}i\sin(N-\psi)[\sin 4\mathrm{f}_1-\sin 2\mathrm{g}_1+\sin 2\mathrm{g}]\text{.........(137)}$$

Now take the case (ii) where the tide-raiser is the sun.

Here we need only consider W_2, but although we may put $\chi=\chi'$, $\psi=\psi'$, $i=i'$, we must not put $j=j'$, $N=N'$, because the tide-raiser is distinct from the moon.

From (131)

$$\frac{d\mathrm{W}_2}{dj'}\Big/\frac{\tau\tau'}{\mathfrak{g}} = \tfrac{1}{2}\mathrm{G}\{i\cos(N'-\psi'-\mathrm{g})+j\cos(N-N'+\mathrm{g})\}$$

Here accented symbols refer to the moon (as perturbed), and unaccented to the sun (as tide-raiser). As we refer the motion to the ecliptic $j=0$, and the last term disappears. Also we want accented symbols to refer to the sun and unaccented to refer to the moon, therefore make τ and τ' interchange their meanings, and drop the accents to N' and ψ'. Thus as far as important

$$\frac{d\mathrm{W}_2}{dj'}\Big/\frac{\tau'\tau}{\mathfrak{g}} = \tfrac{1}{4}i\sin(N-\psi)\sin 2\mathrm{g} \quad\text{.................(138)}$$

This gives the whole effect of the solar tides on the moon.

Collecting results from (137–8), we have by (14)

$$\frac{\xi}{k}\sin j\,\frac{dN}{dt} = \tfrac{1}{4}i\sin(N-\psi)\left[\frac{\tau^2}{\mathfrak{g}}(\sin 4\mathrm{f}_1-\sin 2\mathrm{g}_1+\sin 2\mathrm{g})+\frac{\tau\tau'}{\mathfrak{g}}\sin 2\mathrm{g}\right] \quad\text{.........(139)}$$

This gives the required additional terms due to bodily tides in the equation for dN/dt, viz.: the second of (133).

If the viscosity be small

$$\left.\begin{aligned}\sin 4\mathrm{f}_1-\sin 2\mathrm{g}_1+\sin 2\mathrm{g} &= \sin 4\mathrm{f}\\ \sin 2\mathrm{g} &= \tfrac{1}{2}\sin 4\mathrm{f}\end{aligned}\right\} \quad\text{..............(140)}$$

Next take the secular change of inclination of the lunar orbit.

For this purpose we have to find $d\mathrm{W}/j'dN'+\frac{1}{2}j'd\mathrm{W}/d\epsilon'$, and may drop terms in $i\sin(N-\psi)$.

First take the case (i), where the tide-raiser is the moon.

From (129)

$$\frac{1}{j'}\frac{dW_I}{dN'}\Big/\frac{\tau^2}{\mathfrak{g}} = \tfrac{1}{2}F_1 i \sin(N-\psi+2f_1) = \tfrac{1}{4} i \cos(N-\psi)\sin 4f_1 \quad \ldots(141)$$

$$\tfrac{1}{2}j'\frac{dW_I}{d\epsilon'}\Big/\frac{\tau^2}{\mathfrak{g}} = \tfrac{1}{2}F_1 j \sin 2f_1 \qquad = \tfrac{1}{4} j \sin 4f_1 \quad \ldots(142)$$

From (130)

$$\frac{1}{j'}\frac{dW_1}{dN'}\Big/\frac{\tau^2}{\mathfrak{g}} = -\tfrac{1}{2}G_1\{i\sin(N-\psi+g_1)+j\sin g_1\} = -\tfrac{1}{4}[j+i\cos(N-\psi)]\sin 2g_1 \quad \ldots(143)$$

$$\tfrac{1}{2}j'\frac{dW_1}{d\epsilon'}\Big/\frac{\tau^2}{\mathfrak{g}} = 0 \text{ to present order of approximation} \quad \ldots(144)$$

From (131)

$$\frac{1}{j'}\frac{dW_2}{dN'}\Big/\frac{\tau^2}{\mathfrak{g}} = -\tfrac{1}{2}G\{i\sin(N-\psi-g)-j\sin g\} = \tfrac{1}{4}[j+i\cos(N-\psi)]\sin 2g \quad \ldots(145)$$

$$\tfrac{1}{2}j'\frac{dW_2}{d\epsilon'}\Big/\frac{\tau^2}{\mathfrak{g}} = 0 \text{ absolutely} \quad \ldots(146)$$

Collecting results from the six equations (141–6), we have for the whole perturbation of the moon by the lunar tides

$$\left(\frac{1}{j'}\frac{dW}{dN'}+\tfrac{1}{2}j'\frac{dW}{d\epsilon'}\right)\Big/\frac{\tau^2}{\mathfrak{g}} = \tfrac{1}{4}[j+i\cos(N-\psi)](\sin 4f_1-\sin 2g_1+\sin 2g) \ldots(147)$$

Next take the case (ii), and suppose that the sun is the tide-raiser. Here we need only consider W_2. Noting that $dW_2/d\epsilon'=0$ absolutely, we have from (131)

$$\left(\frac{1}{j'}\frac{dW_2}{dN'}+\tfrac{1}{2}j'\frac{dW_2}{d\epsilon'}\right)\Big/\frac{\tau\tau'}{\mathfrak{g}} = -\tfrac{1}{2}G\{i\sin(N'-\psi'-g)-j\sin(N-N'+g)\}$$

Accented symbols here refer to the moon (as perturbed), unaccented to the sun (as tide-raiser). Therefore $j=0$. Then reverting to the usual notation by shifting accents and dropping useless terms, this expression becomes

$$+\tfrac{1}{4}i\cos(N-\psi)\sin 2g \quad \ldots(148)$$

Collecting results from (147–8), we have by (13)

$$\frac{\xi}{k}\frac{dj}{dt} = -\tfrac{1}{4}[j+i\cos(N-\psi)]\frac{\tau^2}{\mathfrak{g}}(\sin 4f_1-\sin 2g_1+\sin 2g) - \tfrac{1}{4}i\cos(N-\psi)\frac{\tau\tau'}{\mathfrak{g}}\sin 2g \quad \ldots(149)$$

This gives the additional terms due to bodily tides in the equation for dj/dt, viz.: the first of (133).

If the viscosity be small

$$\left.\begin{aligned} \sin 4\mathfrak{f}_1 - \sin 2\mathfrak{g}_1 + \sin 2\mathfrak{g} &= \sin 4\mathfrak{f} \\ \sin 2\mathfrak{g} &= \tfrac{1}{2}\sin 4\mathfrak{f} \end{aligned}\right\} \quad \text{..............(150)}$$

Before proceeding further it may be remarked that to the present order of approximation in case (i)

$$\frac{dW}{d\epsilon'} = \tfrac{1}{2}\sin 4\mathfrak{f}_1$$

and in case (ii) it is zero; thus by (11)

$$\frac{1}{k}\frac{d\xi}{dt} = \tfrac{1}{2}\frac{\tau^2}{\mathfrak{g}}\sin 4\mathfrak{f}_1 \quad \text{..........................(151)}$$

We now turn to the perturbations of the earth's rotation.

Here we have to find dW/di' and

$$\frac{dW}{\tan i\, d\chi'} - \frac{dW}{\sin i\, d\psi'} \quad \text{or} \quad (1 - \tfrac{1}{2}i^2)\frac{dW}{i\, d\chi'} - \frac{dW}{i\, d\psi'}$$

and in the former may drop terms in i and $j\cos(N - \psi)$, and in the latter terms in $j \sin (N - \psi)$.

First take the case (iii), where the moon is tide-raiser and disturber. Here we may take $N = N'$, $\epsilon = \epsilon'$, $j = j'$ throughout, and after differentiation may drop the accents to all the symbols.

From (129)

$$\frac{dW_{\text{I}}}{di'}\Big/\frac{\tau^2}{\mathfrak{g}} = -\tfrac{1}{2}F_1 \{i \cos 2\mathfrak{f}_1 + j\cos (N - \psi + 2\mathfrak{f}_1)\} = \tfrac{1}{4}j\sin(N - \psi)\sin 4\mathfrak{f}_1 \quad \text{......(152)}$$

From (130)

$$\frac{dW_1}{di'}\Big/\frac{\tau^2}{\mathfrak{g}} = \tfrac{1}{2}G_1 \{i\cos \mathfrak{g}_1 + j\cos(N - \psi - \mathfrak{g}_1)\} = \tfrac{1}{4}j\sin(N-\psi)\sin 2\mathfrak{g}_1 \quad \text{......(153)}$$

From (131)

$$\frac{dW_2}{di'}\Big/\frac{\tau^2}{\mathfrak{g}} = \tfrac{1}{2}G\{i\cos\mathfrak{g} + j\cos(N - \psi + \mathfrak{g})\} = -\tfrac{1}{4}j\sin(N-\psi)\sin 2\mathfrak{g} \quad \text{......(154)}$$

Therefore from (152–4) we have for the whole perturbation of the earth, due to attraction of the moon on the lunar tides,

$$\frac{dW}{di'}\Big/\frac{\tau^2}{\mathfrak{g}} = \tfrac{1}{4}j\sin(N-\psi)\,[\sin 4\mathfrak{f}_1 + \sin 2\mathfrak{g}_1 - \sin 2\mathfrak{g}] \quad \text{......(155)}$$

The result for case (iv), where the sun is both tide-raiser and disturber, may be written down by symmetry; and since $j = 0$ here, therefore

$$\frac{dW}{di'}\Big/\frac{\tau'^2}{\mathfrak{g}} = 0 \quad \text{..............................(156)}$$

Next take the cases (v) and (vi), where the tide-raiser and disturber are distinct. Here we need only consider W_2.

From (131) $\dfrac{dW_2}{di'}\Big/\dfrac{\tau\tau'}{\mathfrak{g}} = \frac{1}{2}G\{i\cos g + j\cos(N-\psi+g)\}$

When the moon is tide-raiser and sun disturber, this becomes

$$-\tfrac{1}{4}j\sin(N-\psi)\sin 2g \;\ldots\ldots\ldots\ldots\ldots\ldots\ldots\ldots(157)$$

When sun is tide-raiser and moon disturber it becomes zero.

Collecting results from (155–7), we have by (18)

$$n\sin i\,\frac{d\psi}{dt} = \tfrac{1}{4}j\sin(N-\psi)\left[\frac{\tau^2}{\mathfrak{g}}(\sin 4f_1 + \sin 2g_1 - \sin 2g) - \frac{\tau\tau'}{\mathfrak{g}}\sin 2g\right] \;\ldots\ldots\ldots(158)$$

This gives the additional terms due to bodily tides in the equation for $d\psi/dt$, viz.: the last of (133).

If the viscosity be small

$$\left.\begin{aligned} \sin 4f_1 + \sin 2g_1 - \sin 2g &= \sin 4f(1-2\lambda) \\ \sin 2g &= \tfrac{1}{2}\sin 4f \\ \text{where}\qquad \lambda &= \frac{\Omega}{n} \end{aligned}\right\} \;\ldots\ldots\ldots\ldots(159)$$

Next consider the change in the obliquity of the ecliptic; for this purpose we must find $(1-\frac{1}{2}i^2)\,dW/id\chi' - dW/id\psi'$, and may drop terms involving $j\sin(N-\psi)$.

First take the case (iii), where the moon is both tide-raiser and disturber.

From (129)

$$\frac{dW_I}{d\chi'}\Big/\frac{\tau^2}{\mathfrak{g}} = -F_1\{(1-i^2-j^2)\sin 2f_1 + ij\sin(N-\psi-2f_1) - ij\sin(N-\psi+2f_1)\} \;\ldots(160)$$

$$-\frac{dW_I}{d\psi'}\Big/\frac{\tau^2}{\mathfrak{g}} = F_1\{(1-i^2-j^2)\sin 2f_1 + ij\sin(N-\psi-2f_1) - \tfrac{1}{2}ij\sin(N-\psi+2f_1)\}$$

$$-\tfrac{1}{2}i^2\frac{dW_I}{d\chi'}\Big/\frac{\tau^2}{\mathfrak{g}} = F_1\,\tfrac{1}{2}i^2\sin 2f_1$$

Therefore

$$\left[\frac{1}{i}(1-\tfrac{1}{2}i^2)\frac{dW_I}{d\chi'} - \frac{1}{i}\frac{dW_I}{d\psi'}\right]\Big/\frac{\tau^2}{\mathfrak{g}} = \tfrac{1}{2}F_1\{i\sin 2f_1 + j\sin(N-\psi+2f_1)\}$$

$$= \tfrac{1}{4}[i + j\cos(N-\psi)]\sin 4f_1 \;\ldots\ldots(161)$$

From (130)

$$\frac{dW_1}{d\chi'}\Big/\frac{\tau^2}{\mathfrak{g}} = -\tfrac{1}{2}G_1\{i^2\sin g_1 - ij\sin(N-\psi-g_1) + ij\sin(N-\psi+g_1) + j^2\sin g_1\} \;\ldots(162)$$

$$-\frac{dW_1}{d\psi'}\Big/\frac{\tau^2}{\mathfrak{g}}=\tfrac{1}{2}G_1\{2i^2\sin g_1-2ij\sin(N-\psi-g_1)+ij\sin(N-\psi+g_1)+j^2\sin g_1\}$$

$$-\tfrac{1}{2}i^2\frac{dW_1}{d\chi'}\Big/\frac{\tau^2}{\mathfrak{g}}=0$$

Therefore

$$\left[\frac{1}{i}(1-\tfrac{1}{2}i^2)\frac{dW_1}{d\chi'}-\frac{1}{i}\frac{dW_1}{d\psi'}\right]\Big/\frac{\tau^2}{\mathfrak{g}}=\tfrac{1}{2}G\{i\sin g_1-j\sin(N-\psi-g_1)\}$$
$$=\tfrac{1}{4}[i+j\cos(N-\psi)]\sin 2g_1 \quad\text{......(163)}$$

From (131)

$$\frac{dW_2}{d\chi'}\Big/\frac{\tau^2}{\mathfrak{g}}=-\tfrac{1}{2}G\{i^2\sin g+ij\sin(N-\psi+g)-ij\sin(N-\psi-g)+j^2\sin g\} \quad\text{......(164)}$$

$$-\frac{dW_2}{d\psi'}\Big/\frac{\tau^2}{\mathfrak{g}}=\tfrac{1}{2}G\{-ij\sin(N-\psi-g)+j^2\sin g\}$$

$$-\tfrac{1}{2}i^2\frac{dW_2}{d\chi'}\Big/\frac{\tau^2}{\mathfrak{g}}=0$$

Therefore

$$\left[\frac{1}{i}(1-\tfrac{1}{2}i^2)\frac{dW_2}{d\chi'}-\frac{1}{i}\frac{dW_2}{d\psi'}\right]\Big/\frac{\tau^2}{\mathfrak{g}}=-\tfrac{1}{2}G\{i\sin g+j\sin(N-\psi+g)\}$$
$$=-\tfrac{1}{4}[i+j\cos(N-\psi)]\sin 2g \quad\text{...(165)}$$

Collecting results from (161–3–5), we have for the whole perturbation of the earth due to the attraction of the moon on the lunar tides,

$$\left[\frac{1}{i}(1-\tfrac{1}{2}i^2)\frac{dW}{d\chi'}-\frac{1}{i}\frac{dW}{d\psi'}\right]\Big/\frac{\tau^2}{\mathfrak{g}}=\tfrac{1}{4}[i+j\cos(N-\psi)](\sin 4f_1+\sin 2g_1-\sin 2g) \quad\text{......(166)}$$

The result for case (iv), where the sun is both tide-raiser and disturber, may be written down by symmetry; and since $j=0$ here, therefore

$$\left[\frac{1}{i}(1-\tfrac{1}{2}i^2)\frac{dW}{d\chi'}-\frac{1}{i}\frac{dW}{d\psi'}\right]\Big/\frac{\tau'^2}{\mathfrak{g}}=\tfrac{1}{4}i\sin 4f \quad\text{...........(167)}$$

It is here assumed that the solar slow diurnal tide has the same lag as the sidereal diurnal tide, and that the solar slow semi-diurnal tide has the same lag as the sidereal semi-diurnal tide. This is very nearly true, because Ω' is small compared with n.

Next take the cases (v) and (vi), where the tide-raiser and disturber are distinct. Here we need only consider W_2.

$$\frac{dW_2}{d\chi'}\Big/\frac{\tau\tau'}{\mathfrak{g}}=-\tfrac{1}{2}G\{i^2\sin g+ij\sin(N-\psi+g)-ij'\sin(N'-\psi'-g)+jj'\sin(N-N'+g)\} \quad\text{...(168)}$$

$$-\frac{dW_2}{d\psi'}\Big/\frac{\tau\tau'}{\mathfrak{g}} = \tfrac{1}{2}G\{-ij'\sin(N'-\psi'-g)+jj'\sin(N-N'+g)\}$$

$$-\tfrac{1}{2}i^2\frac{dW_2}{d\chi'}\Big/\frac{\tau\tau'}{\mathfrak{g}} = 0$$

Therefore

$$\left[\frac{1}{i}(1-\tfrac{1}{2}i^2)\frac{dW_2}{d\chi'}-\frac{1}{i}\frac{dW_2}{d\psi'}\right]\Big/\frac{\tau\tau'}{\mathfrak{g}} = -\tfrac{1}{2}G\{i\sin g + j\sin(N-\psi+g)\}$$

When the moon is tide-raiser and the sun disturber, this becomes

$$-\tfrac{1}{4}[i+j\cos(N-\psi)]\sin 2g \ldots\ldots\ldots\ldots\ldots\ldots(169)$$

When the sun is tide-raiser and the moon disturber, this becomes

$$-\tfrac{1}{4}i\sin 2g \ldots\ldots\ldots\ldots\ldots\ldots\ldots\ldots\ldots\ldots(170)$$

Collecting results from (166–7–9, 170), we have by (18),

$$\left.\begin{aligned} n\frac{di}{dt} = \tfrac{1}{4}[i+j\cos(N-\psi)]&\left[\frac{\tau^2}{\mathfrak{g}}(\sin 4f_1+\sin 2g_1-\sin 2g)-\frac{\tau\tau'}{\mathfrak{g}}\sin 2g\right]\\ &+\tfrac{1}{4}i\left[\frac{\tau'^2}{\mathfrak{g}}\sin 4f-\frac{\tau\tau'}{\mathfrak{g}}\sin 2g\right]\end{aligned}\right\}(171)$$

This gives the additional terms due to bodily tides in the equation for di/dt, viz.: the third of (133).

If the viscosity be small

$$\left.\begin{aligned} \sin 4f_1+\sin 2g_1-\sin 2g &= \sin 4f(1-2\lambda)\\ \sin 2g &= \tfrac{1}{2}\sin 4f\\ \text{where}\qquad \lambda &= \frac{\Omega}{n}\end{aligned}\right\}\ldots\ldots\ldots\ldots(172)$$

Also we have from (160–2–4–8) to the present order of approximation,

$$\frac{dW}{d\chi'}\Big/\frac{\tau^2}{\mathfrak{g}} = -\tfrac{1}{2}\sin 4f_1$$

and by symmetry,

$$\frac{dW}{d\chi'}\Big/\frac{\tau'^2}{\mathfrak{g}} = -\tfrac{1}{2}\sin 4f$$

Therefore by (18)

$$-\frac{dn}{dt} = \tfrac{1}{2}\left[\frac{\tau^2}{\mathfrak{g}}\sin 4f_1+\frac{\tau'^2}{\mathfrak{g}}\sin 4f\right]\ldots\ldots\ldots\ldots\ldots\ldots(173)$$

Now let

$$\left.\begin{aligned} \Gamma &= \tfrac{1}{4}\frac{k}{\xi}\frac{\tau^2}{\mathfrak{g}}(\sin 4f_1-\sin 2g_1+\sin 2g)\\ G &= \tfrac{1}{4}\frac{k}{\xi}\left[\frac{\tau^2}{\mathfrak{g}}(\sin 4f_1-\sin 2g_1+\sin 2g)+\frac{\tau\tau'}{\mathfrak{g}}\sin 2g\right]\\ \Delta &= \frac{1}{4n}\left[\frac{\tau^2}{\mathfrak{g}}(\sin 4f_1+\sin 2g_1-\sin 2g)+\frac{\tau'^2}{\mathfrak{g}}\sin 4f-2\frac{\tau\tau'}{\mathfrak{g}}\sin 2g\right]\\ D &= \frac{1}{4n}\left[\frac{\tau^2}{\mathfrak{g}}(\sin 4f_1+\sin 2g_1-\sin 2g)-\frac{\tau\tau'}{\mathfrak{g}}\sin 2g\right]\end{aligned}\right\}\ldots(174)$$

Then the four equations (139), (149), (158), and (171) may be written

$$\left.\begin{aligned}
j\frac{dN}{dt} &= \mathrm{G}i\sin(N-\psi)\\
\frac{dj}{dt} &= -\Gamma j - \mathrm{G}i\cos(N-\psi)\\
i\frac{d\psi}{dt} &= \mathrm{D}j\sin(N-\psi)\\
\frac{di}{dt} &= \Delta i + \mathrm{D}j\cos(N-\psi)
\end{aligned}\right\}\quad\ldots\ldots\ldots\ldots(175)$$

Also from (151) and (173)

$$\left.\begin{aligned}
\frac{1}{k}\frac{d\xi}{dt} &= \tfrac{1}{2}\frac{\tau^2}{\mathfrak{g}}\sin 4\mathrm{f}_1\\
-\frac{dn}{dt} &= \tfrac{1}{2}\frac{\tau^2}{\mathfrak{g}}\sin 4\mathrm{f}_1 + \tfrac{1}{2}\frac{\tau'^2}{\mathfrak{g}}\sin 4\mathrm{f}
\end{aligned}\right\}\quad\ldots\ldots\ldots\ldots(176)$$

These six equations (175–6) contain all the secular inequalities in the motions of the moon and earth, due to the bodily tides raised by the sun and moon, as far as is material for the present investigation. The terms which are omitted only represent very small displacements of the proper planes and of the inclinations of the planes of motion of the two parts of the system to those proper planes.

Reverting to the earlier notation in which

$$\left.\begin{aligned}
y &= j\sin N, \quad \eta = i\sin\psi\\
z &= j\cos N, \quad \zeta = i\cos\psi
\end{aligned}\right\}\quad\ldots\ldots\ldots\ldots(177)$$

we easily find

$$\left.\begin{aligned}
\frac{dz}{dt} &= -\Gamma z - \mathrm{G}\zeta\\
\frac{dy}{dt} &= -\Gamma y - \mathrm{G}\eta\\
\frac{d\zeta}{dt} &= \Delta\zeta + \mathrm{D}z\\
\frac{d\eta}{dt} &= \Delta\eta + \mathrm{D}y
\end{aligned}\right\}\quad\ldots\ldots\ldots\ldots(178)$$

These equations contain the additional terms due to tides, which are to be added to the equations (116), in order to find the secular displacements of the proper planes.

The first application, which will be made hereafter, will be to the case where the viscosity is small, and it will be more convenient to make the transition to that hypothesis at present, although the greater part of what follows in this part will be equally applicable whatever may be the viscosity.

In the case of small viscosity the functions Γ, Δ, G, D will be indicated by the corresponding small letters γ, δ, g, d.

By (140), (150), (159), (172) we have

$$\left.\begin{aligned} &\gamma = \tfrac{1}{4}\frac{k}{\xi}\frac{\sin 4\mathrm{f}}{\mathfrak{g}}[\tau^2], && \mathrm{g} = \tfrac{1}{4}\frac{k}{\xi}\frac{\sin 4\mathrm{f}}{\mathfrak{g}}[\tau^2 + \tfrac{1}{2}\tau\tau'] \\ &\delta = \frac{1}{4n}\frac{\sin 4\mathrm{f}}{\mathfrak{g}}[\tau^2(1-2\lambda) + \tau'^2 - \tau\tau'], && \mathrm{d} = \frac{1}{4n}\frac{\sin 4\mathrm{f}}{\mathfrak{g}}[\tau^2(1-2\lambda) - \tfrac{1}{2}\tau\tau'] \\ &\text{where } \lambda = \frac{\Omega}{n} \end{aligned}\right\} \quad \text{......(179)}$$

In the present case where i and j are small, we have by (112) and (121)

$$\left.\begin{aligned} &\alpha = \frac{k}{\xi}\tau\mathfrak{e} + \tfrac{1}{2}\frac{\tau'}{\Omega}, \quad \beta = \frac{\tau+\tau'}{n}\mathfrak{e} \\ &\mathrm{a} = \frac{k}{\xi}\tau\mathfrak{e}, \qquad \mathrm{b} = \frac{\tau\mathfrak{e}}{n} \\ &\text{where} \quad \mathfrak{e} = \tfrac{1}{2}\frac{n^2}{\mathfrak{g}}, \text{ the permanent ellipticity of the earth} \end{aligned}\right\} \quad \text{......(180)}$$

These equations (180) are the same whether the viscosity be supposed small or not.

The complete equations are

$$\left.\begin{aligned} \frac{dz}{dt} &= \alpha y + \mathrm{a}\eta - (\gamma z + \mathrm{g}\zeta) \\ \frac{dy}{dt} &= -(\alpha z + \mathrm{a}\zeta) - (\gamma y + \mathrm{g}\eta) \\ \frac{d\zeta}{dt} &= \beta\eta + \mathrm{b}y + \delta\zeta + \mathrm{d}z \\ \frac{d\eta}{dt} &= -(\beta\zeta + \mathrm{b}z) + \delta\eta + \mathrm{d}y \end{aligned}\right\} \quad \text{..................(181)}$$

If the viscosity be not small we have Γ, G, Δ, D in place of γ, g, δ, d. As it is more convenient to write small letters than capitals, in the whole of the next section the small letters will be employed, although the same investigation would be equally applicable with Γ, G, &c., in place of γ, g, &c.

The terms in γ, g, δ, d are small compared with those in α, a, β, b, and may be neglected as a first approximation. Also α, a, β, b vary slowly in consequence of tidal reaction, tidal friction, and the consequent change of ellipticity of the earth, but as a first approximation they may be treated as constant.

If we put

$$\left.\begin{aligned} &z_1 = L_1 \cos(\kappa_1 t + \mathrm{m}_1), && z_2 = L_2 \cos(\kappa_2 t + \mathrm{m}_2) \\ &y_1 = L_1 \sin(\kappa_1 t + \mathrm{m}_1), && y_2 = L_2 \sin(\kappa_2 t + \mathrm{m}_2) \\ &\zeta_1 = L_1' \cos(\kappa_1 t + \mathrm{m}_1), && \zeta_2 = L_2' \cos(\kappa_2 t + \mathrm{m}_2) \\ &\eta_1 = L_1' \sin(\kappa_1 t + \mathrm{m}_1), && \eta_2 = L_2' \sin(\kappa_2 t + \mathrm{m}_2) \end{aligned}\right\} \quad \text{.........(182)}$$

by (122) or (118) the first approximation is

$$\left.\begin{aligned} z=z_1+z_2,\quad y=y_1+y_2,\quad \zeta=\zeta_1+\zeta_2,\quad \eta=\eta_1+\eta_2 \\ \text{where}\quad \frac{L_1'}{L_1}=-\frac{\kappa_1+\alpha}{\mathrm{a}}=-\frac{\mathrm{b}}{\kappa_1+\beta},\quad \frac{L_2'}{L_2}=-\frac{\kappa_2+\alpha}{\mathrm{a}}=-\frac{\mathrm{b}}{\kappa_2+\beta} \end{aligned}\right\}\text{......(183)}$$

Before considering the secular changes in the constants L of integration, it will be convenient to take one other step.

The equation of tidal friction (173) may be written approximately

$$-\frac{dn}{dt}=\tfrac{1}{2}\frac{\tau^2+\tau'^2}{\mathfrak{g}}\sin 4\mathrm{f}_1 \text{ ....(184)}$$

because sin 4f will be nearly equal to sin $4\mathrm{f}_1$ as long as τ'^2 is not small compared with τ^2. (See however § 22, Part IV.)

Also the equation of tidal reaction (151) is

$$\frac{1}{k}\frac{d\xi}{dt}=\tfrac{1}{2}\frac{\tau^2}{\mathfrak{g}}\sin 4\mathrm{f}_1 \text{ ....(185)}$$

Dividing one by the other and putting $\tau^2=\tau_0^{\,2}\xi^{-12}$, we have

$$k\frac{dn}{d\xi}=1+\left(\frac{\tau'}{\tau_0}\right)^2\xi^{12}$$

and integrating,

$$\frac{n}{n_0}=1+\frac{1}{kn_0}\left[(1-\xi)+\tfrac{1}{13}\left(\frac{\tau'}{\tau_0}\right)^2(1-\xi^{13})\right] \text{ ....(186)}$$

This is the equation of conservation of moment of momentum of the moon-earth system, as modified by solar tidal friction. From it we obtain n in terms of ξ.

§ 15. *On the secular changes of the constants of integration.*

It is often found difficult on first reading a long analytical investigation to trace the general method amidst the mass of detail, and it is only at the end that the ruling idea is perceived; in such circumstances it has often appeared to me that a preliminary sketch would be of great service to the reader. I shall act on this idea here, and consider some simple equations analogous to those to be treated.

Let the equations be $\dfrac{dz}{dt}=\alpha y,\quad \dfrac{dy}{dt}=-\alpha z$

If α be constant, the solution is obviously

$$z=\mathrm{L}\cos(\alpha t+\mathrm{m}),\quad y=-\mathrm{L}\sin(\alpha t+\mathrm{m})$$

Now suppose α to be slowly varying; put therefore $\alpha+\alpha' t$ for α, and treat α, α' as constants.

Then $$\frac{dz}{dt} = \alpha y + \alpha' t y, \quad \frac{dy}{dt} = -\alpha z - \alpha' t z$$

By differentiation $$\frac{d^2 z}{dt^2} + \alpha^2 z = -\alpha' t\left(\alpha z - \frac{dy}{dt}\right) + \alpha' y$$

$$\frac{d^2 y}{dt^2} + \alpha^2 y = -\alpha' t\left(\alpha y + \frac{dz}{dt}\right) - \alpha' z$$

The terms on the right-hand side of these equations are small, because they involve α', and therefore we may substitute in them from the first approximation.

Hence $$\frac{d^2 z}{dt^2} + \alpha^2 z = -\alpha' L \sin(\alpha t + \mathrm{m}) - 2\alpha' \alpha t L \cos(\alpha t + \mathrm{m})$$

and a similar equation for y.

The solution of this equation is

$$z = L\cos(\alpha t + \mathrm{m}) + \tfrac{1}{2}\frac{\alpha'}{\alpha} Lt\cos(\alpha t + \mathrm{m}) - \tfrac{1}{2}\frac{\alpha'}{\alpha} Lt\cos(\alpha t + \mathrm{m}) - \tfrac{1}{2}\alpha' L t^2 \sin(\alpha t + \mathrm{m})$$

The terms depending on t cut one another out, and

$$z = L\cos(\alpha t + \mathrm{m}) - \tfrac{1}{2}\alpha' L t^2 \sin(\alpha t + \mathrm{m})$$

Similarly we should find

$$y = -L\sin(\alpha t + \mathrm{m}) - \tfrac{1}{2}\alpha' L t^2 \cos(\alpha t + \mathrm{m})$$

The terms in t^2 are obviously equivalent to a change in m, the phase of the oscillation; but the amplitude L is unaffected. We might have arrived at this conclusion about the amplitude if, in solving the differential equations, we had neglected in the solutions the terms depending on t^2, as will be done in considering our equations below. In those equations, however, we shall not find that the terms in t annihilate one another, and thus there will be a change of amplitude.

That this conclusion concerning amplitude is correct, may be seen from the fact that the rigorous solution of the equations

$$\frac{dz}{dt} = \alpha y, \quad \frac{dy}{dt} = -\alpha z$$

is $$z = L\cos(\int \alpha dt + \mathrm{m}_0), \qquad y = -L\sin(\int \alpha dt + \mathrm{m}_0)$$

$$= L\cos(\alpha t + \mathrm{m}_0 - \int \alpha' t dt), \qquad = -L\sin(\alpha t + \mathrm{m}_0 - \int \alpha' t dt)$$

Whence L is unaffected, whilst

$$\mathrm{m} = \mathrm{m}_0 - \int \alpha' t dt$$

So that $$\frac{d\mathrm{m}}{dt} = -t\frac{d\alpha}{dt}$$

Next consider the equations

$$\frac{dz}{dt} = \alpha y - \gamma z, \quad \frac{dy}{dt} = -\alpha z - \gamma y$$

where α is constant, but γ is a very small quantity compared with α, which may vary slowly.

Treat γ as constant, and differentiate, and we have

$$\frac{d^2z}{dt^2} + \alpha^2 z = -\gamma\left(\frac{dz}{dt} + \alpha y\right)$$

$$\frac{d^2y}{dt^2} + \alpha^2 y = -\gamma\left(\frac{dy}{dt} - \alpha z\right)$$

If we neglect γ, we have the first approximation

$$z = L\cos(\alpha t + \mathrm{m}), \quad y = -L\sin(\alpha t + \mathrm{m})$$

Substituting these values for z, y on the right, we have

$$\frac{d^2z}{dt^2} + \alpha^2 z = 2\gamma\alpha L\sin(\alpha t + \mathrm{m})$$

And a similar equation for y.

The solutions are

$$z = \quad L\cos(\alpha t + \mathrm{m}) - \gamma L t\cos(\alpha t + \mathrm{m})$$
$$y = -L\sin(\alpha t + \mathrm{m}) + \gamma L t\sin(\alpha t + \mathrm{m})$$

From this we see that, if we desire to retain the first approximation as the solution, we must have

$$\frac{1}{L}\frac{dL}{dt} = -\gamma \quad \text{..............................(187)}$$

This will be true if γ varies slowly; hence

$$L = L_0 e^{-\int\gamma dt}$$

and the solution is

$$z = \quad L_0 e^{-\int\gamma dt}\cos(\alpha t + \mathrm{m})$$
$$y = -L_0 e^{-\int\gamma dt}\sin(\alpha t + \mathrm{m})$$

It is easy to verify that these are the rigorous solutions of the equations, when α is constant but γ varies.

The equation (187) gives the rate of change of amplitude of oscillation.

The cases which we have now considered, by the method of variation of parameters, are closely analogous to those to be treated below, and have been treated in the same way, so that the reader will be able to trace the process.

They are in fact more than simply analogous, for they are what our equations (181) become if the obliquity of the ecliptic be zero and $\zeta = 0$, $\eta = 0$. In this case $L = j$, and $dj/dt = -j\gamma$.

This shows that the secular change of figure of the earth, and the secular changes in the rate of revolution of the moon's nodes do not affect the rate of alteration of the inclination of the lunar orbit to the ecliptic, so long as the obliquity is zero. This last result contains the implicit assumption that the perturbing influence of the moon on the earth is not so large, but that the obliquity of the equator may always remain small, however the lunar nodes vary. In an exactly similar manner we may show that, if the inclination of the lunar orbit be zero, $di/dt = i\delta$. This is the result of the previous paper "On the Precession of a Viscous Spheroid," when the obliquity is small.

According to the method which has been sketched, the equations to be integrated are given in (181), when we write $\alpha + \alpha' t$ for α, $\mathrm{a} + \mathrm{a}' t$ for a, $\beta + \beta' t$ for β, $\mathrm{b} + \mathrm{b}' t$ for b, and then treat α, a, &c., α', a', &c., γ, g, &c., as constants.

Before proceeding to consider the equations, it will be convenient to find certain relations between the quantities α, a, &c., and the two roots κ_1 and κ_2 of the quadratic $(\kappa + \alpha)(\kappa + \beta) = \mathrm{ab}$.

We have supposed the two roots to be such that

$$\left.\begin{aligned} \kappa_1 + \kappa_2 &= -\alpha - \beta \\ \kappa_1 - \kappa_2 &= -\sqrt{(\alpha - \beta)^2 + 4\mathrm{ab}} \end{aligned}\right\} \quad \text{.....................(188)}$$

Then

$$\kappa_1 \kappa_2 = (\alpha\beta - \mathrm{ab}) \quad \text{.........................(189)}$$

$$\left.\begin{aligned} \kappa_1^2 + \kappa_2^2 &= \alpha^2 + \beta^2 + 2\mathrm{ab} \\ \kappa_1^2 \kappa_2^2 &= (\alpha^2 + \mathrm{ab})(\beta^2 + \mathrm{ab}) - \mathrm{ab}(\alpha + \beta)^2 \end{aligned}\right\} \quad \text{...........(190)}$$

$$\left.\begin{aligned} \beta^2 + \mathrm{ab} - \kappa_1^2 &= (\kappa_1 + \kappa_2)(\kappa_2 + \alpha) \\ \beta^2 + \mathrm{ab} - \kappa_2^2 &= (\kappa_1 + \kappa_2)(\kappa_1 + \alpha) \\ \alpha^2 + \mathrm{ab} - \kappa_1^2 &= (\kappa_1 + \kappa_2)(\kappa_2 + \beta) \\ \alpha^2 + \mathrm{ab} - \kappa_2^2 &= (\kappa_1 + \kappa_2)(\kappa_1 + \beta) \end{aligned}\right\} \quad \text{................(191)}$$

$$\left.\begin{aligned} \kappa_1 + \alpha &= -(\kappa_2 + \beta) \\ \kappa_2 + \alpha &= -(\kappa_1 + \beta) \end{aligned}\right\} \quad \text{.......................(192)}$$

$$\mathrm{ab}(\alpha + \beta) = (\kappa_1 + \alpha)(\kappa_2 + \alpha)(\kappa_1 + \kappa_2) \quad \text{..............(193)}$$

Now suppose our equations (181) to be written as follows:

$$\left.\begin{aligned} \frac{dz}{dt} &= \alpha y + \mathrm{a}\eta + s \\ \frac{dy}{dt} &= -\alpha z - \mathrm{a}\zeta + u \\ \frac{d\zeta}{dt} &= \beta\eta + \mathrm{b}y + \sigma \\ \frac{d\eta}{dt} &= -\beta\zeta - \mathrm{b}z + v \end{aligned}\right\} \quad \text{.......................(194)}$$

where s, u, σ, v comprise all the terms involving α', a', &c., γ, g, &c.

If we write (z) as a type of z, y, ζ, η; (α) as a type of α, a, β, b; (α') as a type of α', a$'$, β', b$'$; (γ) as a type of γ, g, δ, d; and (s) as a type of s, u, σ, υ; it is clear that (s) is $(z)(\alpha')t + (\gamma)(z)$.

Differentiate each of the equations (194), substitute for $\frac{d(z)}{dt}$ after differentiation, and write

$$\left.\begin{aligned} \mathrm{S} &= \frac{ds}{dt} + \alpha u + \mathrm{a}\upsilon \\ \mathrm{U} &= \frac{du}{dt} - \alpha s - \mathrm{a}\sigma \\ \Sigma &= \frac{d\sigma}{dt} + \beta\upsilon + \mathrm{b}u \\ \Upsilon &= \frac{d\upsilon}{dt} - \beta\sigma - \mathrm{b}s \end{aligned}\right\} \qquad (195)$$

The result is

$$\left.\begin{aligned} \frac{d^2 z}{dt^2} &= -(\alpha^2 + \mathrm{ab})\, z - \mathrm{a}(\alpha+\beta)\,\zeta + \mathrm{S} \\ \frac{d^2 y}{dt^2} &= -(\alpha^2 + \mathrm{ab})\, y - \mathrm{a}(\alpha+\beta)\,\eta + \mathrm{U} \\ \frac{d^2 \zeta}{dt^2} &= -(\beta^2 + \mathrm{ab})\,\zeta - \mathrm{b}(\alpha+\beta)\, z + \Sigma \\ \frac{d^2 \eta}{dt^2} &= -(\beta^2 + \mathrm{ab})\,\eta - \mathrm{b}(\alpha+\beta)\, y + \Upsilon \end{aligned}\right\} \qquad (196)$$

From the first of these

$$-(\beta^2 + \mathrm{ab})\,\mathrm{a}(\alpha+\beta)\,\zeta = (\beta^2+\mathrm{ab})\frac{d^2 z}{dt^2} + (\alpha^2+\mathrm{ab})(\beta^2+\mathrm{ab})\,z - \mathrm{S}(\beta^2+\mathrm{ab})$$

Therefore from the third

$$\mathrm{a}(\alpha+\beta)\frac{d^2\zeta}{dt^2} = (\beta^2+\mathrm{ab})\frac{d^2 z}{dt^2} + \{(\alpha^2+\mathrm{ab})(\beta^2+\mathrm{ab}) - \mathrm{ab}(\alpha+\beta)^2\}\,z - \mathrm{S}(\beta^2+\mathrm{ab}) + \Sigma\mathrm{a}(\alpha+\beta)$$

and by (190)

$$\left.\begin{aligned} \mathrm{a}(\alpha+\beta)\frac{d^2\zeta}{dt^2} &= (\beta^2+\mathrm{ab})\frac{d^2 z}{dt^2} + \kappa_1^2\kappa_2^2 z - \mathrm{S}(\beta^2+\mathrm{ab}) + \Sigma\mathrm{a}(\alpha+\beta) \\ &\text{Similarly} \\ \mathrm{a}(\alpha+\beta)\frac{d^2\eta}{dt^2} &= (\beta^2+\mathrm{ab})\frac{d^2 y}{dt^2} + \kappa_1^2\kappa_2^2 y - \mathrm{U}(\beta^2+\mathrm{ab}) + \Upsilon\mathrm{a}(\alpha+\beta) \\ \mathrm{b}(\alpha+\beta)\frac{d^2 z}{dt^2} &= (\alpha^2+\mathrm{ab})\frac{d^2\zeta}{dt^2} + \kappa_1^2\kappa_2^2\zeta - \Sigma(\alpha^2+\mathrm{ab}) + \mathrm{Sb}(\alpha+\beta) \\ \mathrm{b}(\alpha+\beta)\frac{d^2 y}{dt^2} &= (\alpha^2+\mathrm{ab})\frac{d^2\eta}{dt^2} + \kappa_1^2\kappa_2^2\eta - \Upsilon(\alpha^2+\mathrm{ab}) + \mathrm{Ub}(\alpha+\beta) \end{aligned}\right\} \qquad (197)$$

Differentiate the first of (196) twice, using the first of (197), and we have

$$\frac{d^4z}{dt^4} = -(\alpha^2 + \text{ab})\frac{d^2z}{dt^2} - (\beta^2 + \text{ab})\frac{d^2z}{dt^2} - \kappa_1^2\kappa_2^2 z + \left(\beta^2 + \text{ab} + \frac{d^2}{dt^2}\right)\text{S} - \Sigma\text{a}(\alpha + \beta)$$

Therefore by (190)

$$\left[\frac{d^4}{dt^4} + (\kappa_1^2 + \kappa_2^2)\frac{d^2}{dt^2} + \kappa_1^2\kappa_2^2\right] z = \left(\beta^2 + \text{ab} + \frac{d^2}{dt^2}\right)\text{S} - \Sigma\text{a}(\alpha + \beta)$$

Writing (S) as a type of S, Σ, U, Υ,

(S) is of the type $(z)(\alpha)(\alpha')t + (\alpha)(\gamma)(z) + (\alpha')(z) + \frac{d(z)}{dt}(\alpha')t + (\gamma)\frac{d(z)}{dt}$

Hence every term of (S) contains some small term, either (α') or (γ).

Therefore on the right-hand side of the above equation we may substitute for (z) the first approximation, viz.: $(z_1) + (z_2)$ given in (182–3).

When this substitution is carried out, let (S_1), (S_2) be the parts of (S) which contain all terms of the speeds κ_1 and κ_2 respectively.

By (191) and (193) the right-hand side in the above equation may be written

$$(\kappa_1 + \kappa_2)(\kappa_2 + \alpha)\,\text{S}_1 - \frac{\Sigma_1}{\text{b}}(\kappa_1 + \alpha)(\kappa_2 + \alpha)(\kappa_1 + \kappa_2) + \left(\kappa_1^2 + \frac{d^2}{dt^2}\right)\text{S}_1$$

+ the same with 2 and 1 interchanged

Now let D^4 stand for the operation $\frac{d^4}{dt^4} + (\kappa_1^2 + \kappa_2^2)\frac{d^2}{dt^2} + \kappa_1^2\kappa_2^2$, and we have

$$\left.\begin{aligned}
\text{D}^4 z &= (\kappa_1 + \kappa_2)(\kappa_2 + \alpha)\left\{\text{S}_1 - \frac{\kappa_1 + \alpha}{\text{b}}\Sigma_1\right\} + \left(\kappa_1^2 + \frac{d^2}{dt^2}\right)\text{S}_1 \\
&\qquad + \text{the same with 2 and 1 reversed} \\
\text{D}^4 y &= (\kappa_1 + \kappa_2)(\kappa_2 + \alpha)\left\{\text{U}_1 - \frac{\kappa_1 + \alpha}{\text{b}}\Upsilon_1\right\} + \left(\kappa_1^2 + \frac{d^2}{dt^2}\right)\text{U}_1 + \&\text{c.} \\
\text{D}^4 \zeta &= (\kappa_1 + \kappa_2)(\kappa_2 + \beta)\left\{\Sigma_1 - \frac{\kappa_1 + \beta}{\text{a}}\text{S}_1\right\} + \left(\kappa_1^2 + \frac{d^2}{dt^2}\right)\Sigma_1 + \&\text{c.} \\
\text{D}^4 \eta &= (\kappa_1 + \kappa_2)(\kappa_2 + \beta)\left\{\Upsilon_1 - \frac{\kappa_1 + \beta}{\text{a}}\text{U}_1\right\} + \left(\kappa_1^2 + \frac{d^2}{dt^2}\right)\Upsilon_1 + \&\text{c.}
\end{aligned}\right\} \quad (198)$$

The last three of these equations are to be found by a parallel process, or else by symmetry.

If the right-hand sides of (198) be neglected, we clearly obtain, on integration, the first approximation (183) for z, y, ζ, η. This first approximation was originally obtained by mere inspection.

We now have to consider the effects of the small terms on the right on the constants of integration L_1, L_2, L_1', L_2' introduced in the first approximation.

The small terms on the right are, by means of the first approximation, capable of being arranged in one of the alternative forms

$$\left.\begin{matrix}\cos\\ \sin\end{matrix}\right\} \kappa_1 t + t \left.\begin{matrix}\sin\\ \cos\end{matrix}\right\} \kappa_1 t + \text{the same with 2 for 1}$$

Now consider the differential equation

$$\frac{d^4x}{dt^4} + (a^2+b^2)\frac{d^2x}{dt^2} + a^2b^2x = A\cos(at+\eta) + Bt\cos(at+\eta) \ldots (199)$$

First suppose that B is zero, so that the term in A exists alone.

Assume $x = Ct\sin(at+\eta)$ as the solution.

Then $$\frac{d^2x}{dt^2} = C\{-a^2t\sin(at+\eta) + 2a\cos(at+\eta)\}$$

$$\frac{d^4x}{dt^4} = C\{a^4t\sin(at+\eta) - 4a^3\cos(at+\eta)\}$$

By substitution in (199), with $B=0$, we have

$$C\{-4a^3 + 2a(a^2+b^2)\} = A$$

Therefore the solution is

$$x = -\frac{A}{2a(a^2-b^2)}t\sin(at+\eta)$$

By writing $\eta - \frac{1}{2}\pi$ for η, we see that a term $A\sin(at+\eta)$ in the differential equation would generate $\dfrac{A}{2a(a^2-b^2)}t\cos(at+\eta)$ in the solution.

From this theorem it follows that the solution of the equation

$$D^4z = F_1y_1 + F_2y_2$$

is $$z = \frac{tF_1z_1}{2\kappa_1(\kappa_1^2-\kappa_2^2)} + \text{the same with 2 and 1 interchanged}$$

and the solution of $$D^4z = F_1\eta_1 + F_2\eta_2$$

is $$z = \frac{tF_1\zeta_1}{2\kappa_1(\kappa_1^2-\kappa_2^2)} + \text{the same with 2 and 1 interchanged}$$

Also (writing the two alternatives by means of an easily intelligible notation) the solutions of

$$D^4y = F_1\begin{cases}z_1\\ \zeta_1\end{cases} + F_2\begin{cases}z_2\\ \zeta_2\end{cases}$$

are $$y = -\frac{tF_1\begin{cases}y_1\\ \eta_1\end{cases}}{2\kappa_1(\kappa_1^2-\kappa_2^2)} - \text{the same with 2 and 1 interchanged}$$

The similar equations for $D^4\zeta$, $D^4\eta$ may be treated in the same way. The general rule is that:

y and η in the differential equations generate in the solution tz and $t\zeta$ respectively; and z and ζ generate $-ty$ and $-t\eta$ respectively; and the terms are to be divided by $2\kappa_1(\kappa_1^2-\kappa_2^2)$ or $2\kappa_2(\kappa_2^2-\kappa_1^2)$ as the case may be.

Next suppose that $A=0$ in the equation (199), and assume as the solution

$$x = Ct^2 \sin(at+\eta) + Dt\cos(at+\eta)$$

Then

$$\frac{d^2x}{dt^2} = C\{-a^2t^2\sin(at+\eta) + 4at\cos(at+\eta) + 2\sin(at+\eta)\}$$
$$+ D\{-a^2t\cos(at+\eta) - 2a\sin(at+\eta)\}$$

$$\frac{d^4x}{dt^4} = C\{a^4t^2\sin(at+\eta) - 8a^3t\cos(at+\eta) - 12a^2\sin(at+\eta)\}$$
$$+ D\{a^4t\cos(at+\eta) + 4a^3\sin(at+\eta)\}$$

Substituting in (199), we must have

$$4aC(a^2+b^2) - 8a^3C = B$$

and

$$2(C-aD)(a^2+b^2) - 12a^2C + 4a^3D = 0$$

Whence

$$C = -\frac{B}{4a(a^2-b^2)}, \quad D = -\frac{5a^2-b^2}{4a^2(a^2-b^2)^2}B$$

Hence the solution of (199), when $A=0$, is

$$x = -\frac{5a^2-b^2}{4a^2(a^2-b^2)^2}Bt\cos(at+\eta) - \frac{1}{4a(a^2-b^2)}Bt^2\sin(at+\eta)$$

If t be very small, the second of these terms may be neglected.

By writing $\eta-\frac{1}{2}\pi$ for η, we see that a term $Bt\sin(at+\eta)$ in the differential equation, would have given rise in the solution to

$$x = -\frac{5a^2-b^2}{4a^2(a^2-b^2)^2}Bt\sin(at+\eta)$$

t being very small.

By this theorem we see that the solutions of the two alternative differential equations

$$D^4z = tF_1\begin{cases}z_1\\ \zeta_1\end{cases} + tF_2\begin{cases}z_2\\ \zeta_2\end{cases}$$

are, when t is very small,

$$z = -\frac{5\kappa_1^2-\kappa_2^2}{4\kappa_1^2(\kappa_1^2-\kappa_2^2)^2}tF_1\begin{cases}z_1\\ \zeta_1\end{cases} - \text{the same with 2 and 1 interchanged}$$

The similar equations for D^4y, $D^4\eta$, $D^4\zeta$ may be treated similarly. The general rule is that:

tz and $t\zeta$ in the differential equations are reproduced, but with an opposite sign in the solution; and similarly ty and $t\eta$ are reproduced with the opposite sign; and in the solution the terms are to be multiplied by

$$\frac{5\kappa_1^2-\kappa_2^2}{4\kappa_1^2(\kappa_1^2-\kappa_2^2)^2} \text{ or } \frac{5\kappa_2^2-\kappa_1^2}{4\kappa_2^2(\kappa_2^2-\kappa_1^2)^2}$$

For the purpose of future developments it will be more convenient to write these factors in the forms

$$\frac{1}{2\kappa_1(\kappa_1^2-\kappa_2^2)}\left\{\frac{2\kappa_1}{\kappa_1^2-\kappa_2^2}+\frac{1}{2\kappa_1}\right\} \text{ and } \frac{1}{2\kappa_2(\kappa_2^2-\kappa_1^2)}\left\{\frac{2\kappa_2}{\kappa_2^2-\kappa_1^2}+\frac{1}{2\kappa_2}\right\}$$

By means of these two rules we see that the solutions of the two alternative differential equations

$$D^4 z = A_1\begin{Bmatrix} y_1 \\ \eta_1 \end{Bmatrix} + tB_1\begin{Bmatrix} z_1 \\ \zeta_1 \end{Bmatrix} + \text{the same with 2 for 1} \quad \ldots\ldots\ldots(200)$$

are, so long as t is very small,

$$z = z_1 + \frac{tA_1\begin{Bmatrix} z_1 \\ \zeta_1 \end{Bmatrix}}{2\kappa_1(\kappa_1^2-\kappa_2^2)} - \frac{tB_1\begin{Bmatrix} z_1 \\ \zeta_1 \end{Bmatrix}}{2\kappa_1(\kappa_1^2-\kappa_2^2)}\left[\frac{2\kappa_1}{\kappa_1^2-\kappa_2^2}+\frac{1}{2\kappa_1}\right]$$

$$+ \text{ the same with 2 and 1 interchanged} \ldots(201)$$

Putting for z_1, ζ_1, &c., their values from (182), these solutions may be written

$$z = \cos(\kappa_1 t + m_1)\left\{L_1 + \frac{tA_1\begin{Bmatrix} L_1 \\ L_1' \end{Bmatrix}}{2\kappa_1(\kappa_1^2-\kappa_2^2)} - \frac{tB_1\begin{Bmatrix} L_1 \\ L_1' \end{Bmatrix}}{2\kappa_1(\kappa_1^2-\kappa_2^2)}\left[\frac{2\kappa_1}{\kappa_1^2-\kappa_2^2}+\frac{1}{2\kappa_1}\right]\right\}$$

$$+ \text{ the same with 2 for 1} \ldots(202)$$

Hence we may retain the first approximation

$$z = L_1\cos(\kappa_1 t + m_1) + L_2\cos(\kappa_2 t + m_2)$$

as the solution, provided that L_1 and L_2 are no longer constant, but vary in such a way that

$$\left.\begin{array}{l}\dfrac{dL_1}{dt} = \dfrac{A_1\begin{Bmatrix} L_1 \\ L_1' \end{Bmatrix}}{2\kappa_1(\kappa_1^2-\kappa_2^2)} - \dfrac{B_1\begin{Bmatrix} L_1 \\ L_1' \end{Bmatrix}}{2\kappa_1(\kappa_1^2-\kappa_2^2)}\left[\dfrac{2\kappa_1}{\kappa_1^2-\kappa_2^2}+\dfrac{1}{2\kappa_1}\right] \\ \qquad\text{and a similar equation for } L_2\end{array}\right\}\ldots\ldots(203)$$

It will be found, when we come to apply these results, that the solution of the equation for $D^4 y$ will lead to the same equations for the variation of L_1 and L_2 as are derived from the equation for $D^4 z$.

A similar treatment may be applied to the equations for $D^4\zeta$ or $D^4\eta$, and we find similar differential equations for dL_1'/dt and dL_2'/dt.

These equations will be the differential equations for the secular changes in L_1 and L_2', which are the constants of integration in the first approximation.

We will now apply these theorems to the differential equations (181); but as the analysis is rather complex, it will be more convenient to treat the variations of α, a, β, b and the terms in γ, g, δ, d independently.

We will indicate by the symbol Δ the additional terms which arise, and will write the symbol out of which the term arises as a suffix—*e.g.*, we shall write Δz_α for the additional terms in the complete value of z, which arise from the variation of α. Also $(dL/dt)_\alpha$ will be written for the terms in dL/dt which arise from the variation of α.

Terms depending on the variation of α.

We now put for α in (181) $\alpha + \alpha' t$.

Hence in (194)

$$s = \alpha' ty, \quad u = -\alpha' tz, \quad \sigma = 0, \quad v = 0$$

Therefore $$\mathrm{S} = \alpha' \left\{ y - t\left(\alpha z - \frac{dy}{dt}\right)\right\}, \quad \Sigma = -\alpha' \mathrm{b} tz$$

And by substitution from (182–3)

$$\mathrm{S}_1 = \alpha' \{y_1 + tz_1(\kappa_1 - \alpha)\}, \quad \Sigma_1 = -\alpha' \mathrm{b} tz_1$$

S_2, Σ_2 have similar forms with 2 for 1

Then $$\left(\kappa_1^2 + \frac{d^2}{dt^2}\right)\mathrm{S}_1 = 2\alpha'(\kappa_1 - \alpha)\frac{dz_1}{dt} = -2\alpha'\kappa_1(\kappa_1 - \alpha)y_1$$

$$\mathrm{S}_1 - \frac{\kappa_1 + \alpha}{\mathrm{b}}\Sigma_1 = \alpha'\{y_1 + tz_1(\kappa_1 - \alpha)\} + \alpha' t(\kappa_1 + \alpha)z_1$$
$$= \alpha'\{y_1 + 2t\kappa_1 z_1\} \quad \ldots\ldots(204)$$

Thus the equation for z is

$$\frac{1}{\alpha'}\mathrm{D}^4 z = (\kappa_1 + \kappa_2)(\kappa_2 + \alpha)(y_1 + 2t\kappa_1 z_1) - 2\kappa_1(\kappa_1 - \alpha)y_1 + \text{the same with 2 for 1}$$

Hence by the rules found above for the solution of such an equation

$$\frac{1}{\alpha' t}\Delta z_\alpha = \frac{z_1}{2\kappa_1(\kappa_1^2 - \kappa_2^2)}\Big\{(\kappa_1 + \kappa_2)(\kappa_2 + \alpha) - 2\kappa_1(\kappa_1 - \alpha)$$
$$- 2\kappa_1(\kappa_1 + \kappa_2)(\kappa_2 + \alpha)\left[\frac{2\kappa_1}{\kappa_1^2 - \kappa_2^2} + \frac{1}{2\kappa_1}\right]\Big\} + \&\text{c.}$$
$$= -\frac{z_1}{\kappa_1^2 - \kappa_2^2}\left[\kappa_1 - \alpha + \frac{2\kappa_1(\kappa_2 + \alpha)}{(\kappa_1 - \kappa_2)}\right] + \&\text{c.}$$
$$= -z_1\frac{\kappa_1 + \alpha}{(\kappa_1 - \kappa_2)^2} - z_2\frac{\kappa_2 + \alpha}{(\kappa_1 - \kappa_2)^2}$$

Whence

$$\left(\frac{1}{L_1}\frac{dL_1}{dt}\right)_\alpha = -\alpha'\frac{\kappa_1 + \alpha}{(\kappa_1 - \kappa_2)^2}, \quad \left(\frac{1}{L_2}\frac{dL_2}{dt}\right)_\alpha = -\alpha'\frac{\kappa_2 + \alpha}{(\kappa_1 - \kappa_2)^2} \quad \ldots(205)$$

If we form U and Υ, and solve the equation for $\mathrm{D}^4 y$, we obtain the same results.

Again $$\left(\kappa_1^2 + \frac{d^2}{dt^2}\right)\Sigma_1 = -2\alpha'\mathrm{b}\frac{dz_1}{dt} = 2\alpha'\mathrm{b}\kappa_1 y_1$$

$$\Sigma_1 - \frac{\kappa_1 + \beta}{\mathrm{a}}\mathrm{S}_1 = \Sigma_1 - \frac{\mathrm{b}}{\kappa_1 + \alpha}\mathrm{S}_1 = \frac{\mathrm{b}}{\kappa_2 + \beta}\left\{\mathrm{S}_1 - \frac{\kappa_1 + \alpha}{\mathrm{b}}\Sigma_1\right\}$$
$$= \alpha'\frac{\mathrm{b}}{\kappa_2 + \beta}(y_1 + 2t\kappa_1 z_1) \text{ by } (204)$$

Hence the equation for ζ is

$$\frac{1}{\alpha' \mathrm{b}} \mathrm{D}^4\zeta = (\kappa_1 + \kappa_2)(y_1 + 2t\kappa_1 z_1) + 2\kappa_1 y_1 + \text{the same with 2 for 1}$$

And by the rules of solution

$$\frac{1}{\alpha' \mathrm{b}t} \Delta\zeta_a = \frac{z_1}{2\kappa_1(\kappa_1^2 - \kappa_2^2)}\left[\kappa_1 + \kappa_2 + 2\kappa_1 - 2\kappa_1(\kappa_1 + \kappa_2)\left\{\frac{2\kappa_1}{\kappa_1^2 - \kappa_2^2} + \frac{1}{2\kappa_1}\right\}\right] + \&\text{c.}$$

$$= \frac{z_1}{\kappa_1^2 - \kappa_2^2}\left[1 - \frac{2\kappa_1}{\kappa_1 - \kappa_2}\right] + \&\text{c.}$$

$$= -\frac{z_1}{(\kappa_1 - \kappa_2)^2} - \frac{z_2}{(\kappa_1 - \kappa_2)^2}$$

$$= -\frac{1}{\mathrm{b}}\frac{\kappa_2 + \alpha}{(\kappa_1 - \kappa_2)^2}\zeta_1 - \frac{1}{\mathrm{b}}\frac{\kappa_1 + \alpha}{(\kappa_1 - \kappa_2)^2}\zeta_2$$

The last transformation arises from

$$z_1 = \frac{\kappa_2 + \alpha}{\mathrm{b}}\zeta_1, \quad z_2 = \frac{\kappa_1 + \alpha}{\mathrm{b}}\zeta_2$$

Hence

$$\left(\frac{1}{L_1'}\frac{dL_1'}{dt}\right)_a = -\alpha'\frac{\kappa_2 + \alpha}{(\kappa_1 - \kappa_2)^2}, \quad \left(\frac{1}{L_2'}\frac{dL_2'}{dt}\right)_a = -\alpha'\frac{\kappa_1 + \alpha}{(\kappa_1 - \kappa_2)^2} \quad \text{...(206)}$$

If we form U and Υ, and solve the equation for $\mathrm{D}^4\eta$, we obtain the same result.

Terms depending on the variation of β.

The results may be written down by symmetry.

z and y are symmetrical with ζ and η, and therefore unaccented L's are symmetrical with accented ones, and *vice-versâ*; α is symmetrical with β, and *vice-versâ*.

The suffixes 1 and 2 remain unaffected by the symmetry.

By (192) on writing $-(\kappa_1 + \alpha)$ for $\kappa_2 + \beta$, and $-(\kappa_2 + \alpha)$ for $\kappa_1 + \beta$, we have by symmetry with (206),

$$\left(\frac{1}{L_1}\frac{dL_1}{dt}\right)_\beta = \beta'\frac{\kappa_1 + \alpha}{(\kappa_1 - \kappa_2)^2}, \quad \left(\frac{1}{L_2}\frac{dL_2}{dt}\right)_\beta = \beta'\frac{\kappa_2 + \alpha}{(\kappa_1 - \kappa_2)^2} \quad \text{......(207)}$$

And by symmetry with (205),

$$\left(\frac{1}{L_1'}\frac{dL_1'}{dt}\right)_\beta = \beta'\frac{\kappa_2 + \alpha}{(\kappa_1 - \kappa_2)^2}, \quad \left(\frac{1}{L_2'}\frac{dL_2'}{dt}\right)_\beta = \beta'\frac{\kappa_1 + \alpha}{(\kappa_1 - \kappa_2)^2} \quad \text{......(208)}$$

Terms depending on the variation of a.

We now put for a in (181) $\mathrm{a} + \mathrm{a}'t$.

In (194) $\quad s = \mathrm{a}'t\eta, \quad u = -\mathrm{a}'t\zeta, \quad \sigma = 0, \quad \upsilon = 0$

Therefore $\quad S = a'\eta + a't\left(\frac{d\eta}{dt} - \alpha\zeta\right), \quad \Sigma = -a'bt\zeta$

$$S_1 = a'\{\eta_1 + t\zeta_1(\kappa_1 - \alpha)\}, \quad \Sigma_1 = -a'bt\zeta_1$$

S_2, Σ_2 have similar forms with 2 for 1

$$\left(\kappa_1^2 + \frac{d^2}{dt^2}\right) S_1 = 2a'(\kappa_1 - \alpha)\frac{d\zeta_1}{dt} = -2a'\kappa_1(\kappa_1 - \alpha)\eta_1$$

$$S_1 - \frac{\kappa_1 + \alpha}{b}\Sigma_1 = a'[\eta_1 + t\zeta_1(\kappa_1 - \alpha) + t\zeta(\kappa_1 + \alpha)] = a'[\eta_1 + 2\kappa_1 t\zeta_1] \ldots (209)$$

Hence the equation for z is

$$\frac{1}{a'}D^4 z = -2\kappa_1(\kappa_1 - \alpha)\eta_1 + (\kappa_1 + \kappa_2)(\kappa_2 + \alpha)(\eta_1 + 2\kappa_1 t\zeta_1)$$
$$+ \text{ the same with 2 for 1}$$

$$\frac{1}{a't}\Delta z_a = \frac{\zeta_1}{2\kappa_1(\kappa_1^2 - \kappa_2^2)}\left[-2\kappa_1(\kappa_1 - \alpha) + (\kappa_1 + \kappa_2)(\kappa_2 + \alpha) - 2\kappa_1(\kappa_1 + \kappa_2)(\kappa_2 + \alpha)\left(\frac{2\kappa_1}{\kappa_1^2 - \kappa_2^2} + \frac{1}{2\kappa_1}\right)\right] + \&c.$$

$$= -\frac{\zeta_1}{\kappa_1^2 - \kappa_2^2}\left[\kappa_1 - \alpha + \frac{2\kappa_1(\kappa_2 + \alpha)}{\kappa_1 - \kappa_2}\right] - \&c.$$

$$= -\zeta_1\frac{\kappa_1 + \alpha}{(\kappa_1 - \kappa_2)^2} - \zeta_2\frac{\kappa_2 + \alpha}{(\kappa_1 - \kappa_2)^2}$$

$$= -z_1\frac{b}{(\kappa_1 - \kappa_2)^2}\frac{\kappa_1 + \alpha}{\kappa_2 + \alpha} - z_2\frac{b}{(\kappa_1 - \kappa_2)^2}\frac{\kappa_2 + \alpha}{\kappa_1 + \alpha},$$

since $\zeta_1 = z_1\frac{b}{\kappa_2 + \alpha}, \quad \zeta_2 = z_2\frac{b}{\kappa_1 + \alpha}$

Therefore

$$\left(\frac{1}{L_1}\frac{dL_1}{dt}\right)_a = -\frac{a'b}{(\kappa_1 - \kappa_2)^2}\frac{\kappa_1 + \alpha}{\kappa_2 + \alpha}, \quad \left(\frac{1}{L_2}\frac{dL_2}{dt}\right)_a = -\frac{a'b}{(\kappa_1 - \kappa_2)^2}\frac{\kappa_2 + \alpha}{\kappa_1 + \alpha} \ldots (210)$$

Again $\quad \Sigma_1 - \frac{\kappa_1 + \beta}{a}S_1 = \Sigma_1 - \frac{b}{\kappa_1 + \alpha}S_1 = \frac{b}{\kappa_2 + \beta}\left(S_1 - \frac{\kappa_1 + \alpha}{b}\Sigma_1\right)$

$$= \frac{a'b}{\kappa_2 + \beta}(\eta_1 + 2\kappa_1 t\zeta_1) \text{ by } (209)$$

Also $$\left(\kappa_1^2 + \frac{d^2}{dt^2}\right)\Sigma_1 = -2a'b\frac{d\zeta_1}{dt} = 2a'b\kappa_1\eta_1$$

Therefore the equation for ζ is

$$\frac{1}{a'b}D^4\zeta = (\kappa_1 + \kappa_2)(\eta_1 + 2\kappa_1 t\zeta_1) + 2\kappa_1\eta_1 + \text{ the same with 2 for 1}$$

Therefore

$$\frac{1}{a'bt}\Delta\zeta_a = \frac{\zeta_1}{2\kappa_1(\kappa_1^2 - \kappa_2^2)}\left[(\kappa_1 + \kappa_2) + 2\kappa_1 - 2\kappa_1(\kappa_1 + \kappa_2)\left(\frac{2\kappa_1}{\kappa_1^2 - \kappa_2^2} + \frac{1}{2\kappa_1}\right)\right] + \&c.$$

$$= \frac{\zeta_1}{\kappa_1^2 - \kappa_2^2}\left(1 - \frac{2\kappa_1}{\kappa_1 - \kappa_2}\right) + \&c.$$

$$= -\frac{\zeta_1}{(\kappa_1 - \kappa_2)^2} - \frac{\zeta_2}{(\kappa_1 - \kappa_2)^2}$$

Therefore

$$\left(\frac{1}{L_1'}\frac{dL_1'}{dt}\right)_{\mathrm{a}} = -\frac{\mathrm{a}'\mathrm{b}}{(\kappa_1-\kappa_2)^2}, \quad \left(\frac{1}{L_2'}\frac{dL_2'}{dt}\right)_{\mathrm{a}} = -\frac{\mathrm{a}'\mathrm{b}}{(\kappa_1-\kappa_2)^2} \quad \text{......(211)}$$

The same results might have been obtained from the equations to D^4y, $D^4\eta$.

Terms depending on the variation of b.

By symmetry with (211)

$$\left(\frac{1}{L_1}\frac{dL_1}{dt}\right)_{\mathrm{b}} = -\frac{\mathrm{b}'\mathrm{a}}{(\kappa_1-\kappa_2)^2}, \quad \left(\frac{1}{L_2}\frac{dL_2}{dt}\right)_{\mathrm{b}} = -\frac{\mathrm{b}'\mathrm{a}}{(\kappa_1-\kappa_2)^2} \quad \text{........(212)}$$

By symmetry with (210), and putting $-(\kappa_2+\alpha)$ for $(\kappa_1+\beta)$ and $-(\kappa_1+\alpha)$ for $(\kappa_2+\beta)$

$$\left(\frac{1}{L_1'}\frac{dL_1'}{dt}\right)_{\mathrm{b}} = -\frac{\mathrm{b}'\mathrm{a}}{(\kappa_1-\kappa_2)^2}\frac{\kappa_2+\alpha}{\kappa_1+\alpha}, \quad \left(\frac{1}{L_2'}\frac{dL_2'}{dt}\right)_{\mathrm{b}} = -\frac{\mathrm{b}'\mathrm{a}}{(\kappa_1-\kappa_2)^2}\frac{\kappa_1+\alpha}{\kappa_2+\alpha} \quad \text{...(213)}$$

We now come to a different class of terms, viz.: those depending on γ, g, δ, d.

Terms depending on γ.

Here

$$s = -\gamma z, \quad u = -\gamma y, \quad \sigma = 0, \quad v = 0$$

$$\mathrm{S} = -\gamma\left(\frac{dz}{dt} + \alpha y\right), \qquad \Sigma = -\mathrm{b}\gamma y$$

$$\mathrm{S}_1 = \gamma(\kappa_1 - \alpha)y_1, \qquad \Sigma_1 = -\gamma \mathrm{b} y_1$$

S_2, Σ_2 have similar forms with 2 for 1

Obviously

$$\left(\kappa_1^2 + \frac{d^2}{dt^2}\right)\mathrm{S}_1 = 0$$

$$\mathrm{S}_1 - \frac{\kappa_1+\alpha}{\mathrm{b}}\Sigma_1 = 2\gamma\kappa_1 y_1 \quad \text{........................(214)}$$

Hence the equation for z is

$$\frac{1}{\gamma}D^4 z = 2\kappa_1(\kappa_1+\kappa_2)(\kappa_2+\alpha)y_1 + \text{the same with 2 for 1}$$

Therefore

$$\frac{1}{\gamma t}\Delta z_\gamma = z_1\frac{\kappa_2+\alpha}{\kappa_1-\kappa_2} + z_2\frac{\kappa_1+\alpha}{\kappa_2-\kappa_1}$$

And

$$\left(\frac{1}{L_1}\frac{dL_1}{dt}\right)_\gamma = \gamma\frac{\kappa_2+\alpha}{\kappa_1-\kappa_2}, \quad \left(\frac{1}{L_2}\frac{dL_2}{dt}\right)_\gamma = -\gamma\frac{\kappa_1+\alpha}{\kappa_1-\kappa_2} \quad \text{.........(215)}$$

Again

$$\left(\kappa_1^2 + \frac{d^2}{dt^2}\right)\Sigma_1 = 0$$

$$\Sigma_1 - \frac{\kappa_1+\beta}{\mathrm{a}}\mathrm{S}_1 = \frac{\mathrm{b}}{\kappa_2+\beta}\left[\mathrm{S}_1 - \frac{\kappa_1+\alpha}{\mathrm{b}}\Sigma_1\right]$$

$$= \mathrm{b}\gamma\frac{2\kappa_1}{\kappa_2+\beta}y_1$$

And the equation for ζ is

$$\frac{1}{\gamma \mathrm{b}}D^4\zeta = 2\kappa_1(\kappa_1+\kappa_2)y_1 + \text{the same with 2 for 1}$$

Therefore

$$\frac{1}{\gamma \mathrm{b} t}\Delta\zeta_\gamma = \frac{z_1}{\kappa_1-\kappa_2}+\frac{z_2}{\kappa_2-\kappa_1}$$

$$= \frac{\zeta_1}{\mathrm{b}}\frac{\kappa_2+\alpha}{\kappa_1-\kappa_2} - \frac{\zeta_2}{\mathrm{b}}\frac{\kappa_1+\alpha}{\kappa_1-\kappa_2}, \quad \text{since } z_1=\zeta_1\frac{\kappa_2+\alpha}{\mathrm{b}}, \quad z_2=\zeta_2\frac{\kappa_1+\alpha}{\mathrm{b}}$$

Hence
$$\left(\frac{1}{L_1'}\frac{dL_1'}{dt}\right)_\gamma = \gamma\frac{\kappa_2+\alpha}{\kappa_1-\kappa_2}, \quad \left(\frac{1}{L_2'}\frac{dL_2'}{dt}\right)_\gamma = -\gamma\frac{\kappa_1+\alpha}{\kappa_1-\kappa_2} \quad \text{.........(216)}$$

Terms depending on δ.

These may be written down by symmetry.

$-\delta$ is symmetrical with γ. Hence writing $-(\kappa_1+\alpha)$ for $\kappa_2+\beta$, and $-(\kappa_2+\alpha)$ for $(\kappa_1+\beta)$, we have by symmetry with (216)

$$\left(\frac{1}{L_1}\frac{dL_1}{dt}\right)_\delta = \delta\frac{\kappa_1+\alpha}{\kappa_1-\kappa_2}, \quad \left(\frac{1}{L_2}\frac{dL_2}{dt}\right)_\delta = -\delta\frac{\kappa_2+\alpha}{\kappa_1-\kappa_2} \quad \text{.........(217)}$$

And by symmetry with (215)

$$\left(\frac{1}{L_1'}\frac{dL_1'}{dt}\right)_\delta = \delta\frac{\kappa_1+\alpha}{\kappa_1-\kappa_2}, \quad \left(\frac{1}{L_2'}\frac{dL_2'}{dt}\right)_\delta = -\delta\frac{\kappa_2+\alpha}{\kappa_1-\kappa_2} \quad \text{.........(218)}$$

Terms depending on g.

Here
$$s=-\mathrm{g}\zeta, \quad u=-\mathrm{g}\eta, \quad \sigma=0, \quad \upsilon=0$$

$$\mathrm{S}=-\mathrm{g}\left(\frac{d\zeta}{dt}+\alpha\eta\right), \qquad \Sigma=-\mathrm{gb}\eta$$

$$\mathrm{S}_1=\ \mathrm{g}(\kappa_1-\alpha)\eta_1, \qquad \Sigma_1=-\mathrm{gb}\eta_1$$

S_2, Σ_2 have similar forms with 2 for 1

Clearly
$$\left(\kappa_1^2+\frac{d^2}{dt^2}\right)\mathrm{S}_1=0$$

$$\mathrm{S}_1-\frac{\kappa_1+\alpha}{\mathrm{b}}\Sigma_1 = 2\mathrm{g}\kappa_1\eta_1 \quad \text{..........................(219)}$$

Therefore the equation for z is

$$\frac{1}{\mathrm{g}}\mathrm{D}^4 z = 2\kappa_1(\kappa_1+\kappa_2)(\kappa_2+\alpha)\eta_1 + \text{the same with 2 for 1}$$

Thence

$$\frac{1}{\mathrm{g}t}\Delta z_\mathrm{g} = \zeta_1\frac{\kappa_2+\alpha}{\kappa_1-\kappa_2}+\zeta_2\frac{\kappa_1+\alpha}{\kappa_2-\kappa_1}$$

$$= z_1\frac{\mathrm{b}}{\kappa_1-\kappa_2} - z_2\frac{\mathrm{b}}{\kappa_1-\kappa_2}, \quad \text{since } \zeta_1 = z_1\frac{\mathrm{b}}{\kappa_2+\alpha}, \quad \zeta_2=z_2\frac{\mathrm{b}}{\kappa_1+\alpha}$$

Therefore
$$\left(\frac{1}{L_1}\frac{dL_1}{dt}\right)_\mathrm{g} = \mathrm{g}\frac{\mathrm{b}}{\kappa_1-\kappa_2}, \quad \left(\frac{1}{L_2}\frac{dL_2}{dt}\right)_\mathrm{g} = -\mathrm{g}\frac{\mathrm{b}}{\kappa_1-\kappa_2} \quad \text{.........(220)}$$

Again
$$\left(\kappa_1^2+\frac{d^2}{dt^2}\right)\Sigma_1=0$$

and $$\Sigma_1 - \frac{\kappa_1+\beta}{a} S_1 = \frac{b}{\kappa_2+\beta}\left[S_1 - \frac{\kappa_1+\alpha}{b}\Sigma_1\right]$$

$$= g\frac{2b\kappa_1}{\kappa_2+\beta}\eta_1 \text{ by (219)}$$

Therefore the equation for ζ is

$$\frac{1}{gb}D^4\zeta = 2\kappa_1(\kappa_1+\kappa_2)\eta_1 + \text{the same with 2 for 1}$$

Hence $$\frac{1}{gbt}\Delta\zeta_g = \frac{\zeta_1}{\kappa_1-\kappa_2} + \frac{\zeta_2}{\kappa_2-\kappa_1}$$

Therefore $$\left(\frac{1}{L_1'}\frac{dL_1'}{dt}\right)_g = g\frac{b}{\kappa_1-\kappa_2},\quad \left(\frac{1}{L_2'}\frac{dL_2'}{dt}\right)_g = -g\frac{b}{\kappa_1-\kappa_2} \quad\text{.........(221)}$$

The same results may be obtained by means of the equations for $D^4 y$, $D^4\eta$.

Terms depending on d.

These may be written down by symmetry.

$-$d is symmetrical with g. Therefore by symmetry with (221)

$$\left(\frac{1}{L_1}\frac{dL_1}{dt}\right)_d = -d\frac{a}{\kappa_1-\kappa_2},\quad \left(\frac{1}{L_2}\frac{dL_2}{dt}\right)_d = d\frac{a}{\kappa_1-\kappa_2} \quad\text{.........(222)}$$

and by symmetry with (220)

$$\left(\frac{1}{L_1'}\frac{dL_1'}{dt}\right)_d = -d\frac{a}{\kappa_1-\kappa_2},\quad \left(\frac{1}{L_2'}\frac{dL_2'}{dt}\right)_d = d\frac{a}{\kappa_1-\kappa_2} \quad\text{.........(223)}$$

This completes the consideration of the effects on the constants of integration L_1, L_2, L_1', L_2' of all the small terms.

Collecting results from (205–8, 210–13, 215–18, 220–23),

$$\left.\begin{aligned}
\frac{1}{L_1}\frac{dL_1}{dt} &= \frac{1}{(\kappa_1-\kappa_2)^2}\left\{-(\kappa_1+\alpha)(\alpha'-\beta') - a'b\frac{\kappa_1+\alpha}{\kappa_2+\alpha} - b'a\right\} \\
&\quad + \frac{1}{(\kappa_1-\kappa_2)}\{\gamma(\kappa_2+\alpha)+\delta(\kappa_1+\alpha)+gb-da\} \\
\frac{1}{L_2}\frac{dL_2}{dt} &= \frac{1}{(\kappa_1-\kappa_2)^2}\left\{-(\kappa_2+\alpha)(\alpha'-\beta') - a'b\frac{\kappa_2+\alpha}{\kappa_1+\alpha} - b'a\right\} \\
&\quad - \frac{1}{\kappa_1-\kappa_2}\{\gamma(\kappa_1+\alpha)+\delta(\kappa_2+\alpha)+gb-da\} \\
\frac{1}{L_1'}\frac{dL_1'}{dt} &= \frac{1}{(\kappa_1-\kappa_2)^2}\left\{-(\kappa_2+\alpha)(\alpha'-\beta') - a'b - b'a\frac{\kappa_2+\alpha}{\kappa_1+\alpha}\right\} \\
&\quad + \frac{1}{\kappa_1-\kappa_2}\{\gamma(\kappa_2+\alpha)+\delta(\kappa_1+\alpha)+gb-da\} \\
\frac{1}{L_2'}\frac{dL_2'}{dt} &= \frac{1}{(\kappa_1-\kappa_2)^2}\left\{-(\kappa_1+\alpha)(\alpha'-\beta') - a'b - b'a\frac{\kappa_1+\alpha}{\kappa_2+\alpha}\right\} \\
&\quad - \frac{1}{\kappa_1-\kappa_2}\{\gamma(\kappa_1+\alpha)+\delta(\kappa_2+\alpha)+gb-da\}
\end{aligned}\right\} \text{...(224)}$$

We shall now show that these four equations are equivalent to two only, and in showing this shall verify the correctness of the results.

To prove that the four equations (224) *are equivalent to two.*

In (118) we showed that

$$\frac{L_1'}{L_1} = -\frac{\kappa_1 + \alpha}{\mathrm{a}}$$

Therefore we ought to find that

$$\frac{1}{L_1'}\frac{dL_1'}{dt} - \frac{1}{L_1}\frac{dL_1}{dt} = \frac{1}{\kappa_1+\alpha}\frac{d}{dt}(\kappa_1+\alpha) - \frac{\mathrm{a}'}{\mathrm{a}}$$

$$= \frac{\kappa_1+\beta}{\mathrm{ab}}\frac{d}{dt}(\kappa_1+\alpha) - \frac{\mathrm{a}'}{\mathrm{a}}$$

By (188)

$$2(\kappa_1+\alpha) = \alpha - \beta - \sqrt{(\alpha-\beta)^2 + 4\mathrm{ab}}$$

and
$$2\frac{d}{dt}(\kappa_1+\alpha) = \alpha' - \beta' + \frac{(\alpha-\beta)(\alpha'-\beta') + 2(\mathrm{a'b}+\mathrm{ab'})}{\kappa_1-\kappa_2}$$

so that
$$\frac{d}{dt}(\kappa_1+\alpha) = \frac{(\alpha'-\beta')(\kappa_1+\alpha) + \mathrm{a'b}+\mathrm{ab'}}{\kappa_1-\kappa_2}$$

Thus we ought to find that

$$(\kappa_1-\kappa_2)\left[\frac{1}{L_1'}\frac{dL_1'}{dt} - \frac{1}{L_1}\frac{dL_1}{dt}\right] = \alpha' - \beta' - \frac{\mathrm{a}'}{\mathrm{a}}(\kappa_1+\alpha) - \frac{\mathrm{b}'}{\mathrm{b}}(\kappa_2+\alpha)$$

If we subtract the first of equations (224) from the third we shall find this relation to be satisfied. Hence the first and third equations are equivalent to only a single one.

Similarly it may be proved that the second and fourth equations are similarly related.

To prove that the four equations (224) *reduce to those of* § 6, *when the nodes revolve with uniform velocity.*

It appears from § 13 that when a and b are small compared with $\alpha-\beta$, the nodes revolve with approximate uniformity, and the nutations of the system are small.

If this be the case, we have approximately

$$\kappa_1 = -\alpha, \quad \kappa_2 = -\beta$$

It will appear later that $(\alpha'-\beta')/(\alpha-\beta)$, a′/a, b′/b are quantities of the same order of magnitude as γ, g, δ, d.

Now $L_1 = \mathrm{J}$, the inclination of the lunar orbit to its proper plane, and $L_2' = \mathrm{I}$, the inclination of the earth's proper plane to the ecliptic.

Therefore, the first and last of equations (224) become

$$\frac{1}{\mathrm{J}}\frac{d\mathrm{J}}{dt} = -\gamma - \frac{\mathrm{gb}-\mathrm{da}}{\alpha-\beta}$$

$$\frac{1}{\mathrm{I}}\frac{d\mathrm{I}}{dt} = \quad \delta + \frac{\mathrm{gb}-\mathrm{da}}{\alpha-\beta}$$

But since the nodes revolve uniformly, $\mathrm{b}/(\alpha-\beta)$ and $\mathrm{a}/(\alpha-\beta)$ are small, and therefore the latter terms of these equations are negligeable compared with the former.

Hence $$\frac{1}{\mathrm{J}}\frac{d\mathrm{J}}{dt} = -\gamma, \quad \frac{1}{\mathrm{I}}\frac{d\mathrm{I}}{dt} = \delta$$

These results in no way depend on the assumption of the smallness of the viscosity of the planet, and therefore we may substitute Γ and Δ (see (174)) for γ and δ.

A comparison of the expressions for Γ and Δ, with those given in Part II. for dj/dt and in my previous paper for di/dt, will show that our present equations for $d\mathrm{J}/dt$ and $d\mathrm{I}/dt$ are what the previous ones reduce to, when i and j are small. But this comparison shows more than this, for it shows that what the equation (61) § 6 really gives is the rate of change of the inclination of the lunar orbit to its proper plane, and that the equation (66) of the paper on "Precession" really gives the rate of change of the inclination of the earth's proper plane (or mean equator) to the ecliptic.

To show how the equations (224) *reduce to those of* § 10.

We now pass to the other extreme, and suppose the solar influence infinitesimal compared with that of oblateness.

Here $$\alpha = \mathrm{a}, \quad \beta = \mathrm{b}, \quad \gamma = \mathrm{g}, \quad \delta = \mathrm{d}$$

$$\kappa_1 = -(\mathrm{a}+\mathrm{b}), \quad \kappa_2 = 0$$

The equations (224) reduce to

$$\left.\begin{aligned} \frac{1}{L_1}\frac{dL_1}{dt} &= -\mathrm{g}+\mathrm{d}+\frac{\mathrm{a'b}-\mathrm{b'a}}{\mathrm{a}\,(\mathrm{a}+\mathrm{b})} \\ \frac{1}{L_1'}\frac{dL_1'}{dt} &= -\mathrm{g}+\mathrm{d}-\frac{\mathrm{a'b}-\mathrm{b'a}}{\mathrm{b}\,(\mathrm{a}+\mathrm{b})} \\ \frac{1}{L_2}\frac{dL_2}{dt} &= 0, \quad \frac{1}{L_2'}\frac{dL_2'}{dt} = 0 \end{aligned}\right\} \qquad (225)$$

Therefore L_2 and L_2' are constant. Also from the relationship between them

$$\frac{L_2'}{L_2} = -\frac{(\kappa_2+\alpha)}{\mathrm{a}} = -1$$

Hence it follows that the two proper planes are identical with one another, and are fixed in space. They are, in fact, the invariable plane of the system, as appears as follows:

If we use the notation of § 10, $L_1 = j$, $L_1' = i$, and $L_1'/L_1 = -(\kappa_1 + \alpha)/\mathrm{a} = \mathrm{b}/\mathrm{a}$; so that $\mathrm{a}i = \mathrm{b}j$.

Now $\mathrm{a} = k\tau\mathfrak{C}/\xi$, $\mathrm{b} = \tau\mathfrak{C}/n$, and i and j are by hypothesis small, therefore we may write the relationship between a, b, i, j in the form

$$\frac{\xi}{k}\sin j = n \sin i$$

This proves that the two coincident planes fixed in space are identical with the invariable plane of the system (see 108).

But the identity of equations (225) with (71) of § 10 and (29) of the paper on "Precession" remains to be proved.

If i and j be treated as small, those equations are in effect

$$\frac{dj}{dt} = -\mathrm{g}(i+j)$$

$$\frac{di}{dt} = \quad \mathrm{d}(i+j)$$

(or with G and D in place of g and d if the viscosity be not small).

Hence if (225) are identical with (71) and (29) of "Precession," we must have

$$-\mathrm{g}\frac{i}{j} = \mathrm{d} + \frac{\mathrm{a'b} - \mathrm{ab'}}{\mathrm{a}(\mathrm{a}+\mathrm{b})}$$

$$\mathrm{d}\frac{j}{i} = -\mathrm{g} - \frac{\mathrm{a'b} - \mathrm{ab'}}{\mathrm{b}(\mathrm{a}+\mathrm{b})}$$

But $i/j = \mathrm{b}/\mathrm{a}$; therefore the condition for the identity of (225) with (71) and (29) of "Precession" is that

$$(\mathrm{a}+\mathrm{b})(\mathrm{gb}+\mathrm{ad}) + \mathrm{a'b} - \mathrm{ab'} = 0 \quad \ldots\ldots\ldots\ldots\ldots\ldots(226)$$

Or if the viscosity be not small, a similar equation with G and D for g and d.

We cannot prove that this condition is satisfied until a′ and b′ have been evaluated, but it will be proved later in § 16.

This discussion shows that the obliquity of the earth's equator (L_1') to the invariable plane of the moon-earth system, when the solar influence is infinitesimal, degrades into the amplitude of the nineteen-yearly nutation, when the influence of oblateness is infinitesimal. The one quantity is strictly continuous with the other.

This completes the verification of the differential equations (224) in the two extreme cases.

§ 16. *Evaluation of α', a′, &c., in the case of the earth's viscosity.*

The preceding section does not involve any hypothesis as to the constitution of the earth, but it will now be supposed to be viscous, and the various functions, which occur in (224), will be evaluated.

By (184–5) we have

$$\frac{1}{k}\frac{d\xi}{dt} = \frac{1}{2}\frac{\tau^2}{\mathfrak{g}}\sin 4\mathrm{f}_1 \qquad (227)$$

$$-\frac{dn}{dt} = \frac{1}{2}\frac{\tau^2}{\mathfrak{g}}\sin 4\mathrm{f}_1\left\{1 + \left(\frac{\tau'}{\tau}\right)^2\right\} \qquad (228)$$

The last equation is approximate, for by writing it in this form we are neglecting $\tau'^2(\sin 4\mathrm{f} - \sin 4\mathrm{f}_1)/\tau^2 \sin 4\mathrm{f}_1$ compared with unity.

This is legitimate, because when $(\sin 4\mathrm{f} - \sin 4\mathrm{f}_1)/\sin 4\mathrm{f}_1$ is not very small, τ'^2/τ^2 is very small, and *vice-versâ*; see however § 22.

Hence (228) may be written

$$-\frac{dn}{dt} = \left(\frac{1}{k}\frac{d\xi}{dt}\right)\left\{1 + \left(\frac{\tau'}{\tau}\right)^2\right\} \qquad (229)$$

Let

$$\mathfrak{m} = \frac{kn}{\xi} \qquad (230)$$

$\mathfrak{m}$ is the ratio of the moment of momentum of the earth's rotation to that of the orbital motion of moon and earth round their common centre of inertia. (The μ of my paper on "Precession" is equal to the reciprocal of $\mathfrak{m}_0$, where $\mathfrak{m}_0$ is the value of $\mathfrak{m}$ when $t = 0$.)

By (121) and (112) we have

$$\mathrm{a} = \frac{k}{\xi}\tau\mathfrak{e}$$

Now $\mathfrak{e} = \frac{1}{2}n^2/\mathfrak{g}$, the ellipticity of the earth due to rotation; and as $\tau = \frac{3}{2}\mu m/c^3$ and $\xi = \sqrt{c/c_0}$, therefore $\tau = \tau_0/\xi^6$.

Hence

$$\mathrm{a} = \frac{k\tau_0}{2\mathfrak{g}}\frac{n^2}{\xi^7}$$

Differentiating logarithmically

$$\frac{\mathrm{a}'}{\mathrm{a}} = \frac{2}{n}\frac{dn}{dt} - \frac{7}{\xi}\frac{d\xi}{dt} = -\left(\frac{1}{k}\frac{d\xi}{dt}\right)\left\{\frac{2}{n}\left[1 + \left(\frac{\tau'}{\tau}\right)^2\right] + \frac{7k}{\xi}\right\}$$

$$\frac{\mathrm{a}'}{\mathrm{a}} = -\left(\frac{1}{k}\frac{d\xi}{dt}\right)\frac{1}{n}\left\{2\left[1 + \left(\frac{\tau'}{\tau}\right)^2\right] + 7\mathfrak{m}\right\} \qquad (231)$$

Also since

$$\mathrm{a} = \mathfrak{m}\frac{\tau\mathfrak{e}}{n} \qquad (232)$$

$$\mathrm{a}' = -\left(\frac{1}{k}\frac{d\xi}{dt}\right)\left(\frac{\tau\mathfrak{e}}{n^2}\right)\mathfrak{m}\left\{2\left[1 + \left(\frac{\tau'}{\tau}\right)^2\right] + 7\mathfrak{m}\right\} \qquad (233)$$

By (121) and (112)

$$\alpha - \mathrm{a} = \tfrac{1}{2}\frac{\tau'}{\Omega}$$

Since $\Omega = \Omega_0/\xi^3$, and τ' is constant (or at least varies so slowly that we may neglect its variation), we have

$$\frac{\alpha' - \mathrm{a}'}{\alpha - \mathrm{a}} = \frac{3}{\xi}\frac{d\xi}{dt} = \left(\frac{1}{k}\frac{d\xi}{dt}\right)\frac{1}{n}3\mathfrak{m}$$

Now $\dfrac{\Omega}{n} = \lambda$, hence $\qquad \alpha - \mathrm{a} = \dfrac{\tau\mathfrak{e}}{n}\dfrac{\tau'}{\tau}\left(\dfrac{1}{2\lambda\mathfrak{e}}\right)$

Therefore $\qquad \alpha' - \mathrm{a}' = \left(\dfrac{1}{k}\dfrac{d\xi}{dt}\right)\left(\dfrac{\tau\mathfrak{e}}{n^2}\right)\dfrac{\tau'}{\tau}\dfrac{3\mathfrak{m}}{2\lambda\mathfrak{e}}$ ....(234)

From (233–4)

$$\alpha' = \left(\frac{1}{k}\frac{d\xi}{dt}\right)\left(\frac{\tau\mathfrak{e}}{n^2}\right)\mathfrak{m}\left\{\frac{3}{2\lambda\mathfrak{e}}\frac{\tau'}{\tau} - \left[2\left\{1 + \left(\frac{\tau'}{\tau}\right)^2\right\} + 7\mathfrak{m}\right]\right\} \quad \text{......(235)}$$

Also $\qquad \alpha = \dfrac{\tau\mathfrak{e}}{n}\left[\dfrac{\tau'}{\tau}\dfrac{1}{2\lambda\mathfrak{e}} + \mathfrak{m}\right]$ ....(236)

By (121) and (112) $\qquad \mathrm{b} = \dfrac{\tau\mathfrak{e}}{n} = \dfrac{\tau_0}{2\mathfrak{g}}\dfrac{n}{\xi^6}$ ....(237)

Therefore $\qquad \dfrac{\mathrm{b}'}{\mathrm{b}} = \dfrac{1}{n}\dfrac{dn}{dt} - \dfrac{6}{\xi}\dfrac{d\xi}{dt}$

$$= -\left(\frac{1}{k}\frac{d\xi}{dt}\right)\frac{1}{n}\left\{1 + \left(\frac{\tau'}{\tau}\right)^2 + 6\mathfrak{m}\right\} \quad \text{.................(238)}$$

And $\qquad \mathrm{b}' = -\left(\dfrac{1}{k}\dfrac{d\xi}{dt}\right)\left(\dfrac{\tau\mathfrak{e}}{n^2}\right)\left\{1 + \left(\dfrac{\tau'}{\tau}\right)^2 + 6\mathfrak{m}\right\}$ ....(239)

From (231) and (238)

$$\frac{\mathrm{a}'}{\mathrm{a}} - \frac{\mathrm{b}'}{\mathrm{b}} = -\left(\frac{1}{k}\frac{d\xi}{dt}\right)\frac{1}{n}\left\{1 + \left(\frac{\tau'}{\tau}\right)^2 + \mathfrak{m}\right\} \quad \text{..............(240)}$$

By (121) and (112)

$$\beta - \mathrm{b} = \frac{\tau'\mathfrak{e}}{n} = \frac{\tau'}{2\mathfrak{g}}n$$

$$\beta' - \mathrm{b}' = \frac{\tau'}{2\mathfrak{g}}\frac{dn}{dt} = -\left(\frac{1}{k}\frac{d\xi}{dt}\right)\frac{\tau'}{2\mathfrak{g}}\left\{1 + \left(\frac{\tau'}{\tau}\right)^2\right\}$$

$$= -\left(\frac{1}{k}\frac{d\xi}{dt}\right)\left(\frac{\tau\mathfrak{e}}{n^2}\right)\left(\frac{\tau'}{\tau}\right)\left\{1 + \left(\frac{\tau'}{\tau}\right)^2\right\} \quad \text{..............(241)}$$

Therefore

$$\beta' = -\left(\frac{1}{k}\frac{d\xi}{dt}\right)\left(\frac{\tau\mathfrak{e}}{n^2}\right)\left\{1 + \frac{\tau'}{\tau} + \left(\frac{\tau'}{\tau}\right)^2 + \left(\frac{\tau'}{\tau}\right)^3 + 6\mathfrak{m}\right\} \quad \text{........(242)}$$

Lastly $\qquad \beta = \dfrac{\tau\mathfrak{e}}{n}\left(1 + \dfrac{\tau'}{\tau}\right)$ .... (243)

By (174), (227), and (230), when the viscosity is not small, we have

$$\left.\begin{aligned}
\Gamma &= \left(\frac{1}{k}\frac{d\xi}{dt}\right)\frac{\mathfrak{m}}{2n}\frac{(\sin 4\mathrm{f}_1 - \sin 2\mathrm{g}_1 + \sin 2\mathrm{g})}{\sin 4\mathrm{f}_1} \\
\mathrm{G} &= \left(\frac{1}{k}\frac{d\xi}{dt}\right)\frac{\mathfrak{m}}{2n}\frac{(\sin 4\mathrm{f}_1 - \sin 2\mathrm{g}_1 + \sin 2\mathrm{g}) + \frac{\tau'}{\tau}\sin 2\mathrm{g}}{\sin 4\mathrm{f}_1} \\
\Delta &= \left(\frac{1}{k}\frac{d\xi}{dt}\right)\frac{1}{2n}\frac{(\sin 4\mathrm{f}_1 + \sin 2\mathrm{g}_1 - \sin 2\mathrm{g}) - 2\frac{\tau'}{\tau}\sin 2\mathrm{g} + \left(\frac{\tau'}{\tau}\right)^2\sin 4\mathrm{f}}{\sin 4\mathrm{f}_1} \\
\mathrm{D} &= \left(\frac{1}{k}\frac{d\xi}{dt}\right)\frac{1}{2n}\frac{(\sin 4\mathrm{f}_1 + \sin 2\mathrm{g}_1 - \sin 2\mathrm{g}) - \frac{\tau'}{\tau}\sin 2\mathrm{g}}{\sin 4\mathrm{f}_1}
\end{aligned}\right\} \quad (244)$$

If the viscosity be small we have by (179), (227), and (230)

$$\left.\begin{aligned}
\gamma &= \left(\frac{1}{k}\frac{d\xi}{dt}\right)\frac{\mathfrak{m}}{2n}\frac{1}{1-\lambda} \\
\mathrm{g} &= \left(\frac{1}{k}\frac{d\xi}{dt}\right)\frac{\mathfrak{m}}{2n}\frac{1+\frac{1}{2}\frac{\tau'}{\tau}}{1-\lambda} \\
\delta &= \left(\frac{1}{k}\frac{d\xi}{dt}\right)\frac{1}{2n}\frac{1-2\lambda-\frac{\tau'}{\tau}+\left(\frac{\tau'}{\tau}\right)^2}{1-\lambda} \\
\mathrm{d} &= \left(\frac{1}{k}\frac{d\xi}{dt}\right)\frac{1}{2n}\frac{1-2\lambda-\frac{1}{2}\frac{\tau'}{\tau}}{1-\lambda}
\end{aligned}\right\} \quad \text{..............(245)}$$

I think no confusion will arise between the distinct uses made of the symbol g in (244) and (245); in the first it always must occur with a sine, in the second it never can do so.

[If τ' be zero
$$\mathrm{bG} + \mathrm{aD} = -\left(\frac{1}{k}\frac{d\xi}{dt}\right)\left(\frac{\mathfrak{m}}{2n}\right)\left(\frac{\tau\mathfrak{L}}{n}\right)(2)$$

$$\mathrm{a} + \mathrm{b} = \frac{\tau\mathfrak{L}}{n}(1+\mathfrak{m})$$

and by (232), (237), (240)

$$\mathrm{ab}' - \mathrm{ba}' = \left(\frac{1}{k}\frac{d\xi}{dt}\right)\frac{\mathfrak{m}}{n}\left(\frac{\tau\mathfrak{L}}{n}\right)^2(1+\mathfrak{m})$$

Therefore we have

$$(\mathrm{bG} + \mathrm{aD})(\mathrm{a} + \mathrm{b}) + \mathrm{a}'\mathrm{b} - \mathrm{b}'\mathrm{a} = 0$$

This was shown in (226) to be the criterion that the differential equations (224) should reduce to those of (71) and of (29) of "Precession," when the solar influence is evanescent, and the above is the promised proof thereof.]

From (244), (237), and (232) we have

$$\mathrm{bG} - \mathrm{aD} = \left(\frac{1}{k}\frac{d\xi}{dt}\right)\left(\frac{\mathfrak{m}}{2n}\right)\left(\frac{\tau\mathfrak{e}}{n}\right)\frac{2\left(1+\frac{\tau'}{\tau}\right)\sin 2\mathrm{g} - 2\sin 2\mathrm{g}_1}{\sin 4\mathrm{f}_1} \quad \text{...(246)}$$

and similarly

$$\mathrm{bg} - \mathrm{ad} = \left(\frac{1}{k}\frac{d\xi}{dt}\right)\left(\frac{\mathfrak{m}}{2n}\right)\left(\frac{\tau\mathfrak{e}}{n}\right)\frac{2\lambda + \frac{\tau'}{\tau}}{1-\lambda} \quad \text{...............(247)}$$

§ 17. *Change of independent variable, and formation of equations for integration.*

In the equations (224) the time t is the independent variable, but in order to integrate we shall require ξ to be the variable. It has been shown above that these equations are equivalent to only two of them; henceforth therefore we shall only consider the first and last of them. It will also serve to keep before us the physical meaning of the L's, if the notation be changed; the following notation (which has been already used in (127)) will be adopted:

$\mathrm{J} = L_1$ = the inclination of the lunar orbit to the lunar proper plane.
$\mathrm{I} = L_2'$ = the inclination of the earth's proper plane to the ecliptic.
$\mathrm{I}_{,} = L_1'$ = the inclination of the equator to the earth's proper plane.
$\mathrm{J}_{,} = -L_2$ = the inclination of the lunar proper plane to the ecliptic.

Since J, I, &c., are small, we may write

$$\frac{dL_1}{L_1} = d\,.\log\tan\tfrac{1}{2}\mathrm{J}, \quad \frac{dL_2'}{L_2'} = d\,.\log\tan\tfrac{1}{2}\mathrm{I} \quad \text{...........(248)}$$

This particular transformation is chosen because in Part II., where j and i were not small, $dj/\sin j$ seemed to arise naturally.

Also since

$$\frac{L_1'}{L_1} = -\frac{\kappa_1+\alpha}{\mathrm{a}}, \quad \frac{L_2'}{L_2} = -\frac{\kappa_2+\alpha}{\mathrm{a}}$$

we have

$$\left.\begin{aligned}\sin \mathrm{I}_{,} &= -\frac{\kappa_1+\alpha}{\mathrm{a}}\sin \mathrm{J}\\ \sin \mathrm{J}_{,} &= \frac{\mathrm{a}}{\kappa_2+\alpha}\sin \mathrm{I}\end{aligned}\right\} \quad \text{.......................(249)}$$

These equations will give $\mathrm{I}_{,}$ and $\mathrm{J}_{,}$, when J and I are found.

Suppose we divide the first and last of (224) by $d\xi/nkdt$, then their left-hand sides may be written

$$nk\frac{d}{d\xi}\log\tan\tfrac{1}{2}\mathrm{J} \quad \text{and} \quad nk\frac{d}{d\xi}\log\tan\tfrac{1}{2}\mathrm{I}$$

In the last section we have determined the functions α, α', &c., and have them in such a form that Γ, G, Δ, D (or γ, g, δ, d) have a common factor $d\xi/nkdt$.

But this is the expression by which we have to divide the equations in order to change the variable.

Therefore in computing Γ, G, &c. (or γ, g, &c.), we may drop this common factor.

Again α, a, β, b were so written as to have a common factor $\tau\mathfrak{e}/n$; therefore κ_1 and κ_2 also have the same common factor.

Also α', a', β', b' have a common factor $(d\xi/kdt)(\tau\mathfrak{e}/n^2)$.

From this it follows that when the variable is changed, we may drop the factor $\tau\mathfrak{e}/n$ from α, a, β, b, κ_1, κ_2 and the factor $(d\xi/kdt)(\tau\mathfrak{e}/n^2)$ from α', a', β', b'.

Hence the differential equations with the new variable become

$$\left.\begin{aligned} kn\frac{d}{d\xi}\log\tan\tfrac{1}{2}J &= \frac{1}{(\kappa_1-\kappa_2)^2}\left\{-(\kappa_1+\alpha)(\alpha'-\beta')-a'b\frac{\kappa_1+\alpha}{\kappa_2+\alpha}-b'a\right\} \\ &\quad+\frac{1}{(\kappa_1-\kappa_2)}\{\gamma(\kappa_2+\alpha)+\delta(\kappa_1+\alpha)+gb-da\} \\ kn\frac{d}{d\xi}\log\tan\tfrac{1}{2}I &= \frac{1}{(\kappa_1-\kappa_2)^2}\left\{-(\kappa_1+\alpha)(\alpha'-\beta')-a'b-b'a\frac{\kappa_1+\alpha}{\kappa_2+\alpha}\right\} \\ &\quad-\frac{1}{(\kappa_1-\kappa_2)}\{\gamma(\kappa_1+\alpha)+\delta(\kappa_2+\alpha)+gb-da\} \end{aligned}\right\}\quad(250)$$

or similar equations with Γ, G, Δ, D in place of γ, g, δ, d if the viscosity be not small.

But we now have by (232-3-5-6-7-9, 242-3-4-5-6-7)

$$\left.\begin{aligned} &\alpha=\mathfrak{m}+\frac{\tau'}{\tau}\frac{1}{2\lambda\mathfrak{e}},\quad a=\mathfrak{m},\quad \beta=1+\frac{\tau'}{\tau},\quad b=1 \\ &\alpha'=\mathfrak{m}\left\{\frac{\tau'}{\tau}\frac{3}{2\lambda\mathfrak{e}}-\left[2\left\{1+\left(\frac{\tau'}{\tau}\right)^2\right\}+7\mathfrak{m}\right]\right\},\quad a'=-\mathfrak{m}\left\{2\left[1+\left(\frac{\tau'}{\tau}\right)^2\right]+7\mathfrak{m}\right\} \\ &\beta'=-\left\{1+\frac{\tau'}{\tau}+\left(\frac{\tau'}{\tau}\right)^2+\left(\frac{\tau'}{\tau}\right)^3+6\mathfrak{m}\right\},\quad b'=-\left\{1+\left(\frac{\tau'}{\tau}\right)^2+6\mathfrak{m}\right\} \\ &\Gamma=\tfrac{1}{2}\mathfrak{m}\frac{\sin 4f_1-\sin 2g_1+\sin 2g}{\sin 4f_1} \\ &\Delta=\frac{(\sin 4f_1+\sin 2g_1-\sin 2g)-2\dfrac{\tau'}{\tau}\sin 2g+\left(\dfrac{\tau'}{\tau}\right)^2\sin 4f}{2\sin 4f_1} \\ &\gamma=\frac{\mathfrak{m}}{2(1-\lambda)},\quad \delta=\frac{1-2\lambda-\dfrac{\tau'}{\tau}+\left(\dfrac{\tau'}{\tau}\right)^2}{2(1-\lambda)} \\ &bG-aD=\tfrac{1}{2}\mathfrak{m}\frac{2\left(1+\dfrac{\tau'}{\tau}\right)\sin 2g-2\sin 2g_1}{\sin 4f_1} \\ &bg-ad=\frac{\mathfrak{m}\left(2\lambda+\dfrac{\tau'}{\tau}\right)}{2(1-\lambda)} \end{aligned}\right\}\quad\ldots\ldots(251)$$

In these equations we have, recapitulating the notation,

$$\mathfrak{m} = \frac{kn}{\xi}, \quad \lambda = \frac{\Omega}{n}, \quad \mathfrak{e} = \tfrac{1}{2}\frac{n^2}{\mathfrak{g}} \quad \text{......................(252)}$$

Also

$$\left.\begin{aligned} \kappa_1 + \kappa_2 &= -\alpha - \beta \\ \kappa_1 - \kappa_2 &= -\sqrt{(\alpha-\beta)^2 + 4\text{ab}} \end{aligned}\right\} \quad \text{......................(253)}$$

Lastly we have by (186)

$$\frac{n}{n_0} = 1 + \frac{1}{kn_0}\left\{(1-\xi) + \tfrac{1}{13}\left(\frac{\tau'}{\tau_0}\right)^2 (1-\xi^{13})\right\} \quad \text{............(254)}$$

which gives parallel values of n and ξ.

These equations will be solved by quadratures for the case of the moon and earth in Part IV.

If τ'/τ be so small as to be negligeable, and $\tau'/2\lambda\mathfrak{e}\tau$ small compared with unity, the equations (250) admit of reduction to a simple form.

With this hypothesis it is easy to find approximate values of κ_1 and κ_2, and then by some easy, but rather tedious analysis, it may be shown that (250) reduce to the following:

$$\left.\begin{aligned} kn\frac{d}{d\xi}\log\tan\tfrac{1}{2}\text{J} &= -\frac{\mathfrak{m}+1}{\mathfrak{m}}\text{G} + \frac{\tau'}{\tau}\frac{1}{2\lambda\mathfrak{e}}\frac{1+11\mathfrak{m}}{(1+\mathfrak{m})^2} \\ kn\frac{d}{d\xi}\log\tan\tfrac{1}{2}\text{I} &= \frac{\tau'}{\tau}\frac{1}{2\lambda\mathfrak{e}}\frac{1+11\mathfrak{m}}{(1+\mathfrak{m})^2} \end{aligned}\right\} \quad \text{.........(255)}$$

These equations would give the secular changes of J and I, when the solar influence is very small compared with that of the moon. Of course if G be replaced by g, they are applicable to the case of small viscosity.

It is remarkable that the changes of I are independent of the viscosity; they depend in fact solely on the secular change in the permanent ellipticity of the earth.

IV.

Integration of the Differential Equations for Changes in the Inclination of the Orbit and the Obliquity of the Ecliptic.

§ 18. *Integration in the case of small viscosity, where the nodes revolve uniformly.*

It is not, even at the present time, rigorously true that the nodes of the lunar orbit revolve uniformly on the ecliptic and that the inclination of the orbit is constant; but it is very nearly true, and the integration may be carried backwards in time for a long way without an important departure from accuracy.

The integrations will be carried out by the method of quadratures, and the process will be divided into a series of "periods of integration," as explained in § 15 and § 17 of the paper on "Precession." These periods will be the same as those in that paper, and the previous numerical work will be used as far as possible. It will be found, however, that it is not sufficiently accurate to assume the uniform revolution of the nodes beyond the first two periods of integration. For these first two periods the equations of § 7, Part II., will be used; but for the further retrospect we shall have to make the transition to the methods of Part III. It is important to defer the transition as long as possible, because Part III. assumes the smallness of i and j, whilst Part II. does not do so.

By (104) and (86) of Part II. we have, when $j' = 0$, and Ω'/n is neglected,

$$\frac{di}{dt} = \frac{\sin 4\mathfrak{f}}{n\mathfrak{g}} \tfrac{1}{4} \sin i \cos i \left\{\tau^2 (1 - \tfrac{3}{2}\sin^2 j) + \tau'^2 - \frac{2\Omega}{n}\tau^2 \sec i \cos j \right.$$
$$\left. - \tau\tau'(1 - \tfrac{3}{2}\sin^2 j)\right\}$$

$$-\frac{dn}{dt} = \frac{\sin 4\mathfrak{f}}{n\mathfrak{g}} \left\{(1 - \tfrac{1}{2}\sin^2 i)(\tau^2 + \tau'^2) - \tfrac{1}{2}(1 - \tfrac{3}{2}\sin^2 i)\tau^2 \sin^2 j \right.$$
$$\left. - \tau^2 \frac{\Omega}{n} \cos i \cos j + \tfrac{1}{2}\tau\tau' \sin^2 i (1 - \tfrac{3}{2}\sin^2 j)\right\}$$

If we put $1 - \frac{1}{2}\sin^2 i = \cos i$, $1 - \frac{3}{2}\sin^2 j = \cos^3 j$, and neglect $\sin^2 i \sin^2 j$, these may be written

$$\left.\begin{aligned} \frac{di}{dt} &= \frac{\sin 4\mathfrak{f}}{n\mathfrak{g}} \tfrac{1}{4} \sin i \cos i \cos^3 j \left\{\tau^2 + \tau'^2 \sec^3 j - \tau\tau' - \frac{2\Omega}{n}\tau^2 \sec i \sec^2 j\right\} \\ -\frac{dn}{dt} &= \frac{\sin 4\mathfrak{f}}{n\mathfrak{g}} \cos i \cos j \left\{\tau^2 + \tau'^2 \sec j - \tau^2 \frac{\Omega}{n} + \tfrac{1}{2}\tau\tau' \sin i \tan i \cos^2 j\right\} \end{aligned}\right\} \quad (256)$$

If we treat $\sec j$ and $\cos j$ as unity in the small terms in τ'^2, $\tau\tau'$, and Ω/n, (256) only differ from (83) of "Precession" in that di/dt has a factor $\cos^3 j$ and dn/dt has a factor $\cos j$.

Again by (64) and (70)

$$\left.\begin{aligned} -\frac{1}{k}\frac{dj}{dt} &= \frac{1}{\xi}\frac{\sin 4\mathfrak{f}}{\mathfrak{g}} \tau^2 \tfrac{1}{4} \cos i \sin j \\ \frac{1}{k}\frac{d\xi}{dt} &= \frac{\sin 4\mathfrak{f}}{\mathfrak{g}} \tau^2 \tfrac{1}{2} \cos i \cos j \left(1 - \frac{\Omega}{n} \sec i \sec j\right) \end{aligned}\right\} \quad \text{.........(257)}$$

If we divide the second of (256) by the second of (257) we get an equation for $dn/d\xi$, which only differs from (84) of "Precession" in the presence of $\sec j$ in place of unity in certain of the small terms. Now j is small for the lunar orbit; hence the equation (88) of "Precession" for the conservation of moment of momentum is very nearly true. The equation is, with present notation,

$$\frac{n}{n_0} = 1 + \frac{1}{kn_0}\left[1 - \xi + \tfrac{1}{13}\left(\frac{\tau'}{\tau_0}\right)^2 (1 - \xi^{13}) + \tfrac{1}{14}\frac{\tau'}{\tau_0}\sin i \tan i\,(1 - \xi^7)\right]$$
$$+ \tfrac{1}{4}\sin^2 i\,\frac{\Omega_0}{n_0}\,\frac{1}{kn_0+1}\left(\frac{1}{\xi} - 1\right)\left(\frac{1}{\xi} + \frac{kn_0+3}{kn_0+1}\right)$$
$$+ \tfrac{1}{2}\sin^2 i\,\frac{\Omega_0}{n_0}\,\frac{1}{(kn_0+1)^3}\log_e\left(\frac{kn_0+1-\xi}{kn_0\xi}\right) \quad\text{.....................(258)}$$

In this equation we attribute to i, as it occurs on the right-hand side, an average value.

By means of this equation, I had already computed a series of values of n corresponding to equidistant values of ξ.

On dividing the first of (256) by the second of (257) we get an expression which differs from the $d\log\tan^2\frac{1}{2}i/d\xi$ of (84) of "Precession" by the presence of a common factor $\cos^2 j$, and by $\sec j$ occurring in some of the small terms. Hence we may, without much error, accept the results of the integration for i in § 17 of "Precession."

Lastly, dividing the first of (257) by the second, we have

$$\frac{d}{d\xi}\log\sin j = \frac{-1}{2\xi\left(1 - \dfrac{\Omega}{n}\sec i \sec j\right)} \quad\text{..............(259)}$$

This equation has now to be integrated by quadratures.

All the numerical values were already computed for § 17 of "Precession," and only required to be combined.

The present mean inclination of the lunar orbit is 5° 9′, so that $j_0 = 5° 9'$. I then conjecture 5° 12′ as a proper mean value to be assigned to j, as it occurs on the right-hand side of (259) for the first period of integration, which extends from $\xi = 1$ to ·88.

First period of integration.

From $\xi = 1$ to ·88, four equidistant values were computed.

From the computation for § 17 of "Precession" I extract the following:

ξ	$=1$	·96	·92	·88
$\log\dfrac{\Omega}{n}\sec i + 10$	$=8{\cdot}59979$	8·57309	8·56411	8·56746

Introducing $j = 5° 12'$, I find

ξ	$=1$	·96	·92	·88
$\left[2\xi\left(1 - \dfrac{\Omega}{n}\sec i \sec j\right)\right]^{-1}$	$=\cdot5208$	·5412	·5643	·5901

Combining these four values by the rules of the calculus of finite differences, we have

$$\int_{\cdot 88}^{1} \frac{d\xi}{2\xi\left(1-\frac{\Omega}{n}\sec i \sec j\right)} = \cdot 06641$$

This is equal to $\log_e \sin j - \log_e \sin j_0$. Taking $j_0 = 5° \, 9'$, I find $j = 5° \, 30'$.

Second period of integration.

From $\xi = 1$ to ·76, four equidistant values were computed.

From the computation for § 17 "Precession," I extract the following:

ξ	$=1$	·92	·84	·76
$\log \frac{\Omega}{n}\sec i + 10$	$= 8{\cdot}56746$	8·59743	8·65002	8·72318

Assuming 5° 55′ as an average value for j, I find

ξ	$=1$	·92	·84	·76
$\left[2\xi\left(1-\frac{\Omega}{n}\sec i \sec j\right)\right]^{-1}$	$= {\cdot}5193$	·5660	·6232	·6948

Combining these, we have

$$\int_{\cdot 76}^{1} \frac{d\xi}{2\xi\left(1-\frac{\Omega}{n}\sec i \sec j\right)} = \cdot 14345$$

This is equal to $\log_e \sin j - \log_e \sin j_0$. Taking $j_0 = 5° \, 30'$ from the first period, we find $j = 6° \, 21'$.

This completes the integration, so far as it is safe to employ the methods of Part II.

In Part III. it was proved that, in the case where the nodes revolve uniformly, equations (224) reduce to those of Part II. But it was also shown that what the equations of Part II. really give is the change of the inclination of the lunar orbit to the lunar proper plane; also that the equations of "Precession" really give the change of the inclination of the mean equator (that is of the earth's proper plane) to the ecliptic.

The results of the present integration are embodied in the following table, of which the first three columns are taken from the table in § 17 of "Precession."

TABLE I.

Sidereal day in m.s. hours and minutes			Moon's sidereal period in m.s. days	Inclination of mean equator to ecliptic	Inclination of lunar orbit to lunar proper plane
	h.	m.	Days	° ′	° ′
Initial	23	56	27·32	23 28	5 9
	15	28	18·62	20 28	5 30
Final	9	55	8·17	17 4	6 21

We will now consider what amount of oscillation the equator and the plane of the lunar orbit undergo, as the nodes revolve, in the initial and final conditions represented in the above table.

It appears from (119) that $\sin 2j$ oscillates between $\sin 2j_0 \pm \mathrm{a} \sin 2i_0/(\kappa_2+\alpha)$, and that $\sin 2i$ oscillates between $\sin 2i_0 \pm (\kappa_1+\alpha)\sin 2j_0/\mathrm{a}$, where i_0 and j_0 are the mean values of i and j.

With the numerical values corresponding to the initial condition (that is to say in the present configurations of earth, moon, and sun), it will be found on substituting in (115) and (112), with $a_2 = \frac{3}{4}\left(\frac{\Omega'}{\Omega}\right)^2\left(1-\frac{3}{8}\frac{\Omega'}{\Omega}\right)\Omega$ instead of simply $\frac{1}{2}\frac{\tau'}{\Omega}$, that

$$\alpha = \cdot 341251, \quad \beta = \cdot 000318, \quad \mathrm{a} = \cdot 000059, \quad \mathrm{b} = \cdot 000150$$

when the present tropical year is the unit of time.

Since 4ab is very small compared with $(\alpha-\beta)^2$, it follows that we have to a close degree of approximation

$$\kappa_1 = -\alpha, \quad \kappa_2 = -\beta$$

Since $(\kappa_1+\alpha)/\mathrm{a} = \mathrm{b}/(\kappa_1+\beta)$, it follows that $\sin 2j$ oscillates between $\sin 2j_0 \pm \mathrm{a}\sin 2i_0/(\alpha-\beta)$, and $\sin 2i$ between $\sin 2i_0 \mp \mathrm{b}\sin 2j_0/(\alpha-\beta)$.

Let δj and δi be the oscillations of j and i on each side of the mean, then $\delta \sin 2j = \mathrm{a}\sin 2i/(\alpha-\beta)$ and $\delta \sin 2i = \mathrm{b}\sin 2j/(\alpha-\beta)$.

Hence in seconds of arc

$$\left.\begin{aligned} \delta j &= \frac{648000}{\pi}\,\frac{1}{2}\,\frac{\mathrm{a}}{\alpha-\beta}\,\frac{\sin 2i}{\cos 2j} \\ \delta i &= \frac{648000}{\pi}\,\frac{1}{2}\,\frac{\mathrm{b}}{\alpha-\beta}\,\frac{\sin 2j}{\cos 2i} \end{aligned}\right\} \qquad (260)$$

Reducing these to numbers with $j = 5°\ 9'$, $i = 23°\ 28'$, we have $\delta j = 13''\cdot 13$, $\delta i = 11''\cdot 86$*.

* The formulæ here used for the amplitude of the 19-yearly nutation and for the inclination of the lunar proper plane to the ecliptic differ so much from those given by other writers that it will be well to prove their identity.

Hence, if the earth were homogeneous, at the present time we should have δj as the inclination of the proper plane of the lunar orbit to the

---

Laplace (*Méc. Cél.*, liv. vii., chap. 2) gives as the inclination of the proper plane to the ecliptic

$$\frac{a\rho - \frac{1}{2}a\phi}{g-1}\frac{D^2}{a^2}\sin\lambda\cos\lambda$$

Here $a\rho$ is the earth's ellipticity, and is my $\mathfrak{e}$; $a\phi$ is the ratio of equatorial centrifugal force to gravity, and is my n^2a/g, it is therefore $\frac{4}{5}\mathfrak{e}$ when the earth is homogeneous.

Thus his $a\rho - \frac{1}{2}a\phi =$ my $\frac{3}{5}\mathfrak{e}$. His $g-1$ is the ratio of the angular velocity of the nodes to that of the moon, and is therefore my $(\alpha-\beta)/\Omega$. His D is the earth's mean radius, and is my a. His a is the moon's mean distance, and is my c. His λ is the obliquity, and is my i. Thus his formula is $\frac{3}{5}\frac{\mathfrak{e}\Omega}{\alpha-\beta}\frac{a^2}{c^2}\sin i\cos i$ in my notation.

Now my $\tau = 3\mu m/2c^3$, and $\frac{2}{5}a^2 = C/M$.

Therefore the formula becomes

$$\frac{1}{2}\frac{\tau\mathfrak{e}}{\alpha-\beta}(\Omega c)\frac{C}{\mu Mm}\sin 2i$$

But by (5) $C\Omega c/\mu Mm = k$.

Therefore it becomes

$$\frac{1}{2}\frac{k\tau\mathfrak{e}}{\alpha-\beta}\sin 2i$$

By (115) and (112), when $\xi=1$, $\mathrm{a} = k\tau\mathfrak{e}\cos j\cos 2j$.

Therefore in my notation Laplace's result for the inclination of the lunar proper plane to the ecliptic is

$$\frac{1}{2}\frac{\mathrm{a}}{\alpha-\beta}\frac{\sin 2i}{\cos 2j}\sec j$$

This agrees with the result (260) in the text, from which the amount of oscillation of the lunar orbit was computed, save as to the sec j. Since j is small the discrepancy is slight, and I believe my form to be the more accurate.

Laplace states that the inclination is 20″·023 (centesimal) if the earth be heterogeneous, and 41″·470 (centesimal) if homogeneous. Since 41″·470 (centes.) = 13″·44, this result agrees very closely with mine. The difference of Laplace's data explains the discrepancy.

If it be desired to apply my formula to the heterogeneous earth we must take $\frac{5}{6}$ of my k, because the $\frac{2}{5}$ of the formula (6) for s will be replaced by $\frac{1}{3}$ nearly. Also $\mathfrak{e}$, which is $\frac{1}{232}$, must be replaced by the precessional constant, which is ·003272. Hence my previous result in the text must be multiplied by $\frac{5}{6}$ of 232 × ·003272 or ·6326. This factor reduces the 13″·13 of the text to 8″·31. Laplace's result (20″·023 centes.) is 6″·49. Hence there is a small discrepancy in the results; but it must be remembered that Laplace's value of the actual ellipticity (1/334 instead of 1/295) of the earth was considerably in error. The more correct result is I think 8″·31. The amount of this inequality was found by Burg and Burckhardt from the combined observations of Bradley and Maskelyne to be 8″ (Grant's *Hist. Phys. Astr.*, 1852, p. 65).

For the amplitude of the 19-yearly nutation, Airy gives (*Math. Tracts*, 1858, article "On Precession and Nutation," p. 214)

$$\frac{6\pi^2 B}{T'^2\omega\,(n+1)}\frac{\tau}{4\pi}\cos I\sin 2i$$

B is the precess. const. = my $\mathfrak{e}$; his $T' =$ my $2\pi/\Omega$; his $n =$ my ν; his $\omega =$ my n; his I = my i; his $i =$ my j; and his τ is the period of revolution of the nodes, and therefore = my $2\pi/(\alpha-\beta)$.

Then since my $\tau = 3\Omega^2/2\,(1+\nu)$, the above in my notation is

$$\frac{1}{2}\frac{\tau\mathfrak{e}}{n}\frac{1}{\alpha-\beta}\cos i\sin 2j$$

ecliptic, and δi as the amplitude of the 19-yearly nutation. These are very small angles, and therefore initially the method of Part II. was applicable.

Now consider the final condition.

Since the integrations of the two periods have extended from $\xi = 1$ to ·88, and again from $\xi = 1$ to ·76,

$$\tau = \tau_0 (\cdot 88 \times \cdot 76)^{-6}, \quad \Omega = \Omega_0 (\cdot 88 \times \cdot 76)^{-3}, \quad k = k_0 (\cdot 88 \times \cdot 76)^{-1}$$

also the value of n which gives the day of 9 hrs. 55 m. is given by $\log n = 3{\cdot}74451$, and $\log \mathfrak{g} + 10 = 1{\cdot}21217$, when the year is the unit of time.

We now have $i = 17° \, 4'$, $j = 6° \, 21'$.

Using these values in (115) and (112), I find

$$\alpha = \cdot 10872, \quad \beta = \cdot 00627, \quad \mathrm{a} = \cdot 00563, \quad \mathrm{b} = \cdot 00510$$

ab is still small compared with $(\alpha - \beta)$, but not negligeable.

By (117)

$$\kappa_1 - \kappa_2 = -\sqrt{(\alpha - \beta)^2 + 4\mathrm{ab}} = -(\alpha - \beta) - \frac{2\mathrm{ab}}{\alpha - \beta}, \text{ also } \kappa_1 + \kappa_2 = -(\alpha + \beta)$$

Now $2\mathrm{ab}/(\alpha - \beta) = \cdot 00056$.

Hence we have

$$\left.\begin{aligned} \kappa_1 + \kappa_2 &= -\cdot 11499 \\ \kappa_1 - \kappa_2 &= -\cdot 10301 \end{aligned}\right\} \text{ whence } \begin{aligned} \kappa_1 &= -\cdot 10900 \\ \kappa_2 &= -\cdot 00599 \end{aligned}$$

κ_1 and κ_2 have now come to differ a little from $-\alpha$ and $-\beta$, but still not much. With these values I find

$$\log \frac{\mathrm{a}}{\kappa_2 + \alpha} + 10 = 8{\cdot}76472, \quad \log \frac{-\mathrm{b}}{\kappa_1 + \beta} + 10 = 8{\cdot}69606$$

Substituting in the formulæ

$$\delta j = \tfrac{1}{2} \frac{\mathrm{a}}{\kappa_2 + \alpha} \frac{\sin 2i}{\cos 2j}, \quad \delta i = \tfrac{1}{2} \frac{\mathrm{b}}{\kappa_1 + \beta} \frac{\sin 2j}{\cos 2i}$$

I find $\delta j = 57' \, 31''$, $\delta i = 22' \, 42''$

Now by (115) and (112) $\mathrm{b} = \frac{\tau\mathfrak{e}}{n} \cos i \cos 2i$, when $\xi = 1$.

Therefore his result in my notation is

$$\tfrac{1}{2} \frac{\mathrm{b}}{\alpha - \beta} \frac{\sin 2j}{\cos 2i}$$

This is the result used above (in 260) for computing the nutations of the earth.

If my formula is to be used for the heterogeneous earth, $\mathfrak{e}$ must be replaced by the precessional constant, and therefore the result in the text must be multiplied by 232 × ·003272 or ·759. Hence for the heterogeneous earth the 11″·86 must be reduced to 9″·01. Airy computes it as 10″·33, but says the observed amount is 9″·6, but he takes the precessional constant as ·00317, and the moon's mass as 1–70th of that of the earth. I believe that ·00327 and 1–82nd are more in accordance with the now accepted views of astronomers.

Thus the oscillation of the lunar orbit has increased from 13″ to nearly a degree, and that of the equator from 12″ to 23′.

It is clear therefore that we have carried out the integration by the method of Part II., as far back in retrospect as is proper, even for a speculative investigation like the present one.

We shall here then make the transition to the method of Part III.

Henceforth the formulæ used regard the inclination and obliquity as small angles; the obliquity is still however so large that this is not very satisfactory.

§ 19. *Secular changes in the proper planes of the earth and moon when the viscosity is small.*

We now take up the integration, at the point where it stops in the last section, by the method of Part III. The viscosity is still supposed to be small, so that γ, δ, g, d (as defined in (251)) must be taken in place of Γ, Δ, G, D, which refer to any viscosity. The equations are ready for the application of the method of quadratures in (250), and the symbols are defined in (251–4).

The method pursued is to assume a series of equidistant values of ξ, and then to compute all the functions (251–4), substitute them in (250), and combine the equidistant values of the functions to be integrated by the rules of the calculus of finite differences.

The preceding integration terminates where the day is 9 hrs. 55 m., and the moon's sidereal period is 8·17 m.s. days. If the present tropical year be the unit of time, we have, at the beginning of the present integration $\log n_0 = 3{\cdot}74451$, $\log \Omega_0 = 2{\cdot}44836$, and $\log k + 10 = 6{\cdot}20990$, k being $s\Omega_0^{\frac{1}{3}}$ of (7).

The first step is to compute a series of values of n/n_0, by means of (254). As a fact, I had already computed n/n_0 corresponding to $\xi = 1$, ·92, ·84, ·76 for the paper on "Precession," by means of a formula, which took account of the obliquity of the ecliptic; and accordingly I computed n/n_0, by the same formula, for the values of $\xi =$ ·96, ·88, ·80, instead of doing the whole operation by means of (254). The difference between my results here used and those from (254) would be very small.

The following table exhibits some of the stages of the computation. The results are given just as they were found, but it is probable that the last place of decimals, and perhaps the last but one, are of no value. As however we really only require a solution in round numbers, this is of no importance.

TABLE II.

ξ =	1·	·96	·92	·88	·84	·80	·76
$n/n_0=$	1·00000	1·04467	1·08931	1·13392	1·17852	1·22308	1·26763
$\log \mathfrak{e}+10=$	8·40016	8·43812	8·47446	8·50932	8·54284	8·57507	8·60614
$\log \tau'/\tau+10=$	8·61867	8·51230	8·40140	8·28557	8·16435	8·03721	7·90356
$\log \lambda+10=$	8·70384	8·73805	8·77533	8·81581	8·85966	8·90712	8·95841
$\tau'/2\lambda\mathfrak{e}\tau=$	16·3546	10·8418	7·0889	4·5647	2·8895	1·7947	1·0914
$\mathfrak{m}=\mathrm{a}=$	·90035	·97976	1·06603	1·16014	1·26320	1·37648	1·50172
$\log \gamma+10=$	9·67591	9·71452	9·75343	9·79287	9·83307	9·87430	9·91693
$\log \delta+10=$	9·65551	9·65745	9·65824	9·65788	9·65631	9·65341	9·64900
$\log(\mathrm{gb}-\mathrm{ad})+10=$	8·83030	8·86665	8·91307	8·96946	9·03549	9·11080	9·19510
$\alpha'=$	36·696	23·186	12·583	4·144	− 2·747	− 8·605	−13·873
$\mathrm{a}'=$	− 7·4782	− 8·6811	−10·0883	−11·7426	−13·6966	−16·0163	−18·7899
$\beta'=$	− 6·4455	− 6·9122	− 7·4220	− 7·9805	− 8·5940	− 9·2699	−10·0184
$\mathrm{b}'=$	− 6·4038	− 6·8796	− 7·3968	− 7·9612	− 8·5794	− 9·2590	−10·0104
$\log -(\kappa_1+\alpha)+10=$	8·74306	8·95453	9·16587	9·37077	9·55751	9·71146	9·82404
$\log(\kappa_2+\alpha)=$	1·21135	1·03659	·86190	·69374	·54396	·42731	·35255
$\log(\kappa_2-\kappa_1)=$	1·21283	1·04017	·87056	·71393	·58660	·50372	·46520

The further stages in the computation, when these values are used to compute the several terms of the expressions to be integrated, are given in the following table.

TABLE III.

ξ =	1·	·96	·92	·88	·84	·80	·76
$-(\alpha'-\beta')(\kappa_1+\alpha)/kn(\kappa_2-\kappa_1)^2=$	·00995	·02395	·05424	·10413	·13350	·03053	− ·26438
$\mathrm{a}'\mathrm{b}(\kappa_1+\alpha)/kn(\kappa_2+\alpha)(\kappa_2-\kappa_1)^2=$	·00011	·00064	·00376	·02041	·08937	·27505	·57228
$\mathrm{a}'\mathrm{b}/kn(\kappa_2-\kappa_1)^2=$	− ·03117	− ·07671	− ·18670	− ·42945	− ·86628	− 1·42975	− 1·93250
$\mathrm{b}'\mathrm{a}(\kappa_1+\alpha)/kn(\kappa_2+\alpha)(\kappa_2-\kappa_1)^2=$	·00008	·00049	·00294	·01606	·07072	·21887	·45786
$\mathrm{b}'\mathrm{a}/kn(\kappa_2-\kappa_1)^2=$	− ·02403	− ·05970	− ·14593	− ·33778	− ·68546	− 1·13770	− 1·54610
$\gamma(\kappa_1+\alpha)/kn(\kappa_2-\kappa_1)=$	− ·00179	− ·00452	− ·01141	− ·02759	− ·06001	− ·10969	− ·16534
$\gamma(\kappa_2+\alpha)/kn(\kappa_2-\kappa_1)=$	·52483	·54645	·56651	·58035	·58167	·57020	·55832
$\delta(\kappa_1+\alpha)/kn(\kappa_2-\kappa_1)=$	− ·00170	− ·00397	− ·00916	− ·02022	− ·03995	− ·06596	− ·08922
$\delta(\kappa_2+\alpha)/kn(\kappa_2-\kappa_1)=$	·50075	·47916	·45501	·42530	·38719	·34288	·30127
$(\mathrm{bg}-\mathrm{ad})/kn(\kappa_2-\kappa_1)=$	·00460	·00713	·01124	·01764	·02649	·03675	·04704

The method pursued in the integration of the preceding section proceeds virtually on the assumption that the term $\gamma(\kappa_2+\alpha)/kn(\kappa_2-\kappa_1)$ is the only important one in the expression for $d\log\tan\frac{1}{2}J/d\xi$, and that the term $\delta(\kappa_2+\alpha)/kn(\kappa_2-\kappa_1)$ is the only important one in the expression for $d\log\tan\frac{1}{2}I/d\xi$.

Now when $\xi=1$, at the beginning of the present integration, we see from Table III. that the said term in γ is about 22 times as large as any other occurring in $d\log\tan\frac{1}{2}J$, and that the said term in δ is about 16 times as large as any other which occurs in $d\log\tan\frac{1}{2}I$. Hence the preceding integration must have given fairly satisfactory results. But after the first column these terms in γ and δ fail to maintain their relative importance, so that when $\xi=\cdot76$, they have both become considerably less important than other terms—notably $b'a/kn(\kappa_2-\kappa_1)^2$ and $a'b/kn(\kappa_2-\kappa_1)^2$. This is exactly what is to be expected, because the equations are tending towards the form which they would take if the solar influence were nil, and an inspection of (225) shows that these terms would then be prominent.

If we combine these values of the several terms together according to (250), we obtain the seven equidistant values of $d\log\tan\frac{1}{2}J/d\xi$ and $d\log\tan\frac{1}{2}I/d\xi$ exhibited in the following table:

TABLE IV.

ξ =	1·	·96	·92	·88	·84	·80	·76
$d\log\tan\frac{1}{2}J/d\xi$=	−·49386	−·46660	−·37218	−·15627	+·16138	+·35219	+·19330
$d\log\tan\frac{1}{2}I/d\xi$=	+·54460	+·58194	+·69284	+·93287	+1·28273	+1·51135	+1·39323

By interpolation it appears that $dJ/d\xi$ vanishes when $\xi=\cdot8603$. This value of ξ corresponds with 8 hrs. 36 m. for the period of the earth's rotation, and 5·20 m. s. days for the period of the moon's revolution.

Since $d\xi$ is negative in our integration, we see from these values that I, the inclination of the earth's proper plane to the ecliptic, will continue diminishing, and with increasing rapidity. On the other hand, the inclination J of the lunar orbit to its proper plane will increase at first, but at a diminishing rate, and will finally diminish. This is a point of the greatest importance in explaining the present inclination of the lunar orbit to the ecliptic, and we shall recur to it later on.

Now combine the first four values by the rule of finite differences, viz.:

$$[u_0+u_3+3(u_1+u_2)]\tfrac{3}{8}h$$

and all seven by Weddle's rule, viz.:

$$[u_0+u_2+u_3+u_4+u_6+5(u_1+u_3+u_5)]\tfrac{3}{10}h$$

where h is our $d\xi$, and the u's are the several numbers given in the above Table IV.; then we have, on integration from 1 to ·88,

$$\log_e \tan \tfrac{1}{2}\mathrm{J} = \log_e \tan \tfrac{1}{2}\mathrm{J}_0 + \cdot 04750$$
$$\log_e \tan \tfrac{1}{2}\mathrm{I} = \log_e \tan \tfrac{1}{2}\mathrm{I}_0 - \cdot 07953$$

and on integration from 1 to ·76

$$\log_e \tan \tfrac{1}{2}\mathrm{J} = \log_e \tan \tfrac{1}{2}\mathrm{J}_0 + \cdot 02425$$
$$\log_e \tan \tfrac{1}{2}\mathrm{I} = \log_e \tan \tfrac{1}{2}\mathrm{I}_0 - \cdot 23972$$

If we take $\mathrm{J}_0 = 6°$, $\mathrm{I}_0 = 17°$, which are in round numbers the final values of J and I derived from the first method of integration, we easily find,

$$\text{when } \xi = \cdot 88, \quad \mathrm{J} = 6° \, 17', \quad \mathrm{I} = 15° \, 43'$$
$$\text{and when } \xi = \cdot 76, \quad \mathrm{J} = 6° \, 9', \quad \mathrm{I} = 13° \, 25'$$

Next we have by (249)

$$\sin \mathrm{I}_{,} = -\frac{\kappa_1 + \alpha}{\mathrm{a}} \sin \mathrm{J} = \frac{\mathrm{b}}{\kappa_2 + \alpha} \sin \mathrm{J}$$

$$\sin \mathrm{J}_{,} = \frac{\mathrm{a}}{\kappa_2 + \alpha} \sin \mathrm{I} = -\frac{\kappa_1 + \alpha}{\mathrm{b}} \sin \mathrm{I}$$

Now b is always unity, and the logarithms of $(\kappa_2 + \alpha)$ and $-(\kappa_1 + \alpha)$ are given in Table II.; from this we find

$$\text{when } \xi = \cdot 88, \quad \mathrm{I}_{,} = 1° \, 16', \quad \mathrm{J}_{,} = 3° \, 39'$$
$$\text{when } \xi = \cdot 76, \quad \mathrm{I}_{,} = 2° \, 43', \quad \mathrm{J}_{,} = 8° \, 54'$$

By the same formula, when $\xi = 1$ initially, we have $\mathrm{I}_{,} = 22'$, $\mathrm{J}_{,} = 56'$. These two results ought to be identical with the results from (260) of the last section; and they are so very nearly, for at the end of the integration we had $\delta i = 22' \, 42''$, $\delta j = 57' \, 31''$. The small discrepancy which exists is partly due to the assumed smallness of i and j in the present investigation, and also to our having taken the values 6° and 17° for J_0 and I_0 instead of 6° 21′, 17° 4′.

The value $\xi = \cdot 88$ gives the length of day as 8 hrs. 45 m., and the moon's sidereal period as 5·57 m. s. days.

The value $\xi = \cdot 76$ gives the day as 7 hrs. 49 m., and the moon's sidereal period as 3·59 m. s. days. This value of ξ brings us to the point specified as the end of the third period of integration in § 17 of the paper on "Precession."

There is one other point which it will be interesting to determine,—it is to find the rate of the precessional motion of the node of the two proper planes on the ecliptic, and the rate of the motion of the nodes of the equator and orbit upon their respective proper planes. By means of the preceding numerical values, it will be easy to find these quantities at the epochs specified by $\xi = 1$, ·88, ·76.

The period of the precession of the two proper planes is $-2\pi/\kappa_2$, and that of the precession of the two nodes on their proper planes is $2\pi/(\kappa_2-\kappa_1)$.

In the preceding computations we omitted a common factor $\tau\mathfrak{k}/n$ from α, β, a, b, κ_1, κ_2; this factor must now be reintroduced. τ' is a constant and $\log\tau' = 1{\cdot}77242$, then by means of the numerical values given in the first table I find

	$\xi =$	1·	·88	·76
	$\log \tau\mathfrak{k}/n + 10 =$	7·80940	8·19708	8·62750
Also	$\log -\kappa_2 + 10 =$	9·99401	9·89462	9·53295

and $\log(\kappa_2-\kappa_1)$ is given before in Table II.

Introducing the omitted factor $\tau\mathfrak{k}/n$, I find

$\xi =$	1·	·88	·76
$-2\pi/\kappa_2 =$	988 yrs.	509 yrs.	434 yrs.
$2\pi/(\kappa_2-\kappa_1) -$	60 yrs.	77 yrs.	51 yrs.

Thus both precessional movements on the whole increase in rapidity (because of the increasing value of $\tau\mathfrak{k}/n$), but the rate of the precession of the pair of proper planes increases all through, whilst that of the precession on the proper planes diminishes and then increases. It was pointed out towards the end of § 13 that κ_2 is, so to speak, the ancestor of the luni-solar precession, and $\kappa_2-\kappa_1$ the ancestor of the revolution of the moon's nodes. Hence the 988 years has bred (to continue the metaphor) the present 26,000 years of the precessional period, and the 60 years has bred the present 18½ years of the revolution of the moon's nodes.

We see that the $\kappa_2-\kappa_1$ precession attains a minimum at a certain period, being more rapid both earlier and later.

All the above results will be collected and arranged in a tabular form, after further results have been obtained by means of an integration, carrying out the investigation into the more remote past.

The tidal and precessional effects of the sun's influence have now become exceedingly small, and the only way in which the sun continues to exert a sensible effect is in its tendency to make the nodes of the lunar orbit revolve on the ecliptic. In the analysis therefore we may now treat τ' as zero everywhere, except where it occurs in the form $\tau'/\lambda\mathfrak{k}\tau$. Since λ and $\mathfrak{k}$ are both pretty small, these terms in τ'/τ rise in importance.

The equation of conservation of moment of momentum now becomes

$$\frac{n}{n_0} = 1 + \frac{1}{kn_0}(1-\xi)$$

Here kn_0 is equal to the value of $\mathfrak{m}$ in the preceding integration when $\xi = {\cdot}76$; and hence $1/kn_0 = {\cdot}665903$.

We now have $\beta = \mathrm{b}$, $\gamma = \mathrm{g}$, $\delta = \mathrm{d}$, $\beta' = \mathrm{b}'$, $\gamma' = \mathrm{g}'$, $\delta' = \mathrm{d}'$, but α and α' are not equal to a and a′.

It is proposed to carry the new integration over the field defined by $\xi = 1$ to ·88, and to compute four equidistant values.

The following tables give the results of the computation, as in the previous case.

TABLE V.

$\xi =$	1·	·96	·92	·88
$n/n_0 =$	1·00000	1·02664	1·05327	1·07991
$\log \mathfrak{e} + 10 =$	8·60614	8·62898	8·65122	8·67292
$\log \tau'/\tau + 10 =$	7·90356	7·79718	7·68628	7·57045
$\log \lambda + 10 =$	8·95841	9·00018	9·04451	9·09157
$\log \tau'/2\lambda\mathfrak{e}\tau + 10 =$	10·03798	9·86699	9·68952	9·50493
$\mathfrak{m} = \mathrm{a} =$	1·5017	1·6060	1·7193	1·8429
$\log \mathrm{g} + 10 =$	9·91693	9·95049	9·98531	10·02170
$\log \mathrm{d} + 10 =$	9·65322	9·64780	9·64118	9·63303
$\alpha' =$	−13·873	−17·719	−21·607	−25·692
$\mathrm{a}' =$	−18·790	−21·266	−24·130	−27·460
$\beta' = \mathrm{b}' =$	−10·010	−10·636	−11·316	−12·057
$\log -(\kappa_1 + a) + 10 =$	9·82285	9·88247	9·92401	9·95203
$\log (\kappa_2 + a) + 10 =$	·35374	·32327	·31133	·31347
$\log (\kappa_2 - \kappa_1) + 10 =$	·46586	·45758	·46052	·47035

TABLE VI.

$\xi =$	1·	·96	·92	·88
$-(\mathrm{a}' - \beta')(\kappa_1 + a)/kn(\kappa_2 - \kappa_1)^2 =$	− ·1998	− ·4261	− ·6551	− ·8630
$\mathrm{a}'\mathrm{b}(\kappa_1 + a)/kn(\kappa_2 + a)(\kappa_2 - \kappa_1)^2 =$	·4312	·6077	·7500	·8445
$\mathrm{a}'\mathrm{b}/kn(\kappa_2 - \kappa_1)^2 =$	−1·4643	−1·6770	−1·8297	−1·9410
$\mathrm{b}'\mathrm{a}(\kappa_1 + a)/kn(\kappa_2 + a)(\kappa_2 - \kappa_1)^2 =$	·3450	·4882	·6047	·6833
$\mathrm{b}'\mathrm{a}/kn(\kappa_2 - \kappa_1)^2 =$	−1·1714	−1·3469	−1·4752	−1·5706
$\mathrm{g}(\kappa_1 + a)/kn(\kappa_2 - \kappa_1) =$	− ·1251	− ·1540	− ·1777	− ·1965
$\mathrm{g}(\kappa_2 + a)/kn(\kappa_2 - \kappa_1) =$	·4248	·4248	·4335	·4517
$\mathrm{d}(\kappa_1 + a)/kn(\kappa_2 - \kappa_1) =$	− ·0682	− ·0767	− ·0805	− ·0803
$\mathrm{d}(\kappa_2 + a)/kn(\kappa_2 - \kappa_1) =$	·2315	·2116	·1963	·1846
$(\mathrm{bg} - \mathrm{ad})/kn(\kappa_2 - \kappa_1) =$	·0342	·0404	·0469	·0542

Combining these terms according to the formulæ (250), we have

TABLE VII.

ξ =	1·	·96	·92	·88
$d \log \tan \frac{1}{2}J/d\xi=$	+ ·1496	− ·0754	− ·3298	− ·5625
$d \log \tan \frac{1}{2}I/d\xi=$	+1·0601	+·8607	+·6354	+·4370

By interpolation it appears that $dJ/d\xi$ vanishes when $\xi = {}$·9679. This value of ξ corresponds with 7 hrs. 47 m. for the period of the earth's rotation, and 3·25 m. s. days for the period of the moon's revolution.

By the rules of the calculus of finite differences, integrating from $\xi = 1$ to ·88,

$$\log_e \tan \tfrac{1}{2}J = \log_e \tan \tfrac{1}{2}J_0 + \cdot 0244$$

$$\log_e \tan \tfrac{1}{2}I = \log_e \tan \tfrac{1}{2}I_0 - \cdot 0898$$

With $J_0 = 6° \, 9'$, $I_0 = 13° \, 25'$ from the previous integration, we have $J = 6° \, 18'$, $I = 12° \, 16'$.

When $\xi = {}$·88, the length of the day is 7 hrs. 15 m., and the moon's sidereal period is 2·45 m. s. days. Also $I_, = 3° \, 3'$, $J_, = 10° \, 58'$.

Thus we have traced the changes back until the inclination of the proper planes to one another is only $12° \, 16' - 10° \, 58'$ or $1° \, 18'$.

In the same way as before it may be shown that, when $\xi = {}$·88, the period of the precession of the proper planes is 609 years, and the period of the revolution of the two nodes on their moving proper planes is 22 years. The former of the two precessions is therefore at this stage getting slower, whilst the latter goes on increasing in speed.

The physical results of the whole integration of the present section are embodied in the following table.

TABLE VIII. Results of integration in the case of small viscosity.

Day in m. s. hours and min.	Moon's sidereal period in m. s. days	Inclination of earth's proper plane to ecliptic	Inclination of equator to earth's proper plane	Inclination of moon's proper plane to ecliptic	Inclination of lunar orbit to moon's proper plane	Precessional period of the proper planes	Period of revolution of the two nodes on their moving proper planes
h. m.	Days	° ′	° ′	° ′	° ′	Years	Years
9 55	8·17	17 0	0 22	0 57	6 0	988	60
8 45	5·57	15 43	1 16	3 39	6 17	509	77
7 49	3·59	13 25	2 43	8 54	6 9	434	51
7 15	2·45	12 16	3 3	10 58	6 18	609	22

If the integration is to be carried still further back, the solar action may henceforth be neglected, and the motion may be referred to the invariable plane of the system. This plane undergoes a precessional motion due to the sun, which will not interfere with the treatment of it as though fixed. It is inclined to the ecliptic at about 11° 45′, because, at the time when we suppose the solar action to cease, the moment of momentum of the earth's rotation is larger than that of orbital motion, and therefore the earth's proper plane represents the invariable plane of the system more nearly than does the moon's proper plane.

The inclination i of the equator to the invariable plane must be taken as about 3°, and j that of the lunar orbit as something like 5° 30′. The ratio of the two angles 5° 30′ and 3° must be equal to 1·84, which is $\mathfrak{m}$, the ratio of the moment of momentum of the earth's rotation to that of orbital motion, at the point where the preceding integration ceases.

Then in the more remote past the angle i will continue to diminish, until the point is reached where the moon's period is about 12 hours and that of the earth's rotation about 6 hours. The angle j will continue increasing at an accelerating rate.

This may be shown as follows:

The equations of motion are now those of Part II., which may be written

$$kn\frac{dj}{d\xi} = -\mathrm{g}(i+j)$$

$$kn\frac{di}{d\xi} = \quad \mathrm{d}(i+j)$$

But since $i/j = \xi/kn = 1/\mathfrak{m}$, they become

$$kn\frac{d}{d\xi}\log\tan\tfrac{1}{2}j = -\frac{1+\mathfrak{m}}{\mathfrak{m}}\mathrm{g}$$

$$kn\frac{d}{d\xi}\log\tan\tfrac{1}{2}i = \quad (1+\mathfrak{m})\,\mathrm{d}$$

(Compare with the first of equations (255) given in Part III., when $\tau' = 0$.)

These equations are not independent, because of the relationship which must always subsist between i and j.

Substituting from (251) we have for the case of small viscosity

$$kn\frac{d}{d\xi}\log\tan\tfrac{1}{2}j = -\frac{1+\mathfrak{m}}{2(1-\lambda)}$$

$$kn\frac{d}{d\xi}\log\tan\tfrac{1}{2}i = \frac{(1+\mathfrak{m})(1-2\lambda)}{2(1-\lambda)}$$

From this we see that j will always decrease as ξ increases at a rate which tends to become infinite when $\lambda = 1$; and i increases as ξ increases so long as λ is less than ·5, but decreases for values of λ between ·5 and unity at a rate which tends to become infinite when $\lambda = 1$. If we consider the subject retrospectively, ξ decreases, j increases, and i decreases, except for values of λ between ·5 and unity.

This continued increase (in retrospect) of the inclination of the lunar orbit to the invariable plane is certainly not in accordance with what was to be expected, if the moon once formed a part of the earth. For if we continued to trace the changes backwards to the initial condition in which (as shown in "Precession") the two bodies move round one another as parts of a rigid body, we should find the lunar orbit inclined at a considerable angle to the equator; and it is hard to see how a portion detached from the primeval planet could ever have revolved in such an orbit.

These considerations led me to consider whether some other hypothesis than that of infinitely small viscosity of the earth might not modify the above results. I therefore determined to go over the same solution again, but with the hypothesis of very large instead of very small viscosity of the planet.

This investigation is given in the next section, but I shall not retraverse the ground covered by the integration of the first method, but shall merely take up the problem at the point where it was commenced in the present section.

§ 20. *Secular changes in the proper planes of the earth and moon when the viscosity is large.*

Let $\mathfrak{p} = 2gaw/19v$, where v is the coefficient of viscosity of the earth.

By the theory of viscous tides

$$\tan 2\mathrm{f}_1 = \frac{2(n-\Omega)}{\mathfrak{p}}, \quad \tan 2\mathrm{f} = \frac{2n}{\mathfrak{p}}, \quad \tan \mathrm{g}_1 = \frac{n-2\Omega}{\mathfrak{p}}, \quad \tan \mathrm{g} = \frac{n}{\mathfrak{p}} \quad \ldots(261)$$

If the viscosity be very large $\mathfrak{p}$ is very small, and the angles $\frac{1}{2}\pi - 2\mathrm{f}_1$, $\frac{1}{2}\pi - 2\mathrm{f}$, $\frac{1}{2}\pi - \mathrm{g}_1$, $\frac{1}{2}\pi - \mathrm{g}$ are small, so that their cosines are approximately unity and their sines approximately equal to their tangents. Hence

$$\sin 4\mathrm{f}_1 = \frac{\mathfrak{p}}{n-\Omega}, \quad \sin 4\mathrm{f} = \frac{\mathfrak{p}}{n}, \quad \sin 2\mathrm{g}_1 = \frac{2\mathfrak{p}}{n-2\Omega}, \quad \sin 2\mathrm{g} = \frac{2\mathfrak{p}}{n}$$

Introducing $\lambda = \Omega/n$, we have

$$\frac{\sin 4\mathrm{f}}{\sin 4\mathrm{f}_1} = 1-\lambda, \quad \frac{\sin 2\mathrm{g}_1}{\sin 4\mathrm{f}_1} = \frac{2(1-\lambda)}{1-2\lambda}, \quad \frac{\sin 2\mathrm{g}}{\sin 4\mathrm{f}_1} = 2(1-\lambda) \quad \ldots(262)$$

Introducing the transformations (262) into (251), we have

$$\left.\begin{aligned}\Gamma &= \tfrac{1}{2}\mathfrak{m}\left[1 - \frac{4\lambda(1-\lambda)}{1-2\lambda}\right]\\ \Delta &= \tfrac{1}{2}\left[1 + \frac{4\lambda(1-\lambda)}{1-2\lambda} - 4(1-\lambda)\frac{\tau'}{\tau} + (1-\lambda)\left(\frac{\tau'}{\tau}\right)^2\right]\\ \mathrm{bG} - \mathrm{aD} &= -\mathfrak{m}\left[\frac{4\lambda(1-\lambda)}{1-2\lambda} - 2(1-\lambda)\frac{\tau'}{\tau}\right]\end{aligned}\right\}\ldots(263)$$

All the other expressions in (251) remain as they were.

The terms in Γ, Δ, G, D in (250) are the only ones which have to be recomputed, and all the other arithmetical work of the last section will be applicable here. Also all the materials for calculating these new terms are ready to hand.

The results of the computation are embodied in the following tables.

TABLE IX.

ξ =	1·	·96	·92	·88	·84	·80	·76
$\log\Gamma+10=$	9·54901	9·57529	9·59914	9·61994	9·63663	9·64791	9·65092
$\log\Delta=$	·52876	·55517	·58023	·60484	·63005	·65708	·68739
$\log(\mathrm{aD}-\mathrm{bG})+10=$	9·08381	9·22356	9·34416	9·45433	9·55931	9·66259	9·76574

TABLE X.

ξ =	1·	·96	·92	·88	·84	·80	·76
$\Gamma(\kappa_1+a)/kn(\kappa_2-\kappa_1)=$	− ·00133	− ·00328	− ·00800	− ·01853	− ·03818	− ·06513	− ·08961
$\Gamma(\kappa_2+a)/kn(\kappa_2-\kappa_1)=$	·39185	·39657	·39712	·38973	·37003	·33856	·30260
$\Delta(\kappa_1+a)/kn(\kappa_2-\kappa_1)=$	− ·00199	− ·00485	− ·01168	− ·02688	− ·05553	− ·09627	− ·13761
$\Delta(\kappa_2+a)/kn(\kappa_2-\kappa_1)=$	·58529	·58541	·57994	·56554	·53826	·50044	·46468
$(\mathrm{bG}-\mathrm{aD})/kn(\kappa_2-\kappa_1)=$	− ·00825	− ·01622	− ·03034	− ·05388	− ·08850	− ·13092	− ·17504

Combining these terms with those given in Table III., according to the formulæ (250), (with Γ, &c., in place of γ, &c.), we have the following equidistant values.

TABLE XI.

ξ =	1·	·96	·92	·88	·84	·80	·76
$\log\tan\frac{1}{2}\mathrm{J}/d\xi=$	− ·3477	− ·2925	− ·1587	+ ·1125	+ ·5036	+ ·7818	+ ·7195
$\log\tan\frac{1}{2}\mathrm{I}/d\xi=$	+ ·6168	+ ·6661	+ ·7796	+1·0107	+1·3406	+1·5458	+1·4103

By interpolation it appears that $dJ/d\xi$ vanishes when $\xi = \cdot 8966$. This value of ξ corresponds with a period of 8 hrs. 54 m. for the earth's rotation, and 5·89 m. s. days for the moon's revolution.

Integrating as in the last section, from $\xi = 1$ to ·88, we have

$$\log_e \tan \tfrac{1}{2}J = \log_e \tan \tfrac{1}{2}J_0 + \cdot 0238$$
$$\log_e \tan \tfrac{1}{2}I = \log_e \tan \tfrac{1}{2}I_0 - \cdot 0895$$

Taking $I_0 = 6°$, $J_0 = 17°$, we have $I = 15° \, 34'$, $J = 6° \, 9'$.

These values correspond to $I_{,} = 1° \, 15'$, $J_{,} = 3° \, 37'$.

Again integrating from $\xi = 1$ to ·76, we have

$$\log_e \tan \tfrac{1}{2}J = \log_e \tan \tfrac{1}{2}J_0 - \cdot 0461$$
$$\log_e \tan \tfrac{1}{2}I = \log_e \tan \tfrac{1}{2}I_0 - \cdot 2552$$

These give $J = 5° \, 44'$, $I = 13° \, 13'$, which correspond to $I_{,} = 2° \, 33'$, $J_{,} = 8° \, 46'$.

The integration will now be continued over another period, as in the last section. The following are the results of the computations.

TABLE XII.

ξ =	1·	·96	·92	·88
$\log(\Gamma = G) + 10 =$	9·65092	9·64491	9·62783	9·59299
$\log(\Delta = D) + 10 =$	9·84629	9·86040	9·87686	9·89622

TABLE XIII.

ξ =	1·	·96	·92	·88
$G(\kappa_1 + a)/kn(\kappa_2 - \kappa_1) =$	− ·06781	− ·07617	− ·07802	− ·07323
$G(\kappa_2 + a)/kn(\kappa_2 - \kappa_1) =$	·23026	·21018	·19033	·16832
$D(\kappa_1 + a)/kn(\kappa_2 - \kappa_1) =$	− ·10634	− ·12511	− ·13843	− ·14720
$D(\kappa_2 + a)/kn(\kappa_2 - \kappa_1) =$	·36106	·34521	·33771	·33835
$(bG - aD)/kn(\kappa_2 - \kappa_1) =$	− ·13815	− ·16352	− ·19057	− ·35054

Substituting these values in the differential equations (250), we have the following equidistant values:

TABLE XIV.

ξ =	1·	·96	·92	·88
$d \log \tan \tfrac{1}{2}J / d\xi =$	+ ·5547	+ ·3915	+ ·2088	+ ·1925
$d \log \tan \tfrac{1}{2}I / d\xi =$	+ 1·0746	+ ·8682	+ ·6391	+ ·3093

Then integrating from $\xi = 1$ to $\cdot 88$ we have

$$\log_e \tan \tfrac{1}{2}J = \log_e \tan \tfrac{1}{2}J_0 - \cdot 0382$$

$$\log_e \tan \tfrac{1}{2}I = \log_e \tan \tfrac{1}{2}I_0 - \cdot 0886$$

Putting $I_0 = 13° 13'$ and $J_0 = 5° 44'$, from the previous integration, we have $J = 5° 30'$, $I = 12° 6'$.

These values of J and I give $J_, = 10° 49'$, $I_, = 2° 40'$.

The physical meaning of the results of the whole integration is embodied in the following table.

TABLE XV. Results of integration in the case of large viscosity.

Day in m. s. hours and minutes	Moon's sidereal period in m. s. days	Inclination of earth's proper plane to ecliptic	Inclination of equator to earth's proper plane	Inclination of moon's proper plane to ecliptic	Inclination of lunar orbit to moon's proper plane
h. m.	Days				
9 55	8·17	17° 0′	0° 22′	0° 57′	6° 0′
8 45	5·57	15 34	1 15	3 37	6 9
7 49	3·59	13 13	2 33	8 46	5 44
7 15	2·45	12 6	2 40	10 49	5 30

If we compare these results with those in Table VIII. for the case of small viscosity, we see that the inclinations of the two proper planes to one another and to the ecliptic are almost the same as before, but there is here this important distinction, viz.: that the inclinations of the two moving systems to their respective proper planes are less (compare 5° 30′ with 6° 18′, and 2° 40′ with 3° 3′).

And besides, if we had carried the integration, in the case of small viscosity, further back we should have found the inclination of the lunar orbit increasing.

It will now be shown that, in the present case of large viscosity, the inclinations of the equator and the orbit to their proper planes will continue to diminish, as the square root of the moon's distance diminishes, and at an increasing rate.

Suppose that, in continuing the integration, the solar influence be entirely neglected, and the motion referred to the invariable plane of the system. This plane will be in some position intermediate between the two proper planes, but a little nearer to the earth's plane, and will therefore be inclined to the ecliptic at about 11° 45′.

The equations of motion are now those of § 10, Part II., which may be written

$$kn\frac{dj}{d\xi} = -G(i+j)$$

$$kn\frac{di}{d\xi} = \quad D(i+j)$$

But since $i/j = \xi/kn = 1/\mathfrak{m}$, they become

$$kn\frac{d}{d\xi}\log\tan\tfrac{1}{2}j = -\frac{1+\mathfrak{m}}{\mathfrak{m}}G$$

$$kn\frac{d}{d\xi}\log\tan\tfrac{1}{2}i = \quad (1+\mathfrak{m})D$$

(compare with the first of equations (255) given in Part III., when $\tau' = 0$).

These equations are not independent of one another, because of the relationship which must always subsist between i and j.

Substituting from (263) (in which τ' is put zero, and G, D written for Γ, Δ) we have for the case of large viscosity

$$kn\frac{d}{d\xi}\log\tan\tfrac{1}{2}j = -\tfrac{1}{2}(1+\mathfrak{m})\left[1 - \frac{4\lambda(1-\lambda)}{1-2\lambda}\right]$$

$$kn\frac{d}{d\xi}\log\tan\tfrac{1}{2}i = \quad \tfrac{1}{2}(1+\mathfrak{m})\left[1 + \frac{4\lambda(1-\lambda)}{1-2\lambda}\right]$$

When $\lambda = \frac{1}{2}$, $4\lambda(1-\lambda)/(1-2\lambda)$ is infinite, and therefore both $dj/d\xi$ and $di/d\xi$ are infinite. This result is physically absurd.

The absurdity enters by supposing that an infinitely slow tide (viz.: that of speed $n - 2\Omega$) can lag in such a way as to have its angle of lagging nearly equal to 90°. The correct physical hypothesis, for values of λ nearly equal to $\frac{1}{2}$, is to suppose the lag small for the tide $n - 2\Omega$, but large for the other tides. Hence when λ is nearly $= \frac{1}{2}$, we ought to put

$$\sin 4\mathrm{f}_1 = \frac{\mathfrak{p}}{n-\Omega}, \quad \sin 2\mathrm{g} = \frac{2\mathfrak{p}}{n}, \quad \text{but } \sin 2\mathrm{g}_1 = \frac{2(n-2\Omega)}{\mathfrak{p}}$$

Then we should have

$$G = \tfrac{1}{2}\mathfrak{m}\left[1 + 2(1-\lambda) - \frac{2n^2}{\mathfrak{p}^2}(1-\lambda)(1-2\lambda)\right]$$

$$D = \tfrac{1}{2}\left[1 - 2(1-\lambda) + \frac{2n^2}{\mathfrak{p}^2}(1-\lambda)(1-2\lambda)\right]$$

The last term in each of these expressions involves a small factor both in numerator and denominator, viz.: $1-2\lambda$ because $\lambda = \frac{1}{2}$ nearly, and $\mathfrak{p}$, because the viscosity is large. The evaluation of these terms depends on the actual degree of viscosity, but all that we are now concerned with is the fact that when $\lambda = \frac{1}{2}$ the true physical result is that D changes sign by passing through zero and not infinity, and that G does the same for some value of λ not far removed from $\frac{1}{2}$.

Now consider the function $\frac{4\lambda(1-\lambda)}{1-2\lambda}-1$. The following results are not stated retrospectively, and when it is said that i or j increase or decrease, it is meant increase or decrease as t or ξ increases.

(i) From $\lambda=1$ to $\lambda=\cdot 5$ the function is negative.

Hence for these values of λ the inclination j decreases, or zero inclination is dynamically stable.

When $\lambda=\cdot 5$ it is infinite; but we have already remarked on this case.

(ii) From $\lambda=\cdot 5$ to $\lambda=\cdot 191$ it is positive.

Therefore for these values of λ the inclination j increases, or zero inclination is dynamically unstable. It vanishes when $\lambda=\cdot 191$.

(iii) From $\lambda=\cdot 191$ to $\lambda=0$ it is negative.

Therefore for these values of λ the inclination j decreases, or zero inclination is dynamically stable.

Next consider the function $1+\frac{4\lambda(1-\lambda)}{1-2\lambda}$.

(iv) From $\lambda=1$ to $\lambda=\cdot 809$ it is positive.

Therefore for these values of λ the obliquity i increases, or zero obliquity is dynamically unstable. It vanishes when $\lambda=\cdot 809$.

(v) From $\lambda=\cdot 809$ to $\lambda=\cdot 5$ it is negative.

Therefore for these values of λ the obliquity i decreases, or zero obliquity is dynamically stable.

When $\lambda=\cdot 5$ it is infinite; but we have already remarked on this case.

(vi) From $\lambda=\cdot 5$ to $\lambda=0$ it is positive.

Therefore for these values of λ the obliquity i increases, or zero obliquity is dynamically unstable.

Therefore from $\lambda=1$ to $\cdot 809$ the inclination j decreases and the obliquity i increases.

From $\lambda=\cdot 809$ to $\cdot 5$ both inclination and obliquity decrease.

From $\lambda=\cdot 5$ to $\cdot 191$ both inclination and obliquity increase.

From $\lambda=\cdot 191$ to 0 the inclination decreases and the obliquity increases.

At the point where the above retrospective integration stopped, the moon's period was 2·45 days or 59 hours, and the day was 7·25 hours; hence at this point $\lambda = \cdot 123$, which falls between $\cdot 191$ and $\cdot 5$. Hence both inclination and obliquity decrease retrospectively at a rate which tends to become infinite when we approach $\lambda = \cdot 5$, if the viscosity be infinitely great. For large, but not infinite, viscosity the rates become large and then rapidly decrease in the neighbourhood of $\lambda = \cdot 5$.

From this it follows that by supposing the viscosity large enough, the obliquity and inclination may be made as small as we please, when we arrive at the point where $\lambda = \cdot 5$.

It was shown in § 17 of "Precession" that $\lambda = \cdot 5$ corresponds to a month of 12 hours and a day of 6 hours.

Between the values $\lambda = \cdot 5$ and $\cdot 809$ the solutions for both the cases of small and of large viscosity concur in showing zero obliquity and inclination as dynamically stable. But between $\lambda = \cdot 809$ and 1 the obliquity is dynamically unstable for infinitely large, stable for infinitely small viscosity; for these values of λ zero inclination is dynamically stable both for large and small viscosity.

From this it seems probable that for some large but finite viscosity, both zero inclination and zero obliquity would be dynamically stable for values of λ between $\cdot 809$ and unity.

It appears to me therefore that we have only to accept the hypothesis that the viscosity of the earth has always been pretty large, as it certainly is at present, to obtain a satisfactory explanation of the obliquity of the ecliptic and of the inclination of the lunar orbit. This subject will be again discussed in the summary of Part VII.

§ 21. *Graphical illustration of the preceding integrations.*

A graphical illustration will much facilitate the comprehension of the numerical results of the last two sections.

The integrations which have been carried out by quadratures are of course equivalent to finding the areas of certain curves, and these curves will afford a convenient illustration of the nature of those integrations.

In §§ 19, 20 two separate points of departure were taken, the first proceeding from $\xi = 1$ to $\cdot 76$, and the second from $\xi = 1$ to $\cdot 88$. It is obvious that ξ was referred to different initial values c_0 in the two integrations.

In order therefore to illustrate the rates of increase of $\log \tan \frac{1}{2}J$ and $\log \tan \frac{1}{2}I$ from the preceding numerical results, we must either refer the second sets of ξ's to the same initial value c_0 as the first set, or (which will be simpler) we may take $\sqrt{c}$ as the independent variable.

For the values between $\xi = 1$ and ·76, the ordinates of our curves will be the numerical values given in Tables IV. and XI., each divided by $\sqrt{c_0}$. By the choice of a proper scale of length, c_0 may be taken as unity.

For the values in the second integration from $\xi = 1$ to ·88, the $\sqrt{c_0}$ is the final value of $\sqrt{c}$ in the first integration. Hence in order to draw the ordinates in the second part of the curve to the same scale as those of the first, the numbers in Tables VII. and XIV. must be divided by ·76.

Also the second set of ordinates are not spaced out at the same intervals as the first set, for the $d\sqrt{c}$ of the second integration is ·76 of the $d\sqrt{c}$ of the first integration.

Hence the ordinates given in the four Tables, IV., VII., XI., and XIV., are to be drawn corresponding to the abscissæ

$$0, 1, 2, 3, 4, 5, 6, 6{\cdot}76, 7{\cdot}52, 8{\cdot}28$$

In fig. 7 these abscissæ are marked off on the horizontal axis.

The first integration corresponds to the part OO′, and the marked points correspond to the seven values of ξ from 1 to ·76 inclusive. The second integration corresponds to the part O′O″, and the values computed in Tables VII. and XIV. were divided by ·76 to give the ordinates.

The value for $\xi = {\cdot}76$ of the first integration is identical with that for $\xi = 1$ of the second.

The integrations, which have been carried out, correspond to the determination of the areas lying between these curves and the horizontal axis, areas below being esteemed negative.

The two curves for $d \log \tan \frac{1}{2}\mathrm{I}/d\sqrt{c}$ lie very close together, and we thus see that the motion of the earth's proper plane is almost independent of the degree of viscosity.

On the other hand, the two curves for $d \log \tan \frac{1}{2}\mathrm{J}/d\sqrt{c}$ differ considerably. For large viscosity the positive area is much larger than the negative, whilst for small viscosity the positive area is a little smaller than the negative.

If the figure were extended further to the right, the two curves for the variation of I would become identical, and the ordinates would become very small. The two curves for the variation of J would separate widely. That for large viscosity would go upwards in the positive direction, so that its ordinates would be infinite at the point corresponding to $\lambda = \frac{1}{2}$; the curve for small viscosity would go downwards in the negative direction, and the ordinates would be infinite at the point where $\lambda = 1$.

In this figure OO′ is 6 centimetres, OO″ is 8·28 centimetres, and the point corresponding to $\lambda = \frac{1}{2}$ would be 15·2 centimetres from O, and the point corresponding to $\lambda = 1$ would be 17·4 centimetres from O.

We thus see that the degree of viscosity makes an enormous difference in the results.

In the figure, portions of these further parts of the two curves for the variation of J are continued conjecturally by a line of dashes.

The whole figure is to be read from left to right for a retrospective solution, and from right to left if we advance with the time.

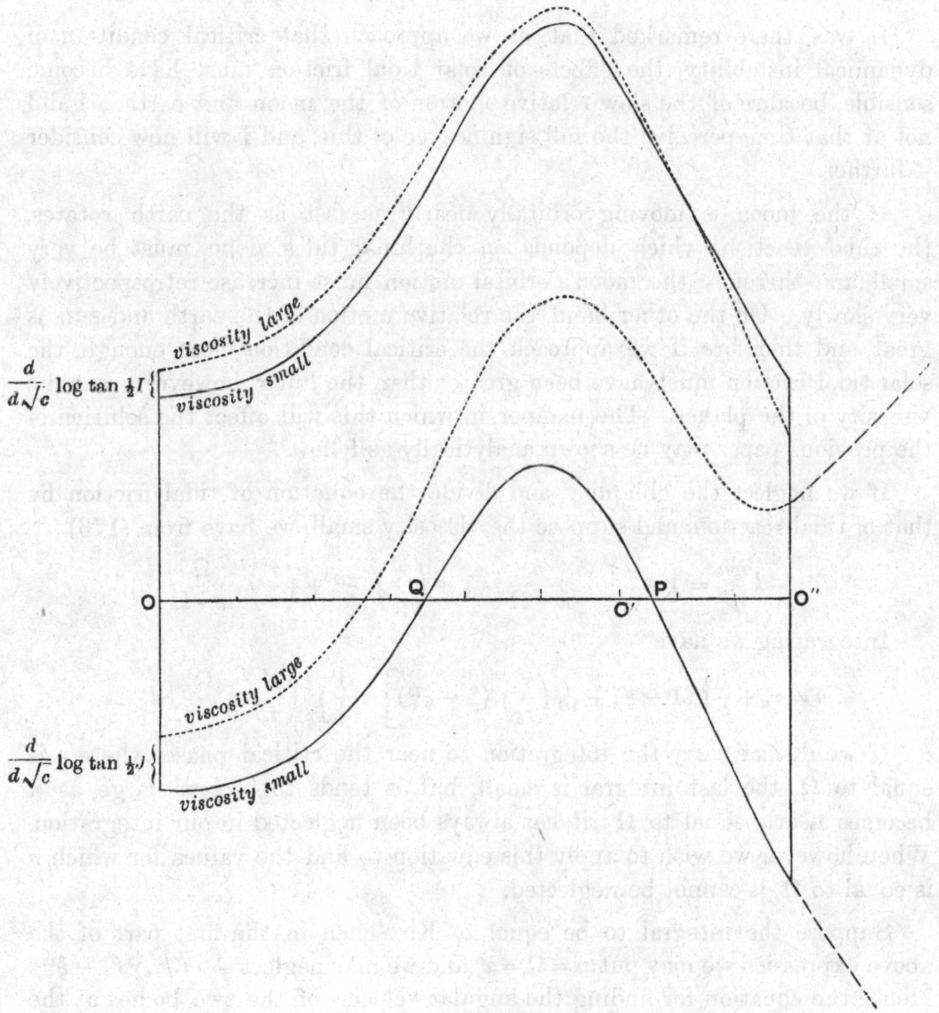

FIG. 7. Diagram to illustrate the motion of the proper planes of the moon and earth.

§ 22. *The effects of solar tidal friction on the primitive condition of the earth and moon.*

In the paper on "Precession," § 16, I found, by the solution of a biquadratic equation, the primitive condition in which the earth and moon moved round together as a rigid body.

Since writing that paper certain additional considerations have occurred to me, which seem to be important in regard to the origin of the moon.

It was there remarked that, as we approach that critical condition of dynamical instability, the effects of solar tidal friction must have become sensible, because of the slow relative motion of the moon and earth. I did not at that time perceive the full significance of this, and I will now consider it further.

If the moon is moving orbitally nearly as fast as the earth rotates, the tidal reaction, which depends on the lunar tides alone, must be very small, and therefore the moon's orbital motion must increase retrospectively very slowly. On the other hand, the relative motion of the earth and sun is great, and therefore if we approach the critical condition close enough, the solar tidal friction must have been greater than the lunar, however great the viscosity of the planet. The manner in which this will affect the solution of the previous paper may be shown analytically as follows.

If we neglect the obliquity, and divide the equation of tidal friction by that of tidal reaction, and suppose the viscosity small, we have from (176)

$$-k\frac{dn}{d\xi} = 1 + \left(\frac{\tau'}{\tau}\right)^2 \frac{n}{n-\Omega} = 1 + \left(\frac{\tau'}{\tau_0}\right)^2 \xi^{12} + \left(\frac{\tau'}{\tau}\right)^2 \frac{\Omega}{n-\Omega}$$

Integrating we have

$$n = n_0 + \frac{1}{k}\left[(1-\xi) + \tfrac{1}{13}\left(\frac{\tau'}{\tau_0}\right)^2 (1-\xi^{13})\right] + \frac{1}{k}\int_\xi^1 \left(\frac{\tau'}{\tau}\right)^2 \frac{\Omega}{n-\Omega}\,d\xi$$

If we do not carry the integration to near the critical phase, where n is equal to Ω, the last integral is small, but it tends to become large as n becomes nearly equal to Ω; it has always been neglected in our integration. When however we wish to apply this equation to find the values for which n is equal to Ω, it cannot be neglected.

Suppose the integral to be equal to K. Then in the first part of the above expression we may put $n = \Omega = x^3$ and we may neglect $\frac{1}{13}(\tau'/\tau_0)^2(1-\xi^{13})$. Hence the equation for finding the angular velocity of the two bodies at the critical phase, when $n = \Omega$, is

$$x^3 = n_0 + \frac{1}{k} - \frac{1}{sx} + \text{K}$$

or

$$x^4 - \left(n_0 + \frac{1}{k} + \text{K}\right)x + \frac{1}{s} = 0$$

The root of this equation, which gives the required phase, is nearly equal to the cube-root of the second coefficient, hence

$$x^3 = n = \Omega = \left(n_0 + \frac{1}{k} + \mathrm{K}\right) \text{ nearly}$$

Now in the paper on "Precession" we found the initial condition, on the hypothesis that K was zero. Hence the effect of solar tidal friction is to increase the angular velocity of the two bodies when their relative motion is zero. Since K may be large, it follows that the disturbance of the solution of § 16 of "Precession" may be considerable.

This therefore shows that it is probable that an accurate solution of our problem would differ considerably from that found in "Precession," and that the common angular velocity of the two bodies might have been great.

If Kepler's law holds good, the periodic time of the moon about the earth, when their centres are 6,000 miles apart, is 2 hrs. 36 m., and when 5,000 miles apart is 1 hr. 57 m.; hence when the two spheroids are just in contact, the time of revolution of the moon would be between 2 hrs. and 2½ hrs.

Now it is a remarkable fact that the most rapid rate of revolution of a mass of fluid, of the same mean density as the earth, which is consistent with an ellipsoidal form of equilibrium, is 2 hrs. 24 m. Is this a mere coincidence, or does it not rather point to the break-up of the primæval planet into two masses in consequence of a too rapid rotation*?

It is not, however, possible to make an adequate consideration of the subject of this section without a treatment of the theory of the tidal friction of a planet attended by a pair of satellites.

It was shown above that if the moon were to move orbitally nearly as fast as the earth rotates, the solar tidal friction would be more important than the lunar, however near the moon might be to the earth. I find that the consequence of this is that the earth's rotation continues to increase retrospectively, and the moon's orbital motion does the same; but the difference between the rotation and the orbital motion continually gets less and less. Meanwhile, the earth's orbital motion round the sun is continually increasing, and the distance from the sun decreasing retrospectively. Theoretically this would go on until the sun and moon (treated as particles) revolve as though rigidly connected with the earth and with one another. This is the configuration of maximum energy of the system.

* [But the instability of a homogeneous ellipsoid sets in for a considerably less rapid rate of rotation. Hence the argument in the text is inexact, and it would appear that the rupture of the primæval planet must have occurred when the rotation was less rapid. The whole subject is full of difficulties, and the conclusions must necessarily remain very speculative.]

The solution is physically absurd, because the distance of the two bodies from the earth would then be very much less than the earth's radius, and *à fortiori* than the sun's radius.

It must be observed, however, that in the retrospect the relative motion of the moon and earth would already have become almost insensible, before the earth's distance from the sun could be sensibly affected.

V.

Secular Changes in the Eccentricity of the Orbit.

§ 23. *Formation of the disturbing function.*

We will now consider the rate of change in the eccentricity and mean distance of the orbit of a satellite, moving in an elliptic orbit, but always remaining in a fixed plane, namely, the ecliptic; and the rate of change of the obliquity of the planet's equator when perturbed by such a satellite will also be found.

Up to the end of Part I. the investigation for the formation of the disturbing function was quite general, and we therefore resume the thread at that point.

In the present problem the inclination of the satellite's orbit to the ecliptic is zero, and we have

$$\varpi = \underline{\varpi} = P = \cos \tfrac{1}{2} i, \quad \kappa = \underline{\kappa} = Q = \sin \tfrac{1}{2} i$$

We thus get rid of the ϖ and κ functions, and henceforth ϖ will indicate the longitude of the perigee.

By equations (24–8),

$$\mathrm{M}_1^2 - \mathrm{M}_2^2 = P^4 \cos 2(\chi - \theta) + 2P^2Q^2 \cos 2\chi + Q^4 \cos 2(\chi + \theta)$$

$$-2\mathrm{M}_1\mathrm{M}_2 = \text{The same with sines for cosines}$$

$$\mathrm{M}_2\mathrm{M}_3 = -P^3Q \cos(\chi - 2\theta) + PQ(P^2 - Q^2)\cos\chi + PQ^3 \cos(\chi + 2\theta)$$

$$\mathrm{M}_1\mathrm{M}_3 = \text{The same with sines for cosines}$$

$$\tfrac{1}{3} - \mathrm{M}_3^2 = \tfrac{1}{3}(P^4 - 4P^2Q^2 + Q^4) + 2P^2Q^2 \cos 2\theta$$

By the definitions (29)

$$\mathrm{X} = \left[\frac{c(1-\mathrm{e}^2)}{r}\right]^{\frac{3}{2}} \mathrm{M}_1, \quad \mathrm{Y} = \left[\frac{c(1-\mathrm{e}^2)}{r}\right]^{\frac{3}{2}} \mathrm{M}_2, \quad \mathrm{Z} = \left[\frac{c(1-\mathrm{e}^2)}{r}\right]^{\frac{3}{2}} \mathrm{M}_3$$

Now let

$$\left.\begin{aligned} \Phi(\alpha) &= \left[\frac{c(1-\mathrm{e}^2)}{r}\right]^3 \cos(2\theta + \alpha) \\ \Psi(\alpha) &= \left[\frac{c(1-\mathrm{e}^2)}{r}\right]^3 \cos\alpha, \quad \mathrm{R} = \left[\frac{c(1-\mathrm{e}^2)}{r}\right]^3 \end{aligned}\right\} \quad \text{.........(264)}$$

Then

$$\left.\begin{aligned}
X^2 - Y^2 &= P^4\Phi(-2\chi) + 2P^2Q^2\Psi(2\chi) + Q^4\Phi(2\chi)\\
2XY &= \text{The same when } \chi + \tfrac{1}{4}\pi \text{ is substituted for } \chi\\
YZ &= -P^3Q\Phi(-\chi) + PQ(P^2 - Q^2)\Psi(\chi) + PQ^3\Phi(\chi)\\
XZ &= \text{The same when } \chi - \tfrac{1}{2}\pi \text{ is substituted for } \chi\\
\tfrac{1}{3}(X^2 + Y^2 - 2Z^2) &= \tfrac{1}{3}(P^4 - P^2Q^2 + Q^4)\,R + 2P^2Q^2\Phi(0)
\end{aligned}\right\} \quad (265)$$

Hence all the terms of the five X-Y-Z functions belong to one of the three types Φ, Ψ, or R.

The equation to the ellipse described by the satellite Diana is

$$\frac{c(1-e^2)}{r} = 1 + e\cos(\theta - \varpi) \quad \text{.....................(266)}$$

Hence

$$\left.\begin{aligned}
R &= 1 + \tfrac{3}{2}e^2 + 3e(1 + \tfrac{1}{4}e^2)\cos(\theta - \varpi) + \tfrac{3}{2}e^2\cos 2(\theta - \varpi) + \tfrac{1}{4}e^3\cos 3(\theta - \varpi)\\
\Phi(\alpha) &= R\cos(2\theta + \alpha) = (1 + \tfrac{3}{2}e^2)\cos(2\theta + \alpha)\\
&\quad + \tfrac{3}{2}e(1 + \tfrac{1}{4}e^2)[\cos(3\theta + \alpha - \varpi) + \cos(\theta + \alpha + \varpi)]\\
&\quad + \tfrac{3}{4}e^2[\cos(4\theta + \alpha - 2\varpi) + \cos(\alpha + 2\varpi)]\\
&\quad + \tfrac{1}{8}e^3[\cos(5\theta + \alpha - 3\varpi) + \cos(\theta - \alpha - 3\varpi)]
\end{aligned}\right\} \quad \text{......(267)}$$

and $\Psi(\alpha) = R\cos\alpha$.

By the theory of elliptic motion, θ the true longitude may be expressed in terms of $\Omega t + \epsilon$ and ϖ, in a series of ascending powers of e the eccentricity. Hence $\Phi(\alpha)$, R, and $\Psi(\alpha)$ may be expressed as the sum of a number of cosines of angles of the form $l(\Omega t + \epsilon) + m\varpi + n\alpha$, and in using these functions we shall require to make α either a multiple of χ or zero, or to differ from a multiple of χ by a constant. Therefore the X-Y-Z functions are expressible as the sums of a number of sines or cosines of angles of the form $l(\Omega t + \epsilon) + m\varpi + n\chi$.

Now χ increases uniformly with the time (being equal to nt + a constant); hence, if we regard the elements of the elliptic orbit as constant, the X-Y-Z functions are expressible as a number of simple time-harmonics. But in § 4, where the state of tidal distortion due to Diana was found, they were assumed to be so expressible; therefore that assumption was justifiable, and the remainder of that section concerning the formation of the disturbing function is applicable.

The problem may now be simplified by the following considerations:— The equation (12) for the rate of variation of the ellipticity of the orbit involves only differentials of the disturbing function with regard to epoch and perigee. It is obvious that in the disturbing function the epoch and perigee will only occur in the argument of trigonometrical functions, therefore after the required differentiations they only occur in the like forms. Now the epoch never occurs except in conjunction with the mean longitude, and the longitude of the perigee increases uniformly with the time (or nearly so), either from the action of other disturbing bodies or from the disturbing action of the permanent oblateness of the planet, which causes a progression of the apses. Hence it follows that the only way in which these differentials of the disturbing function can be non-periodic is when the tide-raiser Diana is identical with the moon. Whence we conclude that—

The tides raised by any one satellite can produce no secular change in the eccentricity of the orbit of any other satellite.

The problem is thus simplified by the consideration that Diana and the moon need only be regarded as distinct as far as regards epoch and perigee, and that they are ultimately to be made identical.

Before carrying out the procedure above sketched, it will be well to consider what sort of approximation is to be made, for the subsequent labour will be thus largely abridged.

From the preceding sketch it is clear that all the terms of the X-Y-Z functions corresponding with Diana's tide-generating potential are of the form

$$(a + be + ce^2 + de^3 + fe^4 + \&c.) \cos [l\chi + m(\Omega t + \epsilon) + n\varpi + \delta]$$

From this it follows that all the terms of the $\mathfrak{X}$-$\mathfrak{Y}$-$\mathfrak{Z}$ functions are of the form

$$F(a + be + ce^2 + de^3 + fe^4 + \&c.) \cos [l\chi + m(\Omega t + \epsilon) + n\varpi + \delta - f]$$

Also by symmetry all the terms of the X′-Y′-Z′ functions are of the form

$$(a + be + ce^2 + de^3 + fe^4 + \&c.) \cos [l\chi' + m(\Omega t + \epsilon') + n\varpi' + \delta]$$

and in the present problem the accent to χ may be omitted.

The products of the $\mathfrak{X}$-$\mathfrak{Y}$-$\mathfrak{Z}$ functions multiplied by the X′-Y′-Z′ functions occur in such a way that when they are added together in the required manner (as for example in $Y'Z'\mathfrak{Y}\mathfrak{Z} + X'Z'\mathfrak{X}\mathfrak{Z}$) only differences of arguments occur, and χ disappears from the disturbing function. Also secular changes can only arise in the satellite's eccentricity and mean distance from such terms in the disturbing function as are independent of $\Omega t + \epsilon$ and ϖ, when we put $\epsilon' = \epsilon$ and $\varpi' = \varpi$. Hence we need only select from the complete products the products of terms of the like argument in the two sets of functions.

Whence it follows that all the part of the disturbing function, which is here important, consists of terms of the form

$$F(a + be + ce^2 + de^3 + fe^4 + \&c.)^2 \cos [m(\epsilon - \epsilon') + n(\varpi - \varpi') - f]$$

or

$$F[a^2 + 2abe + (2ac + b^2)e^2 + (2ad + 2bc)e^3 + (2af + 2bd + c^2)e^4 + \&c.]$$
$$\cos [m(\epsilon - \epsilon') + n(\varpi - \varpi') - f]$$

Now it is intended to develop the disturbing function rigorously with respect to the obliquity of the ecliptic, and as far as the fourth power of the eccentricity.

The question therefore arises, what terms will it be necessary to retain in developing the X-Y-Z functions, so as to obtain the disturbing function correct to e^4.

In the X-Y-Z functions (and in their constituent functions $\Phi(\alpha)$, $\Psi(\alpha)$, R) those terms in which a is not zero will be said to be of the order zero; those in which a is zero, but b not zero, of the first order; those in which $a=b=0$, but c not zero, of the second order, and so on.

By considering the typical term in the disturbing function, we have the following—

Rule of approximation for the development of the X-Y-Z functions and of $\Phi(\alpha)$, $\Psi(\alpha)$, R: develop terms of order zero to e^4; terms of the first order to e^3; terms of the second order to e^2; and drop terms of the third and fourth orders.

To obtain further rules of approximation, and for the subsequent developments, we now require the following theorem.

Expansion of $\cos(k\theta+\beta)$ *in powers of the eccentricity.*

θ is the true longitude of the satellite, $\Omega t+\epsilon$ the mean longitude, and ϖ the longitude of the perigee. For the present I shall write simply Ω in place of $\Omega t+\epsilon$.

By the theory of elliptic motion

$$\Omega=\theta-2e\sin(\theta-\varpi)+\tfrac{3}{4}e^2\left(1+\tfrac{1}{6}e^2\right)\sin 2(\theta-\varpi)-\tfrac{1}{3}e^3\sin 3(\theta-\varpi)+\tfrac{5}{32}e^4\sin 4(\theta-\varpi)$$

If this series be inverted, it will be found that

$$\theta=\Omega+2e\left(1-\tfrac{1}{8}e^2\right)\sin(\Omega-\varpi)+\tfrac{5}{4}e^2\left(1-\tfrac{11}{30}e^2\right)\sin 2(\Omega-\varpi)+\tfrac{13}{12}e^3\sin 3(\Omega-\varpi)+\tfrac{103}{96}e^4\sin 4(\Omega-\varpi)$$

By differentiation we find that, when $e=0$,

$$\frac{d\theta}{de}=2\sin(\Omega-\varpi),\quad \frac{d^2\theta}{de^2}=\tfrac{5}{2}\sin 2(\Omega-\varpi),\quad \frac{d^3\theta}{de^3}=-\tfrac{3}{2}\sin(\Omega-\varpi)+\tfrac{13}{2}\sin 3(\Omega-\varpi)$$

$$\frac{d^4\theta}{de^4}=-11\sin 2(\Omega-\varpi)+\tfrac{103}{4}\sin 4(\Omega-\varpi),\quad \left(\frac{d\theta}{de}\right)^2=2-2\cos(\Omega-\varpi)$$

$$\left(\frac{d\theta}{de}\right)^3=6\sin(\Omega-\varpi)-2\sin 3(\Omega-\varpi),\quad \left(\frac{d\theta}{de}\right)^4=6-8\cos 2(\Omega-\varpi)+2\cos 4(\Omega-\varpi)$$

$$\frac{d\theta}{de}\frac{d^2\theta}{de^2} = \tfrac{5}{2}\cos(\Omega - \varpi) - \tfrac{5}{2}\cos 3(\Omega - \varpi), \quad \left(\frac{d\theta}{de}\right)^2 \frac{d^2\theta}{de^2} = 5\sin 2(\Omega - \varpi) - \tfrac{5}{2}\sin 4(\Omega - \varpi)$$

$$\left(\frac{d^2\theta}{de^2}\right)^2 = \tfrac{25}{8} - \tfrac{25}{8}\cos 4(\Omega - \varpi), \quad \frac{d\theta}{de}\frac{d^3\theta}{de^3} = -\tfrac{3}{2} + 8\cos 2(\Omega - \varpi) - \tfrac{13}{2}\cos 4(\Omega - \varpi)$$

To expand $\cos(k\theta + \beta)$ by means of Maclaurin's theorem, we require the values of the following differentials when $e = 0$ and $\theta = \Omega$:

$$\frac{d}{de}\cos(k\theta + \beta) = -k\sin(k\theta + \beta)\frac{d\theta}{de}$$

$$\frac{d^2}{de^2}\cos(k\theta + \beta) = -k^2\cos(k\theta + \beta)\left(\frac{d\theta}{de}\right)^2 - k\sin(k\theta + \beta)\frac{d^2\theta}{de^2}$$

$$\frac{d^3}{de^3}\cos(k\theta + \beta) = k^3\sin(k\theta + \beta)\left(\frac{d\theta}{de}\right)^3 - 3k^2\cos(k\theta + \beta)\frac{d\theta}{de}\frac{d^2\theta}{de^2} - k\sin(k\theta + \beta)\frac{d^3\theta}{de^3}$$

$$\begin{aligned}\frac{d^4}{de^4}\cos(k\theta + \beta) &= k^4\cos(k\theta + \beta)\left(\frac{d\theta}{de}\right)^4 + 6k^3\sin(k\theta + \beta)\left(\frac{d\theta}{de}\right)^2\frac{d^2\theta}{de^2} \\ &\quad - 3k^2\cos(k\theta + \beta)\left(\frac{d^2\theta}{de^2}\right)^2 - 4k^2\cos(k\theta + \beta)\frac{d\theta}{de}\frac{d^3\theta}{de^3} - k\sin(k\theta + \beta)\frac{d^4\theta}{de^4}\end{aligned}$$

When $e = 0$, $k\theta + \beta = k\Omega + \beta$, and the values of the differentials and functions of differentials of e are given above. If we substitute for these functions their values, and express the products of sines and cosines as the sums of sines and cosines, and introduce an abridged notation in which $k\Omega + \beta + s(\Omega - \varpi)$ is written $(k + s)$, we have

$$\left.\begin{aligned}
\Theta_1 &= \frac{d}{de}\cos(k\theta + \beta) \\
&= -k\cos(k - 1) + k\cos(k + 1) \\
\Theta_2 &= \frac{d^2}{de^2}\cos(k\theta + \beta) \\
&= (k^2 - \tfrac{5}{4}k)\cos(k - 2) - 2k^2\cos k + (k^2 + \tfrac{5}{4}k)\cos(k + 2) \\
\Theta_3 &= \frac{d^3}{de^3}\cos(k\theta + \beta) \\
&= -(k^3 - \tfrac{15}{4}k^2 + \tfrac{13}{4}k)\cos(k - 3) + 3(k^3 - \tfrac{5}{4}k^2 + \tfrac{1}{4}k)\cos(k - 1) \\
&\quad - 3(k^3 + \tfrac{5}{4}k^2 + \tfrac{1}{4}k)\cos(k + 1) + (k^3 + \tfrac{15}{4}k^2 + \tfrac{13}{4}k)\cos(k + 3) \\
\Theta_4 &= \frac{d^4}{de^4}\cos(k\theta + \beta) \\
&= (k^4 - \tfrac{15}{2}k^3 + \tfrac{75}{16}k^2 + 13k^2 - \tfrac{103}{8}k)\cos(k - 4) \\
&\quad - (4k^4 - 15k^3 + 16k^2 - \tfrac{11}{2}k)\cos(k - 2) \\
&\quad + 3(2k^4 - \tfrac{25}{8}k^2 + 2k^2)\cos(k) - (4k^4 + 15k^3 + 16k^2 + \tfrac{11}{2}k)\cos(k + 2) \\
&\quad + (k^4 + \tfrac{15}{2}k^3 + \tfrac{75}{16}k^2 + 13k^2 + \tfrac{103}{8}k)\cos(k + 4)
\end{aligned}\right\} \ldots(268)$$

where the Θ's are merely introduced as an abbreviation.

Then by Maclaurin's theorem

$$\cos(k\theta+\beta)=\cos(k\Omega+\beta)+e\Theta_1+\tfrac{1}{2}e^2\Theta_2+\tfrac{1}{6}e^3\Theta_3+\tfrac{1}{24}e^4\Theta_4\ldots(269)$$

In order to obtain further rules of approximation we will now run through the future developments, merely paying attention to the order of the coefficients and to the factors by which $\Omega t+\epsilon$ will be multiplied in the results. From this point of view we may write

$$\Phi(\alpha)=(e^0)\cos(2\theta)+(e)[\cos(3\theta)+\cos(\theta)]+(e^2)[\cos(4\theta)+\cos(0)]$$
$$+(e^3)[\cos(5\theta)+\cos(\theta)]$$

$$\Psi(\alpha)=R=(e^0)\cos(0)+(e)\cos(\theta)+(e^2)\cos(2\theta)+(e^3)\cos(3\theta)$$

The cosines of the multiples of θ have now to be found by the theorem (269) and substituted in the above equations.

In making the developments the following abridged notation is adopted; a term of the form $\cos[(k+s)\Omega+\beta-s\varpi]$ is written $\{k+s\}$.

Consider the series for $\Phi(\alpha)$ first.

We have by successive applications of (269) with $k=1, 2, 3, 4, 5$:

$$(e^0)\cos(2\theta)=(e^0)\{2\}+(e)[\{1\}+\{3\}]+(e^2)[\{0\}+\{2\}+\{4\}]$$
$$+(e^3)[\{-1\}+\{1\}+\{3\}+\{5\}]+(e^4)[\{-2\}+\{0\}+\{2\}+\{4\}+\{6\}]$$

$$(e)\cos(3\theta)=(e)\{3\}+(e^2)[\{2\}+\{4\}]+(e^3)[\{1\}+\{3\}+\{5\}]$$
$$+(e^4)[\{0\}+\{2\}+\{4\}+\{6\}]$$

$$(e)\cos(\theta)=(e)\{1\}+(e^2)[\{0\}+\{2\}]+(e^3)[\{-1\}+\{1\}+\{3\}]$$
$$+(e^4)[\{-2\}+\{0\}+\{2\}+\{4\}]$$

$$(e^2)\cos(4\theta)=(e^2)\{4\}+(e^3)[\{3\}+\{5\}]+(e^4)[\{2\}+\{4\}+\{6\}]$$

$$(e^2)\cos(0)=(e^2)\{0\}$$

$$(e^3)\cos(5\theta)=(e^3)\{5\}+(e^4)[\{4\}+\{6\}]$$

$$(e^3)\cos(\theta)=(e^3)\{1\}+(e^4)[\{0\}+\{2\}]$$

In these expressions we have no right, as yet, to assume that $\{-2\}$ and $\{-1\}$ are different from $\{2\}$ and $\{1\}$; and in fact we shall find that in the expansion for $\Phi(\alpha)$ they *are* different, but in that for R they are the same.

Adding up these, and rejecting terms of the third and fourth orders by the first rule of approximation, we have

$$\Phi(\alpha)=[(e^0)+(e^2)+(e^4)]\{2\}+[(e)+(e^3)][\{1\}+\{3\}]+[(e^2)+(e^4)][\{0\}+\{4\}]$$
$$+(e^3)\{-1\}+(e^4)\{-2\}$$

It will be observed that $\{5\}$ and $\{6\}$ are wanting, and might have been dropped from the expansions. Also $\{0\}$ and $\{4\}$ are terms of the second order, therefore wherever they are multiplied by (e^4) they might have been dropped. Hence $(e^3)\cos(5\theta)$ need not have been expanded at all. A little further consideration is required to show that $(e^3)\cos(\theta)$ need not have been expanded.

$(e^3)\cos(\theta)$ is an abbreviation for $\frac{1}{8}e^3\cos(\theta-\alpha-3\varpi)$, and therefore in this case $\{1\}=\cos(\Omega-\alpha-3\varpi)$ and $\{2\}=\cos(2\Omega-\alpha-4\varpi)$; but in every other case $\{1\}=\cos(\Omega+\alpha+\varpi)$ and $\{2\}=\cos(2\Omega+\alpha)$. Hence the terms $\{1\}$ and $\{2\}$ in $(e^3)\cos(\theta)$ are of the third and fourth orders and may be dropped, and $\{0\}$ may also be dropped. Thus the whole of $(e^3)\cos(\theta)$ may be dropped.

With respect to $\{-2\}$ and $\{-1\}$, observe that $\{2\}$ in the expansion of $\cos(k_1\theta+\beta_1)$ stands for $\cos[2\Omega+(k_1-2)\varpi+\beta_1]$; and $\{-2\}$ in the expansion of $\cos(k_2\theta+\beta_2)$ stands for $\cos[2\Omega-(k_2+2)\varpi-\beta_2]$; and k_1, k_2 are either 1, 2, 3, or 4; and β_1, β_2 are multiples of $\chi+$ a constant. Hence $\{2\}$ and $\{-2\}$ are necessarily different, but if β_1 and β_2 were multiples of ϖ they might be the same, and indeed in the expansion of R necessarily are the same.

In the same way it may be shown that $\{-1\}$ and $\{1\}$ are necessarily different.

Therefore $\{-1\}$ and $\{-2\}$ being terms of the third and fourth orders may be dropped.

It follows from this discussion that, as far as concerns the present problem,

$$(e^0)\cos(2\theta)=(e^0)\{2\}+(e)[\{1\}+\{3\}]+(e^2)[\{0\}+\{2\}+\{4\}]$$
$$+(e^3)[\{1\}+\{3\}]+(e^4)[\{2\}]$$
$$(e)\cos(3\theta)=(e)\{3\}+(e^2)[\{2\}+\{4\}]+(e^3)[\{1\}+\{3\}]+(e^4)\{2\}$$
$$(e)\cos(\theta)=(e)\{1\}+(e^2)[\{0\}+\{2\}]+(e^3)[\{1\}+\{3\}]+(e^4)\{2\}$$
$$(e^2)\cos(4\theta)=(e^2)\{4\}+(e^3)\{3\}+(e^4)\{2\}$$
$$(e^2)\cos(0)=(e^2)\{0\}$$

And the sum of these expressions is equal to $\Phi(\alpha)$.

We thus get the following rules for the use of the expansion (269) of $\cos(k\theta+\beta)$ for the determination of $\Phi(\alpha)$:

When $k=2$, omit in Θ_3 terms in $\cos(k-3)$, $\cos(k+3)$
in Θ_4 terms in $\cos(k-4)$, $\cos(k-2)$, $\cos(k+2)$, $\cos(k+4)$

When $k=3$, omit in Θ_2 term in $\cos(k+2)$
in Θ_3 terms in $\cos(k-3)$, $\cos(k+1)$, $\cos(k+3)$
all of Θ_4

When $k=1$, omit in Θ_2 term in $\cos(k-2)$
in Θ_3 terms in $\cos(k-3)$, $\cos(k-1)$, $\cos(k+3)$
all of Θ_4

When $k=4$, omit in Θ_1 term in $\cos(k+1)$
in Θ_2 terms in $\cos(k)$, $\cos(k+2)$
all of Θ_3, Θ_4

Following these rules we easily find,

When $k=2$, $\beta=\alpha$

$$\cos(2\theta+\alpha)=(1-4e^2+\tfrac{55}{16}e^4)\cos(2\Omega+\alpha)-2e(1-\tfrac{7}{8}e^2)\cos(\Omega+\alpha+\varpi)$$
$$+2e(1-\tfrac{27}{8}e^2)\cos(3\Omega+\alpha-\varpi)+\tfrac{3}{4}e^2\cos(\alpha+2\varpi)$$
$$+\tfrac{13}{4}e^2\cos(4\Omega+\alpha-2\varpi)\ldots(270)$$

When $k=3$, $\beta=\alpha-\varpi$

$$\cos(3\theta+\alpha-\varpi)=(1-9e^2)\cos(3\Omega+\alpha-\varpi)-3e(1-\tfrac{11}{4}e^2)\cos(2\Omega+\alpha)$$
$$+3e\cos(4\Omega+\alpha-2\varpi)+\tfrac{21}{8}e^2\cos(\Omega+\alpha+\varpi)\ldots(271)$$

When $k=1$, $\beta=\alpha+\varpi$

$$\cos(\theta+\alpha+\varpi)=(1-e^2)\cos(\Omega+\alpha+\varpi)+e(1-\tfrac{5}{4}e^2)\cos(2\Omega+\alpha)$$
$$-e\cos(\alpha+2\varpi)+\tfrac{3}{8}e^2\cos(3\Omega+\alpha-\varpi)\ldots(272)$$

When $k=4$, $\beta=\alpha-2\varpi$

$$\cos(4\theta+\alpha-2\varpi)=\cos(4\Omega+\alpha-2\varpi)-4e\cos(3\Omega+\alpha-\varpi)$$
$$+\tfrac{11}{2}e^2\cos(2\Omega+\alpha)\ldots(273)$$

These are all the series required for the expression of $\Phi(\alpha)$, since $\cos(\alpha+2\varpi)$ does not involve θ, and by what has been shown above $\cos(5\theta+\alpha-3\varpi)$ and $\cos(\theta-\alpha-3\varpi)$ need not be expanded.

We now return again to the series for R or $\Psi(\alpha)$, and consider the nature of the approximations to be adopted there.

With the same notation

$$(e^0)\cos(0)=(e^0)\{0\}$$
$$(e)\cos(\theta)=(e)\{1\}+(e^2)[\{0\}+\{2\}]+(e^3)[\{-1\}+\{1\}+\{3\}]$$
$$+(e^4)[\{-2\}+\{0\}+\{2\}+\{4\}]$$
$$(e^2)\cos(2\theta)=(e^2)\{2\}+(e^3)[\{1\}+\{3\}]+(e^4)[\{0\}+\{2\}+\{4\}]$$
$$(e^3)\cos(3\theta)=(e^3)\{3\}+(e^4)[\{2\}+\{4\}]$$

Since R is a function of $\theta-\varpi$, therefore after expansion it must be a function of $\Omega-\varpi$, and hence $\{1\}$ must be necessarily identical with $\{-1\}$, and $\{2\}$ with $\{-2\}$.

Adding these up, and dropping terms of the third and fourth orders,

$$R=[(e^0)+(e^2)+(e^4)]\{0\}+[(e)+(e^3)]\{1\}+(e^3)\{-1\}$$
$$+[(e^2)+(e^4)]\{2\}+(e^4)\{-2\}$$

Here $\{0\}$ is a term of the order zero, $\{1\}$ of the first order, and $\{2\}$ of the second. Therefore by the first rule of approximation $\{2\}$ and $\{-2\}$ may be dropped when multiplied by (e^4).

Also $\{3\}$ and $\{4\}$ may be dropped.

Hence as far as concerns the present problem

$$(e^0)\cos(0) = (e^0)\{0\}$$

$$(e)\cos(\theta) = (e)\{1\} + (e^2)[\{0\} + \{2\}] + (e^3)[\{-1\} + \{1\}] + (e^4)\{0\}$$

$$(e^2)\cos(2\theta) = (e^2)\{2\} + (e^3)\{1\} + (e^4)\{0\}$$

and $(e^3)\cos(3\theta)$ need not be expanded.

And the sum of these expressions is equal to R.

We thus get the following rules for the use of the expansion of $\cos(k\theta + \beta)$ for the determination of R.

When $k = 1$, omit in Θ_2 term in $\cos(k+2)$
in Θ_3 terms in $\cos(k-3)$, $\cos(k+1)$, $\cos(k+3)$
all of Θ_4

When $k = 2$, omit in Θ_1 term in $\cos(k+1)$
in Θ_2 terms in $\cos(k)$, $\cos(k+2)$
all of Θ_3, Θ_4

Following these rules, we find

When $k = 1$, $\beta = -\varpi$

$$\cos(\theta - \varpi) = (1 - e^2)\cos(\Omega - \varpi) - e + e\cos 2(\Omega - \varpi) \quad \ldots\ldots(274)$$

When $k = 2$, $\beta = -2\varpi$

$$\cos 2(\theta - \varpi) = \cos 2(\Omega - \varpi) - 2e\cos(\Omega - \varpi) + \tfrac{3}{4}e^2 \quad \ldots\ldots(275)$$

These are the only series required for the expansion of R or $\Psi(\alpha)$, since by what is shown above, $\cos 3(\theta - \varpi)$ need not be expanded.

Now multiply (270) by $1 + \frac{3}{2}e^2$; (271) by $\frac{3}{2}e(1 + \frac{1}{4}e^2)$; (272) by $\frac{3}{2}e(1 + \frac{1}{4}e^2)$; and (273) by $\frac{3}{4}e^2$; add the four products together, and add $\frac{3}{4}e^2\cos(\alpha + 2\varpi)$, and we find from (267) after reduction

$$\Phi(\alpha) = (1 - \tfrac{11}{2}e^2 + \tfrac{181}{16}e^4)\cos(2\Omega + \alpha) - \tfrac{1}{2}e(1 - \tfrac{25}{8}e^2)\cos(\Omega + \alpha + \varpi)$$
$$+ \tfrac{7}{2}e(1 - \tfrac{291}{56}e^2)\cos(3\Omega + \alpha - \varpi) + \tfrac{17}{2}e^2\cos(4\Omega + \alpha - 2\varpi) \ldots(276)$$

Next multiply (274) by $3e(1 + \frac{1}{4}e^2)$; (275) by $\frac{3}{2}e^2$; add the two products, and add $1 + \frac{3}{2}e^2$, and we find from (267) after reduction,

$$R = 1 - \tfrac{3}{2}e^2 + \tfrac{3}{8}e^4 + 3e(1 - \tfrac{15}{8}e^2)\cos(\Omega - \varpi) + \tfrac{9}{2}e^2\cos 2(\Omega - \varpi) \ldots(277)$$

Now let

$$\left.\begin{aligned} &E_1 = -\tfrac{1}{2}e(1 - \tfrac{25}{8}e^2); \quad E_2 = 1 - \tfrac{11}{2}e^2 + \tfrac{181}{16}e^4 \\ &E_3 = \tfrac{7}{2}e(1 - \tfrac{291}{56}e^2); \quad E_4 = \tfrac{17}{2}e^2 \\ &J_0 = 1 - \tfrac{3}{2}e^2 + \tfrac{3}{8}e^4; \quad J_1 = \tfrac{3}{2}e(1 - \tfrac{15}{8}e^2); \quad J_2 = \tfrac{9}{4}e^2 \end{aligned}\right\} \ldots\ldots(278)$$

And we have

$$\left.\begin{aligned}\Phi(\alpha) &= \text{E}_1\cos(\Omega+\alpha+\varpi)+\text{E}_2\cos(2\Omega+\alpha)+\text{E}_3\cos(3\Omega+\alpha-\varpi)\\ &\qquad\qquad\qquad\qquad\qquad\qquad +\text{E}_4\cos(4\Omega+\alpha-2\varpi)\\ \text{R} &= \text{J}_0+2\text{J}_1\cos(\Omega-\varpi)+2\text{J}_2\cos 2(\Omega-\varpi)\\ \text{whence}&\\ \Psi(\alpha) &= \text{J}_0\cos\alpha+\text{J}_1[\cos(\Omega+\alpha-\varpi)+\cos(\Omega-\alpha-\varpi)]\\ &\qquad\qquad +\text{J}_2[\cos(2\Omega+\alpha-2\varpi)+\cos(2\Omega-\alpha-2\varpi)]\end{aligned}\right\} \quad \text{......(279)}$$

These three expressions are parts of infinite series which only go as far as terms in e^2, but the terms of the orders e^0 and e have their coefficients developed as far as e^4 and e^3 respectively.

Substituting from (279) for Φ, Ψ, and R their values in the expressions (265), we find

$$\left.\begin{aligned}\text{X}^2-\text{Y}^2 &= P^4[\text{E}_1\cos(2\chi-\Omega-\varpi)+\text{E}_2\cos(2\chi-2\Omega)+\text{E}_3\cos(2\chi-3\Omega+\varpi)\\ &\qquad\qquad +\text{E}_4\cos(2\chi-4\Omega+2\varpi)]\\ &\quad +2P^2Q^2[\text{J}_0\cos 2\chi+\text{J}_1\{\cos(2\chi-\Omega+\varpi)+\cos(2\chi+\Omega-\varpi)\}\\ &\qquad\qquad +\text{J}_2\{\cos(2\chi-2\Omega+2\varpi)+\cos(2\chi+2\Omega-2\varpi)\}]\\ &\quad +Q^4[\text{E}_1\cos(2\chi+\Omega+\varpi)+\text{E}_2\cos(2\chi+2\Omega)+\text{E}_3\cos(2\chi+3\Omega-\varpi)\\ &\qquad\qquad +\text{E}_4\cos(2\chi+4\Omega-2\varpi)]\\ -2\text{XY} &= \text{The same, with sines for cosines}\\ \text{YZ} &= \text{The same as X}^2-\text{Y}^2\text{, but with } -P^3Q \text{ for } P^4,\ PQ(P^2-Q^2) \text{ for}\\ &\qquad 2P^2Q^2,\ PQ^3 \text{ for } Q^4 \text{ and with } \chi \text{ for } 2\chi\\ \text{XZ} &= \text{The same as the last, but with sines for cosines}\\ \tfrac{1}{3}(\text{X}^2+\text{Y}^2-2\text{Z}^2) &= \tfrac{1}{3}(P^4-4P^2Q^2+Q^4)[\text{J}_0+2\text{J}_1\cos(\Omega-\varpi)+2\text{J}_2\cos 2(\Omega-\varpi)]\\ &\quad +2P^2Q^2[\text{E}_1\cos(\Omega+\varpi)+\text{E}_2\cos 2\Omega+\text{E}_3\cos(3\Omega-\varpi)\\ &\qquad\qquad +\text{E}_4\cos(4\Omega-2\varpi)]\end{aligned}\right\} \quad \text{......(280)}$$

If we regard ϖ as constant, and remember that $\chi = nt$, and that Ω stands for $\Omega t+\epsilon$, and if we look through the above functions we see that there are trigonometrical terms of 22 different speeds, viz.: 9 in the first pair all involving $2nt$, 9 in the second pair all involving nt, and 4 in the last.

Since these five functions correspond to Diana's tide-generating potential, we are going to consider the effects of 22 different tides, nine being semi-diurnal, nine diurnal, and the last four may be conveniently called monthly, since their periods are $\frac{1}{4}$, $\frac{1}{3}$, $\frac{1}{2}$ of a month and one month.

We next have to form the 𝔛-𝔜-𝔷 functions. We found that in the X-Y-Z functions there were terms of 22 different speeds; hence we shall now have to introduce 44 symbols indicating the reduction in the height of tide below its equilibrium height, and the retardation of phase. The notation adopted is analogous to that used in the preceding problem, and the following schedule gives the symbols.

Semi-diurnal tides.

speed	$2n-4\Omega$	$2n-3\Omega$	$2n-2\Omega$	$2n-\Omega$	$2n$	$2n+\Omega$	$2n+2\Omega$	$2n+3\Omega$	$2n+4\Omega$
height	F^{iv}	$\mathrm{F}^{\mathrm{iii}}$	F^{ii}	F^{i}	F	F_{i}	F_{ii}	$\mathrm{F}_{\mathrm{iii}}$	F_{iv}
lag	$2\mathrm{f}^{\mathrm{iv}}$	$2\mathrm{f}^{\mathrm{iii}}$	$2\mathrm{f}^{\mathrm{ii}}$	$2\mathrm{f}^{\mathrm{i}}$	$2\mathrm{f}$	$2\mathrm{f}_{\mathrm{i}}$	$2\mathrm{f}_{\mathrm{ii}}$	$2\mathrm{f}_{\mathrm{iii}}$	$2\mathrm{f}_{\mathrm{iv}}$

Diurnal tides.

speed	$n-4\Omega$	$n-3\Omega$	$n-2\Omega$	$n-\Omega$	n	$n+\Omega$	$n+2\Omega$	$n+3\Omega$	$n+4\Omega$
height	G^{iv}	$\mathrm{G}^{\mathrm{iii}}$	G^{ii}	G^{i}	G	G_{i}	G_{ii}	$\mathrm{G}_{\mathrm{iii}}$	G_{iv}
lag	g^{iv}	$\mathrm{g}^{\mathrm{iii}}$	g^{ii}	g^{i}	g	g_{i}	g_{ii}	$\mathrm{g}_{\mathrm{iii}}$	g_{iv}

Monthly tides.*

speed	Ω	2Ω	3Ω	4Ω
height	H^{i}	H^{ii}	$\mathrm{H}^{\mathrm{iii}}$	H^{iv}
lag	h^{i}	$2\mathrm{h}^{\mathrm{ii}}$	$3\mathrm{h}^{\mathrm{iii}}$	$4\mathrm{h}^{\mathrm{iv}}$

The $\mathfrak{X}$-$\mathfrak{Y}$-$\mathfrak{Z}$ functions might now be easily written out; for each term of the X-Y-Z functions is to be multiplied, according to its *speed* by the corresponding *height*, and the corresponding *lag* subtracted from the argument of the trigonometrical term. For example, the first term of $\mathfrak{X}^2-\mathfrak{Y}^2$ is $\mathrm{F}^{\mathrm{i}}\mathrm{E}_1 P^4\cos(2\chi-\Omega-\varpi-2\mathrm{f}^{\mathrm{i}})$. It will however be unnecessary to write out these long expressions.

In order to form the disturbing function W, the $\mathfrak{X}$-$\mathfrak{Y}$-$\mathfrak{Z}$ functions have to be multiplied by the X′-Y′-Z′ functions according to the formula (31). Now the X′-Y′-Z′ functions only differ from the X-Y-Z functions in the accentuation of Ω and ϖ, because Diana is to be ultimately identical with the moon.

In the $\mathfrak{X}$-$\mathfrak{Y}$-$\mathfrak{Z}$ functions Ω is an abbreviation for $\Omega t+\epsilon$, and in the X′-Y′-Z′ functions Ω' for $\Omega t+\epsilon'$; hence wherever in the products we find $\Omega-\Omega'$, we may replace it by $\epsilon-\epsilon'$.

Again, since we are only seeking to find the secular changes in the ellipticity and mean distance, therefore (as before pointed out) we need only multiply together terms whose arguments only differ by the lag. Secular *inequalities*, in the sense in which the term is used in the planetary theory, will indeed arise from the cross-multiplication of certain terms of like *speeds* but of different *arguments*,—for example, the product of the term

$$\mathrm{F}^{\mathrm{ii}}P^4\mathrm{E}_2\cos(2\chi-2\Omega-2\mathrm{f}^{\mathrm{ii}}) \text{ in } \mathfrak{X}^2-\mathfrak{Y}^2$$

multiplied by the term

$$2P^2Q^2\mathrm{J}_2\cos(2\chi-2\Omega'+2\varpi') \text{ in } \mathrm{X}'^2-\mathrm{Y}'^2$$

* With periods of $\frac{1}{4}$, $\frac{1}{3}$, $\frac{1}{2}$, and one month.

when added to the similar cross-product in $4X'Y'\mathfrak{X}\mathfrak{Y}$ (which only differs in having sines for cosines) will give a term

$$2F^{ii}P^6Q^2E_2J_2\cos[2(\epsilon'-\epsilon)-2\varpi'-2f^{ii}]$$

This term in the disturbing function will give a long inequality, but it is of no present interest.

The products may now be written down without writing out in full either the $\mathfrak{X}$-$\mathfrak{Y}$-$\mathfrak{Z}$ functions or the X'-Y'-Z' functions. In order that the results may form the constituent terms of W, the factor $\frac{1}{2}$ is introduced in the first pair of products, the factor 2 in the second pair, and the factor $\frac{3}{2}$ in the last. Then from (280) we have

$$2\,\frac{X'^2-Y'^2}{2}\,\frac{\mathfrak{X}^2-\mathfrak{Y}^2}{2}+2X'Y'\mathfrak{X}\mathfrak{Y}$$

$$\begin{aligned}
=\tfrac{1}{2}P^8\{&F^iE_1^2\cos[(\epsilon'-\epsilon)+(\varpi'-\varpi)-2f^i]+F^{ii}E_2^2\cos[2(\epsilon'-\epsilon)-2f^{ii}]\\
&+F^{iii}E_3^2\cos[3(\epsilon'-\epsilon)-(\varpi'-\varpi)-2f^{iii}]\\
&+F^{iv}E_4^2\cos[4(\epsilon'-\epsilon)-2(\varpi'-\varpi)-2f^{iv}]\}\\
+2P^4Q^4\{&FJ_0^2\cos 2f\\
&+F^iJ_1^2\cos[(\epsilon'-\epsilon)-(\varpi'-\varpi)-2f^i]\\
&+F_iJ_1^2\cos[(\epsilon'-\epsilon)-(\varpi'-\varpi)+2f_i]\\
&+F^{ii}J_2^2\cos[2(\epsilon'-\epsilon)-2(\varpi'-\varpi)-2f^{ii}]\\
&+F_{ii}J_2^2\cos[2(\epsilon'-\epsilon)-2(\varpi'-\varpi)+2f_{ii}]\}\\
+\tfrac{1}{2}Q^8\{&F_iE_1^2\cos[(\epsilon'-\epsilon)+(\varpi'-\varpi)+2f_i]+F_{ii}E_2^2\cos[2(\epsilon'-\epsilon)+2f_{ii}]\\
&+F_{iii}E_3^2\cos[3(\epsilon'-\epsilon)-(\varpi'-\varpi)+2f_{iii}]\\
&+F_{iv}E_4^2\cos[4(\epsilon'-\epsilon)-2(\varpi'-\varpi)+2f_{iv}]\}\ \ldots\ldots\ldots\ldots(281)
\end{aligned}$$

$2Y'Z'\mathfrak{Y}\mathfrak{Z}+2X'Z'\mathfrak{X}\mathfrak{Z}=$ the same, when $2P^6Q^2$ replaces $\frac{1}{2}P^8$; $2P^2Q^2(P^2-Q^2)^2$ replaces $2P^4Q^4$; $2P^2Q^6$ replaces $\frac{1}{2}Q^8$; and G's and g's replace F's and 2f's ....(282)

$$\tfrac{3}{2}\,\frac{X'^2+Y'^2-2Z'^2}{3}\,\frac{\mathfrak{X}^2+\mathfrak{Y}^2-2\mathfrak{Z}^2}{3}$$

$$\begin{aligned}
=\tfrac{1}{6}(P^4-4P^2Q^2+Q^4)^2\{&J_0^2+2H^iJ_1^2\cos[(\epsilon'-\epsilon)-(\varpi'-\varpi)+h^i]\\
&+2H^{ii}J_2^2\cos[2(\epsilon'-\epsilon)-2(\varpi'-\varpi)+2h^{ii}]\}\\
+3P^4Q^4\{&H^iE_1^2\cos[(\epsilon'-\epsilon)+(\varpi'-\varpi)+h^i]+H^{ii}E_2^2\cos[2(\epsilon'-\epsilon)+2h^{ii}]\\
&+H^{iii}E_3^2\cos[3(\epsilon'-\epsilon)-(\varpi'-\varpi)+3h^{iii}]\\
&+H^{iv}E_4^2\cos[4(\epsilon'-\epsilon)-2(\varpi'-\varpi)+4h^{iv}]\}\ \ldots\ldots\ldots\ldots(283)
\end{aligned}$$

The sum of these three last expressions (281–3) when multiplied by $\dfrac{\tau^2}{\mathfrak{g}}\dfrac{1}{(1-e^2)^6}$ is equal to W the disturbing function.

§ 24. *Secular changes in eccentricity and mean distance.*

Before proceeding to the differentiation of W, it is well to note the following coincidences between the coefficients and arguments, viz.: E_1^2 occurs with $(\epsilon'-\epsilon)+(\varpi'-\varpi)$, E_2^2 with $2(\epsilon'-\epsilon)$, E_3^2 with $3(\epsilon'-\epsilon)-(\varpi'-\varpi)$, E_4^2 with $4(\epsilon'-\epsilon)-2(\varpi'-\varpi)$, J_1^2 with $(\epsilon'-\epsilon)-(\varpi'-\varpi)$, J_2^2 with $2(\epsilon'-\epsilon)-2(\varpi'-\varpi)$, and the terms in J_0^2 do not involve $\epsilon, \epsilon', \varpi, \varpi'$. In consequence of these coincidences it will be possible to arrange the results in a highly symmetrical form.

By equations (11) and (12)

$$-\frac{\xi}{k}\frac{d}{dt}\log\eta = \left(\frac{d}{d\epsilon'}+\gamma\frac{d}{d\varpi'}\right)\mathrm{W}, \quad \text{when} \quad \gamma=\frac{1}{\eta}$$

and

$$\frac{1}{k}\frac{d\xi}{dt} = \left(\frac{d}{d\epsilon'}+\gamma\frac{d}{d\varpi'}\right)\mathrm{W}, \quad \text{when} \quad \gamma=0$$

Hence the single operation $d/d\epsilon'+\gamma d/d\varpi'$ will enable us by proper choice the value of γ to find either $\xi d\log\eta/kdt$ or $d\xi/kdt$.

Perform this operation; then putting $\epsilon'=\epsilon$, $\varpi'=\varpi$, and collecting the terms according to their respective E's and J's, we have

$$\begin{aligned}
&\left(\frac{d\mathrm{W}}{d\epsilon'}+\gamma\frac{d\mathrm{W}}{d\varpi'}\right)\div\frac{\tau^2}{\mathfrak{g}}\frac{1}{(1-e^2)^6}\\
&= \mathrm{E}_1^2(1+\gamma)\left\{\tfrac{1}{2}P^8\mathrm{F}^{\mathrm{i}}\sin 2\mathrm{f}^{\mathrm{i}}+2P^6Q^2\mathrm{G}^{\mathrm{i}}\sin \mathrm{g}^{\mathrm{i}}-2P^2Q^6\mathrm{G}_{\mathrm{i}}\sin \mathrm{g}_{\mathrm{i}}\right.\\
&\qquad\qquad\left.-\tfrac{1}{2}Q^8\mathrm{F}_{\mathrm{i}}\sin 2\mathrm{f}_{\mathrm{i}}-3P^4Q^4\mathrm{H}^{\mathrm{i}}\sin \mathrm{h}^{\mathrm{i}}\right\}\\
&\quad+\mathrm{E}_2^2(2)\left\{\text{the same with ii for i, and } 2\mathrm{h}^{\mathrm{ii}} \text{ for } \mathrm{h}^{\mathrm{i}}\right\}\\
&\quad+\mathrm{E}_3^2(3-\gamma)\left\{\text{the same with iii for i, and } 3\mathrm{h}^{\mathrm{iii}} \text{ for } \mathrm{h}^{\mathrm{i}}\right\}\\
&\quad+\mathrm{E}_4^2(4-2\gamma)\left\{\text{the same with iv for i, and } 4\mathrm{h}^{\mathrm{iv}} \text{ for } \mathrm{h}^{\mathrm{i}}\right\}\\
&\quad+\mathrm{J}_1^2(1-\gamma)\left\{2P^4Q^4(\mathrm{F}^{\mathrm{i}}\sin 2\mathrm{f}^{\mathrm{i}}-\mathrm{F}_{\mathrm{i}}\sin 2\mathrm{f}_{\mathrm{i}})\right.\\
&\qquad\left.+2P^2Q^2(P^2-Q^2)^2(\mathrm{G}^{\mathrm{i}}\sin \mathrm{g}^{\mathrm{i}}-\mathrm{G}_{\mathrm{i}}\sin \mathrm{g}_{\mathrm{i}})-\tfrac{1}{3}(P^4-4P^2Q^2+Q^4)^2\mathrm{H}^{\mathrm{i}}\sin \mathrm{h}^{\mathrm{i}}\right\}\\
&\quad+\mathrm{J}_2^2(2-2\gamma)\left\{\text{the same with ii for i, and } 2\mathrm{h}^{\mathrm{ii}} \text{ for } \mathrm{h}^{\mathrm{i}}\right\} \qquad\ldots\ldots\ldots\ldots(284)
\end{aligned}$$

The functions of P and Q, which appear here, will occur hereafter so frequently that it will be convenient to adopt an abridged notation for them. Let x represent either i, ii, iii or iv, and let

$$\left.\begin{aligned}
\phi(\mathrm{x}) &= \tfrac{1}{2}P^8\mathrm{F}^{\mathrm{x}}\sin 2\mathrm{f}^{\mathrm{x}}+2P^6Q^2\mathrm{G}^{\mathrm{x}}\sin \mathrm{g}^{\mathrm{x}}-2P^2Q^6\mathrm{G}_{\mathrm{x}}\sin \mathrm{g}_{\mathrm{x}}-\tfrac{1}{2}Q^8\mathrm{F}_{\mathrm{x}}\sin 2\mathrm{f}_{\mathrm{x}}\\
&\qquad -3P^4Q^4\mathrm{H}^{\mathrm{x}}\sin(\mathrm{xh}^{\mathrm{x}})\\
\psi(\mathrm{x}) &= 2P^4Q^4(\mathrm{F}^{\mathrm{x}}\sin 2\mathrm{f}^{\mathrm{x}}-\mathrm{F}_{\mathrm{x}}\sin 2\mathrm{f}_{\mathrm{x}})+2P^2Q^2(P^2-Q^2)^2(\mathrm{G}^{\mathrm{x}}\sin \mathrm{g}^{\mathrm{x}}-\mathrm{G}_{\mathrm{x}}\sin \mathrm{g}_{\mathrm{x}})\\
&\qquad -\tfrac{1}{3}(P^4-4P^2Q^2+Q^4)^2\mathrm{H}^{\mathrm{x}}\sin(\mathrm{xh}^{\mathrm{x}})
\end{aligned}\right\} \ldots\ldots(285)$$

The generalised definition of the F's, G's, H's, &c., is contained in the following schedule

$$\left.\begin{array}{llllll} \text{speed} & 2n-\mathrm{x}\Omega, & n-\mathrm{x}\Omega, & \mathrm{x}\Omega, & n+\mathrm{x}\Omega, & 2n+\mathrm{x}\Omega \\ \text{height} & \mathrm{F}^{\mathrm{x}} & \mathrm{G}^{\mathrm{x}} & \mathrm{H}^{\mathrm{x}} & \mathrm{G}_{\mathrm{x}} & \mathrm{F}_{\mathrm{x}} \\ \text{lag} & 2\mathrm{f}^{\mathrm{x}} & \mathrm{g}^{\mathrm{x}} & (\mathrm{xh}^{\mathrm{x}}) & \mathrm{g}_{\mathrm{x}} & 2\mathrm{f}_{\mathrm{x}} \end{array}\right\} \quad (286)$$

We must now substitute for the E's and J's their values, and as the ellipticity is chosen as the variable they must be expressed in terms of η instead of e. Also each of the E^2's and J^2's must be divided by $(1-\mathrm{e}^2)^6$.

Since $\sqrt{1-\mathrm{e}^2} = 1-\eta$,

$$\mathrm{e}^2 = 2\eta - \eta^2 \text{ and } (1-\mathrm{e}^2)^{-6} = (1-\eta)^{-12} = 1 + 12\eta + 78\eta^2$$

Then by (278)

$$\left.\begin{array}{ll} \mathrm{E}_1^2 = \tfrac{1}{4}\mathrm{e}^2(1-\tfrac{25}{4}\mathrm{e}^2) = \tfrac{1}{2}\eta(1-13\eta) \quad , \text{ and} & \dfrac{\mathrm{E}_1^2}{(1-\eta)^{12}} = \tfrac{1}{2}\eta(1-\eta) \\ \mathrm{E}_2^2 = 1 - 11\mathrm{e}^2 + \tfrac{423}{8}\mathrm{e}^4 = 1 - 22\eta + \tfrac{445}{2}\eta^2, \text{ and} & \dfrac{\mathrm{E}_2^2}{(1-\eta)^{12}} = 1 - 10\eta + \tfrac{73}{2}\eta^2 \\ \mathrm{E}_3^2 = \tfrac{49}{4}\mathrm{e}^2(1-\tfrac{291}{28}\mathrm{e}^2) = \tfrac{49}{2}\eta(1-\tfrac{149}{7}\eta) \quad , \text{ and} & \dfrac{\mathrm{E}_3^2}{(1-\eta)^{12}} = \tfrac{49}{2}\eta(1-\tfrac{65}{7}\eta) \\ \mathrm{E}_4^2 = \tfrac{289}{4}\mathrm{e}^4 = 289\eta^2 \quad , \text{ and} & \dfrac{\mathrm{E}_4^2}{(1-\eta)^{12}} = 289\eta^2 \\ \mathrm{J}_0^2 = 1 - 3\mathrm{e}^2 + 3\mathrm{e}^4 = 1 - 6\eta + 15\eta^2 \quad , \text{ and} & \dfrac{\mathrm{J}_0^2}{(1-\eta)^{12}} = 1 + 6\eta + 21\eta^2 \\ \mathrm{J}_1^2 = \tfrac{9}{4}\mathrm{e}^2(1-\tfrac{15}{4}\mathrm{e}^2) = \tfrac{9}{2}\eta(1-8\eta) \quad , \text{ and} & \dfrac{\mathrm{J}_1^2}{(1-\eta)^{12}} = \tfrac{9}{2}\eta(1+4\eta) \\ \mathrm{J}_2^2 = \tfrac{81}{16}\mathrm{e}^4 = \tfrac{81}{4}\eta^2 \quad , \text{ and} & \dfrac{\mathrm{J}_2^2}{(1-\eta)^{12}} = \tfrac{81}{4}\eta^2 \end{array}\right\} \quad \ldots\ldots(287)$$

When γ is put equal to $\dfrac{1}{\eta}$ we shall also require the following:

$$\left.\begin{array}{ll} \dfrac{\mathrm{E}_1^2(1+\eta)}{\eta(1-\eta)^{12}} = \tfrac{1}{2} & \dfrac{\mathrm{E}_2^2(2)}{(1-\eta)^{12}} = 2(1-10\eta) \\ \dfrac{\mathrm{E}_3^2(3\eta-1)}{\eta(1-\eta)^{12}} = -\tfrac{49}{2}(1-\tfrac{86}{7}\eta) & \dfrac{\mathrm{E}_4^2(4\eta-2)}{\eta(1-\eta)^{12}} = -578\eta \\ \dfrac{\mathrm{J}_1^2(\eta-1)}{\eta(1-\eta)^{12}} = -\tfrac{9}{2}(1+3\eta) & \dfrac{\mathrm{J}_2^2(2\eta-2)}{\eta(1-\eta)^{12}} = -\tfrac{81}{2}\eta \end{array}\right\} \quad (288)$$

Therefore by putting $\gamma = \dfrac{1}{\eta}$ in equation (284) we have

$$-\frac{\mathfrak{g}}{\tau^2}\frac{\xi}{k}\frac{d}{dt}\log\eta = \tfrac{1}{2}\phi(\mathrm{i}) + 2(1-10\eta)\phi(\mathrm{ii}) - \tfrac{49}{2}(1-\tfrac{86}{7}\eta)\phi(\mathrm{iii}) - 578\eta\phi(\mathrm{iv})$$
$$-\tfrac{9}{2}(1+3\eta)\psi(\mathrm{i}) - \tfrac{81}{2}\eta\psi(\mathrm{ii})$$

and by putting $\gamma = 0$ in (284)

$$\frac{\mathfrak{g}}{\tau^2}\frac{1}{k}\frac{d\xi}{dt} = \tfrac{1}{2}\eta\,(1-\eta)\,\phi\,(\text{i}) + 2\,(1 - 10\eta + \tfrac{73}{2}\eta^2)\,\phi\,(\text{ii}) + \tfrac{147}{2}\eta\,(1 - \tfrac{65}{7}\eta)\,\phi\,(\text{iii})$$
$$+ 1156\eta^2\phi\,(\text{iv}) + \tfrac{9}{2}\eta\,(1+4\eta)\,\psi\,(\text{i}) + \tfrac{81}{2}\eta^2\psi\,(\text{ii})$$

The equations may be also arranged in the following form:

$$-\frac{\mathfrak{g}}{\tau^2}\frac{\xi}{k}\frac{d}{dt}\log\eta = \tfrac{1}{2}\,[\phi\,(\text{i}) + 4\phi\,(\text{ii}) - 49\phi\,(\text{iii}) - 9\psi\,(\text{i})]$$
$$+ \eta\,[-20\phi\,(\text{ii}) + 301\phi\,(\text{iii}) - 578\phi\,(\text{iv}) - \tfrac{27}{2}\psi\,(\text{i}) - \tfrac{81}{2}\psi\,(\text{ii})] \ldots (289)$$

$$\frac{\mathfrak{g}}{\tau^2}\frac{1}{k}\frac{d\xi}{dt} = 2\phi\,(\text{ii})$$
$$+ \eta\,[\tfrac{1}{2}\phi\,(\text{i}) - 20\phi\,(\text{ii}) + \tfrac{147}{2}\phi\,(\text{iii}) + \tfrac{9}{2}\psi\,(\text{i})]$$
$$+ \eta^2\,[-\tfrac{1}{2}\phi\,(\text{i}) + 73\phi\,(\text{ii}) - \tfrac{1365}{2}\phi\,(\text{iii}) + 1156\phi\,(\text{iv}) + 18\psi\,(\text{i}) + \tfrac{81}{2}\psi\,(\text{ii})]$$
$$\ldots\ldots (290)$$

The former of these apparently stops with the first power of η, but it will be observed that we have $d\log\eta/dt$ on the left-hand side so that $d\eta/dt$ is developed as far as η^2.

These equations give the required solutions of the problem.

§ 25. *Application to the case where the planet is viscous.*

If the planet or earth be viscous, we have, as in § 7, $F^x = \cos 2f^x$, $G^x = \cos g^x$, $H^x = \cos(xh^x)$, $G_x = \cos g_x$, $F_x = \cos 2f_x$.

When these values are substituted in (289) we have the equation giving the rate of change of ellipticity in the case of viscosity. The equation is however so long and complex that it does not present to the mind any physical meaning, and I shall therefore illustrate it graphically.

The case taken is the same as that in § 7, where the planet rotates 15 times as fast as the satellite revolves.

The eccentricity or ellipticity is supposed to be small, so that only the first line of (289) is taken.

I took as five several standards of viscosity of the planet, such viscosities as would make the lag f^{ii} of the principal slow semi-diurnal tide, of speed $2n - 2\Omega$, equal to 10°, 20°, 30°, 40°, 44°. (The curves thus correspond to the same cases as in §§ 7 and 10.) Values of $\sin 4f^x$, $\sin 2g^x$, $\sin 2xh^x$, $\sin 2g_x$, $\sin 4f_x$, when $x = $ i, ii, iii were then computed, according to the theory of viscous tides.

These values were then taken for computing values of ϕ (i), ϕ (ii), ϕ (iii), ψ (i) with values of $i = 0°, 15°, 30°, 45°, 60°, 75°, 90°$. The results were then combined so as to give a series of values of $d\log\eta/dt$ or de/edt, and these values are set out graphically in the accompanying fig. 8.

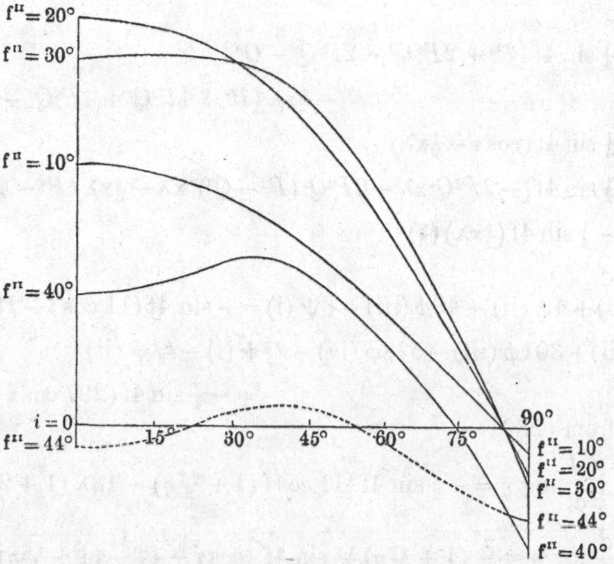

FIG. 8. Diagram showing the rate of change in the eccentricity of the orbit of the satellite for various obliquities and viscosities of the planet $\left(\frac{1}{e}\frac{de}{dt}, \text{ when e is small}\right)$.

In the figure the ordinates are proportional to de/edt, and the abscissæ to i the obliquity; each curve corresponds to one degree of viscosity.

From the figure we see that, unless the viscosity be so great as to approach rigidity (when $f^{ii} = 45°$), the eccentricity will increase for all values of the obliquity, except values approaching 90°.

The rate of increase is greatest for zero obliquity unless the viscosity be very large, and in that case it is a little greater for about 35° of obliquity.

It appears from the paper on "Precession" that if the obliquity be very nearly 90°, the satellite's distance from the planet decreases with the time. Hence it follows from this figure that in general the eccentricity of the orbit increases or diminishes with the mean distance; this is however not true if the viscosity approaches very near rigidity, for then the eccentricity will diminish for zero obliquity, whilst the mean distance will increase.

If the viscosity be very small, the equations (289–90) admit of reduction to very simple forms.

In this case the sines of twice the angles of lagging are proportional to the speeds of the several tides, and we have (as in previous cases)—

$$\frac{\sin 4f^x}{\sin 4f} = 1 - \tfrac{1}{2}x\lambda, \quad \frac{\sin 2g^x}{\sin 4f} = \tfrac{1}{2} - \tfrac{1}{2}x\lambda, \quad \frac{\sin 2xh^x}{\sin 4f} = \tfrac{1}{2}x\lambda$$

$$\frac{\sin 4f_x}{\sin 4f} = 1 + \tfrac{1}{2}x\lambda, \quad \frac{\sin 2g_x}{\sin 4f} = \tfrac{1}{2} + \tfrac{1}{2}x\lambda,$$

Therefore

$$\phi(\mathrm{x}) = \tfrac{1}{4}\sin 4\mathrm{f}\,[P^8 + 2P^6Q^2 - 2P^2Q^6 - Q^8$$
$$- \tfrac{1}{2}\mathrm{x}\lambda\,(P^8 + 4P^6Q^2 + 4P^2Q^6 + Q^8 + 6P^4Q^4)]$$
$$= \tfrac{1}{4}\sin 4\mathrm{f}\,(\cos i - \tfrac{1}{2}\mathrm{x}\lambda)$$
$$\psi(\mathrm{x}) = \tfrac{1}{2}\sin 4\mathrm{f}\,[-2P^4Q^4\mathrm{x}\lambda - 2P^2Q^2(P^2-Q^2)^2\,\mathrm{x}\lambda - \tfrac{1}{6}\mathrm{x}\lambda\,(P^4 - 4P^2Q^2 + Q^4)^2]$$
$$= -\tfrac{1}{4}\sin 4\mathrm{f}\,(\tfrac{1}{2}\mathrm{x}\lambda)\,(\tfrac{2}{3})$$

And

$$\phi(\mathrm{i}) + 4\phi(\mathrm{ii}) - 49\phi(\mathrm{iii}) - 9\psi(\mathrm{i}) = -\sin 4\mathrm{f}\,(11\cos i - 18\lambda)$$
$$-20\phi(\mathrm{ii}) + 301\phi(\mathrm{iii}) - 578\phi(\mathrm{iv}) - \tfrac{27}{2}\psi(\mathrm{i}) - \tfrac{81}{2}\psi(\mathrm{ii})$$
$$= -\tfrac{1}{4}\sin 4\mathrm{f}\,(297\cos i - 756\lambda)$$

Whence from (289)

$$-\frac{\mathfrak{g}}{\tau^2}\frac{\xi}{k}\frac{d}{dt}\log\eta = -\tfrac{1}{2}\sin 4\mathrm{f}\,\{11\cos i\,(1+\tfrac{27}{2}\eta) - 18\lambda\,(1+21\eta)\}$$

or
$$\frac{\xi}{k}\frac{d}{dt}\log\eta = \frac{\tau^2}{\mathfrak{g}}(1+\tfrac{27}{2}\eta)\,\tfrac{11}{2}\sin 4\mathrm{f}\left\{\cos i - \tfrac{18}{11}\frac{\Omega}{n}(1+\tfrac{15}{2}\eta)\right\} \quad\ldots(291)$$

From this we see that, in the case of small viscosity, tidal reaction is in general competent to cause the eccentricity of the orbit of a satellite to increase. But if 18 sidereal days of the planet be greater than 11 sidereal months of the satellite the eccentricity will decrease. Wherefore a circular orbit for the satellite is only dynamically stable provided a period of 18 such days is greater than 11 such months.

Now if we treat the equation (290) for $\frac{d\xi}{dt}$ in the same way, we find—

The first line $= \tfrac{1}{2}\sin 4\mathrm{f}\,(\cos i - \lambda)$.

The second $= \tfrac{1}{2}\eta\sin 4\mathrm{f}\,(27\cos i - 46\lambda)$.

The third $= \tfrac{1}{2}\eta^2\sin 4\mathrm{f}\,(273\cos i - 697\lambda)$.

Therefore

$$\left.\begin{aligned}\frac{\mathfrak{g}}{\tau^2}\frac{1}{k}\frac{d\xi}{dt} &= \tfrac{1}{2}\sin 4\mathrm{f}\,[(1+27\eta+273\eta^2)\cos i - \lambda\,(1+46\eta+697\eta^2)]\\ \text{or}\quad \frac{1}{k}\frac{d\xi}{dt} &= \tfrac{1}{2}\frac{\tau^2}{\mathfrak{g}}(1+27\eta+273\eta^2)\sin 4\mathrm{f}\left[\cos i - \frac{\Omega}{n}(1+19\eta-89\eta^2)\right]\end{aligned}\right\}\ (292)$$

From this it follows that the rate of tidal reaction is greater if the orbit be eccentric than if it be circular. Also for zero obliquity the tidal reaction vanishes when

$$\frac{\Omega}{n} = 1 - 19\eta + 450\eta^2$$

Hence if a satellite were to separate from a planet in such a way that, at the moment after separation, its mean motion were equal to the angular velocity of the planet, if its orbit were eccentric it must fall back into the planet; but if its orbit were circular an infinitesimal disturbance would decide whether it should approach or recede from the planet*.

Now suppose that the viscosity is very large, and that the obliquity is zero.

Then

$$-\frac{\mathfrak{g}}{\tau^2}\frac{\xi}{k}\frac{d}{dt}\log\eta = \tfrac{1}{8}(\sin 4f^{i} + 4\sin 4f^{ii} - 49\sin 4f^{iii} + 6\sin 2h^{i})$$

and the sines are reciprocally proportional to the speeds of the tides, from which they take their origin. As to the term in $\sin 2h^{i}$, which takes its origin from the elliptic monthly tide, the viscosity must make a close approach to absolute rigidity for this term to be reciprocally proportional to the speed of that tide; for the present, therefore, $\sin 2h^{i}$ will be left as it is.

The equation becomes, on this hypothesis,

$$-\frac{\mathfrak{g}}{\tau^2}\frac{\xi}{k}\frac{d}{dt}\log\eta = \tfrac{1}{8}\sin 4f^{ii}\left[\frac{1-\lambda}{1-\frac{1}{2}\lambda} + 4 - \frac{49(1-\lambda)}{1-\frac{3}{2}\lambda}\right] + \tfrac{6}{8}\sin 2h^{i}$$

$$\frac{\mathfrak{g}}{\tau^2}\frac{\xi}{k}\frac{d}{dt}\log\eta = \tfrac{1}{8}\sin 4f^{ii}\frac{44 - 63\lambda + 20\lambda^2}{(1-\frac{1}{2}\lambda)(1-\frac{3}{2}\lambda)} - \tfrac{6}{8}\sin 2h^{i} \quad\text{.............(293)}$$

The numerator of the first term on the right is always positive for values of λ less than unity, and the denominator is always positive if λ be less than $\frac{2}{3}$. Hence if the viscosity be not so great but that the last term is small, the eccentricity always increases if λ lies between zero and $\frac{2}{3}$.

If however λ be not small, then even though the viscosity be not great enough to approach perfect rigidity, we must have $\sin 2h^{i} = 2(1-\lambda)\sin 4f^{ii}/\lambda$. And of course, by supposing the viscosity great enough, this relation may be fulfilled whatever be λ.

Then our equation becomes

$$\frac{\mathfrak{g}}{\tau^2}\frac{\xi}{k}\frac{d}{dt}\log\eta = -\tfrac{1}{8}\sin 4f^{ii}\frac{12 - 80\lambda + 96\lambda^2 - 29\lambda^3}{\lambda(1-\frac{1}{2}\lambda)(1-\frac{3}{2}\lambda)} \quad\text{......(294)}$$

The numerator on the right-hand side is always positive for values of λ less than unity, and the denominator is positive for values of λ less than $\frac{2}{3}$.

Since

$$\frac{1}{k}\frac{d\xi}{dt} = \tfrac{1}{2}\frac{\tau^2}{\mathfrak{g}}\sin 4f^{ii}$$

we have

$$\xi\frac{d}{d\xi}\log\eta = -\tfrac{1}{4}\frac{12 - 80\lambda + 96\lambda^2 - 29\lambda^3}{\lambda(1-\frac{1}{2}\lambda)(1-\frac{3}{2}\lambda)}$$

From this we see that, for very large viscosity,—

For values of λ between 1 and ·6667, the eccentricity increases per unit increase of ξ, and the rate of increase tends to become infinite when $\lambda = $ ·6667.

* See Appendix (p. 374) for further considerations on this subject.

The remarks concerning the physical absurdity of this class of result in § 21 may be repeated in this case.

For values of λ between ·6667 and 0, the eccentricity diminishes.

A similar treatment of the case of small viscosity shows that—

For values of λ between 1 and ·6111 the eccentricity decreases, and for values of λ between ·6111 and 0 the eccentricity increases.

Thus it is only between $\lambda = {}$·6111 and ·6667 that the two cases agree.

Hence in the course of evolution of a satellite revolving about a purely viscous planet:—

For small viscosity the orbit will remain circular until 11 months of the satellite are equal to 18 days of the planet, then the eccentricity will increase until this relationship is again fulfilled, when the eccentricity will again diminish*.

And for very large viscosity the orbit will at once become eccentric, and the eccentricity will increase very rapidly until two months of the satellite are equal to three days of the planet. The eccentricity will then diminish until this relationship is again fulfilled, after which the eccentricity will again increase.

We shall consider later which of these views seems the more probable with regard to the history of the moon.

§ 26. *Secular change in the obliquity and diurnal rotation of the planet, when the satellite moves in an eccentric orbit.*

The method of treating this problem will be the same as that of § 12, to which the reader is referred.

In the complete development of the disturbing function $\chi - \chi'$ would occur wherever the F's and G's occur, but never with the H's.

If we put $\gamma = 1$ in (284), we have

$$\frac{dW}{d\epsilon'} + \frac{dW}{d\varpi'} = \frac{2\tau^2}{\mathfrak{g}(1-\eta)^{12}} \Sigma E_x{}^2 \phi(x) \quad \text{.................(295)}$$

where Σ means summation for i, ii, iii, iv.

This result follows from the fact that in all the E-terms of W, ϵ' and ϖ' enter in the form $l\epsilon' + m\varpi'$, where $l + m = 2$.

In the F^x-terms χ' enters in the form $2\chi'$, and is of the opposite sign from $l+m$; in the F_x-terms it enters in the form $2\chi'$, and is of the same sign as $l+m$; in the G^x-terms it enters in the form χ', and is of the opposite sign

* See "On the Analytical Expressions, &c.," *Proc. Roy. Soc.*, No. 202, 1880. [Paper 7 below.]

from $l+m$; in the G_x-terms it enters in the form χ', and is of the same sign as $l+m$.

Hence as far as regards the E-terms of W, we have

$$\text{in the } F^x\text{-terms } \frac{dW}{d\chi'} = -\left(\frac{dW}{d\epsilon'} + \frac{dW}{d\varpi'}\right)$$

$$\text{in the } F_x\text{-terms } \quad = \frac{dW}{d\epsilon'} + \frac{dW}{d\varpi'}$$

$$\text{in the } G^x\text{-terms } \quad = -\tfrac{1}{2}\left(\frac{dW}{d\epsilon'} + \frac{dW}{d\varpi'}\right)$$

$$\text{in the } G_x\text{-terms } \quad = \tfrac{1}{2}\left(\frac{dW}{d\epsilon'} + \frac{dW}{d\varpi'}\right)$$

$$\text{in the H-terms } \quad = 0$$

In the J-terms of W, χ' enters with coefficient 2 in the F^x- and F_x-terms, and with the coefficient 1 in the G^x- and G_x-terms, and is always of the same sign as the corresponding lag.

Hence for the J-terms

$$\frac{dW}{d\chi'} = \Sigma\left(\frac{dW}{df^x} + \frac{dW}{dg^x}\right)$$

where Σ means summation for the cases where x is zero and both upper and lower i and ii.

From this we have

$$\frac{dn}{dt} = \frac{dW}{d\chi'}$$

$$= -\frac{\tau^2}{\mathfrak{g}}\frac{1}{(1-\eta)^{12}}[\Sigma E_x^2\{P^8F^x\sin 2f^x + 2P^6Q^2G^x\sin g^x + 2P^2Q^6G_x\sin g_x + Q^8F_x\sin 2f_x\} + J_0^2\{4P^4Q^4F\sin 2f + 2P^2Q^2(P^2-Q^2)^2G\sin g\} + \Sigma J_x^2\{4P^4Q^4(F^x\sin 2f^x + F_x\sin 2f_x) + 2P^2Q^2(P^2-Q^2)^2(G^x\sin g^x + G_x\sin g_x)\}] \quad \ldots\ldots(296)$$

the first Σ being from iv to i, and the last only for ii and i.

This is a partial solution for the tidal friction, and corresponds only to the action of the moon on her own tides; that of the sun on his tides may be obtained by symmetry.

It is easy to see that for the joint effect of the two bodies we have

$$\frac{dn}{dt} = -\frac{2\tau\tau'}{\mathfrak{g}}\frac{1}{(1-\eta)^6(1-\eta')^6}J_0J_0'\{4P^4Q^4F\sin 2f + 2P^2Q^2(P^2-Q^2)^2G\sin g\} \quad \ldots\ldots(297)$$

From (296–7) and (287–8) the complete solution may be collected.

In order to find the secular change in the obliquity, we must consider how ψ' would enter in W.

In the development of W, $\Omega' t+\epsilon'$ stands for $\Omega' t+\epsilon'-\psi'$, and ϖ' stands for $\varpi'-\psi'$. Hence from (295)

$$\frac{dW}{d\psi'}=-\left(\frac{dW}{d\epsilon'}+\frac{dW}{d\varpi'}\right)$$
$$=-\frac{2\tau^2}{\mathfrak{g}}\frac{1}{(1-\eta)^{12}}\Sigma E_x{}^2\phi(x) \quad\ldots\ldots\ldots\ldots\ldots\ldots(298)$$

Now by (18) $\qquad n\sin i\frac{di}{dt}=\frac{dW}{d\chi'}\cos i-\frac{dW}{d\psi'}$

Substituting for $\frac{dW}{d\chi'}$ from (296) and for $\frac{dW}{d\psi'}$ from (298), we find

$$\begin{aligned}n\frac{di}{dt}=\frac{\tau^2}{\mathfrak{g}}\frac{1}{(1-\eta)^{12}}\{&\Sigma E_x{}^2[P^7QF^x\sin 2f^x+P^5Q(P^2+3Q^2)G^x\sin g^x\\&-PQ^5(3P^2+Q^2)G_x\sin g_x-PQ^7F_x\sin 2f_x-3P^3Q^3H^x\sin(xh^x)]\\&-J_0{}^2[2P^3Q^3(P^2-Q^2)F\sin 2f+PQ(P^2-Q^2)^3G\sin g]\\&-\Sigma J_x{}^2[2P^3Q^3(P^2-Q^2)(F^x\sin 2f^x+F_x\sin 2f_x)\\&+PQ(P^2-Q^2)^3(G^x\sin g^x+G_x\sin g_x)]\}\ldots(299)\end{aligned}$$

the first Σ being from iv to i, and the last only for ii and i.

This is only a partial solution, and gives the result of the action of the moon on her own tides; that for the sun on his tides may be obtained by symmetry.

It is easy to see that for the joint effect

$$\begin{aligned}n\frac{di}{dt}=-\frac{2\tau\tau'}{\mathfrak{g}}\frac{1}{(1-\eta)^6(1-\eta')^6}J_0J_0'&[2P^3Q^3(P^2-Q^2)F\sin 2f\\&+PQ(P^2-Q^2)^3G\sin g]\ldots\ldots(300)\end{aligned}$$

From (299, 300) and (287–8) the complete solution may be collected.

If these solutions be applied to the case where the earth is viscous and where the viscosity is small, it will be found after reduction as in previous cases that

$$\begin{aligned}-\frac{dn}{dt}=\frac{\sin 4f}{2\mathfrak{g}}\Big[&\tau^2(1-\tfrac{1}{2}\sin^2 i)(1+15\eta+\tfrac{195}{2}\eta^2)\\&+\tau'^2(1-\tfrac{1}{2}\sin^2 i)(1+15\eta'+\tfrac{195}{2}\eta'^2)\\&-\tau^2\frac{\Omega}{n}\cos i(1+27\eta+273\eta^2)-\tau'^2\frac{\Omega'}{n}\cos i(1+27\eta'+273\eta'^2)\\&+\tau\tau'\tfrac{1}{2}\sin^2 i(1+3\eta+3\eta'+6\eta^2+9\eta\eta'+6\eta'^2)\Big]\ldots\ldots\ldots(301)\end{aligned}$$

$$\begin{aligned}n\frac{di}{dt}=\frac{\sin 4f}{4\mathfrak{g}}\sin i\cos i\Big[&\tau^2(1+15\eta+\tfrac{195}{2}\eta^2)+\tau'^2(1+15\eta'+\tfrac{195}{2}\eta'^2)\\&-2\tau^2\frac{\Omega}{n}\sec i(1+27\eta+273\eta^2)-2\tau'^2\frac{\Omega'}{n}\sec i(1+27\eta'+273\eta'^2)\\&-\tau\tau'(1+3\eta+3\eta'+6\eta^2+9\eta\eta'+6\eta'^2)\Big]\ldots\ldots(302)\end{aligned}$$

These results give the tidal friction and rate of change of obliquity due both to the sun and moon; η is the ellipticity of the lunar orbit, and η' of the solar (or terrestrial) orbit.

If η and η' be put equal to zero they agree with the results obtained in the paper on "Precession."

§ 27. *Verification of analysis, and effect of evectional tides.*

The analysis of this part of the paper has been long and complex, and therefore a verification is valuable.

The moment of momentum of the orbital motion of the moon and earth round their common centre of inertia is proportional to the square root of the *latus rectum* of the orbit, according to the ordinary theory of elliptic motion. In the present notation this moment of momentum is equal to $C\xi(1-\eta)/k$. If we suppose the obliquity of the ecliptic to be zero, the whole moment of momentum of the system (supposing only one satellite to exist) is

$$C\left\{\frac{\xi}{k}(1-\eta)+n\right\}$$

Therefore we ought to find, if the analysis has been correctly worked, that

$$\frac{\xi}{k}\frac{d\eta}{dt}=(1-\eta)\frac{1}{k}\frac{d\xi}{dt}+\frac{dn}{dt}$$

This test will be only applied in the case where the viscosity is small, because the analysis is pretty short; but it may also be applied in the general case.

When $i=0$, we have from (292), after multiplying both sides by $1-\eta$,

$$\frac{2}{\sin 4\mathrm{f}}\frac{\mathfrak{g}}{\tau^2}(1-\eta)\frac{1}{k}\frac{d\xi}{dt}=1+26\eta+246\eta^2-\lambda(1+45\eta+651\eta^2)$$

And when $i=0$ and $\tau'=0$, from (301)

$$-\frac{2}{\sin 4\mathrm{f}}\frac{\mathfrak{g}}{\tau^2}\frac{dn}{dt}=1+15\eta+\tfrac{195}{2}\eta^2-\lambda(1+27\eta+273\eta^2)$$

Hence

$$(1-\eta)\frac{1}{k}\frac{d\xi}{dt}+\frac{dn}{dt}=\tfrac{1}{2}\sin 4\mathrm{f}\,\frac{\tau^2}{\mathfrak{g}}\left[11\eta\left(1+\tfrac{27}{2}\eta\right)-18\lambda\eta(1+21\eta)\right]$$

$$=\frac{\xi}{k}\frac{d\eta}{dt}\text{ from (291)}$$

Thus the above formulæ satisfy the condition of the constancy of the moment of momentum of the system.

The most important lunar inequality after the Equation of the Centre is the Evection. The effects of lagging evectional tides may be worked out on the same plan as that pursued above for the Equation of the Centre.

I will not give the analysis, but will merely state that, in the case of small viscosity of the earth, the equation for the rate of change of ellipticity, inclusive of the evectional terms, becomes

$$\frac{\xi}{k}\frac{d}{dt}\log\eta = \tfrac{11}{2}\left(1+\tfrac{27}{2}\eta\right)\sin 4f\,\frac{\tau^2}{\mathfrak{g}}\left\{1-\tfrac{18}{11}\frac{\Omega}{n}-\tfrac{675}{352}\left(\frac{\Omega'}{\Omega}\right)^2\right\}$$

where Ω' is the earth's mean motion in its orbit round the sun.

From this we see that, even at the present time, the evectional tides will only reduce the rate of increase of the ellipticity by $\frac{1}{88}$th part of the whole. In the integrations to be carried out in Part VI. this term will sink in importance, and therefore it will be entirely neglected.

The Variation is another lunar inequality of slightly less importance than the Evection; but it may be observed that the Evection was only of any importance because its argument involved the lunar perigee, and its coefficient the eccentricity. Now neither of these conditions are fulfilled in the case of the Variation. Moreover in the retrospective integration the coefficients will degrade far more rapidly than those of the evectional terms, because they will depend on $(\Omega'/\Omega)^4$. Hence the secular effects of the variational tides will not be given, though of course it would be easy to find them if they were required.

VI.

Integration for Changes in the Eccentricity of the Orbit.

§ 28. *Integration in the case of small viscosity.*

By (291–2), we have approximately

$$\frac{\xi}{k}\frac{d}{dt}\log\eta = \tfrac{11}{2}\sin 4f\,\frac{\tau^2}{\mathfrak{g}}\left(1+\tfrac{27}{2}\eta\right)\left[\cos i-\tfrac{18}{11}\lambda\right]$$

$$\frac{1}{k}\frac{d\xi}{dt} = \tfrac{1}{2}\sin 4f\,\frac{\tau^2}{\mathfrak{g}}\left(1+27\eta\right)\left[\cos i-\lambda\right]$$

Therefore

$$\left(1+\tfrac{27}{2}\eta\right)\frac{d}{d\xi}\log\eta = \frac{11}{\xi}\cdot\frac{1-\frac{18}{11}\lambda\sec i}{1-\lambda\sec i}$$

$$=\frac{11}{\xi}-7\frac{\Omega}{\xi n}\sec i \text{ approximately}$$

The last transformation assumes that λ or Ω/n is small compared with unity; this will be the case in the retrospective integration for a long way back.

As a first approximation we have

$$\eta=\eta_0\xi^{11}$$

Therefore

$$\int_1^\xi \tfrac{27}{2}\eta d\log\eta = \tfrac{27}{2}(\eta-\eta_0) = -\tfrac{27}{2}\eta_0(1-\xi^{11}) \text{ approximately}$$

And for a second approximation

$$\log_e\left(\frac{\eta}{\eta_0\xi^{11}}\right) = \tfrac{27}{2}\eta_0(1-\xi^{11}) - 7\Omega_0\int_1^\xi \frac{\sec i}{\xi^4 n}d\xi \text{ ....(303)}$$

The integral in this expression is very small, and therefore to evaluate it we may assign to i an average value, say I, and neglect the solar tidal friction in assigning a value to n, so that we take

$$n = n_0 + \frac{1}{k}(1-\xi)$$

Let $kn_0 + 1 = \kappa$; whence $n = \frac{1}{k}(\kappa - \xi)$

Hence the last term in (303) is approximately equal to

$$-7k\Omega_0 \sec \mathrm{I}\int_1^\xi \frac{d\xi}{\xi^4(\kappa-\xi)}$$

$$= 7k\Omega_0\sec\mathrm{I}\left[\frac{1}{3\kappa}\left(\frac{1}{\xi^3}-1\right)+\frac{1}{2\kappa^2}\left(\frac{1}{\xi^2}-1\right)+\frac{1}{\kappa^3}\left(\frac{1}{\xi}-1\right)\right]-\frac{7k\Omega_0}{\kappa^4}\sec\mathrm{I}\log\left(\frac{\xi n_0}{n}\right)$$

In the last term n has been written for $(\kappa-\xi)/k$.

If we write

$$\mathrm{K} = \left[\frac{1}{3\kappa}\left(\frac{1}{\xi^3}-1\right)+\frac{1}{2\kappa^2}\left(\frac{1}{\xi^2}-1\right)+\frac{1}{\kappa^3}\left(\frac{1}{\xi}-1\right)\right]7k\Omega_0\sec\mathrm{I} + \tfrac{27}{2}\eta_0(1-\xi^{11})$$

then $$\eta = \eta_0\xi^{11}\left(\frac{n}{n_0\xi}\right)^{\frac{7k\Omega_0}{\kappa^4}\sec\mathrm{I}}e^{\mathrm{K}} \text{ ....(304)}$$

This formula will now be applied to trace the changes in the eccentricity of the lunar orbit.

The integration will be made over a series of "periods" which cover the same ground as those in the paper on "Precession"; and the numerical results of that paper will be used for assigning the values to n and I.

kn_0 is equal to $1/\mu$ of that paper, and therefore κ is $(1+\mu)/\mu$.

First period of integration.

From $\xi = 1$ to ·88.

I is taken as 22°. In "Precession" μ was 4·0074, therefore $kn_0 = \cdot 24954$ and $\kappa = 1\cdot 24954$. Also $k\Omega_0 = kn_0\Omega_0/n_0$, and $\Omega_0/n_0 = 1/27\cdot 32$.

In computing for § 17 of "Precession" I found at the end of the period $\log n/n_0 = \cdot 18971$.

Using these values I find

$$\log_{10}\left(\frac{n}{\xi n_0}\right)^{\frac{7k}{\kappa^4}\Omega_0 \sec I} = \cdot 00692$$

Also $$K = \cdot 01980 + \tfrac{27}{2}\eta_0(1-\xi^{11})$$

Now e_0, the present eccentricity of the lunar orbit, is ·054908.

Whence $\eta_0 = 1 - \sqrt{1-e_0^2} = \cdot 001509$, and $\frac{27}{2}\eta_0(1-\xi^{11}) = \cdot 015375$

Using these values I find

$\log_{10}\eta = 6{\cdot}59007 - 10$, and the first approximation gave $\log_{10}\eta = 6{\cdot}56788 - 10$

Therefore $\eta = \cdot 00038911$, and $e = \cdot 02789$, at the end of the first period of integration.

Second period of integration.

From $\xi = 1$ to ·76. I was taken as 18° 45′.

A similar calculation gives

$$\log\left(\frac{n}{n_0\xi}\right)^{\frac{7k\Omega_0}{\kappa^4}\sec I} = \cdot 00817$$

The first part of $K = \cdot 06998$, and $\frac{27}{2}\eta_0(1-\xi^{11}) = \cdot 00500$

Whence

$\log\eta = 5{\cdot}31758 - 10$, and the first approximation gave $\log\eta = 5{\cdot}27902 - 10$

Therefore $\eta = \cdot 000020777$ and $e = \cdot 006446$, at the end of the second period of integration.

Third period of integration.

From $\xi = 1$ to ·76. I was taken as 16° 13′.

Then a similar calculation gave

$$\log\left(\frac{n}{n_0\xi}\right)^{\frac{7k}{\kappa^4}\Omega_0 \sec I} = \cdot 00566$$

The first part of $K = \cdot 12355$, and $\frac{27}{2}\eta_0(1-\xi^{11}) = \cdot 00027$

Whence

$\log\eta = 4{\cdot}06584 - 10$, and the first approximation gave $\log\eta = 4{\cdot}00653 - 10$

Therefore $\eta = \cdot 0000011637$, and $e = \cdot 001526$ at the end of the third period of integration.

Fourth period of integration.

The procedure is now changed in the same way, and for the same reason, as in the fourth period of § 17 of "Precession."

Let $N = \frac{n}{n_0}$ (as in that paper). Then the equation of tidal friction is

$$-\frac{dN}{dt} = \tfrac{1}{2} \sin 4f \frac{\tau^2}{\mathfrak{g} n_0} (1 - \lambda)$$

and the equation for the change in η may be written approximately

$$-\frac{d}{dN} \log \eta = \frac{n_0 k}{\xi} \frac{11 - 18\lambda}{1 - \lambda}$$

Since λ or Ω/n is no longer small, this expression will be integrated by quadratures.

Using the numerical values given in § 17 of "Precession," I find the following corresponding values:

$$N = 1{\cdot}000 \qquad 1{\cdot}107 \qquad 1{\cdot}214 \qquad 1{\cdot}321$$

$$\frac{kn_0}{\xi} \frac{11 - 18\lambda}{1 - \lambda} = 15{\cdot}469 \qquad 17{\cdot}665 \qquad 19{\cdot}465 \qquad 11{\cdot}994$$

Integrating by quadratures with the common difference dN equal to ·107, we find the integral equal to 5·5715.

Whence $\eta = 44{\cdot}273 \times 10^{-10}$, and $e = {\cdot}00009411$.

The results of the whole integration are given in the following table, of which the first two columns are taken from the paper on "Precession."

TABLE XVI.

Days in m. s. hours and minutes	Moon's sidereal period in m. s. days	Eccentricity of lunar orbit
h. m.	Days	
23 56	27·32	·054908
15 28	18·62	·027894
9 55	8·17	·006446
7 49	3·59	·001526
5 55	12 hours	·000094

Beyond this the eccentricity would decrease very little more, because this integration stops where λ is about $\frac{1}{2}$, and the eccentricity ceases to diminish when λ is $\frac{11}{18}$.

The final eccentricity in the above table is only $\frac{1}{580}$th of the initial eccentricity, and the orbit is very nearly circular.

§ 29. *The change of eccentricity when the viscosity is large.*

I shall not integrate the equations in the case where the viscosity is large, because the solution depends so largely on the exact degree of viscosity.

If the viscosity were infinitely large, then in the retrospective integration the eccentricity would be found getting larger and larger and finally would become infinite, when λ is equal to $\frac{2}{3}$. This result is of course physically absurd. If on the other hand the viscosity were large, we might find the eccentricity diminishing, then stationary, and finally increasing until $\lambda = \frac{2}{3}$, after which it would diminish again. Thus by varying the viscosity, supposed always large, we might get a considerable diversity of results.

VII.

Summary and Discussion of Results.

§ 30. *Explanation of problem.—Summary of Parts I. and II.*

In considering the changes in the orbit of a satellite due to frictional tides, very little interest attaches to those elements of the orbit which are to be specified, in order to assign the position which the satellite would occupy at a given instant of time. We are rather here merely concerned with those elements which contain a description of the nature of the orbit.

These elements are the mean distance, inclination, and eccentricity. Moreover all those inequalities in these three elements, which are periodic in time, whether they fall into the class of "secular" or "periodic" inequalities, have no interest for us, and what we require is to trace their *secular changes.*

Similarly, in the case of the planet we are only concerned to discover the secular changes in the period of its rotation, and in the obliquity of its equator to a fixed plane.

It has unfortunately been found impossible to direct the investigation strictly according to these considerations. Amongst the ignored elements are the longitudes of the nodes of the orbit and equator upon the fixed plane, and it was found in one part of the investigation, viz.: Part III., that secular inequalities (in the ordinary acceptance of the term) had to be taken into consideration both in the five elements which define the nature of the orbit, and the planet's mode of motion, and also in the motion of the two nodes.

In the paper on "Precession" I considered the secular changes in the mean distance of the satellite, and the obliquity and rotation-period of the planet, but the satellite's orbit was there assumed to be circular and confined to the fixed plane. In the present paper the inclination and eccentricity are specially considered, but the introduction of these elements has occasioned a modification of the results attained in the previous paper. For convenience of diction I shall henceforth speak of the planet as the earth, and of the satellites as the moon and sun; for, as far as regards tides, the sun may be treated as a satellite of the earth. The investigation has been kept as far as possible general, so as to be applicable to any system of tides in the earth; but it has been directed more especially towards the conception of a bodily distortion of the earth's mass, and all the actual applications are made on the hypothesis that the earth is a viscous body. A very slight modification would however make the results applicable to frictional oceanic tides on a rigid nucleus (see § 1 immediately after (15)).

I thought it sufficient to consider the problem as divisible into the two following cases:—

1st. Where the moon's orbit is circular, but inclined to the ecliptic. (Parts I., II., III., IV.)

2nd. Where the orbit is eccentric, but always coincident with the ecliptic. (Parts I., V., VI.)

Now that these problems are solved, it would not be difficult, although laborious, to unite the two investigations into a single one; but the additional interest of the results would hardly repay one for the great labour, and besides this division of the problem makes the formulæ considerably shorter, and this conduces to intelligibility.

For the present I only refer to the first of the above problems.

It appears that the problem requires still further subdivision, for the following reasons:—

It is a well-known result of the theory of perturbed elliptic motion, that the orbit of a satellite, revolving about an oblate planet and perturbed by a second satellite, always maintains a constant inclination to a certain plane, which is said to be *proper* to the orbit; the nodes also of the orbit revolve with a uniform motion on that plane, apart from "periodic" inequalities.

If then the moon's proper plane be inclined at a very small angle to the ecliptic, the nodes revolve very nearly uniformly on the ecliptic, and the orbit is inclined at very nearly a constant angle thereto. In this case the equinoctial line revolves also nearly uniformly, and the equator is inclined at nearly a constant angle to the ecliptic.

Here then any inequalities in the motion of the earth and moon, which depend on the longitudes of the nodes or of the equinoctial line, are harmonically periodic in time (although they are "secular inequalities"), and cannot lead to any cumulative effects which will alter the elements of the earth or moon.

Again, suppose that the moon and earth are the only bodies in existence. Here the axis of resultant moment of momentum of the system, or the normal to the invariable plane, remains fixed in space. The component moments of momentum are those of the earth's rotation, and of the moon's and earth's orbital revolution round their common centre of inertia. Hence the earth's axis and the normal to the lunar orbit must always be coplanar with the normal to the invariable plane, and therefore the orbit and equator must have a common node on the invariable plane. This node revolves with a uniform precessional motion, and (so long as the earth is rigid) the inclinations of the orbit and equator to the invariable plane remain constant.

Here also inequalities, which depend on the longitude of the common node, are harmonically periodic in time, and can lead to no cumulative effects.

But if the lunar proper plane be not inclined at a small angle to the ecliptic, the nodes of the orbit may either revolve with much irregularity, or may oscillate about a mean position* on the ecliptic. In this case the inclinations of the orbit and equator to the ecliptic may oscillate considerably.

Here then inequalities, which depend on the longitudes of the node and of the equinoctial line, are not simply periodic in time, and may and will lead to cumulative effects.

This explains what was stated above, namely, that we cannot entirely ignore the motion of the two nodes.

Our problem is thus divisible into three cases :—

(i) Where the nodes revolve uniformly on the ecliptic, and where there is a second disturbing satellite, viz. : the sun.

(ii) Where the earth and moon are the only two bodies in existence.

(iii) Where the nodes either oscillate, or do not revolve uniformly.

The cases (i) and (ii) are distinguished by our being able to ignore the nodes. They afford the subject matter for the whole of Part II.

It is proved in §5 that the tides raised by any one satellite can produce directly no secular change in the mean distance of any other satellite. This is true for all three of the above cases.

It is also shown that, in cases (i) and (ii), the tides raised by any one satellite can produce directly no secular change in the inclination of the orbit of any other satellite to the plane of reference. This is not true for case (iii).

The change of inclination of the moon's orbit in case (i) is considered in §6. The equation expressive of the rate of change of inclination is given in (61) and (62). In §7 this is applied in the case where the earth is viscous. Fig. 4 illustrates the physical meaning of the equation, and the reader is

* It is true that this mean position will itself have a slow precessional motion.

referred to §7 for an explanation of the figure. From this figure we learn that the effect of the frictional tides is in general to diminish the inclination of the lunar orbit to the ecliptic, unless the obliquity of the ecliptic be large, when the inclination will increase. The curves also show that for moderate viscosities the rate of decrease of inclination is most rapid when the obliquity of the ecliptic is zero, but for larger viscosities the rate of decrease has a maximum value, when the obliquity is between 30° and 40°.

If the viscosity be small the equation for the rate of decrease of inclination is reducible to a very simple form; this is given in (64) § 7.

In §§ 8, 9, is found the law of increase of the square root of the moon's distance from the earth under the influence of tidal reaction. The law differs but little from that found and discussed in the paper on "Precession," where the plane of the lunar orbit was supposed to be coincident with the ecliptic. If the viscosity be small the equation reduces to a very simple form; this is given in (70). In § 10 I pass to case (ii), where the earth and moon are the only bodies. The equation expressive of the rate of change of inclination of the lunar orbit to the invariable plane is given in (71). Fig. 5 illustrates the physical meaning of the equation, and an explanation of it is given in § 10. From it we learn that the effect of the tides is always to cause a diminution of the inclination—at least so long as the periodic time of the satellite, as measured in rotations of the planet, is pretty long. The following considerations show that this must generally be the case. It appears from the paper on "Precession" that the effect of tidal friction is to cause a continual transference of moment of momentum from that of terrestrial rotation to that of orbital motion; hence it follows that the normal to the lunar orbit must continually approach the normal to the invariable plane. It is true that the rate of this approach will be to some extent counteracted by a parallel increase in the inclination of the earth's axis to the same normal. It will appear later that if the moon were to revolve very rapidly round the earth, and if the viscosity of the earth were great, then this counteracting influence might be sufficiently great to cause the inclination to increase*. This possible increase of inclination is not exhibited in fig. 5, because it illustrates the case where the sidereal month is 15 days long.

In § 11 it is shown that, for case (ii), the rate of variation of the mean distance, obliquity, and terrestrial rotation follow the laws investigated in "Precession," but that the angle, there called the obliquity of the ecliptic, must be interpreted as the angle between the plane of the lunar orbit and the equator.

In § 12 I return again to case (i) and find the laws governing the rate of increase of the obliquity of the ecliptic, and of decrease of the diurnal rotation

* See the abstract of this paper, *Proc. Roy. Soc.*, No. 200, 1879 [Appendix A, p. 380 below], for certain general considerations bearing on this case.

of the earth. The results differ so little from those discussed in "Precession" that they need not be further referred to here.

Up to this point no approximation has been admitted with regard to smallness either in the obliquity or the inclination of the orbit, but mathematical difficulties have rendered it expedient to assume their smallness in the following part of the paper.

§31. *Summary of Part III.*

Part III. is devoted to case (iii) of our first problem. It was found necessary in the first instance to consider the theory of the secular inequalities in the motion of a moon revolving about an oblate rigid earth, and perturbed by a second satellite, the sun. The sun being large and distant, the ecliptic is deemed sensibly unaffected, and is taken as the fixed plane of reference.

The proper plane of the lunar orbit has been already referred to, but I was here led to introduce a new conception, viz.: that of a second proper plane to which the motion of the earth is referred. It is proved that the motion of the system may then be defined as follows:—

The two proper planes intersect one another on the ecliptic, and their common node regredes on the ecliptic with a slow precessional motion. The lunar orbit and the equator are respectively inclined at constant angles to their proper planes, and their nodes on their respective planes also regrede uniformly and at the same speed. The motions are timed in such a way that when the inclination of the orbit to the ecliptic is at the maximum, the obliquity of the equator to the ecliptic is at the minimum, and *vice versâ*.

Now let us call the angular velocity with which the nodes of the orbit would regrede on the ecliptic, if the earth were spherical, *the nodal velocity.*

And let us call the angular velocity with which the common node of the orbit and equator would regrede on the invariable plane of the system, if the sun did not exist, *the precessional velocity.*

If the various obliquities and inclinations be not large, the precessional velocity is in fact the purely lunar precession.

Then if the nodal velocity be large compared with the precessional velocity, the lunar proper plane is inclined at a small angle to the ecliptic, and the equator is inclined at a small angle to the earth's proper plane.

This is the case with the earth, moon, and sun at present, because the nodal period is about $18\frac{1}{2}$ years, and the purely lunar precession would have a period of between 20,000 and 30,000 years. It is not usual to speak of a proper plane of the earth, because it is more simple to conceive a mean equator, about which the true equator nutates with a period of about $18\frac{1}{2}$ years.

Here the precessional motion of the two proper planes is the whole luni-solar precession, and the regression of the nodes on the proper planes is practically the same as the regression of the lunar nodes on the ecliptic.

A comparison of my result with the formula ordinarily given will be found at the end of § 13, and in a note to § 18.

Secondly, if the nodal velocity be small compared with the precessional velocity, the lunar proper plane is inclined at a small angle to the earth's proper plane.

Also the inclination of the equator to the earth's proper plane bears very nearly the same ratio to the inclination of the orbit to the moon's proper plane as the orbital moment of momentum of the two bodies bears to that of the rotation of the earth.

In the planets of the solar system, on account of the immense mass of the sun, the nodal velocity is never small compared with the precessional velocity, unless the satellite moves with a very short periodic time round its planet, or unless the satellite be very small; and if either of these be the case the ratio of the two moments of momentum is small.

Hence it follows that in our system, if the nodal velocity be small compared with the precessional velocity, the proper plane of the satellite is inclined at a small angle to the equator of the planet. The rapidity of motion of the satellites of Mars, Jupiter, and of some of the satellites of Saturn, and their smallness compared with their planets, necessitates that their proper planes should be inclined at small angles to the equators of the planets. A system may, however, be conceived in which the two proper planes are inclined at a small angle to one another, but where the satellite's proper plane is not inclined at a small angle to the planet's equator.

In the case now before us the regression of the common node of the two proper planes is a sort of compound solar precession of the planet with its attendant moon, and the regression of the two nodes on their respective proper planes is very nearly the same as the purely lunar precession on the invariable plane of the system. Thus there are two precessions, the first of the system as a whole, and the second going on within the system, almost as though the external precession did not exist.

If the nodal velocity be of nearly equal speed with the precessional velocity, the regression of the proper planes and that of the nodes on those planes are each a compound phenomenon, which it is rather hard to disentangle without the aid of analysis. Here none of the angles are necessarily small.

It appears from the investigation in "Precession" that the effect of tidal friction is that, on tracing the changes of the system backwards in time, we find the moon getting nearer and nearer to the earth. The result of this is that the ratio of the nodal velocity to the precessional velocity continually diminishes retrospectively; it is initially very large, it decreases, then becomes equal to unity, and finally is very small. Hence it follows that a retrospective solution will show us the lunar proper plane departing from its present close proximity to the ecliptic, and gradually passing over until it becomes inclined at a small angle to the earth's proper plane.

Therefore the problem, involved in the history of the obliquity of the ecliptic and in the inclination of the lunar orbit, is to trace the secular changes in the pair of proper planes, and in the inclinations of the orbit and equator to their respective proper planes.

The four angles involved in this system are however so inter-related, that it is only necessary to consider the inclination of one proper plane to the ecliptic, and of one plane of motion to its proper plane, and afterwards the other two may be deduced. I chose as the two, whose motions were to be traced, the inclination of the lunar orbit to its proper plane, and the inclination of the earth's proper plane to the ecliptic; and afterwards deduced the inclination of the moon's proper plane to the ecliptic, and the inclination of the equator to the earth's proper plane.

The next subject to be considered (§14 to end of Part III.) was the rate of change of these two inclinations, when both moon and sun raise frictional tides in the earth. The change takes place from two sets of causes:—

First because of the secular changes in the moon's distance and periodic time, and in the earth's rotation and ellipticity of figure—for the earth must always remain a figure of equilibrium.

The nodal velocity varies directly as the moon's periodic time, and it will decrease as we look backwards in time.

The precessional velocity varies directly as the ellipticity of the earth's figure (the earth being homogeneous) and inversely as the cube of the moon's distance, and inversely as the earth's diurnal rotation; it will therefore increase retrospectively. The ratio of these two velocities is the quantity on which the position of the proper planes principally depends.

The *second* cause of disturbance is due directly to the tidal interaction of the three bodies.

The most prominent result of this interaction is, that the inclination of the lunar orbit to its proper plane in general diminishes as the time increases, or increases retrospectively. This statement may be compared with the results of Part II., where the ecliptic was in effect the proper plane. The retrospective increase of inclination may be reversed however, under special conditions of tidal disturbance and lunar periodic time.

Also the inclination of the earth's proper plane to the ecliptic in general increases with the time, or diminishes retrospectively. This is exemplified by the results of the paper on "Precession," where the obliquity of the ecliptic was found to diminish retrospectively. This retrospective decrease may be reversed under special conditions.

It is in determining the effects of this second set of causes, that we have to take account of the effects of tidal disturbance on the motions of the nodes of the orbit and equator on the ecliptic.

After a long analytical investigation, equations are found in (224), which give the rate of change of the positions of the proper planes, and of the inclinations thereto.

It is interesting to note how these equations degrade into those of case (i) when the nodal velocity is very large compared with the precessional velocity, and into those of case (ii) when the same ratio is very small.

In order completely to define the rate of change of the configuration of the system, there are two other equations, one of which gives the rate of increase of the square root of the moon's distance (which I called in a previous paper the equation of tidal reaction), and the other gives the rate of retardation of the earth's diurnal rotation (which I called before the equation of tidal friction). For the latter of these we may however substitute another equation, in which the time is not involved, and which gives a relationship between the diurnal rotation and the square root of the moon's distance. It is in fact the equation of conservation of moment of momentum of the moon-earth system, as modified by the solar tidal friction. This is the equation which was extensively used in the paper on "Precession."

Except for the solar tidal friction and for the obliquity of the orbit and equator, this equation would be rigorously independent of the kind of frictional tides existing in the earth. If the obliquities are taken as small, they do not enter in the equation, and in the present case the degree of viscosity of the earth only enters to an imperceptible degree, at least when the day is not very nearly equal to the sidereal month. When that relation between the day and month is very nearly fulfilled, the equation may become largely affected by the viscosity; and I shall return to this point later, while for the present I shall assume the equation to give satisfactory results.

This equation of conservation of moment of momentum enables us to compute as many parallel values of the day and month as may be desired.

Now we have got the time-rates of change of the inclinations of the lunar orbit to its proper plane, and of the earth's proper plane to the ecliptic, and we have also the time-rate of change of the square root of the moon's distance. Hence we may obtain the square-root-of-moon's-distance-rate (or shortly the distance-rate) of change of the two inclinations.

The element of time is thus entirely eliminated; and as the period of time required for the changes has been adequately considered in the paper on "Precession," no further reference will here be made to time.

In a precisely similar manner the equations giving the time-rate in the cases (i) and (ii) of our first problem, may be replaced by equations of distance-rate.

Up to this point terrestrial phraseology has been used, but there is nothing which confines the applicability of the results to our own planet and satellite.

§ 32. *Summary of Part IV.*

We now, however, pass to Part IV., which contains a retrospective integration of the differential equations, with special reference to the earth, moon, and sun. The mathematical difficulties were so great that a numerical solution was the only one found practicable*. The computations made for the paper on "Precession" were used as far as possible.

The general plan followed was closely similar to that of the previous paper, and consists in arbitrarily choosing a number of values for the distance of the moon from the earth (or what amounts to the same thing for the sidereal month), and then computing all the other elements of the system by the method of quadratures.

The first case considered is where the earth has a small viscosity. And here it may be remarked that although the solution is only rigorous for infinitely small viscosity, yet it gives results which are very nearly true over a considerable range of viscosity. This may be seen to be true by a comparison of the results of the integrations in §§ 15 and 17 of "Precession," in the first of which the viscosity was not at all small; also by observing that the curves in fig. 2 of "Precession" do not differ materially from the curve of sines until ϵ (the f of this paper) is greater than 25°; also by noting a similar peculiarity in figs. 4 and 5 of this paper. The hypothesis of large viscosity does not cover nearly so wide a field.

That which we here call a small viscosity is, when estimated by terrestrial standards, very great (see the summary of "Precession").

To return, however, to the case in hand:—We begin with the present configuration of the three bodies, when the moon's proper plane is almost identical with the ecliptic, and when the inclination of the equator to its proper plane is very small. This is the case (i) of the first problem:—

It appears that the solution of "Precession" is sufficiently accurate for this stage of the solution, and accordingly the parallel values of the day,

* An analytical solution in the case of a single satellite, where the viscosity of the planet is small, is given in *Proc. Roy. Soc.*, No. 202, 1880. [Paper 7, below.]

month, and obliquity of the earth's proper plane (or mean equator) are taken from § 17 of that paper; but the change in the new element, the inclination of the lunar orbit, has to be computed.

The results of the solution are given in Table I., § 18, to which the reader is referred.

This method of solution is not applicable unless the lunar proper plane is inclined at a small angle to the ecliptic, and unless the equator is inclined at a small angle to its proper plane. Now at the beginning of the integration, that is to say with a homogeneous earth, and with the moon and sun in their present configuration, the moon's proper plane is inclined to the ecliptic at 13″, and the equator is inclined to the earth's proper plane at 12″ (for the heterogeneous earth these angles are about 8″·3 and 9″·0); and at the end of this integration, when the day is 9 hrs. 55 m. and the month 8·17 m.s. days, the former angle has increased to 57′ 31″, and the latter to 22′ 42″. These last results show that the nutations of the system have already become considerable, and although subsequent considerations show that this method of solution has not been overstrained, yet it here becomes advisable to carry out the solution into the more remote past by the methods of Part III.

It was desirable to postpone the transition as long as possible, because the method used up to this point does not postulate the smallness of the inclinations, whereas the subsequent procedure does make that supposition.

In § 19 the solution is continued by the new method, the viscosity of the earth still being supposed to be small. After laborious computations results are obtained, the physical meaning of which is embodied in Table VIII. The last two columns give the periods of the two precessional motions by which the system is affected. The precession of the pair of proper planes is, as it were, the ancestor of the actual luni-solar precession, and the revolution of the two nodes on their proper planes is the ancestor of the present revolution of the lunar nodes on the ecliptic, and of the 19-yearly nutation of the earth's axis.

This table exhibits a continued approach of the two proper planes to one another, so that at the point where the integration is stopped they are only separated by 1° 18′; at the present time they are of course separated by 23° 28′.

The most remarkable feature in this table is that (speaking retrospectively) the inclination of the lunar orbit to its proper plane first increases, then diminishes, and then increases again.

If it were desired to carry the solution still further back, we might without much error here make the transition to the method of case (ii) of the first problem, and neglecting the solar influence entirely, refer the motion to the invariable plane of the moon-earth system. This invariable plane would have to be taken as somewhere between the two proper planes, and therefore inclined to the ecliptic at about 11° 45′; the invariable plane would then really continue to have a precessional motion due to the solar influence on the system formed by the earth and moon together, but this would not much affect the treatment of the plane as though it were fixed in space.

We should then have to take the obliquity of the equator to the invariable plane as about 3°, and the inclination of the lunar orbit to the same plane as about 5° 30′.

In the more remote past the obliquity of the equator to the invariable plane would go on diminishing, but at a slower and slower rate, until the moon's period is 12 hours and the day is 6 hours, when it would no longer diminish; and the inclination of the orbit to the invariable plane would go on increasing, until the day and month come to an identity, and at an ever increasing rate.

It follows from this, that if we continued to trace the changes backwards, until the day and month are identical, we should find the lunar orbit inclined at a considerable angle to the equator. If this were necessarily the case, it would be difficult to believe that the moon is a portion of the primeval planet detached by rapid rotation, or by other causes. But the previous results are based on the hypothesis that the viscosity of the earth is small, and it therefore now became important to consider how a different hypothesis concerning the constitution of the earth might modify the results.

In § 20 the solution of the problem is resumed, at the point where the methods of Part III. were first applied, but with the hypothesis that the viscosity of the earth is very large, instead of very small. The results for any intermediate degree of viscosity must certainly lie between those found before and those to be found now.

Then having retraversed the same ground, but with the new hypothesis, I found the results given in Table XV.

The inclinations of the two proper planes to the ecliptic are found to be very nearly the same as in the case of small viscosity. But the inclination of the lunar orbit to its proper plane increases at first and then continues diminishing, without the subsequent reversal of motion found in the previous solution.

If the solution were carried back into the more remote past, the motion being referred to the invariable plane, we should find both the obliquity of the equator and the inclination of the orbit diminishing at a rate which tends to become *infinite*, if the viscosity is *infinitely* great. Infinite viscosity is of course the same as perfect rigidity, and if the earth were perfectly rigid the system would not change at all. The true interpretation to put on this result is that the rate of change of inclination becomes large, if the viscosity be large. This diminution would continue until the day was 6 hours and the month 12 hours. For an analysis of the state of things further back than this, the reader is referred to § 20.

From this it follows, that by supposing the viscosity large enough we may make the obliquity and inclination to the invariable plane as small as we please, by the time that state is reached in which the month is equal to twice the day.

Hence, on the present hypothesis, we trace the system back until the lunar orbit is sensibly coincident with the equator, and the equator is inclined to the ecliptic at an angle of 11° or 12°.

It is probable that in the still more remote past the plane of the lunar orbit would not have a tendency to depart from that of the equator. It is not, however, expedient to attempt any detailed analysis of the changes further back, for the following reason. Suppose a system to be unstable, and that some infinitesimal disturbance causes the equilibrium to break down; then after some time it is moving in a certain way. Now suppose that from a knowledge of the system we endeavour to compute backwards from the observed mode of its motion at that time, and so find the condition from which the observed state of motion originated. Our solution will carry us back to a state very near to that of instability, from which the system really departed, but as the calculation can take no account of the infinitesimal disturbance, which caused the equilibrium to break down, it can never bring us back to the state which the system really had. And if we go on computing the preceding state of affairs, the solution will continue to lead us further and further astray from the truth. Now this, I take it, is likely to have been the case with the earth and moon; at a certain period in the evolution (viz.: when the month was twice the day) the system probably became dynamically unstable, and the equilibrium broke down. Thus it seems more likely that we have got to the truth, if we cease the solution at the point where the lunar orbit is nearly coincident with the equator, than by going still further back.

In § 21, fig. 7, is given a graphical illustration of the distance-rate of change in the inclinations of the lunar orbit to its proper plane, and of the earth's proper plane to the ecliptic; the dotted curves refer to the hypothesis of large viscosity, and the firm-curves to that of small viscosity.

The figure is explained and discussed in that section; I will here only draw attention to the wideness apart of the two curves illustrative of the rate of change of the inclination of the lunar orbit. This shows how much influence the degree of viscosity of the earth must have had on the present inclination of the lunar orbit to the ecliptic.

It is particularly interesting to observe that in the case of small viscosity this curve rises above the horizontal axis. If this figure is to be interpreted retrospectively, along with our solution, it must be read from left to right, but if we go with the time, instead of against it, from right to left.

Now if the earth had had in its earlier history infinitely small viscosity, and if the moon had moved primitively in the equator, then until the evolution had reached the point represented by P, the lunar orbit would have always remained sensibly coincident with its proper plane. In passing from P to Q the inclination of the orbit to its proper plane would have increased, but the whole increase could not have amounted to more than a few minutes of arc. At the point P the day is 7 hrs. 47 m. in length, and the month 3·25 m. s. days in length; at the point Q the day is 8 hrs. 36 m., and the month 5·20 m. s. days. From Q down to the present state this small inclination would have always decreased.

If then the earth had had small viscosity throughout its evolution, the lunar orbit would at present be only inclined at a very small angle to the ecliptic. But it is actually inclined at about 5° 9′, hence it follows that while the hypothesis of small viscosity is competent to explain *some* inclination, it cannot explain the actually existing inclination.

It was shown in the papers on "Tides" and "Precession" that, if the earth be not at present perfectly rigid or perfectly elastic, its viscosity must be very large. And it was shown in "Precession" that if the viscosity be large, the obliquity of the ecliptic must at present be decreasing. Now it will be observed that in resuming the integration with the hypothesis of large viscosity, the solution of the first method with the hypothesis of small viscosity was accepted as the basis for continuing the integration with large viscosity. This appears at first sight somewhat illogical, and to be strictly correct, we ought to have taken as the initial inclination of the earth's proper plane to the ecliptic, at the beginning of the application of the methods of Part III. to the hypothesis of large viscosity, some angle probably a little less than 23½°* instead of 17°. This would certainly disturb the results, but I have not thought it advisable to take this course for the following reasons.

It is probable that at the present time the greater part, if not the whole of the tidal friction is due to oceanic tides, and not to bodily tides. If the ocean were frictionless, it would be low tide under the moon; consequently the effects of fluid friction must be to accelerate, not retard, the ocean tides†. In order to apply our present analysis to the case of oceanic tidal

* In the present configuration of the earth, moon, and sun, the obliquity will decrease, if the viscosity be very large. But if we integrate backwards this retrospective increase of obliquity would soon be converted into a decrease. Thus at the end of "the first period of integration," the obliquity would be a little greater than 23½°, but by the end of the "second period" it would probably be a little less than 23½°. It is at the end of the "second period" that the method of Part III. is first applied.

† Otherwise the lunar attraction on the tides would accelerate the earth's rotation—a clear violation of the principles of energy.

friction, that angle which has been called the lag of the tide must be interpreted as the acceleration of the tide.

We know that the actual friction in water is small, and hence the tides of long period will be less affected by friction than those of short period; thus the effects of fluid tidal friction will probably be closely analogous to those resulting from the hypothesis of small viscosity of the whole earth and bodily tides. On the other hand, it is probable that the earth was once more plastic than at present, either superficially or throughout its mass, and therefore it seems probable that the bodily tides, even if small at present, were once more considerable. I think therefore that on the whole we shall be more nearly correct in supposing that the terrestrial nucleus possessed a high degree of stiffness in the earliest times, and that it will be best to apply the hypothesis of small viscosity to the more modern stages of the evolution, and that of large viscosity to the more ancient.

At any rate this appears to be a not improbable theory, and one which accords very well with the present values of the obliquity of the ecliptic, and of the inclination of the lunar orbit.

§ 33. *On the initial condition of the earth and moon.*

It was remarked above that the equation of conservation of moment of momentum, as modified by the effects of solar tidal friction, could only be regarded as practically independent of the degree of viscosity of the earth, so long as the moon's sidereal period was not nearly equal to the day; and that if this relationship were nearly satisfied, the equation which we have used throughout might be considerably in error.

Now in the paper on "Precession" the system was traced backwards, in much the same way as has been done here, until the moon's tide-generating influence was very large compared with that of the sun; the solar influence was then entirely neglected, and the equation of conservation of moment of momentum was used for determining that initial condition, where the month and day were identical, from which the system started its course of development*. The period of revolution of the system in its initial configuration was found to be about $5\frac{1}{2}$ hours. I now however see reason to believe that the solar tidal friction will make the numerical value assigned to this period of revolution considerably in error, whilst the general principle remains almost unaffected. This subject is considered in § 22.

The necessity of correction arises from the assumption that because the moon is retrospectively getting nearer and nearer to the earth, therefore the

* See also a paper on "The Determination of the Secular Effects of Tidal Friction by a Graphical Method," *Proc. Roy. Soc.*, No. 197, 1879. [Paper 5.]

effects of lunar tidal friction must more and more preponderate over those of solar tidal friction, so that if the solar tidal friction were once negligeable it would always remain so. But tidal friction depends on two elements, viz.: the magnitude of the tide-generating influence, and the relative motion of the two bodies. Now whilst the tide-generating influence of the moon *does* become larger and larger, as we approach the critical state, yet the relative motion of the moon and earth becomes smaller and smaller; on the other hand the tide-generating influence of the sun remains sensibly constant, whilst the relative motion of the earth and sun slightly increases*.

From this it follows that the solar tidal friction must ultimately become actually more important than the lunar, notwithstanding the close proximity of the moon to the earth.

The complete investigation of this subject involves considerations which will require special treatment. In § 22 it is only so far considered as to show that, when there is identity of the periods of revolution of the moon and earth, the angular velocity of the system must be greater than that given by the solution in § 18 of "Precession."

When the earth rotates in $5\frac{1}{2}$ hours, the motion of the moon relatively to the earth's surface would already be pretty slow. If the system were traced into the more remote past, the earth's rotation would be found getting more and more rapid, and the moon's orbital angular velocity also continually increasing, but ever approximating to identity with the earth's rotation.

When the surfaces of the two bodies are almost in contact, the motion of the moon relatively to the earth's surface would be almost insensible. This appears to point to the break-up of the primeval planet into two parts, in consequence of a rotation so rapid as to be inconsistent with an ellipsoidal form of equilibrium†.

§ 34. *Summary of Parts V. and VI.*

I now come to the second of the two problems, where the moon moves in an eccentric orbit, always coincident with the ecliptic.

In § 23 it is shown that the tides raised by any one satellite can produce no secular change in the eccentricity of the orbit of any other satellite; thus the eccentricity and the mean distance are in this respect on the same footing.

It was found to be more convenient to consider the ellipticity of the orbit instead of the eccentricity. In § 24 (289) and (290), are given the time-rates

* In the paper on "Precession" it was stated in Section 18 that this must be the case, but I did not at that time perceive the importance of this consideration.

† [See a footnote to § 22 on p. 323.]

of increase of the ellipticity and of the square root of mean distance. In § 25 the result for the ellipticity is applied to the case where the earth is viscous, and its physical meaning is graphically illustrated in fig. 8.

This figure shows that in general the ellipticity will increase with the time; but if the obliquity of the ecliptic be nearly 90°, or if the viscosity be so great that the earth is very nearly rigid, the ellipticity will diminish. This last result is due to the rising into prominence of the effects of the elliptic monthly tide.

If the viscosity be very small the equation is reducible to a very simple form, which is given in (291). From (291) we see that if the obliquity of the ecliptic be zero, the ellipticity will either increase or diminish, according as 18 rotations of the planet take a shorter or a longer time than 11 revolutions of the satellite. From this it follows that in the history of a satellite revolving about a planet of small viscosity, the circular orbit is dynamically stable until 11 months of the satellite have become longer than 18 days of the planet. Since the day and month start from equality and end in equality, it follows that the eccentricity will rise to a maximum and ultimately diminish again.

It is also shown that if a satellite be started to move in a circular orbit with the same periodic time as that of the planet's rotation (with maximum energy for given moment of momentum), then if infinitesimal eccentricity be given to the orbit the satellite will ultimately fall into the planet; and if, the orbit being circular, infinitesimal decrease of distance be given the satellite will fall in, whilst if infinitesimal increase of distance be given the satellite will recede from the planet. Thus this configuration, in which the planet and satellite move as parts of a single rigid body, has a complex instability; for there are two sorts of disturbance which cause the satellite to fall in, and one which causes it to recede from the planet*.

If the planet have very large viscosity the case is much more complex, and it is examined in detail in § 25.

It will here only be stated that the eccentricity will diminish if 2 months of the satellite be longer than 3 days of the planet, but will increase if the 2 months be shorter than 3 days; also the rate of increase of eccentricity tends to become infinite, for infinitely great viscosity, if the 2 months are equal to the 3 days.

These results are largely due to the influence of the elliptic monthly tide, and with most of the satellites of the solar system, this is a very slow tide

* This passage appeared to the referee, requested by the R. S. to report on this paper, to be rather obscure, and it has therefore been somewhat modified. To further elucidate the point I have added in an appendix [p. 374] a graphical illustration of the effects of eccentricity, similar to those given in No. 197 of *Proc. Roy. Soc.*, 1879. [Paper 5.]

See also the abstract of this paper in the *Proc. Roy. Soc.*, No. 200, 1879 [Appendix A, p. 380], for certain general considerations bearing on the problem of the eccentricity.

compared with the semi-diurnal tides; therefore it must in general be supposed that the viscosity of the planet makes a close approximation to perfect rigidity, in order that this statement may be true.

The infinite value of the rate of change of eccentricity is due to the speed of the slower elliptic semi-diurnal tide being infinitely slow, when 2 months are equal to 3 days. The result is physically absurd, and its true meaning is commented on in § 25.

In § 26 the time-rates of change of the obliquity of the planet's equator, and of the diurnal rotation are investigated, when the orbits of the tide-raising satellites are eccentric; the only point of general interest in the result is, that the rate of change of obliquity and the tidal friction are both augmented by the eccentricity of the orbit, as was foreseen in the paper on "Precession."

In § 27 it is stated that the effect of the evectional tides is such as to diminish the eccentricity of the orbit, but the formula given shows that the effect cannot have much importance, unless the moon be very distant from the earth.

In Part VI. the equations giving the rate of change of eccentricity are integrated, on the hypothesis that the earth has small viscosity.

The first step is to convert the time-rates of change into distance-rates, and thus to eliminate the time, as in the previous integrations.

The computations made for the paper on "Precession" were here made use of, as far as possible.

The results of the retrospective integration are given in Table XVI., § 28. This table exhibits the eccentricity falling from its present value of $\frac{1}{18}$th down to about $\frac{1}{10800}$th, so that at the end the orbit is very nearly circular.

The integration in the case of large viscosity is not carried out, because the actual degree of viscosity will exercise so very large an influence on the result.

If the viscosity were *infinitely* large, we should find the eccentricity getting larger and larger retrospectively, and ultimately becoming *infinite*, when 2 months were equal to 3 days. This result is of course absurd, and merely represents that the larger the viscosity, the larger would be the eccentricity. On the other hand, if the viscosity were merely large, we might find the eccentricity decreasing at first, then stationary, then increasing until 2 months were equal to 3 days, and then decreasing again.

It follows therefore that various interpretations may be put to the present eccentricity of the lunar orbit.

If, as is not improbable, the more recent changes in the configuration of our system have been chiefly brought about by oceanic tidal friction, whilst the earlier changes were due to bodily tidal friction, with considerable viscosity of the planet, then, supposing the orbit to have been primevally circular, the history of the eccentricity must have been as follows: first an increase to a maximum, then a decrease to a minimum, and finally an increase to the present value. There seems nothing to tell us how large the early maximum, or how small the subsequent minimum of eccentricity may have been.

VIII.

Review of the Tidal Theory of Evolution as applied to the Earth and the other Members of the Solar System.

I will now collect the various results so as to form a sketch of what the previous investigations show as the most probable history of the earth and moon, and in order to indicate how far this history is the result of calculation, references will be given to the parts of my several papers in which each point is especially considered.

We begin with a planet, not very much more than 8,000 miles in diameter*, and probably partly solid, partly fluid, and partly gaseous. This planet is rotating about an axis inclined at about 11° or 12° to the normal to the ecliptic†, with a period of from 2 to 4 hours‡, and is revolving about the sun with a period not very much shorter than our present year§.

The rapidity of the planet's rotation causes so great a compression of its figure that it cannot continue to exist in an ellipsoidal form‖ with stability; or else it is so nearly unstable that complete instability is induced by the solar tides¶.

The planet then separates into two masses, the larger being the earth and the smaller the moon. I do not attempt to define the mode of separation, or to say whether the moon was initially more or less annular. At any rate it must be assumed that the smaller mass became more or less conglomerated, and finally fused into a spheroid—perhaps in consequence of impacts between its constituent meteorites, which were once part of the primeval planet. Up to this point the history is largely speculative, [for although we know the limit of stability of a homogeneous mass of rotating liquid, yet it surpasses the power of mathematical analysis to follow the manner of rupture when the limiting velocity of rotation is surpassed.]

* "Precession," Section 24 [p. 115].

† This at least appears to be the obliquity at the earliest stage to which the system has been traced back in detail, but the effect of solar tidal friction would make the obliquity primevally less than this, to an uncertain and perhaps considerable amount.

‡ "Precession," Section 18 [p. 101], and Part IV., Section 22 [p. 322; but see footnote on p. 323].

§ "Precession," Section 19 [p. 105].

‖ "Precession," Section 18 [p. 101], and Part IV., Section 22 [p. 322, and footnote on p. 323].

¶ Summary of "Precession" [p. 132].

We now have the earth and the moon nearly in contact with one another, and rotating nearly as though they were parts of one rigid body.

This is the system which has been made the subject of the present dynamical investigation.

As the two masses are not rigid, the attraction of each distorts the other; and if they do not move rigorously with the same periodic time, each raises a tide in the other. Also the sun raises tides in both.

In consequence of the frictional resistance to these tidal motions, such a system is dynamically unstable*. If the moon had moved orbitally a little faster than the earth rotates she must have fallen back into the earth; thus the existence of the moon compels us to believe that the equilibrium broke down by the moon revolving orbitally a little slower than the earth rotates. Perhaps the actual rupture into two masses was the cause of this slower motion; for if the detached mass retained the same moment of momentum as it had initially, when it formed a part of the primeval planet, this would, I think, necessarily be the case.

In consequence of the tidal friction the periodic time of the moon (or the month) increases in length, and that of the earth's rotation (or the day) also increases; but the month increases in length at a much greater rate than the day.

At some early stage in the history of the system, the moon has conglomerated into a spheroidal form, and has acquired a rotation about an axis nearly parallel with that of the earth. We will now follow the moon itself for a time.

The axial rotation of the moon is retarded by the attraction of the earth on the tides raised in the moon, and this retardation takes place at a far greater rate than the similar retardation of the earth's rotation†. As soon as the moon rotates round her axis with twice the angular velocity with which she revolves in her orbit, the position of her axis of rotation (parallel with the earth's axis) becomes dynamically unstable‡. The obliquity of the lunar equator to the plane of the orbit increases, attains a maximum, and then diminishes. Meanwhile the lunar axial rotation is being reduced towards identity with the orbital motion.

Finally her equator is nearly coincident with the plane of her orbit, and the attraction of the earth on a tide, which degenerates into a permanent ellipticity of the lunar equator, causes her always to show the same face to

* "Secular Effects, &c.," *Proc. Roy. Soc.*, No. 197, 1879 [Paper 5, p. 204]; and "Precession," Section 18 [p. 101].

† "Precession," Section 23 [p. 113].

‡ "Precession," Section 17 [p. 93]. It is of course possible that the lunar rotation was very rapidly reduced by the earth's attraction on the lagging tides, and was never permitted to be more than twice the orbital motion. In this case the lunar equator has never deviated much from the plane of the orbit.

the earth*. Laplace has shown that this is a necessary consequence of the elliptic form of the lunar equator.

All this must have taken place early in the history of the earth, to which I now return.

As the month increases in length the lunar orbit becomes eccentric, and the eccentricity reaches a maximum when the month occupies about a rotation and a half of the earth. The maximum of eccentricity is probably not large. After this the eccentricity diminishes†.

The plane of the lunar orbit is at first practically identical with the earth's equator, but as the moon recedes from the earth the sun's attraction begins to make itself felt. Here then we must introduce the conception of the two ideal planes (here called the proper planes), to which the motion of the earth and moon must be referred‡. The lunar proper plane is at first inclined at a very small angle to the earth's proper plane, and the orbit and equator coincide with their respective proper planes.

As soon as the earth rotates with twice the angular velocity with which the moon revolves in her orbit, a new instability sets in. The month is then about 12 of our present hours, and the day is about 6 of our present hours in length.

The inclinations of the lunar orbit and of the equator to their respective proper planes increase. The inclination of the lunar orbit to its proper plane increases to a maximum of 6° or 7°§, and ever after diminishes; the inclination of the equator to its proper plane increases to a maximum of about 2° 45′||, and ever after diminishes. The maximum inclination of the lunar orbit to its proper plane takes place when the day is a little less than 9 of our present hours, and the month a little less than 6 of our present days. The maximum inclination of the equator to its proper plane takes place earlier than this.

Whilst these changes have been going on, the proper planes have been themselves changing in their positions relatively to one another and to the ecliptic. At first they were nearly coincident with one another and with the earth's equator, but they then open out, and the inclination of the lunar proper plane to the ecliptic continually diminishes, whilst that of the terrestrial proper plane continually increases.

* [At the time when this was written I thought that Helmholtz had been the first to suggest the reduction of the moon's axial rotation by means of tidal friction. But the same idea had been advanced both by Kant and Laplace, independently of one another, at much earlier dates. See Chapter 16 of *The Tides and Kindred Phenomena in the Solar System*, by G. H. Darwin.]

† Parts V. and VI. The exact history of the eccentricity is somewhat uncertain, because of the uncertainty as to the degree of viscosity of the earth.

‡ See Parts III. and IV. (and the summaries thereof in Part VII.) for this and what follows about proper planes.

§ Table XV., Part IV. [p. 316].

|| Found from the values in Table XV. [p. 316], and by a graphical construction.

At some stage the earth has become more rigid, and oceans have been formed, so that it is probable that oceanic tidal friction has come to play a more important part than bodily tidal friction*. If this be the case the eccentricity of the orbit, after passing through a stationary phase, begins to increase again.

We have now traced the system to a state in which the day and month are increasing, but at unequal rates; the inclinations of the lunar proper plane to the ecliptic and of the orbit to its proper plane are diminishing; the inclination of the terrestrial proper plane to the ecliptic is increasing, and of the equator to its proper plane is diminishing; and the eccentricity of the orbit is increasing.

No new phase now supervenes†, and at length we have the system in its present configuration. The minimum time in which the changes from first to last can have taken place is 54,000,000 years‡.

In a previous paper it was shown that there are other collateral results of the viscosity of the earth; for during this course of evolution the earth's mass must have suffered a screwing motion, so that the polar regions have travelled a little from west to east relatively to the equator. This affords a possible explanation of the north and south trend of our great continents§. Also a large amount of heat has been generated by friction deep down in the earth, and some very small part of the observed increase of temperature in underground borings may be attributable to this cause||.

The preceding history might vary a little in detail, according to the degree of viscosity which we attribute to the earth's mass, and according as oceanic tidal friction is or is not, now and in the more recent past, a more powerful cause of change than bodily tidal friction.

The argument reposes on the imperfect rigidity of solids, and on the internal friction of semi-solids and fluids; these are *veræ causæ*. Thus changes of the kind here discussed must be going on, and must have gone on in the past. And for this history of the earth and moon to be true throughout, it is only necessary to postulate a sufficient lapse of time, and that there is not enough matter diffused through space to resist materially the motions of the moon and earth in perhaps several hundred million years.

It hardly seems too much to say that granting these two postulates, and the existence of a primeval planet, such as that above described, then a system

* Compare with "Precession," Section 14 [p. 69], where the present secular acceleration of the moon's mean motion is considered.

† Unless the earth's proper plane (or mean equator) be now slowly diminishing in obliquity, as would be the case if the bodily tides are more potent than the oceanic ones. In any case this diminution must ultimately take place in the far future.

‡ "Precession," end of Section 18 [p. 105].

§ "Problems," Part I. [p. 151 and p. 188].

|| "Problems," Part II. [p. 155 and p. 193].

would necessarily be developed which would bear a strong resemblance to our own.

A theory, reposing on *veræ causæ*, which brings into quantitative correlation the lengths of the present day and month, the obliquity of the ecliptic, and the inclination and eccentricity of the lunar orbit, must, I think, have strong claims to acceptance.

But if this has been the evolution of the earth and moon, a similar process must have been going on elsewhere. The present investigation has only dealt with a single satellite and the sun, but the theory may of course be extended, with some modification, to planets attended by several satellites. I will now therefore consider some of the other members of the solar system.

A large planet has much more energy of rotation to be destroyed, and moment of momentum to be redistributed than a small one, and therefore a large planet ought to proceed in its evolution more slowly than a small one. Therefore we ought to find the larger planets less advanced that the smaller ones.

The masses of such of the planets as have satellites are, in terms of the earth's mass, as follows: Mars $= \frac{1}{10}$; Jupiter $= 301$; Saturn $= 90$; Uranus $= 14$; Neptune $= 16$.

Mars should therefore be furthest advanced in its evolution, and it is here alone in the whole system that we find a satellite moving orbitally faster than the planet rotates. This will also be the ultimate fate of our moon, because, after the moon's orbital motion has been reduced to identity with that of the earth's rotation, solar tidal friction will further reduce the earth's angular velocity, the tidal reaction on the moon will be reversed, and the moon's orbital velocity will increase, and her distance from the earth will diminish. But since the moon's mass is very large, the moon must recede to an enormous distance from the earth, before this reversal will take place. Now the satellites of Mars are very small, and therefore they need only to recede a short distance from the planet before the reversal of tidal reaction*.

The periodic time of the satellite Deimos is 30 hrs. 18 m.†, and as the period of rotation of Mars is 24 hrs. 37 m.‡, Deimos must be still receding from Mars, but very slowly.

The periodic time of the satellite Phobos is 7 hrs. 39 m.; therefore Phobos must be approaching Mars. It does not seem likely that it has ever been remote from the planet.

* In the graphical method of treating the subject, "the line of momentum" will only just intersect "the curve of rigidity." See *Proc. Roy. Soc.*, No. 197, 1879. [Paper 5, p. 201.]

† *Observations and Orbits of the Satellites of Mars*, by Asaph Hall. Washington Government Printing Office, 1878.

‡ According to Kaiser, as quoted by Schmidt. *Ast. Nach.*, Vol. LXXXII. p. 333.

The eccentricities of the orbits of both satellites are small, though somewhat uncertain. The eccentricity of the orbit of Phobos appears however to be the larger of the two.

If the viscosity of the planet be small, or if oceanic tidal friction be the principal cause of change, both eccentricities are diminishing; but if the viscosity be large, both are increasing. In any case the rate of change must be excessively slow. As we have no means of knowing whether the eccentricities are increasing or diminishing this larger eccentricity of the orbit of Phobos cannot be a fact of much importance either for or against the present views. But it must be admitted that it is a slightly unfavourable indication.

The position of the proper plane of a satellite is determined by the periodic time of the satellite, the oblateness of the planet, and the sun's distance. The inclination of the orbit of a satellite to its proper plane is not determined by anything in the system. Hence it is only the inclination of the orbit which can afford any argument for or against the theory.

The proper planes of both satellites are necessarily nearly coincident with the equator of the planet; but it is in accordance with the theory that the inclinations of the orbits to their respective proper planes should be small*.

Any change in the obliquity of the equator of Mars to the plane of his orbit must be entirely due to solar tides. The present obliquity is about 27°, and this points also to an advanced stage of evolution—at least if the axis of the planet was primitively at all nearly perpendicular to the ecliptic.

We now come to the system of Jupiter.

This enormous planet is still rotating in about 10 hours, its axis is nearly perpendicular to the ecliptic, and three of its satellites revolve in 7 days or less, whilst the fourth has a period of 16 days 16 hrs. This system is obviously far less advanced than our own.

The inclinations of the proper planes to Jupiter's equator are necessarily small, but the inclinations of the orbits to the proper planes appear to be very interesting from a theoretical point of view. They are as follows†:—

Satellite	Inclination of orbit to proper plane
	° ′ ″
First	0 0 0
Second	0 27 50
Third	0 12 20
Fourth	0 14 58

* For the details of the Martian system, see the paper by Professor Asaph Hall, above quoted. With regard to the proper planes, see a paper by Professor J. C. Adams read before the R. Ast. Soc. on Nov. 14, 1879, *R. A. S. Monthly Not.* There is also a paper by Mr Marth, *Ast. Nach.*, No. 2280, Vol. xcv., Oct. 1879.

† Herschel's *Astron.*, Synoptic Tables in Appendix.

Now we have shown above that the orbit of a satellite is at first coincident with its proper plane, that the inclination afterwards rises to a maximum, and finally declines. If then we may assume, as seems reasonable, that the satellites are in stages of evolution corresponding to their distances from the planet, these inclinations accord well with the theory.

The eccentricities of the orbits of the two inner satellites are insensible, those of the outer two small. This does not tell strongly either for or against the theory, because the history of the eccentricity depends considerably on the degree of viscosity of the planet; yet it on the whole agrees with the theory that the eccentricity should be greater in the more remote satellites. It appears that the satellites of Jupiter always present the same face to the planet, just as does our moon*. This was to be expected.

The case of Saturn is not altogether so favourable to the theory. The extremely rapid rotation, the ring, and the short periodic time of the inner satellites point to an early stage of development; whilst the longer periodic time of the three outer satellites, and the high obliquity of the equator indicate a later stage. Perhaps both views may be more or less correct, for successive shedding of satellites would impart a modern appearance to the system. It may be hoped that the investigation of the effects of tidal friction in a planet surrounded by a number of satellites may throw some light on the subject. This I have not yet undertaken, and it appears to have peculiar difficulties. It has probably been previously remarked, that the Saturnian system bears a strong analogy with the solar system, Titan being analogous to Jupiter, Hyperion and Iapetus to Uranus and Neptune, and the inner satellites being analogous to the inner planets. Thus anything which aids us in forming a theory of the one system will throw light on the other†.

The details of the Saturnian system seem more or less favourable to the theory.

The proper planes of the orbits (except that of Iapetus) are nearly in the plane of the ring, and the inclinations of all the orbits to their proper planes appear not to be large.

Herschel gives the following eccentricities of orbit:—

Tethys ·04 (?), Dione ·02 (?), Rhea ·02 (?), Titan ·029314, Hyperion "rather large"; and he says nothing of the eccentricities of the orbits of the remaining three satellites. If the dubious eccentricities for the first three of the above are of any value, we seem to have some indication of the early maximum of eccentricity to which the analysis points; but perhaps this is pushing the

* Herschel's *Astron.*, 9th ed., § 546.

† Another investigation [Paper 8] seems to show pretty conclusively that tidal friction cannot be in all cases the most important feature in the evolution of such systems as that of Saturn and his satellites, and the solar system itself. I am not however led to reject the views maintained in this paper.

argument too far. The satellite Iapetus appears always to present the same face to the planet*.

Concerning Uranus and Neptune there is not much to be said, as their systems are very little known; but their masses are much larger than that of the earth, and their satellites revolve with a short periodic time. The retrograde motion and high inclination of the satellites of Uranus are, if thoroughly established, very remarkable.

The above theory of the inclination of the orbit has been based on an assumed smallness of inclination, and it is not very easy to see to what results investigation might lead, if the inclination were large. It must be admitted however that the Uranian system points to the possibility of the existence of a primitive planet, with either retrograde rotation, or at least with a very large obliquity of equator.

It appears from this review that the other members of the solar system present some phenomena which are strikingly favourable to the tidal theory of evolution, and none which are absolutely condemnatory. Perhaps by further investigations some light may be thrown on points which remain obscure.

APPENDIX.

(Added July, 1880.)

A graphical illustration of the effects of tidal friction when the orbit of the satellite is eccentric.

In a previous paper (Paper 5†) a graphical illustration of the effects of tidal friction was given for the case of a circular orbit. As this method makes the subject more easily intelligible than the purely analytical method of the present paper, I propose to add an illustration for the case of the eccentric orbit.

Consider the case of a single satellite, treated as a particle, moving in an elliptic orbit, which is co-planar with the equator of the planet.

Let Ch be the resultant moment of momentum of the system. Then with the notation of the present paper, by § 27 the equation of conservation of moment of momentum is

$$n + \frac{\xi}{k}(1-\eta) = h$$

* Herschel's *Astron.*, 9th ed., § 547.

† The last sentence of [Paper 5, p. 207] contains an erroneous statement [which I unfortunately omitted to correct in passing this reprint for the press]. It will be seen from the figure on p. 378 that the line of zero eccentricity on the energy surface is not a ridge as there stated.

Here Cn is the moment of momentum of the planet's rotation, and $C\xi(1-\eta)/k$ is the moment of momentum of the orbital motion; and the whole moment of momentum is the sum of the two.

By the definitions of ξ and k in § 2, $C\dfrac{\xi}{k} = \dfrac{\mu Mm}{\sqrt{\mu(M+m)}}\sqrt{c}$, where μ is the attraction between unit masses at unit distance.

By a proper choice of units we may make $\mu Mm/\sqrt{\mu(M+m)}$ and C equal to unity*.

Let x be equal to the square root of the satellite's mean distance c, and the equation of conservation of moment of momentum becomes

$$n + x(1-\eta) = h \quad\text{..........................}(\alpha)$$

If in (α), the *ellipticity* of the orbit η, be zero, we have equation (3) of Paper 5, p. 197.

It is well known that the sum of the potential and kinetic energies in elliptic motion is independent of the eccentricity of the orbit, and depends only on the mean distance.

Hence if CE be the whole energy of the system, we have (as in equations (2) and (4) of Paper 5), with the present units

$$2E = n^2 - \frac{1}{x^2}$$

If z be written for 2E, and if the value of n be substituted from (α), we have

$$z = \{h - x(1-\eta)\}^2 - \frac{1}{x^2} \quad\text{..........................}(\beta)$$

This is the equation of energy of the system.

* In the paper above referred to, and in Paper 7, below, the physical meaning of the units adopted is scarcely adequately explained.

The units are such that C, the planet's moment of inertia, is unity, that $\mu(M+m)$ is unity, and that a quantity called s and defined in (6) of this paper is unity.

From this it may be deduced that the unit length is such a distance that the moment of inertia of planet and satellite, treated as particles, when at this distance apart about their common centre of inertia is equal to the moment of inertia of the planet about its own axis. If γ be this unit of length, this condition gives $\dfrac{Mm}{M+m}\gamma^2 = C$, or $\gamma = \sqrt{\dfrac{C(M+m)}{Mm}}$.

The unit of time is the time taken by the satellite to describe an arc of 57°·3 in a circular orbit at distance γ; it is therefore $\left(\dfrac{C}{\mu Mm}\right)^{\frac{1}{2}}\left(C\dfrac{M+m}{Mm}\right)^{\frac{1}{4}}$. The unit of mass is $\dfrac{Mm}{M+m}$.

From this it follows that the unit of moment of momentum is the moment of momentum of orbital motion when the satellite moves in a circular orbit at distance γ. The critical moment of momentum of the system, referred to in those two papers and below in this appendix, is $4/3^{\frac{3}{4}}$ of this unit of moment of momentum.

In whatever manner the two bodies may interact on one another, the resultant moment of momentum h must remain constant, and therefore (α) will always give one relation between n, x, and η; a second relation would be given by a knowledge of the nature of the interaction between the two bodies.

The equation (α) might be illustrated by taking n, x, η as the three rectangular co-ordinates of a point, and the resulting surface might be called the surface of momentum, in analogy with the "line of momentum" in the above paper.

This surface is obviously a hyperboloid, which cuts the plane of nx in the straight line $n + x = h$; it cuts the planes of $n\eta$ and $\eta = 1$ in the straight line determined by $n = h$; and the plane of $x\eta$ in the rectangular hyperbola $x(1-\eta) = h$.

The contour lines of this surface for various values of n are a family of rectangular hyperbolas with common asymptotes, viz.: $\eta = 1$ and $x = 0$. It does not however seem worth while to give a figure of them.

If the satellite raises frictional tides of any kind in the planet, the system is non-conservative of energy, and therefore in equation (β) x and η must so vary that z may always diminish.

Suppose that equation (β) be represented by a surface the points on which have co-ordinates x, η, z, and suppose that the axis of z is vertical. Then each point on the surface represents by the co-ordinates x and η one configuration of the system, with given moment of momentum h. Since the energy must diminish, it follows that the point which represents the configuration of the system must always move down hill. To determine the exact path pursued by the point it would be necessary to take into consideration the nature of the frictional tides which are being raised by the satellite.

I will now consider the nature of the surface of energy.

It is clear that it is only necessary to consider positive values of η lying between zero and unity, because values of η greater than unity correspond to a hyperbolic orbit; and the more interesting part of the surface is that for which η is a pretty small fraction.

The curves, formed on the surface by the intersection of vertical planes parallel to x, have maxima and minima points determined by $dz/dx = 0$.

This condition gives by differentiation of (β)

$$x^4 - \frac{h}{1-\eta}x^3 + \frac{1}{(1-\eta)^2} = 0 \quad \ldots\ldots\ldots\ldots\ldots\ldots\ldots(\gamma)$$

From the considerations adduced in Paper 5 and in the next following Paper 7, it follows that this equation has either two real roots or no real roots.

When $\eta = 0$ the equation has real roots provided h be greater than $4/3^{\frac{3}{4}}$, and since this case corresponds to that of all but one of the satellites of the solar system, I shall henceforth suppose that h is greater than $4/3^{\frac{3}{4}}$. It will be seen presently that in this case every section parallel to x has a maximum and minimum point, and the nature of the sections is exhibited in the curves of energy in the two other papers.

Now consider the condition $n = \Omega$, which expresses that the planet rotates in the same period as that in which the satellite revolves, so that if the orbit be circular the two bodies revolve like a single rigid body.

With the present units $\Omega = 1/x^3$, and by (α), $n = h - x(1-\eta)$.

Hence the condition $n = \Omega$ leads to the biquadratic

$$x^4 - \frac{h}{1-\eta}x^3 + \frac{1}{1-\eta} = 0 \qquad (\delta)$$

If η be zero this equation is identical with (γ), which gives the maxima and minima of energy.

Hence if the orbit be circular the maximum and minimum of energy correspond to two cases in which the system moves as a rigid body. If however the orbit be elliptical, and if $n = \Omega$, there is still relative motion during revolution of the satellite, and the energy must be capable of degradation. The principal object of the present note is to investigate the stability of the circular orbit in these cases, and this question involves a determination of the nature of the degradation when the orbit is elliptical.

In Part V. of the present paper it has been shown that if the planet be a fluid of small viscosity the ellipticity of the satellite's orbit will increase if 18 rotations of the planet be less than 11 revolutions of the satellite, and *vice versâ*. Hence the critical relation between n and Ω is $n = \frac{18}{11}\Omega$. This leads to the biquadratic

$$x^4 - \frac{h}{1-\eta}x^3 + \frac{18}{11}\frac{1}{1-\eta} = 0 \qquad (\epsilon)$$

This is an equation with two real roots, and when it is illustrated graphically it will lead to a pair of curves. For configurations of the system represented by points lying between these curves the eccentricity increases, and outside it diminishes,—supposing the viscosity of the planet and the eccentricity of the satellite's orbit to be small.

In order to illustrate the surface of energy (β) and the three biquadratics (γ), (δ), and (ϵ), I chose $h = 3$, which is greater than $4/3^{\frac{3}{4}}$.

By means of a series of solutions, for several values of η, of the equations (γ), (δ), (ϵ), and a method of graphical interpolation, I have drawn the accompanying figure.

The horizontal axis is that of x, the square root of the satellite's distance, and the numbers written along it are the various values of x. The vertical axis is that of η, and it comprises values of η between 0 and 1. The axis of z is perpendicular to the plane of the paper, but the contour lines for various values of z are projected on to the plane of the paper.

The numbers written on the curves represent the values of z, viz., $z=0, 1, 2, 3, 4, 5$.

The ends of the contour lines on the right are joined by dotted lines, because it would be impossible to draw the curves completely without a very large extension of the figure.

The broken lines (---) marked "line of maxima," terminating at A, and "line of minima," terminating at B, represent the two roots of the biquadratic (γ).

The lines marked $n=\Omega$ represent the two roots of (δ), but computation showed that the right-hand branch fell so very near the line of minima, that it was necessary somewhat to exaggerate the divergence in order to show it on the figure.

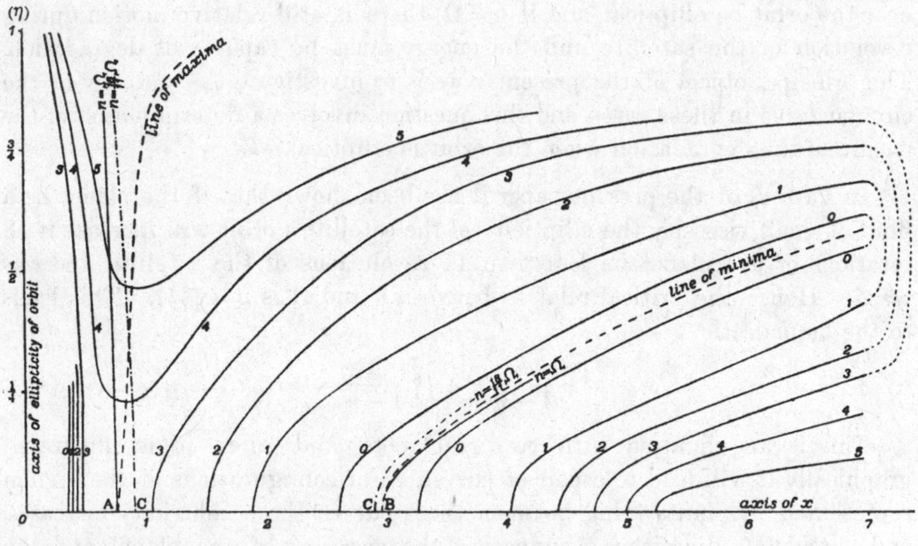

Fig. 9. Contour lines of surface of energy.

The chain-dot lines (–·–·–) C, C, marked $n=\frac{18}{11}\Omega$, represent the two roots of (ϵ). For configurations of the system represented by points lying between these two curves, the ellipticity of orbit will increase; for the regions outside it will decrease. This statement only applies to cases of small ellipticity, and small viscosity of the planet.

Inspection of the figure shows that the line of minima is an infinitely long valley of a hyperbolic sort of shape, with gently sloping hills on each side, and the bed of the valley gently slopes up as we travel away from B.

The line of maxima is a ridge running up from A with an infinitely deep ravine on the left, and the gentle slopes of the valley of minima on the right.

Thus the point B is a true minimum on the surface, whilst the point A is a maximum-minimum, being situated on a saddle-shaped part of the surface.

The lines $n = \Omega$ start from A and B, but one deviates from the ridge of maxima towards the ravine; and the other branch deviates from the valley of minima by going up the slope on the side remote from the origin.

This surface enables us to determine perfectly the stabilities of the circular orbit, when planet and satellite are moving as parts of a rigid body.

The configuration B is obviously dynamically stable in all respects; for any configuration represented by a point near B must degrade down to B.

It is also clear that the configuration A is dynamically unstable, but the nature of the instability is complex. A displacement on the right-hand side of the ridge of maxima will cause the satellite to recede from the planet, because x must increase when the point slides down hill.

If the viscosity be small, the ellipticity given to the orbit will diminish, because A is not comprised between the two chain-dot curves. Thus for this class of tide the *circularity* is stable, whilst the configuration is unstable.

A displacement on the left-hand side of the ridge of maxima will cause the satellite to fall into the planet, because the point will slide down into the ravine. But the circularity of the orbit is again stable.

This figure at once shows that if planet and satellite be revolving with maximum energy as parts of a rigid body, and if, without altering the total moment of momentum, or the equality of the two periods, we impart infinitesimal ellipticity to the orbit, the satellite will fall into the planet. This follows from the fact that the line $n = \Omega$ runs on to the slope of the ravine.

If on the other hand without affecting the moment of momentum, or the circularity, we infinitesimally disturb the relation $n = \Omega$, then the satellite will either recede from or approach towards the planet according to the nature of the disturbance.

These two statements are independent of the nature of the frictional interaction of the two bodies.

The only parts of this figure which postulate anything about the nature of the interaction are the curves $n = \frac{18}{11}\Omega$.

I have not thought it worth while to illustrate the case where h is less than $4/3^{\frac{3}{4}}$, or the negative side of the surface of energy; but both illustrations may easily be carried out.

APPENDIX A.

An extract from the abstract of the foregoing paper, Proc. Roy. Soc., Vol. XXX. (1880), pp. 1—10.

. . . .

The following considerations (in substitution for the analytical treatment of the paper) will throw some light on the general effects of tidal friction:—

Suppose the motions of the planet and of its solitary satellite to be referred to the invariable plane of the system. The axis of resultant moment of momentum is normal to this plane, and the component rotations are that of the planet's rotation about its axis of figure, and that of the orbital motion of the planet and satellite round their common centre of inertia; the axis of this latter rotation is clearly the normal to the satellite's orbit. Hence the normal to the orbit, the axis of resultant moment of momentum, and the planet's axis of rotation, must always lie in one plane. From this it follows that the orbit and the planet's equator must necessarily have a common node on the invariable plane.

If either of the component rotations alters in amount or direction, a corresponding change must take place in the other, such as will keep the resultant moment of momentum constant in direction and magnitude.

It appears from the previous papers that the effect of tidal friction is to increase the distance of the satellite from the planet, and to transfer moment of momentum from that of planetary rotation to that of orbital motion.

If then the direction of the planet's axis of rotation does not change, it follows that the normal to the lunar orbit must approach the axis of resultant moment of momentum. By drawing a series of parallelograms on the same diameter and keeping one side constant in direction, this may be easily seen to be true.

The above statement is equivalent to saying that the inclination of the satellite's orbit will decrease.

But this decrease of inclination does not always necessarily take place, for the previous investigations show that another effect of tidal friction may be to increase the obliquity of the planet's equator to the invariable plane, or in other words to increase the inclination of the planet's axis to the axis of resultant moment of momentum.

Now if a parallelogram be drawn with a constant diameter, it will easily be seen that by increasing the inclination of one of the sides to the diameter (and even decreasing its length), the inclination of the other side to the diameter may also be increased.

The most favourable case for such a change is when the side whose inclination is increased is nearly as long as the diameter. From this it follows that the inclination of the satellite's orbit to the invariable plane may increase, and also that the case when it is most likely to increase is when the moment of momentum of planetary rotation is large compared with that of the orbital motion. The analytical solution of the problem agrees with these results, for it shows that if the viscosity of the planet be small the inclination of the orbit always diminishes, but if the viscosity be large, and if the satellite moves with a short periodic time (as estimated in rotations of the planet), then the inclination of the orbit will increase.

These results serve to give some idea of the physical causes which, according to the memoir, gave rise to the present inclination of the lunar orbit to the ecliptic. For the analytical investigation shows that the inclination of the lunar orbit to its proper plane (which replaces the invariable plane when the solar attraction is introduced) was initially small, that it then increased to a maximum, and finally diminished, and that it is still diminishing.

. . . .

The following considerations (in substitution for the analytical treatment of the paper) throw some light on the physical causes of these results [as to the eccentricity of the orbit].

Consider a satellite revolving about a planet in an elliptic orbit, with a periodic time which is long compared with the period of rotation of the planet; and suppose that frictional tides are raised in the planet.

The major axis of the tidal spheroid always points in advance of the satellite, and exercises a force on the satellite which tends to accelerate its linear velocity.

When the satellite is in perigee the tides are higher, and this disturbing force is greater than when the satellite is in apogee.

The disturbing force may, therefore, be represented as a constant force, always tending to accelerate the motion of the satellite, and a periodic force which accelerates in perigee and retards in apogee. The constant force causes a secular increase of the satellite's mean distance and a retardation of its mean motion.

The accelerating force in perigee causes the satellite to swing out further than it would otherwise have done, so that when it comes round to apogee it is more remote from the planet. The retarding force in apogee acts exactly inversely, and diminishes the perigeean distance. Thus, the apogeean distance increases and the perigeean distance diminishes, or in other words, the eccentricity of the orbit increases.

Now consider another case, and suppose the satellite's periodic time to be identical with that of the planet's rotation. When the satellite is in perigee it is moving faster than the planet rotates, and when in apogee it is moving slower; hence at apogee the tides lag, and at perigee they are accelerated. Now the lagging apogeean tides give rise to an accelerating force on the satellite, and increase the perigeean distance, whilst the accelerated perigeean tides give rise to a retarding force, and decrease the apogeean distance. Hence in this case the eccentricity of the orbit will diminish.

It follows from these two results that there must be some intermediate periodic time of the satellite, for which the eccentricity does not tend to vary*.

But the preceding general explanations are in reality somewhat less satisfactory than they seem, because they do not make clear the existence of certain antagonistic influences.

Imagine a satellite revolving about a planet, and subject to a constant accelerating force, which we saw above would result from tidal reaction.

In a circular orbit a constant tangential force makes the satellite's distance increase, but the larger the orbit the less does the given force increase the mean distance. Now the satellite, moving in the eccentric orbit, is in the apogeean part of its orbit like a satellite moving in a circular orbit at a certain mean distance, but in the perigeean part of the orbit it is like a satellite moving in a circular orbit but at a smaller mean distance; in both parts of the orbit it is subject to the same tangential force. Then the distance at the perigeean part of the orbit increases more rapidly than the distance at the apogeean part. Hence the constant tangential force on the satellite in the eccentric orbit will make the eccentricity diminish. It is not clear from the preceding general explanation, when this cause for decreasing eccentricity will be less important than the previous cause for increasing eccentricity.

. . . .

* The substance of the preceding general explanation was suggested to me in conversation by Sir William Thomson, when I mentioned to him the results at which I had arrived.

7.

ON THE ANALYTICAL EXPRESSIONS WHICH GIVE THE HISTORY OF A FLUID PLANET OF SMALL VISCOSITY, ATTENDED BY A SINGLE SATELLITE.

[*Proceedings of the Royal Society*, Vol. XXX. (1880), pp. 255—278.]

IN a series of papers read from time to time during the past two years before the Royal Society*, I have investigated the theory of the tides raised in a rotating viscous spheroid, or planet, by an attendant satellite, and have also considered the secular changes in the rotation of the planet, and in the revolution of the satellite. Those investigations were intended to be especially applicable to the case of the earth and moon, but the friction of the solar tides was found to be a factor of importance, so that in a large part of those papers it became necessary to conceive the planet as attended by two satellites.

The differential equations which gave the secular changes in the system were rendered very complex by the introduction of solar disturbance, and I was unable to integrate them analytically; the equations were accordingly treated by a method of numerical quadratures, in which all the data were taken from the earth, moon, and sun. This numerical treatment did not permit an insight into all the various effects which might result from frictional tides, and an analytical solution, applicable to any planet and satellite, is desirable.

In the present paper such an analytical solution is found, and is interpreted graphically. But the problem is considered from a point of view which is at once more special and more general than that of the previous papers.

The point of view is more general in that the planet may here be conceived to have any density and mass whatever, and to be rotating with any angular velocity, provided that the ellipticity of figure is not large, and that the satellite may have any mass, and may be revolving about its planet, either consentaneously with or adversely to the planetary rotation. On the other hand, the problem here considered is more special in that the planet is

* [The previous papers of the present volume.]

supposed to be a spheroid of fluid of small viscosity; that the obliquity of the planet's equator, the inclination and the eccentricity of the satellite's orbit to the plane of reference are treated as being small, and, lastly, it is supposed that the planet is only attended by a single satellite.

The satellite itself is treated as an attractive particle, and the planet is supposed to be homogeneous.

The notation adopted is made to agree as far as possible with that of Paper 5, in which the subject was treated from a similarly general point of view, but where it was supposed that the equator and orbit were co-planar, and the orbit necessarily circular*.

The motion of the system is referred to the invariable plane, that is, to the plane of maximum moment of momentum.

The following is the notation adopted:—

For the planet:—

M = mass; a = mean radius; g = mean pure gravity; C = moment of inertia (neglecting ellipticity of figure); n = angular velocity of rotation; i = obliquity of equator to invariable plane, considered as small; $\mathfrak{g} = \frac{2}{5}g/a$.

For the satellite:—

m = mass; c = mean distance; Ω = mean motion; e = eccentricity of orbit, considered as small; j = inclination of orbit, considered as small; $\tau = \frac{3}{2}m/c^3$, where m is measured in the astronomical unit.

For both together:—

$\nu = M/m$, the ratio of the masses; $s = \frac{2}{5}[(a\nu/g)^2(1+\nu)]^{\frac{1}{3}}$; h = the resultant moment of momentum of the whole system; E = the whole energy, both kinetic and potential, of the system.

By a proper choice of the units of length, mass, and time, the notation may be considerably simplified.

Let the unit of length be such that $M + m$, when measured in the astronomical unit, may be equal to unity.

Let the unit of time be such that s or $\frac{2}{5}[(a\nu/g)^2(1+\nu)]^{\frac{1}{3}}$ may be unity.

Let the unit of mass be such that C, the planet's moment of inertia, may be unity†.

Then we have $$\Omega^2 c^3 = M + m = 1 \quad\text{..............................(1)}$$

Now, if we put for g its value M/a^2, and for ν its value M/m, we have

$$s = \tfrac{2}{5}\left\{\left[\frac{aM}{m}\cdot\frac{a^2}{M}\right]^2 \frac{M+m}{m}\right\}^{\frac{1}{3}} = \tfrac{2}{5}\frac{a^2}{m}, \text{ since } M+m \text{ is unity,}$$

and since s is unity, $m = \frac{2}{5}a^2$, when m is estimated in the astronomical unit.

* "Determination of the Secular Effects of Tidal Friction by a Graphical Method," *Proc. Roy. Soc.*, No. 197, 1879. [Paper 5.]

† [See a footnote on p. 375 for a consideration of the nature of these units.]

Again, since $C = \frac{2}{5}Ma^2$, and since C is unity, therefore $M = \frac{5}{2}/a^2$, where M is estimated in the mass unit.

Therefore $Mm/(M+m)$ is unity, when M and m are estimated in the mass unit, with the proposed units of length, time, and mass.

According to the theory of elliptic motion, the moment of momentum of the orbital motion of the planet and satellite about their common centre of inertia is $\frac{Mm}{M+m}\Omega c^2 \sqrt{1-e^2}$. Now it has been shown that the factor involving M is unity, and by (1) $\Omega c^2 = \Omega^{-\frac{1}{3}} = c^{\frac{1}{2}}$.

Hence, if we neglect the square of the eccentricity e, the moment of momentum of orbital motion is numerically equal to $\Omega^{-\frac{1}{3}}$ or $c^{\frac{1}{2}}$.

Let $x = \Omega^{-\frac{1}{3}} = c^{\frac{1}{2}}$.

In this paper x, the moment of momentum of orbital motion, will be taken as the independent variable. In interpreting the figures given below it will be useful to remember that it is also equal to the square root of the mean distance.

The moment of momentum of the planet's rotation is equal to Cn; and since C is unity, n will be either the moment of momentum of the planet's rotation, or the angular velocity of rotation itself.

With the proposed units $\tau = \frac{3}{2}m/c^3 = \frac{3}{5}a^2x^{-6}$, since $m = \frac{2}{5}a^2$; and

$$\mathfrak{g} = \frac{2}{5}g/a = \frac{2}{5}M/a^3 = \frac{2}{5}m, \quad M/ma^3 = \frac{4}{25}\nu/a$$

Also $\tau^2/\mathfrak{g}$ (a quantity which occurs below) is equal to $\frac{9}{4}a^5/\nu x^{12}$.

Let t be the time, and let $2f$ be the phase-retardation of the tide which I have elsewhere called the sidereal semi-diurnal tide of speed $2n$, which tide is known in the British Association Report on Tides as the faster of the two K tides.

If the planet be a fluid of small viscosity, the following are the differential equations which give the secular changes in the elements of the system:

$$\frac{dn}{dt} = -\tfrac{1}{2}\frac{\tau^2}{\mathfrak{g}}\sin 4f\left(1-\frac{\Omega}{n}\right) \qquad (2)$$

$$\frac{dx}{dt} = \tfrac{1}{2}\frac{\tau^2}{\mathfrak{g}}\sin 4f\left(1-\frac{\Omega}{n}\right) \qquad (3)$$

$$\frac{di}{dt} = \tfrac{1}{4}\frac{\tau^2}{\mathfrak{g}}\sin 4f\left(\frac{i+j}{n}\right)\left(1-\frac{2\Omega}{n}\right) \qquad (4)$$

$$\frac{dj}{dt} = -\tfrac{1}{4}\frac{\tau^2}{\mathfrak{g}}\sin 4f\left(\frac{i+j}{x}\right) \qquad (5)$$

$$\frac{1}{e}\frac{de}{dt} = \tfrac{1}{4}\frac{\tau^2}{\mathfrak{g}}\sin 4f \cdot \frac{1}{x}\left(11-\frac{18\Omega}{n}\right) \qquad (6)$$

The first three of these equations are in effect established in equation (80) of Paper 3, p. 91. The suffix m^2 to the symbols i and N there indicates that the equations (80) only refer to the action of the moon, and as here we only have a single satellite, they are the complete equations. N is equal to n/n_0, so that n_0 disappears from the first and second of (80); also $\mu = 1/sn_0\Omega_0^{\frac{1}{3}}$, and thus n_0 disappears from the third equation. $P = \cos i$, $Q = \sin i$, and, since we are treating i the obliquity as small, $P = 1$, $Q = i$; also $\lambda = \Omega/n$; the ϵ of that paper is identical with the f of the present one; lastly ξ is equal to $\Omega_0^{\frac{1}{3}}\Omega^{-\frac{1}{3}}$, and since with our present units $s = 1$, therefore $\mu d\xi/dt = d\Omega^{-\frac{1}{3}}/n_0 dt = dx/n_0 dt$.

With regard to the transformation of the first of (80) into (4) of the present paper, I remark that treating i as small $\frac{1}{4}PQ - \frac{1}{2}\lambda Q = \frac{1}{4}i(1 - 2\Omega/n)$, and introducing this transformation into the first of (80), equation (4) is obtained, except that i occurs in place of $(i + j)$. Now in Paper 3 the inclination of the orbit of the satellite to the plane of reference was treated as zero, and hence j was zero; but I have proved on pp. 292, 295 of Paper 6 that when we take into account the inclination of the orbit of the satellite, the P and Q on the right-hand sides of equation (80) of "Precession" must be taken as the cosine and sine of $i + j$ instead of i. Equations (5) and (6) are proved in § 10, Part II, p. 238 and § 25, Part V, p. 338, of Paper 6.

The integrals of this system of equations will give the secular changes in the motion of the system under the influence of the frictional tides. The object of the present paper is to find an analytical expression for the solution, and to interpret that solution geometrically.

From equations (2) and (4) we have

$$i\frac{dn}{dt} + n\frac{di}{dt} = \frac{1}{4}\frac{\tau^2}{\mathfrak{g}}\sin 4f\left[-(i+j) + 2j\left(1 - \frac{\Omega}{n}\right)\right]$$

But from (3) and (5) $x\,dj/dt + j\,dx/dt$ is equal to the same expression; hence

$$i\frac{dn}{dt} + n\frac{di}{dt} = x\frac{dj}{dt} + j\frac{dx}{dt}$$

The integral of this equation is $in = jx$,

or $$\frac{i}{j} = \frac{x}{n} \qquad \text{.................................(7)}$$

Equation (7) may also be obtained by the principle of conservation of moment of momentum. The motion is referred to the invariable plane of the system, and however the planet and satellite may interact on one another, the resultant moment of momentum must remain constant in direction and magnitude. Hence if we draw a parallelogram of which the diagonal is h

(the resultant moment of momentum of the system), and of which the sides are n and x, inclined respectively to the diagonal at the angles i and j, we see at once that

$$\frac{\sin i}{\sin j} = \frac{x}{n}$$

If i and j be treated as small this reduces to (7).

Again the consideration of this parallelogram shows that

$$h^2 = n^2 + x^2 + 2nx \cos(i + j)$$

which expresses the constancy of moment of momentum. If the squares and higher powers of $i + j$ be neglected, this becomes

$$h = n + x \quad \text{.................................(8)}$$

Equation (8) may also be obtained by observing that $dn/dt + dx/dt = 0$, and therefore on integration $n + x$ is constant. It is obvious from the principle of moment of momentum that the planet's equator and the plane of the satellite's orbit have a common node on the invariable plane of the system.

If we divide equations (4) and (6) by (3), we have the following results:—

$$\frac{1}{i}\frac{di}{dx} = \frac{1}{2n}\left(1 + \frac{j}{i}\right)\frac{n - 2\Omega}{n - \Omega} \quad \text{.......................(9)}$$

$$\frac{1}{e}\frac{de}{dx} = \frac{1}{2x}\,\frac{11n - 18\Omega}{n - \Omega} \quad \text{.......................(10)}$$

But from (7) and (8)

$$1 + \frac{j}{i} = 1 + \frac{n}{x} = \frac{h}{x}$$

Also $\Omega = x^{-3}$, and $n = h - x$.

Hence (9) and (10) may be written

$$\left.\begin{aligned} \frac{d}{dx}\log i &= \frac{\frac{1}{2}h}{x(h-x)}\,\frac{x^3(h-x) - 2}{x^3(h-x) - 1} \\ \frac{d}{dx}\log e &= \frac{1}{2x}\cdot\frac{11x^3(h-x) - 18}{x^3(h-x) - 1} \end{aligned}\right\} \quad \text{...............(11)}$$

Now $$\frac{h\{x^3(h-x) - 2\}}{2x(h-x)\{x^3(h-x) - 1\}} = \frac{h}{x(h-x)} + \frac{\frac{1}{2}hx^2}{x^4 - hx^3 + 1}$$

Therefore $$\frac{d}{dx}\log i = \frac{1}{x} + \frac{1}{h-x} + \frac{\frac{1}{2}hx^2}{x^4 - hx^3 + 1} \quad \text{................(12)}$$

Also $$\frac{11x^3(h-x) - 18}{2x\{x^3(h-x) - 1\}} = \frac{9}{x} - \frac{\frac{7}{2}x^2(x-h)}{x^4 - hx^3 + 1}$$

Therefore $$\frac{d}{dx}\log e = \frac{9}{x} - \frac{\frac{7}{2}x^2(x-h)}{x^4 - hx^3 + 1} \quad \text{.....................(13)}$$

These two equations are integrable as they stand, except as regards the last term in each of them.

It was shown in (4) of Paper 5, p. 197 that the whole energy of the system, both kinetic and potential, was equal to $\frac{1}{2}[n^2 - x^{-2}]$.

Then integrating (12) and (13), and writing down (7) and (8) again, and the expression for the energy, we have the following equations, which give the variations of the elements of the system in terms of the square root of the satellite's distance, and independently of the time:—

$$\left.\begin{aligned} \log i &= \log \frac{x}{h-x} + \tfrac{1}{2}h \int \frac{x^2 dx}{x^4 - hx^3 + 1} + \text{const.} \\ \log e &= \log x^9 - \tfrac{7}{2} \int \frac{x^2 (x-h)\, dx}{x^4 - hx^3 + 1} + \text{const.} \\ j &= \frac{h-x}{x} i \\ n &= h - x \\ 2E &= (h-x)^2 - \frac{1}{x^2} \end{aligned}\right\} \quad \ldots\ldots\ldots\ldots(14)$$

When the integration of these equations is completed, we shall have the means of tracing the history of a fluid planet of small viscosity, attended by a single satellite, when the system is started with any given moment of momentum h, and with any mean distance and (small) inclination and (small) eccentricity of the satellite's orbit, and (small) obliquity of the planet's equator. It may be remarked that h is to be taken as essentially positive, because the sign of h merely depends on the convention which we choose to adopt as to positive and negative rotations.

These equations do not involve the time, but it will be shown later how the time may be also found as a function of x. It is not, however, necessary to find the expression for the time in order to know the sequence of events in the history of the system.

Since the fluid which forms the planet is subject to friction, the system is non-conservative of energy, and therefore x must change in such a way that E may diminish.

If the expression for E be illustrated by a curve in which E is the vertical ordinate and x the horizontal abscissa, then any point on this "curve of energy" may be taken to represent one configuration of the system, as far as regards the mean distance of the satellite. Such a point must always slide down a slope of energy, and we shall see which way x must vary for any given configuration. This consideration will enable us to determine the sequence of events, when we come to consider the expressions for i, e, j, n in terms of x.

We have now to consider the further steps towards the complete solution of the problem.

The only difficulty remaining is the integration of the two expressions involved in the first and second of (14). From the forms of the expressions to be integrated, it is clear that they must be split up into partial fractions. The forms which these fractions will assume will of course depend on the nature of the roots of the equation $x^4 - hx^3 + 1 = 0$.

Some of the properties of this biquadratic were discussed in a previous paper, but it will now be necessary to consider the subject in more detail.

It will be found by Ferrari's method that

$$x^4 - hx^3 + 1 = \left\{x^2 + 2x\frac{\lambda^{\frac{3}{2}} - h}{4} + \frac{\lambda^{\frac{3}{2}} - h}{2\lambda^{\frac{1}{2}}}\right\}\left\{x^2 - 2x\frac{\lambda^{\frac{3}{2}} + h}{4} + \frac{\lambda^{\frac{3}{2}} + h}{2\lambda^{\frac{1}{2}}}\right\}$$

where $\lambda^3 - 4\lambda - h^2 = 0$.

By using the property $(\lambda^{\frac{3}{2}} - h)(\lambda^{\frac{3}{2}} + h) = 4\lambda$, this expression may be written in the form

$$[\{x + \tfrac{1}{4}(\lambda^{\frac{3}{2}} - h)\}^2 + \{\tfrac{1}{4}(\lambda^{\frac{3}{2}} - h)\sqrt{1 + 2h\lambda^{-\frac{3}{2}}}\}^2]$$
$$\times [\{x - \tfrac{1}{4}(\lambda^{\frac{3}{2}} + h)\}^2 + \{\tfrac{1}{4}(\lambda^{\frac{3}{2}} + h)\sqrt{1 - 2h\lambda^{-\frac{3}{2}}}\}^2]$$

which is of course equivalent to finding all the roots of the biquadratic in terms of h and λ.

Now let a curve be drawn of which h^2 is the ordinate (negative values of h^2 being admissible) and λ the abscissa; it is shown in fig. 1. Its equation is $h^2 = \lambda(\lambda^2 - 4)$.

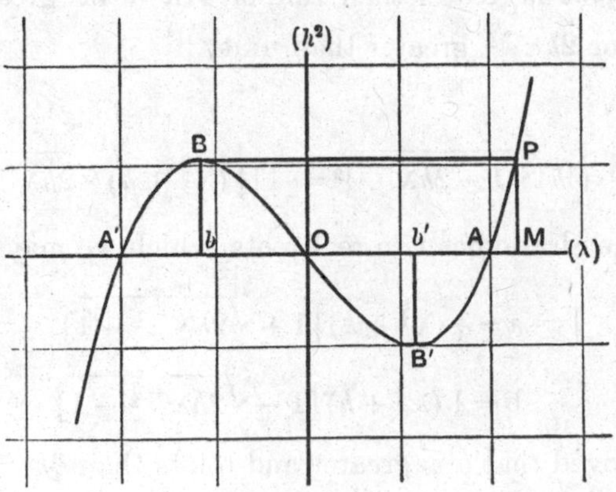

FIG. 1.

N.B. The ordinates are drawn to one-third of the scale to which the abscissæ are drawn.

It is obvious that $OA = OA' = 2$.

The maximum and minimum values of h^2 (viz., Bb, $B'b'$) are given by $3\lambda^2 = 4$ or $\lambda = \pm 2/3^{\frac{1}{2}}$.

Then $Bb = B'b' = -2^3/3^{\frac{3}{2}} + 4.2/3^{\frac{1}{2}} = (4/3^{\frac{3}{4}})^2$.

Since in the cubic, on which the solution of the biquadratic depends, h^2 is necessarily positive, it follows that if h be greater than $4/3^{\frac{3}{4}}$ the cubic has one real positive root greater than OM, and if h be less than $4/3^{\frac{3}{4}}$, it has two real negative roots lying between O and OA′, and one real positive root lying between OA and OM.

To find OM we observe that since h^2 is equal to $(4/3^{\frac{3}{4}})^2$, and since the root of $\lambda^3 - 4\lambda - h^2 = 0$ which is equal to $-2/3^{\frac{1}{2}}$ is repeated twice, therefore, if ϵ be the third root (or OM) we must have

$$\left(\lambda + \frac{2}{3^{\frac{1}{2}}}\right)^2 (\lambda - \epsilon) = \lambda^3 - 4\lambda - \frac{16}{3^{\frac{3}{2}}}$$

whence $(2/3^{\frac{1}{2}})^2 \epsilon = (4/3^{\frac{3}{4}})^2$, and ϵ or $OM = 4/3^{\frac{1}{2}}$.

Now $OA = 2$; hence, if h be less than $4/3^{\frac{3}{4}}$, the cubic has a positive root between 2 and $4/3^{\frac{1}{2}}$, and if h be greater than $4/3^{\frac{3}{4}}$, the cubic has a positive root between $4/3^{\frac{1}{2}}$ and infinity.

It will only be necessary to consider the positive root of the cubic.

Suppose h to be greater than $4/3^{\frac{3}{4}}$.

Then it has just been shown that λ is greater than $4/3^{\frac{1}{2}}$, and hence (λ being positive) $3\lambda^3$ is greater than 16λ, or $4(\lambda^3 - 4\lambda)$ greater than λ^3, or $4h^2$ greater than λ^3, or $2h\lambda^{-\frac{3}{2}}$ greater than unity.

Therefore

$$\{\tfrac{1}{4}(\lambda^{\frac{3}{2}} + h)\sqrt{1 - 2h\lambda^{-\frac{3}{2}}}\}^2 = -\{\tfrac{1}{4}(\lambda^{\frac{3}{2}} + h)\sqrt{2h\lambda^{-\frac{3}{2}} - 1}\}^2$$

Thus the biquadratic has two real roots, which we may call a and b,

where
$$a = \tfrac{1}{4}(\lambda^{\frac{3}{2}} + h)\left[1 + \sqrt{2h\lambda^{-\frac{3}{2}} - 1}\right]$$
$$b = \tfrac{1}{4}(\lambda^{\frac{3}{2}} + h)\left[1 - \sqrt{2h\lambda^{-\frac{3}{2}} - 1}\right]$$

It will be proved that a is greater and b less than $\frac{3}{4}h$.

Now
$$a > \text{ or } < \tfrac{3}{4}h$$
as
$$(\lambda^{\frac{3}{2}} + h)\left[1 + \sqrt{2h\lambda^{-\frac{3}{2}} - 1}\right] > \text{ or } < 3h$$

as $$\frac{(\lambda^{\frac{3}{2}}+h)}{\lambda^{\frac{3}{4}}}\sqrt{2h-\lambda^{\frac{3}{2}}} > \text{ or } < 2h-\lambda^{\frac{3}{2}}$$

as $$\lambda^{\frac{3}{2}}+h > \text{ or } < \lambda^{\frac{3}{4}}\sqrt{2h-\lambda^{\frac{3}{2}}}$$

as $$\lambda^3+2h\lambda^{\frac{3}{2}}+h^2 > \text{ or } < 2h\lambda^{\frac{3}{2}}-\lambda^3$$

as $$2\lambda^3+h^2 > \text{ or } < 0$$

Since the left-hand side is essentially positive, a is greater than $\frac{3}{4}h$.

Again $$\text{b} > \text{ or } < \tfrac{3}{4}h$$

as $$(\lambda^{\frac{3}{2}}+h)\left[1-\sqrt{2h\lambda^{-\frac{3}{2}}-1}\right] > \text{ or } < 3h$$

as $$-\frac{(\lambda^{\frac{3}{2}}+h)}{\lambda^{\frac{3}{4}}}\sqrt{2h-\lambda^{\frac{3}{2}}} > \text{ or } < 2h-\lambda^{\frac{3}{2}}$$

Since the left-hand side is negative and the right positive, the left is less than the right, and therefore b is less than $\frac{3}{4}h$.

If, therefore, h be greater than $4/3^{\frac{3}{4}}$, we may write

$$x^4-hx^3+1=(x-\text{a})(x-\text{b})[(x-\alpha)^2+\beta^2]$$

where a $-\frac{3}{4}h$, $\frac{3}{4}h-$ b are positive, and where α is negative.

We now turn to the other case and suppose h less than $4/3^{\frac{3}{4}}$. All the roots of the biquadratic are now imaginary, and we may put

$$x^4-hx^3+1=[(x-\alpha)^2+\beta^2][(x-\gamma)^2+\delta^2]$$

If α be taken as $-\frac{1}{4}(\lambda^{\frac{3}{2}}-h)$, then γ is $\frac{1}{4}(\lambda^{\frac{3}{2}}+h)$.

It only remains to prove that γ is greater than $\frac{3}{4}h$.

Now $$\gamma > \text{ or } < \tfrac{3}{4}h$$

as $$\lambda^{\frac{3}{2}} > \text{ or } < 2h$$

as $$\lambda^3 > \text{ or } < 4h^2 = 4(\lambda^3-4\lambda)$$

as $$16 > \text{ or } < 3\lambda^2$$

as $$4/3^{\frac{1}{2}} > \text{ or } < \lambda$$

but it has been already shown that in this case, λ is less than $4/3^{\frac{1}{2}}$, wherefore γ is greater than $\frac{3}{4}h$.

We may now proceed to the required integrations.

First case where h is greater than $4/3^{\frac{3}{4}}$.

Let $$x^4-hx^3+1=(x-\text{a})(x-\text{b})[(x-\alpha)^2+\beta^2]$$

so that the roots are a, b, $\alpha\pm\beta\iota$.

Also let a be the root which is greater than $\frac{3}{4}h$, b that which is less, and let

$$\mathrm{a} = \mathrm{a}_1 + \tfrac{3}{4}h, \quad \mathrm{b} = \tfrac{3}{4}h - \mathrm{b}_1, \quad \alpha = \tfrac{3}{4}h - \alpha_1$$

To find the expression for i we have to integrate $\dfrac{x^2}{x^4 - hx^3 + 1}$.

Let $f(x) = (x - \mathrm{a})\,\psi(x)$, and let $x^2/f(x) = A/(x - \mathrm{a}) + \phi(x)/\psi(x)$.

Then $$x^2 (x - \mathrm{a}) = Af(x) + (x - \mathrm{a})^2 \phi(x)$$

Hence $$A = \mathrm{a}^2/f'(\mathrm{a})$$

If, therefore, $$f(x) = x^4 - hx^3 + 1, \quad A = 1/(4\mathrm{a} - 3h) = 1/4\mathrm{a}_1$$

Thus the partial fractions corresponding to the roots a and b are

$$\frac{1}{4\mathrm{a}_1}\frac{1}{x - \mathrm{a}} - \frac{1}{4\mathrm{b}_1}\frac{1}{x - \mathrm{b}} \quad \text{.........................(15)}$$

If the pair of fractions corresponding to the roots $\alpha \pm \beta\iota$ be formed and added together, we find

$$\frac{1}{2(\alpha_1^2 + \beta^2)}\frac{-\alpha_1 (x - \alpha) + \beta^2}{[(x - \alpha)^2 + \beta^2]} \quad \text{.......................(16)}$$

The sum of (15) and (16) is equal to $\dfrac{x^2}{x^4 - hx^3 + 1}$, and

$$\int \frac{x^2 dx}{x^4 - hx^3 + 1} = \frac{1}{4\mathrm{a}_1}\log(x \sim \mathrm{a}) - \frac{1}{4\mathrm{b}_1}\log(x \sim \mathrm{b}) - \frac{\alpha_1}{4(\alpha_1^2 + \beta^2)}\log[(x - \alpha)^2 + \beta^2]$$

$$+ \frac{\beta}{2(\alpha_1^2 + \beta^2)}\arctan\frac{x - \alpha}{\beta} \quad \text{......(17)}$$

Substituting in the first of (14) we have

$$i = A\,\frac{x}{h - x}\cdot\frac{(x \sim \mathrm{a})^{\frac{h}{8\mathrm{a}_1}} \exp.\left[\dfrac{h\beta}{4(\alpha_1^2 + \beta^2)}\arctan\dfrac{x - \alpha}{\beta}\right]}{(x \sim \mathrm{b})^{\frac{h}{8\mathrm{b}_1}}\,[(x - \alpha)^2 + \beta^2]^{\frac{h\alpha_1}{8(\alpha_1^2 + \beta^2)}}} \quad \text{......(18)}$$

where A is a constant to be determined by the value of i, which corresponds with a particular value of x.

From the third of (14) we see that by omitting the factor $x/(h - x)$ from the above, we obtain the expression for j.

To find the expression for e we have to integrate $\dfrac{x^2 (x - h)}{x^4 - hx^3 + 1}$.

Now $$x^2 (x - h) = \tfrac{1}{4}(4x^3 - 3hx^2) - \tfrac{1}{4}hx^2$$

and therefore

$$\int \frac{x^2 (x - h)\,dx}{x^4 - hx^3 + 1} = \tfrac{1}{4}\log(x^4 - hx^3 + 1) - \tfrac{1}{4}h\int\frac{x^2 dx}{x^4 - hx^3 + 1}$$

The integral remaining on the right hand has been already determined in (17). Substituting in the second of (14), we have

$$e = \frac{Bx^9}{(x^4 - hx^3 + 1)^{\frac{7}{8}}} \left\{ \frac{(x \sim \text{a})^{\frac{h}{8\text{a}_1}} \exp. \left[\frac{h\beta}{4(\alpha_1^2 + \beta^2)} \arc\tan \frac{x - \alpha}{\beta} \right]}{(x \sim \text{b})^{\frac{h}{8\text{b}_1}} [(x - \alpha)^2 + \beta^2]^{\frac{h\alpha_1}{8(\alpha_1^2 + \beta^2)}}} \right\}^{\frac{7}{4}} \quad \text{...(19)}$$

where B is a constant to be determined by the value of e, corresponding to some particular value of x.

From this equation we get the curious relationship

$$e = \frac{B}{A^{\frac{7}{4}}} \cdot \frac{x^9}{(x^4 - hx^3 + 1)^{\frac{7}{8}}} j^{\frac{7}{4}} \text{......................(20)}$$

This last result will obviously be equally true even if all the roots of $x^4 - hx^3 + 1 = 0$ are imaginary.

In the present case the complete solution of the problem is comprised in the following equations:—

$$\left. \begin{aligned} j &= A \frac{(x \sim \text{a})^{\frac{h}{8\text{a}_1}} \exp. \left[\frac{h\beta}{4(\alpha_1^2 + \beta^2)} \arc\tan \frac{x - \alpha}{\beta} \right]}{(x \sim \text{b})^{\frac{h}{8\text{b}_1}} [(x - \alpha)^2 + \beta^2]^{\frac{h\alpha_1}{8(\alpha_1^2 + \beta^2)}}} \\ i &= \frac{x}{h - x} j \\ e &= \frac{B}{A^{\frac{7}{4}}} \frac{x^9}{(x^4 - hx^3 + 1)^{\frac{7}{8}}} j^{\frac{7}{4}} \\ n &= h - x \\ 2E &= (h - x)^2 - \frac{1}{x^2} \end{aligned} \right\} \text{........(21)}$$

It is obvious that the system can never degrade in such a way that x should pass through one of the roots of the biquadratic $x^4 - hx^3 + 1 = 0$. Hence the solution is divided into three fields, viz., (i) $x = +\infty$ to $x = \text{a}$; here we must write $x - \text{a}$, $x - \text{b}$ for the $x \sim \text{a}$, $x \sim \text{b}$ in the above solution; (ii) $x = \text{a}$ to $x = \text{b}$; here we must write $\text{a} - x$, $x - \text{b}$ (this is the part which has most interest in application to actual planets and satellites); (iii) $x = \text{b}$ to $x = -\infty$; here we must write $\text{a} - x$, $\text{b} - x$. When x is negative the physical meaning is that the revolution of the satellite is adverse to the planet's rotation.

By referring to (4) and (6), we see that i must be a maximum or minimum when $n = 2\Omega$, and e a maximum or minimum when $n = \frac{18}{11}\Omega$. Hence the corresponding values of x are the roots of the equations $x^4 - hx^3 + 2 = 0$, and $x^4 - hx^3 + \frac{18}{11} = 0$ respectively.

Now

$$\frac{x^2}{x^4 - hx^3 + 1} = \frac{1}{4\mathrm{a}_1}\frac{1}{x-\mathrm{a}} - \frac{1}{4\mathrm{b}_1}\frac{1}{x-\mathrm{b}} + \frac{1}{2(\alpha_1^2+\beta^2)}\frac{-\alpha_1(x-\alpha)+\beta^2}{[(x-\alpha)^2+\beta^2]}$$

therefore

$$x^2 = \frac{1}{4\mathrm{a}_1}(x-\mathrm{b})[(x-\alpha)^2+\beta^2] - \frac{1}{4\mathrm{b}_1}(x-\mathrm{a})[(x-\alpha)^2+\beta^2]$$

$$+ \frac{1}{2(\alpha_1^2+\beta^2)}[-\alpha_1(x-\alpha)+\beta^2](x-\mathrm{a})(x-\mathrm{b})$$

Hence the coefficient of x^3 on the right-hand side must be zero, and therefore

$$\frac{1}{4\mathrm{a}_1} - \frac{1}{4\mathrm{b}_1} - \frac{\alpha_1}{2(\alpha_1^2+\beta^2)} = 0$$

And

$$\frac{h}{8\mathrm{a}_1} = \frac{h}{8\mathrm{b}_1} + \frac{h\alpha_1}{4(\alpha_1^2+\beta^2)}$$

Now when $x = +\infty$, arc tan $\frac{x-\alpha}{\beta} = \frac{1}{2}\pi$, and when $x = -\infty$, it is equal to $-\frac{1}{2}\pi$.

Hence when $x = \pm\infty$, $j = A \exp. [\pm \pi h\beta/8(\alpha_1^2+\beta^2)]$, $i = -j$; the upper sign being taken for $+\infty$ and the lower for $-\infty$.

Then since j tends to become constant when $x = \pm\infty$, and since $9 - \frac{7}{2} = \frac{11}{2}$, therefore when x is very large e tends to vary as $x^{\frac{11}{2}}$.

If x be very small j has a finite value, and i varies as x, and e varies as x^9.

j, i, and e all become infinite when $x = \mathrm{b}$, and i also becomes infinite when $x = h$.

This analytical solution is so complex that it is not easy to understand its physical meaning; a geometrical illustration will, however, make it intelligible.

The method adopted for this end is to draw a series of curves, the points on which have x as abscissa and i, j, e, n, E as ordinates. The figure would hardly be intelligible if all the curves were drawn at once, and therefore a separate figure is drawn for i, j, and e; but in each figure the straight line which represents n is drawn, and the energy curve is also introduced in order to determine which way the figure is to be read. The zero of energy is of course arbitrary, and therefore the origin of the energy curve is in each case shifted along the vertical axis, in such a way that the energy curve may clash as little as possible with the others.

It is not very easy to select a value of h which shall be suitable for drawing these curves within a moderate compass, but after some consideration I chose $h = 2{\cdot}6$, and figs. 2, 3, and 4 are drawn to illustrate this value of h. If the cubic $\lambda^3 - 4\lambda - (2{\cdot}6)^2 = 0$, be solved by Cardan's method, it will be found that

$\lambda = 2{\cdot}5741$, and using this value in the formula for the roots of the biquadratic we have

$$x^4 - 2{\cdot}6x^3 + 1 = (x - 2{\cdot}539)(x - {\cdot}826)[(x + {\cdot}382)^2 + ({\cdot}575)^2]$$

Hence $\mathrm{a} = 2{\cdot}539$, $\mathrm{b} = {\cdot}826$, $\alpha = -{\cdot}382$, $\beta = {\cdot}575$, $\frac{3}{4}h = 1{\cdot}95$, and $4\mathrm{a}_1 = 2{\cdot}356$, $4\mathrm{b}_1 = 4{\cdot}496$, $\alpha_1 = 2{\cdot}332$, $\alpha_1^2 + \beta^2 = 5{\cdot}771$.

Then we have

$$\left.\begin{aligned} j &= A\,\frac{(2{\cdot}539 \sim x)^{{\cdot}552}\,\exp.\,[{\cdot}062\,\text{arc tan}\,(1{\cdot}740x + {\cdot}665)]}{(x \sim {\cdot}826)^{{\cdot}289}\,(x^2 + {\cdot}765x + {\cdot}477)^{{\cdot}1314}} \\ i &= \frac{x}{2{\cdot}6 - x}\,j \\ e &= \frac{B}{A^{\frac{7}{4}}}\,\frac{x^9}{(x^4 - 2{\cdot}6x^3 + 1)^{\frac{7}{8}}}\,j^{\frac{7}{4}} \\ n &= 2{\cdot}6 - x \\ 2E &= (2{\cdot}6 - x)^2 - \frac{1}{x^2} \end{aligned}\right\}\quad\ldots\ldots(22)$$

The maximum and minimum values of i are given by the roots of the equation $x^4 - 2{\cdot}6x^3 + 2 = 0$, viz., $x = 2{\cdot}467$ and $x = 1{\cdot}103$. The maximum and minimum values of e are given by the roots of the equation $x^4 - 2{\cdot}6x^3 + \frac{18}{11} = 0$, viz., $x = 2{\cdot}495$ and $x = 1{\cdot}0095$. The horizontal asymptotes for i/A and j/A are at distances from the axis of x equal to $\exp.\,({\cdot}062 \times \frac{1}{2}\pi)$ and $\exp.\,(-{\cdot}062 \times \frac{1}{2}\pi)$, which are equal to $1{\cdot}102$ and ${\cdot}908$ respectively.

Fig. 2 shows the curve illustrating the changes of i, the obliquity of the equator to the invariable plane.

The asymptotes are indicated by broken lines; that at A is given by $x = {\cdot}826$, and is the ordinate of maximum energy; that at B is given by $x = 2{\cdot}6$, and gives the configuration of the system for which the planet has no rotation. The point C is given by $x = 2{\cdot}539$, and lies on the ordinate of minimum energy. Geometrically the curve is divided into three parts by the vertical asymptotes, but it is further divided physically.

The curve of energy has four slopes, and since the energy must degrade, there are four methods in which the system may change, according to the way in which it was started. The arrows marked on the curve of obliquity show the direction in which the curve must be read.

Since none of these four methods can ever pass into another, this figure really contains four figures, and the following parts of the figure are quite independent of one another, viz.: (i) from $-\infty$ to O; (ii) from A to O; (iii) from A to C; (iv) from $+\infty$ to C. The figures 3 and 4 are also similarly in reality four figures combined. For each of these parts the constant A must be chosen with appropriate sign; but in order to permit the curves in fig. 2 to be geometrically continuous the obliquity is allowed to change sign.

The actual numerical interpretation of this figure depends on the value of A. Thus if for any value of x in any of the four fields the obliquity has an assigned value, then the ordinate corresponding to that value of x will give a scale of obliquity from which all the other ordinates within that field may be estimated.

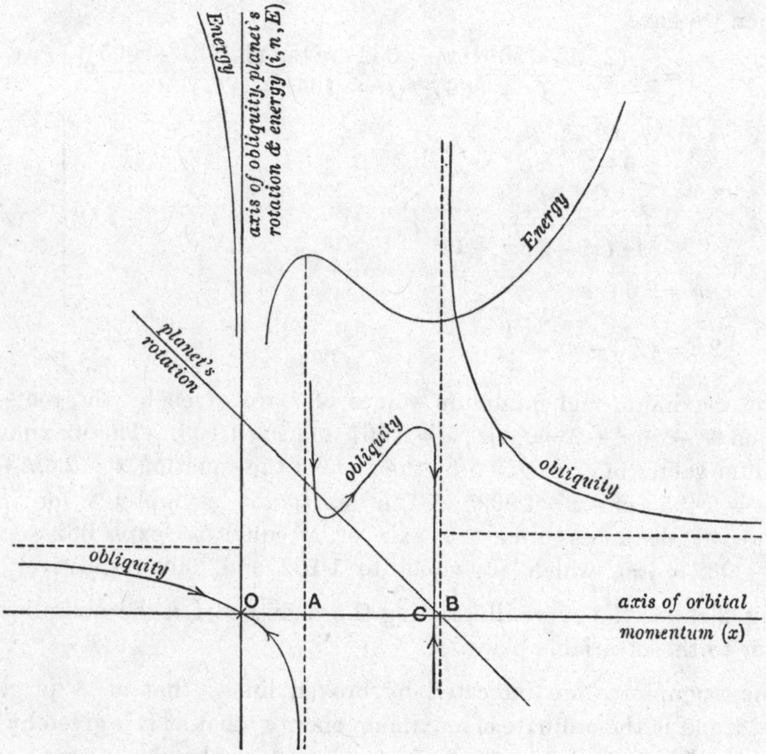

Fig. 2. Diagram for Obliquity of Planet's Equator.—First case.

As a special example of this we see that, if the obliquity be zero at any point, a consideration of the curve will determine whether zero obliquity be dynamically stable or not; for if the arrows on the curve of obliquity be approaching the axis of x, zero obliquity is dynamically stable, and if receding from the axis of x, dynamically unstable.

Hence from $x = +\infty$ to B, zero obliquity is dynamically unstable, from $-\infty$ to O and A to O dynamically stable, and from A to B, first stable, then unstable, and finally stable.

The infinite value of the obliquity at the point B has a peculiar significance, for at B the planet has no rotation, and being thus free from what Sir William Thomson calls "gyroscopic domination," the obliquity changes with infinite ease. In fact at B the term equator loses its meaning. The infinite value at A has a different meaning. The configuration A is one of maximum energy and of dynamical equilibrium, but is unstable as regards mean distance and planetary rotation; at this point the system changes infinitely slowly as regards time, and therefore the infinite value of the obliquity does not indicate an infinite rate of change of obliquity. In fact if we put $n=\Omega$ in (1) we see that $di/idt = -\frac{1}{4}(\tau^2/\mathfrak{g})\sin 4f$. However, to consider this case adequately we should have to take into account the obliquity in the equations for dn/dt and dx/dt, because the principal semi-diurnal tide vanishes when $n=\Omega$.

Similarly at the minimum of energy the system changes infinitely slowly, and thus the obliquity would take an infinite time to vanish.

We may now state the physical meaning of fig. 2, and this interpretation may be compared with a similar interpretation in [Paper 5].

A fluid planet of small viscosity is attended by a single satellite, and the system is started with an amount of positive moment of momentum which is greater than $4/3^{\frac{3}{4}}$, with our present units of length, mass and time.

The part of the figure on the negative side of the origin indicates a negative revolution of the satellite and a positive rotation of the planet, but the moment of momentum of planetary rotation is greater (by an amount h) than the moment of momentum of orbital motion. Then the satellite approaches the planet and ultimately falls into it, and the obliquity always diminishes slowly. The part from O to A indicates positive rotation of both parts of the system, but the satellite is very close to the planet and revolves round the planet quicker than the planet rotates, as in the case of the inner satellite of Mars. Here again the satellite approaches and ultimately falls in, and the obliquity always diminishes.

The part from A to C indicates positive rotation of both parts, but the satellite revolves slower than the planet rotates. This is the case which has most interest for application to the solar system. The satellite recedes from the planet, and the system ceases its changes when the satellite and planet revolve slowly as parts of a rigid body—that is to say, when the energy is a minimum. The obliquity first decreases, then increases to a maximum, and ultimately decreases to zero*.

The part from infinity to C indicates a positive revolution of the satellite, and from infinity to B a negative rotation of the planet, but from B to C a positive rotation of the planet, which is slower than the revolution of the satellite. In either of these cases the satellite approaches the planet, but the changes cease when the satellite and planet move slowly round as parts of a

* According to the present theory, the moon, considered as being attended by the earth as a satellite, has gone through these changes.

rigid body—that is to say, when the energy is a minimum. If the rotation of the planet be positive, the obliquity diminishes, if negative it increases. If the rotation of the planet be *nil*, the term obliquity ceases to have any meaning, since there is no longer an equator.

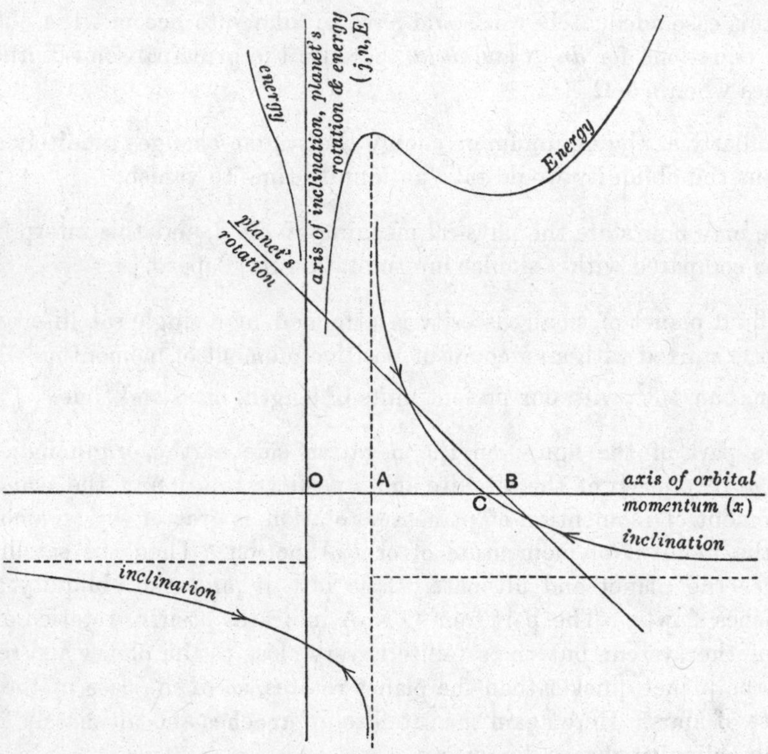

FIG. 3. Diagram for Inclination of Satellite's Orbit.—First case.

Fig. 3 illustrates the changes of inclination of the satellite's orbit, and may be interpreted in the same way as fig. 2. It appears from the part of the figure for which x is negative, that if the revolution of the satellite be negative, and the rotation of the planet positive, but the moment of momentum of planetary rotation greater than that of orbital motion, then, as the satellite approaches the planet, the inclination of the orbit increases, or zero inclination is dynamically unstable. In every other case the inclination will decrease, or zero inclination is dynamically stable.

This result undergoes an important modification when a second satellite is introduced, as appeared in Paper 6.

Fig. 4 shows a similar curve for the eccentricity of the orbit. The variations of the eccentricity are very much larger than those of the obliquity and inclination, so that it was here necessary to draw the ordinates on a much reduced scale. It was not possible to extend the figure far in either direction, because for large values of x, e varies as a high power of x (viz., $\frac{11}{2}$). The curve presents a resemblance to that of obliquity, for in the field comprised between the two roots of the biquadratic (viz., between A and C) the eccentricity diminishes to a minimum, increases to a maximum, and ultimately vanishes at C. This field represents a positive rotation both of the planet

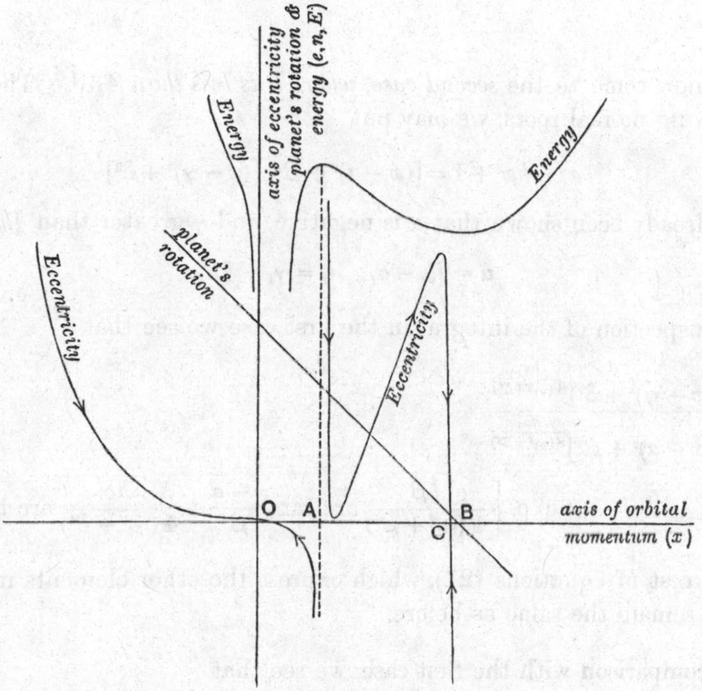

Fig. 4. Diagram for Eccentricity of Satellite's Orbit.—First case.

and satellite, but the satellite revolves slower than the planet rotates. This part represents the degradation of the system from the configuration of maximum energy to that of minimum energy, and the satellite recedes from the planet, until the two move round slowly like the parts of a rigid body.

In every other case the eccentricity degrades rapidly, whilst the satellite approaches the planet.

The very rapid rate of variation of the eccentricity, compared with that of the obliquity would lead one to expect that the eccentricity of the orbit of a satellite should become very large in the course of its evolution, whilst the obliquity should not increase to any very large extent. But it must be remembered that we are here only treating a planet of small viscosity, and it appeared, in Paper 6, that the rate of increase or diminution of the eccentricity is very much less rapid (per unit increase of x) if the viscosity be not small, whilst the rate of increase or diminution of obliquity (per unit increase of x) is slightly increased with increase of viscosity. Thus the observed eccentricities of the orbits of satellites and of obliquities of their planets cannot be said to agree in amount with the theory that the planets were primitively fluids of small viscosity, though I believe they *do* agree with the theory that the planets were fluids or quasi-solids of large viscosity.

We now come to the *second case, where h is less than* $4/3^{\frac{3}{4}}$. The biquadratic having no real roots, we may put

$$x^4 - hx^3 + 1 = [(x-\alpha)^2 + \beta^2][(x-\gamma)^2 + \delta^2]$$

It has already been shown that α is negative, and γ greater than $\frac{3}{4}h$.

Let
$$\alpha = \tfrac{3}{4}h - \alpha_1, \quad \gamma = \gamma_1 + \tfrac{3}{4}h$$

By inspection of the integral in the first case we see that

$$j = A\frac{[(x-\gamma)^2+\delta^2]^{\frac{h\gamma_1}{8(\gamma_1^2+\delta^2)}}}{[(x-\alpha)^2+\beta^2]^{\frac{h\alpha_1}{8(\alpha_1^2+\beta^2)}}}$$
$$\times \exp.\left[\frac{h\beta}{4(\alpha_1^2+\beta^2)}\text{arc tan}\frac{x-\alpha}{\beta} + \frac{h\delta}{4(\gamma_1^2+\delta^2)}\text{arc tan}\frac{x-\gamma}{\delta}\right]$$

The rest of equations (21), which express the other elements in terms of j and x, remain the same as before.

By comparison with the first case, we see that

$$\frac{x^2}{x^4-hx^3+1} = \frac{1}{2(\alpha_1^2+\beta^2)}\,\frac{-\alpha_1(x-\alpha)+\beta^2}{(x-\alpha)^2+\beta^2} + \frac{1}{2(\gamma_1^2+\delta^2)}\,\frac{\gamma_1(x-\gamma)+\delta^2}{(x-\gamma)^2+\delta^2}$$

On multiplying both sides of this identity by $x^4 - hx^3 + 1$, and equating the coefficients of x^3, we find

$$0 = \frac{-\alpha_1}{2(\alpha_1^2+\beta^2)} + \frac{\gamma_1}{2(\gamma_1^2+\delta^2)}$$

Therefore
$$\frac{h\alpha_1}{8(\alpha_1^2+\beta^2)} = \frac{h\gamma_1}{8(\gamma_1^2+\delta^2)}$$

Thus when x is equal to $\pm\infty$

$$j = A\exp.\left[\pm\frac{\pi h\beta}{8(\alpha_1^2+\beta^2)} \pm \frac{\pi h\delta}{8(\gamma_1^2+\delta^2)}\right]$$

the upper sign being taken for $+\infty$, and the lower for $-\infty$. This expression gives the horizontal asymptotes for j and i.

In order to illustrate this solution, I chose $h=1$, and found by trigonometrical solution of the cubic $\lambda^3 - 4\lambda - 1 = 0$, $\lambda = 2{\cdot}1149$, and thence

$$\left.\begin{aligned}
j &= A\left(\frac{x^2 - 2{\cdot}038x + 1{\cdot}401}{x^2 + 1{\cdot}038x + {\cdot}714}\right)^{{\cdot}077} \quad \text{exp.}\,[{\cdot}081 \text{ arc tan}\,(1{\cdot}500x + {\cdot}778) \\
&\qquad\qquad\qquad\qquad + {\cdot}346 \text{ arc tan}\,(1{\cdot}659x - 1{\cdot}691)] \\
i &= \frac{x}{1-x}\,j \\
e &= \frac{B}{A^{\frac{7}{4}}}\,\frac{x^9}{(x^4 - x^3 + 1)^{\frac{7}{8}}}\,j^{\frac{7}{4}} \\
n &= 1 - x \\
2E &= (1-x)^2 - \frac{1}{x^2}
\end{aligned}\right\} \quad (23)$$

When $\quad x = +\infty, \quad j/A = 1{\cdot}956 = -i/A$

and when $\quad x = -\infty, \quad j/A = \ {\cdot}512 = -i/A$

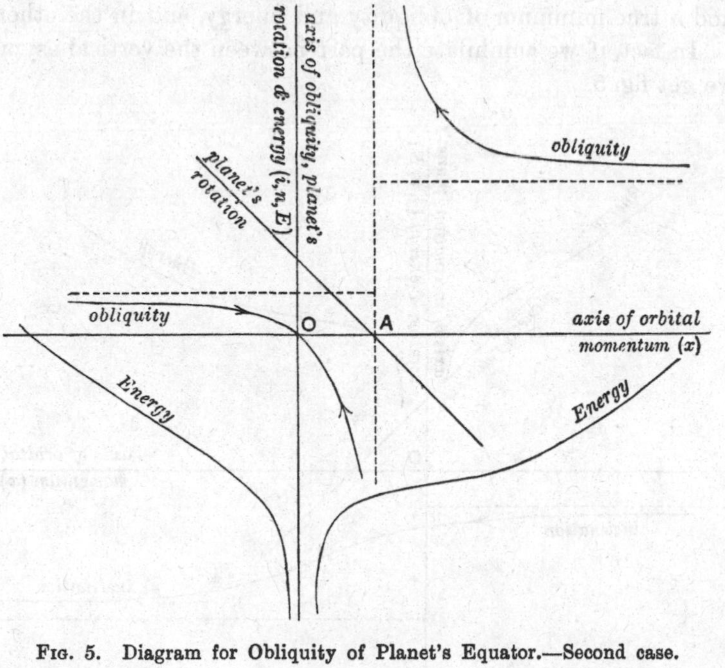

FIG. 5. Diagram for Obliquity of Planet's Equator.—Second case.

These solutions are illustrated as in the previous case by the three figures 5, 6, 7. There are here only two slopes of energy, and hence these figures each of them only contain two separate figures.

Fig. 5 illustrates the changes of i, the obliquity of the equator to the invariable plane.

In this figure there is only one vertical asymptote, viz., that corresponding to $x=1$. For this value of x the planet has no rotation, is free from "gyroscopic domination," and the term equator loses its meaning.

The figure shows that if the rotation of the planet be negative, but the moment of momentum of planetary rotation less than that of orbital motion, then the obliquity increases, whilst the satellite approaches the planet.

This increase of obliquity only continues so long as the rotation of the planet is negative. The rotation becomes positive after a time, and the obliquity then diminishes, whilst the satellite falls into the planet. In the corresponding part of fig. 2 the satellite did not fall into the planet, but the two finally moved slowly round together as the parts of a rigid body.

If the revolution of the satellite be negative, and the rotation of the planet positive, but the moment of momentum of rotation greater than that of revolution, the obliquity always diminishes as the satellite falls towards the planet.

Figs. 2 and 5 only differ in the fact that in the one there is a true maximum and a true minimum of obliquity and energy, and in the other there is not so. In fact, if we annihilate the part between the vertical asymptotes of fig. 2 we get fig. 5.

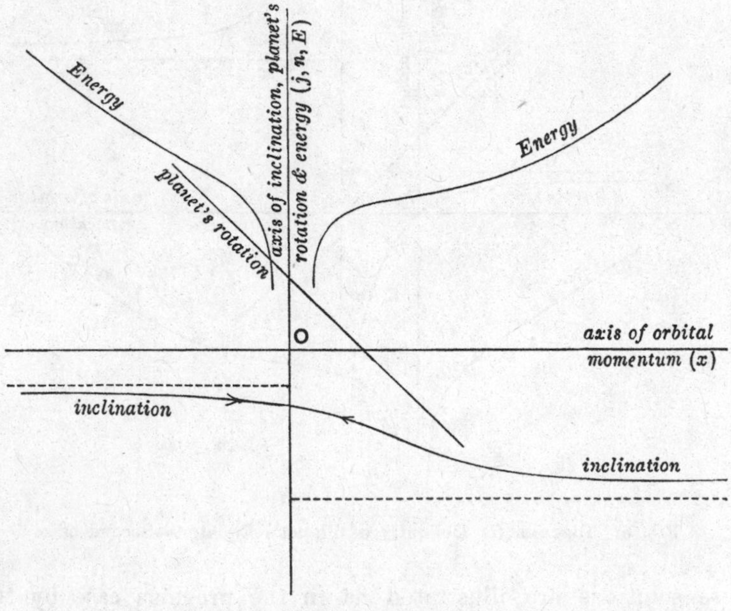

Fig. 6. Diagram for Inclination of Satellite's Orbit.—Second case.

Fig. 6 illustrates the changes of inclination of the orbit. It does not possess very much interest, since it simply shows that however the system be started with positive revolution of the satellite, whether the rotation of the planet be positive or not, the inclination of the orbit slightly diminishes as the satellite falls in.

And however the system be started with negative revolution of the satellite, and therefore necessarily positive rotation of the planet, the inclination of the orbit slightly increases. Fig. 6 again corresponds to fig. 3, if in the latter the part lying between the maximum and minimum of energy be annihilated.

Fig. 7 illustrates the changes of eccentricity, and shows that it always diminishes rapidly however the system is started, as the satellite falls towards the planet. This figure again corresponds with fig. 4, if in the latter the parts between the maximum and minimum of energy be annihilated.

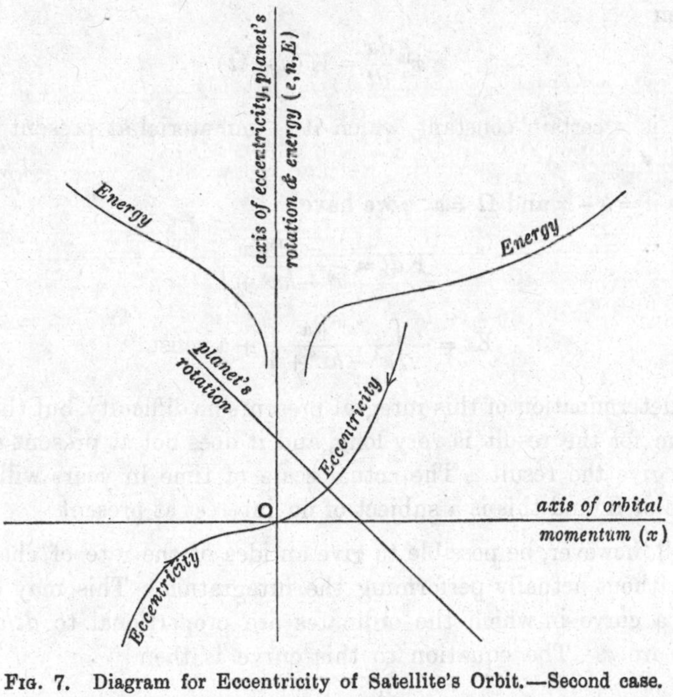

FIG. 7. Diagram for Eccentricity of Satellite's Orbit.—Second case.

These three figures may be interpreted as giving the various stabilities and instabilities of the system, just as was done in the first case.

The solution of the problem, which has been given and discussed above, gives merely the sequence of events, and does not show the rate at which the changes in the system take place. It will now be shown how the time may be found as a function of x.

Consider the equation

$$\frac{dx}{dt} = \frac{1}{2}\frac{\tau^2}{\mathfrak{g}} \sin 4f \left(1 - \frac{\Omega}{n}\right)$$

f is here the angle of lag of the sidereal semi-diurnal tide of speed $2n$. By the theory of the tides of a viscous spheroid, $\tan 2f = 2n/\mathfrak{p}$, where $\mathfrak{p}$ is a certain function of the radius of the planet and its density, and which varies inversely as the coefficient of viscosity of the spheroid*.

Since by hypothesis the viscosity is small, f is a small angle, so that $\sin 4f$ may be taken as equal to $2 \tan 2f$. Thus, $\sin 4f/n$ is a constant, depending on the dimensions, density, and viscosity of the planet.

It has already been shown that τ^2 varies as x^{-12}, and $\mathfrak{g}$ is a constant, which depends only on the density of the planet. Hence, the above equation may be written

$$x^{12}\frac{dx}{dt} = \mathrm{K}(n - \Omega)$$

where K is a certain constant, which it is immaterial at present to evaluate precisely.

Since $n = h - x$ and $\Omega = x^{-3}$, we have

$$\mathrm{K}dt = \frac{-x^{15}dx}{x^4 - hx^3 + 1}$$

or

$$\mathrm{K}t = -\int \frac{x^{15}dx}{x^4 - hx^3 + 1} + \text{a const.}$$

The determination of this integral presents no difficulty, but the analytical expression for the result is very long, and it does not at present seem worth while to give the result. The actual scale of time in years will depend on the value of K, and this is a subject of no interest at present.

It will, however, be possible to give an idea of the rate of change of the system without actually performing the integration. This may be done by drawing a curve in which the ordinates are proportional to dt/dx, and the abscissæ are x. The equation to this curve is then

$$\mathrm{K}\frac{dt}{dx} = \frac{-x^{15}}{x^4 - hx^3 + 1}$$

The maximum and minimum values (if any) of dt/dx are given by the real roots of the equation

$$11x^4 - 12hx^3 + 15 = 0$$

One of such roots will be found to be intermediate between a and b, and the other greater than a.

* "On the Bodily Tides of Viscous and semi-elastic Spheroids," &c. [Paper 1, § 5.]

Fig. 8 shows the nature of the curve when drawn with the free hand. It was not found possible to draw this figure to scale, because when $h=2{\cdot}6$ it was found that the minimum M was equal to ·85, and could not be made distinguishable from a point on the asymptote A, whilst the minimum m was equal to about 900,000, and could not be made distinguishable from a point on the asymptote C.

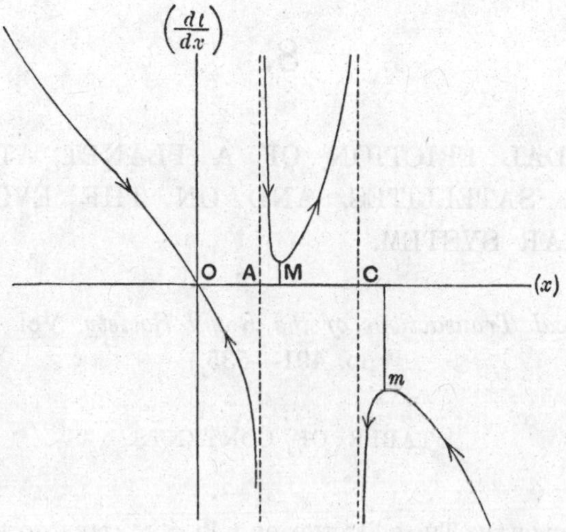

FIG. 8. Diagram illustrating the Rate of Change of the System.

The area intercepted between this curve, the axis of x, and any pair of ordinates corresponding to two values of x, will be proportional to the time required to pass from the one configuration to the other.

When dt/dx is negative, that is to say, when the satellite is falling into the planet, the areas fall below the axis of x. This is clearly necessary in order to have geometrical continuity in the curve.

The figure shows that the rate of alteration in the system becomes very slow when the satellite is far from the planet; this must indeed obviously be the case, because the tidal effects vary as the inverse sixth power of the satellite's mean distance.

8.

ON THE TIDAL FRICTION OF A PLANET ATTENDED BY SEVERAL SATELLITES, AND ON THE EVOLUTION OF THE SOLAR SYSTEM.

[*Philosophical Transactions of the Royal Society*, Vol. 172 (1881), pp. 491—535.]

TABLE OF CONTENTS.

PAGE

Introduction 406

I. THE THEORY OF THE TIDAL FRICTION OF A PLANET ATTENDED BY SEVERAL SATELLITES.

§ 1. Statement and limitation of the problem 407

§ 2. Formation and transformation of the differential equations . 408

§ 3. Sketch of method for solution of the equations by series . 415

§ 4. Graphical solution in the case where there are not more than three satellites 416

§ 5. The graphical method in the case of two satellites . . . 418

II. A DISCUSSION OF THE EFFECTS OF TIDAL FRICTION WITH REFERENCE TO THE EVOLUTION OF THE SOLAR SYSTEM.

§ 6. General consideration of the problem presented by the solar system 433

§ 7. Numerical data and deductions therefrom 437

§ 8. On the part played by tidal friction in the evolution of planetary masses 447

§ 9. General discussion and summary 452

Introduction.

IN previous papers on the subject of tidal friction* I have confined my attention principally to the case of a planet attended by a single satellite. But in order to make the investigation applicable to the history of the earth and moon it was necessary to take notice of the perturbation of the sun. In

* [The previous papers in the present volume.]

consequence of the largeness of the sun's mass it was not there requisite to make a complete investigation of the theory of a planet attended by a pair of satellites.

In the first part of this paper the theory of the tidal friction of a central body attended by any number of satellites is considered.

In the second part I discuss the degree of importance to be attached to tidal friction as an element in the evolution of the solar system and of the several planetary sub-systems.

The last paragraph contains a discussion of the evidence adduced in this part of the paper, and a short recapitulation of the observed facts in the solar system which bear on the subject. This is probably the only portion which will have any interest for others than mathematicians.

I.

THE THEORY OF THE TIDAL FRICTION OF A PLANET ATTENDED BY ANY NUMBER OF SATELLITES.

§ 1. *Statement and limitation of the problem.*

Suppose there be a planet attended by any number of satellites, all moving in circular orbits, the planes of which coincide with the equator of the planet; and suppose that all the satellites raise tides in the planet. Then the problem proposed for solution is to investigate the gradual changes in the configuration of the system under the influence of tidal friction.

This problem is only here treated under certain restrictions as to the nature of the tidal friction and in other respects. These limitations however will afford sufficient insight into the more general problem. The planet is supposed to be a homogeneous spheroid formed of viscous fluid, and the only case considered in detail is that where the viscosity is small; moreover, in the tidal theory adopted the effects of inertia are neglected. I have however shown elsewhere that this neglect is not such as to vitiate the theory materially*. The satellites are treated as attractive particles which have the power of attracting and being attracted by the planet, but have no influence upon one another. A consequence of this is that each satellite only raises a single tide in the planet, and that it is not necessary to take into consideration the actual distribution of the satellites at any instant of time. We are thus only concerned in determining the changes in the distances of the satellites and in the rotation of the planet.

If the mutual perturbation of the satellites were taken into account the problem would become one of the extremest complication. We should have

* [Paper 4.]

all the difficulties of the planetary theory in determining the various inequalities, and, besides this, it would be necessary to investigate an indefinitely long series of tidal disturbances induced by these inequalities of motion, and afterwards to find the secular disturbances due to the friction of these tides.

It is however tolerably certain that in general these inequality-tides will exercise a very small influence compared with that of the primary tide. Supposing a relationship between the mean motions of two, three, or more satellites, like that which holds good in the Jovian system, to exist at any epoch, it is not credible but that such relationship should be broken down in time by tidal friction. General considerations would lead one to believe that the first effect of tidal friction would be to set up amongst the satellites in question an oscillation of mean motions about the average values which satisfy the supposed definite relationship; afterwards this oscillation would go on increasing indefinitely until a critical state was reached in which the average mean motions would break loose from the relationship, and the oscillation would subsequently die away. It seems probable therefore that in the history of such a system there would be a series of periods during which the mutual perturbations of the satellites would exercise a considerable but temporary effect, but that on the whole the system would change nearly as though the satellites exercised no mutually perturbing power.

There is however one case in which mutual perturbation would probably exercise a lasting effect on the system. Suppose that in the course of the changes two satellites came to have nearly the same mean distance, then these two bodies might either come ultimately into collision or might coalesce so as to form a double system like that of the earth and moon, which revolve round the sun in the same period. In this paper I do not make any attempt to trace such a case, and it is supposed that any satellite may pass freely through a configuration in which its distance is equal to that of any other satellite.

§ 2. *Formation and transformation of the differential equations.*

In this paper I shall have occasion to make frequent use of the idea of moment of momentum. This phrase is so cumbrous that I shall abridge it and speak generally of angular momentum, and in particular of rotational momentum and orbital momentum when meaning moment of momentum of a planet's rotation and moment of momentum of the orbital motion of a satellite. I shall also refer to the principle of conservation of moment of momentum as that of conservation of momentum.

The notation here adopted is almost identical with that of previous papers on the case of the single satellite and planet; it is as follows:—

For the planet let:

M = mass; a = mean radius; g = mean pure gravity; w = mass per unit volume; υ = viscosity; $\mathfrak{g} = \frac{2}{5}g/a$; n = angular velocity of rotation; C = moment of inertia about the axis of rotation, and therefore, neglecting the ellipticity of figure, equal to $\frac{2}{5}Ma^2$.

For any particular one of the system of satellites, let:

m = mass; c = distance from planet's centre; Ω = orbital angular velocity.

Also μ being the attraction between unit masses at unit distance, let $\tau = \frac{3}{2}\mu m/c^3$; and let $\nu = M/m$.

These same symbols will be used with suffixes 1, 2, 3, &c., when it is desired to refer to the 1st, 2nd, 3rd, &c., satellite, but when (as will be usually the case) it is desired simply to refer to any satellite, no suffixes will be used.

Where it is necessary to express a summation of similar terms, each corresponding to one satellite, the symbol Σ will be used; *e.g.*, $\Sigma\kappa c^{\frac{1}{2}}$ will mean $\kappa_1 c_1^{\frac{1}{2}} + \kappa_2 c_2^{\frac{1}{2}}$ + &c.

Now consider the single satellite m, c, Ω, &c.

If this satellite alone were to raise a tide in the planet, the planet would be distorted into an ellipsoid with three unequal axes, and in consequence of the postulated internal friction, the major axis of the equatorial section of the planet would be directed to a point somewhat in advance of the satellite in its orbit.

Let f be the angle made by this major axis with the satellite's radius vector; f is then a symbol subject to suffixes 1, 2, 3, &c., because it will be different for each satellite of the system.

It is proved in (22) of my paper on the "Precession of a Viscous Spheroid*," that the tidal frictional couple due to this satellite's attraction is $C\frac{1}{2}\frac{\tau^2}{\mathfrak{g}}\sin 4\text{f}$.

Now it appears from Sec. 14 of the same paper that the tidal reaction, which affects the motion of each satellite, is independent of the tides raised by all the other satellites.

Hence the principle of conservation of momentum enables us to state, that the rate of increase of the orbital momentum of any satellite is equal to the rate of the loss of rotational momentum of the planet which is caused by that satellite alone. The rate of loss of this latter momentum is of course equal to the above tidal frictional couple.

When the planet is reduced to rest the orbital momentum of the satellite

* [Paper 3, p. 49.]

in the circular orbit is $\Omega c^2 Mm/(M+m)$. Hence the equation of tidal reaction, which gives the rate of change in the satellite's distance, is

$$\frac{d}{dt}\left[\frac{Mm}{M+m}\Omega c^2\right] = C\tfrac{1}{2}\frac{\tau^2}{\mathfrak{g}}\sin 4\mathrm{f} \quad \text{.....................(1)}$$

A similar equation will hold true for each satellite of the system.

This equation will now be transformed.

By Kepler's law $\Omega^2 c^3 = \mu(M+m)$ and therefore

$$\frac{Mm}{M+m}\Omega c^2 = \mu^{\frac{1}{2}}\frac{Mm}{(M+m)^{\frac{1}{2}}}c^{\frac{1}{2}}$$

By the theory of the tides of a viscous spheroid (Paper 1)

$$\tan 2\mathrm{f} = \frac{2(n-\Omega)}{2\mathfrak{p}}, \quad \text{where } 2\mathfrak{p} = \frac{2gaw}{19v}$$

Hence $$\sin 4\mathrm{f} = \frac{2(n-\Omega)/\mathfrak{p}}{1+(n-\Omega)^2/\mathfrak{p}^2}, \quad \text{also } \tau^2 = (\tfrac{3}{2})^2\frac{\mu^2 m^2}{c^6}$$

Hence (1) becomes

$$\frac{\mu^{\frac{1}{2}}Mm}{(M+m)^{\frac{1}{2}}}\frac{dc^{\frac{1}{2}}}{dt} = (\tfrac{3}{2})^2\frac{C}{\mathfrak{g}}\frac{(\mu m)^2}{c^6}\frac{(n-\Omega)/\mathfrak{p}}{1+(n-\Omega)^2/\mathfrak{p}^2} \quad \text{..............(2)}$$

Now let Ch be the angular momentum of the whole system, namely that due to the planet's rotation and to the orbital motion of all the satellites. And let CE be the whole energy, both kinetic and potential, of the system. Then h is the angular velocity with which the planet would have to rotate in order that the rotational momentum might be equal to that of the whole system; and E is twice the square of the angular velocity with which the planet would have to rotate in order that the kinetic energy of planetary rotation might be equal to the whole energy of the system. By the principle of conservation of momentum h is constant, and since the system is non-conservative of energy E is variable, and must diminish with the time.

The kinetic energy of the orbital motion of the satellite m is $\frac{1}{2}\mu Mm/c$, and the potential energy of position of the planet and satellite is $-\mu Mm/c$; the kinetic energy of the planet's rotation is $\frac{1}{2}Cn^2$. Thus we have,

$$Ch = Cn + \Sigma\frac{\mu^{\frac{1}{2}}Mm}{(M+m)^{\frac{1}{2}}}c^{\frac{1}{2}} \quad \text{.........................(3)}$$

$$2CE = Cn^2 - \Sigma\frac{\mu Mm}{c} \quad \text{..................................(4)}$$

In the equations (3) and (4) we may regard C as a constant, provided we neglect the change of ellipticity of the planet's figure as its rotation slackens.

Let the symbol ∂ indicate partial differentiation; then from (3) and (4)

$$\frac{\partial n}{\partial (c^{\frac{1}{2}})} = -\frac{1}{C}\frac{\mu^{\frac{1}{2}} Mm}{(M+m)^{\frac{1}{2}}}$$

$$-\frac{\partial E}{\partial (c^{\frac{1}{2}})} = \frac{1}{C}\frac{\mu^{\frac{1}{2}} Mm}{(M+m)^{\frac{1}{2}}} n - \frac{1}{C}\frac{\mu Mm}{c^{\frac{3}{2}}}$$

But
$$\frac{1}{C}\frac{\mu Mm}{c^{\frac{3}{2}}} = \frac{1}{C}\frac{\mu^{\frac{1}{2}} Mm}{(M+m)^{\frac{1}{2}}}\Omega$$

and therefore
$$-\frac{C(M+m)^{\frac{1}{2}}}{\mu^{\frac{1}{2}} Mm}\frac{\partial E}{\partial (c^{\frac{1}{2}})} = n - \Omega \quad \text{.......................(5)}$$

From equations (2) and (5) we may express the rate of increase of the square root of any satellite's distance in terms of the energy of the whole system, in the general case where the planet has any degree of viscosity. A good many transformations, analogous to those below, may be made in this general case, but as I shall only examine in detail the special case in which the viscosity is small, it will be convenient to make the transition thereto at once.

When the viscosity is small, $\mathfrak{p}$, which varies inversely as the viscosity, is large. Then, unless $n - \Omega$ be very large, $(n - \Omega)/\mathfrak{p}$ is small compared with unity. Thus in (2) we may neglect $(n - \Omega)^2/\mathfrak{p}^2$ in the denominator compared with unity.

Substituting from (5) in (2), and making this approximation, we have

$$\frac{\mu^{\frac{1}{2}} Mm}{(M+m)^{\frac{1}{2}}}\frac{dc^{\frac{1}{2}}}{dt} = -(\tfrac{3}{2})^2 \frac{C}{\mathfrak{g}}\frac{(m\mu)^2}{c^6}\frac{C(M+m)^{\frac{1}{2}}}{\mu^{\frac{1}{2}} Mm\mathfrak{p}}\frac{\partial E}{\partial (c^{\frac{1}{2}})} \quad \text{...........(6)}$$

Now let
$$\xi = \left(\frac{M}{M+m}\right)^{\frac{1}{2}}\left(\frac{c}{a}\right)^{\frac{7}{2}} \quad \text{..........................(7)}$$

where a is any constant length, which it may be convenient to take either as equal to the mean radius of the planet, or as the distance of some one of the satellites at some fixed epoch. ξ is different for each satellite and is subject to the suffixes 1, 2, 3, &c.

The equation (6) may be written

$$\left(\frac{M}{M+m}\right)^{\frac{1}{2}} 7c^3\frac{dc^{\frac{1}{2}}}{dt} = -(\tfrac{3}{2})^2 \times 49 \times \frac{\mu C^2}{M\mathfrak{g}\mathfrak{p}} \times \left(\frac{M+m}{M}\right)^{\frac{1}{2}}\frac{1}{7c^3}\frac{\partial E}{\partial (c^{\frac{1}{2}})}$$

Now let
$$A = (\tfrac{3}{2})^2 \times 49 \times \frac{\mu C^2}{Ma^7\mathfrak{g}\mathfrak{p}} \quad \text{.........................(8)}$$

And we have
$$\frac{d\xi}{dt} = -A\frac{\partial E}{\partial \xi} \quad \text{..............................(9)}$$

In order to calculate A it may be convenient to develop its expression further.

$$M\mathfrak{g}\alpha^7 = \tfrac{2}{5}\frac{\mu M}{a^3} \cdot M\alpha^7 = \tfrac{5}{2}\mu C^2 \left(\frac{\alpha}{a}\right)^7, \quad \text{so that } \frac{\mu C^2}{M\mathfrak{g}\alpha^7} = \tfrac{2}{5}\left(\frac{a}{\alpha}\right)^7$$

and
$$A = (\tfrac{3}{2})^2 (\tfrac{2}{5})\, 49\, \frac{(a/\alpha)^7}{\mathfrak{p}}, \quad \text{where } \mathfrak{p} = \frac{gaw}{19\upsilon} \quad \text{.............(10)}$$

Since $\mathfrak{p}$ is an angular velocity A is a period of time, and A is the same for all the satellites.

In (9) ξ is the variable, but it will be convenient to introduce an auxiliary variable x, such that

$$\left.\begin{aligned} x^7 &= \xi \\ \text{or} \quad x &= \left(\frac{M}{M+m}\right)^{\frac{1}{14}} \left(\frac{c}{\alpha}\right)^{\frac{1}{2}} \end{aligned}\right\} \quad \text{..............................(11)}$$

Then
$$\frac{\mu^{\frac{1}{2}} Mm}{(M+m)^{\frac{1}{2}}}\, c^{\frac{1}{2}} = \frac{\mu^{\frac{1}{2}} M^{\frac{13}{14}} m\alpha^{\frac{1}{2}}}{(M+m)^{\frac{3}{7}}}\, x$$

Let
$$\kappa = \frac{\mu^{\frac{1}{2}} M^{\frac{13}{14}} m\alpha^{\frac{1}{2}}}{C(M+m)^{\frac{3}{7}}} \quad \text{..............................(12)}$$

κ is different for each satellite and is subject to suffixes 1, 2, 3, &c.

Thus (3) may be written

$$h = n + \Sigma\kappa x \quad \text{..................................(13)}$$

Again
$$\frac{\mu Mm}{c} = \frac{\mu M^{\frac{8}{7}} m}{(M+m)^{\frac{1}{7}}\alpha}\, \frac{1}{x^2}$$

Let
$$\lambda = \frac{\mu M^{\frac{8}{7}} m}{C(M+m)^{\frac{1}{7}}\alpha} \quad \text{............... ....(14)}$$

λ is different for each satellite and is subject to suffixes 1, 2, 3, &c.

On comparing (12) and (14) we see that

$$\frac{\lambda}{\kappa} = \Omega x^3 \quad \text{..................................(15)}$$

This is of course merely a form of writing the equation

$$\mu(M+m) = \Omega^2 c^3$$

Then (4) may be written

$$2E = n^2 - \Sigma\frac{\lambda}{x^2} \quad \text{............................(16)}$$

In order to compute κ and λ we may pursue two different methods.

First, suppose $\alpha = a$, the planet's mean radius.

Then $\dfrac{m\alpha^{\frac{1}{2}}}{C} = \dfrac{5}{2\nu a^{\frac{3}{2}}}$; $M^{\frac{7}{14}}\mu^{\frac{1}{2}} = (ga^2)^{\frac{1}{2}}$; $\dfrac{M^{\frac{6}{14}}}{(M+m)^{\frac{3}{7}}} = \left(\dfrac{\nu}{1+\nu}\right)^{\frac{3}{7}}$

$\kappa = \frac{5}{2}[\nu^4(1+\nu)^3]^{-\frac{1}{7}}\left(\dfrac{g}{a}\right)^{\frac{1}{2}}$, of same dimensions as an angular velocity.

$\lambda = \frac{5}{2}[\nu^6(1+\nu)]^{-\frac{1}{7}}\left(\dfrac{g}{a}\right)$, of same dimensions as the square of an angular velocity.

If ν be large compared with unity, as is generally the case, the expressions become

$$\kappa = \frac{5m}{2M}\sqrt{\frac{g}{a}}, \quad \lambda = \frac{5m}{2M}\left(\frac{g}{a}\right) \quad \ldots\ldots\ldots\ldots\ldots\ldots\ldots(17)$$

Secondly, suppose M large compared with all the m's, and suppose for example that the solar system as a whole is the subject of investigation. Then take α as the earth's present radius vector, and ω as its present mean motion, and

$$\kappa = \frac{m}{C}\sqrt{\mu M\alpha}, \quad \text{and } \lambda = \frac{\mu M m}{C\alpha}$$

or

$$\kappa = m\left(\frac{\omega\alpha^2}{C}\right), \quad \lambda = m^2\left(\frac{\omega^2\alpha^2}{C}\right) \quad \ldots\ldots\ldots\ldots\ldots(18)$$

C is here the sun's moment of inertia.

Collecting results from (9), (13), (16), the equations which determine the changes in the system are

$$\left.\begin{array}{l} \dfrac{d\xi}{dt} = -A\dfrac{\partial E}{\partial \xi} \\ \text{and a similar equation for each satellite} \\ n = h - \Sigma\kappa x \\ 2E = n^2 - \Sigma\dfrac{\lambda}{x^2} \end{array}\right\} \quad \ldots\ldots\ldots\ldots\ldots\ldots\ldots\ldots(19)$$

where $x^7 = \xi$; A is a certain time to be computed as above shown in (10); κ an angular velocity to be computed as above shown in (17) and (18); and λ the square of an angular velocity to be computed as above in (17) and (18).

Also
$$\xi = \left(\frac{M}{M+m}\right)^{\frac{1}{2}}\left(\frac{c}{\alpha}\right)^{\frac{7}{2}} = \left(\frac{\nu}{1+\nu}\right)^{\frac{1}{2}}\left(\frac{c}{\alpha}\right)^{\frac{7}{2}}$$

If ν be large compared with unity, ξ is very approximately proportional to the seventh power of the square root of the satellite's distance.

The solution of this system of simultaneous differential equations would give each of the ξ's in terms of the time; afterwards we might obtain n and E in terms of the time from the last two of (19).

These differential equations possess a remarkable analogy with those which represent Hamilton's principle of varying action (Thomson and Tait's *Nat. Phil.*, 1879, § 330 (14)).

The rate of loss of energy of the system may be put into a very simple form. This function has been called by Lord Rayleigh (*Theory of Sound*, Vol. I. § 81) the Dissipation Function, and the name is useful, because this function plays an important part in non-conservative systems.

In the present problem the Dissipation Function or Dissipativity, as it is called by Sir William Thomson, is $-C\dfrac{dE}{dt}$.

Now
$$\frac{dE}{dt} = \Sigma \frac{\partial E}{\partial \xi}\frac{d\xi}{dt}$$

From (19) the dissipativity is therefore either

$$CA\Sigma\left(\frac{\partial E}{\partial \xi}\right)^2 \text{ or } \frac{C}{A}\Sigma\left(\frac{d\xi}{dt}\right)^2$$

This quantity is of course essentially positive.

It is easy to show that $\dfrac{\partial E}{\partial \xi} = -\dfrac{\kappa}{7x^6}(n-\Omega)$

On substituting for the various symbols in the expression for the dissipativity their values in terms of the original notation, we have

$$-\frac{dE}{dt} = \Sigma \frac{\tau^2}{\mathfrak{gp}}(n-\Omega)^2$$

Or if N be the tidal frictional couple corresponding to the satellite m,

$$-C\frac{dE}{dt} = \Sigma N(n-\Omega)$$

This last result would be equally true whatever were the viscosity of the planetary spheroid.

The dissipativity, converted into heat by Joule's equivalent, expresses the amount of heat generated per unit time within the planetary spheroid. This result has been already obtained in a different manner for the case of a single satellite in a previous paper [Paper 4, p. 158].

§ 3. *Sketch of method for solution of the equations by series.*

It does not seem easy to obtain a rigorous analytical solution of the system (19) of differential equations. I have however solved the equations by series, so as to obtain analytical expressions for the ξ's, as far as the fourth power of the time. This solution is not well adapted for the purposes of the present paper, because the series are not rapidly convergent, and therefore cannot express those large changes in the configuration of the system which it is the object of the present paper to trace.

As no subsequent use is made of this solution, and as the analysis is rather long, I will only sketch the method pursued.

If $\frac{1}{49}A$ be taken as the unit of time $\dfrac{dE}{dt} = -\frac{1}{49}\Sigma\left(\dfrac{\partial E}{\partial \xi}\right)^2$.

Differentiating again and again with regard to the time, and making continued use of this equation, we find d^2E/dt^2, d^3E/dt^3, &c., in terms of $\partial E/\partial \xi$.

It is then necessary to develop these expressions by performing the differentiations with regard to ξ.

An abridged notation was used in which $\left[\dfrac{\text{a, b}}{\text{p}}\right]^{\text{k}}$ represented

$$\left(\frac{\text{a}\lambda x^{-3} - \text{b}n\kappa}{x^{\text{p}}}\right)^{\text{k}} \text{ or } \left[\frac{\kappa(\text{a}\Omega - \text{b}n)}{x^{\text{p}}}\right]^{\text{k}}$$

With this notation the whole operation may be shown to depend on the performance of $\partial/\partial\xi$ on expressions of the form

$$\Sigma\gamma\left[\frac{\text{a}_1,\ \text{b}_1}{\text{p}_1}\right]^{\text{k}_1}\left[\frac{\text{a}_2,\ \text{b}_2}{\text{p}_2}\right]^{\text{k}_2}\cdots\left[\frac{\text{a}_r,\ \text{b}_r}{\text{p}_r}\right]^{\text{k}_r}$$

where γ is independent of ξ, but may be a function of the mass of each satellite.

Having evaluated the successive differentials of E we have

$$E = E_0 + t\left(\frac{dE}{dt}\right)_0 + \frac{t^2}{1\,.\,2}\left(\frac{d^2E}{dt^2}\right)_0 + \frac{t^3}{1\,.\,2\,.\,3}\left(\frac{d^3E}{dt^3}\right)_0 + \text{\&c.}$$

where the suffix 0 indicates that the value, corresponding to $t = 0$, is to be taken.

It is also necessary to evaluate the successive differentials of $\partial E/\partial\xi$ with regard to the time, and then we have

$$\xi = \xi_0 - t\left(\frac{\partial E}{\partial \xi}\right)_0 - \frac{t^2}{1\,.\,2}\left(\frac{\partial}{\partial\xi}\frac{dE}{dt}\right)_0 - \frac{t^3}{1\,.\,2\,.\,3}\left(\frac{\partial}{\partial\xi}\frac{d^2E}{dt^2}\right)_0 - \text{\&c.}$$

The coefficient of t^4 was found to be very long even with the abridged notation, and involved squares and products of Σ's.

§ 4. *Graphical solution in the case where there are not more than three satellites.*

Although a general analytical solution does not seem attainable, yet the equations have a geometrical or quasi-geometrical meaning, which makes a complete graphical solution possible, at least in the case where there are not more than three satellites.

To explain this I take the case of two satellites only, and to keep the geometrical method in view I change the notation, and write z for E, x for ξ_1, and y for ξ_2, also I write Ω_x for Ω_1, and Ω_y for Ω_2. The unit of time is chosen so that $A = 1$.

Then the equations (19) become

$$\frac{dx}{dt} = -\frac{\partial z}{\partial x}, \quad \frac{dy}{dt} = -\frac{\partial z}{\partial y} \qquad \text{..........................(20)}$$

and

$$2z = (h - \kappa_1 x^{\frac{1}{7}} - \kappa_2 y^{\frac{1}{7}})^2 - \frac{\lambda_1}{x^{\frac{2}{7}}} - \frac{\lambda_2}{y^{\frac{2}{7}}} \qquad \text{.................(21)}$$

Suppose a surface constructed to illustrate (21), x, y, z being the co ordinates of any point on it. Let the axes of x and y be drawn horizontally, and that of z vertically upwards. The z ordinate of course gives the energy of the system corresponding to any values of x and y which are consistent with the given angular momentum h.

We have for the dissipativity of the system

$$-\frac{dz}{dt} = \left(\frac{\partial z}{\partial x}\right)^2 + \left(\frac{\partial z}{\partial y}\right)^2$$

Whence

$$\frac{dx}{dz} = \frac{\frac{\partial z}{\partial x}}{\left(\frac{\partial z}{\partial x}\right)^2 + \left(\frac{\partial z}{\partial y}\right)^2}, \quad \frac{dy}{dz} = \frac{\frac{\partial z}{\partial y}}{\left(\frac{\partial z}{\partial x}\right)^2 + \left(\frac{\partial z}{\partial y}\right)^2} \qquad \text{...........(22)}$$

Let $(X - x)/\lambda = (Y - y)/\mu = Z - z$ be the equations to a straight line through a point x, y, z on the surface. If this line lies in the tangent plane at that point

$$\lambda \frac{\partial z}{\partial x} + \mu \frac{\partial z}{\partial y} - 1 = 0$$

The inclination of this line to the axis of z will be a maximum or minimum when $\lambda^2 + \mu^2$ is a maximum or minimum. In other words if this straight line is a tangent line to the steepest path through x, y, z on the surface, $\lambda^2 + \mu^2$ must be a maximum or minimum.

Hence for this condition to be fulfilled we must have

$$\lambda\delta\lambda + \mu\delta\mu = 0$$

$$\frac{\partial z}{\partial x}\delta\lambda + \frac{\partial z}{\partial y}\delta\mu = 0$$

And therefore $\lambda\Big/\frac{\partial z}{\partial x} = \mu\Big/\frac{\partial z}{\partial y}$, and these are both equal to

$$1\Big/\left\{\left(\frac{\partial z}{\partial x}\right)^2 + \left(\frac{\partial z}{\partial y}\right)^2\right\}$$

Therefore the equation to the tangent to the steepest path is

$$\frac{X-x}{\partial z/\partial x} = \frac{Y-y}{\partial z/\partial y} = \frac{Z-z}{(\partial z/\partial x)^2 + (\partial z/\partial y)^2} \quad\text{.................(23)}$$

If this steepest path on the energy surface is the path actually pursued by the point which represents the configuration of the system, equation (23) must be satisfied by

$$X = x + \frac{dx}{dz}\delta z, \quad Y = y + \frac{dy}{dz}\delta z, \quad Z = z + \delta z$$

And therefore we must have

$$\frac{dx}{dz} = \frac{\frac{\partial z}{\partial x}}{\left(\frac{\partial z}{\partial x}\right)^2 + \left(\frac{\partial z}{\partial y}\right)^2}, \quad \frac{dy}{dz} = \frac{\frac{\partial z}{\partial y}}{\left(\frac{\partial z}{\partial x}\right)^2 + \left(\frac{\partial z}{\partial y}\right)^2}$$

But these are the values already found in (22) for dx/dz and dy/dz.

Therefore we conclude that the representative point always slides down a steepest path on the energy surface. Hence it only remains to draw the surface, and to mark out the lines of steepest slope in order to obtain a complete graphical solution of the problem. Since the lines of greatest slope cut the contours at right angles, if we project the contours orthogonally on to the plane of xy, and draw the system of orthogonal trajectories of the contours, we obtain a solution in two dimensions. This solution will be exhibited below, but for the present I pass on to more general considerations.

A precisely similar argument might be applied to the case where there are any number of satellites, but as space has only three dimensions, a geometrical solution is not possible. If there be r satellites, the problem to be solved may be stated in geometrical language thus:—

It is required to find the path which is inclined at the least angle to the axis of E on the locus

$$2E = (h - \Sigma\kappa\xi^{\frac{1}{7}})^2 - \Sigma\frac{\lambda}{\xi^{\frac{2}{7}}}$$

This locus is described in space of $r+1$ dimensions. One axis is that of E, and the remaining r axes are the axes of the r different ξ's. The solution may be depressed so as to merely require space of r dimensions, for we may, in space of r dimensions, construct the orthogonal trajectories of the contour loci found by attributing various values to E.

Thus we might actually solve geometrically the case of three satellites. The energy locus here involves space of four dimensions, but the contour loci are a family of surfaces in three dimensions. If such a system of surfaces were actually constructed, it would be possible to pass through them a number of wires or threads which should be a good approximation to the orthogonal trajectories. The trouble of execution would however be hardly repaid by the results, because most of the interesting general conclusions may be drawn from the case of two satellites, where we have only to deal with curves.

If the case of a single satellite be considered, we see that the energy locus is a curve, and the transit along the steepest path degenerates merely into travelling down hill. Now as the slopes of the energy curve are not altered in direction, but merely in steepness, by taking the abscissæ of points on the curve as any power of ξ, the solution may still be obtained if we take x (or $\xi^{\frac{1}{7}}$) as the abscissa instead of ξ. This reduces the solution to exactly that which was given in a previous paper, where the graphical method was applied to the case of a single satellite*.

§ 5. *The graphical method in the case of two satellites.*

I now return to the special case in which there are only two satellites. The equation to the surface of energy is given in (21). The maxima and minima values of z (if any) are given by equating $\partial z/\partial x$ and $\partial z/\partial y$ to zero. This gives

$$\left.\begin{aligned} h - \kappa_1 x^{\frac{1}{7}} - \kappa_2 y^{\frac{1}{7}} &= \frac{\lambda_1}{\kappa_1 x^{\frac{3}{7}}} \\ h - \kappa_1 x^{\frac{1}{7}} - \kappa_2 y^{\frac{1}{7}} &= \frac{\lambda_2}{\kappa_2 y^{\frac{3}{7}}} \end{aligned}\right\} \qquad (24)$$

By (15) and (19) we see that these equations may be written

$$\left.\begin{aligned} n &= \Omega_x \\ n &= \Omega_y \end{aligned}\right\} \qquad (25)$$

They also lead to the equations

$$\left.\begin{aligned} (x^{\frac{1}{7}})^4 - \frac{h - \kappa_2 y^{\frac{1}{7}}}{\kappa_1} (x^{\frac{1}{7}})^3 + \frac{\lambda_1}{\kappa_1^2} &= 0 \\ (y^{\frac{1}{7}})^4 - \frac{h - \kappa_1 x^{\frac{1}{7}}}{\kappa_2} (y^{\frac{1}{7}})^3 + \frac{\lambda_2}{\kappa_2^2} &= 0 \end{aligned}\right\} \qquad (26)$$

* [Paper 5, p. 195.]

Now an equation of the form $Y^4 - \alpha Y^3 + \beta = 0$ may be written

$$(Y\beta^{-\frac{1}{4}})^4 - \alpha\beta^{-\frac{1}{4}}(Y\beta^{-\frac{1}{4}})^3 + 1 = 0$$

And I have proved in a previous paper* that an equation $x^4 - hx^3 + 1 = 0$ has two real roots, if h be greater than $4/3^{\frac{3}{4}}$, but has no real roots if h be less than $4/3^{\frac{3}{4}}$. Hence it follows that this equation in Y has two real roots, if α be greater than $4\beta^{\frac{1}{4}}/3^{\frac{3}{4}}$, but no real roots if it be less.

If we consider the two equations (26) as biquadratics for $x^{\frac{1}{7}}$ and $y^{\frac{1}{7}}$ respectively, we see that the first has, or has not, a pair of real roots, according as

$$h - \kappa_2 y^{\frac{1}{7}} \text{ is greater or less than } \left(\frac{4}{3^{\frac{3}{4}}}\right)\lambda_1^{\frac{1}{4}}\kappa_1^{\frac{1}{2}}$$

and the second has, or has not, a pair of real roots, according as

$$h - \kappa_1 x^{\frac{1}{7}} \text{ is greater or less than } \left(\frac{4}{3^{\frac{3}{4}}}\right)\lambda_2^{\frac{1}{4}}\kappa_2^{\frac{1}{2}}$$

If we substitute for the λ's and κ's their values, we find that

$$\lambda_1^{\frac{1}{4}}\kappa_1^{\frac{1}{2}} = \frac{\mu^{\frac{1}{2}}(Mm_1)^{\frac{3}{4}}}{C^{\frac{3}{4}}(M+m_1)^{\frac{1}{4}}}$$

$$\lambda_2^{\frac{1}{4}}\kappa_2^{\frac{1}{2}} = \text{the same with } m_2 \text{ in place of } m_1$$

Now let γ_1 and γ_2 be two lengths determined by the equations

$$\frac{Mm_1}{M+m_1}\gamma_1^2 = \frac{Mm_2}{M+m_2}\gamma_2^2 = C$$

Or in words—let γ_1 be such a distance that the moment of inertia of the planet (concentrated at its centre) and the first satellite about their common centre of inertia may be equal to the planet's moment of inertia about its axis of rotation; and let γ_2 be a similar distance involving the second satellite instead of the first.

And let ω_1, ω_2 be two angular velocities determined by the equations

$$\omega_1^2\gamma_1^3 = \mu(M+m_1), \quad \omega_2^2\gamma_2^3 = \mu(M+m_2)$$

Or in words—let ω_1 be the angular velocity of the first satellite when revolving in a circular orbit at distance γ_1, and ω_2 a similar angular velocity for the second satellite when revolving at distance γ_2.

$$\text{Now} \quad \left[\frac{\mu Mm_1}{C}\right]^{\frac{1}{2}} = \omega_1\gamma_1^{\frac{1}{2}} \text{ and } \left[\frac{Mm_1}{C(M+m_1)}\right]^{\frac{1}{4}} = \frac{Mm_1}{C(M+m_1)}\gamma_1^{\frac{3}{2}}$$

$$\text{so that} \quad \lambda_1^{\frac{1}{4}}\kappa_1^{\frac{1}{2}} = \frac{Mm_1}{C(M+m_1)}\omega_1\gamma_1^2$$

* [Paper 6, pp. 390–1.]

and similarly $\lambda_2^{\frac{1}{4}}\kappa_2^{\frac{1}{2}}$ = the same with the suffix 2 in place of 1. Hence the first of the two equations (26) has, or has not, a pair of real roots, according as

$$C(h-\kappa_2 y^{\frac{1}{4}}) \text{ is greater or less than } \frac{4}{3^{\frac{3}{4}}}\frac{Mm_1}{M+m_1}\omega_1\gamma_1^2$$

and the second has, or has not, a pair of real roots, according as

$$C(h-\kappa_1 x^{\frac{1}{4}}) \text{ is greater or less than } \frac{4}{3^{\frac{3}{4}}}\frac{Mm_2}{M+m_2}\omega_2\gamma_2^2$$

It is obvious that $Mm_1\omega_1\gamma_1^2/(M+m_1)$ is the orbital momentum of the first satellite when revolving at distance γ_1, and similarly $Mm_2\omega_2\gamma_2^2/(M+m_2)$ is the orbital momentum of the second satellite when revolving at distance γ_2.

If the second or y-satellite be larger than the first or x-satellite the latter of these momenta is larger than the first.

Now Ch is the whole angular momentum of the system, and in order that there may be maxima and minima determined by the equations $\partial z/\partial x = 0$, $\partial z/\partial y = 0$, the equations (26) must have real roots. Then on putting y equal to zero in the first of the above conditions, and x equal to zero in the second we get the following results:—

First, there are no maxima and minima points for sections of the energy surface either parallel to x or y, if the whole momentum of the system be less than $4/3^{\frac{3}{4}}$ times the orbital momentum of the smaller or x-satellite when moving at distance γ_1.

Second, there are maxima and minima points for sections parallel to x, but not for sections parallel to y, if the whole momentum be greater than $4/3^{\frac{3}{4}}$ times the orbital momentum of the smaller or x-satellite when moving at distance γ_1, but less than $4/3^{\frac{3}{4}}$ times the orbital momentum of the larger or y-satellite when moving at distance γ_2.

Third, there are maxima and minima for both sections, if the whole momentum be greater than $4/3^{\frac{3}{4}}$ times the orbital momentum of the larger or y-satellite when moving at distance γ_2.

This third case now requires further subdivision, according as whether there are not or are absolute maximum or minimum points on the surface.

If there are such points the two equations (24) or (25) must be simultaneously satisfied.

Hence we must have $n = \Omega_x = \Omega_y$, in order that there may be a maximum or minimum point on the surface.

But in this case the two satellites revolve in the same periodic time, and may be deemed to be rigidly connected together, and also rigidly connected with the planet. Hence the configurations of maximum or minimum energy are such that all three bodies move as though rigidly connected together.

The simultaneous satisfaction of (24) necessitates that

$$x^{\frac{3}{7}} = \frac{\lambda_1\kappa_2}{\kappa_1\lambda_2} y^{\frac{3}{7}} \text{ or } y^{\frac{3}{7}} = \frac{\lambda_2\kappa_1}{\kappa_2\lambda_1} x^{\frac{3}{7}}$$

Hence the equations (24) become

$$h - \left[\kappa_1 + \kappa_2 \left(\frac{\lambda_2\kappa_1}{\lambda_1\kappa_2}\right)^{\frac{1}{3}}\right] x^{\frac{1}{7}} = \frac{\lambda_1}{\kappa_1 x^{\frac{3}{7}}}$$

$$h - \left[\kappa_1 \left(\frac{\lambda_1\kappa_2}{\lambda_2\kappa_1}\right)^{\frac{1}{3}} + \kappa_2\right] y^{\frac{1}{7}} = \frac{\lambda_2}{\kappa_2 y^{\frac{3}{7}}}$$

These equations may be written

$$\left.\begin{aligned} (x^{\frac{1}{7}})^4 - \frac{h}{\kappa_1 + \kappa_2 (\lambda_2\kappa_1/\lambda_1\kappa_2)^{\frac{1}{3}}} (x^{\frac{1}{7}})^3 + \frac{\lambda_1/\kappa_1}{\kappa_1 + \kappa_2 (\lambda_2\kappa_1/\lambda_1\kappa_2)^{\frac{1}{3}}} = 0 \\ (y^{\frac{1}{7}})^4 - \frac{h}{\kappa_1 (\lambda_1\kappa_2/\lambda_2\kappa_1)^{\frac{1}{3}} + \kappa_2} (y^{\frac{1}{7}})^3 + \frac{\lambda_2/\kappa_2}{\kappa_1 (\lambda_1\kappa_2/\lambda_2\kappa_1)^{\frac{1}{3}} + \kappa_2} = 0 \end{aligned}\right\} \quad \ldots(27)$$

Treating these biquadratics in the same way as before, we find that they have, or have not, two real roots, according as h is greater or less than

$$\frac{4}{3^{\frac{3}{4}}} [(\lambda_1\kappa_1^2)^{\frac{1}{3}} + (\lambda_2\kappa_2^2)^{\frac{1}{3}}]^{\frac{3}{4}}$$

Now $$(\lambda_1\kappa_1^2)^{\frac{1}{3}} + (\lambda_2\kappa_2^2)^{\frac{1}{3}} = \frac{\mu^{\frac{2}{3}} M}{C} \left[\frac{m_1}{(M + m_1)^{\frac{1}{3}}} + \frac{m_2}{(M + m_2)^{\frac{1}{3}}}\right]$$

Therefore there is, or there is not, a pair of real solutions of the equations $n = \Omega_x = \Omega_y$, according as the total momentum of the system is, or is not, greater than

$$\frac{4}{3^{\frac{3}{4}}} \mu^{\frac{1}{2}} C^{\frac{1}{4}} M^{\frac{3}{4}} \left[\frac{m_1}{(M + m_1)^{\frac{1}{3}}} + \frac{m_2}{(M + m_2)^{\frac{1}{3}}}\right]^{\frac{3}{4}}$$

And this is also the criterion whether or not there is a maximum, or minimum, or maximum-minimum point on the energy surface.

In the case where the masses of the satellites are small compared with the mass of the planet, we may express the critical value of the momentum of the system in the form

$$\frac{4}{3^{\frac{3}{4}}} \mu^{\frac{1}{2}} C^{\frac{1}{4}} \frac{[M (m_1 + m_2)]^{\frac{3}{4}}}{(M + m_1 + m_2)^{\frac{1}{4}}}$$

A comparison of this critical value with the two previous ones shows that if the two satellites be fused together, and if γ be such that

$$\frac{M (m_1 + m_2)}{M + m_1 + m_2} \gamma^2 = C$$

and if ω be the orbital angular velocity of the compound satellite when moving at distance γ, then the above critical value of the momentum of the whole system is

$$\frac{4}{3^{\frac{3}{4}}} \frac{M(m_1+m_2)}{M+m_1+m_2} \omega\gamma^2$$

and this is $4/3^{\frac{3}{4}}$ times the orbital momentum of the compound satellite when revolving at distance γ.

Hence if the masses of the satellites are small compared with that of the planet, there are, or are not, maximum or minimum or maximum-minimum points on the surface of energy, according as the total momentum of the system is greater or less than $4/3^{\frac{3}{4}}$ times the orbital momentum of the compound satellite when moving at distance γ.

In the case where the masses of the satellites are not small compared with that of the planet, I leave the criterion in its analytical form.

There are thus three critical values of the momentum of the whole system, and the actual value of the momentum determines the character of the surface of energy according to its position with reference to these critical values.

In proceeding to consider the graphical method of solution by means of the contour lines of the energy surface, I shall choose the total momentum of the system to be greater than this third critical value, and the surface will have a maximum point. From the nature of the surface in this case we shall be able to see how it would differ if the total momentum bore any other position with reference to the three critical values. It will be sufficient if we only consider the case where the masses of the two satellites are small compared with that of the planet.

By (17) we have, with an easily intelligible alternative notation,

$$\left.\begin{matrix}\kappa_1\\ \kappa_2\end{matrix}\right\} = \tfrac{5}{2}\frac{\left.\begin{matrix}m_1\\ m_2\end{matrix}\right\}}{M}\sqrt{\frac{g}{a}}, \qquad \left.\begin{matrix}\lambda_1\\ \lambda_2\end{matrix}\right\} = \tfrac{5}{2}\frac{\left.\begin{matrix}m_1\\ m_2\end{matrix}\right\}}{M}\frac{g}{a}$$

Now κ_1 is an angular velocity, and if we choose $1/\kappa_1$ as the unit of time, we have

$$\kappa_1 = 1, \quad \kappa_2 = \frac{m_2}{m_1}$$

also

$$\lambda_1 = \tfrac{2}{5}\frac{M}{m_1}\kappa_1^2, \quad \lambda_2 = \tfrac{2}{5}\frac{M}{m_2}\kappa_2^2$$

If we choose the mass of the first satellite as unit of mass, then $m_1 = 1$, and we have

$$\kappa_1 = 1, \quad \kappa_2 = m_2, \quad \lambda_1 = \tfrac{2}{5}M, \quad \lambda_2 = \tfrac{2}{5}Mm_2$$

The unit of length has been already chosen as equal to the mean radius of the planet.

Substituting in (21) we have as the equation to the energy surface

$$2z = (h - x^{\frac{1}{7}} - m_2 y^{\frac{1}{7}})^2 - \tfrac{2}{5} M \left(\frac{1}{x^{\frac{2}{7}}} + \frac{m_2}{y^{\frac{2}{7}}} \right)$$

Since we suppose m_1 and m_2 to be small compared with M, we have

$$x = \left(\frac{c_1}{a}\right)^{\frac{7}{2}}, \quad y = \left(\frac{c_2}{a}\right)^{\frac{7}{2}}$$

On account of the abruptness of the curvatures, this surface is extremely difficult to illustrate unless the figure be of very large size, and it is therefore difficult to choose appropriate values of h, M, m_2, so as to bring the figure within a moderate compass.

In order to exhibit the influence of unequal masses in the satellites, I choose $m_2 = 2$, the mass of the first satellite being unity. I take $M = 50$, so that $\frac{2}{5}M = 20$.

With these values for M and m, the first critical value for h is 3·711, the second is 6·241, and the third is 8·459.

I accordingly take $h = 9$, which is greater than the third critical value. The surface to be illustrated then has the equation

$$2z = (9 - x^{\frac{1}{7}} - 2y^{\frac{1}{7}})^2 - 20 \left(\frac{1}{x^{\frac{2}{7}}} + \frac{2}{y^{\frac{2}{7}}} \right)$$

There is also another surface to be considered, namely

$$n = 9 - x^{\frac{1}{7}} - 2y^{\frac{1}{7}}$$

which gives the rotation of the planet corresponding to any values of x and y.

The equations $n = \Omega_x, \quad n = \Omega_y$

have also to be exhibited.

The computations requisite for the illustration were laborious, as I had to calculate values of z and n corresponding to a large number of values of x and y, and then by graphical interpolation to find the values of x and y, corresponding to exact values of z and n.

The surface of energy will be considered first.

Fig. 1 shows the contour-lines (that is to say, lines of equal energy) in the positive quadrant, z being either positive or negative.

I speak below as though the paper were held horizontally, and as though positive z were drawn vertically upwards.

The numbers written along the axes give the numerical values of x and y.

The numbers written along the curves are the corresponding values of $-2z$. Since all the numbers happen to be negative, smaller numbers indicate greater energy than larger ones; and, accordingly, in going down hill we pass from smaller to greater numbers.

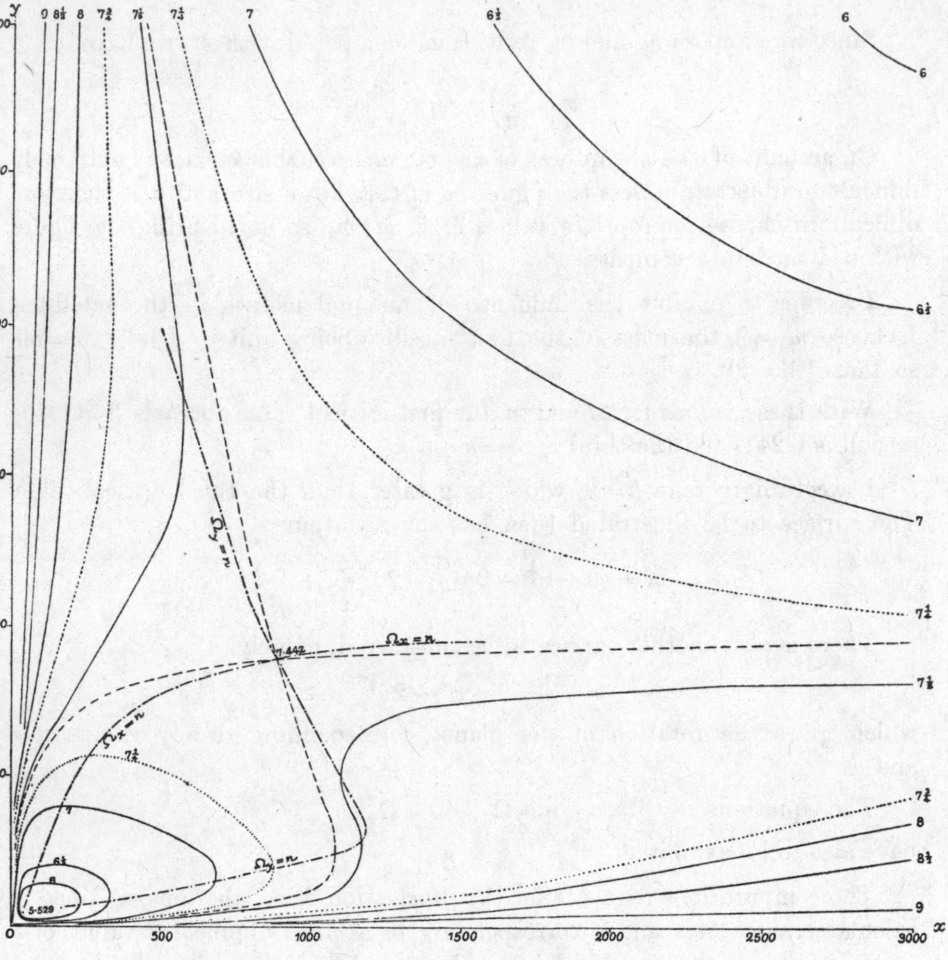

FIG. 1. Contours of the surface $2z = (9 - x^{\frac{1}{7}} - 2y^{\frac{1}{7}})^2 - 20\left(\frac{1}{x^{\frac{2}{7}}} + \frac{2}{y^{\frac{2}{7}}}\right)$ when x and y are both positive.

N.B. The values of $2z$ indicated by the numbers on the contour-lines are all negative, so that the smaller numbers indicate higher contours.

The full-line contours are equidistant, and correspond to the values 9, $8\frac{1}{2}$, 8, $7\frac{1}{2}$, 7, $6\frac{1}{2}$, and 6 of $-2z$; but since the slopes of the surface are very gentle in the central part, dotted lines (....) are drawn for the contours $7\frac{3}{4}$ and $7\frac{1}{4}$.

The points marked 5·529 and 7·442 are equidistant from x and y, and therefore correspond to the case where the two satellites have the same distance from the planet, or, which amounts to the same thing, are fused together. The former is a maximum point on the surface, the latter a maximum-minimum.

The dashed line (- - -) through 7·442 is the contour corresponding to that value of $-2z$.

The chain-dot lines (-·-·-) through the same point will be explained below.

An inspection of these contours shows that along the axes of x and y the surface has infinitely deep ravines; but the steepness of the cliffs diminishes as we recede from the origin.

The maximum point 5·529 is at the top of a hill bounded towards the ravines by very steep cliffs, but sloping more gradually in the other directions.

The maximum-minimum point 7·442 is on a saddle-shaped part of the surface, for we go up hill, whether proceeding towards O or away from O, and we go down hill in either direction perpendicular to the line towards O.

If the total angular momentum of the system had been less than the smallest critical value, the contour lines would all have been something like rectangular hyperbolas with the axes of x and y as asymptotes, like the outer curves marked 6, $6\frac{1}{2}$, 7 in fig. 1. In this case the whole surface would have sloped towards the axes.

If the momentum had been greater than the smallest, and less than the second critical value, the outer contours would have still been like rectangular hyperbolas, and the branches which run upwards, more or less parallel to y, would still have preserved that character nearer to the axes, whilst the branches more or less parallel to x would have had a curve of contrary reflexure, somewhat like that exhibited by the curve $7\frac{1}{2}$ in fig. 1, but less pronounced. In this case all the lines of steepest slope would approach the axis of x, but some of them in some part of their course would recede from the axis of y.

If the momentum had been greater than the second, but less than the third critical value, the contours would still all have been continuous curves, but for some of the inner ones there would have been contrary reflexure in both branches, somewhat like the curve marked $7\frac{1}{2}$ in fig. 1. There would still have been no closed curves amongst the contours. Here some of the lines of greatest slope would in part of their course have receded from the axis of x, and some from the axis of y, but the same line of greatest slope would never have receded from both axes.

Finally, if the momentum be greater than the third critical value, we have the case exhibited in fig. 1.

Fig. 2 exhibits the lines of greatest slope on the surface. It was constructed by making a tracing of fig. 1, and then drawing by eye the

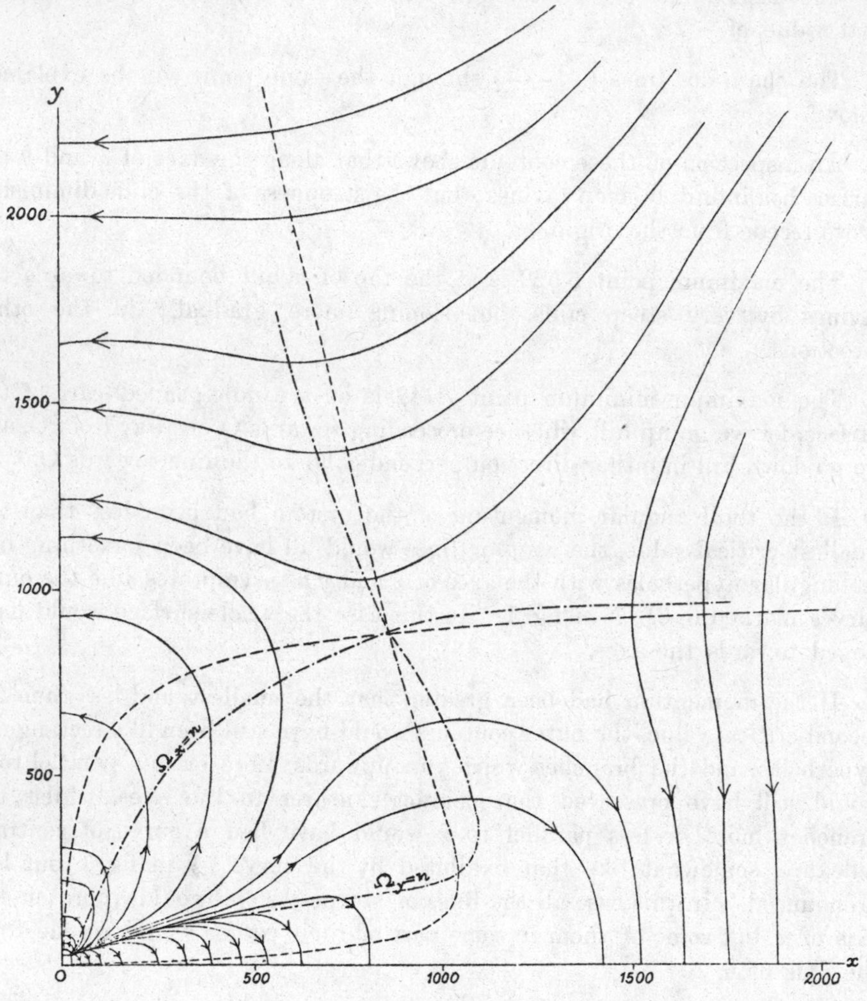

FIG. 2. Lines of greatest slope on the surface $2z = (9 - x^{\frac{1}{7}} - 2y^{\frac{1}{7}})^2 - 20\left(\frac{1}{x^{\frac{2}{7}}} + \frac{2}{y^{\frac{2}{7}}}\right)$.

orthogonal trajectories of the contours of equal energy. The dashed line (– – –) is the contour corresponding to the maximum-minimum point 7·442 of fig. 1. The chain-dot line (– · – · –) will be explained later.

One set of lines all radiate from the maximum point 5·529 of fig. 1. The arrows on the curves indicate the downward direction. It is easy to see how these lines would have differed, had the momentum of the system had various smaller values.

Fig. 3 exhibits the contour lines of the surface

$$n = 9 - x^{\frac{1}{7}} - 2y^{\frac{1}{7}}$$

It is drawn on nearly the same scale as fig. 1, but on a smaller scale than fig. 2.

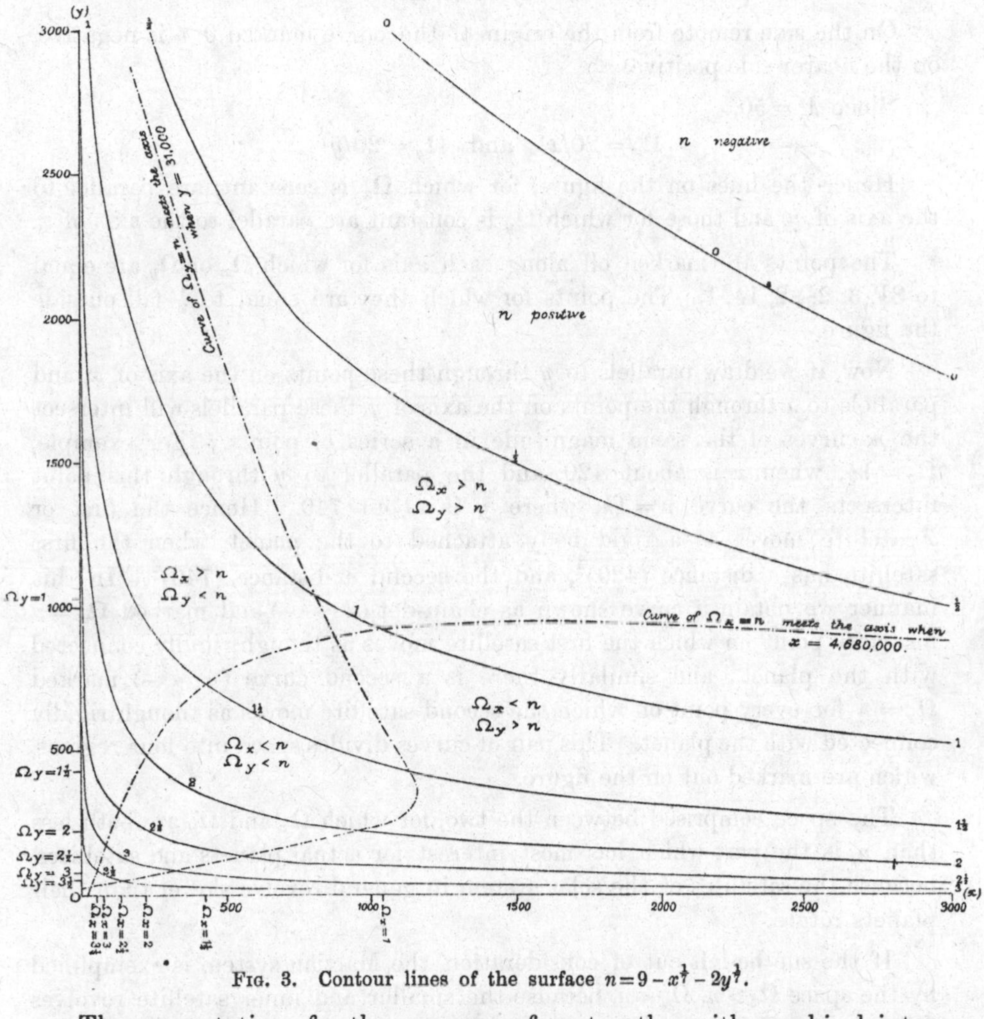

FIG. 3. Contour lines of the surface $n=9-x^{\frac{1}{7}}-2y^{\frac{1}{7}}$.

The computations for the energy surface, together with graphical interpolation, gave values of x and y corresponding to exact values of n.

The axis of n is perpendicular to the paper, and the numbers written on the curves indicate the various values of n.

These curves are *not* asymptotic to the axes, for they all cut both axes. The angles, however, at which they cut the axes are so acute that it is impossible to exhibit the intersections.

None of the curves meet the axis of x within the limits of the figure.

The curve $n=3$ meets the axis of y when $y=2150$, and that for $n=3\frac{1}{2}$ when $y=1200$, but for values of n smaller than 3 the intersections with the axis of y do not fall within the figure. The thickness, which it is necessary to give to the lines in drawing, obviously prevents the possibility of showing these facts, except in a figure of very large size.

On the side remote from the origin of the curve marked 0, n is negative, on the nearer side positive.

Since $M=50$,

$$\Omega_x = 20/x^{\frac{3}{7}} \quad \text{and} \quad \Omega_y = 20/y^{\frac{3}{7}}$$

Hence the lines on the figure, for which Ω_x is constant, are parallel to the axis of y, and those for which Ω_y is constant are parallel to the axis of x.

The points are marked off along each axis for which Ω_x or Ω_y are equal to $3\frac{1}{2}$, 3, $2\frac{1}{2}$, 2, $1\frac{1}{2}$, 1. The points for which they are equal to $\frac{1}{2}$ fall outside the figure.

Now, if we draw parallels to y through these points on the axis of x, and parallels to x through the points on the axis of y, these parallels will intersect the n curves of the same magnitude in a series of points. For example, $\Omega_x = 1\frac{1}{2}$, when x is about 420, and the parallel to y through this point intersects the curve $n=1\frac{1}{2}$, where y is about 740. Hence the first or x-satellite moves as a rigid body attached to the planet, when the first satellite has a distance $(420)^{\frac{2}{7}}$, and the second a distance $(740)^{\frac{2}{7}}$. In this manner we obtain a curve shown as chain-dot (–·–·–) and marked $\Omega_x = n$ for every point on which the first satellite moves as though rigidly connected with the planet; and similarly there is a second curve (–·–·–) marked $\Omega_y = n$ for every point on which the second satellite moves as though rigidly connected with the planet. This pair of curves divides space into four regions, which are marked out on the figure.

The space comprised between the two, for which Ω_x and Ω_y are both less than n, is the part which has most interest for actual planets and satellites, because the satellites of the solar system in general revolve slower than their planets rotate.

If the sun be left out of consideration, the Martian system is exemplified by the space $\Omega_x > n$, $\Omega_y < n$, because the smaller and inner satellite revolves quicker than the planet rotates, and the larger and outer one revolves slower.

The little quadrilateral space near O is of the same character as the external space $\Omega_x > n$, $\Omega_y > n$, but there is not room to write this on the figure.

These chain-dot curves are marked also on figs. 1 and 2. In fig. 1 the line $\Omega_x = n$ passes through all those points on the contours of energy whose tangents are parallel to x, and the line $\Omega_y = n$ passes through points whose tangents are parallel to y.

The tangents to the lines of greatest slope are perpendicular to the tangents to the contours of energy; hence in fig. 2 $\Omega_x = n$ passes through points whose tangents are parallel to y, and $\Omega_y = n$ through points whose tangents are parallel to x.

Within each of the four regions into which space is thus divided the lines of slope preserve the same character; so that if, for example, at any part of the region they are receding from x and y, they do so throughout.

This is correct, because dx/dt changes sign with $n - \Omega_x$ and dy/dt with $n - \Omega_y$; also either $n - \Omega_x$ or $n - \Omega_y$ changes sign in passing from one region to another. In these figures a line drawn at 45° to the axes through the origin divides the space into two parts; in the upper region y is greater than x, and in the lower x is greater than y. Hence configurations, for which the greater or y-satellite is exterior to the lesser or x-satellite, are represented by points in the upper space and those in which the lesser satellite is exterior by the lower space.

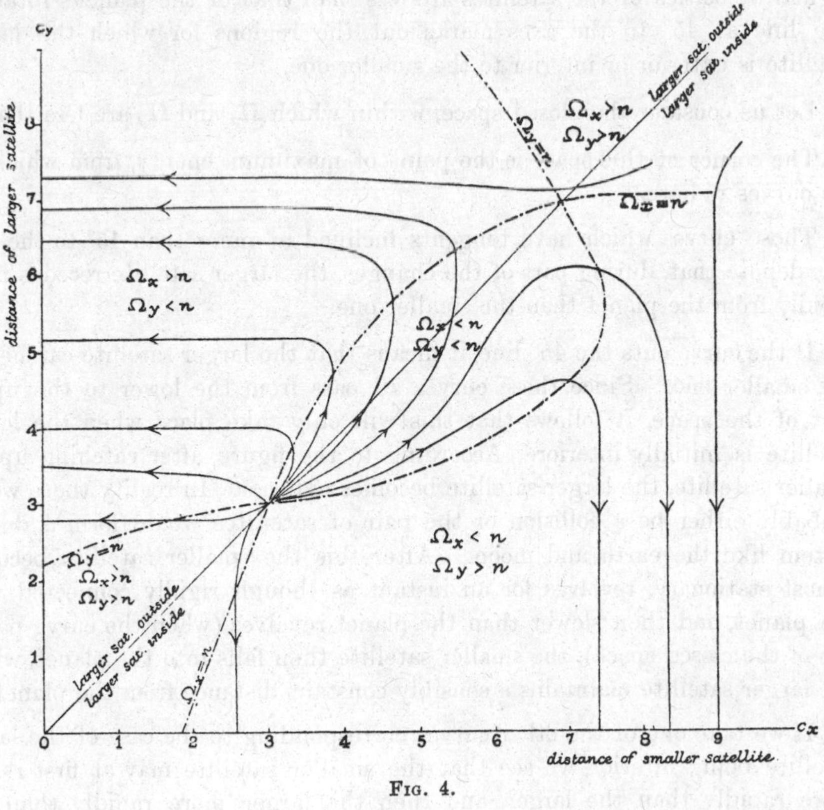

FIG. 4.

In the figures of which I have been speaking hitherto the abscissæ and ordinates are the $\frac{7}{2}$ power of the distances of the two satellites; now this is an inconveniently high power, and it is not very easy to understand the physical meaning of the result. I have therefore prepared another figure in which the abscissæ and ordinates are the actual distances. In fig. 4 the curves are no longer lines of steepest slope.

The reduction from fig. 2 to fig. 4 involved the raising of all the ordinates and abscissæ of the former one to the $\frac{2}{7}$ power. This process was rather troublesome, and fig. 4 cannot claim to be drawn with rigorous accuracy; it is, however, sufficiently exact for the hypothetical case under consideration. If we had to treat any actual case, it would only be necessary to travel along a single line of change, and for that purpose special methods of approximation might be found for giving more accurate results.

In this figure the numbers written along the axes denote the distances of the satellites in mean radii of the planet—the radius of the planet having been chosen as the unit of length.

The chain-dot curves, as before, enclose the region for which the orbital angular velocities of the satellites are less than that of the planet's rotation. The line at 45° to the axes marks out the regions for which the larger satellite is exterior or interior to the smaller one.

Let us consider the closed space, within which Ω_x and Ω_y are less than n.

The corner of this space is the point of maximum energy, from which all the curves radiate.

Those curves which have tangents inclined at more than 45° to the axis of x denote that, during part of the changes, the larger satellite recedes more rapidly from the planet than the smaller one.

If the curve cuts the 45° line, it means that the larger satellite catches up the smaller one. Since these curves all pass from the lower to the upper part of the space, it follows that this will only take place when the larger satellite is initially interior. According to the figure, after catching up the smaller satellite, the larger satellite becomes exterior. In reality there would probably either be a collision or the pair of satellites would form a double system like the earth and moon. After this the smaller satellite becomes almost stationary, revolves for an instant as though rigidly connected with the planet, and then slower than the planet revolves (when the curve passes out of the closed space); the smaller satellite then falls into the planet, whilst the larger satellite maintains a sensibly constant distance from the planet.

If we take one of the other curves corresponding to the case of the larger satellite being interior, we see that the smaller satellite may at first recede more rapidly than the larger, and then the larger more rapidly than the smaller, but not so as to catch it up. The larger one then becomes nearly stationary, whilst the smaller one still recedes. The larger one then falls in, whilst the smaller one is nearly stationary.

If we now consider those curves which are from the beginning in the upper half of the closed space, we see that if the larger satellite is initially exterior, it recedes at first rapidly, whilst the smaller one recedes slowly. The smaller and inner satellite then comes to revolve as though rigidly connected with the planet, and afterwards falls into the planet, whilst the distance of the larger one remains nearly unaltered.

Either satellite comes into collision with the planet when its distance therefrom is unity. When this takes place the colliding satellite becomes fused with the planet, and the system becomes one where there is only a single satellite; this case might then be treated as in previous papers.

The divergence of the curves from the point of maximum energy shows that a very small difference of initial configuration in a pair of satellites may in time lead to very wide differences of configuration. Accordingly tidal friction alone will not tend to arrange satellites in any determinate order. It cannot, therefore, be definitely asserted that tidal friction has not operated to arrange satellites in any order which may be observed.

I have hitherto only considered the positive quadrant of the energy surface, in which both satellites revolve positively about the planet. There are, however, three other cases, viz.: where both revolve negatively (in which case the planet necessarily revolves positively, so as to make up the positive angular momentum), or where one revolves negatively and the other positively.

These cases will not be discussed at length, since they do not possess much interest.

Fig. 5 exhibits the contours of energy for that quadrant in which the smaller or x-satellite revolves positively and the larger or y-satellite negatively. This figure may be conceived as joined on to fig. 1, so that the x-axes coincide*. The numbers written on the contours are the values of $2z$; they are positive and pretty large. Whence it follows that these contours are enormously higher than those shown in fig. 1, where all the numbers on the contours were negative.

The contours explain the nature of the surface. It may, however, be well to remark that, although the contours appear to recede from the x-axis for ever, this is not the case; for, after receding from the axis for a long way, they ultimately approach it again, and the axis is asymptotic to each of them. The point, at which the tangent to each contour is parallel to the axis of x, becomes more and more remote the higher the contour.

The lines of steepest slope on this surface give, as before, the solution of the problem.

* [The reduction of the figures from the originals for the purpose of this reprint has unfortunately not been made on exactly the same scale. Fig. 5 would have to be increased in linear scale by the fraction $\frac{2}{1}\frac{0}{9}$ths in order to fit exactly on to Fig. 1.]

If we hold this figure upside down, and read x for y and y for x, we get a figure which represents the general nature of the surface for the case where the x-satellite revolves negatively and the y-satellite positively. But of course the figure would not be drawn correctly to scale.

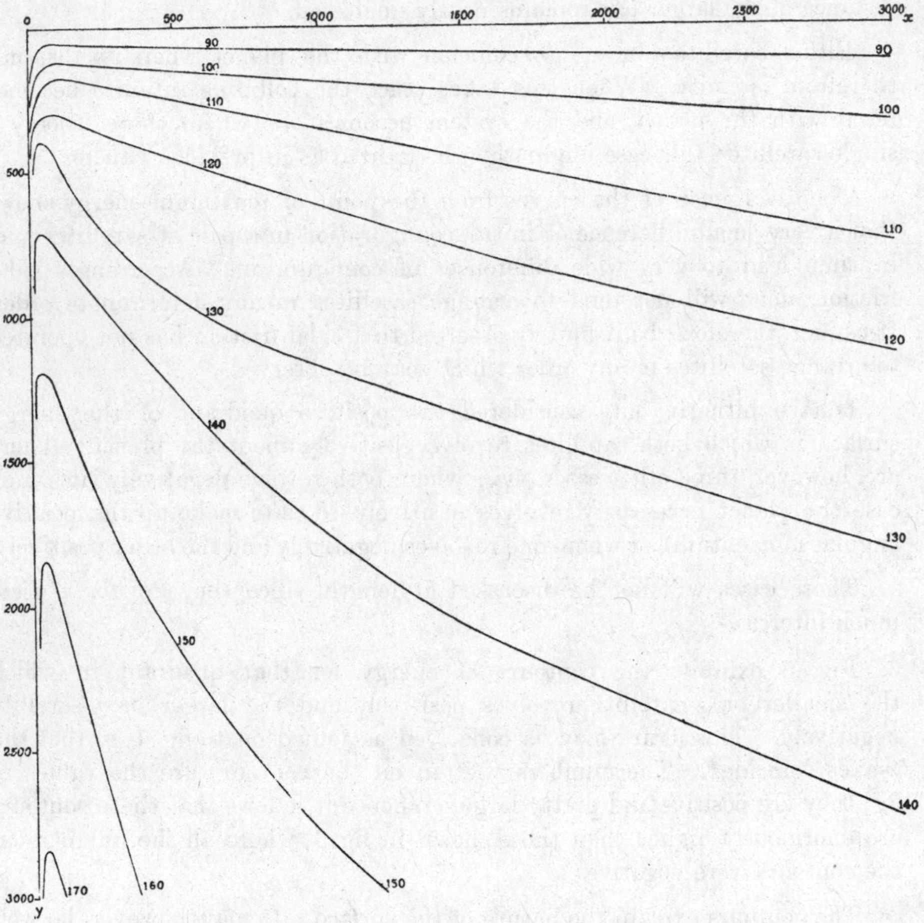

FIG. 5. Contour lines of the surface $2z=(9-x^{\frac{1}{7}}-2y^{\frac{1}{7}})^2-20\left(\frac{1}{x^{\frac{2}{7}}}+\frac{2}{y^{\frac{2}{7}}}\right)$ when x is positive and y negative.

The contours for the remaining quadrant, in which both satellites revolve negatively, would somewhat resemble a family of rectangular hyperbolas with the axes as asymptotes. I have not thought it worth while to construct them, but the physical interpretation is obviously that both satellites always must approach the planet.

II.

A Discussion of the Effects of Tidal Friction with reference to the Evolution of the Solar System.

§ 6. *General consideration of the problem presented by the solar system.*

In a series of previous papers I have traced out the changes in the manner of motion of the earth and moon which must have been caused by tidal friction. By adopting the hypothesis that tidal friction has been the most important element in the history of those bodies, we are led to coordinate together all the elements in their motions in a manner so remarkable, that the conclusion can hardly be avoided that the hypothesis contains a great amount of truth.

Under these circumstances it is natural to inquire whether the same agency may not have been equally important in the evolution of the other planetary sub-systems, and of the solar system as a whole.

This inquiry necessarily leads on to wide speculations, but I shall endeavour to derive as much guidance as possible from numerical data.

In the first part of the present paper the theory of the tidal friction of a planet, attended by several satellites, has been treated.

It would, at first sight, seem natural to replace this planet by the sun, and the satellites by the planets, and to obtain an approximate numerical solution. We might suppose that such a solution would afford indications as to whether tidal friction has or has not been a largely efficient cause in modifying the solar system.

The problem here suggested for solution differs, however, in certain points from that actually presented by the solar system, and it will now be shown that these differences are such as would render the solution of no avail.

The planets are not particles, as the suggested problem would suppose them to be, but they are rotating spheroids in which tides are being raised both by their own satellites and by the sun. They are, therefore, subject to a complicated tidal friction; the reaction of the tides raised by the satellites goes to expand the orbits of the satellites, but the reaction of the tide raised in the planet by the sun, and that raised in the sun by the planet both go towards expanding the orbit of the planet. It is this latter effect with which we are at present concerned.

I propose then to consider the probable relative importance of these two causes of change in the planetary orbits.

But before doing so it will be well as a preliminary to consider another point.

In considering the effects of tidal friction the theory has been throughout adopted that the tidally-disturbed body is homogeneous and viscous. Now we know that the planets are not homogeneous, and it seems not improbable that the tidally-disturbed parts will be principally more or less superficial—as indeed we know that they are in the case of terrestrial oceans. The question then arises as to the extent of error introduced by the hypothesis of homogeneity.

For a homogeneous viscous planet we have shown that the tidal frictional couple is approximately equal* to

$$C\frac{\tau^2}{\mathfrak{g}}\frac{n-\Omega}{\mathfrak{p}}, \text{ where } \mathfrak{p}=\frac{gaw}{19v}$$

Now how will this expression be modified, if the tidally-disturbed parts are more or less superficial, and of less than the mean density of the planet?

To answer this query we must refer back to the manner in which the expression was built up.

By reference to my paper "On the Tides of a Viscous Spheroid" (Paper 1, p. 13), it will be seen that $\mathfrak{p}$ is really $(\frac{5}{2}gaw-\frac{3}{2}gaw)/19v$, and that in both of these terms w represents the density of the tidally-disturbed matter, but that in the former g represents the gravitation of the planet and in the second it is equal to $\frac{4}{3}\pi\mu aw$, where w is the density of the tidally-disturbed matter. Now let f be the ratio of the mean density of the spheroid to the density of the tidally-disturbed matter.

Then in the former term

$$gaw=\frac{3}{4\pi\mu}\cdot g\cdot\tfrac{4}{3}\pi\mu aw=\frac{3}{4\pi\mu}g^2\times\frac{1}{f}$$

And in the latter

$$gaw=\frac{3}{4\pi\mu}g^2\times\frac{1}{f^2}$$

Hence if the planet be heterogeneous and the tidally-disturbed matter superficial, $\mathfrak{p}$ must be a coefficient of the form

$$\frac{3}{4\pi\mu\times 19}\frac{g^2}{vf}(\tfrac{5}{2}-\tfrac{3}{2}/f)$$

If f be unity this reduces to the form $gaw/19v$, as it ought; but if the tidally-disturbed matter be superficial and of less than the mean density, then $\mathfrak{p}$ must be a coefficient which varies as $\frac{g^2}{\mu vf}(1-\frac{3}{5}/f)$. The exact form of the coefficient will of course depend upon the nature of the tides. If

* I leave out of account the case of "large" viscosity, because as shown in a previous paper that could only be true of a planet which in ordinary parlance would be called a solid of great rigidity.—See Paper 3, p. 126.

f be large the term $3/5f$ will be negligeable compared with unity. Again, if we refer to Paper 3, p. 46, it appears that the C in the expression for the tidal frictional couple represents $\frac{2}{5}(\frac{4}{3}\pi a^3 w) a^2$, where w is the density of the tidally-disturbed matter; hence C should be replaced by C/f.

If we reconstruct the expression for the tidal frictional couple, we see that it is to be divided by f, because of the true meaning to be assigned to C, but it is to be multiplied by f on account of the true meaning to be assigned to $\mathfrak{p}$.

From this it follows that for a given viscosity it is, roughly speaking, probable that the tidal frictional couple will be nearly the same as though the planet were homogeneous. The above has been stated in an analytical form, but in physical language the reason is because the lagging of the tide will be augmented by the deficiency of density of the tidally-disturbed matter in about the same proportion as the frictional couple is diminished by the deficiency of density of the tide-wave upon which the disturbing satellite has to act.

This discussion appeared necessary in order to show that the tidal frictional couple is of the same order of magnitude whether the planet be homogeneous or heterogeneous, and that we shall not be led into grave errors by discussing the theory of tidal friction on the hypothesis of the homogeneity of the tidally-disturbed bodies.

We may now proceed to consider the double tidal action of a planet and the sun.

Let us consider the particular homogeneous planet whose mass, distance from sun, and orbital angular velocity are m, c, Ω. For this planet, let C' = moment of inertia; a' = mean radius; w' = density; g' = gravity; $\mathfrak{g}' = \frac{2}{5}g'/a'$; v' = the viscosity; $\mathfrak{p}' = g'a'w'/19v' = \frac{3}{4\times19}g'^2/\pi\mu v'$; and n' = angular velocity of diurnal rotation.

The same symbols when unaccented are to represent the parallel quantities for the sun.

Suppose the sun to be either perfectly rigid, perfectly elastic, or perfectly fluid. Then *mutatis mutandis*, equation (2) gives the rate of increase of the planet's distance from the sun under the influence of the tidal friction in the planet. It becomes

$$\mu^{\frac{1}{2}} \frac{Mm}{(M+m)^{\frac{1}{2}}} \frac{dc^{\frac{1}{2}}}{dt} = (\tfrac{3}{2})^2 \frac{C'}{\mathfrak{g}'} \frac{(\mu M)^2}{c^6} \frac{n'-\Omega}{\mathfrak{p}'}$$

If the planet have no satellite the right-hand side is equal to $-C'dn'/dt$, because the equation was formed from the expression for the tidal frictional couple.

Hence, if none of the planets had satellites we should have a series of equations of the form

$$C'n' + \mu^{\frac{1}{2}} \frac{Mm}{(M+m)^{\frac{1}{2}}} c^{\frac{1}{2}} = h$$

with different h's corresponding to each planet.

We may here remark that the secular effects of tidal friction in the case of a rigid sun attended by tidally-disturbed planets, with no satellites, may easily be determined. For if we put $c^{\frac{1}{2}} = x$, and note that Ω varies as x^{-3}, and that n' has the form $(h - kx)/C'$, we see that it would only be necessary to evaluate a series of integrals of the form $\int_{x_0}^{x} \frac{\epsilon x^{15} dx}{\alpha - \beta x^3 + \gamma x^4}$. This integral is in fact merely the time which elapses whilst x changes from x_0 to x, and the time scale is the same for all the planets. It is not at present worth while to pursue this hypothetical case further.

Now if we suppose the planet to raise frictional tides in the sun, as well as the sun to raise tides in the planets, we easily see by a double application of (2) that

$$\mu^{\frac{1}{2}} \frac{Mm}{(M+m)^{\frac{1}{2}}} \frac{dc^{\frac{1}{2}}}{dt} = (\tfrac{3}{2})^2 \frac{1}{c^6} \left[(\mu m)^2 \frac{C}{\mathfrak{gp}} (n - \Omega) + (\mu M)^2 \frac{C'}{\mathfrak{g}'\mathfrak{p}'} (n' - \Omega) \right] \quad \text{...(28)}$$

The tides raised in the planet by its satellites do not occur explicitly in this equation, but they do occur implicitly, because n', the planet's rotation, is affected by these tides.

The question which we now have to ask is whether in the equation (28) the solar term (without accents) or the planetary term (with accents) is the more important.

In the solar system the rotations of the sun and planets are rapid compared with the orbital motions, so that Ω may be neglected compared with both n and n'.

Hence the planetary term bears to the solar term approximately the ratio $\frac{M^2 C' n' \mathfrak{gp}}{m^2 C n \mathfrak{g}'\mathfrak{p}'}$.

Now $\left(\frac{M}{m}\right)^2 \frac{C'}{C} \frac{\mathfrak{g}}{\mathfrak{g}'} = \frac{M}{m} \left(\frac{a'}{a}\right)^2 \frac{g}{g'} \frac{a'}{a} = \left(\frac{g}{g'}\right)^2 \frac{a'}{a}$. Also $\frac{\mathfrak{p}}{\mathfrak{p}'} = \left(\frac{g}{g'}\right)^2 \frac{v'}{v}$

Therefore the ratio is $\left(\frac{g}{g'}\right)^4 \frac{a'}{a} \frac{n'}{n} \frac{v'}{v}$

Solar gravity is about 26·4 times that of the earth and about 10·4 times that of Jupiter. The solar radius is about 109 times that of the earth and about 10 times that of Jupiter. The earth's rotation is about 25·4 times that of the sun, and Jupiter's rotation is about 61 times that of the sun.

Combining these data I find that the effect of solar tides in the earth is about 113,000 v'/v times as great as the effect of terrestrial tides in the sun, and the effect of solar tides in Jupiter is about 70,000 v'/v times as great as the effect of Jovian tides in the sun. It is not worth while to make a similar comparison for any of the other planets.

It seems reasonable to suppose that the coefficient of tidal friction in the planets is of the same order of magnitude as in the sun, so that it is improbable that v'/v should be either a large number or a small fraction.

We may conclude then from this comparison that the effects of tides raised in the sun by the planets are quite insignificant in comparison with those of tides raised in the planets by the sun.

It appears therefore that we may fairly leave out of account the tides raised in the sun in studying the possible changes in the planetary orbits as resulting from tidal friction.

But the difference of physical condition in the several planets is probably considerable, and this would lead to differences in the coefficients of tidal friction to which there is no apparent means of approximating. It therefore seems inexpedient at present to devote time to the numerical solution of the problem of the rigid sun and the tidally-disturbed planets.

§ 7. *Numerical data and deductions therefrom.*

Although we are thus brought to admit that it is difficult to construct any problem which shall adequately represent the actual case, yet a discussion of certain numerical values involved in the solar system and in the planetary sub-systems will, I think, lead to some interesting results.

The fundamental fact with regard to the theory of tidal friction is the transformation of the rotational momentum of the planet as it is destroyed by tidal friction into orbital momentum of the tide-raising body.

Hence we may derive information concerning the effects of tidal friction by the evaluation of the various momenta of the several parts of the solar system.

Professor J. C. Adams has kindly given me a table of values of the planetary masses, each with its attendant satellites. The authorities were as follows: for Mercury, Encke; for Venus, Le Verrier; for the Earth, Hansen; for Mars, Hall; for Jupiter, Bessel; for Saturn, Bessel; for Uranus, Von Asten; for Neptune, Newcomb.

The masses were expressed as fractions of the sun. The results, when earth plus moon is taken as unity, are given in the table below. The mean distances, taken from Herschel's *Astronomy*, are given in a second column.

The unit of mass is earth plus moon, the unit of length is the earth's mean distance from the sun, and the unit of time will be taken as the mean solar day.

	Masses (m)	Mean Distances (c)
Sun . . .	315,511·	
Mercury . .	·06484	·387098
Venus . .	·78829	·723332
Earth . . .	1·00000	1·000000
Mars . . .	·10199	1·523692
Jupiter . .	301·0971	5·202776
Saturn . .	90·1048	9·538786
Uranus . .	14·3414	19·18239
Neptune . .	16·0158	30·05660

Then μ being the attraction between unit masses at unit distance, M being sun's mass, and 365·25 being the earth's periodic time, we have

$$\sqrt{\mu M} = \frac{2\pi}{365 \cdot 25} = \frac{10^{8 \cdot 23558}}{10^{10}}$$

The momentum of orbital motion of any one of the planets round the sun is given by $m \,.\, \sqrt{\mu M} \,.\, \sqrt{c}$.

With the above data I find the following results*.

TABLE I.

Planet	Orbital momentum
Mercury	·00079
Venus	·01309
Earth	·01720
Mars	·00253
Jupiter	13·469
Saturn	5·456
Uranus	1·323
Neptune	1·806
Total	22·088

We must now make an estimate of the rotational momentum of the sun, so as to compare it with the total orbital momentum of the planets.

It seems probable that the sun is much more dense in the central portion, than near the surface†. Now if the Laplacian law of internal density were

* These values are of course not rigorously accurate, because the attraction of Jupiter and Saturn on the internal planets is equivalent to a diminution of the sun's mass for them, and the attraction of the internal planets on the external ones is equivalent to an increase of the sun's mass.

† I have elsewhere shown that there is a strong probability that this is the case with Jupiter, and that planet probably resembles the sun more nearly than does the earth.—See *Ast. Soc. Month. Not.*, Dec., 1876. [See Vol. III.]

to hold with the sun, but with the surface density infinitely small compared with the mean density, we should have

$$C=\tfrac{2}{3}\left[1-\frac{6}{\pi^2}\right]Ma^2$$

If on the other hand the sun were of uniform density we should have $C=\tfrac{2}{5}Ma^2$*.

* These considerations lead me to remark that in previous papers, where the tidal theory was applied numerically to the case of the earth and moon, I might have chosen more satisfactory numerical values with which to begin the computations.

It was desirable to use a consistent theory of frictional tides, and that founded on the hypothesis of a homogeneous viscous planet was adopted.

The earth had therefore to be treated as homogeneous, and since tidal friction depends on relative motion, the rotation of the homogeneous planet had to be made identical with that of the real earth. A consequence of this is that the rotational momentum of the earth in my problem bore a larger ratio to the orbital momentum of the moon than is the case in reality. Since the consequence of tidal friction is to transfer momentum from one part of the system to the other, this treatment somewhat vitiated subsequent results, although not to such an extent as could make any important difference in a speculative investigation of that kind.

If it had occurred to me, however, it would have been just as easy to replace the actual heterogeneous earth by a homogeneous planet mechanically equivalent thereto. The mechanical equivalence referred to lies in the identity of mass, moment of inertia, and rotation between the homogeneous substitute and the real earth. These identities of course involve identity of rotational momentum and of rotational energy, and, as will be seen presently, other identities are approximately satisfied at the same time.

Suppose that roman letters apply to the real earth and italic letters to the homogeneous substitute.

By Laplace's theory of the earth's figure, with Thomson and Tait's notation (*Natural Philosophy*, § 824)

$$\mathrm{C}=\tfrac{2}{3}\left\{1-\frac{6(f-1)}{f\theta^2}\right\}\mathrm{Ma}^2$$

where f is the ratio of mean to surface density, and θ is a certain angle.

Also
$$C=\tfrac{2}{5}Ma^2(1+\tfrac{2}{3}e)$$

where e is the ellipticity of the homogeneous planet's figure.

By the above conditions of mechanical identity

$$\mathrm{M}=M \text{ and } \mathrm{C}=C$$

whence
$$\left(\frac{a}{\mathrm{a}}\right)^2=\tfrac{5}{3}(1-\tfrac{2}{3}e)\left\{1-\frac{6(f-1)}{f\theta^2}\right\}$$

Now put $\mathrm{m}=\mathrm{n}^2\mathrm{a}/\mathrm{g}$, $m=n^2a/g$; where g, g are mean pure gravity in the two cases. Then the remaining condition gives $\mathrm{n}=n$.

Therefore
$$\frac{m}{\mathrm{m}}=\frac{\mathrm{g}a}{g\mathrm{a}}=\left(\frac{a}{\mathrm{a}}\right)^3$$

But
$$e=\tfrac{5}{4}m=\tfrac{5}{4}\mathrm{m}\left(\frac{a}{\mathrm{a}}\right)^3$$

Hence
$$\left(\frac{a}{\mathrm{a}}\right)^2=\tfrac{5}{3}\left\{1-\tfrac{5}{6}\mathrm{m}\left(\frac{a}{\mathrm{a}}\right)^3\right\}\left\{1-\frac{6(f-1)}{f\theta^2}\right\}$$

This is an equation which gives the radius of the homogeneous substituted planet in terms of

The former of these two suppositions seems more likely to be near the truth than the latter.

Now $\frac{2}{3}(1 - 6\pi^{-2}) = \cdot 26138$, so that C may lie between $(\cdot 26138)\,Ma^2$ and $(\cdot 4)\,Ma^2$.

The sun's apparent radius is 961″·82, therefore the unit of distance being the present distance of the earth from the sun, $a = 961\cdot 82\pi/648{,}000$; also $M = 315{,}511$.

Lastly the sun's period of rotation is about 25·38 m.s. days, so that $n = 2\pi/25\cdot 38$.

Combining these numerical values I find that Cn (the solar rotational momentum) may lie between ·444 and ·679. The former of these values seems however likely to be far nearer the truth than the latter.

It follows therefore that the total orbital momentum of the planetary system, found above to be 22, is about 50 times that of the solar rotation.

In discussing the various planetary sub-systems I take most of the numerical values from the excellent tables of astronomical constants in Professor Ball's *Astronomy**, and from the table of masses given above.

---

that of the earth. It may be solved approximately by first neglecting $\frac{2}{5}\mathrm{m}\,(a/\mathrm{a})^3$, and afterwards using the approximate value of a/a for determining that quantity.

The density of the homogeneous planet is found from

$$w = \mathrm{w}\left(\frac{\mathrm{a}}{a}\right)^3$$

where w is the earth's mean density.

To apply these considerations to the earth, we take $\theta = 142°\ 30'$, $f = 2\cdot 057$, which give $\frac{1}{295}$ as the ellipticity of the earth's surface.

With these values (Thomson and Tait's *Natural Philosophy*, § 824, table, col. vii., they give however ·835)

$$\tfrac{5}{3}\left\{1 - \frac{6(f-1)}{f\theta^2}\right\} = \cdot 83595$$

The first approximation gives $\dfrac{a}{\mathrm{a}} = \cdot 9143$, and the second $\dfrac{a}{\mathrm{a}} = \cdot 9133$.

Hence the radius of the actual earth 6,370,000 metres becomes, in the homogeneous substitute, 5,817,000 metres.

Taking 5·67 as the earth's mean specific gravity, that of the homogeneous planet is 7·44.

The ellipticity of the homogeneous planet is ·00329 or $\frac{1}{303}$, which differs but little from that of the real earth, viz.: $\frac{1}{295}$.

The precessional constant of the homogeneous planet is equal to the ellipticity, and is therefore ·00329. If this be compared with the precessional constant ·00327 of the earth, we see that the homogeneous substitute has sensibly the same precession as has the earth.

If a similar treatment be applied to Jupiter, then (with the numerical values given in a previous paper, *Ast. Soc. Month. Not.*, Dec. 1876 [see Vol. III.]) the homogeneous planet has a radius equal to ·8 of the actual one; its density is about half that of the earth, and its ellipticity is $\frac{1}{19}$.

* *Text Book of Science: Elements of Astronomy.* Longmans. 1880.

Mercury.

The diameter at distance unity is about 6″·5; the diurnal period is $24^h\ 0^m\ 50^s$ (?). The value of the mass seems very uncertain, but I take Encke's value given above. Assuming that the law of internal density is the same as in the earth (see below), we have $C = \cdot 33438ma^2$, and $n = 2\pi$ very nearly. Whence I find for the rotational momentum

$$Cn = \frac{\cdot 34}{10^{10}}.$$

Venus.

The diameter at distance unity is about 16″·9; the diurnal period is $23^h\ 21^m\ 22^s$ (?). Assuming the same law of internal density as for the earth, I find

$$Cn = \frac{28{\cdot}6}{10^{10}}.$$

Herschel remarks (*Outlines of Astronomy*, § 509) that "both Mercury and Venus have been concluded to revolve on their axes in about the same time as the Earth, though in the case of Venus, Bianchini and other more recent observers have contended for a period of twenty-four times that length." He evidently places little reliance on the observations [and now, in 1908, it seems perhaps more probable that the periods of rotation of both planets are the same as their periods about the sun].

The Earth.

I adopt Laplace's theory of internal density (with Thomson and Tait's notation), and take, according to Colonel Clarke, the ellipticity of surface to be $\frac{1}{295}$. This value corresponds with the value 2·057 for the ratio of mean to surface density (the f of Thomson and Tait), and to 142° 30′ for the auxiliary angle θ.

The moment of inertia is given by the formula

$$C = \tfrac{2}{3}\left[1 - \frac{6(f-1)}{f\theta^2}\right] ma^2$$

These values of θ and f give $\tfrac{5}{3}\left[1 - \frac{6(f-1)}{f\theta^2}\right] = \cdot 83595$

Whence $$C = \cdot 33438ma^2$$

The numerical coefficient is the same as that already used in the case of the two previous planets.

The moon's mass being $\frac{1}{82}$nd of the earth's, the earth's mass is $\frac{82}{83}$ in the chosen unit of mass.

With sun's parallax 8″·8, and unit of length equal to earth's mean distance

$$a = \frac{8{\cdot}8\pi}{648000}$$

The angular velocity of diurnal rotation, with unit of time equal to the mean solar day,

$$n = \frac{2\pi}{\cdot 99727}$$

Combining these values I find for the earth's rotational momentum

$$Cn = \frac{37\cdot 88}{10^{10}}$$

Writing m' for the moon's mass, and neglecting the eccentricity of the lunar orbit, the moon's orbital momentum* is

$$\frac{mm'}{m+m'}\,\Omega c^2$$

Taking the moon's parallax as 3422″·3 (which gives a distance of 60·27 earth's radii), and the sun's parallax as 8″·8, we have

$$c = \frac{8\cdot 8}{3422\cdot 3}$$

Taking the lunar period as 27·3217 m.s. days we have

$$\Omega = \frac{2\pi}{27\cdot 3217}$$

As above stated, m is $\frac{82}{83}$, and m' is $\frac{1}{83}$; whence it will be found that the moon's orbital momentum is $\frac{181}{10^{10}}$.

This is 4·78 times the earth's rotational momentum.

The resultant angular momentum of the system, with obliquity of ecliptic 23° 28′, is 5·71 times the earth's rotational momentum, and is $\frac{216}{10^{10}}$.

Mars.

The polar diameter at distance unity is 9″·352 (Hartwig, *Nature*, June 3, 1880). With an ellipticity $\frac{1}{300}$ this gives 4″·686 as the mean radius. The diurnal period is $24^h\ 37^m\ 23^s$. Assuming the law of internal density to be the same as in the earth I find

$$Cn = \frac{1\cdot 08}{10^{10}}$$

The masses of the satellites are very small, and their orbital momentum must also be very small.

* If we determine μ from the formula

$$\mu = \Omega^2 c^3 = \left(\frac{2\pi}{27\cdot 3217}\right)^2 \left(\frac{8\cdot 8}{3422\cdot 3}\right)^3$$

and observe that $\mu M = (2\pi/365\cdot 25)^2$, we obtain 329,000 as the sun's mass. This disagrees with the value 315,511 used elsewhere. The discrepancy arises from the neglect of solar perturbation of the moon, and of planetary perturbation of the earth.

Jupiter.

The polar and equatorial diameters at the planet's mean distance from the sun are 35″·170 and 37″·563 (Kaiser and Bessel, *Ast. Nach.* vol. 48, p. 111). These values give a mean radius 5·2028 × 18″·383 at distance unity.

The period of rotation is $9^h\ 55\frac{1}{2}^m$, or ·4136 m.s. day.

I have elsewhere shown reason to believe that the surface density of Jupiter is very small compared with the mean density. It appears that we have approximately

$$C = \tfrac{2}{3} \times \frac{1}{2{\cdot}528}\, ma^2 = {\cdot}2637\, ma^2 *$$

The numerical coefficient differs but little from that which we should have, if the Laplacian law of internal density were true, with infinitely small surface density (f infinite, $\theta = 180°$); for, as appeared in considering the sun's moment of inertia, the factor would be in that case ·26138.

With these values I find

$$Cn = \frac{2{,}594{,}000}{10^{10}} = {\cdot}0002594$$

The distances of the satellites referred to the mean distance of Jupiter from the sun are

I.	II.	III.	IV.
111″·74	177″·80	283″·61	498″·87

Taking Jupiter's mean distance to be 5·20278, the logarithms of the distances in terms of the earth's distance from the sun are

I.	II.	III.	IV.
7·45002 − 10	7·65174 − 10	7·85453 − 10	8·09980 − 10

The periodic times are in m.s. days (Herschel's *Astronomy*, Appendix)

I.	II.	III.	IV.
1·76914	3·55181	7·15455	16·6888

The masses given me by Professor Adams† from a revision of Damoiseau's work are in terms of Jupiter's mass

I.	II.	III.	IV.
$\frac{2{\cdot}8311}{10^5}$	$\frac{2{\cdot}3236}{10^5}$	$\frac{8{\cdot}1245}{10^5}$	$\frac{2{\cdot}1488}{10^5}$

Combining these data according to the formula $m\Omega c^2$, where m is the

* *Ast. Soc. Month. Not.*, Dec. 1876, p. 83. [See Vol. III.]

† He kindly gave me these data for another purpose.—See *Ast. Soc. Month. Not.*, Dec., 1876, p. 81.

mass of the satellite, I find for the orbital momenta of the satellites expressed in terms of the chosen units—

I.	II.	III.	IV.
$\frac{2406}{10^{10}}$	$\frac{2489}{10^{10}}$	$\frac{10993}{10^{10}}$	$\frac{3857}{10^{10}}$

The sum of these is $19745/10^{10}$ and is the total orbital momentum of the satellites. It is $\frac{1}{130}$th of the rotational momentum of the planet as found above.

The whole angular momentum of the Jovian system is $\frac{2,614,000}{10^{10}}$

Saturn.

There seems to be much doubt as to the diameter of the planet.

The values of the *mean* radius at distance unity given by Bessel, De La Rue, and Main (with ellipticities 1/10·2, 1/11 (?), 1/9·227 respectively) are 79″, 82″, and 94″ respectively*.

The period of rotation is $10^h\ 29\frac{1}{4}^m$ or ·437 m.s. day.

Assuming (as with the sun) that the surface density is infinitely small compared with the mean density, we have $C = {\cdot}2614ma^2$. I find then that these three values give respectively,

$$Cn = \frac{497,000}{10^{10}} \text{ and } \frac{535,000}{10^{10}} \text{ and } \frac{703,000}{10^{10}}$$

The masses of the satellites are unknown, but Herschel thinks that Titan is nearly as large as Mercury.

If we take its mass as ·06 in terms of the earth's mass, its distance as 176″·755 at the planet's mean distance from the sun, and its periodic time as 15·95 m.s. days, we find the orbital momentum to be $16,000/10^{10}$. The whole orbital momentum of the satellites and the ring is likely to be greater than this, for the ring has been variously estimated to have a mass equal to $\frac{1}{120}$th to $\frac{1}{620}$th of the planet.

It is probable therefore that orbital momentum of the system is $\frac{1}{30}$th, or thereabouts, of the rotational momentum of the planet.

Nothing is known concerning the rotation of Uranus and Neptune, and but little of their satellites.

The results of this numerical survey of the planets are collected in the following table.

* Deduced from values of the equatorial diameter found by these observers, referred to the planet's mean distance from the sun, as given by Ball.

TABLE II.

Planet	i Rotational momentum of planet $\times 10^{10}$	ii Orbital momentum of satellites $\times 10^{10}$	iii Ratio of ii to i	iv Total momentum of each planet's system $\times 10^{10}$
Mercury . .	·34 ?	...	...	·34 ?
Venus . . .	28·6 ?	...	...	28·6 ?
Earth . . .	37·88	181	4·78	216
Mars . . .	1·08	very small	very small	1·08
Jupiter . .	2,594,000	20,000	$\frac{1}{130}$	2,614,000
Saturn . .	500,000 to 700,000	16,000 or more	$\frac{1}{30}$ or more	520,000 to 720,000

The numbers marked with queries are open to much doubt.

If the numbers given in column iv. of this table be compared with those given in Table I., it will be seen that the total internal momentum of each of the planetary sub-systems is very small compared with the orbital momentum of the planet in its motion round the sun. This ratio is largest in the case of Jupiter, and here the internal momentum is ·00026 whilst the orbital momentum is 13; hence in the case of Jupiter the orbital momentum is 50000 times the sum of the rotational momentum of the planet and the orbital momentum of its satellites. From this it follows that if the whole of the momentum of Jupiter and his satellites were destroyed by solar tidal friction, the mean distance of Jupiter from the sun would only be increased by $\frac{1}{25000}$th part. The effect of the destruction of the internal momentum of any of the other planets would be very much less.

If therefore the orbits of the planets round the sun have been considerably enlarged, during the evolution of the system, by the friction of the tides raised in the planets by the sun, the primitive rotational momentum of the planetary bodies must have been thousands of times greater than at present. If this were the case then the enlargement of the orbits must simultaneously have been increased, to some extent, by the reaction of the tides raised in the sun by the planets.

But it does not seem probable that the planetary masses ever possessed such an enormous amount of rotational momentum, and therefore it is not probable that tidal friction has affected the dimensions of the planetary orbits considerably.

It is difficult to estimate the degree of attention which should be paid to Bode's empirical law concerning the mean distances of the planets, but it may perhaps be supposed that that law (although violated in the case of Neptune, and only partially satisfied by the asteroids) is the outcome of the laws governing the successive epochs of instability in the history of a rotating and contracting nebula. Now if, after the genesis of the planets, tidal friction had affected the planetary distances considerably, all appearance of such primitive law in the distances would be thereby obliterated. If therefore there be now observable a sort of law of mean distances, it to some extent falls in with the conclusion arrived at by the preceding numerical comparisons.

The extreme relative smallness of the masses of the Martian and Jovian satellites tends to show the improbability of very large changes in the dimensions of the orbits of those satellites; although the argument has not nearly equal force in these cases, because the distances of the satellites from these planets are small.

The numbers given in column iii. of Table II. show in a striking manner the great difference between the present physical conditions of the terrestrial system and those of Mars, Jupiter, and Saturn. These numbers may perhaps be taken as representing the amount of effect which the tidal friction due to the satellites has had in their evolution, and confirms the conclusion that, whilst tidal friction may have been (and according to previous investigations certainly appears to have been) the great factor in the evolution of the earth and moon, yet with the satellites of the other planets it has not had such important effects.

In previous papers the expansion of the lunar orbit under the influence of terrestrial tidal friction was examined, and the moon was traced back to an origin close to the present surface of the earth. The preceding numerical comparisons suggest that the contraction of the planetary masses has been the more important factor elsewhere, and that the genesis of satellites occurred elsewhere earlier in the evolution.

It has been shown that the case of the earth and moon does actually differ widely from that of the other planets, and we may therefore reasonably suppose that the history has also differed considerably.

Although we might perhaps leave the subject at this point, yet, after arriving at the above conclusions, it seems natural to inquire in what manner the simultaneous action of the contraction of a planetary mass and of tidal friction is likely to have operated.

The subject is necessarily speculative, but the conclusions at which I arrive are, I think, worthy of notice, for although they involve much of mere conjectural assumption in respect to the quantities and amounts assumed, yet they are deduced from the rigorous dynamical principles of angular momentum and of energy.

§ 8. *On the part played by tidal friction in the evolution of planetary masses.*

To consider the subject of this section, we require—

(α) Some measure of the relative efficiency of solar tidal friction in reducing the rotational momentum and the rotation of the several planets.

(β) We have to consider the manner in which the simultaneous action of the contraction of the planetary mass and of solar tidal friction co-operates.

(γ) We have to discuss how the separation of a satellite from the contracting mass is likely to affect the course of evolution.

It is not possible to treat these questions rigorously, but without some guidance on these points further discussion would be fruitless.

The probable influence of the heterogeneity of the planetary mass on tidal friction has been already discussed, and it has been shown that the case of homogeneity will probably give good indications of the result in the true case. I therefore adhere here also to the hypothesis of homogeneity.

I will begin with (α) and consider—

The relative efficiency of solar tidal friction.

The rate at which the rotation of any one of the planets is being reduced is $\tau^2(n-\Omega)/\mathfrak{g}\mathfrak{p}$, where n, $\mathfrak{g}$, $\mathfrak{p}$ refer to the planet, and are the quantities which were previously indicated by the same symbols accented.

τ is $\frac{3}{2}M/c^3$, and therefore varies as Ω^2. With all the planets (excepting, perhaps, Mercury and Venus, according to Herschel) Ω is small compared with n, and we may write n for $n-\Omega$.

It has been already shown that $\mathfrak{p} = \dfrac{3}{19 \times 4\pi}\dfrac{g^2}{\mu v}$, and $\mathfrak{g} = \frac{2}{5}\dfrac{g}{a}$.

$$\text{Hence} \qquad \mathfrak{p}\mathfrak{g} = \frac{2 \times 3}{5 \times 19 \times 4\pi}\frac{g^3}{\mu a v} = \frac{2 \times 3}{5 \times 19 \times 4\pi}\frac{\mu^2 m^3}{a^7 v}$$

Therefore the rate of reduction of planetary rotation is proportional to $\Omega^4 a^7 n m^{-3} v$.

The coefficient of friction v is quite unknown, but we shall obtain indications of the relative importance of tidal retardation in the several planets by supposing v to be the same in all. If we multiply this expression by ma^2, we obtain an expression to which the rate of reduction of rotational momentum is proportional. By means of the data used in the preceding section I find the following results.

TABLE III.

Planet	Number to which tidal retardation is proportional	Number to which rate of destruction of rotational momentum is proportional
Mercury . . .	1000· (?)	9·1 (?)
Venus . . .	11· (?)	8·1 (?)
Earth . . .	1·	1·0
Mars	·89	·026
Jupiter . . .	·00005	2·3
Saturn . . .	·000020 to ·000066	·11 to ·54

This table only refers to solar tidal friction, and the numbers are computed on the hypothesis of the identity of the viscosity for all the planets.

The figures attached to Mercury and Venus are open to much doubt. Perhaps the most interesting point in this table is that the rate of solar tidal retardation of Mars is nearly equal to that of the earth, notwithstanding the comparative closeness of the latter to the sun. The significance of these figures will be commented on below.

I shall now consider—

(β) *The manner in which solar tidal friction and the contraction of the planetary nebula work together.*

It will be supposed that the contraction is the more important feature, so that the acceleration of rotation due to contraction is greater than the retardation due to tidal friction.

Let h be the rotational momentum of the planet at any time; then

$$Cn = h \text{ or } n = \tfrac{5}{2}\frac{h}{ma^2} \quad\text{.............................}(29)$$

In accordance with the above supposition h is a quantity which diminishes slowly in consequence of tidal friction, and a diminishes in consequence of contraction, at such a rate that dn/dt is positive.

We also have
$$\mathfrak{pg} = \frac{2\times 3}{19\times 5\times 4\pi}\frac{g^3}{\mu v a}$$

The rate of change in the dimensions of the planet's orbit about the sun remains insensible, so that τ and Ω may be treated as constant.

Then the rate of loss of rotational momentum of the planetary mass is $Cn\frac{\tau^2}{\mathfrak{gp}}\left(1-\frac{\Omega}{n}\right)$. By the above transformations we see that this expression varies as $\frac{hva}{g^3}\left(1-\frac{2}{5}\frac{\Omega ma^2}{h}\right)$.

But $m=\frac{4}{3}\pi wa^3$, and therefore $a=\left(\frac{3m}{4\pi}\right)^{\frac{1}{3}}w^{-\frac{1}{3}}$. Also $g=\mu m/a^2$.

On substituting this expression for g, and then replacing a throughout by its expression in terms of w, we see that, on omitting constant factors, the rate of loss of rotational momentum varies as $\frac{hv}{w^{\frac{7}{3}}}-\frac{kv}{w^3}$, where $k=\frac{2}{5}\left(\frac{3}{4\pi}\right)^{\frac{2}{3}}\Omega m^{\frac{5}{3}}$, a constant.

From (29) we see that if h varies as a^2, or as $w^{-\frac{2}{3}}$, n the angular velocity of rotation remains constant.

If therefore we suppose h to vary as some power of a less than 2 (which power may vary from time to time) we represent the hypothesis that the contraction causes more acceleration of rotation than tidal friction causes retardation. Let us suppose then that $h=\mathrm{H}w^{-\frac{2}{3}+\beta}$ where β is less than $\frac{2}{3}$, and varies from time to time.

Then the rate of loss of momentum varies as

$$vw^{-3}(\mathrm{H}w^{\beta}-k)$$

In order to determine the rate of loss of rotation we must divide this expression by C, which varies as $w^{-\frac{2}{3}}$.

Therefore the rate of loss of angular velocity of rotation varies as

$$vw^{-\frac{7}{3}}(\mathrm{H}w^{\beta}-k)$$

In order to determine how tidal friction and contraction co-operate it is necessary to adopt some hypothesis concerning the coefficient of friction v.

So long as the tides consist of a bodily distortion, the coefficient of friction must be some function of the density, and will certainly increase as the density increases.

Now if, as regards the loss of momentum, v varies as a power less than the cube of the density, the first factor vw^{-3} diminishes as the density increases; and if, as regards the loss of rotation, v varies as a power less than $\frac{7}{3}$ of the density, the first factor $vw^{-\frac{7}{3}}$ diminishes as the density increases.

As the cube and $\frac{7}{3}$ powers both represent very rapid increments of the coefficient of friction with increase of density, it is probable that the first factor in both expressions diminishes as the contraction of the planetary mass proceeds.

Now consider the second factor $Hw^\beta - k$, which corresponds to the factor $n - \Omega$ in the expression for the rate of loss of rotational momentum; Hw^β is large compared with k so long as the rotation of the planet is fast compared with the orbital motion of the planet about the sun, and since this factor is always positive, it always increases as the contraction increases.

For planets remote from the sun, where contraction has played by far the more important part, β will be very nearly equal to $\frac{2}{3}$, and for those nearer to the sun β will be small (or it might be negative if the tidal retardation exceeds the contractional acceleration).

We thus have one factor always increasing and the other always diminishing, and the importance of the increasing factor is greater for planets remote from than for those near to the sun.

If β be small it is difficult to say how the two rates will vary as the contraction proceeds. But if β does not differ very much from $\frac{2}{3}$ both rates are probably initially small, rise to a maximum and then diminish.

Hence it may be concluded as probable that in the history of a contracting planetary mass, which is sufficiently far from the sun to allow contraction to be a more important factor than tidal friction, both the rate of loss of rotational momentum and of loss of rotation, due to solar tidal friction, were initially small, rose to a maximum and then diminished.

These considerations are important as showing that the efficiency of solar tidal friction was probably greater in the past than at present.

We now come to (γ)—

The effect of the genesis of a satellite on the evolution.

This subject is necessarily in part obscure, and the conclusions must be, in so far, open to doubt.

When a satellite separates from a planetary mass, it seems probable that that part of the planetary mass, which before the change had the greatest angular momentum, is lost by the planet. Hence the rotational momentum of the planet suffers a diminution, and the mass is also diminished. An inspection of the expressions in the last paragraphs shows that it is probable that the loss of a satellite diminishes the rate of loss of planetary rotational momentum, but slightly increases the rate of loss of rotation due to solar tidal friction.

Now if the satellite be large the effect of the tides raised by the satellite in the planet is to cause a much more powerful reduction of planetary rotation than was effected by the sun. The rotational momentum thus removed from the planet reappears in the orbital momentum of its satellite. And the reduction of rotation of the planet causes a reduction of rate of solar tidal effects, by diminishing the angular velocity of the planet's rotation relatively to the sun.

The first and immediate effect of the separation of a satellite is no doubt highly speculative, but the second effect seems to follow undoubtedly, whatever be the mode of separation of the satellite.

From these considerations we may conclude that the effect of the separation of a satellite is to destroy planetary rotation, but to preserve angular momentum within the planetary sub-system.

Hence we ought to find that those planets which have large satellites have a slow rotation, but have a relatively large amount of angular momentum within their systems.

A proper method of comparison between the several planets is difficult of attainment, but these ideas seem to agree with the fact that the earth, which is large compared with Mars, rotates in the same time, but that the whole angular momentum of earth and moon is large*.

* A method of comparing the various members of the solar system has occurred to me, but it is not founded on rigorous argument.

It seems probable that the small density of the larger planets is due to their not being so far advanced in their evolution as the smaller ones, and it is likely that they are continuing to contract and will some day be as dense as the earth.

The proposed method of comparison is to estimate how fast each of the planets must rotate if, with their actual rotational momenta, they were as condensed as the earth, and had the same law of internal density.

The period of this rotation may be called the "effective period."

With the data used above, taking the earth's mean density as unity, the mean density of Mars is ·675, that of Jupiter ·235, that of Saturn ·125 or ·111 or ·074, according to the data used.

To condense these planets we must reduce their radii in the proportion of the cube-roots of these numbers.

Their actual moments of inertia must be reduced by multiplying by the $\frac{2}{3}$rd power of these numbers, and as we suppose the law of internal density to be the same as in the earth, the moments of inertia of Jupiter and Saturn must be also increased in the proportion ·33438 to ·26138.

Then the "effective period" will be the actual period reduced by the same factors as have been given for reducing the moments of inertia.

In this way I find that the Martian day is to be divided by 1·3; the Jovian day by 2; and the Saturnian day by 3·14 to 4·44 according to the data adopted. The earth's day of course remains unchanged.

The following table gives the results.

TABLE IV.

Planet	Actual period of rotation	Effective period of rotation
Earth	$23^h\ 56^m$	$23^h\ 56^m$
Mars	$24^h\ 37^m$	19^h
Jupiter	$9^h\ 55^m$	5^h
Saturn	$10^h\ 29^m$	$3^h\ 20^m$ to $2^h\ 20^m$

§ 9. *General discussion and summary.*

According to the nebular hypothesis the planets and the satellites are portions detached from contracting nebulous masses. In the following discussion I shall accept that hypothesis in its main outline, and shall examine what modifications are necessitated by the influence of tidal friction*.

In § 7 it is shown that the reaction of the tides raised in the sun by the planets must have had a very small influence in changing the dimensions of the planetary orbits round the sun, compared with the influence of the tides raised in the planets by the sun.

From a consideration of numerical data with regard to the solar system and the planetary sub-systems, it appears improbable that the planetary orbits have been much enlarged by tidal friction, since the origin of the several planets. But it is possible that part of the eccentricities of the planetary orbits is due to this cause.

We must therefore examine the several planetary sub-systems for the effects of tidal friction.

From arguments similar to those advanced with regard to the solar system as a whole, it appears unlikely that the satellites of Mars, Jupiter, and Saturn originated very much nearer the present surfaces of the planets than we now observe them. But the data being insufficient, we cannot feel sure that the alteration in the dimensions of the orbits of these satellites has not been considerable. It remains, however, nearly certain that they cannot have first

---

This seems to me to illustrate the arguments used above. For there should in general be a diminution of effective period as we recede from the sun.

It will be noted that the earth, although ten times larger than Mars, has a longer effective period. The larger masses should proceed in their evolution slower than the smaller ones, and therefore the greater proximity of the earth to the sun does not seem sufficient to account for this, more especially as it is shown above that the efficiency of solar tidal friction is of about the same magnitude for the two planets. It is explicable however by the considerations in the text, for it was there shown that a large satellite was destructive of planetary rotation.

If we estimate how fast the earth must rotate in order that the whole internal momentum of moon and earth should exist in the form of rotational momentum, then we find an effective period for the earth of $4^h\ 12^m$. This again illustrates what was stated above, viz.: that a large satellite is preservative of the internal momentum of the planet's system.

The orbital momentum of the satellites of the other planets is so small, that an effective period for the other planets, analogous to the $4^h\ 12^m$ of the earth, would scarcely differ sensibly from the periods given in the table.

If Jupiter and Saturn will ultimately be as condensed as the earth, then it must be admitted as possible or even probable that Saturn (and perhaps Jupiter) will at some future time shed another satellite; for the efficiency of solar tidal friction at the distance of Saturn is small, and a period of two or three hours gives a very rapid rotation.

* [Since the date of this paper the nebular hypothesis has been much criticised, and it is now hardly possible to accept it in the form which was formerly held to be satisfactory. It has not, however, been thought expedient to modify this discussion.]

originated almost in contact with the present surfaces of the planets, in the same way as, in previous papers, has been shown to be probable with regard to the moon and earth.

The numerical data in Table II., § 7, exhibit so striking a difference between the terrestrial system and those of the other planets, that, even apart from the considerations adduced in this and previous papers, we should have grounds for believing that the modes of evolution have been considerably different.

This series of investigations shows that the difference lies in the genesis of the moon close to the present surface of the planet, and we shall see below that solar tidal friction may be assigned as a reason to explain how it happened that the terrestrial planet had contracted to nearly its present dimensions before the genesis of a satellite, but that this was not the case with the exterior planets.

The numbers given in Table III., § 8, show that the efficiency of solar tidal friction is very much greater in its action on the nearer planets than on the further ones. But the total amount of rotation of the various planetary masses destroyed from the beginning cannot be at all nearly proportional to the numbers given in that table, for the more remote planets must be much older than the nearer ones, and the time occupied by the contraction of the solar nebula from the dimensions of the orbit of Saturn down to those of the orbit of Mercury must be very long. Hence the time during which solar tidal friction has been operating on the external planets must be very much longer than the period of its efficiency for the interior ones, and a series of numbers proportional to the total amount of rotation destroyed in the several planets would present a far less rapid decrease, as we recede from the sun, than do the numbers given in Table III. Nevertheless the disproportion between these numbers is so great that it must be admitted that the effect produced by solar tidal friction on Jupiter and Saturn has not been nearly so great as on the interior planets.

In § 8 it has been shown to be probable that, as a planetary mass contracts, the rate of tidal retardation of rotation, and of destruction of rotational momentum increases, rises to a maximum, and then diminishes. This at least is so, when the acceleration of rotation due to contraction exceeds the retardation due to tidal friction; and this must in general have been the case. Thus we may suppose that the rate at which solar tidal friction has retarded the planetary rotations in past ages was greater than the present rate of retardation, and indeed there seems no reason why many times the present rotational momenta of the planets should not have been destroyed by solar tidal friction. But it remains very improbable that so large an amount of momentum should have been destroyed as to affect the orbits of the planets round the sun materially.

I will now proceed to examine how the differences of distance from the sun would be likely to affect the histories of the several planetary masses.

According to the nebular hypothesis a planetary nebula contracts, and rotates quicker as it contracts. The rapidity of the revolution causes its form to become unstable, or, perhaps a portion gradually detaches itself; it is immaterial which of these two really takes place. In either case the separation of that part of the mass, which before the change had the greatest angular momentum, permits the central portion to resume a planetary shape. The contraction and increase of rotation proceed continually until another portion is detached, and so on. There thus recur at intervals a series of epochs of instability or of abnormal change.

Now tidal friction must diminish the rate of increase of rotation due to contraction, and therefore if tidal friction and contraction are at work together, the epochs of instability must recur more rarely than if contraction acted alone.

If the tidal retardation is sufficiently great, the increase of rotation due to contraction will be so far counteracted as never to permit an epoch of instability to occur.

Now the rate of solar tidal frictional retardation decreases rapidly as we recede from the sun, and therefore these considerations accord with what we observe in the solar system.

For Mercury and Venus have no satellites, and there is a progressive increase in the number of satellites as we recede from the sun. Moreover, the number of satellites is not directly connected with the mass of the planet, for Venus has nearly the same mass as the earth and has no satellite, and the earth has relatively by far the largest satellite of the whole system. Whether this be the true cause of the observed distribution of satellites amongst the planets or not, it is remarkable that the same cause also affords an explanation, as I shall now show, of that difference between the earth with the moon, and the other planets with their satellites, which has caused tidal friction to be the principal agent of change with the former but not with the latter.

In the case of the contracting terrestrial mass we may suppose that there was for a long time nearly a balance between the retardation due to solar tidal friction and the acceleration due to contraction, and that it was not until the planetary mass had contracted to nearly its present dimensions that an epoch of instability could occur.

It may also be noted that if there be two equal planetary masses which generate satellites, but under very different conditions as to the degree of condensation of the masses, then the two satellites so generated would be likely to differ in mass; we cannot of course tell which of the two planets would generate the larger satellite. Thus if the genesis of the moon was deferred until a late epoch in the history of the terrestrial mass, the mass of the moon relatively to the earth, would be likely to differ from the mass of other satellites relatively to their planets.

If the contraction of the planetary mass be almost completed before the genesis of the satellite, tidal friction, due jointly to the satellite and to the sun, will thereafter be the great cause of change in the system, and thus the hypothesis that it is the sole cause of change will give an approximately accurate explanation of the motion of the planet and satellite at any subsequent time.

That this condition is fulfilled in the case of the earth and moon, I have endeavoured to show in the previous papers of this series.

At the end of the last of those papers the systems of the several planets were reviewed from the point of view of the present theory. It will be well to recapitulate shortly what was there stated and to add a few remarks on the modifications and additions introduced by the present investigation.

The previous papers were principally directed to the case of the earth and moon, and it was there found that the primitive condition of those bodies was as follows:—the earth was rotating, with a period of from two to four hours, about an axis inclined at 11° or 12° to the normal to the ecliptic, and the moon was revolving, nearly in contact with the earth, in a circular orbit coincident with the earth's equator, and with a periodic time only slightly exceeding that of the earth's rotation.

Then it was proved that lunar and solar tidal friction would reduce the system from this primitive condition down to the state which now exists by causing a retardation of terrestrial rotation, an increase of lunar period, an increase of obliquity of ecliptic, an increase of eccentricity of lunar orbit, and a modification in the plane of the lunar orbit too complex to admit of being stated shortly.

It was also found that the friction of the tides raised by the earth in the moon would explain the present motion of the moon about her axis, both as regards the identity of the axial and orbital revolutions, and as regards the direction of her polar axis.

Thus the theory that tidal friction has been the ruling power in the evolution of the earth and moon completely coordinates the present motions of the two bodies, and leads us back to an initial state when the moon first had a separate existence as a satellite.

This initial configuration of the two bodies is such that we are almost compelled to believe that the moon is a portion of the primitive earth detached by rapid rotation or other causes.

There may be some reason to suppose that the earliest form in which the moon had a separate existence was in the shape of a ring, but this annular condition precedes the condition to which the dynamical investigation leads back.

The present investigation shows, in confirmation of preceding ones*, that at this origin of the moon the earth had a period of revolution about the sun shorter than at present by perhaps only a minute or two, and it also shows that since the terrestrial planet itself first had a separate existence the length of the year can have increased but very little—almost certainly by not so much as an hour, and probably by not more than five minutes†.

With regard to the 11° or 12° of obliquity which still remains when the moon and earth are in their primitive condition, it may undoubtedly be partly explained by the friction of the solar tides before the origin of the moon, and perhaps partly also by the simultaneous action of the ordinary precession and the contraction and change of ellipticity of the nebulous mass‡.

In the review referred to I examined the eccentricities and inclinations of the orbits of the several other satellites, and found them to present indications favourable to the theory. In the present paper I have given reasons for supposing that the tidal friction arising from the action of the other satellites on their planets cannot have had so much effect as in the case of the earth. That those indications were not more marked, and yet seemed to exist, agrees well with this last conclusion.

The various obliquities of the planets' equators to their orbits were also considered, and I was led to conclude that the axes of the planets from Jupiter inwards were primitively much more nearly perpendicular to their orbits than at present. But the case of Saturn and still more that of Uranus (as inferred from its satellites) seem to indicate that there was a primitive obliquity at the time of the genesis of the planets, arising from causes other than those here considered.

The satellites of the larger planets revolve with short periodic times; this admits of a simple explanation, for the smallness of the masses of these satellites would have prevented tidal friction from being a very efficient cause of change in the dimensions of their orbits, and the largeness of the planets' masses would have caused them to proceed slowly in their evolution.

* "Precession," § 19 [p. 105].

† If the change has been as much as an hour the rotational momentum of the earth destroyed by solar tidal friction must have been 33 times the present total internal momentum of moon and earth. For the orbital momentum of a planet varies as the cube root of its periodic time, and if we differentiate logarithmically we obtain the increment of periodic time in terms of the increment of orbital momentum. Taking the numerical data from Tables I. and II. we see that this statement is proved by the fact that 3×33 times $[216 \div \cdot 01720 \times 10^{10}] \times 365{\cdot}25 \times 24$ is very nearly equal to unity.

‡ See a paper "On a Suggested Explanation of the Obliquity of Planets to their Orbits," *Phil. Mag.*, March, 1877. [See Vol. III.] See however § 21 "Precession" [p. 110].

If the planets be formed from chains of meteorites or of nebulous matter the rotation of the planets has arisen from the excess of orbital momentum of the exterior over that of the interior matter. As we have no means of knowing how broad the chain may have been in any case, nor how much it may have closed in on the sun in course of concentration, we have no means of computing the primitive angular momentum of a planet. A rigorous method of comparison of the primitive rotations of the several planets is thus wanting.

If however the planets were formed under similar conditions, then, according to the present theory, we should expect to find the exterior planets now rotating more rapidly than the interior ones. It has been shown above (see Table IV., note to § 8) that, on making allowance for the different degrees of concentration of the planets, this is the case.

That the interior satellite of Mars revolves with a period of less than a third of its planet's rotation is perhaps the most remarkable fact in the solar system. The theory of tidal friction explains this perfectly*, and we find that this will be the ultimate fate of all satellites, because the solar tidal friction retards the planetary rotation without directly affecting the satellite's orbital motion.

The numerical comparison in Table III. shows that the efficiency of solar tidal friction in retarding the terrestrial and Martian rotations is of about the same degree of importance, notwithstanding the much greater distance of the planet Mars.

From the discussion in this paper it will have been apparent that the earth and moon do actually differ from the other planets in such a way as to permit tidal friction to have been the most important factor in their history.

By an examination of the probable effects of solar tidal friction on a contracting planetary mass, we have been led to assign a cause for the observed distribution of satellites in the solar system, and this again has itself afforded an explanation of how it happened that the moon so originated that the tidal friction of the lunar tides in the earth should have been able to exercise so large an influence.

* It is proper to remark that the rapid revolution of this satellite might perhaps be referred to another cause, although the explanation appears very inadequate.

It has been pointed out above that the formation of a satellite out of a chain or ring of matter must be accompanied by a diminution of periodic time and of distance. Thus a satellite might after formation have a shorter periodic time than its planet.

If this, however, were the explanation, we should expect to find other instances elsewhere, but the case of the Martian satellite stands quite alone.

[Since the date of this paper the work of several investigators seems to indicate that the earliest form of a satellite may not be annular. The papers which will be reproduced in Vol. III. tend also in this direction.]

In this summary I have endeavoured not only to set forth the influence which tidal friction may, and probably has had in the history of the system, but also to point out what effects it cannot have produced.

The present investigations afford no grounds for the rejection of the nebular hypothesis, but while they present evidence in favour of the main outlines of that theory, they introduce modifications of considerable importance.

Tidal friction is a cause of change of which Laplace's theory took no account*, and although the activity of that cause is to be regarded as mainly belonging to a later period than the events described in the nebular hypothesis, yet its influence has been of great, and in one instance of even paramount importance in determining the present condition of the planets and their satellites.

* Note added on July 28, 1881.

Dr T. R. Mayer appears to have been amongst the first to draw attention to the effects of tidal friction. I have recently had my attention called to his paper on "Celestial Dynamics" [Translation, *Phil. Mag.*, 1863, Vol. xxv., pp. 241, 387, 417], in which he has preceded me in some of the remarks made above. He points out that, as the joint result of contraction and tidal friction, "the whole life of the earth therefore may be divided into three periods—youth with increasing, middle age with uniform, and old age with decreasing velocity of rotation."

9.

ON THE STRESSES CAUSED IN THE INTERIOR OF THE EARTH BY THE WEIGHT OF CONTINENTS AND MOUNTAINS.

[*Philosophical Transactions of the Royal Society*, Vol. 173 (1882), pp. 187—230, with which is incorporated "Note on a previous paper," *Proc. Roy. Soc.* Vol. 38 (1885), pp. 322—328.]

[ABOUT a year after the publication of this paper in the *Transactions*, an essay was submitted to me at Cambridge by Mr Charles Chree, now Director of the Kew Observatory, in which an error in my procedure was pointed out. As Mr Chree's treatment of the problem was quite different from mine I wrote a "Note" for the *Proceedings* to correct my mistake in accordance with his criticism. The investigation and the correction are now fused together. This has necessitated the re-writing of certain portions of the paper, and the new matter is indicated by square parentheses. Parts of the "Note" are also inserted in their proper places so as to make good the defects in the original investigation.]

TABLE OF CONTENTS.

	PAGE
Introduction	460
Part I. THE MATHEMATICAL INVESTIGATION.	
§ 1. On the state of internal stress of a strained elastic sphere	460
§ 2. The determination of the stresses when the disturbing potential is an even zonal harmonic	465
§ 3. On the direction and magnitude of the principal stresses in a strained elastic solid	472
§ 4. The application of the previous analysis to the determination of the stresses produced by the weight of superficial inequalities	474
§ 5. The state of stress due to ellipticity of figure or to tide-generating forces	476
§ 6. On the stresses due to a series of parallel mountain chains	481
§ 7. On the stresses due to the even zonal harmonic inequalities	484
§ 8. On the stresses due to the weight of an equatorial continent	490
§ 9. On the strength of various substances	494
§ 10. On the case when the elastic solid is compressible	497
Part II. SUMMARY AND DISCUSSION	501

In this paper I have considered the subject of the solidity and strength of the materials of which the earth is formed, from a point of view from which it does not seem to have been hitherto discussed.

The first part of the paper is entirely devoted to a mathematical investigation, based upon a well-known paper of Sir William Thomson's. The second part consists of a summary and discussion of the preceding work. In this I have tried, as far as possible, to avoid mathematics, and I hope that a considerable part of it may prove intelligible to the non-mathematical reader.

I.

Mathematical Investigation.

[§ 1. *On the state of internal stress of a strained elastic sphere.*

In this section it is proposed to find the solutions of two closely analogous problems.

(i) Consider a homogeneous elastic sphere of density w and radius a, for which $\omega - \frac{1}{3}\upsilon$ is the modulus of incompressibility and υ the modulus of rigidity*. Let the system be referred to rectangular axes x, y, z with origin at centre, and let r be radius vector. It is required to find the state of strain of the sphere when it is subject to superficial normal traction defined by $W_i a^i/r^i$, where W_i is a solid harmonic of degree i.

(ii) Consider the same sphere, and let it be free from any superficial forces but subject to forces acting throughout the whole mass such that the force acting on unit volume is that due to a gravitation potential W_i. It is required to determine the state of strain.

I begin with the first problem.

Lord Kelvin (Sir William Thomson) has shown that the component strains α, β, γ of such a sphere, when subject to superficial tractions whose three components are A_i, B_i, C_i (being surface harmonics of order i), are given by

$$\alpha = \frac{1}{\upsilon a^{i-1}} \left\{ \frac{\omega}{2K}(a^2 - r^2)\frac{d\Psi_{i-1}}{dx} + \frac{(i+2)\,\omega - (2i-1)\,\upsilon}{(i-1)(2i+1)\,K} r^{2i+1} \frac{d}{dx}(\Psi_{i-1} r^{-2i+1}) + \frac{1}{2i(i-1)(2i+1)} \frac{d\Phi_{i+1}}{dx} + \frac{1}{i-1} A_i r^i \right\}\dagger \quad \text{..........(1)}$$

* The phraseology adopted by Thomson and Tait (*Natural Philosophy*, first edition) and others seems a little unfortunate. One might be inclined to suppose that compressibility and rigidity were things of the same nature; but rigidity and the reciprocal of compressibility are of the same kind. If one may give exact meanings to old words of somewhat general meanings, then one may pair together compressibility and "pliancy," and call the moduli for the two sorts of elasticity the "incompressibility" and rigidity.

† Thomson and Tait's *Natural Philosophy*, § 737 (52).

where for brevity we write K for $(2i^2+1)\,\omega - (2i-1)\,\upsilon$, and where Ψ, Φ are auxiliary functions defined by

$$\Psi_{i-1} = \frac{d}{dx}\,(A_i r^i) + \frac{d}{dy}\,(B_i r^i) + \frac{d}{dz}\,(C_i r^i)$$

$$\Phi_{i+1} = r^{2i+3}\left\{\frac{d}{dx}\,(A_i r^{-i-1}) + \frac{d}{dy}\,(B_i r^{-i-1}) + \frac{d}{dz}\,(C_i r^{-i-1})\right\}$$

The components β, γ are found from α by cyclical changes of letters.

In the case of problem (i) the three components of superficial force are $xW_i a^{i-1}/r^i$, $yW_i a^{i-1}/r^i$, $zW_i a^{i-1}/r^i$.

Since

$$\frac{xW_i}{r^{i+1}} = \frac{1}{2i+1}\left[r^{-i+1}\frac{dW_i}{dx}\right] - \frac{1}{2i+1}\left[r^{i+2}\frac{d}{dx}\,(W_i r^{-2i-1})\right]$$

and since these two terms are surface harmonics of orders $i-1$ and $i+1$, and since x, y, z is a point on the surface of the sphere, we have

$$xW_i \frac{a^{i-1}}{r^i} = A_{i-1} + A_{i+1}$$

where

$$A_{i-1} = \frac{a^i}{2i+1}\left[r^{-i+1}\frac{dW_i}{dx}\right], \quad A_{i+1} = -\frac{a^i}{2i+1}\left[r^{i+2}\frac{d}{dx}\,(W_i r^{-2i-1})\right]$$

with similar expressions for B_{i-1}, B_{i+1}, C_{i-1}, C_{i+1}.

The corresponding auxiliary functions are easily found to be

$$\Psi_{i-2} = 0, \quad \Phi_i = -\frac{i(2i-1)}{2i+1}\,a^i W_i, \quad \Psi_i = \frac{(i+1)(2i+3)}{2i+1}\,a^i W_i, \quad \Phi_{i+2} = 0$$

These must be substituted in (1), and α is the sum of the two values found when i of the formula becomes $i-1$ and $i+1$. It will be noticed that the function K will only occur in the form in which the i of (1) becomes $i+1$; thus henceforth I write

$$K = [2(i+1)^2+1]\,\omega - (2i+1)\,\upsilon$$

and as a further abbreviation write

$$I = 2(i+1)^2 + 1$$

$$\left.\right\}\quad \dots\dots\dots\dots(2)$$

On completing the substitutions indicated, we find the solution to be

$$\alpha = (Ea^2 - Fr^2)\frac{dW_i}{dx} - Gr^{2i+3}\frac{d}{dx}\,(W_i r^{-2i-1})$$

where

$$E = \frac{i(i+2)\,\omega - \upsilon}{2(i-1)\,K\upsilon}, \quad F = \frac{(i+1)(2i+3)\,\omega}{2(2i+1)\,K\upsilon}, \quad G = \frac{i\omega + (2i+1)\,\upsilon}{(2i+1)\,K\upsilon}$$

$$\left.\right\}\quad \dots\dots\dots(3)$$

Throughout the greater part of the present paper the elastic solid will be treated as incompressible, and it will be convenient to proceed at once to arrange this solution in such a form that the portion of the whole in which compressibility plays a part shall be separated from the rest. This may be effected as follows :—

By means of (2) we have

$$\omega = \frac{1}{I}[K + (2i + 1)\upsilon]$$

Thus ω may be eliminated, except in so far as it is involved in K.

On substitution in (3) we find

$$\alpha = \frac{1}{2I\upsilon}\left[\frac{i(i+2)}{(i-1)}a^2 - \frac{(i+1)(2i+3)}{(2i+1)}r^2\right]\frac{dW_i}{dx} - \frac{i}{(2i+1)I\upsilon}r^{2i+3}\frac{d}{dx}(W_i r^{-2i-1})$$
$$+ \frac{1}{2K}\left(1 + \frac{i}{I}\right)\left[(a^2 - r^2)\frac{dW_i}{dx} - 2r^{2i+3}\frac{d}{dx}(W_i r^{-2i-1})\right] \quad\dots\dots\dots\dots(4)$$

Another and more useful form may be obtained from (4) by completing the differentiations in the second terms of each line; thus we find

$$\alpha = \frac{1}{2I\upsilon}\left[\frac{i(i+2)}{i-1}a^2 - (i+3)r^2\right]\frac{dW_i}{dx} + \frac{i}{I\upsilon}xW_i$$
$$+ \frac{1}{2K}\left(1 + \frac{i}{I}\right)\left[(a^2 - 3r^2)\frac{dW_i}{dx} + 2(2i+1)xW_i\right]\dots\dots\dots(5)$$

When the solid is incompressible ω becomes infinite compared with υ, and K also becomes infinite. Thus the second line, which represents the effect of compressibility, will disappear.

Let P, Q, R, S, T, U be the six stresses, across three planes mutually at right angles to one another at the point x, y, z, estimated as is usual in works on the theory of elasticity; and let P, Q, R be tractions and not pressures.

Then if we write $$\delta = \frac{d\alpha}{dx} + \frac{d\beta}{dy} + \frac{d\gamma}{dz} \dots\dots\dots\dots\dots\dots(6)$$

so that δ is the dilatation, we have by the usual formula*

$$\left.\begin{aligned} P &= (\omega - \upsilon)\delta + 2\upsilon\frac{d\alpha}{dx} \\ T &= \upsilon\left(\frac{d\alpha}{dz} + \frac{d\gamma}{dx}\right) \end{aligned}\right\} \dots\dots\dots\dots\dots(7)$$

and the other four stresses are expressed from these by the proper cyclic changes of letters.

---

* Thomson and Tait's *Natural Philosophy*, § 693.

In order to determine the stresses it is necessary to find the differential coefficients of the displacements with respect to x, y, z, and thence to determine δ.

By means of the properties of harmonic and of homogeneous functions we find

$$\delta = \frac{d\alpha}{dx} + \frac{d\beta}{dy} + \frac{d\gamma}{dz}$$

$$= \frac{(i+1)(2i+3)}{K} W_i \quad \text{.......................(8)}$$

We also have $$\omega - \upsilon = \frac{K}{I} - \frac{2(i^2+i+1)}{I}\upsilon$$

The final results of the processes indicated are as follows:—

$$P = \frac{1}{I}\left\{\left[\frac{i(i+2)}{i-1}a^2 - (i+3)r^2\right]\frac{d^2W_i}{dx^2} - 6x\frac{dW_i}{dx} + (3i+I)W_i\right\}$$

$$+ \frac{\upsilon}{K}\left(1+\frac{i}{I}\right)\left\{(a^2-3r^2)\frac{d^2W_i}{dx^2} + 4(i-1)x\frac{dW_i}{dx} - 2i(i-1)W_i\right\}$$

$$T = \frac{1}{I}\left\{\left[\frac{i(i+2)}{i-1}a^2 - (i+3)r^2\right]\frac{d^2W_i}{dxdz} - 3\left(x\frac{dW_i}{dz} + z\frac{dW_i}{dx}\right)\right\}$$

$$+ \frac{\upsilon}{K}\left(1+\frac{i}{I}\right)\left\{(a^2-3r^2)\frac{d^2W_i}{dxdz} + 2(i-1)\left(x\frac{dW_i}{dz} + z\frac{dW_i}{dx}\right)\right\} \quad \text{......(9)}$$

The other four stresses are determinable by cyclic changes of letters.

When the solid is incompressible the second lines in these formulæ disappear.

This is the required solution when the sphere is subject only to superficial normal stresses defined by $W_i a^i/r^i$.

We now turn to the second problem where there are no superficial stresses, but where there is a force acting throughout the whole sphere defined by the potential W_i.

This problem has been solved by Lord Kelvin*, and the result is

$$\alpha = (La^2 - Mr^2)\frac{d^2W_i}{dx^2} - Nr^{2i+3}\frac{d}{dx}(W_i r^{-2i-1})$$

where

$$L = \frac{i(i+2)\omega - i\upsilon}{2(i-1)K\upsilon}, \quad M = \frac{(i+1)(2i+3)\omega - (2i+1)\upsilon}{2(2i+1)K\upsilon}, \quad N = \frac{i\omega}{(2i+1)K\upsilon}$$

* Thomson and Tait's *Natural Philosophy*, § 834, (8) and (9).

If this solution be transformed in the same way as the last, and if we form δ and P, T as before, we find

$$\left.\begin{aligned}
\alpha &= \frac{1}{2Iv}\left[\frac{i(i+2)}{i-1}a^2-(i+3)r^2\right]\frac{dW_i}{dx}+\frac{i}{Iv}xW_i \\
&\quad+\frac{1}{2K}\left(\frac{i}{I}\right)\left[(a^2-3r^2)\frac{dW_i}{dx}+2(2i+1)xW_i\right] \\
\delta &= \frac{i}{K}W_i;\quad I=2(i+1)^2+1;\quad K=I\omega-(2i+1)v \\
\mathrm{P} &= \frac{1}{I}\left\{\left[\frac{i(i+2)}{i-1}a^2-(i+3)r^2\right]\frac{d^2W_i}{dx^2}-6x\frac{dW_i}{dx}+3iW_i\right\} \\
&\quad+\frac{v}{K}\left(\frac{i}{I}\right)\left\{(a^2-3r^2)\frac{d^2W_i}{dx^2}+4(i-1)x\frac{dW_i}{dx}-2i(i-1)W_i\right\} \\
\mathrm{T} &= \frac{1}{I}\left\{\left[\frac{i(i+2)}{i-1}a^2-(i+3)r^2\right]\frac{d^2W_i}{dxdz}-3\left(x\frac{dW_i}{dz}+z\frac{dW_i}{dx}\right)\right\} \\
&\quad+\frac{v}{K}\left(\frac{i}{I}\right)\left\{(a^2-3r^2)\frac{d^2W_i}{dxdz}+2(i-1)\left(x\frac{dW_i}{dz}+z\frac{dW_i}{dx}\right)\right\}
\end{aligned}\right\} \quad \ldots\ldots(10)$$

As before when the solid is incompressible the second lines in the expressions for α, P and T vanish.

On comparison between (5), (9) and (10) we see that for an incompressible solid α, β, γ and S, T, U are the same for both problems, and $\mathrm{P}-W_i$, $\mathrm{Q}-W_i$, $\mathrm{R}-W_i$ of the first problem are the same as P, Q, R of the second. In other words the solution of the first problem may be derived from that of the second by the addition of a hydrostatic pressure $-W_i$.

For the immediate object in view we adopt the solution of the second problem when the solid is incompressible, and we then have

$$\left.\begin{aligned}
I\mathrm{P} &= \left\{\frac{i(i+2)}{i-1}a^2-(i+3)r^2\right\}\frac{d^2W_i}{dx^2}-6x\frac{dW_i}{dx}+3iW_i \\
I\mathrm{T} &= \left\{\frac{i(i+2)}{i-1}a^2-(i+3)r^2\right\}\frac{d^2W_i}{dxdz}-3\left(x\frac{dW_i}{dz}+z\frac{dW_i}{dx}\right)
\end{aligned}\right\} \ldots\ldots(11)$$

The expressions for Q, R, S, U may be written down from these by means of cyclic changes of the symbols.

In order to find the magnitudes and directions of the principal stresses at any point it would be necessary to solve a cubic equation. The solution of this equation appears to be difficult, but the special case in which it reduces to a quadratic equation will fortunately give adequate results. It may be seen from considerations of symmetry that if W_i be a zonal harmonic, two of the principal stress-axes lie in a meridional plane and the third is perpendicular thereto.

I shall therefore take W_i to be a zonal harmonic, and as the future developments will be by means of series (which though finite will be long for the higher orders of harmonics) I shall attend more especially to the equatorial regions of the sphere.]

§ 2. *The determination of the stresses when the disturbing potential is an even zonal harmonic.*

[In this section the stresses are found from the solution of the second problem of § 1. As pointed out, we are at the same time determining the stresses arising from the solution of the first problem.]

If θ be colatitude the expression for a zonal surface harmonic or Legendre's function of order i is

$$\cos^i\theta - \frac{i(i-1)}{4.(1!)^2}\cos^{i-2}\theta\sin^2\theta + \frac{i(i-1)(i-2)(i-3)}{4^2(2!)^2}\cos^{i-4}\theta\sin^4\theta - \dots$$

or if we begin by the other end of the series, and take i as an even number, the expression is

$$(-)^{\frac{1}{2}i}\frac{i!}{2^i\{\frac{1}{2}i!\}^2}\left[\sin^i\theta - \frac{i^2}{2!}\sin^{i-2}\theta\cos^2\theta + \frac{i^2(i-2)^2}{4!}\sin^{i-4}\theta\cos^4\theta - \dots\right]$$

This latter is the appropriate form when we wish to consider especially the equatorial regions, because $\cos\theta$ is small for that part of the sphere.

There is of course a similar formula when i is odd, but of this I shall make no use.

Let $\rho^2 = x^2 + y^2$, so that $\sin\theta = \rho/r$, $\cos\theta = z/r$.

Then we may put

$$W_i = \rho^i - \frac{i^2}{2!}\rho^{i-2}z^2 + \frac{i^2(i-2)^2}{4!}\rho^{i-4}z^4 - \frac{i^2(i-2)^2(i-4)^2}{6!}\rho^{i-6}z^6 + \dots \quad (12)$$

W_i is a solid zonal harmonic of degree i; but $r^{-i}W_i$ requires multiplication by a factor $(-)^{\frac{1}{2}i}i!/2^i\{\frac{1}{2}i!\}^2$ in order to make it a Legendre's function.

The factors by which W_i must be deemed to be multiplied in order that it may be a potential, will be dropped for the present, to be inserted later. Or we may, if we like, suppose that the units of length or of time are so chosen as to make the factor equal to unity.

Let

$$\beta_0 = 1,\quad \beta_2 = \frac{i^2}{2!},\quad \beta_4 = \frac{i^2(i-2)^2}{4!},\quad \beta_6 = \frac{i^2(i-2)^2(i-4)^2}{6!}\ \text{\&c.} \quad \dots(13)$$

Then, dropping the suffix to W for brevity, we may write

$$W = \beta_0\rho^i - \beta_2\rho^{i-2}z^2 + \beta_4\rho^{i-4}z^4 - \beta_6\rho^{i-6}z^6 + \dots \quad \dots\dots\dots(14)$$

I shall now find P, Q, R, T at any point in the meridional plane which is determined by $y=0$.

In evaluating the first differential coefficients of W we must not put $y=0$, in as far as these coefficients are a first step towards the determination of the second differential coefficients. But in as far as these first coefficients are directly involved in the expressions for P, Q, R, and T, and in the second coefficients in the same expressions, we may put $y=0$, and thus write x in place of ρ. We have further

$$\rho\frac{d\rho}{dx}=x,\quad \rho\frac{d\rho}{dy}=y,\quad \frac{d\rho}{dz}=0,\quad \text{since } \rho^2=x^2+y^2.$$

Then

$$\frac{dW}{dx}=x\,[i\beta_0\rho^{i-2}-(i-2)\,\beta_2\rho^{i-4}z^2+(i-4)\,\beta_4\rho^{i-6}z^4-\ldots]$$

$$\frac{dW}{dy}=y\,[\text{same series}\qquad\qquad\qquad\qquad\qquad]$$

$$\frac{dW}{dz}=z\,[\qquad\quad -2\beta_2\rho^{i-2}+4\beta_4\rho^{i-4}z^2-6\beta_6\rho^{i-6}z^4+\ldots]$$

In differentiating a second time we may treat ρ as identical with x, because y is to be put equal to zero. Thus

$$\left.\begin{aligned}
\frac{d^2W}{dx^2}&=i(i-1)\,\beta_0x^{i-2}-(i-2)(i-3)\,\beta_2x^{i-4}z^2+(i-4)(i-5)\,\beta_4x^{i-6}z^4-\ldots\\
\frac{d^2W}{dy^2}&=i\beta_0x^{i-2}-(i-2)\,\beta_2x^{i-4}z^2+(i-4)\,\beta_4x^{i-6}z^4-\ldots\\
\frac{d^2W}{dz^2}&=-1.2\beta_2x^{i-2}+3.4\beta_4x^{i-4}z^2-5.6\beta_6x^{i-6}z^4+\ldots
\end{aligned}\right\}\quad\ldots\ldots(15)$$

$$\left.\begin{aligned}
\frac{d^2W}{dx\,dz}&=xz\,[-2(i-2)\,\beta_2x^{i-4}+4(i-4)\,\beta_4x^{i-6}z^2-6(i-6)\,\beta_6x^{i-8}z^4+\ldots]\\
\frac{d^2W}{dx\,dy}&=0,\qquad \frac{d^2W}{dy\,dz}=0
\end{aligned}\right\}\quad\ldots\ldots(16)$$

Also treating ρ as identical with x, and putting $y=0$,

$$\left.\begin{aligned}
x\frac{dW}{dx}&=i\beta_0x^i-(i-2)\,\beta_2x^{i-2}z^2+(i-4)\,\beta_4x^{i-4}z^4-\ldots\\
y\frac{dW}{dy}&=0\\
z\frac{dW}{dz}&=-2\beta_2x^{i-2}z^2+4\beta_4x^{i-4}z^4-6\beta_6x^{i-6}z^6+\ldots
\end{aligned}\right\}\quad\ldots\ldots(17)$$

$$\left.\begin{aligned}
\left(z\frac{dW}{dx}+x\frac{dW}{dz}\right)&=xz\,\{(i\beta_0-2\beta_2)\,x^{i-2}-[(i-2)\,\beta_2-4\beta_4]\,x^{i-4}z^2\\
&\qquad\qquad+[(i-4)\,\beta_4-6\beta_6]\,x^{i-6}z^4-\ldots\}\\
\left(y\frac{dW}{dx}+x\frac{dW}{dy}\right)&=0,\qquad \left(y\frac{dW}{dz}+z\frac{dW}{dy}\right)=0
\end{aligned}\right\}\quad\ldots(18)$$

These various results have now to be introduced into the expressions (11) for P, Q, R, S, T, U.

In performing these operations it will be convenient to write J for $i(i+2)/(i-1)$. Also $r^2 = \rho^2 + z^2 = x^2 + z^2$, when $y = 0$.

From these formulæ we see that $S = 0$, $U = 0$; which shows that a meridional plane is one of the three principal planes, a result already observed from principles of symmetry.

Now

$$\left.\begin{aligned} r^2 \frac{d^2 W}{dx^2} &= i(i-1)\beta_0 x^i + [i(i-1)\beta_0 - (i-2)(i-3)\beta_2] x^{i-2} z^2 \\ &\qquad - [(i-2)(i-3)\beta_2 - (i-4)(i-5)\beta_4] x^{i-4} z^4 + \dots \\ r^2 \frac{d^2 W}{dy^2} &= i\beta_0 x^i + [i\beta_0 - (i-2)\beta_2] x^{i-2} z^2 - [(i-2)\beta_2 - (i-4)\beta_4] x^{i-4} z^4 + \dots \\ r^2 \frac{d^2 W}{dz^2} &= -1.2\beta_2 x^i - [1.2\beta_2 - 3.4\beta_4] x^{i-2} z^2 + [3.4\beta_4 - 5.6\beta_6] x^{i-4} z^4 - \dots \end{aligned}\right\} \quad \dots\dots(19)$$

$$\left.\begin{aligned} -2x \frac{dW}{dx} + iW &= -i\beta_0 x^i + (i-4)\beta_2 x^{i-2} z^2 - (i-8)\beta_4 x^{i-4} z^4 + \dots \\ -2y \frac{dW}{dy} + iW &= i\beta_0 x^i - i\beta_2 x^{i-2} z^2 + i\beta_4 x^{i-4} z^4 - \dots \\ -2z \frac{dW}{dz} + iW &= i\beta_0 x^i - (i-4)\beta_2 x^{i-2} z^2 + (i-8)\beta_4 x^{i-4} z^4 - \dots \end{aligned}\right\} \dots(20)$$

$$\begin{aligned} r^2 \frac{d^2 W}{dx\,dz} = xz\,\{&-2(i-2)\beta_2 x^{i-2} - [2(i-2)\beta_2 - 4(i-4)\beta_4] x^{i-4} z^2 \\ &+ [4(i-4)\beta_4 - 6(i-6)\beta_6] x^{i-6} z^4 - \dots\} \quad \dots(21) \end{aligned}$$

Then multiplying (19) by $-(i+3)$, (20) by 3, and (15) by Ja^2, and adding them each to each, we get the expressions for P, Q, R.

Also multiplying (21) by $-(i+3)$, (18) by -3, and (16) by Ja^2 and adding, we get the expression for T. The results are

$$\begin{aligned} I\mathrm{P} = &- [(i+3)i(i-1) + 3i]\beta_0 x^i \\ &+ [\{(i+3)(i-2)(i-3) + 3(i-4)\}\beta_2 - (i+3)i(i-1)\beta_0] x^{i-2} z^2 \\ &- [\{(i+3)(i-4)(i-5) + 3(i-8)\}\beta_4 - (i+3)(i-2)(i-3)\beta_2] x^{i-4} z^4 \\ &+ [\{(i+3)(i-6)(i-7) + 3(i-12)\}\beta_6 \\ &\qquad\qquad - (i+3)(i-4)(i-5)\beta_4] x^{i-6} z^6 - \dots \\ &+ Ja^2 [i(i-1)\beta_0 x^{i-2} - (i-2)(i-3)\beta_2 x^{i-4} z^2 + (i-4)(i-5)\beta_4 x^{i-6} z^4 - \dots] \end{aligned}$$

$$\begin{aligned} I\mathrm{R} = &[(i+3).1.2\beta_2 + 3i\beta_0] x^i - [(i+3).3.4\beta_4 - \{(i+3).1.2 - 3(i-4)\}\beta_2] x^{i-2} z^2 \\ &+ [(i+3).5.6\beta_6 - \{(i+3).3.4 - 3(i-8)\}\beta_4] x^{i-4} z^4 \\ &- [(i+3).7.8\beta_8 - \{(i+3).5.6 - 3(i-12)\}\beta_6] x^{i-6} z^6 + \dots \\ &- Ja^2 [1.2\beta_2 x^{i-2} - 3.4\beta_4 x^{i-4} z^2 + 5.6\beta_6 x^{i-6} z^4 - \dots] \end{aligned}$$

$$IQ = -[(i+3)i - 3i]\beta_0 x^i + [\{(i+3)(i-2) - 3i\}\beta_2 - (i+3)i\beta_0]x^{i-2}z^2$$
$$-[\{(i+3)(i-4) - 3i\}\beta_4 - (i+3)(i-2)\beta_2]x^{i-4}z^4$$
$$+[\{(i+3)(i-6) - 3i\}\beta_6 - (i+3)(i-4)\beta_4]x^{i-6}z^6 - \ldots$$
$$+ Ja^2[ix^{i-2} - (i-2)\beta_2 x^{i-4}z^2 + (i-4)\beta_4 x^{i-6}z^4 - \ldots]$$

$$\frac{IT}{xz} = [\{(i+3)2(i-2) + 3.2\}\beta_2 - 3i\beta_0]x^{i-2}$$
$$-[\{(i+3)4(i-4) + 3.4\}\beta_4 - \{(i+3)2(i-2) + 3(i-2)\}\beta_2]x^{i-4}z^2$$
$$+[\{(i+3)6(i-6) + 3.6\}\beta_6 - \{(i+3)4(i-4) + 3(i-4)\}\beta_4]x^{i-6}z^4$$
$$-[\{(i+3)8(i-8) + 3.8\}\beta_8 - \{(i+3)6(i-6) + 3(i-6)\}\beta_6]x^{i-8}z^6 + \ldots$$
$$- Ja^2[2(i-2)\beta_2 x^{i-4} - 4(i-4)\beta_4 x^{i-6}z^2 + 6(i-6)\beta_6 x^{i-8}z^4 - \ldots]$$

The general law of formation of the successive coefficients is obvious, and it is easy to write down the general term in each of the eight series involved in these four expressions; the best way indeed of obtaining the formulæ given below is to write down and transform the general term.

The semi-polar coordinates used hitherto are not so convenient as true polar coordinates; I therefore substitute r, radius vector, and l, latitude, for the x, z system, and putting $x = r\cos l$, $z = r\sin l$ write

$$\left.\begin{aligned}
P &= r^i\cos^i l\,(A_0 + A_2\tan^2 l + A_4\tan^4 l + \ldots) \\
&\qquad + a^2 r^{i-2}\cos^{i-2} l\,(B_0 + B_2\tan^2 l + B_4\tan^4 l + \ldots) \\
R &= r^i\cos^i l\,(C_0 + C_2\tan^2 l + C_4\tan^4 l + \ldots) \\
&\qquad + a^2 r^{i-2}\cos^{i-2} l\,(D_0 + D_2\tan^2 l + D_4\tan^4 l + \ldots) \\
T &= r^i\sin l\cos^{i-1} l\,(E_0 + E_2\tan^2 l + E_4\tan^4 l + \ldots) \\
&\qquad + a^2 r^{i-2}\sin l\cos^{i-3} l\,(F_0 + F_2\tan^2 l + F_4\tan^4 l + \ldots) \\
Q &= r^i\cos^i l\,(G_0 + G_2\tan^2 l + G_4\tan^4 l + \ldots) \\
&\qquad + a^2 r^{i-2}\cos^{i-2} l\,(H_0 + H_2\tan^2 l + H_4\tan^4 l + \ldots)
\end{aligned}\right\}\ldots(22)$$

Introducing for J and for the β's their values in terms of i, I find that the coefficients A, B, &c., are reducible to the forms given in the following equations:—

$$\left.\begin{aligned}
IA_0 &= -\frac{1}{0\,!}\{i(i+2)(i-0)-3.0.(i+1)\}+0.(i+3)(i+2)(i+1)\\
&= -i^2(i+2)\\
IA_2 &= \frac{i^2}{2\,!}\{i(i-0)(i-2)-3.2(i-1)\}-\frac{1}{0\,!}(i+3)(i+0)(i-1)\\
IA_4 &= -\frac{i^2(i-2)^2}{4\,!}\{i(i-2)(i-4)-3.4(i-3)\}\\
&\quad+\frac{i^2}{2\,!}(i+3)(i-2)(i-3)\\
IA_6 &= \frac{i^2(i-2)^2(i-6)^2}{6\,!}\{i(i-4)(i-6)-3.6(i-5)\}\\
&\quad-\frac{i^2(i-2)^2}{4\,!}(i+3)(i-4)(i-5)\\
\&c. &= \&c.\\
IB_0 &= \frac{i(i+2)}{i-1}\frac{1}{0\,!}i(i-1)\\
IB_2 &= -\frac{i(i+2)}{i-1}\frac{i^2}{2\,!}(i-2)(i-3)\\
IB_4 &= \frac{i(i+2)}{i-1}\frac{i^2(i-2)^2}{4\,!}(i-4)(i-5)\\
\&c. &= \&c.
\end{aligned}\right\}\text{......(23)}$$

$$\left.\begin{aligned}
IC_0 &= \frac{1}{0\,!}\{(i+3)i^2-[i(1\{-2\}-1)+3.0.1]\}=i[(i+1)(i+2)+1]\\
IC_2 &= -\frac{i^2}{2\,!}\{(i+3)(i-2)^2-[i(3.0-1)+3.2.3]\}\\
IC_4 &= \frac{i^2(i-2)^2}{4\,!}\{(i+3)(i-4)^2-[i(5.2-1)+3.4.5]\}\\
IC_6 &= -\frac{i^2(i-2)^2(i-4)^2}{6\,!}\{(i+3)(i-6)^2-[i(7.4-1)+3.6.7]\}\\
\&c. &= \&c.\\
ID_0 &= -\frac{i(i+2)}{i-1}\frac{i^2}{0\,!}\\
ID_2 &= \frac{i(i+2)}{i-1}\frac{i^2(i-2)^2}{2\,!}\\
ID_4 &= -\frac{i(i+2)}{i-1}\frac{i^2(i-2)^2(i-4)^2}{4\,!}\\
\&c. &= \&c.
\end{aligned}\right\}\text{...(24)}$$

$$\left.\begin{aligned}
IE_0 &= -\frac{1}{1\,!}\,i\,[1\,(3i+3)-i\,(i-0)\,(i+1)] = i\,(i+1)\,(i^2-3)\\
IE_2 &= \frac{i^2}{3\,!}\,(i-2)\,[3\,(5i+3)-i\,(i-2)\,(i-1)]\\
IE_4 &= -\frac{i^2\,(i-2)^2}{5\,!}\,(i-4)\,[5\,(7i+3)-i\,(i-4)\,(i-3)]\\
IE_6 &= \frac{i^2\,(i-2)^2\,(i-4)^2}{7\,!}\,(i-6)\,[7\,(9i+3)-i\,(i-6)\,(i-5)]\\
\&c. &= \&c.\\
IF_0 &= -\frac{i^3\,(i^2-4)}{i-1}\\
IF_2 &= \frac{i^3\,(i^2-4)}{i-1}\,\frac{(i-2)\,(i-4)}{3\,!}\\
IF_4 &= -\frac{i^3\,(i^2-4)}{i-1}\,\frac{(i-2)\,(i-4)^2\,(i-6)}{5\,!}\\
IF_6 &= \frac{i^3\,(i^2-4)}{i-1}\,\frac{(i-2)\,(i-4)^2\,(i-6)^2\,(i-8)}{7\,!}\\
\&c. &= \&c.
\end{aligned}\right\}\quad\ldots(25)$$

$$\left.\begin{aligned}
IG_0 &= -\frac{1}{0\,!}\,\{i\,(i-0)-3\,.\,0\}+0\,.\,(i+3)\,(i+2) = -\,i^2\\
IG_2 &= \frac{i^2}{2\,!}\,\{i\,(i-2)-3\,.\,2\}-\frac{1}{0\,!}\,(i+3)\,i\\
IG_4 &= -\frac{i^2\,(i-2)^2}{4\,!}\,\{i\,(i-4)-3\,.\,4\}+\frac{i^2}{2\,!}\,(i+3)\,(i-2)\\
\&c. &= \&c.\\
IH_0 &= \frac{i\,(i+2)}{i-1}\,\frac{1}{0\,!}\,i\\
IH_2 &= -\frac{i\,(i+2)}{i-1}\,\frac{i^2}{2\,!}\,(i-2)\\
IH_4 &= \frac{i\,(i+2)}{i-1}\,\frac{i^2\,(i-2)^2}{4\,!}\,(i-4)\\
\&c. &= \&c.
\end{aligned}\right\}\quad\ldots\ldots(26)$$

These sets of coefficients are all written down in such a form that the laws of their formation are obvious, and the general terms may easily be found. I have computed their values from these formulæ for the even zonal harmonics of orders 2, 4, 6, 8, 10, 12; the results are given in the following tables both in the form of fractions and of decimals approximately equal to those fractions.

The G's and H's were not computed because their values were not required for subsequent operations.

TABLE I. The coefficients for expressing the stress P.

i	A_0	A_2	A_4	A_6	B_0	B_2	B_4	B_6
2	$-\frac{16}{19}$ -·8421	$-\frac{22}{19}$ -1·1579			$+\frac{16}{19}$ +·8421			
4	$-\frac{32}{17}$ -1·8824	$+\frac{28}{51}$ +·5490	$+\frac{48}{17}$ +2·8235		$+\frac{32}{17}$ +1·8824	$-\frac{128}{51}$ -2·5098		
6	$-\frac{32}{11}$ -2·9091	+18 +18·0000	$+\frac{184}{11}$ +16·7273	$-\frac{272}{55}$ -4·9455	$+\frac{32}{11}$ +2·9091	$-\frac{1152}{55}$ -20·9455	$+\frac{256}{55}$ +4·6545	
8	$-\frac{640}{163}$ -3·9264	$+\frac{10328}{163}$ +63·3620	$-\frac{2112}{163}$ -12·9571	$-\frac{12160}{163}$ -74·6012	$+\frac{640}{163}$ +3·9264	$-\frac{76800}{1141}$ -67·3094	$+\frac{92160}{1141}$ +80·7713	$-\frac{8192}{1141}$ -7·1797
10	$-\frac{400}{81}$ -4·9383	$+\frac{36130}{243}$ +148·6831	$-\frac{69200}{243}$ -284·7737	$-\frac{56000}{243}$ -230·4527	$+\frac{400}{81}$ +4·9383	$-\frac{112000}{729}$ -153·6352	$+\frac{320000}{729}$ +438·9561	$-\frac{51200}{243}$ -210·6996
12	$-\frac{672}{113}$ -5·9469	$+\frac{32316}{113}$ +285·9823	$-\frac{138000}{113}$ -1221·2389	$+\frac{24000}{113}$ +212·3894	$+\frac{672}{113}$ +5·9469	$-\frac{362880}{1243}$ -291·9389	$+\frac{1881600}{1243}$ +1513·7570	$-\frac{2150400}{1243}$ -1730·0080

TABLE II. The coefficients for expressing the stress R.

i	C_0	C_2	C_4	C_6	D_0	D_2	D_4	D_6
2	$+\frac{26}{19}$ +1·3684	$+\frac{32}{19}$ +1·6842			$-\frac{32}{19}$ -1·6842			
4	$+\frac{124}{51}$ +2·4314	$-\frac{112}{51}$ -2·1961	$-\frac{256}{51}$ -5·0196		$-\frac{128}{51}$ -2·5098	$+\frac{256}{51}$ +5·0196		
6	$+\frac{38}{11}$ +3·4545	-24 -24·0000	$-\frac{208}{11}$ -18·9091	$+\frac{512}{55}$ +9·3091	$-\frac{192}{55}$ -3·4909	$+\frac{1536}{55}$ +27·9273	$-\frac{512}{55}$ -9·3091	
8	$+\frac{728}{163}$ +4·4663	$-\frac{12352}{163}$ -75·7791	$+\frac{4224}{163}$ +25·9141	$+\frac{76288}{815}$ +93·6049	$-\frac{5120}{1141}$ -4·4873	$+\frac{92160}{1141}$ +80·7713	$-\frac{122880}{1141}$ -107·6950	$+\frac{16384}{1141}$ +14·3593
10	$+\frac{1330}{243}$ +5·4733	$-\frac{41200}{243}$ -169·5473	$+\frac{84800}{243}$ +348·9712	$+\frac{60160}{243}$ +247·5720	$-\frac{4000}{729}$ -5·4870	$+\frac{128000}{729}$ +175·5830	$-\frac{384000}{729}$ -526·7490	$+\frac{204800}{729}$ +280·9328
12	$+\frac{732}{113}$ +6·4779	$-\frac{35856}{113}$ -317·3097	$+\frac{158400}{113}$ +1401·7700	$-\frac{38400}{113}$ -339·8230	$-\frac{8064}{1243}$ -6·4875	$+\frac{403200}{1243}$ +324·3765	$-\frac{2150400}{1243}$ -1730·0080	$+\frac{2580480}{1243}$ +2076·0097

TABLE III. The coefficients for expressing the stress T.

i	E_0	E_2	E_4	E_6	F_0	F_2	F_4	F_6
2	$+\frac{6}{19}$ $+\cdot3158$							
4	$+\frac{260}{51}$ $+5\cdot0980$	$+\frac{80}{17}$ $+4\cdot7059$			$-\frac{256}{51}$ $-5\cdot0196$			
6	$+14$ $+14\cdot0000$	$-\frac{56}{11}$ $-5\cdot0909$	$-\frac{1008}{55}$ $-18\cdot3273$		$-\frac{768}{55}$ $-13\cdot9636$	$+\frac{1024}{55}$ $+18\cdot6182$		
8	$+\frac{4392}{163}$ $+26\cdot9448$	$-\frac{13248}{163}$ $-81\cdot2761$	$-\frac{10368}{163}$ $-63\cdot6074$	$+\frac{244224}{5705}$ $+42\cdot8088$	$-\frac{30720}{1141}$ $-26\cdot9238$	$+\frac{122880}{1141}$ $+107\cdot6950$	$-\frac{49152}{1141}$ $-43\cdot0780$	
10	$+\frac{10670}{243}$ $+43\cdot9095$	$-\frac{74800}{243}$ $-307\cdot8189$	$+\frac{17600}{243}$ $+72\cdot4280$	$+\frac{577280}{1701}$ $+339\cdot3769$	$-\frac{32000}{729}$ $-43\cdot8958$	$+\frac{256000}{729}$ $+351\cdot1660$	$-\frac{102400}{243}$ $-421\cdot3992$	$+\frac{409600}{5103}$ $+80\cdot2664$
12	$+\frac{7332}{113}$ $+64\cdot8850$	$-\frac{90480}{113}$ $-800\cdot7089$	$+\frac{137280}{113}$ $+1214\cdot8673$	$+\frac{99840}{113}$ $+883\cdot5398$	$-\frac{80640}{1243}$ $-64\cdot8753$	$+\frac{1075200}{1243}$ $+865\cdot0040$	$-\frac{2580480}{1243}$ $-2076\cdot0097$	$+\frac{1474560}{1243}$ $+1186\cdot2912$

If W be a 2nd, 4th, or 6th harmonic these tables give the complete expressions for P, R, and T; if W be an 8th harmonic the only further coefficients required are A_8 and C_8.

For the cases of the 10th and 12th harmonics the values in the tables are sufficient to give the stresses approximately over a wide equatorial belt, because the series for P, R, T proceed by powers of the tangent of the latitude, and the omitted terms involve high powers of that tangent. It would hardly be safe however to apply the formula—at least as regards the 12th harmonic—for latitudes greater than 15°, because the coefficients are large.

§ 3. *On the direction and magnitude of the principal stresses in a strained elastic solid.*

Let P, Q, R, S, T, U specify the stresses in a homogeneously stressed and strained elastic solid. Let l, m, n be the direction cosines of a principal stress axis.

The consideration, that at the extremity of a principal axis the normal to the stress quadric is coincident with the radius vector, gives the equations

$$(\mathrm{P}-\lambda)\,l+\mathrm{U}m+\mathrm{T}n=0$$
$$\mathrm{U}l+(\mathrm{Q}-\lambda)\,m+\mathrm{S}n=0$$
$$\mathrm{T}l+\mathrm{S}m+(\mathrm{R}-\lambda)\,n=0$$

These equations lead to the discriminating cubic for the determination of λ, and the solution for l, m, n is then

$$\frac{l^2}{(Q-\lambda)(R-\lambda)-S^2}=\frac{m^2}{(P-\lambda)(R-\lambda)-T^2}=\frac{n^2}{(P-\lambda)(Q-\lambda)-U^2}$$

In the case considered in the preceding sections S and U vanish, and the cubic reduces to the quadratic

$$(P-\lambda)(R-\lambda)-T^2=0$$

of which the solution is

$$2\lambda=P+R\pm\sqrt{(P-R)^2+4T^2}$$

m is obviously zero and l, n are determinable from

$$l^2(P-\lambda)=n^2(R-\lambda)$$

Let $\qquad l=\cos\vartheta,\quad n=\sin\vartheta$

Then it is easily proved that

$$\cot 2\vartheta=\frac{P-R}{2T}\quad\text{..............................(27)}$$

This equation gives the directions of the principal stress-axes.

The two principal stresses N_1, N_3 are the two values of λ, so that

$$\left.\begin{aligned}N_1&=\tfrac{1}{2}(P+R)+\tfrac{1}{2}\sqrt{(P-R)^2+4T^2}\\N_3&=\tfrac{1}{2}(P+R)-\tfrac{1}{2}\sqrt{(P-R)^2+4T^2}\end{aligned}\right\}\quad\text{.................(28)}$$

and the third principal stress is of course Q.

When an elastic solid is in a state of stress it is supposed, in all probability with justice, that the tendency of the solid to rupture at any point is to be estimated by the form of the stress quadric. At any rate the hypothesis is here adopted that the tendency to break is to be estimated by the difference between the greatest and least principal stresses. For the sake of brevity I shall refer to the difference between the greatest and least principal stresses as "the stress-difference." This quantity I shall find it convenient to indicate by Δ.

We may also look at the subject from another point of view:—It is a well-known theorem in the theory of elastic solids that the greatest shearing stress at any point is equal to a half of the stress-difference. It is difficult to conceive any mode in which an elastic solid can rupture except by shearing, and hence it appears that the greatest shearing stress is a proper measure of the tendency to break. This measure of tendency to break is exactly one-half of the stress-difference, and it is therefore a matter of indifference whether we take greatest shearing stress or stress-difference. For the sake of comparison with experimental results as to the stresses under which wires and rods of various materials will break and crush, I have found it more convenient to use stress-difference throughout; but the results may all be reduced to shearing stresses by merely halving the numbers given.

[In the case where N_2 is the mean of the principal stresses we have from (28)]

$$\Delta = \sqrt{(P-R)^2 + 4T^2} \qquad \text{(29)}$$

[If however N_2 is the least of the stresses we have

$$\Delta = N_1 - N_2 \qquad \text{(30)]}$$

§ 4. *The application of previous analysis to the determination of the stresses produced by the weight of superficial inequalities.*

[In this section I shall in the first instance consider the formulæ for the stresses when the solid is supposed to be compressible, and shall afterwards proceed to the limit when the solid is treated as incompressible. This is the plan already followed above in § 1.

Suppose that $r = a + h\varsigma_i$ is the equation to an harmonic spheroid of the ith order, forming inequalities on the surface of the sphere, whose density is w.

There are two causes of stress in the interior of the sphere; the first of these is the weight of the inequalities, acting only on the surface; the second is the attraction of the inequalities acting throughout the whole sphere. These two causes correspond to the two problems solved in § 1.

The weight of the inequalities gives rise to a superficial normal traction equal to $-gwh\varsigma_i$. Hence in using the solutions (5) and (9) of the first problem we must put $-gwhr^i\varsigma_i/a^i$ for W_i.

As regards the second cause (the attraction of the inequalities), the potential of the layer of matter $h\varsigma_i$ in the interior of the sphere is $3gwhr^i\varsigma_i/(2i+1)\,a^i$; this is the value which must be attributed to W_i when we use the solution (10) of the second problem.

We see then that the potential to be used in the solution of the second problem is $-3/(2i+1)$ times the potential to be used in the solution of the first problem.

The solution of the first problem for α, P, T is the same as that for the second with certain additional terms. In so far as the two solutions are the same the two problems may be fused together, when the two causes cooperate, by using the solution of the second problem with

$$W_i = -gwh\frac{r^i\varsigma_i}{a^i}\left(1 - \frac{3}{2i+1}\right) = -\frac{2(i-1)}{2i+1}\,gwh\,\frac{r^i\varsigma_i}{a^i}$$

In as far as concerns the additional terms arising from the solution of the problem in its first form, if we adopt the definition that

$$W_i = -\frac{2(i-1)}{2i+1} gwh \frac{r^i \varsigma_i}{a^i} \qquad \text{.......................(31)}$$

the value of W_i to be used in these additional terms is

$$-gwh\frac{r^i \varsigma_i}{a^i} = \frac{2i+1}{2(i-1)} W_i$$

In this way we see that the complete solution of the problem is

$$\begin{aligned}
\alpha &= \frac{1}{2Iv}\left[\frac{i(i+2)}{i-1}a^2 - (i+3)r^2\right]\frac{dW_i}{dx} + \frac{i}{Iv}xW_i \\
&\quad + \frac{1}{2K}\left[\frac{2i+1}{2(i-1)} + \frac{i}{I}\right]\left[(a^2 - 3r^2)\frac{dW_i}{dx} + 2(2i+1)xW_i\right] \\
\mathrm{P} &= \frac{1}{I}\left\{\left[\frac{i(i+2)}{i-1}a^2 - (i+3)r^2\right]\frac{d^2W_i}{dx^2} - 6x\frac{dW_i}{dx} + 3iW_i\right\} + \frac{2i+1}{2(i-1)}W_i \\
&\quad + \frac{v}{K}\left[\frac{2i+1}{2(i-1)} + \frac{i}{I}\right]\left\{(a^2 - 3r^2)\frac{d^2W_i}{dx^2} + 4(i-1)x\frac{dW_i}{dx} - 2i(i-1)W_i\right\} \\
\mathrm{T} &= \frac{i}{I}\left\{\left[\frac{i(i+2)}{i-1}a^2 - (i+3)r^2\right]\frac{d^2W}{dx\,dz} - 3\left(x\frac{dW_i}{dz} + z\frac{dW_i}{dx}\right)\right\} \\
&\quad + \frac{v}{K}\left[\frac{2i+1}{2(i-1)} + \frac{i}{I}\right]\left\{(a^2 - 3r^2)\frac{d^2W}{dx\,dz} + 2(i-1)\left(x\frac{dW_i}{dz} + z\frac{dW_i}{dx}\right)\right\}
\end{aligned} \qquad \text{......(32)}$$

where W_i has the value defined above in (31), and where the other components of displacement and the other stresses are to be found by cyclic changes of letters.

If now we proceed to the limit when the solid is incompressible, K becomes infinite and (32) becomes

$$\left.\begin{aligned}
I\mathrm{P} - \frac{2i+1}{2(i-1)}W_i &= \left[\frac{i(i+2)}{i-1}a^2 - (i+3)r^2\right]\frac{d^2W_i}{dx^2} - 6x\frac{dW_i}{dx} + 3iW_i \\
I\mathrm{T} &= \left[\frac{i(i+2)}{i-1}a^2 - (i+3)r^2\right]\frac{d^2W_i}{dx\,dz} - 3\left(x\frac{dW_i}{dz} + z\frac{dW_i}{dx}\right)
\end{aligned}\right\} \qquad \text{......(33)}$$

These are the quantities which are tabulated in § 2 under the headings P, Q, R, T.

If we write $\qquad \varsigma_i = \sin^i\theta - \frac{i^2}{2!}\sin^{i-2}\theta\cos^2\theta + \text{\&c.}$

where θ is the colatitude, h is the height above the mean sphere of the elevation at the equator.

Now W_i was put equal to $r^i\varsigma_i$ in § 2.]

Thus in order to apply the preceding results to finding the stresses caused in a sphere, possessing the power of gravitation, by the weight and attraction of surface inequalities expressed by

$$r = a + h\varsigma_i \qquad (34)$$

we must multiply the preceding results for P, R, T, Q by

$$-\frac{2(i-1)}{2i+1}\frac{gwh}{a^i} \qquad (35)$$

§ 5*. *The state of stress due to ellipticity of figure or to tide-generating forces.*

When the potential W_i is a solid harmonic of the second degree, the solution found will give the stresses caused by oblateness or prolateness of the spheroid. It will also serve for the case of a rotating spheroid with more or less oblateness than is appropriate for the equilibrium figure. When an elastic sphere is under the action of tide-generating forces, the disturbing potential is a solid harmonic of the second degree, and therefore the present solution will apply to this case also.

If we extract the case $i = 2$ from Tables I, II, III, and put $i = 2$ in (33), and substitute colatitude θ for latitude l, we have after some simple reductions

$$\left.\begin{aligned} 19\left(\mathrm{P} - \tfrac{5}{2}W_2\right) &= 16a^2 - (19 + 3\cos 2\theta)\,r^2 \\ 19\left(\mathrm{R} - \tfrac{5}{2}W_2\right) &= -32a^2 + (29 + 3\cos 2\theta)\,r^2 \\ 19\left(\mathrm{Q} - \tfrac{5}{2}W_2\right) &= 16a^2 - (13 + 9\cos 2\theta)\,r^2 \\ 19\mathrm{T} &= 3\sin 2\theta r^2 \end{aligned}\right\}$$

Let N_1, N_2, N_3, be the three principal stresses, each diminished by $\frac{5}{2}W_2$, so that

$$\left.\begin{aligned} \left.\begin{matrix}\mathrm{N}_1 + \frac{5}{2}W_2 \\ \mathrm{N}_3 + \frac{5}{2}W_2\end{matrix}\right\} &= \tfrac{1}{2}(\mathrm{P} + \mathrm{R}) \pm \tfrac{1}{2}\sqrt{\{(\mathrm{P} - \mathrm{R})^2 + 4\mathrm{T}^2\}} \\ \mathrm{N}_2 + \tfrac{5}{2}W_2 &= \mathrm{Q} \end{aligned}\right\}$$

Then

$$\left.\begin{aligned} \left.\begin{matrix}19\mathrm{N}_1 \\ 19\mathrm{N}_3\end{matrix}\right\} &= -8a^2 + 5r^2 \pm 3\sqrt{\{64(a^2 - r^2)^2 + r^4 - 16r^2(a^2 - r^2)\cos 2\theta\}} \\ 19\mathrm{N}_2 &= 16a^2 - 13r^2 - 9r^2\cos 2\theta \end{aligned}\right\}$$

Now let us find the surfaces, if any, over which $\mathrm{N}_2 = \mathrm{N}_1$ or N_3. They are obviously given by

$$24a^2 - 18r^2 - 9r^2\cos 2\theta = \pm 3\sqrt{\{64(a^2 - r^2)^2 + \dots \&c.\}}$$

* This section in its present form is extracted from "Note on a previous paper," *Proc. Roy. Soc.*, 1885.

This easily reduces to

$$r^2(1-\cos 2\theta)\,[32a^2-20r^2-9r^2(1+\cos 2\theta)]=0$$

Thus the solutions are

$$\left.\begin{aligned} r&=0\\ \theta&=0 \text{ and } \pi\\ \text{and}\qquad 32a^2-20(x^2+y^2)-38z^2&=0\end{aligned}\right\}\text{.....................(36)}$$

By trial it is easy to see that at the centre and all along the polar axis $N_2=N_1$, and that inside the ellipsoid $10(x^2+y^2)+19z^2=16a^2$, N_2 is greater than N_1, and outside it is less.

Hence inside the ellipsoid N_2-N_3 and outside it N_1-N_3 is the stress-difference. N_2-N_3 vanishes nowhere so long as N_2 is not equal to N_1, and N_1-N_3 vanishes where $r=\frac{2}{3}\sqrt{2}\,.\,a=\cdot9428a$ and $\theta=0$, which is inside the region for which N_1-N_3 is the stress-difference. This is the only point in the whole sphere for which the stress-difference vanishes.

The ellipsoid of separation cuts the sphere in colatitude $\sin^{-1}\frac{1}{3}$ or 35° 16′.

If we put Δ for stress-difference, between the centre and the ellipsoid

$$19\Delta=24a^2-18r^2-9r^2\cos 2\theta$$
$$+3\sqrt{\{64(a^2-r^2)^2+r^4-16(a^2-r^2)r^2\cos 2\theta\}}\text{ ......(37)}$$

and between the polar surface regions and the ellipsoid

$$19\Delta=6\sqrt{\{64(a^2-r^2)^2+r^4-16(a^2-r^2)r^2\cos 2\theta\}}\text{ ....(38)}$$

This last also holds for the whole polar axis, along which

$$19\Delta=6(8a^2-9r^2)\text{ or }6(9r^2-8a^2)$$

In order to find the actual value of Δ in any special case, we have to multiply the expression for Δ by appropriate factors, to be determined hereafter. For the present it will be convenient to omit these factors.

We may now from (37) and (38) determine the distribution of stress-difference throughout the sphere.

By computation and graphical interpolation I have drawn the figure (1), showing the curves of equal stress-difference throughout a meridional section of the sphere. The numbers written on the curves give the values of 19Δ, when the radius of the sphere is unity. The point marked 0 is that where Δ vanishes.

The dotted curve is the ellipse of separation cutting the circle in colatitude 35° 16′.

Over the polar cap and at the surface 19Δ is constant and equal to 6; at the surface from colatitude 35° 16′ to the equator 19Δ increases from 6 to 18, varying as the square of the sine of the colatitude.

At the centre 19Δ is 48, being eight times the polar superficial value.

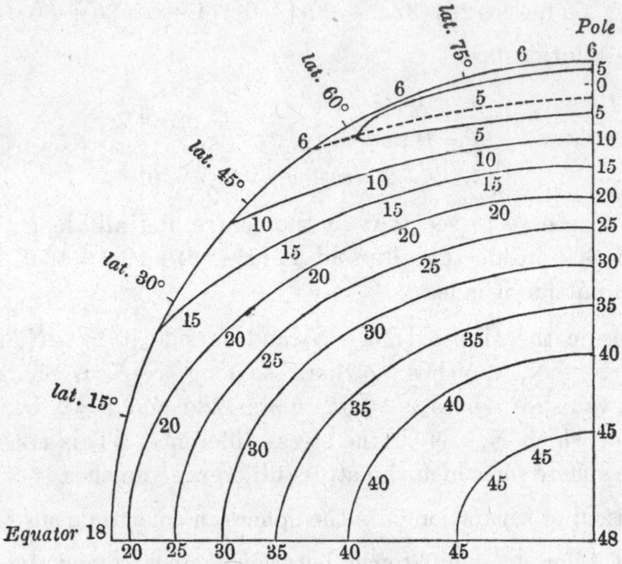

FIG. 1. Diagram showing surfaces of equal stress-difference due to the weight of second harmonic inequalities or to tide-generating force.

If this figure be made to rotate about the polar axis, the several curves will of course generate the surfaces of equal stress-difference throughout the sphere.

Writing ϑ for the inclination of one of the principal axes to the equator, we have by means of the formula (27)

$$\cot 2\vartheta = \frac{\mathrm{P}-\mathrm{R}}{2\mathrm{T}} = 8\left\{\frac{a^2}{r^2}-1\right\} \operatorname{cosec} 2\theta - \cot 2\theta$$

It would be easy to trace out the changes of direction of the principal stress-axes throughout the sphere, but I will only now make the observation that all over the surface they are parallel to and perpendicular to the surface, and that at the centre they are polar and equatorial, the stress-quadric being of course an ellipsoid of revolution.

We have next to find the actual amount of stress-difference which arises from given ellipticity of form of the spheroid. Putting $i=2$ we have

$$\varsigma_i = \sin^2\theta - 2\cos^2\theta = 3\left[\tfrac{1}{3} - \cos^2\theta\right]$$

The equation to the spheroid is

$$r = a + h\varsigma_i$$

$$= a\left[1 + 3\frac{h}{a}\left(\tfrac{1}{3} - \cos^2\theta\right)\right] = a\left[1 + e\left(\tfrac{1}{3} - \cos^2\theta\right)\right]$$

Thus $3h/a$ is the ellipticity of the spheroid, which we may put equal to e.

[It was shown in (31) § 4 that the results for $P - \frac{5}{2} W_2$, $Q - \frac{5}{2} W_2$, $R - \frac{5}{2} W_2$, T are to be multiplied by $-\frac{2}{5}\frac{gwh}{a^2}$, and this will of course also be the factor for the stress-difference Δ.

Substituting e for $3h/a$, and introducing the factor $\frac{1}{19}$, which has been omitted in considering the distribution of stress-difference throughout the sphere, we see that ellipticity e gives a stress-difference represented by the greater of the two expressions

$$\left.\begin{aligned} \Delta &= -\tfrac{2}{95}\, egw \left[8 - 6\frac{r^2}{a^2} - 3\frac{r^2}{a^2}\cos 2\theta \right. \\ &\qquad \left. + \sqrt{\left\{ 64\left(1 - \frac{r^2}{a^2}\right)^2 + \frac{r^4}{a^4} - 16\left(1 - \frac{r^2}{a^2}\right)\frac{r^2}{a^2}\cos 2\theta \right\}} \right] \\ \Delta &= -\tfrac{4}{95}\, egw \sqrt{\left\{ 64\left(1 - \frac{r^2}{a^2}\right)^2 + \frac{r^4}{a^4} - 16\left(1 - \frac{r^2}{a^2}\right)\frac{r^2}{a^2}\cos 2\theta \right\}} \end{aligned}\right\} \ldots(39)$$

Throughout the polar caps the latter is greater, and throughout the rest of the ellipsoid the former is greater.

If we estimate the forces in gravitation units the factor g must be omitted.

Over the polar cap at the surface, the stress-difference

$$\Delta = -\frac{4ewa}{95}, \text{ a constant}$$

Over the rest of the surface it increases as the square of the sine of the colatitude and is given by the formula

$$\Delta = -\frac{12wa}{95}\sin^2\theta$$

The stress-difference at the centre is

$$\Delta = -\frac{32wa}{95}$$

In these formulæ the negative sign has clearly no significance and may be omitted.]

To apply this to the case of the earth, take $a = 637 \times 10^6$ cm., and $w = 5{\cdot}66$, and we find the polar and equatorial stress-differences at the surface to be respectively $152e$ and $456e$ metric tonnes per square centimetre; the central stress-difference is $1214e$ metric tonnes per square centimetre.

If these numbers be multiplied by 6·34, we get the same results expressed in tons per square inch. Thus in British units these three stress-differences are $962e$, $2887e$ and $7698e$.

If then the ellipticity e be $\frac{1}{1000}$th, at the pole and equator on the surface and at the centre, the stress-differences will be nearly 1 ton, 3 tons, and nearly 8 tons to the square inch respectively.

From the Table VII. in § 9 it will appear that cast brass ruptures with a stress-difference of about 8 tons to the square inch.

Thus a spheroid, made of material as strong as brass, and of the same dimensions and density as the earth, would only just support an excess or deficiency of ellipticity equal to $\frac{1}{1000}$th, above or below the equilibrium ellipticity adapted for its rotation.

The following is a second example:—If the homogeneous earth (with ellipticity $\frac{1}{232}$) were to stop rotating, the stress-difference at the centre would be 33 tons per square inch.

Now suppose the cause of internal stress to be the moon's tide-generating influence, and let m = moon's mass, and c = moon's distance.

Then the potential under which the earth is stressed is

$$-\tfrac{3}{2}\frac{m}{c^3}\left(\tfrac{1}{3}-\cos^2\theta\right)wr^2$$

or according to the notation of § 4

$$-\tfrac{1}{2}\frac{m}{c^3}wr^2\varsigma_2$$

If we took into account the elastic yielding of the earth and the weight and attraction of the tidal protuberance, this potential would have to be diminished. To estimate the diminution we must of course know the amount of elastic yielding, but as there is no means of approximating thereto, it will be left out of account.

Then it is obvious that the factor by which Δ, as given in (39), must be multiplied in order to give the stress-difference is $\frac{1}{2}mw/c^3$. Thus the surface stress-difference at the polar cap is $\frac{6}{19}\cdot\frac{1}{2}(m/c^3)\,wa^2$ in absolute force units.

Putting M for the earth's mass, and dividing by gravity, we have $\frac{3}{19}(ma^3/Mc^3)\,wa$ as the polar surface value of Δ in gravitation units. The central value of Δ is of course eight times as great.

With the numerical data used above, $wa = 3605$ metric tonnes per square cm., and $m/M = \frac{1}{82}$, $a/c = \frac{1}{60}$, whence the polar surface stress-difference is 32 grammes, and the central stress-difference 257 grammes per square centimetre.

But this conclusion is erroneous for the following reason. If we suppose the moon to revolve in the terrestrial equator, and imagine that the meridian from which longitudes are measured is the meridian in which the moon stands at the instant under consideration, the tide-generating potential is

$$-\tfrac{3}{2}\frac{m}{c^3}r^2\left[\tfrac{1}{3}-\sin^2\theta\cos^2\phi\right]$$

This expression may be written

$$\tfrac{3}{4}\frac{m}{c^3}r^2\left(\tfrac{1}{3}-\cos^2\theta+\sin^2\theta\cos 2\phi\right)$$

The former of these terms produces a permanent increase of the earth's ellipticity, and is confused and lost in the ellipticity due to terrestrial rotation, and can produce no stress in the earth. The second term is the true tide-generating potential, but it is a sectorial harmonic, and I have failed to treat such cases. Now the first of these terms causes ellipticity in a homogeneous earth equal to $(\frac{5}{2}a/g)(\frac{3}{4}m/c^3)$ according to the equilibrium-tide theory. This ellipticity is equal to $\cdot 1039 \times 10^{-6}$, an excessively small quantity. If however this permanent ellipticity does not exist (and the above investigation in reality presumes it not to exist), then there will be a superficial stress-difference at the poles equal to $152 \times \cdot 1039 \times 10^{-6}$ metric tonnes per square centimetre, and a central stress-difference of eight times as much.

Since a metric tonne is a million grammes this polar surface stress-difference is 16 grammes, and the central 128 grammes per square centimetre. These stress-differences are exactly the halves of those which have been computed above. Thus the remaining stress-difference which is due to the moon's tide-generating influence is 16 grammes at the surface and 128 grammes at the centre per square centimetre.

A flaw in this reasoning is that stress-difference is a non-linear function of the stresses, and therefore the stress-difference arising from the sum of two sets of bodily stresses is not the sum of their separate stress-differences.

I conceive however that the above conclusion is not likely to be much wrong.

These stresses are very small compared with those arising from the weights of mountains and continents as computed below, nevertheless they are so considerable that we can understand the enormous rigidity which Sir William Thomson has shown that the earth must possess in order to resist considerable tidal deformations of its mass.

§ 6. *On the stresses due to a series of parallel mountain chains.*

Having considered the case of the second harmonic, I now pass to the other extreme and suppose the order of harmonics i to be infinitely great, whilst the radius of the sphere is also infinitely great.

The equatorial belt now becomes infinitely wide, and the surface inequalities consist of a number of parallel simple harmonic mountains and valleys.

If i be infinitely large, we have from (12)

$$W_i = \rho^i \left[1 - \frac{1}{2\,!}\left(\frac{iz}{\rho}\right)^2 + \frac{1}{4\,!}\left(\frac{iz}{\rho}\right)^4 - \&c.\right]$$

Now let ξ be the depth below the surface of the point indicated in the sphere (now infinitely large) by x, y, z.

As the formulæ given above apply to the meridional plane for which $y = 0$, we have $\rho = a - \xi$.

Let $b=a/i$, then when both i and a are infinite

$$\rho^i = a^i\left(1-\frac{\xi}{a}\right)^i = a^i\epsilon^{-\xi/b}$$

and since in the limit $\rho/i = a/i = b$,

$$W_i = a^i\epsilon^{-\xi/b}\left(1-\frac{1}{2\,!}\frac{z^2}{b^2}+\frac{1}{4\,!}\frac{z^4}{b^4}-\&c.\right)$$

$$=(a^i)\,\epsilon^{-\xi/b}\cos\frac{z}{b}$$

This expression for W involves the infinite factor a^i, and in order to get rid of it we must now consider the factor by which it is to be multiplied, in introducing the height of the mountains and gravity.

This factor is computed in § 4; it is there shown that if $r=a+h\varsigma_i$ be a harmonic spheroid, the factor is $-2(i-1)\,gwh/(2i+1)\,a^i$.

Now if the harmonic i be of an infinitely high order, ς_i becomes simply $\cos z/b$, and the equation to the surface is

$$\xi = -h\cos\frac{z}{b}$$

ξ being measured downwards. Thus the harmonic spheroid $h\varsigma_i$ now represents a series of parallel harmonic mountains and valleys of height and depth h, and wave-length $2\pi b$.

The factor becomes $-gwh/a^i$, when i is infinite.

Thus the effective disturbing potential W, which is competent to produce the same state of stress and strain as the weight of the mountains and valleys, is given by

$$W=-gwh\epsilon^{-\xi/b}\cos\frac{z}{b}\ \ldots\ldots\ldots\ldots\ldots\ldots\ldots\ldots(40)$$

Now revert to the expressions (33) for the stresses.

When i is infinite $I=2i^2$, and they become, on changing x into $(a-\xi)$

$$\mathrm{P}-\frac{1}{2i^2}W=\frac{1}{2i}(a^2-r^2)\frac{d^2W}{d\xi^2}+\frac{6(a-\xi)}{2i^2}\frac{dW}{d\xi}+\frac{3}{2i}W_i$$

$$\mathrm{T}=-\frac{1}{2i}(a^2-r^2)\frac{d^2W}{d\xi dz}-\frac{3}{2i^2}\left[(a-\xi)\frac{dW}{dz}-z\frac{dW}{d\xi}\right]$$

As shown above $a^2-r^2=2a\xi$, and $a/i=b$ in the limit; making these substitutions, and dropping the terms which become infinitely small when i is infinite, we have

$$\left.\begin{aligned}&\mathrm{P}=\xi b\frac{d^2W}{d\xi^2}, \qquad \mathrm{T}=-\xi b\frac{d^2W}{d\xi dz}\\ &\text{and by a similar process}\\ &\mathrm{R}=\xi b\frac{d^2W}{dz^2}, \qquad \mathrm{Q}=0\end{aligned}\right\}\ \ldots\ldots\ldots\ldots\ldots\ldots(41)$$

Then from (40) and (41) we have

$$\left.\begin{aligned} P &= -gwh\frac{\xi}{b}\epsilon^{-\xi/b}\cos\frac{z}{b} \\ R &= gwh\frac{\xi}{b}\epsilon^{-\xi/b}\cos\frac{z}{b} \\ T &= gwh\frac{\xi}{b}\epsilon^{-\xi/b}\sin\frac{z}{b} \end{aligned}\right\} \qquad (42)$$

Since the stress-difference

$$\Delta = \sqrt{(P-R)^2 + 4T^2}$$

we have

$$\Delta = 2gwh\frac{\xi}{b}\epsilon^{-\xi/b} \qquad (43)$$

The directions of the stress-axes are given by

$$\cot 2\vartheta = \frac{P-R}{2T} = \cot\frac{z}{b}$$

so that

$$\vartheta = \frac{1}{2}\frac{z}{b} \qquad (44)$$

Equation (43) gives the stress-difference at a depth ξ below the mean surface, and is very remarkable as showing that the stress-difference depends on depth below the mean horizontal surface and not at all on the position of the point considered with reference to the ridges and furrows.

Equation (44) shows that if we travel uniformly and horizontally through the solid perpendicular to the ridges, the stress-axes revolve with a uniform angular velocity.

They are vertical and horizontal when we are under a ridge, and they have turned through a right angle and are again vertical and horizontal by the time we have arrived under a furrow.

Since the function $x\epsilon^{-x}$ is a maximum when $x = 1$, the stress-difference Δ is a maximum when $\xi = b$,—that is to say, at a depth equal to $1/2\pi$ of the wave-length—and is then equal to $2gwh\epsilon^{-1}$ or in gravitation units of force to $\cdot 736wh$. It is interesting to notice that the value of this maximum depends only on the height and density of the mountains, and does not involve the distance from crest to crest. The depth at which this maximum is reached depends of course on the wave-length.

Fig. 2 shows the distribution of stress-difference, the vertical ordinates representing stress-difference, and the horizontal ones depth below the surface. The numbers written on the horizontal axis are multiples of b; the distance OL on this scale is equal to 6·28, and is therefore equal to the wave-length from crest to crest, and the distance OH is the semi-wave-length from crest to furrow.

In the case of terrestrial mountains w is about 2·8, and if we suppose h to be 2000 metres, or a little over 6000 feet, we have the case of a series of lofty

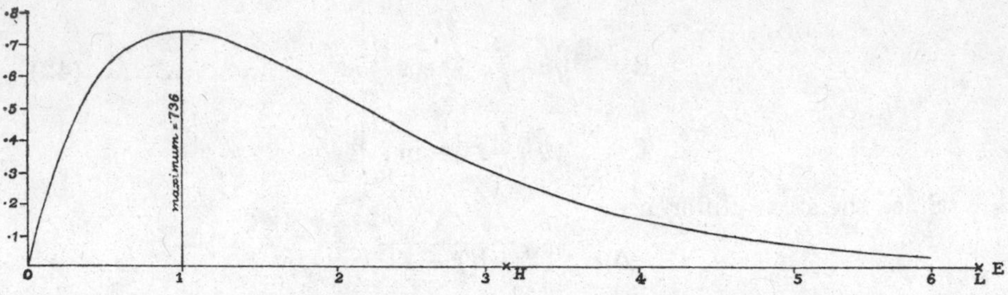

FIG. 2. Diagram showing the difference between the principal stresses due to harmonic mountains and valleys on a horizontal plane.

mountain chains—for it must be remembered that the valleys run down to 2000 metres below the mean surface, so that the mountains are some 13,000 feet above the valley-bottoms.

Then $h = 2 \times 10^5$, $w = 2{\cdot}8$, and the maximum stress-difference is

$$736 \times 2{\cdot}8 \times 2 \times 10^5 = {\cdot}412 \times 10^6 \text{ grammes per square centimetre.}$$

This stress-difference is, in British measure, 2·6 tons per square inch.

If we suppose (as is not unreasonable) that it is 314 miles from crest to crest of the mountains, the maximum stress will be reached at 50 miles below the surface.

From Table VII., § 9, it will be seen that if the materials of the earth at this depth of 50 miles had only as much tenacity as sheet lead, the mountain chains would sink down, but they would just be supported if the tenacity were equal to that of cast tin.

§ 7. *On the stresses due to the even zonal harmonic inequalities.*

Having considered the two extreme cases where i is 2, and infinity, I pass now to the intermediate ones. As the odd zonal harmonics were not required for the investigation in the following section I have only worked out in detail the even ones.

The surface of the sphere is now corrugated by a series of undulations parallel to the equator, and the altitude of the corrugations increases towards the poles. The form of the undulation in the neighbourhood of the equator is exhibited in Fig. 3.

As in the case of the second harmonic there are regions within which $N_2 - N_3$ is the proper measure of stress-difference, and others in which $N_1 - N_3$ is the proper measure. The complete determination of these regions might be difficult, but as these harmonics are only used for the determination of stress-difference in the equatorial regions, it is sufficient to find the boundary of the regions for that part of the sphere.

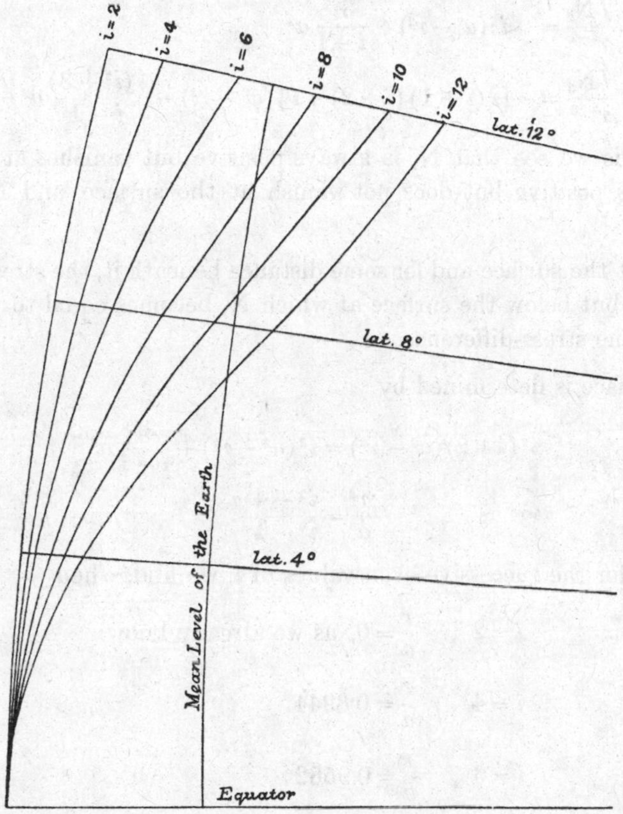

FIG. 3. Diagram showing the profile of the even harmonics near the Equator. Radius of sphere 18 inches.

We see from (22) that $\sqrt{\{(P-R)^2+4T^2\}}$ only differs from $P-R$ by terms which depend on the square of the sine of the latitude.

Hence as far as the first power of $\sin l$ we have

$$N_1 = P - \frac{2i+1}{2(i-1)} W_i, \quad N_2 = Q - \frac{2i+1}{2(i-1)} W_i, \quad N_3 = R - \frac{2i+1}{2(i-1)} W_i$$

Therefore if we neglect terms depending on the square of the sine of the latitude, we have from (22),

$$\frac{IN_1}{r^{i-2}} = A_0 r^2 + B_0 a^2, \quad \frac{IN_2}{r^{i-2}} = G_0 r^2 + H_0 a^2, \quad \frac{IN_3}{r^{i-2}} = C_0 r^2 + D_0 a^2$$

Substituting, for A_0, B_0, &c., their values from (23), (24), (26), and effecting some easy reductions, we find,

$$\frac{IN_1}{r^{i-2}} = i^2(i+2)(a^2-r^2)$$

$$\frac{IN_2}{r^{i-2}} = i^2(a^2-r^2)+\frac{3i^2}{i-1}a^2$$

$$\frac{IN_3}{r^{i-2}} = -[i(i+1)(i+2)+i](a^2-r^2)-\frac{i(i^2+3)}{i-1}a^2$$

From this we see that N_1 is always positive but vanishes at the surface, N_2 is always positive but does not vanish at the surface, and N_3 is always negative.

Hence at the surface and for some distance beneath it, the stress-difference is $N_2 - N_3$; but below the surface at which N_1 becomes equal to N_2, we have $N_1 - N_3$ as the stress-difference.

This surface is determined by

$$i^2(i+2)(a^2-r^2) = i^2(a^2-r^2)+\frac{3i^2}{i-1}a^2$$

whence
$$\frac{r^2}{a^2} = \frac{i^2-4}{i^2-1}$$

Solving for the successive even values of i, we find, when

$i=2$, $\frac{r}{a}=0$, as we already know

$i=4$, $\frac{r}{a}=0{\cdot}8944$

$i=6$, $\frac{r}{a}=0{\cdot}9562$

$i=8$, $\frac{r}{a}=0{\cdot}9759$

$i=10$, $\frac{r}{a}=0{\cdot}9847$

We see that even when $i=4$, the region is very thin in which $N_2 - N_3$ is the proper measure. For the higher harmonics it soon becomes negligeable*.

[I shall then only consider the stress-difference $N_1 - N_3$, and this] is as before given by

$$\Delta = \sqrt{(P-R)^2+4T^2}$$

To form this expression the series in (22) for R must be subtracted from the series for P. Since the C's and D's of Table II. have always the opposite

* As far as this point, this section is taken from a "Note on a previous paper" referred to above.

signs from the A's and B's of Table I., this algebraic subtraction becomes a numerical addition of the numbers in these two tables.

The results are given in the following table.

TABLE IV. The coefficients for expressing P − R.

i	$A_0 - C_0$	$A_2 - C_2$	$A_4 - C_4$	$A_6 - C_6$	$B_0 - D_0$	$B_2 - D_2$	$B_4 - D_4$	$B_6 - D_6$
2	− 2·2105	− 2·8421			+ 2·5263			
4	− 4·3137	+ 2·7451	+ 7·8431		+ 4·3922	− 7·5294		
6	− 6·3636	+ 42	+ 35·6364	− 14·2545	+ 6·4	− 48·8727	+ 13·9636	
8	− 8·3926	+ 139·1411	− 38·8712	− 168·2061	+ 8·4137	− 148·0806	+ 188·4663	− 21·5390
10	− 10·4115	+ 318·2304	− 633·7449	− 478·0247	+ 10·4252	− 329·2182	+ 965·7051	− 491·6324
12	− 12·425	+ 603·292	− 2623·009	+ 552·212	+ 12·434	− 616·315	+ 3243·765	− 3806·018

Then we have

$$\mathrm{P} - \mathrm{R} = r^i \cos^i l\,[(A_0 - C_0) + (A_2 - C_2)\tan^2 l + \ldots]$$
$$+ a^2 r^{i-2} \cos^{i-2} l\,[(B_0 - D_0) + (B_2 - D_2)\tan^2 l + \ldots]$$

The materials for computing T have been already given in Table III.

The series for P − R and for 2T should now be squared and added together, but the result would be so complex that it is not worth while to proceed algebraically.

At the equator $\mathrm{T} = 0$, and $\Delta = \mathrm{P} - \mathrm{R}$, and P − R reduces to only two terms, whatever be the order of harmonic.

By reference to (23) and (24) we see that at the equator

$$\Delta = \frac{i r^{i-2}}{2(i+1)^2 + 1}\left[\frac{i(i+2)(2i-1)}{i-1}a^2 - (i+1)(2i+3)r^2\right]$$

or

$$\Delta = \frac{i(i+1)(2i+3)a^2 r^{i-2}}{(i+1)(2i+3) - i}\left[1 - \frac{r^2}{a^2} + \frac{3}{(i^2-1)(2i+3)}\right] \quad \ldots\ldots\ldots(45)$$

This value for Δ requires of course multiplication by appropriate factors involving the height of the continents and gravity.

Even when i is no larger than 6, (45) differs but little from $i r^{i-2}(a^2 - r^2)$, at least for values of r not very nearly equal to a.

Δ clearly reaches a maximum when

$$\frac{r^2}{a^2} = \frac{i-2}{i}\left\{1 + \frac{3}{(i^2-1)(2i+3)}\right\}$$

For large values of i this maximum is nearly equal to $2\{(i-2)/i\}^{\frac{1}{2}i-1} a^i$.

From these formulæ the following results are easily obtained.

TABLE V (*a*).

$i=$	2	4	6	8	10	12
Maximum value of Δ	2·526	1·118	·959	·894	·859	·836
Value of r/a when Δ is max[m] .	0	·714	·819	·867	·895	·913

[If we compare these values of r/a with the values found for the limit of the region in which $N_1 - N_3$ is the proper measure of stress-difference, we see that it always falls far within that region. It appears then that in these cases it suffices to use only the form for Δ which has furnished these numbers.]

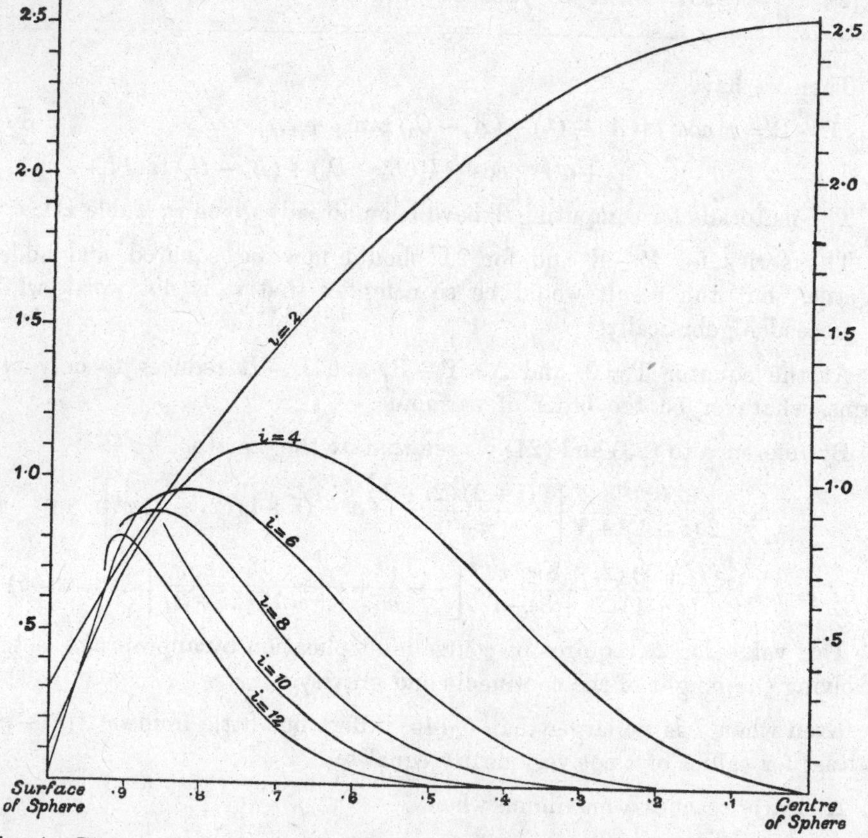

FIG. 4. Diagram showing the difference of the principal stresses at the equator, due to inequalities represented by the even zonal harmonics.

Fig. 4 shows graphically the law of diminution of the stress-difference $N_1 - N_3$ for the even zonal harmonics, the vertical ordinates representing stress-difference and the horizontal ones the distances from the surface or from the centre of the globe.

In order to find a numerical value of these maximum stress-differences which shall be intelligible according to ordinary mechanical ideas, I will suppose the height of each of the harmonics at the equator to be 1500 metres. On account of the small density of the superficial layers in the earth, this is very nearly the same as supposing that in the earth the maximum height of the continents above, and the maximum depth of the depressions below the mean level of the earth are each about 3000 metres. In the summary at the end I shall show that there is reason to believe that this is about the magnitude of terrestrial inequalities.

Then by (35) we have to multiply the maximum stress-differences in the above table by $2(i-1)\,wh/(2i+1)$, in order to obtain the stress-differences for the supposed continents in grammes or tonnes per square centimetre.

Now $w = 5{\cdot}66$ and $h = 1{\cdot}5 \times 10^5$ according to the above hypothesis as to height of continent; and the coefficient in i is of course different for each harmonic.

By performing these multiplications I find the following results.

TABLE V (*b*). Maximum stress-differences due to harmonic continents and seas.

Order of harmonic	2	4	6	8	10	12
Max. stress-difference, in metric tonnes per sq. cm. due to continents 1500 metres high	·858	·633	·626	·625	·625	·625
Ditto in British tons per sq. inch, for same continents	5·43	4·01	3·97	3·96	3·96	3·96
Depth in British miles at which this stress is attained	{ Centre of earth }	1146	725	532	420	347

N.B.—The continents referred to are supposed to be of the earth's mean density and are equivalent to actual continents of double the height.

Thus far we have only considered the stress-differences at the equator immediately underneath the centres of the continents, but we must now see how they differ as the latitude of the place of observation increases. In order to attain this result a good deal of computation was necessary.

For this purpose I calculated P − R and 2T for a number of points and found the square root of the sum of these squares. As the computations were laborious, and as the results given in the following table are amply sufficient for the purpose in hand, I did not think it worth while to trace the changes to a greater depth than $r = \cdot 7$. Moreover the correctness of the last significant figures given cannot be guaranteed, although I believe that it is correct in most cases.

TABLE VI. Showing the stress-difference $N_1 - N_3$ due to the several harmonic inequalities at various depths and in various latitudes.

i		Equator	Lat. 6°	Lat. 12°	i		Equator	Lat. 6°	Lat. 12°
2	$r=1.$	·316	·316	·316	8	$r=1.$	·021	·015	·000
	$r=\cdot 9$	·736	·732	·721		$r=\cdot 9$	·859	·853	·853
	$r=\cdot 8$	1·112	1·108	1·097		$r=\cdot 8$	·798	·795	·797
	$r=\cdot 7$	1·443	1·440	1·431		$r=\cdot 7$	·506	·505	·507
4	$r=1.$	·079	·074	·061	10	$r=1.$	·014	·008	·007
	$r=\cdot 9$	·727	·719	·700		$r=\cdot 9$	·857	·854	·860
	$r=\cdot 8$	1·044	1·038	1·025		$r=\cdot 8$	·631	·630	·635
	$r=\cdot 7$	1·116	1·113	1·104		$r=\cdot 7$	·307	·307	·309
6	$r=1.$	·036	·031	·016	12	$r=1.$	·010	·003	·019
	$r=\cdot 9$	·817	·810	·800		$r=\cdot 9$	·827	·824	·835
	$r=\cdot 8$	·953	·949	·945		$r=\cdot 8$	·481	·481	·486
	$r=\cdot 7$	·788	·786	·785		$r=\cdot 7$	·179	·179	·181

The numbers given in the column marked "equator" might be computed from (45), and are those exhibited graphically in fig. 4; they are here given as a means of comparison with the numbers corresponding to latitudes 6° and 12°.

The result to be deduced from this table is that the lines of equal stress-difference are very nearly parallel with the surface, and that it is for all practical purposes sufficient to know the stress-difference immediately under the centre of the continents.

We have already seen in § 6 that for harmonics of infinitely high orders the lines of equal stress-difference are rigorously parallel with the mean surface.

§ 8. *On the stresses due to the weight of an equatorial continent.*

The actual continents and seas on the earth's surface have not quite the regular wavy character of the elevations and depressions which have been treated hitherto. The subject of the present section possesses therefore a peculiar interest for the purpose of application to the earth. Had I however foreseen, at the beginning, the direction which the results of this whole investigation would take, it is probable that I might not have carried out the long computations which were required for discussing the case of an isolated continent. But now that that end has been reached, it seems worth while to place the results on record.

The function exp. $[-\cos^2\theta/\sin^2\alpha]$ (where θ is colatitude) obviously represents an equatorial belt of elevation, and I therefore chose it as the form of the required equatorial continent. This function has to be expanded in a series of zonal harmonics in order to apply the analytical solutions for the stresses produced by the weight of the continent.

It is obvious by inspection that the odd zonal harmonics can take no part in the representation of the function, and it was on this account that I have above only worked out the cases of the even zonal harmonics.

The multiplication of this function by the successive Legendre's functions, and integration over the surface of the sphere, are operations algebraically tedious, and wholly uninteresting, and I therefore simply give the results.

I find that if $\alpha = 10°$, and

$$\varsigma_i = \sin^i\theta - \frac{i^2}{2\,!}\sin^{i-2}\theta\cos^2\theta + \frac{i^2(i-2)^2}{4\,!}\sin^{i-4}\theta\cos^4\theta - \&c.$$

then

$$\left.\begin{aligned}&2\epsilon^{-\cos^2\theta/\sin^2\alpha} - \beta_0 = \beta_2\varsigma_2 + \beta_4\varsigma_4 + \beta_6\varsigma_6 + \beta_8\varsigma_8 + \beta_{10}\varsigma_{10} + \beta_{12}\varsigma_{12} + \dots\\ &\text{where}\\ &\beta_0 = \cdot 3078,\quad \beta_2 = \cdot 3673,\quad \beta_4 = \cdot 3339,\quad \beta_6 = \cdot 2829,\quad \beta_8 = \cdot 2252,\\ &\qquad\qquad \beta_{10} = \cdot 1688,\quad \beta_{12} = \cdot 1193\end{aligned}\right\}\quad (46)$$

β_0 is put on the left-hand side in order that the mean value of the function may be zero. The β's obviously decrease very slowly, and as I stop with the 12th harmonic, the representation of the function is very imperfect.

Fig. 5 illustrates the results of the representation, the portion of a circle marked "mean level of earth" represents a meridional section of the earth; the dotted curve marked "inequality to be represented" shows the true value of the function 2 exp. $[-\cos^2\theta/\sin^2\alpha] - \beta_0$; the curve marked "representation" shows the right-hand side of (46) stopping with the 12th harmonic; the second curve is the same without the 2nd harmonic constituent $\beta_2\varsigma_2$, and it is introduced for the reasons explained in the discussion and summary at the end.

The equatorial value of the exponential function is 1·792, that of the "representation" is 1·497, and that of "the representation without the 2nd harmonic" is 1·130.

The polar value of the exponential is $-$ ·3078, that of the "representation" is $-$ ·0830, and that of "the representation without 2nd harmonic" is + ·6516. This latter function thus gives us an equatorial continent and a pair of polar continents of less height.

The extreme difference of height in the "representation" between the elevation at the equator and the depression at the pole is $(1{\cdot}497 + {\cdot}083)\,h$ or $1{\cdot}58h$. I do not know exactly the extreme difference in the case where the 2nd harmonic is omitted, because I have not traced the inequality throughout, but it is probably about $1{\cdot}3$ or $1{\cdot}4h$.

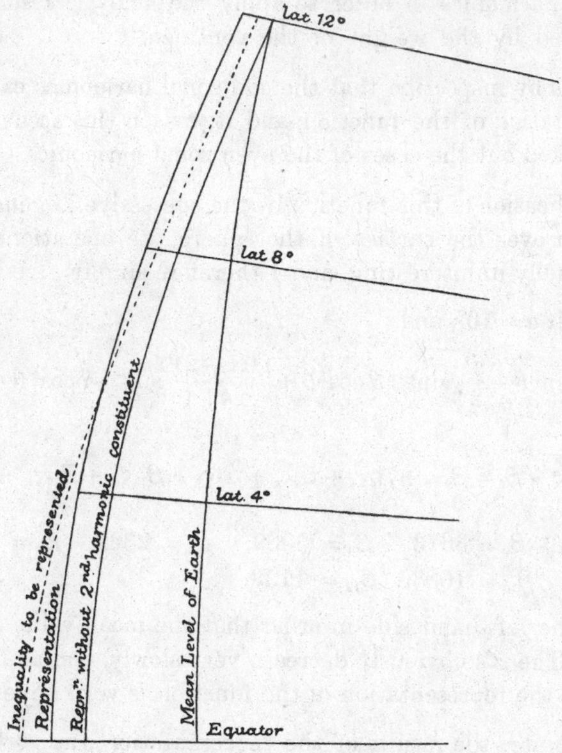

Fig. 5. Diagram showing the profile of isolated equatorial continents. Radius of sphere 18 inches.

Let Δ_i be the numerical value, as computed for § 7, of the stress-difference $N_1 - N_3$ due to the harmonic spheroid ς_i. Then it has been shown above that the stress-difference due to the spheroid whose equation is $r = a + h\varsigma_i$ is $-2(i-1)\,gwh\Delta_i/(2i+1)$.

Now stress-difference is a non-linear function of the component stresses P, R, T, and therefore the stress-difference due to a compound harmonic spheroid is not in general the sum of the stress-differences due to the constituent harmonic spheroids. At any point, however, where the principal stress-axes are all coincident in direction and where all the greater stress-axes coincide, and not a greater with a less, and where $T = 0$, the stress-difference is linear and is the sum of the constituent stress-differences. This is the case at the equator for the present equatorial continent.

Hence, if Δ be the stress-difference $N_1 - N_3$ at the equator due to the spheroid,

$$r = a + h(\beta_2 \varsigma_2 + \beta_4 \varsigma_4 + \ldots + \beta_{12} \varsigma_{12})$$

We have

$$\Delta = -gwh\left[\tfrac{2}{5}\beta_2\Delta_2 + \tfrac{6}{9}\beta_4\Delta_4 + \tfrac{10}{13}\beta_6\Delta_6 + \tfrac{14}{17}\beta_8\Delta_8 + \tfrac{18}{21}\beta_{10}\Delta_{10} + \tfrac{22}{25}\beta_{12}\Delta_{12}\right] \ldots (47)$$

In this formula the Δ_i's are the numbers which were computed for drawing fig. 4, from the formula (45), namely

$$\Delta_i = -\frac{i(i+1)(2i+3)}{2(i+1)^2+1}\left(\frac{r}{a}\right)^{i-2}\left[1 - \frac{r^2}{a^2} + \frac{3}{(i^2-1)(2i+3)}\right]$$

By using these computations I have drawn Fig. 6. The vertical ordinates are $-\Delta \div gwh$, and the horizontal are the distances from surface or centre of the sphere.

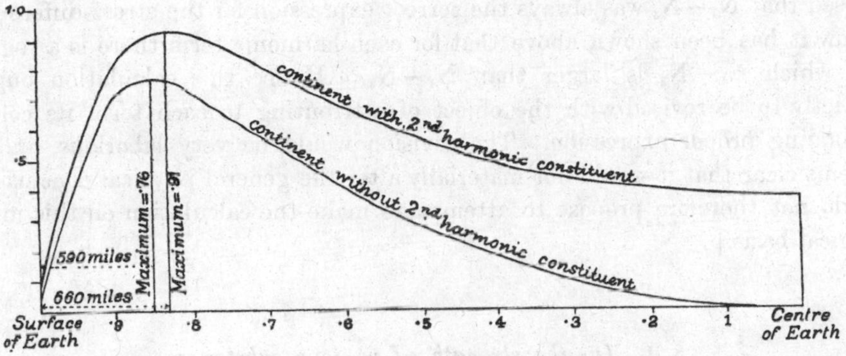

FIG. 6. Diagram showing the difference of principal stresses at the equator due to isolated equatorial continents.

The maxima in the two curves are merely found graphically, and the distances where the maxima are reached (viz.: 660 and 590 miles from the surface) are written down on the supposition that the radius of the sphere is 4000 miles.

In the discussion in the second part of this paper, I have endeavoured to make an estimate of the proper elevation to attribute to these isolated continents; so as to make the case, as nearly as may be, analogous to the earth.

Although it appears impossible to make an accurate estimate, I conclude that it will not be excessive if we assume that the greatest difference of height, between the highest point in the equatorial elevation and the approximately spherical remainder of the globe, is 2000 metres.

Accordingly for the representation I put $1{\cdot}58h = 2000$, and for the second curve $1{\cdot}4h = 2000$; these give $h = 1{\cdot}27 \times 10^5$ cm. and $h = 1{\cdot}4 \times 10^5$ cm. respectively.

Taking $w = 5{\cdot}66$, then for the representation we have $wh = {\cdot}72$ tonnes per square centimetre, and for the other curve $wh = {\cdot}79$ of the same units. The maximum stress-differences are $\cdot 91wh$ and $\cdot 76wh$ respectively.

Therefore for the equatorial table-land (called above the representation) we have a maximum stress-difference of ·66 metric tonne per square cm. or 4·1 British tons per square inch; and for the equatorial table-land balanced by a pair of polar continents (2nd harmonic omitted) we have a maximum stress-difference of ·60 tonne per square cm. or 3·8 tons per square inch.

I therefore conclude that our great continental plateaus produce a stress-difference of about 4 tons per square inch at a depth of 600 or 700 miles from the earth's surface.

[The whole of this calculation was made at a time when I wrongly supposed that $N_1 - N_3$ was always the correct expression for the stress-difference. Now it has been shown above that for each harmonic term there is a region in which $N_2 - N_3$ is larger than $N_1 - N_3$. Hence this calculation ought strictly to be revised with the object of attributing to each term its corresponding proper expression. The revision would be very laborious, and it seems clear that it would not materially alter the general physical conclusion. I do not therefore propose to attempt to make the calculation on this more logical basis.]

§ 9. *On the strength of various substances.*

In order to have a proper comprehension of the strength which the earth's mass must possess in order to resist the tendency to rupture produced by the unequal distribution of weights on the surface, it is necessary to consider the results of experiment.

Rankine* gives a large number of results obtained by various experimenters, and Sir William Thomson also gives similar tables in his article on "Elasticity†."

Amongst other constants Sir William gives Young's modulus and the greatest elastic extension. If the materials of a wire remain perfectly elastic when the wire is just on the point of breaking under tension, then the product of Young's modulus into the greatest elastic extension should be equal to what is called the tenacity, which is defined as the breaking tension

* *Useful Rules and Tables:* Griffin, London, 1873, p. 191, *et seq.*

† "Elasticity": Black, Edinburgh, 1878. This is the article from the *Encyclopedia Britannica*, but it is also published as a separate work.

per square centimetre of the area of the wire. If however a permanent set begins to take place before the wire breaks, this product should be less than the tenacity. I do not see how it can ever be greater, unless there be a marked departure from Hooke's law "ut tensio sic vis"; or different sets of experiments with the same class of material might make it seem greater. In some of the results given by Sir William Thomson the product of modulus and elastic extension is however greater than tenacity.

Ordinary experience would lead one to suppose that such materials as lead and copper would undergo a considerable stress beyond the limits of perfect elasticity, before breaking. It is surprising therefore to see how nearly identical this product is to the tenacity—indeed in the case of lead absolutely identical, as may be seen in the table below.

With regard to the earth we require to know what is the limiting stress-difference under which a material takes permanent set or begins to flow, rather than the stress-difference under which it breaks; for if the materials of the earth were to begin to flow, the continents would sink down and the sea bottoms rise up.

It will be seen from the definition of tenacity given above that it is the rupturing stress-difference for tensional stresses. There is no word specially applied to rupturing stress-difference under pressure.

I am inclined to think that for the purposes of this investigation these tables in most cases rate the strength of materials somewhat too highly; for it seems probable that a permanent set would be taken, if a material were subjected for a long time to a stress-difference, which is a considerable fraction of the limiting value. We are likely to know more on this point in some years' time when the wires hung by Sir William Thomson in the tower of Glasgow University have been subjected to several years of tension. However this may be I give the results of some of the experiments as collected and quoted by Sir William Thomson and the late Professor Rankine. The first table of tenacity, except the results denoted by the letter R, are taken from Sir William Thomson. The second table of crushing stress-difference is taken entirely from Rankine. The multiplications and reductions to different units I have done myself.

TABLE VII. Limiting stress-difference.

Produced by tension				Produced by crushing		
Material	Breaking stress-difference in metric tonnes per square centimetre	Stress-difference at which permanent set begins in—		Material	Breaking stress-difference in—	
		Metric tonnes per square centimetre	British tons per square inch		Metric tonnes per square centimetre	British tons per square inch
Sheet lead	·23	·23	1·46	Strong red brick	·077	·49
Cast tin	·416	·417	2·64	Strong sandstone	·39	2·45
(R) „	·325	...	2·06 (R)	(F) Strong limestone	·60	3·80
Wood (ash)	1·20	1·20	7·61	Marble	·39	2·45
Cast brass	1·27	1·27	8·05	Granite	·39 to ·77	2·45 to 4·91
„ iron	·94 to 2·04	1·14 to 1·87	7·23 to 11·86	(F) Granite (Mount Sorrel)	·905	5·74
Drawn copper	4·10	4·00	25·36	(F) Grauwacke	1·19	7·54
English steel pianoforte wire	23·62	23·56	149·6	Ash (along the grain)	·63	4·02
(R) Brick, cement	·020 to ·021	...	·125 to ·134	Cast brass	·73	4·60
(R) Glass	·66	...	4·20	Wrought iron	2·52 to 2·84	16 to 18
(R) Slate	·68 to ·90	...	4·3 to 5·7			

NOTE.—The second and third columns give the product of Young's modulus into greatest elastic extension, and this should give the stress-difference when permanent set begins. Rankine does not give the data for this quantity, but the breaking stress-difference is given in both metric and British units, the latter being in the third column.

In the second half of the table the results marked F are from Sir William Fairbairn's experiments.

The only cases in these two tables in which we have the opportunity of comparing the strength for resisting the stress-difference, when produced in the two manners, are for the materials cast brass and ash; in both cases we see that the substance is considerably weaker for crushing than for tension.

I should be inclined to suppose that the crushing strength is more nearly the datum we require for the case of the stresses in the earth.

In the first half of the table we probably see the effect of permanent set in the cases of copper and pianoforte wire (compare 4·00 with 4·10, and 23·56 with 23·62), but it is surprising that the contrast between the two columns is not more marked.

§ 10. *On the case when the elastic solid is compressible.*

It appears desirable to know how far the results of the preceding investigation may differ, if the elastic solid be compressible. According to the views of Dr Ritter, referred to in the summary, this may be the case, perhaps to a large extent.

[In the paper as originally presented to the Royal Society I had failed to notice that, as pointed out in § 1 as revised for the present volume, there are really two distinct problems under discussion. Accordingly the present section has been rewritten.

The effects of compressibility will now be considered in the two limiting cases of harmonics of the second and of infinite orders. Preparations for this have already been made in the revision of § 4.

For the case of the second harmonic, we have to put $i = 2$ in the definition (2) of § 1 and in equation (9). The analysis will be much abbreviated if we write

$$\lambda = \frac{99}{38\left(\frac{\omega}{\upsilon} - \frac{5}{19}\right)}$$

Then since $K = 19\omega - 5\upsilon$, we have

$$\left(\frac{2i+1}{2(i-1)} + \frac{i}{I}\right)\frac{\upsilon}{K} = \tfrac{1}{19}\lambda$$

Thus (9) becomes

$$P = \tfrac{5}{2}W_2 + \tfrac{1}{19}\left[(8a^2 - 5r^2)\frac{d^2 W_2}{dx^2} - 6x\frac{dW_2}{dx} + 6W_2\right]$$
$$+ \tfrac{1}{19}\lambda\left[(a^2 - 3r^2)\frac{d^2 W_2}{dx^2} + 4x\frac{dW_2}{dx} - 4W_2\right]$$

Also T has a similar form.

We thus obtain

$$19P - \tfrac{1}{2}(107 - 8\lambda) W_2 = [(8+\lambda) a^2 - (5+3\lambda) r^2] \frac{d^2 W_2}{dx^2} - 2(3 - 2\lambda) x \frac{dW_2}{dx}$$

$$19T = [(8+\lambda) a^2 - (5+3\lambda) r^2] \frac{d^2 W_2}{dx\,dz} - (3 - 2\lambda)\left(x \frac{dW_2}{dz} - z \frac{dW_2}{dx}\right)$$

These formulæ, and others derived from them by cyclical changes of letters, are to be applied in the case where

$$W_2 = -\tfrac{2}{5} gwh \frac{r^2}{a^2} \varsigma_2$$

$$r^2 \varsigma_2 = x^2 + y^2 - 2z^2$$

As in § 5 it is more convenient to substitute e, the ellipticity of the ellipsoid, for h, so that h has the value $\tfrac{1}{3}ea$.

Thus we are to take

$$W_2 = -\tfrac{2}{15} \frac{gwe}{a} (x^2 + y^2 - 2z^2)$$

For the present the factor $-\tfrac{2}{15} gwe/a$ may be omitted, to be reintroduced at a later stage.

We find then

$$\tfrac{19}{2}P - \tfrac{1}{4}(107 - 8\lambda) W_2 = (8+\lambda) a^2 - (5+3\lambda) r^2 - 2(3-2\lambda) x^2$$

$$\tfrac{19}{2}Q - \tfrac{1}{4}(107 - 8\lambda) W_2 = (8+\lambda) a^2 - (5+3\lambda) r^2 - 2(3-2\lambda) y^2$$

$$\tfrac{19}{2}R - \tfrac{1}{4}(107 - 8\lambda) W_2 = -2(8+\lambda) a^2 + 2(5+3\lambda) r^2 + 4(3-2\lambda) z^2$$

$$\tfrac{19}{2}T = (3-2\lambda) xz; \quad \tfrac{19}{2}S = (3-2\lambda) yz; \quad \tfrac{19}{2}U = (3-2\lambda) xy$$

It will suffice to consider only the case when $y = 0$, and we may use polar coordinates with $x = r\sin\theta$, $y = r\cos\theta$.

Thus

$$\tfrac{19}{2}P - \tfrac{1}{4}(107 - 8\lambda) W_2 = (8+\lambda)(a^2 - r^2) + (3-2\lambda) r^2 \cos 2\theta$$

$$\tfrac{19}{2}Q - \tfrac{1}{4}(107 - 8\lambda) W_2 = (8+\lambda)(a^2 - r^2) + (3-2\lambda) r^2$$

$$\tfrac{19}{2}R - \tfrac{1}{4}(107 - 8\lambda) W_2 = -2(8+\lambda)(a^2 - r^2) + 2(3-2\lambda) r^2 \cos 2\theta$$

$$\tfrac{19}{2} . 2T = (3-2\lambda) r^2 \sin 2\theta; \quad S = U = 0$$

It may be well, before proceeding further, to consider what range of values is legitimately attributable to λ.

Navier and Poisson maintained that the modulus of rigidity in a solid was $\tfrac{3}{5}$ of that of incompressibility, but Stokes showed that this was very far from being the case for many solids*. Although then Poisson's contention is not well founded, yet it gives a value for the ratio which is more or less correct for some solids and perhaps for many; and it is curious that in the case now under consideration Poisson's theory gives the dividing line which separates solutions of one kind from those of another kind.

According to Navier and Poisson the modulus of rigidity v is $\frac{3}{5}$ of that of compressibility, which is $\omega - \frac{1}{3}v$; hence $\omega = 2v$. This makes $\lambda = \frac{3}{2}$, and $3 - 2\lambda$ vanishes.

Thus according to Poisson's theory, with $\lambda = \frac{3}{2}$, we obtain the following curious result:—

$$\mathrm{P} - \tfrac{5}{2}W_2 = \mathrm{Q} - \tfrac{5}{2}W_2 = \quad a^2 - r^2$$
$$\mathrm{R} - \tfrac{5}{2}W_2 = -2(a^2 - r^2)$$
$$\mathrm{T} = 0$$

Our coordinate axes are therefore principal stress-axes; one being parallel to the polar axis of the globe, and the other two (being equal *inter se*) anywhere mutually at right angles in a small circle of latitude.

The stress-difference is $3(a^2 - r^2)$.

Thus on introducing the omitted factors and changing the sign (as is clearly permissible) the stress-difference is given throughout the sphere by

$$\Delta = \tfrac{2}{5}gwea\left(1 - \frac{r^2}{a^2}\right)$$

For a homogeneous sphere of the size and mean density of the earth this gives

$$\Delta = 1442e\left(1 - \frac{r^2}{a^2}\right) \text{ metric tonnes per square cm.}$$

It vanishes at the surface and at the centre is $1442e$. In the case of incompressibility we found in § 5 the central stress-difference to be $1214e$ in the same units. Thus this degree of compressibility entirely relieves the superficial stress-difference and only augments the central stress-difference by a sixth part.

Returning to the consideration of the range of values of λ:—

We know that $\lambda = 0$ corresponds to complete incompressibility. The modulus of incompressibility $\omega - \frac{1}{3}v$ vanishes when $\omega = \frac{1}{3}v$, and this furnishes the other limit. In this case $\lambda = 37\frac{1}{3}$. It would not be very interesting to obtain numerical results for these large values of λ, but it is well to determine the limit of values permissible.

In the general case the procedure to be followed is exactly parallel to that of § 5. It will be found that, if for brevity we write

$$\mathfrak{R}^2 = 64(1 + \tfrac{1}{3}\lambda)^2(a^2 - r^2)^2 + (1 - \tfrac{2}{3}\lambda)^2 r^4 - 16(1 + \tfrac{1}{8}\lambda)(1 - \tfrac{2}{3}\lambda)(a^2 - r^2)r^2\cos 2\theta$$

* Thomson and Tait's *Natural Philosophy*, § 684.

The stress-difference Δ is equal to the greater of the two following expressions:

(i) $\Delta = \frac{6}{19}\,\mathfrak{R}$

(ii) $\Delta = \frac{3}{19}\left[8\left(1+\frac{1}{8}\lambda\right)(a^2-r^2)+2\left(1-\frac{2}{3}\lambda\right)r^2-3\left(1-\frac{2}{3}\lambda\right)r^2\cos 2\theta+\mathfrak{R}\right]$

When $\lambda = 0$, these of course reduce to the results of § 5.

When $\lambda = \frac{3}{2}$ the forms (i) and (ii) become identical, and we come back on the simple case corresponding to Poisson's hypothesis.

The equation to the ellipsoid, which separates the regions in which the forms (i) and (ii) are respectively applicable, is

$$(10-2\lambda)(x^2+y^2)+19z^2=(16+2\lambda)a^2$$

When λ is greater than $\frac{3}{2}$ this ellipsoid entirely encloses the sphere, and the separation becomes nugatory. It appears then that—

When λ is less than $\frac{3}{2}$, the form (ii) holds inside the ellipsoid, and the form (i) outside.

When λ is greater than $\frac{3}{2}$, the form (i) is applicable everywhere.

I have not reduced these results to numbers in any specific cases, because it suffices to learn that compressibility affects the result largely in the case of the second harmonic, although for moderate values of λ the maximum value of Δ is not changed very much in amount.

It will now be shown that whatever may be the compressibility of the solid, we get the same solution in the case when the harmonic is of an infinitely high order. If δP, δT denote the additional terms introduced by compressibility, when i is very large we have from (32)

$$\delta \mathrm{P} = \frac{v}{2i^2\omega}\left(1+\frac{1}{2i}\right)\left\{(a^2-3r^2)\frac{d^2W_i}{dx^2}+4ix\frac{dW_i}{dx}-2i^2W_i\right\}$$

$$= -\frac{v}{\omega}W_2$$

In a similar manner we find $\delta\mathrm{T} = 0$.

It is clear then that $\delta\mathrm{P} = \delta\mathrm{Q} = \delta\mathrm{R} = -\dfrac{v}{\omega}W_2$. It follows that the difference of stresses is unaffected, and our former result is unaffected.

It may be concluded that except for the lower harmonic inequalities compressibility introduces but little change in our results.]

II.

Summary and Discussion.

The existence of dry land proves that the earth's surface is not a figure of equilibrium appropriate for the diurnal rotation.

Hence the interior of the earth must be in a state of stress, and as the land does not sink in, nor the sea-bed rise up, the materials of which the earth is made must be strong enough to bear this stress.

We are thus led to inquire how the stresses are distributed in the earth's mass, and what are their magnitudes. These points cannot be discussed without an hypothesis as to the interior constitution of the earth.

In this paper I have solved a problem of the kind indicated for the case of a homogeneous incompressible elastic sphere, and have applied the results to the case of the earth.

It may of course be urged that the earth is not such as this treatment postulates.

The view which was formerly generally held was that the earth consists of a solid crust floating on a molten nucleus. It has also been lately maintained by Dr August Ritter in a series of interesting papers that the interior of the earth is gaseous*. A third opinion, contended for by Sir William Thomson, and of which I am myself an adherent, is that the earth is throughout a solid of great rigidity; he explains the flow of lava from volcanoes either by the existence of liquid vesicles in the interior, or by the melting of solid matter, existing at high temperature and pressure, at points where diminution of pressure occurs.

There is another consideration, which is consistent with Sir William Thomson's view, and which was pointed out to me by Professor Stokes. It may be that underneath each continent there is a region of deficient density; then underneath this region there would be no excess of pressure†.

For the present investigation it is to some extent a matter of indifference as to which of these views is correct, for if it is only the crust of the earth which possesses rigidity, or if Professor Stokes's suggestion of the regions of

* "Anwendung der mechanischen Wärmetheorie auf kosmologische Probleme." Carl Rümpler, Hannover, 1879. This is a reprint of six papers in Wiedemann's *Annalen*.

Dr Ritter contends that the temperature in the interior of the planet is above the critical temperature and that of dissociation for all the constituents, so that they can only exist as gas. Data are wanting with regard to the mechanical properties of matter at, say 10,000° Fahr., and a pressure of many tons to the square inch. Is it not possible that such "gas" may have the density of mercury and the rigidity and tenacity of granite? Although such a conjectural "gaseous" solid might possess high rigidity, it would [presumably] have great compressibility.

† [We may regard this view as now (1906) established by the modern results of geodesy.]

deficient density be correct, then the stresses in the crust or in the parts near the surface must be greater than those here computed—enormously greater if the crust be thin*, or if the region of deficient density be of no great thickness.

With regard to the property of incompressibility which is here attributed to the elastic sphere, it appears from § 10 that even if we suppose the elastic solid to be compressible, yet [for inequalities of moderate extension on the surface of the globe] the results with regard to the internal stresses are almost the same as though it were incompressible; [but this is not so for such an inequality as is represented by a harmonic of the second order]. I think the hypothesis of great incompressibility is likely to be much nearer to the truth than is that of great compressibility; I shall therefore adhere to the supposition of infinite incompressibility.

I take then a homogeneous incompressible elastic sphere, and suppose it to have the power of gravitation and to be superficially corrugated. In consequence of mathematical difficulties the problem is here only solved for the particular class of surface inequalities called zonal harmonics, the nature of which will be explained below.

Before discussing the state of stress produced by these inequalities, it will be convenient to explain the proper mode of estimating the strength of an elastic solid under stress.

At any point in the interior of a stressed elastic solid there are three lines mutually at right angles, which are called the principal stress-axes. Inside the solid at the point in question imagine a small plane (say a square centimetre or inch) drawn perpendicular to one of the stress-axes; such a small plane will be called an inter-face†. The matter on one side of the ideal inter-face might be removed without disturbing the equilibrium of the elastic solid, provided some proper force be applied to the inter-face; in other words, the matter on one side of an inter-face exerts a force on the matter on the other side. Now a stress-axis has the property that this force is parallel to the stress-axis to which the inter-face is perpendicular. Thus along a stress-axis the internal force is either purely a traction or purely a pressure. Treating pressures as negative tractions, we may say that at any point of a stressed elastic solid, there are three mutually perpendicular directions along which the stresses are purely tractional. The traction which must be applied to an inter-face of a square centimetre in area, in order to maintain equilibrium when the matter on one side of the inter-face is removed, is called a principal stress, and is of course to be measured by grammes weight per square centimetre.

* The evaluation of the stresses in a crust, with fluid beneath, would be tedious, but not more difficult than the present investigation. I may perhaps undertake this at some future time.

† This term is due to Professor James Thomson.

If the three stresses be equal and negative, the matter at the point in question is simply squeezed by hydrostatic pressure, and it is not likely that in a homogeneous solid any simple hydrostatic pressure, *absolutely* equal in all directions, would ever rupture the solid. The effect of the equality of the three stresses when they are positive and tractional is obscure, but at least physicists do not in general suppose that this is the cause of rupture when a solid breaks.

If the three principal stresses be unequal, one must of course be greatest and one least, and there is reason to suppose that tendency of the solid to rupture is to be measured by the difference between these principal stresses.

In one very simple case we know that this is so, for if we imagine a square bar, of which the section is a square centimetre, to be submitted to simple longitudinal tension, then two of the principal stresses are zero (namely, the stresses perpendicular to the faces of the rod), and the third is equal to the longitudinal traction. The traction under which the rod breaks is a measure of its strength, and this is equal to the difference of principal stresses.

If at the same time the rod were subjected to great hydrostatic pressure, the breaking load would be very little, if at all affected; now the hydrostatic pressure subtracts the same quantity from all three principal stresses, but leaves the difference between the greatest and least principal stresses the same as before.

Difference of principal stresses may also be produced by crushing.

In this paper I call the difference between the greatest and least principal stresses the "stress-difference," and I say that, if calculation shows that the weight of a certain inequality on the surface of the earth will produce such and such stress-difference at such and such a place, then the matter at that place must be at least as strong as matter which will break when an equal stress-difference is produced by traction or crushing.

I shall usually estimate stress-difference by metric tonnes (a million grammes) per square centimetre, or by British tons per square inch.

In Table VII., § 9, are given the experimentally determined values of the breaking stress-difference for various substances. The table is divided into two parts, in the former of which the stress-difference was produced by tension, and in the latter by crushing. It is not necessary here to advert to the difference in meaning of the numbers given in the first column and those given in the two latter columns in the first half of the table.

The cases of wood and cast brass are the only ones where a comparison is possible between the two breaking stress-differences, as differently produced. It will be seen that the material is weaker for crushing than for tension. For the reasons given in that section, I am inclined to think that these tables rate the strength of the materials somewhat too highly for the purposes of this investigation. I conceive that the results derived from crushing are more appropriate for the present purpose than those derived from tension; and fortunately the results for various kinds of rocks seem to have been principally derived from crushing stresses.

This table will serve as a means of comparison with the numerical results derived below, so that we shall see, for example, whether or not at 500 miles from the surface the materials of the earth are as strong as granite.

We may now pass to the mathematical investigation. It appears therefrom that the distribution of stress-difference is quite independent of the absolute heights and depths of the inequalities. Although the questions of distribution and magnitude of the stresses are thus independent, it will in general be convenient to discuss them more or less simultaneously.

The problem has only been solved for the class of superficial inequalities called zonal harmonics, and their nature will now be explained.

A zonal harmonic consists of a series of undulations corrugating the surface in parallels of latitude with reference to some equator on the globe; the number of the undulations is estimated by the order of the harmonic. The harmonic of the second order is the most fundamental kind, and consists of a single undulation forming an elevation round the equator, and a pair of depressions at the poles of that equator; it may also be defined as an elliptic spheroid of revolution, and the absolute magnitude is measured by the ellipticity of the spheroid.

If the order of the harmonic be high, say 30 or 40, we have a regular series of mountain chains and intervening valleys running round the sphere in parallels of latitude.

For the sake of convenience I shall always speak as though the equator were a region of elevation, but the only effect of changing elevations into depressions, and *vice versâ*, is to reverse diametrically the directions of all the stresses.

The harmonics of the orders 2, 6, 10, &c., have depressions at the poles of the sphere; those of orders 4, 8, 12, &c., have elevations at the poles.

The harmonic of the fourth order consists of an equatorial continent and a pair of circular polar continents, with an intervening depression. That of the sixth order consists of an equatorial continent and a pair of annular continents in latitudes (about) 60° on one and the other side of the equator. The 8th harmonic brings down these new annular continents to about latitude 45°, and adds a pair of polar continents; and so on.

By a continuation of this process the transition to the mountain chains and valleys is obvious.

In § 5 the case of the 2nd harmonic is considered. As above explained the sphere is deformed into a spheroid of revolution. The investigation also applies to the case of a rotating spheroid, such as the earth, with either more or less oblateness than is appropriate for the figure of equilibrium.

The lines throughout a meridional section of the spheroid along which the stress-difference is constant are shown in fig. 1, and the numbers written on the curves give the relative magnitudes of the stress-difference.

[At the surface the stress-difference is constant over two polar caps which extend southward from the north pole and northward from the south pole as far as latitudes 54° 44′ north and south. Between these latitudes and the equator the stress-difference increases until at the equator it is three times as great as over the polar caps.] In the polar regions the stress-difference diminishes as we descend into the spheroid and then increases again; in the equatorial regions it always increases as we descend. The maximum value is at the centre, and there the stress-difference is eight times as great as at the [poles].

If the elastic solid be compressible the stress-difference [at the surface is much less than] on the hypothesis of incompressibility. [For a certain value of the compressibility the superficial stress-difference vanishes all over the surface. In this case the stress-difference at the centre of the globe is only greater than that found on the hypothesis of complete incompressibility by one-sixth part of itself. It would thus seem as if compressibility would not make a very great difference in the actual strength of the globe.]

On evaluating the stress-difference, on the hypothesis of incompressibility, arising from given ellipticity in a spheroid of the size and density of the earth, it appears that if the excess or defect of ellipticity above or below the equilibrium value (namely $\frac{1}{232}$ for the homogeneous earth) were $\frac{1}{1000}$, then the stress-difference at the centre would be 8 tons per square inch, and accordingly, if the sphere were made of material as strong as brass (see Table VII.), it would be just on the point of rupture. Again if the homogeneous earth, with ellipticity $\frac{1}{232}$, were to stop rotating, the central stress-difference would be 33 tons per square inch, and it would rupture if made of any material excepting the finest steel.

A rough calculation* will show that if the planet Mars has ellipticity $\frac{1}{80}$ (about twice the ellipticity on the hypothesis of homogeneity) the central stress-difference must be 6 tons per square inch. It was formerly supposed that the ellipticity of the planet was even greater than $\frac{1}{80}$, and even if the

* The data for the calculation are: Ratio of terrestrial radius to Martian radius 1·878. Ratio of Martian mass to terrestrial mass ·1020. Whence ratio of Martian gravity to terrestrial gravity ·3596. Central stress-difference, due to ellipticity e, $996e$ tons per square inch. "Homogeneous" ellipticity of Mars $\frac{1}{166}$; and $\frac{996}{166}$ equal to 6.

latest telescopic evidence had not been adverse to such a conclusion, we should feel bound to regard such supposed ellipticity with the greatest suspicion, in the face of the result just stated.

The state of internal stress of an elastic sphere under tide-generating forces is identical with that caused by ellipticity of figure*. Hence the investigation of § 5 gives the distribution of stress-difference caused in the earth by the moon's attraction. In fig. 1, the point called "the pole" is the point where the moon is in the zenith.

Computation shows that the stress-difference at the [polar] surfaces, due to the lunar tide-generating forces, is 16 grammes per square centimetre, and at the centre eight times as much. These stresses are considerable, although very small compared with those due to terrestrial inequalities, as will appear below.

In § 6 the stresses produced by harmonic inequalities of high orders are considered. This is in effect the case of a series of parallel mountains and valleys, corrugating a mean level surface with an infinite series of parallel ridges and furrows. [In this case compressibility makes absolutely no difference in the result, as shown in § 10.]

It is found that the stress-difference depends only on the depth below the mean surface, and is independent of the position of the point considered with regard to ridge and furrow; the direction of the stresses does however depend on this latter consideration.

In fig. 2 is shown the law by which the stress-difference increases and then diminishes as we go below the surface. The vertical ordinates of the curve indicate the relative magnitude of the stress-difference, and the horizontal ones the depth below the surface. The depth OL on the figure is equal to the distance between adjacent ridges, and the figure shows that the stress-difference is greatest at a depth equal to $\frac{1}{6{\cdot}28}$ of OL.

The greatest stress-difference depends merely on the height and density of the mountains, and the depth at which it is reached merely on the distance from ridge to ridge.

Numerical calculation shows that if we suppose a series of mountains, whose crests are 4000 metres or about 13,000 feet above the intermediate valley-bottoms, formed of rock of specific gravity 2·8, then the maximum stress-difference is 2·6 tons per square inch (about the tenacity of cast tin); also if the mountain chains are 314 miles apart the maximum stress-difference is reached at 50 miles below the mean surface.

It may be necessary to warn the geologist that this investigation is approximate in a certain sense, for the results do not give the state of stress

* This is subject to certain qualifications noticed in § 5.

actually within the mountain prominences or near the surface in the valley-bottoms. The solution will however be very nearly accurate at some five or six miles below the valley-bottoms. The solution shows that the stress-difference is *nil* at the mean surface, but it is obvious that both the mountain masses and the valley-bottoms are in some state of stress.

The mathematician will easily see that this imperfection arises, because the problem really treated is that of an infinite elastic plane, subjected to simple harmonic tractions and pressures.

To find the state of stress actually within the mountain masses would probably be difficult.

The maximum stress-difference just found for the mountains and valleys obviously cannot be so great as that at the base of a vertical column of this rock, which has a section of a square inch and is 4000 metres high. The weight of such a column is 7·1 tons, and therefore the stress-difference at the base would be 7·1 tons per square inch. The maximum stress-difference computed above is 2·6, which is about three-eighths of 7·1 tons per square inch. Thus the support of the contiguous masses of rock, in the case just considered, serves as a relief to the rock to the extent of about five-eighths of the greatest possible stress-difference. This computation also gives a rough estimate of the stress-differences which must exist if the crust of the earth be thin. It is shown below that there is reason to suppose that the height from the crest to the bottom of the depression in such large undulations as those formed by Africa and America is about 6000 metres. The weight of a similar column 6000 metres high is nearly 11 tons.

In § 7 I take the cases of the even zonal harmonics from the 2nd to the 12th, but for all except the 2nd harmonic only the equatorial region of the sphere is considered.

Fig. 3 shows an exaggerated outline of the equatorial portion of the inequalities; it only extends far enough to show half of the most southerly depression, even for the 12th harmonic. It did not seem worth while to trace the surfaces of equal stress-difference throughout the spheroid, but the laborious computations are carried far enough to show that these surfaces must be approximately parallel to the surface of the mean sphere. It is accordingly sufficient to find the law for the variation of stress-difference immediately underneath the equatorial belt of elevation. It requires comparatively little computation to obtain the results numerically, and the results of the computation are exhibited graphically in fig. 4.

Table V. (*b*), § 7, gives the maximum stress-differences, resulting from these several inequalities, computed under conditions adequately noted in the table itself. It will be convenient to postpone the discussion of the results.

In § 8 I build up out of these six harmonics an isolated equatorial continent. The nature of the elevation is exhibited in fig. 5, in the curve marked "representation"; no notice need be now taken of the dotted curve. This curve exhibits a belt of elevation of about 15° of latitude in semi-breadth, and the rest of the spheroid is approximately spherical. This kind of elevation requires the 2nd as one of its harmonic constituents, and this harmonic means ellipticity of the whole globe. Now it may perhaps be fairly contended that on the earth we have no such continent as would require a perceptible 2nd harmonic constituent. I therefore give in fig. 5, a second curve which represents an equatorial belt of elevation counterbalanced by a pair of polar continents in such a manner that there is no second harmonic constituent.

I have not attempted to trace the curves of equal stress-difference arising from these two kinds of elevation, but I believe that they will consist of a series of much elongated ovals, whose longer sides are approximately parallel with the surface of the globe, drawn about the maximum point in the interior of the sphere at the equator. The surfaces of equal stress-difference in the solid figure will thus be a number of flattened tubular surfaces one within the other.

At the equator however the law of variation of stress-difference is easy to evaluate, and fig. 6 shows the results graphically, the vertical ordinates representing stress-difference and the horizontal the depths below the surface. The upper curve in fig. 6 corresponds with the "representation curve" of fig. 5, and the lower curve with the case where there is no 2nd harmonic constituent.

The central stress-difference, which may be observed in the upper curve, results entirely from the presence of the 2nd harmonic constituent in the corresponding equatorial belt of elevation.

The maximum stress-differences in these two cases occur at about 660 and 590 miles from the surface respectively.

We now come to perhaps the most difficult question with regard to the whole subject—namely, how to apply these results most justly to the case of the earth.

The question to a great extent turns on the magnitude and extent of the superficial inequalities in the earth. As the investigation deals with the larger inequalities, it will be proper to suppose the more accentuated features of ridges, peaks and holes to be smoothed out.

The stresses caused in the earth by deficiency of matter over the sea-beds are the same as though the seas were replaced by a layer of rock, having everywhere a thickness of about $\frac{1{\cdot}02}{2{\cdot}75}$ or nearly $\frac{4}{11}$ of the actual depths of sea.

The surface being partially smoothed and dried in this manner, we require to find an ellipsoid of revolution which shall intersect the corrugations in such a manner that the total volume above it shall be equal to the total volume below it.

Such a spheroid may be assumed to be the figure of equilibrium appropriate to the earth's diurnal rotation; if it departs from the equilibrium form by even a little, then we shall much underestimate the stress in the earth's interior by supposing it to be a form of equilibrium.

Professor Bruns has introduced the term "geoid" to express any one of the "level" surfaces in the neighbourhood of the earth's surface, and he endeavours to form an estimate of the departure of the continental masses and sea-bottoms from some mean geoid*. From the geodesic point of view the conception is valuable, but such an estimate is scarcely what we require in the present case. The mean geoid itself will necessarily partake of the contortions of the solid earth's surface, even apart from disturbances caused by local inequalities of density, and thus it cannot be a figure of equilibrium.

Thus, even if we were to suppose that the solid earth were everywhere coincident with a geoid—which is far from being the case—a state of stress would still be produced in the interior of the earth.

An example of this sort of consideration is afforded by the geodesic results arrived at by Colonel Clarke, R.E.†, who finds that the ellipsoid which best satisfies geodesic measurement, has three unequal axes, and that one equatorial semi-axis is 1524 feet longer than the other. Now such an ellipsoid as this, although not exactly one of Bruns' geoids, must be more nearly so than any spheroid of revolution; and yet this inequality (if really existent, and Colonel Clarke's own words do not express any very great confidence) must produce stress in the earth. Colonel Clarke's results show an ellipticity of the equator equal to $\frac{1}{13731}$, and this in the homogeneous elastic earth will be about equivalent to ellipticity $\frac{1}{27000}$; such ellipticity would produce a central stress-difference of $\frac{7698}{27000}$ or nearly one-third of a British ton per square inch.

From this discussion it may, I think, be fairly concluded that if we assume the sea-level as being the figure of equilibrium and estimate the departures therefrom, we shall be well within the mark.

The average height of the continents is about 350 metres (1150 feet), and the average depth of the great oceans is in round numbers 5000 metres (16,000 feet); but the latter datum is open to much uncertainty‡. When the sea is solidified into rock the 5000 metres of depth is reduced to 3200 metres below the actual sea-level. Thus the average effective depression of sea-bed is about nine times as great as the average height of the land.

* *Die Figur der Erde*. Von Dr H. Bruns. Berlin: Stankiewicz, 1878.

† *Phil. Mag.*, Aug., 1878.

‡ In a previous paper, "Geological Changes, &c." *Phil. Trans.*, Vol. 167, Part I., p. 295, I have endeavoured to discuss this subject, and references to a few authorities will be found there. [This paper will probably be included in Vol. III. of this collection.]

I shall take it as exactly nine times as great, and put the depth as 3150 metres; but it is of course to be admitted that perhaps eight and perhaps ten might be more correct factors.

In the analytical investigation of this paper the outlines of the vertical section of the continents and depressions are always sweeping curves of the harmonic type, and the magnitudes of the elevations and depressions are estimated by the greatest heights and depths, measured from a mean surface which equally divides the two.

We have already supposed the outlines of continents and sea-beds to have been smoothed down into sweeping curves, which we may take as being, roughly speaking, of the harmonic type. The smoothing will have left the averages unaffected.

The averages are not however estimated from a mean spheroidal surface, but from one which is far distant from the mean.

The questions now to be determined are as follows:—What is the proper greatest height and depression, estimated from a mean spheroid, which will bring out the above averages estimated from present sea-level, and what is the position of the mean spheroid with reference to the sea-level.

From the solution of the problem considered in the note below*, it

* Conceive a series of straight harmonic undulations corrugating a mean horizontal surface, and suppose them to be flooded with water. This will represent fairly well the undulations on the dried earth, and the water-level will represent the sea-level.

Suppose that the average heights and depths of the parts above and below water are known, and that it is required to find the position of the mean horizontal surface with reference to the water-level, and the height of the undulations measured from that mean surface.

Take an origin of coordinates in the water-level, the axis of x in the water-level and perpendicular to the undulations, and the axis of y measured upwards.

Let
$$y = h(\cos x - \cos a)$$
be the equation to the undulations.

The average height of the dry parts is clearly $\frac{1}{2a}\int_{-a}^{+a} y\,dx$ or $\frac{h}{a}(\sin a - a\cos a)$. Similarly the average depth below water is $\frac{h}{\pi - a}[\sin(\pi - a) - (\pi - a)\cos(\pi - a)]$ or $\frac{h}{\pi - a}[\sin a + (\pi - a)\cos a]$.

If the latter average be p times as great as the former
$$ph\cos a\left(\frac{1}{a}\tan a - 1\right) = h\cos a\left(\frac{1}{\pi - a}\tan a + 1\right).$$
This is an equation for determining a.

Now I find that $a = 34° 30'$ gives $p = 8·983$, which corresponds very nearly with $p = 9$ of the text above.

This value of a corresponds with an average equal to $·1165h$ for the height above water, and $1·0469h$ for the depth below water. Now if we put
$$1·0469h = 3150 \text{ metres}$$
which gives
$$·1165h = 350 \text{ metres very nearly,}$$
we have $h = 3009$ metres.

appears that, if the continents and sea-beds have sections which are harmonic curves, then if we take,—

The mean level bisecting elevations and depressions as 2480 metres (8150 feet) below the sea-level, and the greatest elevation and depression from that mean level as 3009 metres (9840 feet), it results that the *average* height of the land above sea-level is 350 metres and the *average* depression of dried sea-bed is 3150 metres.

It thus appears that 3000 metres would be a proper greatest elevation and depression to assume for the harmonic analysis of this paper, if the earth were homogeneous. But as the density of superficial rocks is only a half of the mean density of the earth, I shall take 1500 metres as the greatest elevation and depression from the mean equilibrium spheroid of revolution.

It is proper here to note that the height of the undulations of elevation and depression in the zonal harmonic inequalities is considerably greater towards the poles than it is about the equator; it might therefore be maintained that by making 1500 metres the equatorial height, we are taking too high an estimate. But the state of stress caused in the sphere at any point depends very much more on the height of the inequality in the neighbourhood of a superficial point immediately over the point considered, than it does on the inequalities in remote parts of the sphere.

Now in all the inequalities, except the 2nd harmonic, I have considered the state of stress in the equatorial region, and it will therefore I think be proper to adhere to the 1500 metres for the greatest height and depression.

We have next to consider, what order of harmonic inequalities is most nearly analogous to the great terrestrial continents and oceans. The most obvious case to take is that of the two Americas and Africa with Europe. The average longitude of the Americas is between 60° and 80° W., and the average longitude of Africa is about 25° E., hence there is a difference of longitude of about a right-angle between the two masses. These two great continents would be more nearly represented by an harmonic of the sectorial

---

The depth below water-level of the mean level is $h \cos 34° 30'$ or 2480 metres.

The greatest height of the dry part above the water-level is $3009 - 2480$ or 429 metres, and the greatest depth of the submerged part below water-level is $3009 + 2480$ or 5489 metres.

After the proof-sheets of this paper had been corrected, Professor Stokes pointed out to me that, according to Rigaud (*Cam. Phil. Soc.*, vol. 6), the area of land is about four-fifteenths of the whole area of the earth's surface. Now, in the ideal undulations we are here considering the area above water is about one-tenth of the whole area; hence in this respect the analogy is not satisfactory between these undulations and the terrestrial continents. If I have not considerably over-estimated the average depth of the sea (and I do not think that I have done so), the discrepancy must arise from the fact that actual continents and sea-beds do not present in section curves which conform to the harmonic type; there must also be a difference between corrugated spherical and plane surfaces.

The geological denudation of the land must, to some extent, render our continents flat-topped.

class*, rather than by a zonal harmonic, nevertheless I think the solution for the zonal harmonic will be adequate for the present purpose.

Now it has been explained above that the harmonic of the fourth order represents an equatorial continent and a pair of polar continents. In the case of the 4th harmonic therefore there is a right angle of a great circle between contiguous continents. We may conclude from this that the large terrestrial inequalities are about equivalent to the harmonic of the fourth order.

Table V. (*b*), § 7, gives the maximum stress-differences under the centre of the equatorial elevation of the several zonal harmonics, the height of each being 1500 metres. The point at which this maximum is reached is given in each case, and fig. 4 illustrates graphically the law of variation of stress-difference.

The second harmonic cannot be said to represent a continent, and the table shows that in each of the other cases the maximum stress-difference is very nearly 4 tons per square inch. The depths of the maximum point are of course very different in each case.

We have concluded above that Africa and America are about equivalent to an harmonic of the fourth order, hence it may be concluded that the stress-difference under those continents is at a maximum at more than 1100 miles from the earth's surface, and there amounts to about 4 tons per square inch. A comparison with Table VII. shows that marble would break under this stress, but that *strong* granite would stand.

The case of the isolated continent investigated in § 8 appeared likely to prove the most interesting one, for the purpose of application to the case of the earth. But unfortunately I have found it difficult to arrive at a satisfactory conclusion as to the proper height to attribute to the continent.

The average height of the American continent is about 1100 feet above the sea, and the average depth of the Pacific Ocean about 15,000 feet. If the water of the Pacific be congealed into rock, it will have an effective depth of 10,000 feet. The greatest height of the American continent above the bed of the dried Pacific when smoothed down must be fully 12,000 feet or 3700 metres. The height of the great central Asian plateau above the average bed of the southern ocean (after drying) must be considerably more than this.

Now in the application to the homogeneous planet the heights are to be halved to allow for the smaller density of surface rock.

* The sectorial harmonic of the fourth order $\sin^4\theta\cos 4\phi$ would represent these two great continents well. It would represent China and Australia fairly; but would annihilate the Himalayan plateau, and place another great continent in mid-Pacific. It is not at all difficult to find the stress-difference under the centre of a sectorial inequality, but to find it generally involves the solution of a cubic equation.

I therefore take 2000 metres as the height of the top of the equatorial table-land above the remaining approximately spherical portion of the sphere.

The investigation of § 8 then shows that the equatorial table-land will give rise to a stress-difference of 4·1 tons per square inch at a depth of 660 miles; and that the equatorial table-land counterbalanced by the pair of polar continents (the second harmonic constituent being absent) gives a stress-difference of about 3·8 tons per square inch at a depth of 590 miles.

This estimate of stress-difference agrees in amount, with singular exactness, with that just found from the case of the 4th zonal harmonic, but the maximum is reached 400 or 500 miles nearer to the earth's surface.

I think there can be no doubt but that there *are* terrestrial inequalities of much greater breadth than that of my isolated continent; thus this investigation for the isolated continent will give a position for the maximum stress-difference too near the surface to correspond with the largest continents. On the other hand, I do not feel at all sure that I have not considerably underestimated the height of such a comparatively narrow plateau.

In the present paper it has been impossible to take any notice of the stresses produced by the most fundamental inequality on the earth's surface, because it depends essentially on heterogeneity of density.

It is well known that the earth may be divided into two hemispheres, one of which consists almost entirely of land, and the other of sea. If the south of England be taken as the pole of a hemisphere, it will be found that almost the whole of the land, excepting Australia, lies in that hemisphere, whilst the antipodal hemisphere consists almost entirely of sea. This proves that the centre of gravity of the earth's mass is more remote from England, than the centre of figure of the solid globe.

A deformation of this kind is expressed by a surface harmonic of the first order, for such an harmonic is equivalent to a small displacement of the sphere as a whole, without true deformation. Now if we consider the surface forces produced by such a deformation in a homogeneous sphere, we find of course that there is an unbalanced resultant force acting on the whole sphere in the direction diametrically opposed to that of the equivalent displacement of the whole sphere.

The fact that in the homogeneous sphere such an unbalanced force exists shows that in this case the problem is meaningless; it is in fact merely equivalent to a mischoice in the origin for the coordinates. But in the case of the earth such an inequality does exist, and the force referred to must of course be counterbalanced somehow. The balance can only be maintained by inequalities of density, which are necessarily unknown. The problem therefore apparently eludes mathematical treatment.

It is certain that so wide-spreading an inequality, even if not great in amount, must produce great stress within the globe. And just as the 2nd harmonic produces a more even distribution of stress than the 4th, so it is likely that the first would produce a more even distribution than the 2nd.

It is difficult to avoid the conclusion that the whole of the solid portion of the earth is in a sensible state of stress.

I would not however lay very much emphasis on this point, because we are in such complete ignorance as to the manner in which the equilibrium of the solid part of the earth is maintained.

From this discussion it appears that if the earth be solid throughout, then at a thousand miles from the surface the material must be as strong as granite. If it be fluid or gaseous inside, and the crust a thousand miles thick that crust must be stronger than granite, and if only two or three hundred miles in thickness much stronger than granite. This conclusion is obviously strongly confirmatory of Sir William Thomson's view that the earth is solid throughout.

INDEX TO VOLUME II.

A

Acceleration, secular—of the moon, 73, 122
Airy, Sir G.—fortnightly nutation of earth, 303

B

Bromwich, T. J.—vibration of elastic sphere, 1
Bruns, H.—definition of geoid, 509
Bryan, G. H.—precession of liquid spheroid, 54
Burckhardt and Burg—lunar inequality due to oblateness of earth, 303
Butcher, J. G.—equations of equilibrium of elastico-viscous material, 17

C

Carret, J.—sliding of earth's crust on liquid interior, 190
Chree, C.—stresses in earth due to weight of continents and mountains, 459
Clarke, A. R.—numerical values for figure of the earth, 31, 509
Compressible elastic sphere—stresses in, 497
Continents—configuration of, 190; stresses due to weight of (table of contents), 459; average elevation of, 510
Cosmogony—numerical results giving history of earth and moon, 87, 98, 302, 311, 316, 349; general review, 367

D

Distortion, secular—of a viscous spheroid, 141, 189
Disturbing function—theory of as applicable to tidal problem, 212

E

Earth, Clarke's values for figure of, 31, 509; remote history of, 36; rate of secular change of rotation of, 59; initial condition of earth and moon, 101, 130, 322, 363; secular contraction of, 115; lunar inequality due to oblateness of, 250; plane proper to motion of, or mean equator, 259, 303; stresses due to weights of continents and mountains (table of contents), 459; solidity of, 514
Eccentricity of orbit of satellite, rate of change of, 324 et seq., 374
Elastic spheroid, effects of inertia in oscillations of, 167 et seq.; stresses in, due to weights of continents and mountains (table of contents), 459
Elastic and semi-elastic spheroid—tides, strain, and vibrations of, 1, 34, 460
Elasticity, rate of degradation of, 30
Elastico-viscous spheroid—tides of, 17
Energy, curves and surfaces of, for finding secular effects of tidal friction, 195, 374
Evans, Sir J., sliding of earth's crust on liquid interior, 190

F

Fairbairn, Sir W., on strength of materials, 496
Fortnightly tide, 35, 123

G

Graphical determination of effects of tidal friction, 195, 374, 416
Graphical results of theory of tidal friction for planet and single satellite, 396 et seq.

H

Heat generated by internal tidal friction, 155
Hecker, O.—elastic oscillations of solid earth, vi, x
Helmholtz, G. von, on reduction of lunar rotation by tidal friction, 132, 360
Herglotz, G., elastic tides of heterogeneous spheroid, 31
History of earth and moon, numerical results giving, 87, 98, 302, 311, 316, 349; general review of, 367

I

Inclination of orbit of a satellite (table of contents), 209
Inertia, effects of—in oscillations of elastic, viscous and fluid spheroids, 167

J

Joly, J.—geological time, x
Jupiter, system of, reviewed, 372, 488

K

Kant, I.—on reduction of lunar rotation by tidal friction, ix, 132, 369
Kelvin, Lord—on geological time, viii; tides and strain of elastic sphere, 1, 34, 460; precession of liquid spheroid, 54; oscillations of liquid sphere, 132, 181; experiments on prolonged tension of wires, 495; solidity of the earth, 514
Kohlrausch, on the rate of degradation of elasticity, 29

L

Lamb, Horace — oscillations of an elastic sphere, 1, 185
Lamé, G.—on stresses and strains of an elastic sphere, 1
Land, proportion of, on earth, 511
Laplace—fictitious satellites producing true tide-generating force, 55; reduction of lunar rotation by tidal friction, ix, 132, 369; formula for plane proper to moon's motion, 303

Liouville—equations of motion of a body which is changing its shape, 37
Liquid spheroid—effects of inertia in oscillations of, 167 et seq.
Love, A. E. H.—on the birth of the moon, 132
Lunar inequality due to oblateness of earth considered conjointly with precession, 250

M

Mars—satellites of, 133, 371, 457; Schiapparelli's map of, 190; ellipticity of, 505
Maxwell, J. C.—'modulus of time of relaxation of rigidity' defined, 17
Mayer, T. R.—on tidal friction, 458
Moment of momentum of the several planets and their systems, 445
Moon—secular acceleration of, 73, 122; rate of secular change of mean distance and mean motion of, 69; initial condition of earth and, 101, 130, 322, 363; plane proper to the motion of, 259, 303; rate of change of eccentricity of orbit, 324 et seq.
Mountains and continents—stresses due to weights of (table of contents), 459

O

Obliquity of ecliptic—secular change of, 58, 129; E. J. Stone on, 133
Oscillations of viscous, elastic and fluid spheroids, 1 et seq., 167 et seq.

P

Perturbing function—theory of as applicable to tidal problem, 212
Planes proper to the motions of the earth and moon, 259, 303, 321
Planetary systems—survey of, in connection with tidal friction, 133, 371, 433
Plasticity of matter—Tresca and St Vénant on, 29
Precession, of viscous spheroid, 36; of elastic and liquid spheroids, 53; and lunar inequality due to oblateness treated as a single problem, 250

R

Rankine, J. M.—on strength of materials, 496
Rayleigh, Lord—on the fortnightly tide, 123
Rigaud—on the proportion of land and sea, 511
Ritter, A.—theory of gaseous stars, 501

S

St Vénant—on plasticity of matter, 29
Satellite, system of fictitious satellites giving true tide-generating force, 55; secular changes in elements of (table of contents), 208
Saturn—system of, reviewed, 372
Schiapparelli's map of Mars, 190
Schweydar, W.—fortnightly tide on heterogeneous globe, 35, 123
Sea—proportion of, on earth, 511
Secular, contraction of earth, 115; distortion of viscous spheroid, 141, 189; changes in elements of orbit of satellite (table of contents), 208
See, T. J. J.—orbits of double stars, ix
Solar system, survey of, with reference to tidal friction, 433
Stokes, Sir G. G.—on deficiency of density underlying the continents, 501
Stone, E. J.—tidal friction and the obliquity of the ecliptic, 133
Strength of various materials, 494
Stresses in earth due to weights of continents and mountains (table of contents), 459
Strutt, Hon. R.—on geological time, ix
Sun—moment of momentum of rotation of, 439

T

Thomson, Sir W.—see Kelvin, Lord
Tidal friction—rate of change of obliquity, earth's rotation, moon's distance, 58–78, 113; see also table of contents, 208; tables giving total changes in elements of motion of earth and moon, 87, 98, 302, 311, 316, 349; survey of planetary system in connection with, 133, 371; graphical determination of effects of, 195, 374, 416; analytical solution for planet and single satellite, 383; of planet attended by several satellites (table of contents), 406; survey of solar system, 433; T. R. Mayer on, 458
Tide-generating force—specification of by means of fictitious satellites, 55
Tides—of a viscous spheroid, 1; amount of reduction of oceanic tides on a yielding nucleus, 13, 19, 22; of elastico-viscous spheroid, 17; of heterogeneous spheroid, 31; problems connected with tides of a viscous spheroid (table of contents), 140
Time, geological—viii, ix, x
Time occupied in changes of configuration of earth and moon, 105, 125
Tresca—on plasticity of matter, 29

V

Viscous spheroid—rate of subsidence of inequalities on, 11; heat generated in, by tidal friction, 155; effects of inertia in oscillations of, 167; see also various other titles.

CAMBRIDGE: PRINTED BY JOHN CLAY, M.A. AT THE UNIVERSITY PRESS.